International Symposium on Materials and Electrical Engineering (ISMEE 2017)

IOP Conference Series: Materials Science and Engineering
Volume 384

Bandung, Indonesia
16 November 2017

Editors:

Ade Gafar Abdullah
Iwan Kustiawan
Didin Wahyudin

ISBN: 978-1-5108-6753-6
ISSN: 1757-8981

Printed from e-media with permission by:

Curran Associates, Inc.
57 Morehouse Lane
Red Hook, NY 12571

Printed by Curran Associates, Inc. (2018)

For permission requests, please contact the Institute of Physics
at the address below.

Institute of Physics
Dirac House, Temple Back
Bristol BS1 6BE UK

Phone: 44 1 17 929 7481
Fax: 44 1 17 920 0979

techtracking@iop.org

Additional copies of this publication are available from:

Curran Associates, Inc.
57 Morehouse Lane
Red Hook, NY 12571 USA
Phone: 845-758-0400
Fax: 845-758-2633
Email: curran@proceedings.com
Web: www.proceedings.com

TABLE OF CONTENTS

IOP Conf. Series: Materials Science and Engineering **384** (2018) 011001 doi:10.1088/1757-899X/384/1/011001

Preface

The International Symposium on Materials and Electrical Engineering (ISMEE) 2017 is a fundamental stage of annual symposium encompassing a broad range of issues related to material, electrical and computer/information technology. It is intended to be used as a scientific forum for accommodating discussion among researchers in the field of applied science and engineering, and engineering education.

The ISMEE symposium is organized by Department of Electrical Engineering Education, Universitas Pendidikan Indonesia. This year, the first ISMEE, being held in 16 November at Bandung, Indonesia. ISMEE 2017 is structured as a meta-conference to admit researchers in the to connect with international research communities and with each other for the worldwide dissemination and sharing of research, development, and deployment ideas that span the field of material, electrical and computer/information technology.

There are 169 papers from various topics including material engineering and education, electrical engineering and education, and computer engineering and education. More than 70 papers are selected to go through a strict peer reviewed process. Selected papers will be published in IOP Conference Series: Material Science and Engineering.

We would like to thank the organizing committee and the members of reviewers for their kind assistance in reviewing the papers. We would also extend our best gratitude for Professor Budi Mulyanti, Professor Kuang-Hui Chi, and Professor Shinobu Hasegawa for their invaluable contribution and worthwhile ideas shared in the conference.

The Editors
Dr. Ade Gafar Abdullah
Didin Wahyudin, Ph.D
Iwan Kustiawan, Ph.D

IOP Conf. Series: Materials Science and Engineering **384** (2018) 011001 doi:10.1088/1757-899X/384/1/011001

LIST OF COMMITTEE

Conference Chair:
Didin Wahyudin, Ph.D – Universitas Pendidikan Indonesia

Co- Conference Chair:
Iwan Kustiawan, Ph.D – Universitas Pendidikan Indonesia

International Advisory Boards:

Prof. Peter Blanchfield – Nottingham University, UK
Assoc. Prof. Hiroyuki Mitsuhara – Tokushima University, Japan
Assoc. Prof. Shinobu Hasegawa – JAIST, Japan
Assoc. Prof. Kuang-Hui Chi – National Yunlin University, Taiwan.
Dr. Syarif Hidayat – Institut Teknologi Bandung, Indonesia
Prof. Dr. Khairrurijal – Institut Teknologi Bandung, Indonesia
Dr. Suryono – Universitas Diponegoro, Indonesia
Dr. Yohandri – Universitas Negeri Padang, Indonesia
Dr. Eng. Muhammad Anshar – Universitas Negeri Malang, Indonesia
Dr. Mahardhika Prasetya Adji – Universitas Negeri Semarang, Indonesia
Dr. Eng. Khoeruddin – Universitas Negeri Yogyakarta, Indonesia

Scientific Committee

Prof. Sumarto – Universitas Pendidikan Indonesia
Prof. Bachtiar Hasan – Universitas Pendidikan Indonesia
Prof. Janulis Purba – Universitas Pendidikan Indonesia
Prof. Mukhidin – Universitas Pendidikan Indonesia
Dr. Enjang Akhmad Djuanda – Universitas Pendidikan Indonesia
Dr. Tasma Sucita – Universitas Pendidikan Indonesia

Organizing Committee

Dr. Ade Gafar Abdullah – Universitas Pendidikan Indonesia
Dr. Aip Saripudin – Universitas Pendidikan Indonesia
Dr. Isma Widiaty – Universitas Pendidikan Indonesia
Dr. Dadang Lukman Hakim – Universitas Pendidikan Indonesia
Agus Juhana, S.Pd – Universitas Pendidikan Indonesia
Diky Zakaria, S.Pd – Universitas Pendidikan Indonesia
Nia Amelia, S.Pd – Universitas Pendidikan Indonesia
Wildan Arasid, S.Pd – Universitas Pendidikan Indonesia

IOP Conf. Series: Materials Science and Engineering **384** (2018) 011001 doi:10.1088/1757-899X/384/1/011001

CONFERENCE PHOTOGRAPH

ISMEE
CONSTRUCTING THE ADVANCED
TECHNOLOGY FOR
FUTURE

Economic Dispatch (ED)
2017
ISMEE
CONSTRUCTING THE ADVANCED TECHNOLOGY FOR THE FUTURE

IOP Conf. Series: Materials Science and Engineering **384** (2018) 011001 doi:10.1088/1757-899X/384/1/011001

.bdg
for
2017
ISMEE
International Symposium on
Material and
Electrical Engineering
November 16th 2017
IOP Publishing
Conference Series
IET Insp
Compendex

IOP Conf. Series: Materials Science and Engineering **384** (2018) 011002 doi:10.1088/1757-899X/384/1/011002

Peer review statement

All papers published in this volume of *IOP Conference Series: Materials Science and Engineering* have been peer reviewed through processes administered by the proceedings Editors. Reviews were conducted by expert referees to the professional and scientific standards expected of a proceedings journal published by IOP Publishing.

IOP Conf. Series: Materials Science and Engineering **384** (2018) 012001 doi:10.1088/1757-899X/384/1/012001

Analysis of the Coverage Area of the Access Point Using Netspot Simulation

A T Suryani* and A B Pantjawati

Departemen of Electrical Engineering Education, Universitas Pendidikan Indonesia, Jl. Dr. Setiabudhi No. 229. Bandung, Indonesia

*atiasuryani@student.upi.edu

Abstract. Currently the ease of accessing the internet has become a fundamental requirements primarily a college surroundings. The requirement to always be connected to the internet network is seen with the growing number of access points that are installed to meet these needs. Installation of the access point is often not structured properly so that the area of scoped in the area of coverage is uneven. On the research of the simulation is done against the signal strength of the access point toward the receiver Wi-Fi in the room and the room of the student learning lecturer electrical engineering building FPTK UPI 3 and 4 floors. This research aims to map the territory of blankspot on that location, so that it can be used as a reference for optimizing the placement of access points. The research method used was the experiment through the drive test using Nestpot software. The analysis performed includes: Signal level, Signal-to-Interference ratio, and Quantity of Access point. Experimental results indicate the quality of the reception on location observed is not the same, some point receive very strong signals from more than one access point, while some other point still entered into the realm of blankspot. Of research results also gained some room located near some of the access point are having Signal to Interference Ratio-(SIR).

1. Introduction

The current utilization of access to internet network is already widely used. Request student and lecturer for the always connected to the internet network on the rise. The number of access point installed should be able to meet the needs, but the reality that there are currently mounting point of the access point was not arranged properly, so that the laying on of access point at some locations are outside the range of the area.

In line with the development of many current technologies utilized for communication through the medium of the internet. The current wireless network has been developing very rapidly with the latest technology refers to the IEEE 802. 11n [1]. At this time there have been many schools, colleges and agencies that utilize wireless networks. In planning a wireless network we need to know the condition of the room as well as a place to placing the acces point to work optimally, and every client in every room can capture the signals properly.

A study been done by A.R. Sandeep, y. Shreyas, Shivam Seth, Rajat Agarwal, and g. Sadashivappa [2], who is researching about the comparison Log Distance Path Loss Model and Empirical Indoor propagation Model (IEPM) in layout optimization Free Wi-Fi in the R.V. college of Engineering campus in Bangalore. The simulation is done by laying the Wi-Fi network anywhere within the campus.

IOP Conf. Series: Materials Science and Engineering **384** (2018) 012001 doi:10.1088/1757-899X/384/1/012001

This research was conducted to analyze wireless network in the building of the lecture with attention to parameters: Signal level, Signal-to-interference ratio, Quantity and access point.

2. Research method

The methods used in this research is a method of experimentation through a test drive using Netspot PRO Software. Nespot PRO serves as the scanning software used for Wi-Fi networks with the primary SSID parameter and can keep track of the strength of the signal, the Netspot PRO can scan an area and identify the quantity and strength of all access points in range.

Area Map – each survey site Netspot PRO is based on the location of that want scanned. The map must be floor plans, with all of the interior walls and other obstructions are marked. The map must be drawn to the actual scale so that accurate scanning. Map area can be loaded into the Netspot PRO from scanner or image program outside of, or a map can be drawn directly on the Netspot PRO, using Illustrator map function.

Access point serves as a transmitter that emits radio wave signals. AP (Access Point)-the wireless access points (such as a wireless router) that allows other wireless devices (such as a computer) to connect to a local area network using radio waves.

The purpose of the research method of the experiment was to investigate the room blankspot include Signal level, signal-to-Interference ratio and Quantity access point, the author did this experiment on space DPTE lecturer who was on the 3rd floor FPTK UPI as well as lecture room of electrical students are on the 4th floor FPTK UPI. These stages can be seen in figure 1.

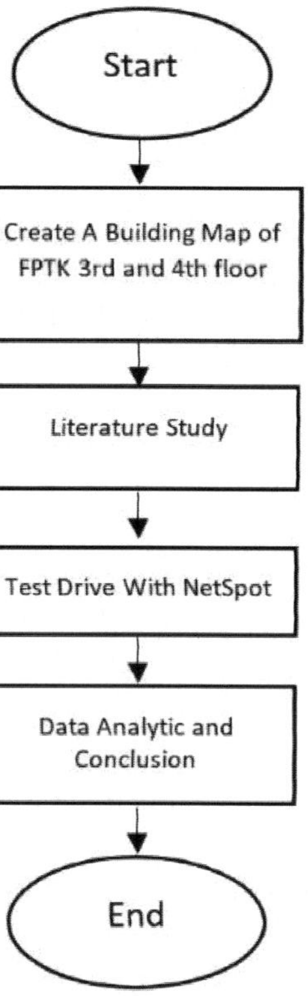

Figure 1. Flowchart.

The first step to be prepared is the scanning location plan. To be used on Nestpot PRO, the plan should be in .dwg or .jpg format.

The numbering the room on the floor plan is sequentially from numbers 1 to 9. The floor plan includes all the rooms, with all interior walls and other obstructions. Floor 3th is used for lecturers and administrative rooms. The floor plan can be seen in figure 2.

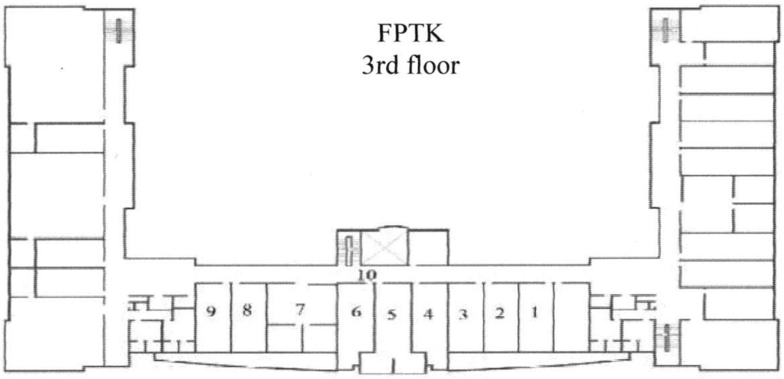

Figure 2. Room lecturer FPTK 4th floor UPI.

Floor 4th is used for electrical engineering student classrooms. It consists of four rooms and the space in front of the rooms. The Floor plan can be seen in figure 3.

The parameters examined:

Signal level is the value of the size of a computer connection to connect to the internet through a Wi-Fi network, strong signal effect with the speed of the network connection to the internet and are affected by the presence of a barrier that would keep the signal experienced the downturn.

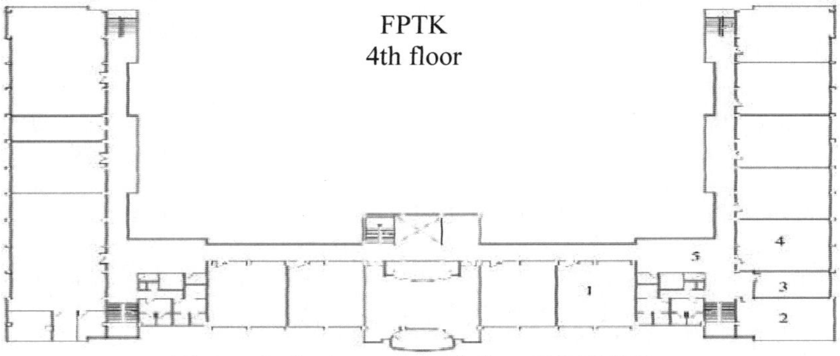

Figure 3. Student room 4th floor FPTK UPI.

Signal to Noise Ratio (SIR) is defined as the ratio between the power of the desired signal power noise (noise), and is widely used as a standard measure of the quality of the signals for communication systems [3]. A signal information as communication tools will experience a lot of disturbance by noise (noise), so that the signal can spoil it. Signal misbehaving suffered a decline in its quality. The quality of this signal can be determined from the value of the Signal to Noise Ratio (SIR) that is measured in units of decibels (dB). The quality of the signal can be distorted due to SIR there are many objects such as buildings, barrier walls etc.

Quantity of Access Point is the visualization to indicate the number of the access point is detected at each measurement point on the floor plan. One of the visualizations on Netspot PRO or Enterprise.

2.1. Test Drive using Netspot

A test drive is done using Netspot PRO software that has been installed on a laptop with runs on any room to the other and in a State ON to the software netspot, when doing a test drive on one of the room consists of a 6-7 point Heatmap, the author just took some SSID or that can be used by faculty and students of electro.

2.2. Data selection

After doing a test drive next the retrieved value signal in dBm or dB of some parameters obtained from netspot software including Signal level, Signal to Interference Ratio, Quantity, Frequency band Access Point coverage, PHY mode coverage, Download Speed, Upload Speed and the Wireless transmit rate. Not all parameters taken by the author so that the parameters taken by the writer there are only three parameters later related to the topics that are covered include: Signal level, Signal-to-Interference ratio, Quantity and Access Point.

2.3. Data analysis

The next step after the author specifies the parameter that is doing the analysis by means of entering a value signals into m. excel graph is then created to be easy to analyze equity SSID on the room.

3. Results and discussion

3.1. Signal Level

Signal strength can make quality reliable network, the signal level on a wireless reliability can be measured by knowing the value of a quantity specified dBm when scanning for Wi-Fi. The more positive approach the dBm value then the signal is getting good and reliable. Data retrieval is performed for 3 times and the average was taken. The results signal level from the SSID on every space lecturer FPTK electro 3rd floor can be seen in figure 4.

Based on Figure 4 the fourth access point can be seen gathered in Room 6, so only the nearest covered room i.e. Room 4, 5, 6 and 7. In an area of 1, 2, 3, 8, 9, and 10 not covered due to its distance away with 6. Should have been placed one access point between the rooms 7, 8, 9 so that the signal level is spread evenly. It is visible when a software simulation using Netspot default range set -96 dBm (minimum) and -10 dBm (maximum). If the signal level is approaching -10 dBm (maximum) then the signal level is said to be good as it approaches the maximum value. But the value of the average signal level is obtained by space 1.2 .7, 8, 9, and 10 i.e. above -60 dBm maximum value away from. The results of the signal level of each space lecturer FPTK electro floors 3 to a different SSID can be seen in figure 5.

Based on Figure 5 it can be seen that the uneven Division of the SSID on each room. The get the SSID with the value of the signal level above -62 dBm is room 3, 4, 5, and 6. Room 6 is a gathering place for the access point so that it looks to have a good level of signal and reliable. In Room 8 and 9 need to be placed in the room because of the access point is in the region of blankspot and no one of the SSID can be covered.

In accordance with the previous explanation, a good level of signal it is nearing its maximum point is -10 dBm. Future experiments conducted on FPTK 4th floor. The results signal level from the SSID on every Lecture Hall electro FPTK 4th floor can be seen in figure 6.

Based on Figure 6 can be seen there are five access points where the value of the signal level high above-60 dBm to each room, no one room is all covered by the access point. Therefore the lecture room on the 4th floor of electrical students could be said to blankspot area.

Need for the transfer of the position of the access point in the lecture room on the 4th floor of the students so that students can access the internet network. The results of the signal level of each lecture hall electro FPTK 4th floor to a different SSID can be seen in figure 7.

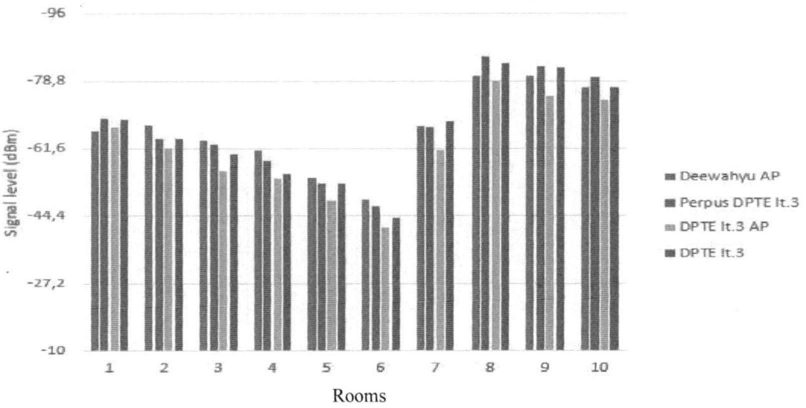

Figure 4. Comparison of the Signal level from the SSID on every space Lecturer FPTK Electro floor 3.

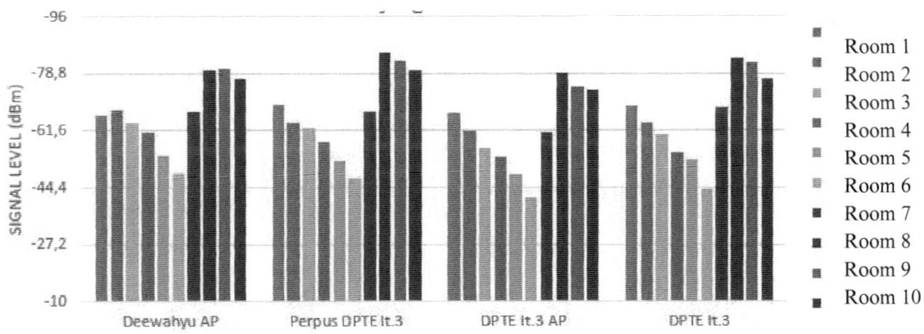

Figure 5. Comparison of the Signal Level of the Lecturer FPTK Spaces Every third floor to a different SSID.

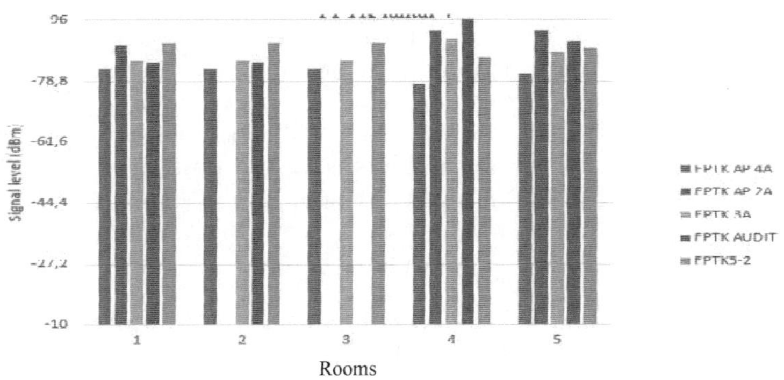

Figure 6. Comparison of the Signal Level of the SSID on the Lecture Hall Electro FPTK 4th floor.

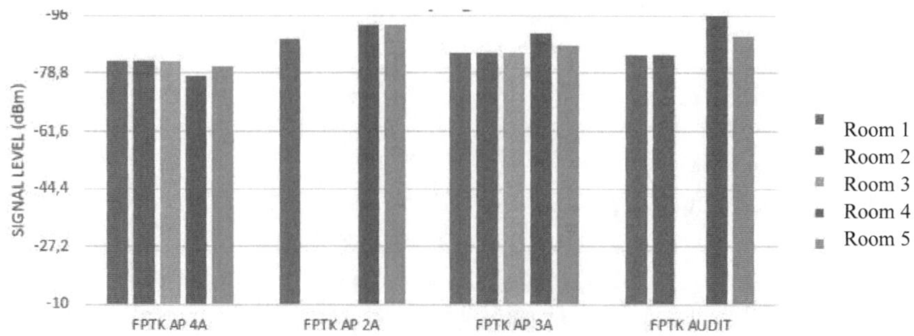

Figure 7. Comparison of the Signal Level of each lecture hall 4th floor FPTK Electro to a different SSID.

Based on figure 7 shows the placement of acces point here not near student lecture room electro. On the SSID FPTK AP AP FPTK 4A, 2A, 3A, and AP FPTK FPTK Audit, has the most low level signal under -70dBm.

3.2. Signal to-Interference ratio

Signal-to-Interference ratio (SIR) is a comparison between strong signals with interference signal strong total. Influence of interference is calculated using the parameters of SIR in a point with a certain distance from the access point. The value obtained from the comparison of SIR Received Signal Level (RSL) received from the access point (S) with a total of interference is received on a particular observation point [4]. The results of the signal-to-interference ratio of the SSID on every space lecturer FPTK electro 3rd floor can be seen in Figure 8, and lecturer FPTK electro space each to different SSID can be seen in Figure 9.

Obtained information that can establish a connection to SIR when SIR positive tend to approach the value, the ratio is the minimum value of the dB 10-15 received to establish the connection. Based on Figure 8 and 9 have a value, value tends to be negative SIR tends to be negative is influenced by the transmission of a mutual "collided, resulting in almost a nuisance. This transmission caused by collected SSID in Room 6. 3 Lt DPTE SSID on the AP has value high and at least SIR room 2 and 3 is positive, for SSID of DPTE library on 3rd floor value is at least SIR Room 8 and 9. The values in Figure 9 is still less than the ratio of the minimum acceptable for establishing the connection.

The results of the signal-to-interference ratio of the SSID on every student room electro FPTK 4th floor can be seen in Figure 10, and every student room electro FPTK UPI for a different SSID can be seen in Figure 11.

Based on Figure 10 and 11 have the value low, which tend to be SIR on the values FPTK AP SSID 2A low-value SIR 96 dB is located in an area of 2 was caused by obstructions and position the SSID of spaces lecture, for SIR on SSID FPTK 4A tend to be among the other SSID but its value is still less than the value of the minumun SIR reliable.

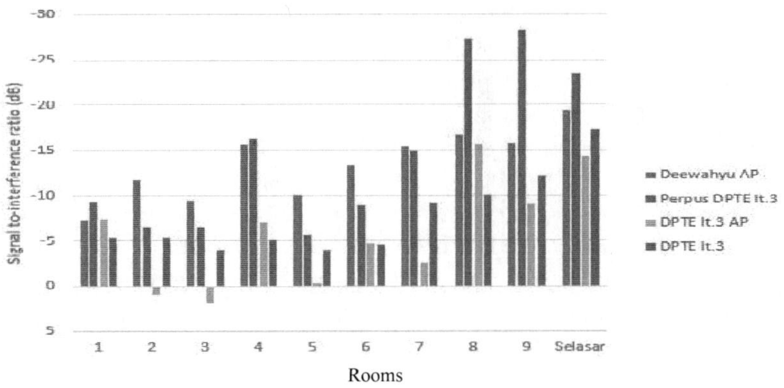

Figure 8. Comparison of SIR from the SSID on every space Lecturer FPTK Electro floor 3.

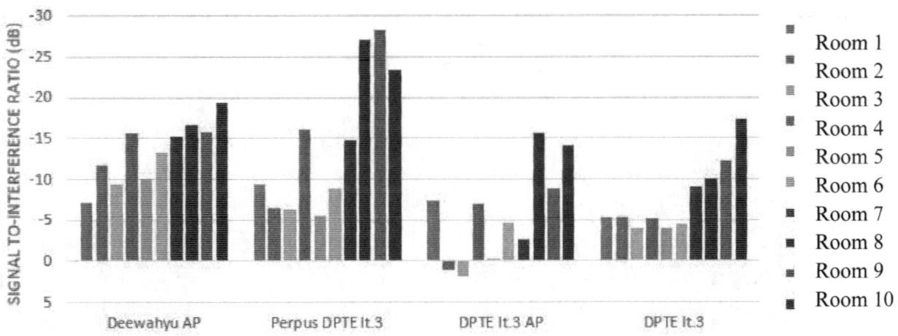

Fig 9. Comparison of SIR from every space Lecturer FPTK Electro floors 3 to a different SSID.

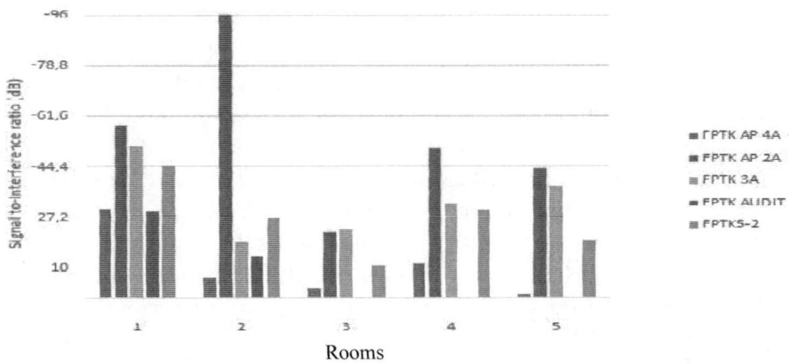

Figure 10. Comparison of SIR from the SSID on every space Lecturer FPTK Electro floor 4.

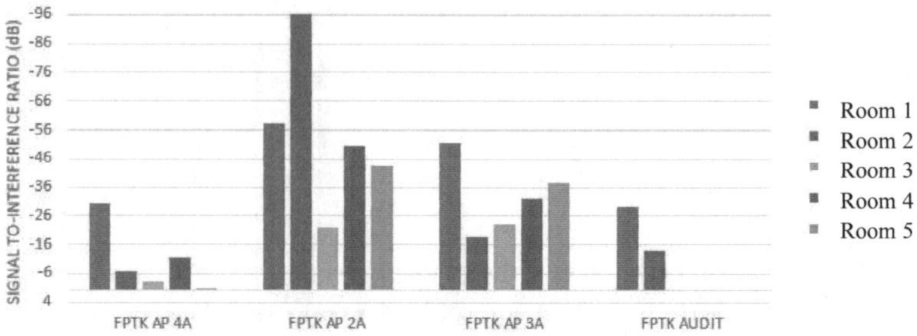

Fig 11. Comparison of SIR from the SSID on every Lecture Hall 4th floor FPTK Electro.

3.3. Quantity of Access Point

On this parameter to indicate the number of visualization is an access point that were detected at each measurement point on the floor plan. One of the visulisasi available on NetSpot PRO or Enterprise. The results of the Quantity of Access Point 3rd floor space electro lecturer can be seen in figure 12, and to lecture room of electrical 4 floors can be seen in figure 13.

Figure 12. Visualization of QAP 3rd floor.

Based on Figure 12 room professors tend to be orange which means it is rated 4, on Quantity this access point has a range from 1 to 5 with sampe description purple mean 0, light blue color has the value 1, green 2, yellow 3, orange 4, and the best is Red 5. On the 8th and 9th spaces tend to be blue are worth 0. Indicates the SSID is not covered on the Room 8 and 9 as well as the front verandah room lecturer FPTK electro UPI. In addition to Room 8 and 9, in room 1 and 2 its not everything tinted color orange because it is affected by the other four SSID SSID, in an area of 3, 4, 5, 6, 7 room that has a value of 4 indicates that the room can be covered by the four SSID.

Based on Figure 13 spaces and associated electrical students, tend to be colored light blue and purple signifies the value 0 and 1. There is 1 room on 2 SSID and yet still not be said good, room lectures students of electrical is entirely to be said in the blankspot by looking at the visuals in figure 13 [5][6].

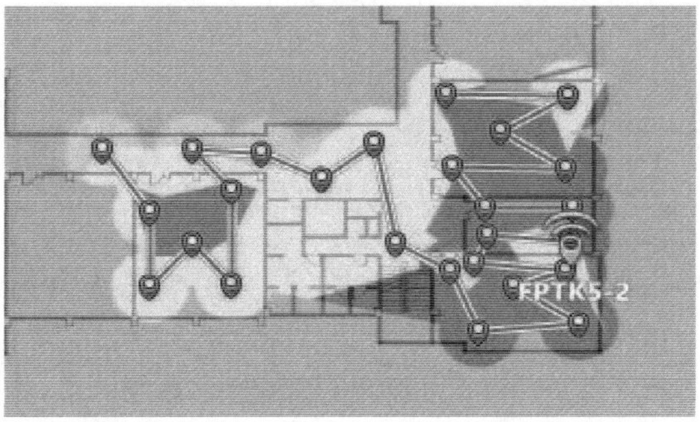

Figure 13. Visualization of QAP 4th floor.

4. Conclusion

Based on testing and analysis that has been done on this research, the following conclusions can be drawn: Based on the results of the simulation are represented on the graph can be known that the signal level on the 3rd floor is uneven for the entire space, there is a buildup of electrical lecturer SSID on the one room that causes beberapan other blankspot. On the 4th floor has a low level signal values above-70 dBm causes all the room in the region blankspot. On the 3rd floor space lecturer electro tends to be negative due to gather in an area 6 SSID so that transmission of mutual "collide" yet on SSID DPTE Lt. 3 AP value is high among others SIR. On the 4th floor student space electro has the value low at all SIR due to the placement of the access points that are too far away and eclipsed by room causes all of the room can not be covered by the Access Point.

References

[1] Crow BP, Widjaja I, Kim JG, Sakai PT. 1997. IEEE 802.11 wireless local area networks. *IEEE Communications magazine.* 1997 Sep;35(9):116-26.

[2] Sandeep AR, Shreyas Y, Seth S, Agarwal R, Sadashivappa G. 2008. Wireless network visualization and indoor empirical propagation model for a campus wi-fi network. *World Academy of Science, Engineering and Technology.* 2008;42:730-4.

[3] Lo YS. *Signal-to-Noise Ratio (SNR) estimation in additive Gaussian mixture noise channel* (Doctoral dissertation, Multimedia University).

[4] Henderson T, Kotz D, Abyzov I. 2008. The changing usage of a mature campus-wide wireless network. *Computer Networks.* 2008 Oct 9;52(14):2690-712.

[5] Battiti R, Brunato M, Delai A. 2003. *Optimal wireless access point placement for location-dependent services.* University of Trento.

[6] Köbben B. 2007. Wireless campus LBS: a test bed for WiFi positioning and location based services. *Cartography and Geographic Information Science.* 2007 Jan 1;34(4):285-92.

IOP Conf. Series: Materials Science and Engineering **384** (2018) 012002 doi:10.1088/1757-899X/384/1/012002

Design of a High Gain Circularly Polarized Microstrip Antenna for GPS Receiver

A Rozana* and T Hariyadi

Departement of Electrical Engineering, Universitas Pendidikan Indonesia, Jl. Dr. Setiabudhi No. 207, Bandung, Indonesia

*alyarozana@student.upi.edu

Abstract. Advances in communication technology using radio wave continue to increase including in Global Positioning System (GPS). This triggered the emergence of various kinds of research and development in the field of radio wave communication. One of the most widely developed devices today is the antenna. Various innovations and antenna technologies are emerging. In this paper, we will design a high gain microstrip patch antenna with circular polarization characteristics that can be used for GPS applications operating in the L1 1.575 GHz frequency region. The microstrip antenna design consists of a patch radiator with a dual-feed line in order to produce a circular polarization. The design process of this antenna starts with the determination of the antenna specification, calculate the antenna dimension, designs and simulation the antenna using CST Studio Suite software, realizes the antenna, and analyzes the simulation results and the measurement of the antenna. The simulation results show gain of our propose antenna design simulation on the L1 frequency ,that is: gain is 6.254 dB, the bandwidth of antenna is 290 Mhz from 1.366 GHz to 1.655 GHz for return loss less than -10 dB with return loss -20.77 dB and axial ratio is 2.38.

1. Introduction

Circular polarization is widely used in Satellite Communications Systems [1], and Global Positioning System (GPS) [2], because circularly polarized characteristics can reduce multipath effects and provide flexibility in the orientation angle between a transmitter and a receiver. There are a number of types of circularly polarized printed antennas. Most of them are reported as various patch antennas, which operate by using small perturbations providing the degenerate modes and 90 degree phase difference. Various techniques for broadband circular polarization has been reported using a radiating element, a parasitic element and a stacked structure [3]. Recently, various printed dipole antennas [4] have been realized due to several attractive features, such as broad bandwidth, lower profile, and lighter weight. In some references, a printed dipole antenna with integrated microstrip balun is used, and broadband characteristics are obtained. The dual-band printed dipole antenna is also reported, which is provided by two pairs of combined parallel arms. With a transverse dipole using rotated configuration then circularly polarized characteristics are met.

In this paper, a circularly-polarized printed dipole antenna is employed to provide circular polarization characteristics. The method used in this simulation is antenna cross-dipole. The effects of the parameters are discussed in the next sections.

2. Antenna Design

In the design of an antenna microstrip dipole, it must first be calculated the dimensions of the antenna to be made, especially the length of the dipole arm. To obtain the dimension of dipole microstrip antenna, we must know the parameters of the material to be used, that is dielectric thickness (h), and dielectric constant (ε_r). To calculate the length of the dipole microstrip arm, we must first calculate the effective Dielectric constant (ε_{eff}) of the microstrip using Eq. (1) [5]:

$$\varepsilon_{eff} = \frac{\varepsilon_r+1}{2} + \frac{\varepsilon_r-1}{2}\left[\frac{1}{\sqrt{1+12\frac{d}{w}}}\right] \quad (1)$$

$$W = \left(\frac{c}{2f\sqrt{\frac{\varepsilon_r+1}{2}}}\right) \times 0.1 \quad (2)$$

Where,

ε_r = dielectric Constant
W = width of dipole arm (mm)
f = center frequency
c = 3 x10^8 (m/s)
d = substrate thickness (mm)

Then the obtain the total length of the dipole arm using Eq. (3) [5]:

$$L = \frac{c}{2f\sqrt{\varepsilon_{eff}}} \quad (3)$$

Where,

L = length of dipole arm
ε_{eff} = effective dielectric Constant

First, we design a single dipole antenna, secondly after the results of a good single antenna simulation has met with the specification, we design the cross-dipole antenna. Figure 1 shows the geometry of the single dipole antenna. Figure 2 shows the geometry of the dipole-cross antenna. It can be seen from Figure 2(a) that the antenna element is simple and easy to manufacture, and also has good characteristics of circular polarization [6].

The antenna substrate is using FR-4 with permittivity 4.5 and thickness 1.6 mm. It is found, in the design process, that the permittivity has no significant effect on the antenna dipole cross performance. The location and size of the patch shown in Figure 1 and 2 will affect the resonant frequency of the antenna.

The dimension of the proposed single dipole antenna can be seen in Figure 1. The dimension of the patch single antenna dipole shown in table 1. And the proposed crossed-dipole antenna can be seen in Figure 2. The feeding network of the crossed-dipole antenna can be seen in Figure 2(d). The feeding network consists of microstrip structure, and the radius of the diameter is 90 mm. This feeding network that we used can create circular polarization, for which a 90° phase shift is used. A coax port with 50 Ω is used to feed the cross dipole antenna dipole with the ports of the feeding network just as shown in figure 2 (d).

Table 1. Dimension of the proposed single dipole antenna

Symbol	Value	Description
Hs	49.51 mm	Width of the substrate antenna
Wd	17.5 mm	Width of single antenna dipole
Ld	40.5 mm	Length of single antenna dipole
Wd2	33.6 mm	Length of antenna
Ld2	11.5 mm	Width of antenna
LF11	22.3 mm	Length of feedline network
LF12	14.5 mm	Length of feedline network
LF13	5.00 mm	Length of feedline network

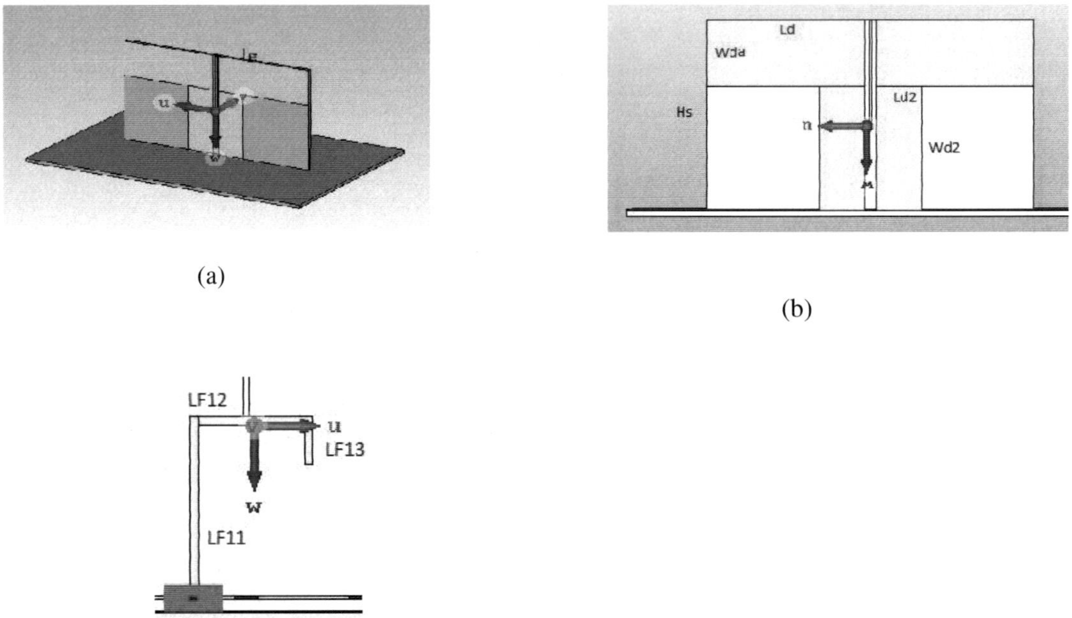

(a)

(b)

(c)

Figure 1. (a) Layout of the proposed single dipole antenna. (b) proposed printed single dipole antenna. (c) feedline network.

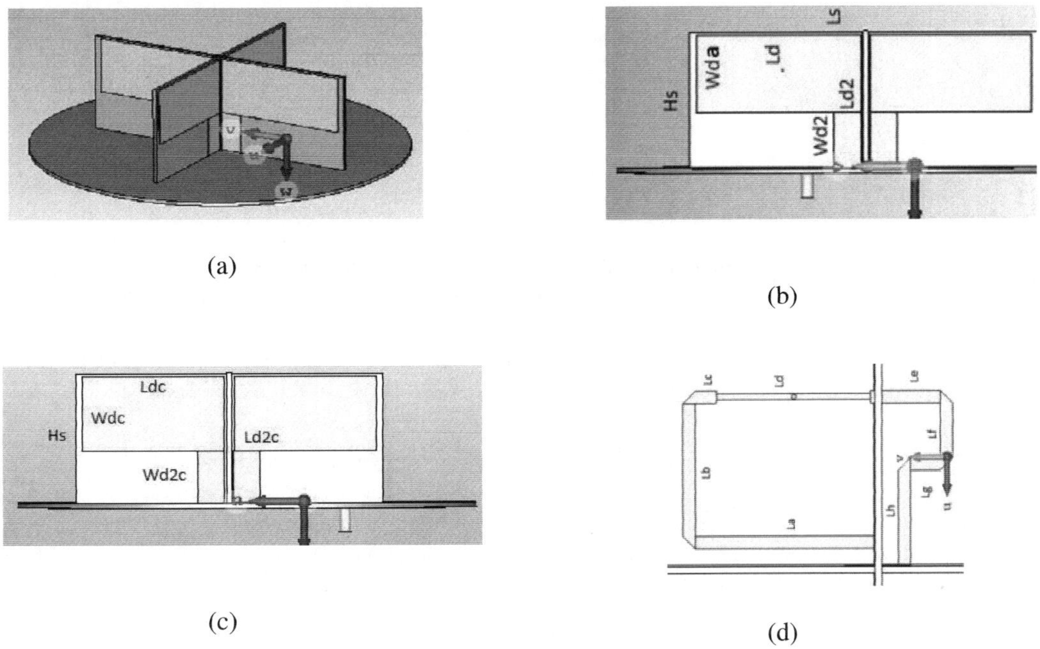

Figure 2. (a) Layout of the proposed crossed-dipole antenna. (b) Proposed printed dipole antenna 1. (c) Proposed printed dipole antenna 2. (d) Feedline network.

In order to create circular polarization in the axial direction of the antenna, it should have the same feed current amplitude with 90 degree phase of different and in the other direction could be elliptical polarization. The antenna height and tilt angle dipole were adjusted during the design process. During the design, simulation was performed on the proposed structure in order to optimize some important parameters.

3. Results and discussion

According to the optimized results using CST Studio Suite, the parameters of the proposed antenna can be seen in table 2.

Table 2. Dimension of the proposes crossed-dipole antenna

Symbol	Value	Description
Hs	47 mm	Width of the substrate antenna
Ls	122 mm	Length of the substrate antenna
Wda	27.1 mm	Width of antenna dipole 1
Ld	56.7 mm	Length of antenna dipole 1
Wd2	20.6 mm	Width of antenna dipole 1
Ld2	9.5 mm	Length of antenna dipole 1
Lsc	109 mm	Length of the substrate antenna
Wdc	27.1 mm	Width of antenna dipole 2
Ldc	50.7 mm	Length of antenna dipole 2
Wd2c	20.6 mm	Width of antenna dipole 2
Ld2c	9.5 mm	Length of antenna dipole 2
La	42.96 mm	Length of feedline network
Lb	35.2 mm	Length of feedline network

Table 2. Cont.

Lc	8.1 mm	Length of feedline network
Ld	37 mm	Length of feedline network
Le	17.2 mm	Length of feedline network
Lf	16.2 mm	Length of feedline network
Lg	10.4 mm	Length of feedline network
Lg	25.76 mm	Length of feedline network

Figure 3 shows the simulated return loss of the proposed antenna. The results show that the impedance bandwidth for return loss below -10 dB is from 1.366 GHz – 1.655 GHz. The VSWR of the antenna design in this simulation averaged less than 2 in the frequency range 1.366 GHz – 1.655 GHz, with the lowest value at the working frequency of 1.575 GHz is 1.2, as shown in Figure 4. The radiation pattern of proposed antenna is directional that can be observed as in Figure 5.

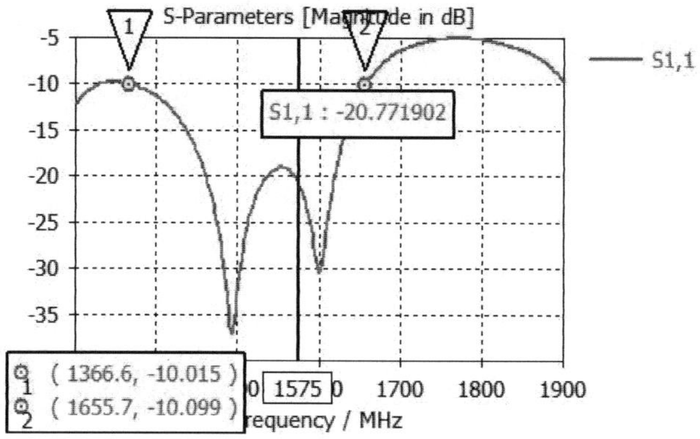

Figure 3. The simulation result of return loss for the proposed printed dipole antenna.

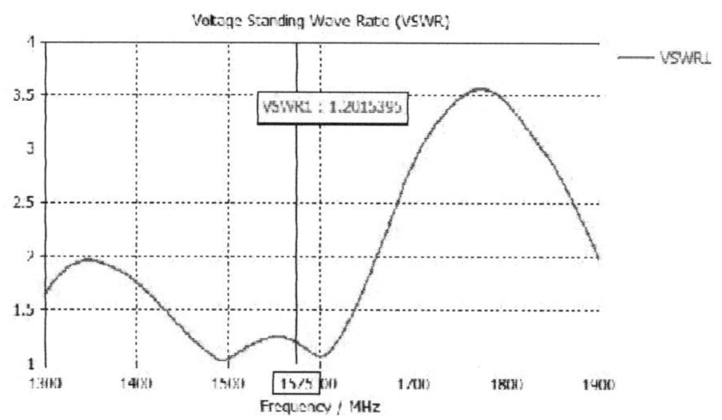

Figure 4. The simulation result of VSWR.

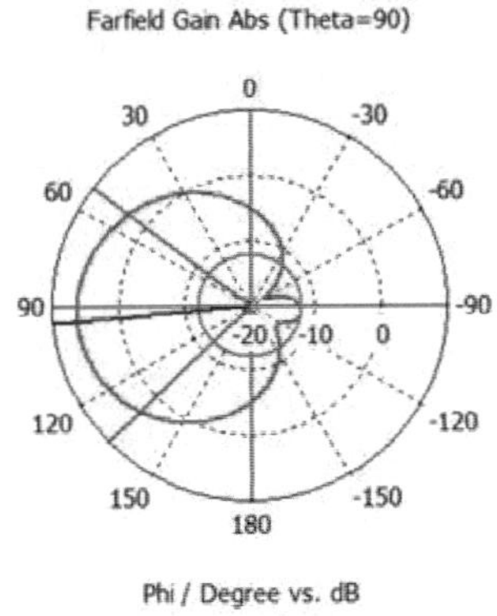

Figure 5. The Simulation result of radiation patterns for the proposed antenna at 1.575 GHz

In satellite communications, the receiving antenna generally uses circular polarization to be able to receive signals from satellites at each position, so that the design of the receiving antenna needs to be observed as well as the axial ratio bandwidth value, as in Figure. 6. To ensure circular polarization, the axial ratio is less than 3, if seen of the main beam, and in the range between 0 degree until 60 degree and the range 0 degree until -60 degree is less than 6 dB [7]. Figure 7 shows the radiation pattern and gain of dipole antenna. It can be seen from Figure 7 that this antenna has directional radiation pattern and 6.254 of gain at 1.575 GHz.

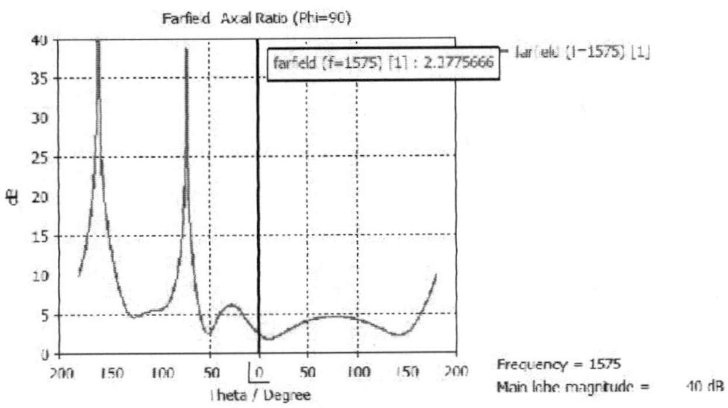

Figure 6. The simulation result of axial ratio for the proposed printed dipole antenna at 1.575 GHz

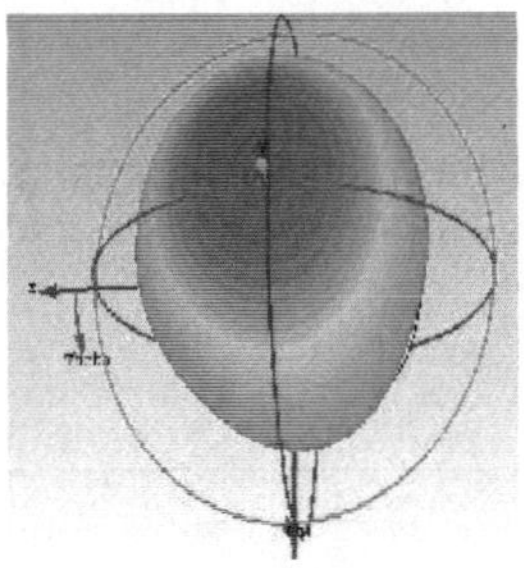

Type	Farfield
Approximation	enabled (kR >> 1)
Monitor	farfield (f=1575) [1]
Component	Abs
Output	Gain
Frequency	1575
Rad. effic.	-1.634 dB
Tot. effic.	-1.671 dB
Gain	6.254 dB

Figure 7. The simulation result of gain for the proposed printed dipole

The effects of some important antenna parameters are studied and discussed using CST Studio Suite. Several key parameters which affect the return loss of the printed dipole antenna have been studied as follows.

3.1. The width of the printed arms between frequency

The variation of the width of the dipole antenna arm effects the bandwidth impedance and the frequency range. When the width of the arm becomes wide, the changes in frequency and return loss is not very significant and bandwidth becomes wider as shown in Figure 8.

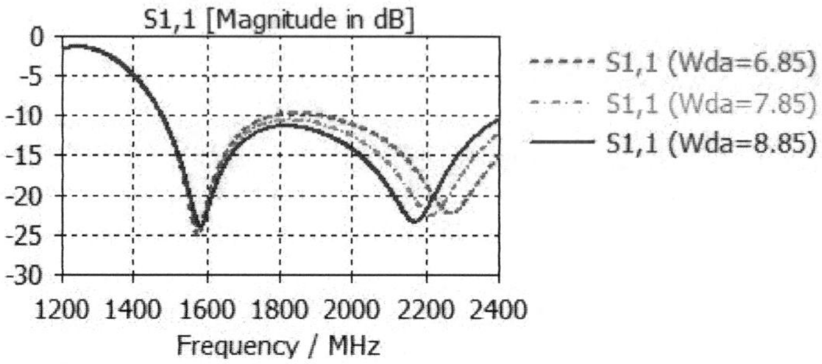

Figure 8. The simulation results of return loss with different arm width.

3.2. Length of feedline network before in cross

The Variations of the length of the network feedline affect the frequency range and bandwidth. When the network feedline becomes wide, the frequency will decrease slightly, and the bandwidth becomes narrower, as shown in Figure 9.

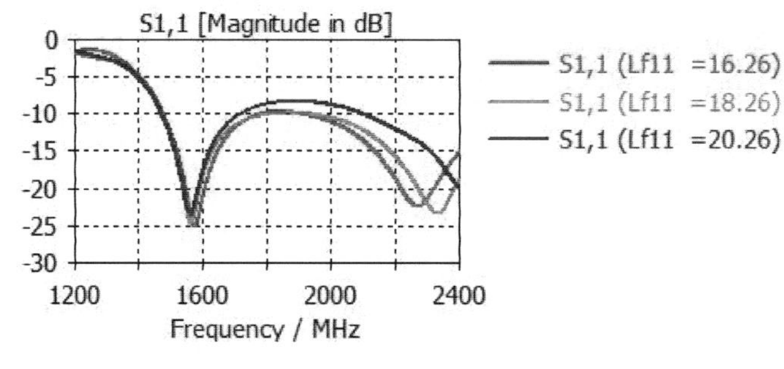

Figure 9. The simulation resuts of return loss with different length of feedline network

3.3. The width of the dipole antenna arm after the cross

The Variations of dipole antenna width affect bandwidth impedance and work frequency. When the width of the arm becomes wide, the frequency decreases and the bandwidth becomes narrower, as shown in Figure 10

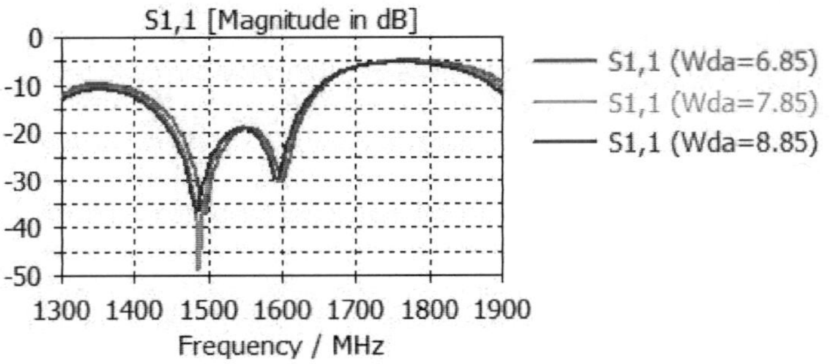

Figure 10. The Simulation results of return loss with different arm width after cross.

4. Conclusion

In this paper, we designed a cross-dipole antenna microstrip antenna that works at a frequency of 1.575 GHz for GPS receiver applications. Simulation results at 1.575 GHz has a return loss of -20.77 dB and VSWR 1.2. In addition, the bandwidth is 290 MHz with frequency range 1.366 GHz – 1.655 GHz for return loss less than -10 dB. The antenna gain is 6.254 dBi with Axial ratio less than 3 dB. The Axial ratio from -60 degree to 60 degree from main beam is less than 6 dB.

References

[1] F.Ferrero, C.Luxy, G.Jacquemod, and R.Staraj, DualBand Circularly Polarized Microstrip Antenna for Satellite Applications, IEEE Antennasand Wireless Propagation Letters, Vol.4, 2005, pp.13-15.

[2] X.Bao, M.J.Ammann, “A Printed Dipole Antenna for Wideband Circular Polarization Operation” IEEE Antennas and Wireless Propagation, Vol.5, 2006, pp.323-326.

[3] L.R.Li, J.Laskar, and M.M.Tentzeris, Wideband Probe-Fed Circularly Polarised Circular Loop Antenna, Electronics Letters, 2005, Vol.41, No.18, pp.997-999.

[4] B.Edward, and D.Rees, A Broadband Printed Dipole with Integrated Balun, Microwave Journal, May 1987, pp.339-344

[5] Jamaluddin, M.H.; Rahim, M.K. A.; Aziz, M. Z. A. Abd.; Asrokin, A. “Microstrip dipole antenna For Wlan Application”. Ieee Conference Publications, Publication Year: 2005, Page(S): 30 – 33.

[6] L. Wang, H. C. Yang, And Y. Li, “A Novel Printed Dipole Antenna Using In High Latitudes For Inmarsat”, Progress In Electromagnetics Research Letters, Jan 2011, Vol. 20, 37- 44.

[7] Cobham Antenna Systems, Microwave Antennas. 2012. Specialist Antenna Design and Manufacture Antennas for Thuraya Satellite/GPS/GSM. Online. http://www.cobham.com/communications-and-connectivity/antenna-systems/antennas-and-terminals-for-satcom/uhfls-band-satellite-antennas/uhf-l-and-s-band-antennas-for-satellite-applications/antennas-for-thuraya-satellite-gpsgsm/

IOP Conf. Series: Materials Science and Engineering **384** (2018) 012003 doi:10.1088/1757-899X/384/1/012003

Bottle Filling Simulation System Using Programmable Logic Control

A W Pradana and M Somantri*

Department of Electrical Engineering Education, Universitas Pendidikan Indonesia, Jl. Dr. Setiabudhi 207, Bandung 40154, Indonesia.

*msomantri@upi.edu

Abstract. The use of manual systems in the process of bottle filling is still often used, when using the manual process, then make the production is not maximal, and the volume is often different. The process is also done of course repeatedly, of course, takes a long time. The proposed tools are used to make bottle filling automatically to make the work faster and more efficient. In the process of filling the bottle on this prototype, the user can choose the option to fill the bottle. This prototype is PLC-based, PLC is considered capable to replace hundreds of relays, time etc., then PLC is considered mostly for the automation process. In this research, the method used is direct experimental method, which is done by research planning, control design, circuit, tool design, up to test phase, when tested in experiment 1 for 7 times experiment, with 1 second filling time, bottles have not been consistent, if averaged the result is 39,571 ml. then in the second experiment for 7 experiments, with a 2 second filling time, the water that filled the bottle has not been consistent, if the average yield is 108 ml.

1. Introduction

Technological developments are increasingly advanced. Industry changes its system from manual to automatic. Currently charging bottle in the industry are in desperate need of automation so that when charging can be faster and more efficient. Automation is the use of control systems that can reduce human needs in terms of work [1]. Automating tasks in the industry will increase productivity and reduce the possibility of errors in the system [2]. Some automatic bottle fill prototypes have many features already in use, such as a number of volumes and user-defined volume selection [3]. To make a process or a good system can use a variety of applied bias automation techniques [4]. Current automation techniques are used to improve systems, most industries use PLCs for industrial automation [4].

PLC is the easiest solution for the process automation system because of its less complexity than the relay logic system [4]. PLC is widely used in the automation industry because its functionality is able to replace hundreds of relays, time etc. [2]. With the automatic system all the process runs well, we just monitor whether the poses run in accordance with the desired. The prototype made is very suitable for home industry, and easy when in operation. The purpose of making this prototype is:

1. To help manual process be automatic
2. Help overcome the problem of the household industry to produce more products.
3. Helping the home industry to be able to compete with large industry.

2. Method

2.1. Research Planning
Based on some problems in the conventional industry regarding the bottle filling still using the manual system, it takes a tool that can simplify the work process in the bottle filling. In addition, there are several things to note namely, the estimation of needs, price estimates, then the application of the tool.

2.2. Program Design
At this stage, the software used to create the program on this prototype is Cx - Programmer, with programming languages using ladder diagrams.

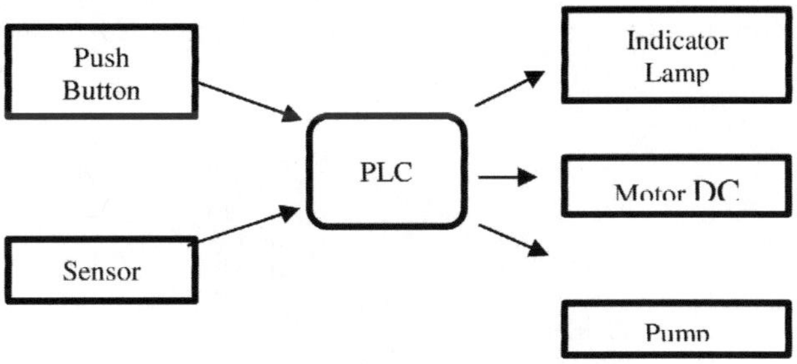

Figure 1. Flowchart Process Control

The process is done the selection of electrical components, the function of each component affects the working system of the control circuit used to operate this prototype.

- Electrical component working system
- The working capability of electrical components

From the figure 1 shows that the input section that is using push button and proximity sensor and for the processing is done by PLC, then at the output of the indicator light as indicator standby state, where this situation PLC get voltage source, then there is indicator light ready, this condition indicate that motor, and pump.

2.3. Tool Design
At this stage the process is to determine the components of equipment needed to form this prototype, at this stage, there are some things in a note such as:

- The working function of each tool
- Tool Characteristics

2.3.1. A subsubsection. In designing this prototype used a conveyor as shown in figure 2 is able to move the bottle to the place of charging, this conveyor is driven by motor 12 VDC, while the direction of this conveyor motion is horizontal. In order to move this dc motor in supply with a voltage of 12 VDC connected to the relay is set by the PLC to ON / OFF it.

Figure 2. Conveyor

IOP Conf. Series: Materials Science and Engineering **384** (2018) 012003 doi:10.1088/1757-899X/384/1/012003

2.3.2. Push Button. The push button is functioning as a circuit breaker or electric current from the power source, in this prototype, there is 3 push button function that is the first function as a stop button, where this button serves to stop the system being worked. Then the "A" and "B" buttons, these two buttons are the work command buttons for the system to run, only that distinguishes that when the bottle filling.

2.3.3. PLC (Programmable Logic Control). The PLC used in this study is the Omron CP1L - L20DR - A PLC, with 12 inputs and 8 outputs, where the voltage source can be 220 VAC and 24VDC. Omron type PLC is able to communicate with the program CX-Programmer.

2.3.4. Photoelectric Sensor. Photoelectric sensors as shown in figure 3 work based on the presence or absence of objects by using their capacitance change values. Sensors are used is able to detect objects both types of metals and non-metallic type, the working voltage of this sensor ranges from 10 VDC - 30 VDC.

Figure 3. Photoelectric Sensor

2.3.5. DC Motor. DC motor as shown in figure 4 is an electrical machine that requires electrical energy to work, where electrical energy is converted into mechanical energy. This mechanical energy can be used to rotate the fan, and also the conveyor.

Figure 4. DC Motor

2.3.6. Brushless DC Pump. This type of pump must be installed in the exposed state of water in the suction. This motor works at a voltage of 12 VDC, then the positive cable is marked in red and the negative cable is marked in blue as shown in figure 5.

Figure 5. Brushless DC Pump

2.3.7. MCB (Mini Circuit Breaker). MCB is used as a security when there is a short from the outside that can endanger PLC, MCB working principle that if getting excessive heat then bimetal in MCB will expand and break MCB. On this prototype used MCB 2 A.

2.3.8. Relay. The relay used is the type of Omron type MY2N - J, where the working voltage of the coil is used 24 VDC, which is capable of passing current 5A at 220 VAC relay working voltage, and 5A at a working voltage of 28 VDC.

2.3.9. Functional Test. This process is an early stage test process, where testing is done from every design that has been created and synchronizes with controls and tools that have been made. This is necessary to reduce the occurrence of errors, from the side of the program, system control and tools.

2.3.10. Testing Tool. This whole test process is a process whereby the incorporation of a program that has been designed in the application cx - programmer, then connected to the CP1L PLC that has been connected with the circuit controls and tools that have been made, below is a flow diagram of the work process of the tool made.

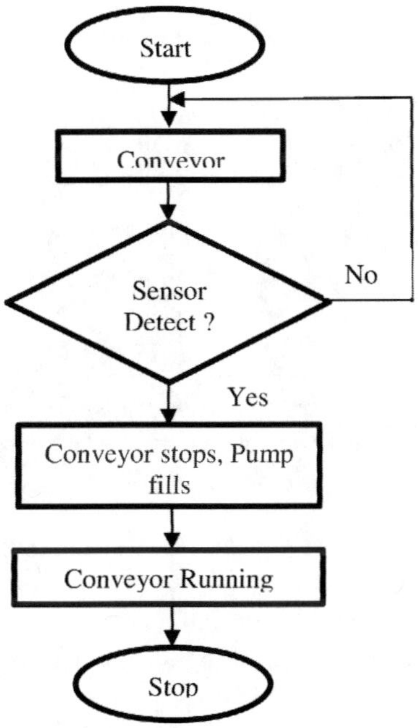

Figure 6. Flowchart Process

In Figure 6 it is explained that the bottle filling process takes place, where when starting the filling process, the bottle already placed on the conveyor will go to the charging area, when the photoelectric sensor detects an object then the conveyor will stop, then the pump will fill the bottle during the period which is set, when it is finished filling, the conveyor will re-run, this system uses a system loop (continuous).

This tool is designed to be able to perform two types of charging process, in process 1 working time bottle filling is set for 1 second, then for process 2 working time bottle filling is set for 2 seconds. As for the steps-steps to perform testing tools as follows:

Step 1: Connect the required supply voltage already installed either for PLC, or other equipment.

Step 2: Connect the input and output parts with PLC, other hardware must also be connected with PLC such as dc motor, sensors and brushless DC Pump.

Step 3: Select the option to adjust how much water will fill into the bottle. Make a selection by pressing a button that has been labelled.

Step 4: After we press the button then the conveyor will run, where on the top of the conveyor there is an empty bottle.

Step 5: When the bottle covers the proximity sensor the dc motor will stop. Then brushless DC Pump will work and fill the water for the time already set in the program.

Step 6: When the brushless working time of DC Pump is complete, Brushless DC Pump will stop then the dc motor will move back and rotate the conveyor belt.

Step 7: To stop the tool press the stop button.

3. Result and Discussion

From the design of the automatic bottle filling device, the tool works well, the system works in accordance with the already designed. This tool has 2 processes where the first process of the bottle runs over the conveyor, then after about the dead conveyor sensor and brushless DC pump will work and fill the bottle that is right under the charger.

Figure 7. Bottle Filling In Process 2

Figure 8. Bottle Filling In Process 1

Figure 7 explains that a bottle filling process is taking place, in this process the time spent to fill the bottle for 1 second. Figure 8 shows the test in process 2 where the time used to fill the bottle for 2 seconds.

Figure 9. Control Circuit

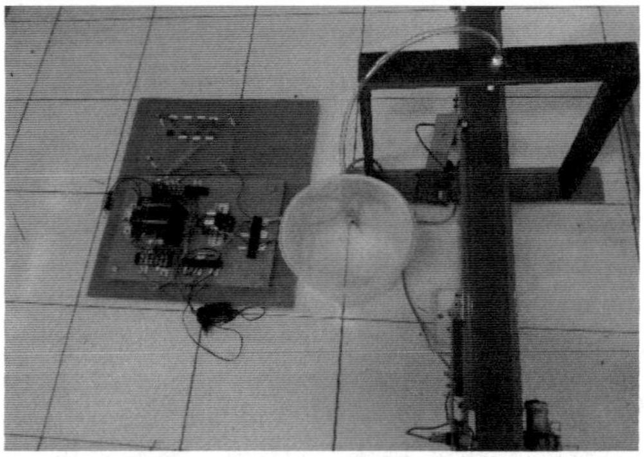

Figure 10. Look At the Bottle Filling Tool

Figure 9 is a control set made in accordance with the plan, while figure 10 is a top view of a tool already made. After the finished tool is made, it is done several times experiments on 2 charging process system. The following table results from experiment 1:

Table 1. Test Results during 1 Second

No Experience	Volume
1	40 ml
2	40 ml
3	39 ml
4	40 ml
5	39 ml
6	39 ml
7	40 ml

In table 1 is done 7 times, with charging time for 1 second. Seen from the existing data that there is volume inequality during charging, for experiments 1 and 2 volumes obtained that is 40 ml, then in 3 volumes experiments obtained by 39 ml, then in experiment 4 volumes obtained for 40 ml, then on experiments 5 and 6 volumes obtained 39 ml, and in the 7th experiment again get a volume of 40 ml. From the data obtained also that the volume produced during the 7 experiments ranged from 39 ml - 40 ml. if averaged - average during the experimental process then the average is 39,571 ml.

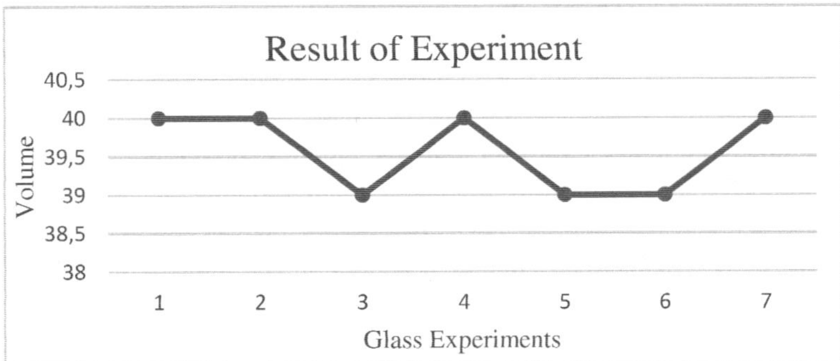

Figure 11. Graphic Result of Experiment during 1 Second

In Figure 11 showed that in glass experiment 1 and 2 were obtained the same data that were as much as 40 ml. Then, in glass experiment 3 underwent a decreasing amount of volume that was only about 39 ml. then on glass experiment 4 had increased again that was equal to 40 ml, then at experiment of glass number 5 and 6 again got the decreasing result about 39 ml, and for the last data taken the volume of water obtained in glass 7 that is equal to 40 ml.

Then at the time of the experiment, there were several bottles not detected by the sensor just because the bottle was far out of the distance range of the sensor. Furthermore, when a bottle had a larger size, many times the sensor read out the object 2 times. So that for 1 bottle, can charge 2 times. In addition the photoelectric sensor, the test also uses a capacitive probe type, when the experiment is often filling failure due to the distance of the bottle with the sensor far enough, the ability of the sensor to detect the object only about 4mm, therefore the sensor selection is needed to be able to fill the bottle with maximum.

4. Conclusion

Based on the experiments conducted and from the data obtained, in experiment 1 where this experiment was conducted at the time of filling the bottle for 1 second, that the data obtained was inconsistent, where the data ranged in the number 39 ml - 40 ml. then in go back to experiment 2 where the setting time for the process of bottle filling for 2 seconds is also obtained result so variation, that is 105 ml range. 109 ml, and 110 ml.

In the experimental execution several times there is an error where the sensor does not detect the coming of the bottle, so there is a bottle that is not filled with water, this proves that the photoelectric sensor used during the manufacture of this tool is still experiencing miss in its use. For that better sensor selection is needed, in order to reduce the error rate and every bottle that passes can be filled.

The results of the reading in this experiment can be in some inequality, this is because when testing the bottle that has been filled with water, the reading using measuring glass as a measure, thus causing inaccuracies in the reading. Therefore, the use of digital measuring instruments is necessary to reduce the error rate when reading the results of each time the experiment. Overall the tool works well, the system control works well, and for the components used

References

[1] N. V. Sabnis, 2017. *Improvement in Productivity of Bottle Filling Operation by using Multi-Nozzle PLC System,* vol. 10, no. 1, pp. 724–727, 2017.

[2] M. Sreejeth and S. Chouhan, 2017. PLC based automated liquid mixing and bottle filling system, *1st IEEE Int. Conf. Power Electron. Intell. Control Energy Syst. ICPEICES 2016*, pp. 1–5, 2017.

[3] K. Thiyagarajan, R. Meenakshi, and P. Suganya, 2016. Vision based bottle classification and automatic bottle filling system, *2016 Int. Conf. Adv. Hum. Mach. Interact. HMI 2016*, pp. 40–42, 2016.

[4] J. Pannu, R. Kulkarni, and M. S. B. Ranjana, 2016. On the automated multiple liquid bottle filling system, *Proc. IEEE Int. Conf. Circuit, Power Comput. Technol. ICCPCT 2016*, pp. 1–3, 2016.

IOP Conf. Series: Materials Science and Engineering **384** (2018) 012004 doi:10.1088/1757-899X/384/1/012004

Android Mobile Application for Heart Rate Measurement and Monitoring

A I S Hamdani, W Purnama and D Wahyudin*

Department of Electrical Engineering Education, Universitas Pendidikan Indonesia, Jl. Dr. Setiabudhi 207, Bandung 40154, Indonesia.

*deewahyu@upi.edu

Abstract. It is an undeniable fact that technological advance in health aspect holds a very important role body because health is one thing that humans should be concerned about for living a proper life. One of the most vital parts of the human body is the heart. A heart condition can be measured by heart rate. Heart rate is the number of times of a human heart beats per minute that is usually called bpm (beats per minute). Due to the importance of knowing the heart condition every time, it is Necessary for the human to have a helping device to check the heart rate every time without going to a hospital or seeing the doctor first. This device and application is supposed to be easy to use and effective for daily life. In addition, this heart rate measuring device is designed using a microcontroller and Android phones.

1. Introduction

Development of the advanced technology has been growing rapidly as it affects many aspects of our lives, especially in the health sector. It can be seen by numerous tools, machines and even software that can be used by humans to check, examine and maintain their health [1], [2].

Technological progress in the health sector is inevitable. Body health is something that is essential for humans to survive in this world. One important part of the human body is the heart which is a vital organ to pump the blood that will be distributed throughout the body. The heart works continuously and repeatedly. Heart's condition can be detected by measuring the heart rate. [3]–[5]. The heart rate measurement is done by measuring the number of times the heart beats per minute which is usually shortened by bpm. The heart rate can vary according to the body's physical needs, gender, age, body mass, activity, etc. Sudden change in heartbeat can indicate heart failure that needs a medical treatment [6], [7].

However, a lot of research and projects have been carried out either nationally or internationally to invent something related to heart rate measurement. Several previous research has succeeded to measure heart rate using the pulse sensor. These studies mostly use The Arduino as a microcontroller and Bluetooth as data communication. In this project, the ESP 8266 is used as a microcontroller, and In this project, the ESP 8266 is used as a microcontroller as well as data-communicating tool using Wifi.

Based on the background and previous research review, it is necessary to have a tool to monitor the heartbeat at any time without having to go to a hospital or doctor. This tool should be simple and effective for everyday use like a smartphone that is always used by humans. Based on the exposure of the problem and review of research described above, a simple and effective tool for heart rate measurement has been created in this project. This tool uses pulse sensor, ESP 8266 as a microcontroller and data-communicating Wifi as well. After the trial, the error rate of this tool is 2,37 %.

2. Design and Methods

2.1. Basic theory

2.1.1. Heart. The heart is one of the most important organs of the human body. Heart plays a crucial role in performing daily activities. The heart functions as a pump in the circulatory system to provide a continuous flow of blood throughout the body and to transport O^2 and other elements needed by the muscle in order to actively move in daily basis [4].

2.1.2. Heart rate. The heart rate measures the number of times the heart beats per minute which is usually shortened by bpm. The heart rate can vary according to the body's physical needs, gender, age, body mass, activity, as shown in Table 1 and Table 2 [6].

Normal heartbeat ranges from 60 to 100 BPM This table is a classification of heart rate based on frequency and age and gender [8].

Table 1. Heart Rate for Male

Resting heart rate chart for Men						
Age	Heartbeats Per Minute					
	Excellent	Good	Above Average	Average	Below Average	Poor
18-25	56-61	62-65	66-69	70-73	74-81	82+
25-35	55-61	62-65	66-70	71-74	75-81	82+
36-45	57-62	63-66	67-70	71-75	76-82	83+
46-55	58-63	64-67	68-71	72-76	77-83	84+
56-65	57-61	62-67	68-71	72-75	76-81	82+
65+	56-61	62-65	66-69	70-73	74-79	80+

Table 2. Heart Rate for Female

Resting heart rate chart for Women						
Age	Heartbeats Per Minute					
	Excellent	Good	Above Average	Average	Below Average	Poor
18-25	56-61	62-65	66-69	70-73	74-81	82+
25-35	60-64	65-68	69-72	73-76	77-82	83+
36-45	60-64	65-69	70-73	74-78	79-84	85+
46-55	61-65	66-69	70-73	74-77	78-83	84+
56-65	60-64	65-68	69-73	74-77	78-83	84+
65+	60-64	65-68	69-72	73-76	77-84	84+

2.2. Component

2.2.1. ESP 8266

Figure 1. ESP 8266 NodeMCU

ESP8266 is a microcontroller designed by Espressif Systems. As shown in Figure 1, ESP 8266 is a solution for independent Wifi network which becomes a bridge to connect the available microcontroller to the Wifi and to run its own application. ESP8266 modules which have a variety of input and output pins are also equipped with a USB connector. Just like Arduino, ESP8266 can be easily connected to a laptop using a USB cable [9], [10].

2.2.2. Pulse Sensor

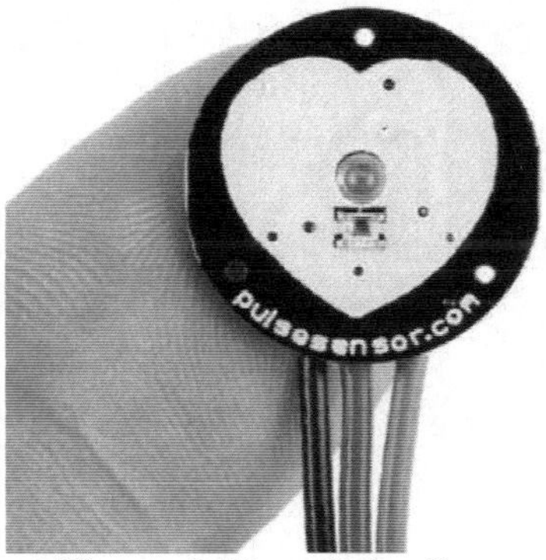

Figure 2. Heart Pulse Sensor

Pulse Sensor is a plug-and-play detector designed to be applied in Arduino. To analyze the heart rate, the sensor is placed in a certain body part, such as a fingertip, earlobes, and temples. Once it is programmed, the pulse sensor will soon scan the heart rate in the real time.

Pulse sensor has three legs named Pulse, Negative, and Positive Supply. Three to five volt is said to be the ideal voltage level for this sensor. This sensor, as shown in Figure 2, can only absorb the power up to 4 mA. That is why this sensor is suitable for mobile device [11].

2.2.3. OLED LCD

OLED is a Light-Emitting Diode which it's electromagnetic layer is made by organic compounds that emit light in response to an electric current. The organic semiconductor layer is located between two electrodes, one of which is a transparent electrode. OLED is used to create a digital display on a device, such as television screen, computer monitor, and portable systems, for example, mobile phone, handheld game console, and PDA. In contrast to LCD, OLED does not need back light; it glows on its own. OLED's waste is much lower than LCD's, making it very suitable for portable device [12].

2.2.4. Battery / Power Supply

The battery is a device that can convert the chemical energy contained in the active ingredient components of the battery into electrical energy through an electrochemical reaction of reduction and oxidation [13].

2.3. Design

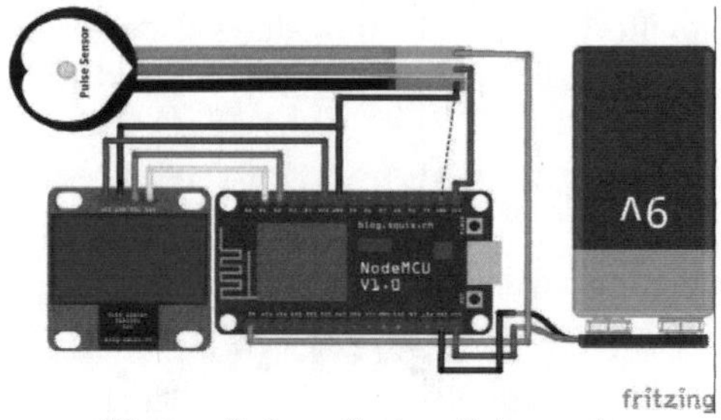

Figure 3. Installation Schematic

Figure 3 shows a schematic of the circuit design and measurement of heart rate is. Table 3 is a scheme of the installation of such equipment.

Table 3. Installation Schematic

ESP 8266	Pulse Sensor	OLED LCD	Battery
Vin			+
Gnd	-	GND	-
3V	+	VCC	
analog 0	Signal		
digital 1		SDA	
digital 2		SCL	

2.4. Block program

Block diagram of the heart rate measuring device is shown in Figure 4. The block diagram of parts and their functions can be seen below:

1. Power supply serves as the power supply to all parts except android smartphone.
2. Screen LCD is used to display the heart rate measurement.
3. The pulse sensor is a heartbeat sensor.
4. The esp8266 microcontroller is used as sensor data processing and data transmission intermediaries.
5. Android smartphone as the heart rate measurement data viewer.

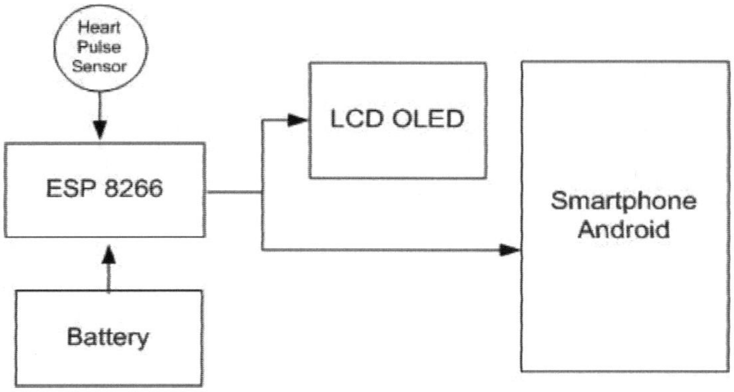

Figure 4. Block Diagram

2.5. Flow chart

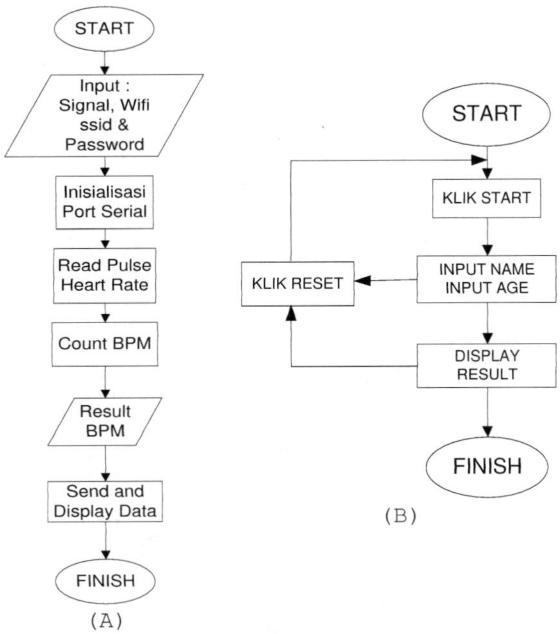

Figure 5. Flow chart

Figure 5 (a) shows the flowchart for the process of reading and measurement of the pulse sensor. Figure 9 (b) shows the flowchart for the usage of the application on the Android smartphone.

2.6. Programming

Programming device using Arduino IDE. Arduino IDE (Integrated Development Environment) is software developed by the Arduino is used to design various processes associated with the Arduino programming[15].

Arduino app was created using App Inventor. App Inventor is an open source web app originally developed by Google and is currently managed by the Massachusetts Institute of Technology (MIT). App Inventor allows new users to create programs on the computer to create software applications for the Android operating system [16].

2.7. Data collection

The process of data completion is done by a test method. This tool is tested to ten people at random sampling. They were asked to use this tool and see the results of heart rate measurement. Sensors on the tool are mounted on the index finger. The comparison is done by conventional heart rate, by calculating the pulse next to the wrist.

3. Results and Discussion

3.1. Android Application Test

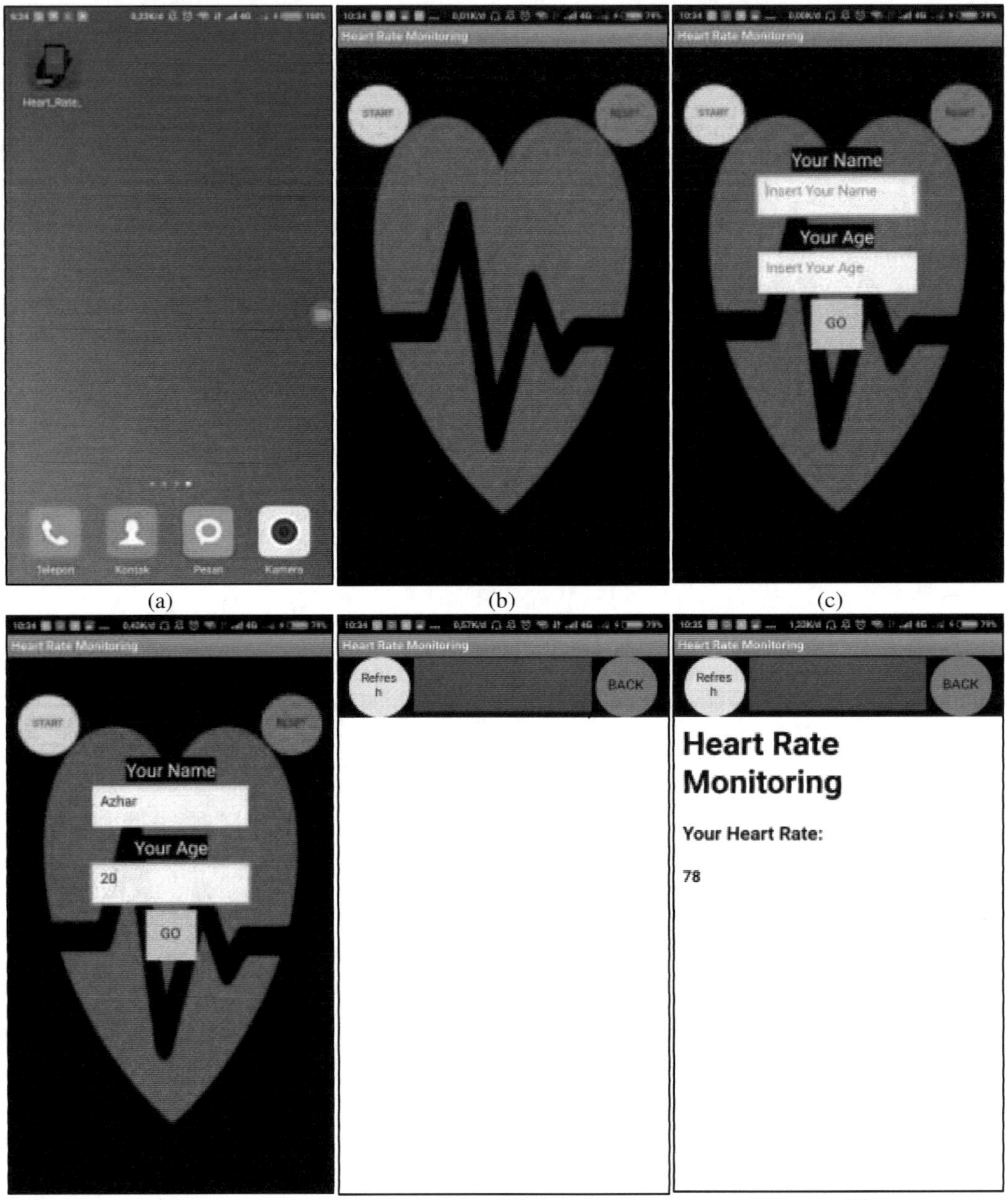

(d) (e) (f)

Figure 6. Android Application Display

Figure 6 is an android app view of this heartbeat measurement and monitoring tool. Figure 6 (a) is an icon display of an application called Heart Rate Measurement. Figure 6 (b) is the initial display when the application is opened, aka two key tones i.e., Start to start and Reset to reset. Figure 6 (c) is the view when the Start button is pressed; the user is prompted to enter the name and age of the column provided.

Figure 6 (d) is the view when the name and age have been entered, and the Go button can be pressed. When the Go button is pressed, it will go to the next screen like picture 6 (e). Figure 6 (f) is the display when the Refresh button is pressed which will bring up the measurement result of a heartbeat. When the Back button is pressed, then the display will return to view in Figure 6 (b).

3.2. Tool Measuremetn Result

In Table 4, it can be seen the results of varying heart rate measurements. Measurements are done by putting the sensor on the user's index finger; then the sensor will start reading and sending it to the smartphone. The sensitivity of the sensor is very high, so it takes some time for the sensor reading value to be stable. Due to the rapidly changing value of sensor readings, the heart rate taken is of minimal value and the maximum value that often appears at the time of measurement. Heart rate from tool measurements ranges from 70 to 95 BPM. This means that the measured heart rate can be said to be normal because the normal BPM ranges from 60 to 100 BPM. The variation in heart rate is influenced by various things, such as gender, age, and activity performed before or during measurement.

Table 5 and Figure 7 show the comparison between the measurement results of the tool and the measurement of the heartbeat manually. Measurements are done manually by measuring the pulse on the wrist for 1 minute. It can be seen that the difference between the two measurements is ranging from 0 - 4. From the difference is obtained error for each measurement. The average error of 2.37%, this means that the heart rate measurement tool is quite effective because it has a relatively small error. Figure 8 is a signal form of the heartbeat read by the tool that has been created. The signal on Serial Plotter software in Arduino IDE.

Table 4. Measurement Device Result

NO	Gender	Age	BPM		Mean
			Min	Max	
1	Female	22	91	99	95
2	Female	22	79	85	82
3	Female	21	79	85	82
4	Male	21	85	91	88
5	Female	21	91	99	95
6	Male	22	66	74	70
7	Female	22	85	91	88
8	Male	12	70	74	72
9	Female	17	71	85	78
10	Male	24	71	79	75

Table 5. Comparison Result

NO	Device Measurement Result	Manual Measurement	Difference Measurement	Error (%)
1	95	96	1	1.05
2	82	80	2	2.43
3	82	84	2	2.43
4	88	84	4	4.54
5	95	96	1	1.05
6	70	72	2	2.85
7	88	88	0	0
8	72	74	2	2.77
9	82	80	2	2.56
10	75	78	3	4

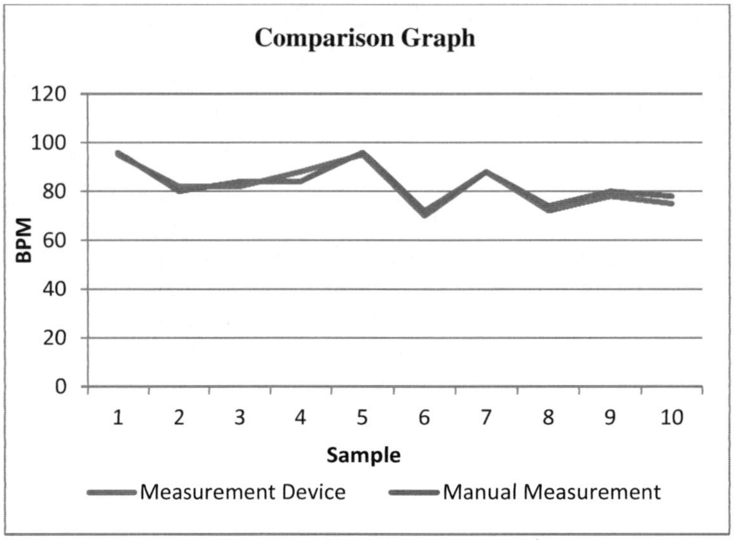

Figure 7. Comparison Graph

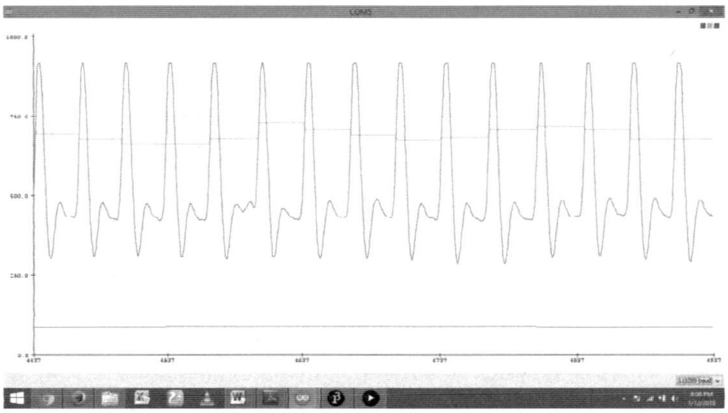

Figure 8. Heartpulse signal

4. Conclusion

Pulse sensor was had a high sensitivity and takes a long time for the reader can be stable. In addition, NodeMCU used in this system serves as a data transmission with using Wifi connection is fast and easy to use. The android applications can display the heart rate of the device, in contras it cannot use for displaying heartbeat signals. Based on the results of the evaluation, this application is not in accordance with the results of manual measurement because there were some errors.

References

[1] G. W. Wohingati, A. Subari, F. Teknik, and U. Diponegoro, "ARDUINO UNO R3 YANG DIINTEGRASIKAN DENGAN BLUETOOTH," vol. 17, no. 2, pp. 65–71, 2013.

[2] H. Heruryanto, W. B. Nurdin, and B. Armynah, "Sistem Pengukuran Denyut Jantung Berbasis Mikrokontroler ATMega 8535." Makassar, Indonesia.

[3] Y. Ostchega, D. Ph, K. S. Porter, J. Hughes, and C. F. Dillon, "Resting Pulse Rate Reference Data for Children , Adolescents , and Adults : United States , 1999 – 2008," no. 41, pp. 1999–2008, 2011.

[4] A. Karya, T. Ilmiah, and I. E. M. S, "Perubahan denyut nadi pada mahasiswa setelah aktivitas naik turun tangga."
[5] N. Chirakanphaisarn, T. Thongkanluang, and Y. Chiwpreechar, "Heart Rate Measurement and Electrical Pulse Signal Analysis for Subjects Span of 20-80 Years," pp. 70–74, 2016.
[6] A. H. Association, "All About Heart Rate (Pulse)," All About heart Rate (Pulse), 2015. .
[7] H. Mansor, S. S. Meskam, and N. Sakinah, "Portable Heart Rate Measurement for Remote Health Monitoring System," no. June 2013, pp. 0–4, 2015.
[8] D. V. Marchione, "Healthy Resting Heart Rate By Age for Men and Women," 2016. [Online]. Available: https://www.belmarrahealth.com/resting-heart-rate-chart-factors-influence-heart-rate-elderly/.
[9] H. Connections, "ESP8266 WiFi Module Quick Start Guide."
[10] E. Systems and I. O. T. Team, "ESP8266EX Datasheet," 2015.
[11] P. Sensor, T. Pulse, and S. Kit, "Pulse Sensor Getting Started Guide."
[12] D. Matrix, "HTDS Series Single Color OLED Display."
[13] A. Satriady, W. Alamsyah, A. H. I. Saad, and S. Hidayat, "Pengaruh Luas Elektroda Terhadap Karakteristik Baterai Lifepo 4," vol. 6, no. 2, pp. 43–48, 2016.
[14] Android Overview _ Open Handset Alliance.
[15] Arduino, "Arduino - Environment." 2017.
[16] M. I. of Technology, "About Us _ Explore MIT App Inventor." .

IOP Conf. Series: Materials Science and Engineering **384** (2018) 012005 doi:10.1088/1757-899X/384/1/012005

UAV Vision System for Rescue Payload Delivery

D Mardiansyah* and A H S Budi

Department of Electrical Engineering Education, Universitas Pendidikan Indonesia, Jl. Dr. Setiabudhi 207, Bandung 40154, Indonesia.

*dicky.mardiansyah8@gmail.com

Abstract. Disaster can happen anytime and anywhere. First aid should be done as soon as possible, but not all disaster locations can be accessed easily. Therefore, it needs a vehicle that can deliver a rescue payload quickly and accurately to the location that difficult to access by humans. This research aims to create a multirotor Unmanned Aerial Vehicle (UAV) that can deliver rescue payload or another payload automatically to a specific location. The multirotor type used is a hexacopter. It can carry a maximum payload of 2 Kg. The flight control (FC) used in the UAV is Pixhawk. FC used to control the UAV automatically based on the coordinate input and the specific direction. To determine the target of payload shipment, the UAV equipped with a webcam that can track the shipping target by capturing colour and shape. The UAV use the raspberry pi to process the image then send instructions to the pixhawk by using Mavlink and Mavproxy communication to drop the payload in the desired place.

1. Introduction

Nowadays the use of Unmanned Aerial Vehicle (UAV) has grown rapidly. The UAV, also known as Drone used in various needs such as military, Search and Rescue (SAR), photography, agriculture, courier, and others. Autonomous flight of the UAV has attracted much research efforts during the past few years [1]. In the field of SAR that concerns on human safety is required to always standby in overcoming the disasters. Speed and accuracy are the main concern for SAR. However, some disaster locations are very difficult and risky to access by humans. U.S. Coast Guard has expressed their interest in the use of UAVs to aid its helicopter in maritime or SAR missions [2]. In addition, UAVs can increase the searching area using the onboard infrared camera, by reducing the risk of human safety with a low-cost budget.

Autonomous required many abilities such as evaluate global positions, altitudes and movements to control their stability [3]. However, autonomous systems still have some lacks, especially in position systems that only use GPS sensor [4]. Therefore it needs a system that can handle the lacks of autonomous UAVs that only use GPS sensor. One of the methods to reduce the GPS weakness is to use a computer vision system. Research about this vision system has been developed on UAV to improve that technology at this time.

Determined position system with Image-Based Visual Servoing (IBVS), which tracks certain features on the landing platform. This method depends on adaptive control rule, moreover by considering failsafe rule flight mode that currently in use is also affected. The result is satisfying, they landed with a smooth position on moving landing[5].

The research on [6] presents the determining of position system with image-based by using quadrotor equipped with facing down GoPro Camera. This UAV is using vision algorithm. Images that captured by the UAV processed by the laptop with long-distance control, then the data is resubmitted to the onboard controller. The Laptops will give command periodically with new coordinates to the UAV so it can be landing in the right position.

This research will explain about computer vision system that will install on UAV as rescue payload sender to increase the precision of determining the payload delivering target. This system consists of Raspberry Pi 3 as a Single Board Computer (SBC) and webcam Logitech C920 Pro HD 1080p as a vision sensor that can see the bottom area of UAV. This System is using image processing technique with OpenCV library and python programming. The System will read the Hue Saturation Value (HSV) of the target that will be used as delivering target or landing target. UAV will still use GPS as coordinate determinants. However, when the UAV is on destination's coordinate the Vision system will work and give a feedback to Flight control (FC) for going to the centre of the target.

2. Unmanned Aerial Vehicle

2.1. Design

Basically, the UAV vision system is a general drone that equipped with a vision system to add the target tracking capabilities. This is the component used in this UAV (table 1).

Table 1. Hardware Component

Component	Type	Category
Pixhawk	2.4.8	Controller
Raspberry Pi	V.3	Controller
Battery Lipo	4S 5500 mah	Power
Telemetry	433 Mhz	Communication
Radio Transmitter	Turnigy 9XR	Comunication
GPS	M8N	Sensor
Webcam	Logitech C922	Sensor
ESC	40A	Speed Control
Brushless Motor	DJI 920Kv	Actuator
Servo	Towerpro mg90s	Actuator

This UAV as shown in figure 1 is using frame kit S550 with plastic injection module material that light enough to be combined with PBC on that main part.

Figure 1. UAV Construction

2.2. Block Diagram

On Figure 2. It explains how the UAV vision system are connected with each part. All the components are integrated on 1 UAV.

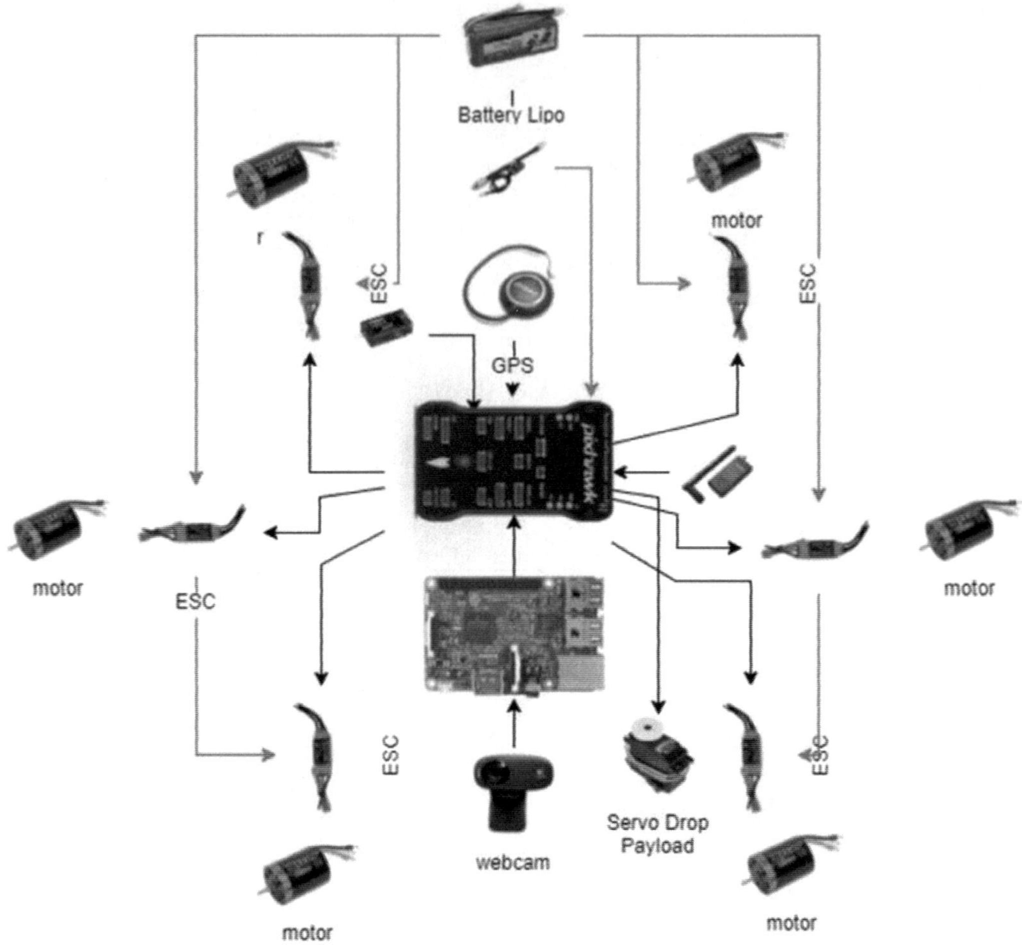

Figure 2. Block Diagram UAV Vision System

2.3. Flowchart

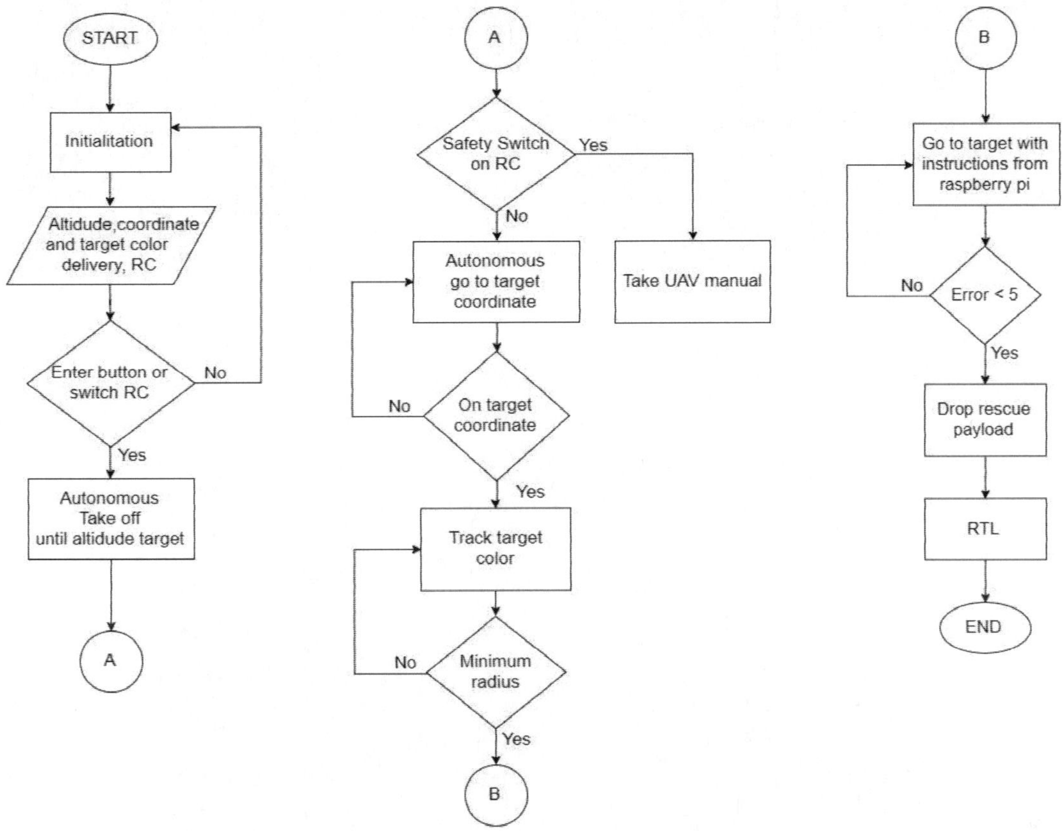

Figure 3. Flowchart UAV Vision System

Figure 3 will be described as followed. First initialization, then the UAV will wait for input from the user in the form of altitude, coordinate and target colour delivery, and RC. After that, if the enter key or the take-off switch button on the RC is pressed, the UAV will autonomously take off until the altitude target is reached. If the safety switch button on the RC is pressed the UAV will be piloted manually using RC. The safety switch button is for emergency only. If the switch button is not pressed the UAV will go towards the target's coordinate in auto-pilot mode. If the target coordinate has been reached, the UAV will detect the target colour (by using HSV). To minimize the colour noises, a minimum radius measurement is given so that the camera could detect target accurately. After detected the colour, the Raspberry will give instructions to Pixhawk through MAVlink to get to the centre point of the target. If the error < 5 (minimum point to get the centre) the UAV will go down to the specified altitude then drop the rescue payload. After that, the UAV will come back to the station and land automatically or manually.

3. Vision System

This UAV used Vision System composed of Raspberry Pi as its controller and Logitech C920 Pro HD webcam as its vision sensor. The webcam is connected to the controller through the USB port. It is faced down so that it could detect the delivery target (Figure 4).

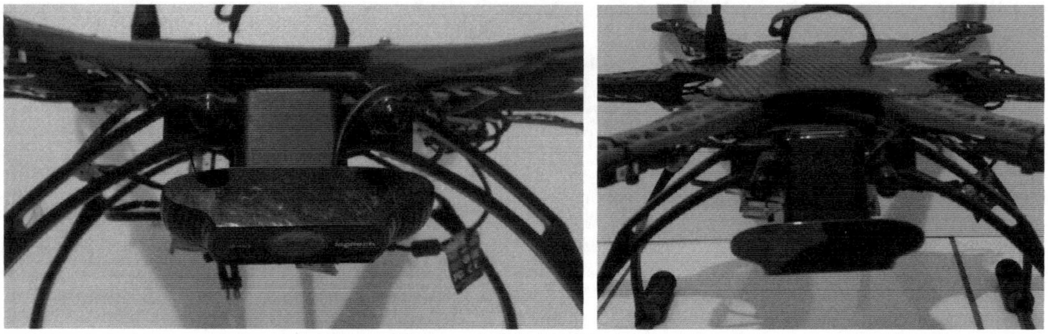

Figure 4. Vision System

Raspberry Pi has its limitation on data processing as it only uses a 1.2 GHz quad-core processor and 1 GB of RAM. To reduce the possibility of hanging up, the webcam resolution couldn't be used to its fullest, only 480x480 is used.

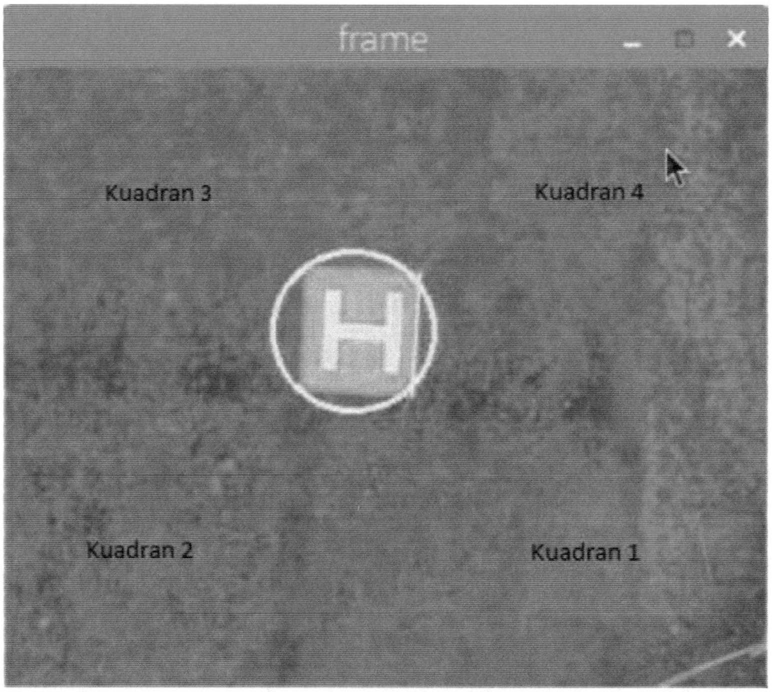

Figure 5. Quadrant Division

The controller used OpenCV 2.4.8 for image processing. The method used for delivery target detection is by detecting the colour using the HSV filter. To anticipate the colour noise, size limitation is given. Image processing will provide the coordinate (x, y) position of the delivery target. But as the coordinate only use 8-bit resolution, it needs to be processed first to find the centre point and divide the image into a 4 quadrant (Figure 5) using:

$$X = X - 125 \quad (1)$$

$$Y = Y - 125 \quad (2)$$

In addition, for more accurate target positioning the angle is required. In determining the target position angle used the formula:

$$\theta = \arctan\frac{y}{x} \tag{3}$$

From the data (x, y) and the angle obtained it will be converted into a UAV movement to get to the target delivery with accurate. In the change used formula Proportional Integral Derivative (PID). The Raspberry Pi will send instructions to the FC to make the movement in accordance with the direction of the target position. The communication used the serial port using MAVLink protocol.

4. Experimental Result

The result of determining the target using vision system can be seen in figure 6 shows that to simplify controller UAV image divided into 4 area based on quadrant which can be seen in figure 7.

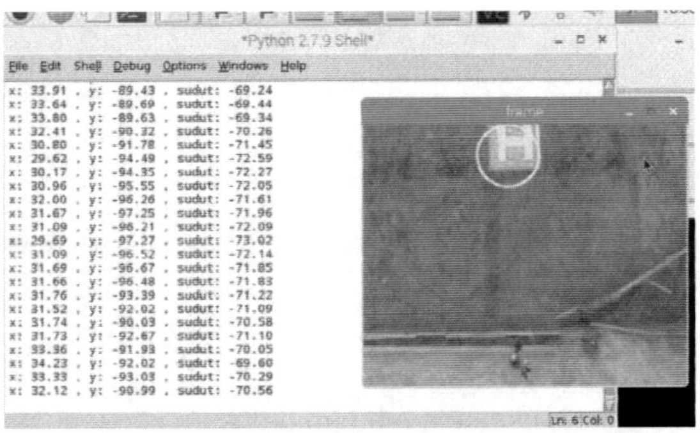

(a)

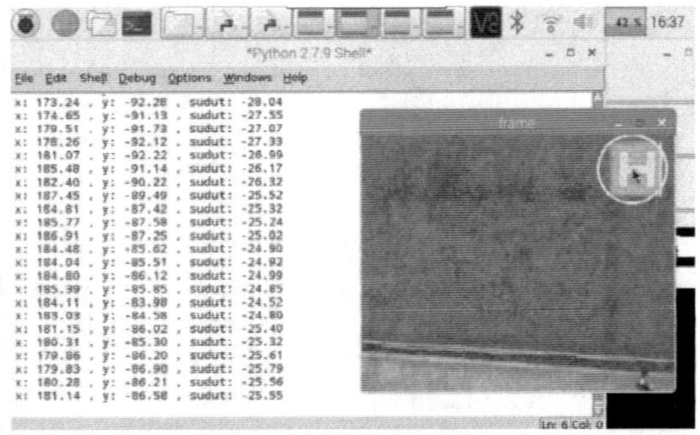

(b)

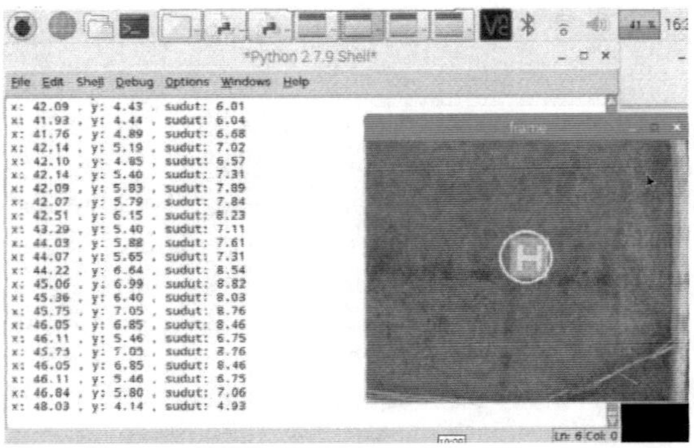

(c)

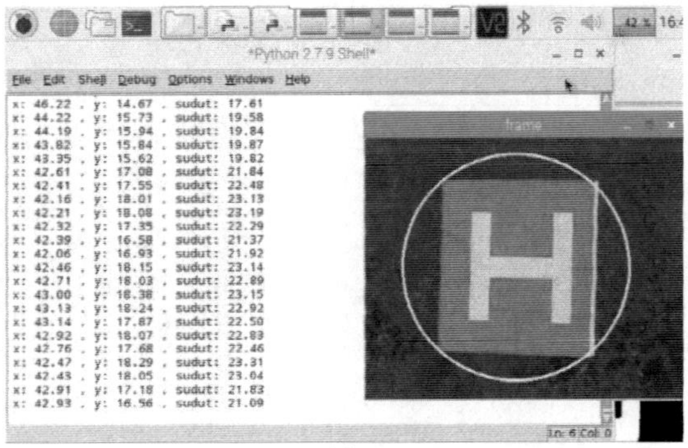

(d)

Figure 6. Trial Results

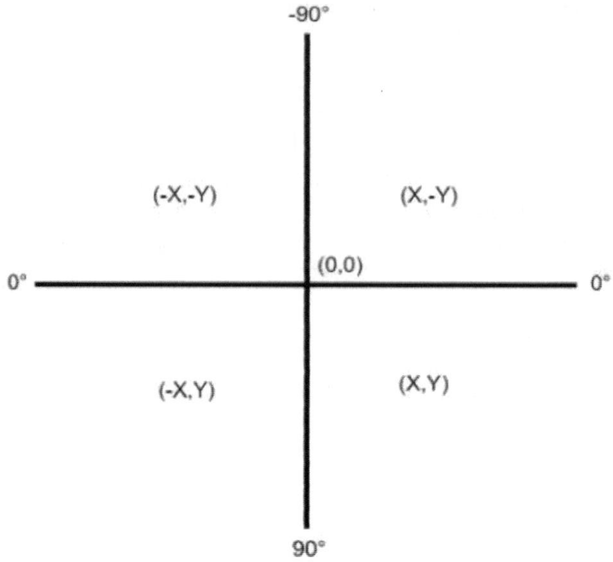

Figure 7. Cartesian Coordinate

From the above result, UAV will be able to position itself according to coordinates (x, y, Θ). For example when the value obtained (x, y,) then UAV will do yaw as much as Θ.

5. Conclusion and Future Work

A UAV with a vision system will make delivering payload on a target more accurate than just using GPS to determine the destination coordinates. But in the implementation, it will take longer because it will receive feedback from the camera for precision with the target. For future research is expected to find a new algorithm to reduce camera noise, so the target will be faster.

References

[1] Gautam A, Sujit P B and Saripalli S 2014 A survey of autonomous landing techniques for UAVs *2014 International Conference on Unmanned Aircraft Systems (ICUAS)* (IEEE) pp 1210–8
[2] Ryan A and Hedrick J K 2005 A mode-switching path planner for UAV-assisted search and rescue *Proceedings of the 44th IEEE Conference on Decision and Control* vol 2005(IEEE)pp 1471–6
[3] Demonceaux C, Vasseur P and Pégard C 2006 Omnidirectional vision on UAV for attitude computation *Proceedings - IEEE International Conference on Robotics and Automation* vol 2006pp 2842–7
[4] Jin S, Zhang J, Shen L and Li T 2016 On-board vision autonomous landing techniques for quadrotor: A survey *Chinese Control Conference, CCC* vol 2016–Augus (IEEE) pp 10284–9
[5] Lee D, Ryan T and Kim H J 2012 Autonomous landing of a VTOL UAV on a moving platform using image-based visual servoing *Proceedings - IEEE International Conference on Robotics and Automation* pp 971–6
[6] Vidal V F, Honório L M, Santos M F, Silva M F, Cerqueira A S and Oliveira E J 2017 UAV vision aided positioning system for location and landing *2017 18th International Carpathian Control Conference, ICCC 2017* pp 228–33

IOP Conf. Series: Materials Science and Engineering **384** (2018) 012006 doi:10.1088/1757-899X/384/1/012006

Design of a Directional Microstrip Antenna at UHF-Band for Passive Radar Application

D L N Fauzi* and T Hariyadi

Department of Electrical Engineering Education, Universitas Pendidikan Indonesia, Jl. Dr. Setiabudhi No.207, Bandung, Indonesia

*denilistyanto@student.upi.edu

Abstract. Passive radar is an electromagnetic system that can receive the frequency signals from transmitted source such as FM radio, TV broadcasting station, or mobile communication base station to detect the presence of an object by utilizing its frequency reflection. One of the components of a passive radar is an antenna. An antenna in a passive radar is to receive an information signal from the reflection of an object and then the data that already received will be processed into the form of signal before it was displayed on the monitor. This paper aims to create a microstrip antenna with directional pattern for a passive radar. Microstrip antenna is a type of antenna that has a light weight, easy to fabricate, has a simple shape and lower cost. The design and simulation of antenna are using CST Studio Suite software. If the simulation results fit with the requirement, we fabricate and measure it and then compare it with the simulation results. The result of antenna has an ultrahigh frequency (UHF), directional pattern, gain 6dBi and range bandwidth minimum 470Mhz – 780Mhz.

1. Introduction

Passive radar is an electromagnetic system that can receive the frequency signals from transmitted source such as radio station, TV broadcasting station, or mobile communication base station to detect and track target by utilizing its frequency reflection [1]. Recently, passive radar can attract considerable interest of people due to its several advantages if compared with active system. Specifically, because a passive radar does not transmit, it is naturally convert and offers high tolerance to electronic countermeasure (ECM), it does not require additional spectrum and has been widely considered as a unique sensing capability to manage the radio spectrum congestion problem[2-3]. Passive radar can be generally classified into two categories, which are reference channel (RC) and surveillance channel (SC). In the first category, an RC is utilized at the receiver to collect the direct-path (transmitter-to receiver) signal, while a separate surveillance channel (SC) is used to collect the target echo [1], [4], [5], Work Principle passive radar can be seen in Figure 1.

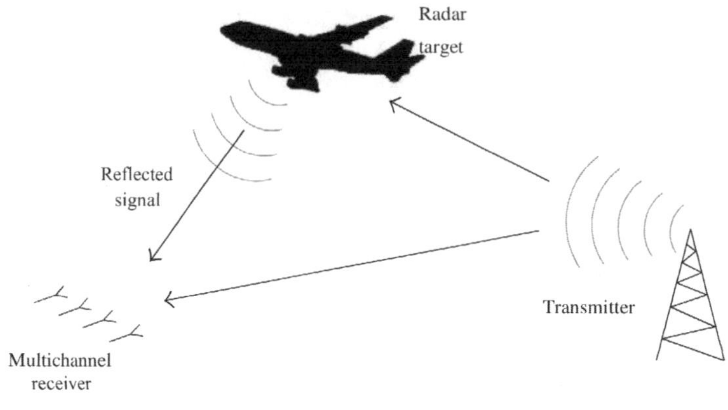

Figure 1. Passive radar principle

Antenna is one of the important components in a passive radar that acts as receiving a frequency signal. Passive radar antennas are generally large with microwave antenna type to be placed in certain positions. In this study the author will make a passive radar antenna using a type of microstrip antenna [1].

Microstrip antennas have become quite popular types of antennas in various applications. This is because the microstrip antenna has several advantages compared to conventional antennas. The advantages include: light, small, thin, and low-cost manufacturing. Because of the excess the microstrip antenna is called a low-profile antenna [6].

In this research, passive radar antennas utilize sources from TV transmitters. Frequency used is UHF (476-806 MHz) [7], in accordance with the regulation of TV broadcast spectrum in Indonesia. Antenna microstrip for passive radar applications have been made by Peter Knott [2], but uses an array method to obtain a large gain while in this research uses director and reflector to obtain a large gain.

2. Antenna design

Before designing of microstrip antenna to determine a specification and materials used for which antenna, In this research we use FR-4 as a substrate with dielectric constant of 4.5 and thickness 1.6 mm. To obtain the dimension of microstrip patch antenna we use equation (1) while to obtain patch length can be obtain from equation (2) – (10).

The width of dimension a patch microstrip antenna using the following equation [8].

$$W = \frac{c}{2f}\sqrt{\frac{2}{\varepsilon+1}} \tag{1}$$

where,

W = Width of the patch

c = Speed of light in free space ($3x10^8$ m/s)

ε_r = dielectric constant

$$L = \frac{c}{2fo\sqrt{\varepsilon_r}} - 2\Delta L \tag{2}$$

where,

$$\Delta L = 0{,}412h\ \frac{\varepsilon_{re} \quad +0{,}3 \quad w/h+0{,}2}{\varepsilon_{re} \quad -0{,}2 \quad w/h-0{,}8} \tag{3}$$

where,

h = height of the substrate and ε_r:

$$\varepsilon_r = \frac{\varepsilon_r + 1}{2} + \frac{\varepsilon_r - 1}{2}[1 + 12\frac{h}{w}] - 1/2 \quad (4)$$

The dimension of ground plane length (lg) and width (wg) of microstrip antenna can be obtain from equation (5) and (6) [9].

$$Lg = 6h + L \quad (5)$$

$$Wg = 6h + W \quad (6)$$

The calculation of feed width (Wr) to achieve 50 Ω characteristic impedance.

$$\frac{W_0}{h} = \frac{2}{\pi}[B - 1 - ln(2B - 1) + \frac{\varepsilon_r - 1}{2\varepsilon_r}(ln(B - 1) + 0{,}39 - \frac{0.61}{\varepsilon_r}] \quad (7)$$

Where,

$$B = \frac{3\ \pi}{2Z_{0\sqrt{\varepsilon_r}}} \quad (8)$$

And formula of the length a feed antenna

$$L = \frac{\lambda_g}{4} \quad (9)$$

Where,

$$\lambda_g = \frac{\lambda_U}{\sqrt{\varepsilon_r}} \quad (10)$$

The result of design antenna used that formula L x W (117.18 mm x 150 mm) and Lg x Wg (126.78 mm x159.6 mm) have a 58.603 mm for a length of feed and 7.93 mm for width of feed antenna for specification frequency 586 MHz and used FR-4 epoxy of dielectric constant 4.5 for bandwidth 216 MHz. In the Figure 2 and Figure 3 is the result design antenna and result of return loss before modify.

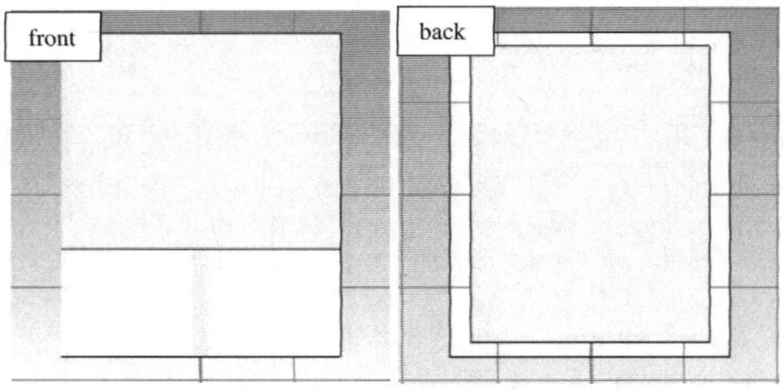

Figure 2. Design of antenna before modify.

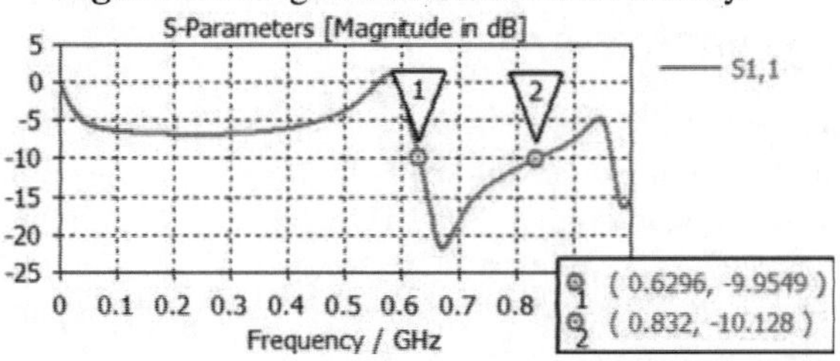

Figure 3. The result of return loss and bandwidth before modify design of antenna

The modify antenna used substrate FR4 epoxy of dielectric constant (ε_r) 4.5 and thickness (h) 1.6 mm. The proposed antenna design is shown in Figure 4 and the dimensions of the proposed design are listed in Table 1. The design of antenna inspirited by antenna DVB-T [6].

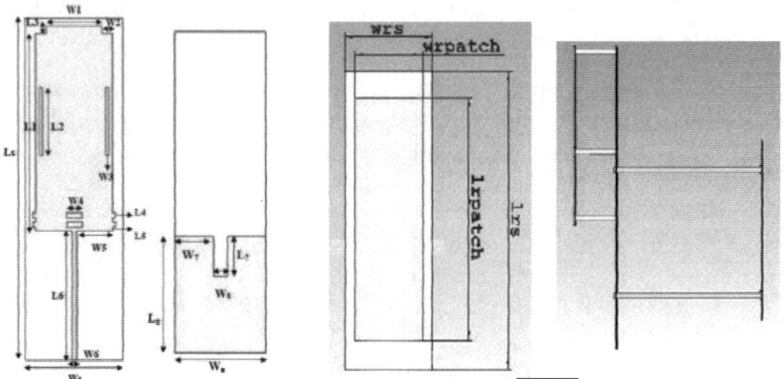

A B

Figure 4. A. Microstrip patch antenna

B. Design of director, reflector and antenna in side view

Table 1. Dimension list of antenna design

Parameter	Value (mm)	Parameter	Value (mm)
W1	36	L1	130
W2	7	L2	44.5
W5	11	L5	3.5
W6	3	L6	85
W7	32	L7	60
W8	7	Ls	225
Ws	64	Lg	82
Wg	64	Lrs	200
Wrs	64	Lrpatch	162.5
Wrpatch	50		

The substrate dimension of antenna is 225 mm x 64 mm x 1.6 mm. The impedance that was obtained is 50 Ω, this value is gotten after setting W6 with 3 mm. This director design has a function to direct the radiation pattern become previously directional omni-directional with the size of director substrate about 200 mm x 64 mm x 1.6 mm and the patch size is 162.5 mm x 50 mm x 0.035 mm. The patch design greatly influences with radiation pattern.

The distance between reflector to driver, and vice versa is using the equation of 0.35λ for reflector and 0.14λ for director [10]. The size is adjusting the antenna size and using parameter sweep for simulation in order to analyze the expected result.

3. Results and discussion

This antenna is designed and simulated by using CST Studio Suite software with 50 Ω impedance. The return loss value and the wide of bandwidth is 300 MHz. Figure 5 and Figure 6 is the simulation result from antenna before it was added by reflector and director. From Figure 5 it can be seen that the badhwidth is narrower before adding director and reflector. The radiation pattern is omnidirectional and the antenna gain is 2.16 dBi as shown in Figure 6.

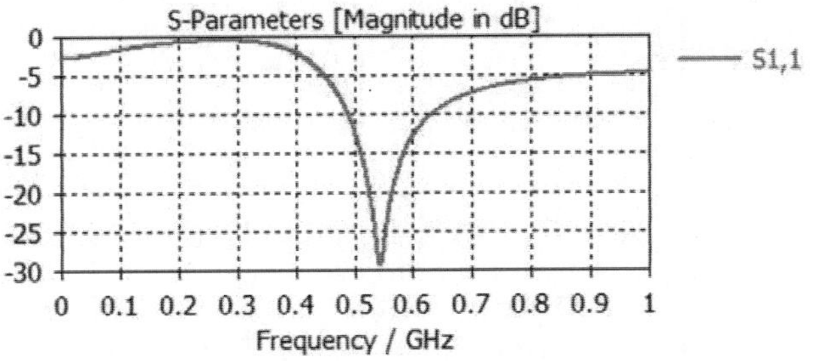

Figure 5. Return loss before adding the reflector and director.

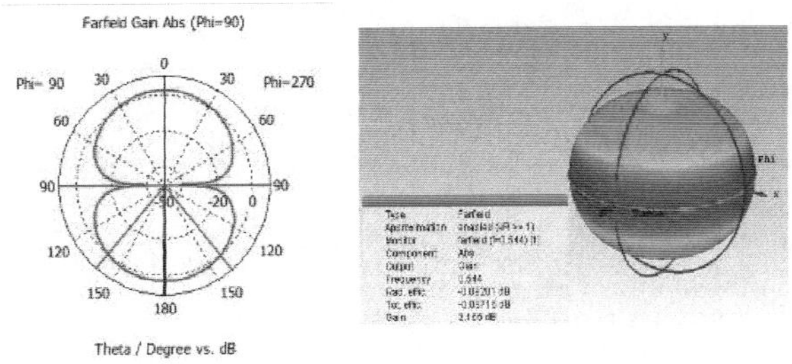

Figure 6. Radiation Pattern before adding the reflector and director.

After simulating the antenna before it is added by director and reflector, then add director with design that has been made. In the Figure 7 to 10 is the parameter sweep for optimization to matching the antenna specifications, the result of simulation after added director only can be seen in the Figure 11 until Figure 13.

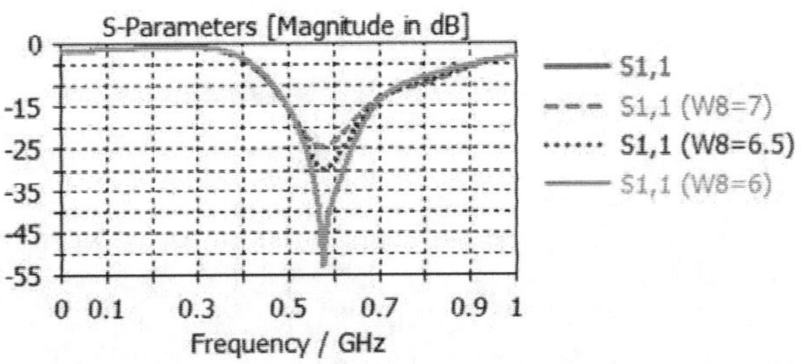

Figure 7. Simulation result of return loss with different w8 at l7 = 50 mm.

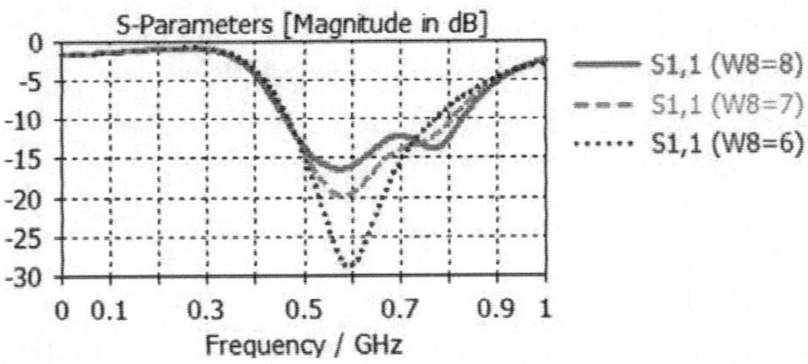

Figure 8. Simulation result of return loss with different w8 at l7 = 60 mm

IOP Conf. Series: Materials Science and Engineering **384** (2018) 012006 doi:10.1088/1757-899X/384/1/012006

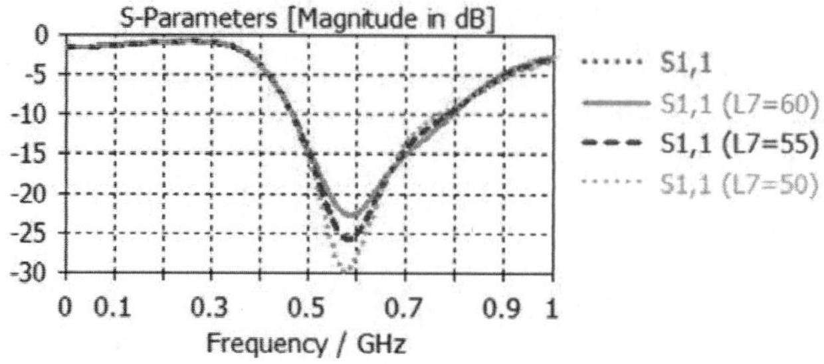

Figure 9. Simulation result of return loss with different l7 at w8 = 6 mm

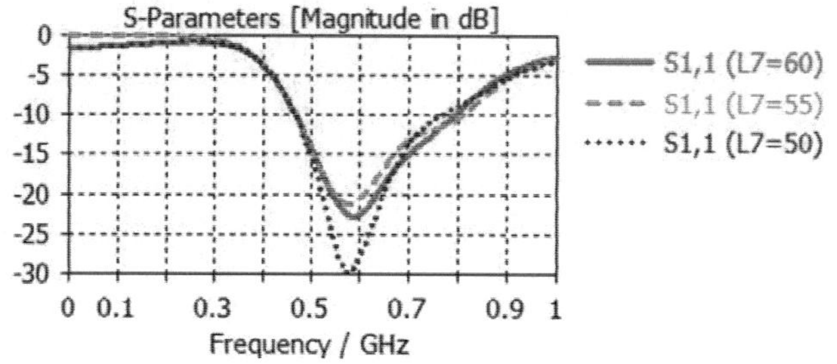

Figure 10. Simulation result of return loss with different l7 at w8 = 6.5 mm

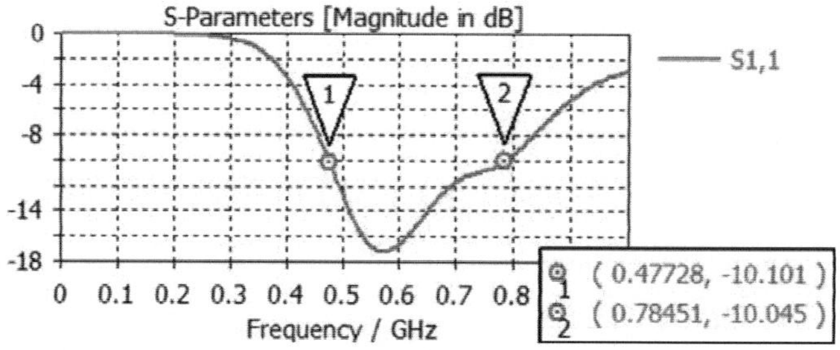

Figure 11. Optimization of bandwidth result before added reflector

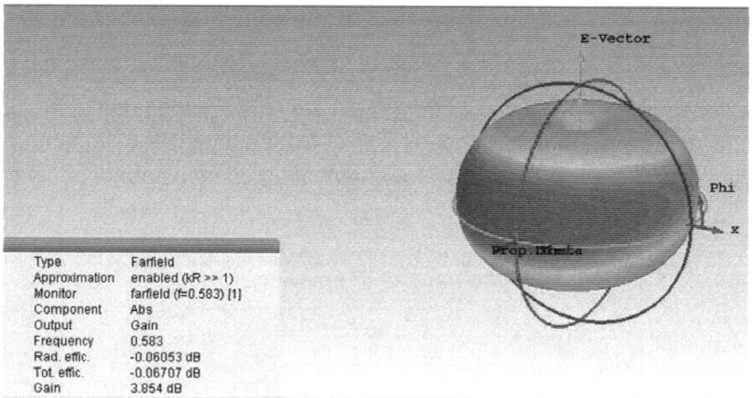

Figure 12. Radiation pattern and gain result after added reflector

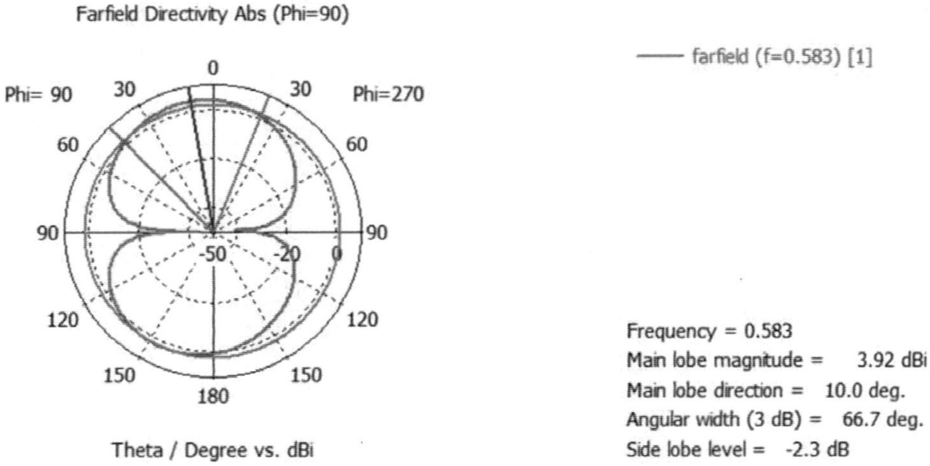

Figure 13. Radiation pattern result after added reflector

After adding the director, this antenna is added with reflector to obtain the better gain, but not reduce the bandwidth, as in the Figure 14 and Figure 15. From Figure 14 we can see that the return loss is better than before added the reflector. The antenna gain also increase after adding the reflector as shown in Figure 15.

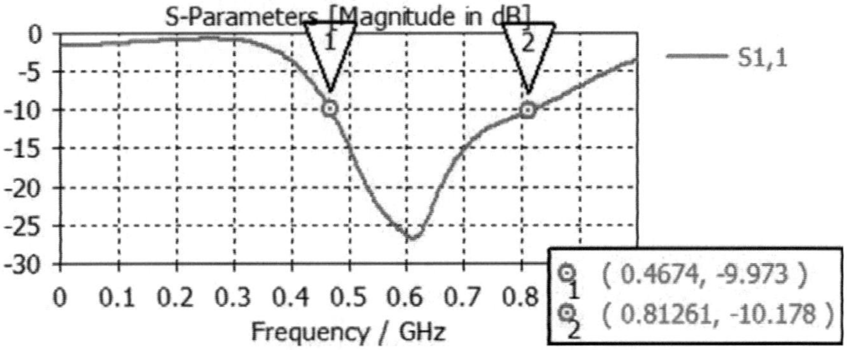

Figure 14. Bandwidth result after added reflector

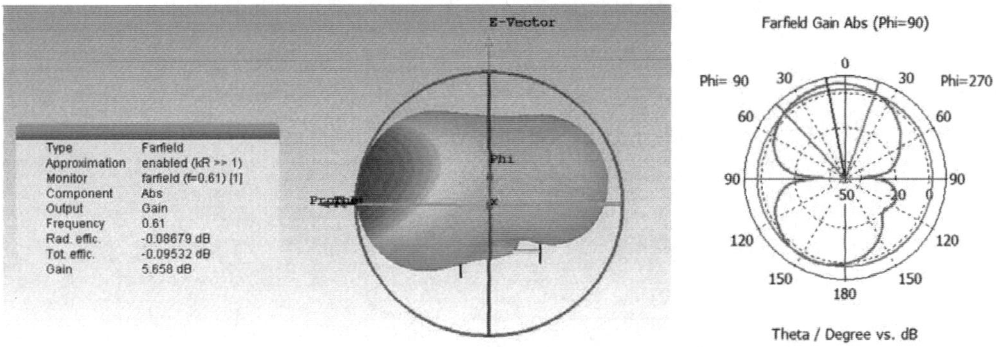

Figure 15. Gain result after added reflector

The gain can optimization applying by enlarge the dimension of reflector 225 mm x 64 mm to 225 mm x 225 mm with the same range, as in Figure 16 design of enlarge the reflector. In the Figure 17 is the result the gain after added reflector has been enlarge, and Figure 18 is the result radiation pattern.

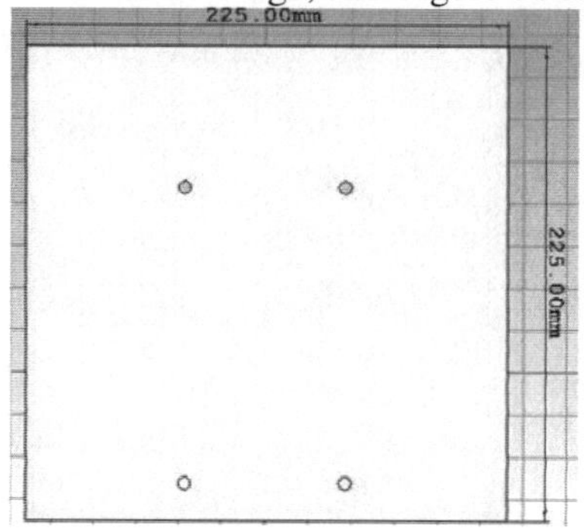

Figure 16. The design reflector 225 mm x 225 mm to optimization the gain and magnitude of frequency

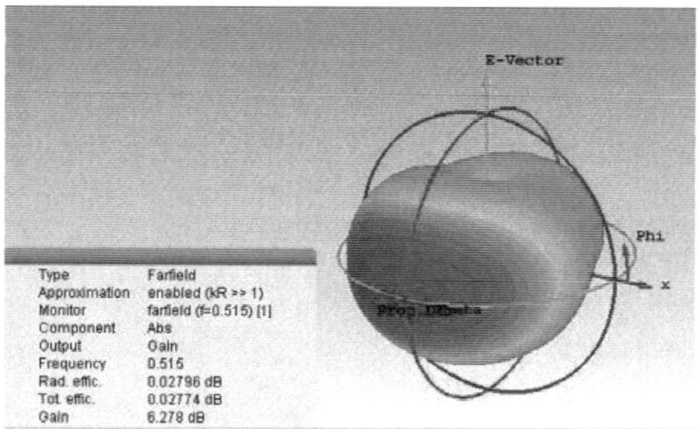

Figure 17. The result of gain after added reflector 225 mm x 225 mm

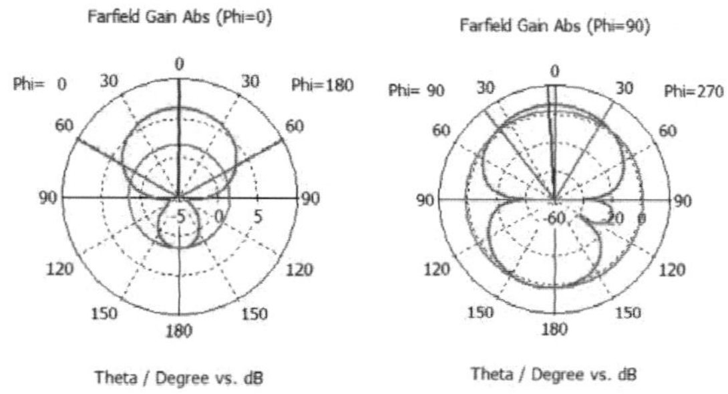

Figure 18. The result of radiation pattern with cos phi 0° and 90° after added reflector 225 mm x 225 mm.

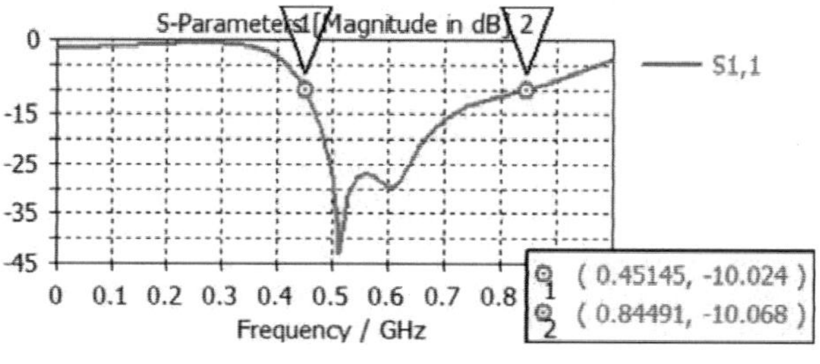

Figure 19. The result of return loss after added reflector 225 mm x 225 mm.

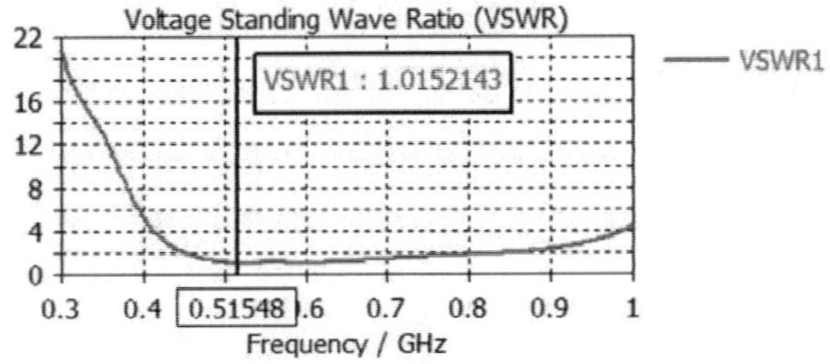

Figure 20. VSWR of the optimal design

The final result of return loss can be seen in Figure 19 with 393 MHz of bandwidth and the result of the VSWR can be seen Figure 20.

4. Conclusions

Design of antenna microstrip with directional radiation pattern for radar passive system that operate in Indonesia has UHF-band frequency range correspond with the frequency of television transmitter. This antenna used director and reflector to have gain that was expected, with 0.14λ for the distance of director to driver and 0.35λ for the distance of reflector to driver. The substrate materials that used are FR-4 with 4.5 dielectric constant, while the size is 255 mm and 64 mm. The dimension of director is 200 mm x 64 mm and the dimension of reflector is 225 mm x 225 mm in order to the result of gain is 6.27 dBi and bandwidth 393 MHz from 451 – 844 MHz for return loss less than -10 dB.

References

[1] H. Kuschel, J. Heckenbach, and R. Appel 2010 Countering stealth with passive, multi-static, low frequency radars *IEEE Aerosp. Electron. Syst. Mag.*, vol. **25**, no. 9, pp. 11–17.

[2] P. Knott, 2013 Design of a printed dipole antenna array for a passive radar system *Int. J. Antennas Propag.*, vol. **2013**.

[3] Q. He and R. S. Blum, 2014 The significant gains from optimally processed multiple signals of opportunity and multiple receive stations in passive radar *IEEE Signal Process. Lett.*, vol. **21**, no. 2, pp. 180–184.

[4] X. Zhang, H. Li, and B. Himed 2017 Multistatic Detection for Passive Radar with Direct-Path Interference *IEEE Trans. Aerosp. Electron. Syst.*, vol. **53**, no. 2, pp. 915–925

[5] X. Zhang, H. Li, J. Liu, and B. Himed 2016 Joint delay and Doppler estimation for passive sensing with direct-path interference *IEEE Trans. Signal Process.*, vol. **64**, no. 3, pp. 630–640

[6] R. Nurvitasari, T. Hariyadi, B. Mulyanti, D. Pendidikan, T. Elektro, and U. P. Indonesia 2014 Antena Mikrostrip Monopole Pita Lebar Segi Empat untuk Aplikasi DVB-T vol. **13**, no. 2, pp. 161–166

[7] M. Perhubungan and R. Indonesia 2003 Keputusan menteri perhubungan nomor : km. 76 tahun 2003 tentang rencana induk (," pp. 1–89, 2003.

[8] G. V. P. Pranathi, N. D. Rani, M. Satyanarayana, and P. G. T. Rao 2015 *Patch Antenna Parameters Variation with* pp. 7344–7350, 2015.

[9] S. Kumar and H. Gupta 2013 Design and Study of Compact and Wideband Microstrip U-Slot Patch Antenna for Wi-Max Application *IOSR J. Electron. Commun. Eng.*, vol. **5**, no. 2, pp. 2278–8735

[10] C. Run-Nan, Y. Ming-Chuan, L. Shu, Z. Xing-Qi, Z. Xin-Yue, and L. Xiao-Feng 2012 Design and analysis of printed yagi-uda antenna and two-element array for WLAN applications," *Int. J. Antennas Propag.*, vol. **2012.**

IOP Conf. Series: Materials Science and Engineering **384** (2018) 012007 doi:10.1088/1757-899X/384/1/012007

Automation and Monitoring Smart Kitchen Based on Internet of Things (IoT)

F Nugroho* and A B Pantjawati

Department of Electrical Engineering Education, Universitas Pendidikan Indonesia, Jalan Dr. Setiabudi no. 207, Bandung 40154 Indonesia

*fauzinugroho@student.upi.edu

Abstract. The kitchen is one of the important places in a house. Safety factor is the main aspect that must be taken into account during the activity in the kitchen. The existence of gas leakage, uncontrolled fire and excessive temperatures must be quickly identified and addressed. The purpose of this research is to make prototype of kitchen security system using Internet of Things. The system is designed using 4 types of sensors and Arduino UNO. DHT 11 sensor is used to monitor temperature and humidity, IR Flames sensor is used to detect fire, MQ-135 sensors are used to detect gas leakage, and PIR sensors are used to detect human activities in the kitchen. The sensors output are then connected to the Arduino which will control the relay. The relay acts as a fan switch in the event of a gas leak, uncontrolled fire and excessive temperature increase. Under these conditions, Arduino will also turn on the alarm and the led, and send information to the server. The results show that the system can work according to the desired specifications.

1. Introduction

1.1. Background

Nowadays the existence of the Internet of Things (IoT) has already changed the human life. IoT is a concept that allows objects around us to communicate with each other [1]. This capability will make objects understand what humans need without any command, including providing important information wherever we are [2] [3].

In everyday life, people perform various activities. One of the places where human activity is the kitchen. With a variety of activities conducted in the kitchen, will certainly make the kitchen temperature conditions quickly changed. In addition to temperature, the use of gas stoves has a high fire risk. Based on that, the use of IoT in the kitchen is necessary to keep the kitchen air condition always be comfortable and to reduce risk due from the use of gas stoves.

1.2. Journal Review

Many research and projects have been made regarding this smart kitchen. As research by Vu Trieu Minh and Riva Khannadari from Tallinn University of Technology. And Mutsalklisana Chaiyaorn from Northeastern University, Boston, USA.

From some previous research, these research mostly only reveal a concept, some research that succeeded in making a system that can work but still use ATmega32 as a microcontroller with a limited sensor. In this paper, Explain about designing an IoT based automation and monitoring system using

Arduino UNO as microcontroller and ESP8266-01 as data communication device using wireless fidelity (Wi-Fi) network.

1.3. Proposed

Based on background and review from research that has been described, we need a system that can monitor and send various information about the condition of the kitchen continuously, even though we are in a place far from the kitchen.

This project created a system which can detect changes in temperature and fire caused by the use of gas stoves in the kitchen. In the system, the DHT11 sensor is mounted to detect temperature changes, an MQ-135 gas sensor to detect Liquefied Petroleum Gas (LPG) leak, IR flame sensor to detect fire, and passive infrared (PIR) sensors to detect human activity in the kitchen. In this system, there is also a relay to control the fan that serves to control the temperature and blow out of gas in the event of a gas leak or smoke from the kitchen in case of fire. This system can be controlled and monitored via the internet directly from laptops or smartphones anytime, even from a place far from the kitchen. In case of fire or gas leak, this system will provide a warning in the form of alarm, as well as information to be sent via short message service (SMS), e-mail, or in-app notification on a smartphone.

2. Method

The project of this system is built by some architecture, starting from hardware architecture, software architecture and system architecture.

2.1. Hardware Architecture

Figure 1 shows the design and schematic of this system. The main components used in this circuit are Arduino Uno, ESP8266-01 (wifi module), DHT11, MQ-135, IR Flame sensor, PIR sensors, buzzer, three pieces LED, Relay module and A DC Fan.

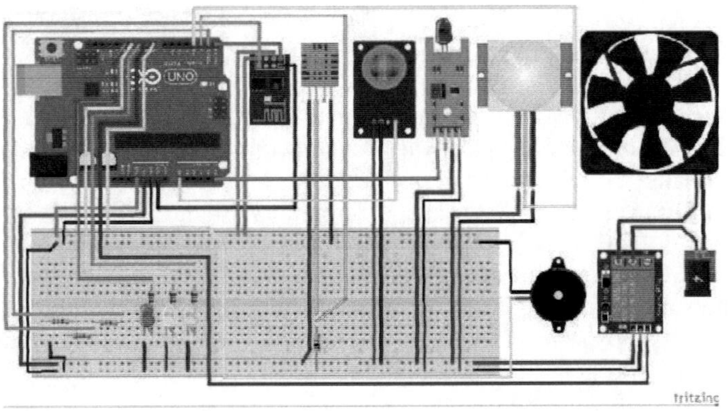

Figure 1. Schematic Installation

2.1.1. Arduino Uno. Arduino Uno is a microcontroller that is currently popular. Arduino Uno uses ATMEGA328 as its microcontroller, has 14 digital I / O pins and 6 analog inputs. To program it, the Arduino is connected to the computer using USB type A to type B or what we often call the printer cable [4].

2.1.2. ESP8266. ESP8266 is a wifi module that serves as a microcontroller enhancement such as Arduino to connect directly to wifi and create TCP / IP connections. This module requires about 3.3v power with three modes of wifi namely Station, Access Point and Both (tools). This module is also equipped with processor, memory, and GPIO where the number of pins depends on the type of ESP8266 we use. On this system, This device is used as Wi-Fi shield for Arduino which will connect to Wi-Fi network from available hotspot router. This device will send the collected data from Arduino to the server.

2.1.3. Temperature and Humidity Sensor (DHT11). DHT11 uses a capacitive humidity sensor and a thermistor to measure ambient air. DHT11 sends output data in the form of digital output without requiring analog pins. This sensor is very easy to use but has a measurement limit of only up to 60 ° C [5]. In this system, DHT11 is used to detect temperature and humidity.

2.1.4. Gas Leakage Sensor (MQ-135). Gas sensor MQ-135 is very useful to detect a gas leak in the kitchen. This sensor can detect flammable gases such as LPG, CH4, CO, Propane to detect smoke from fire. It has a very quick response to gas leakage, but it takes a long time when it first starts up. In this sensor, there is a potentiometer that serves to regulate the sensitivity of the sensor. When the sensor detects a gas leak, the sensor's conductivity will increase depending on the amount of gas concentration detected [6]. In this system, MQ-135 is used to detect leaks on LPG gas.

2.1.5. IR Flame Sensor. In this system, IR flame sensor functions as a fire detector. This sensor will detect fires based on the infrared wave spectrum generated by the fire.

2.1.6. Passive Infrared (PIR) Sensor. In this system, the PIR will serve to determine the presence of human activities in the kitchen. This sensor determines the presence of human activities based on the infrared emission from the human.

2.1.7. Power Supply. This system uses an adapter that will work on 100 - 240V. The resulting output is 5V, 1A. This power supply will be connected directly to the microcontroller. Figure 8 shows the power supply.

2.2. Software Architecture

Software that is used in this system is Arduino IDE, Cayenne web server.

2.2.1. Arduino IDE. Arduino Uno is programmed using the Arduino Integrated Development Environment (IDE). Arduino IDE is a software developed by Arduino. This program is used to design various processes related to Arduino programming [4]. Not only the Arduino device, ESP 8266-01 is also programmed using the serial monitor available on the Arduino IDE.

2.2.2. Cayenne web server. Cayenne is an application and web server to create project internet of things. Cayenne has features for monitoring and controlling the objects connected to it. Using Cayenne is very easy to use because it is only with visual drag and drop. On this system, Cayenne use to show all data from the sensor, control the relay fan and send warning notices to the mobile device via E-mail and SMS.

2.3. System Architecture

The system architecture of the smart kitchen system is shown in figure 2.

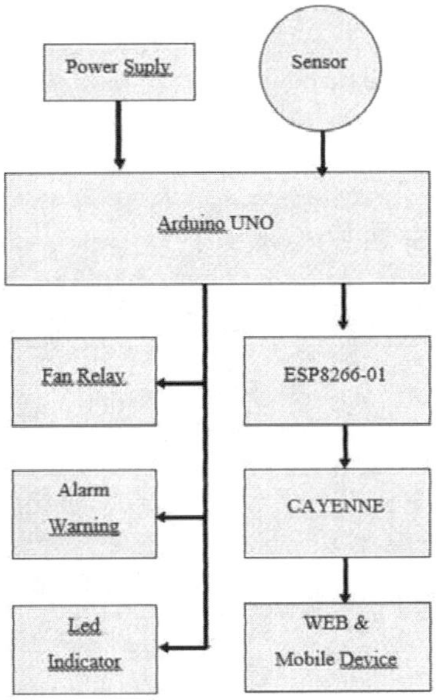

Figure 2. Architecture of Smart Kitchen System

- The power supply is connected directly to Arduino. The sensor powered from pin 5v (VCC) and pin 3.3 V in Arduino.
- The sensor used in this system is DHT11 to detect temperature change, Mq-135 to detect LPG gas leak, IR flame sensor to detect fire, PIR sensor to detect human activity in the kitchen. Data received by the sensor will be sent directly to the Arduino via digital or analog pins.
- Arduino Uno serves as the controlling center in this system, all data sent by the sensor will be processed to become a reference in the next action. Data received by Arduino will also be sent to the Cayenne server.
- Esp8266 is used as Wi-Fi shield that will connect the Arduino to the server via Wi-Fi.
- Fan Relay functions as a fan controller. This relay will work according to Arduino command. The Arduino will command the Fan to turn on when the DHT11 sensor detects air temperature> 25 ° C, MQ-135 sensor detects LPG gas leakage, IRflame sensors detects a fire, and when a command from a mobile device passes through the Cayenne server.
- Alarm works when there is a gas leak or a fire occurs.
- The indicator lights consist of three different lamps, each light indicator having different functions. A red light indicates a fire, yellow lights indicate a gas leak, green lights indicate the kitchen safely.
- Cayyene serves as a web server that will accommodate various information transmitted by Arduino via Wi-Fi connection. Cayenne will process the information and will be displayed on the dashboard. The dashboard is a view of the page or application that shows the information received by the Cayenne server. In the dashboard, there is also a section that can control the components connected to the Cayenne server. In this system, which can be controlled by the Cayenne is fan relay. Cayenne also has a trigger function. Trigger function will send notifications in the form of SMS and Email based on sensor readings that have been determined.
- The mobile device serves as the receiver and information viewer of the smart kitchen system through the Cayenne server.

3. Result and Discussion

To determine the performance of this system, the smart kitchen prototype that will be placed in a miniature kitchen, as shown in figure 3. The layout of each indicator and sensor is adjusted to the actual system condition. Prior to use in this system, the sensor was independently tested to determine its eligibility.

Figure 3. Kitchen Miniature

The test results show that the system can work properly. Each sensor can work correctly. All data sent by the sensor can be accepted by Arduino to determine an action for the output device. Output devices such as Fan, alarm, and indicator lights can work according to their function. The data that received by Arduino is also sent to the Cayenne server via Wi-Fi. Data received by the Cayenne instantly displayed on the dashboard. Data displayed on the dashboard can be seen in figure 4.

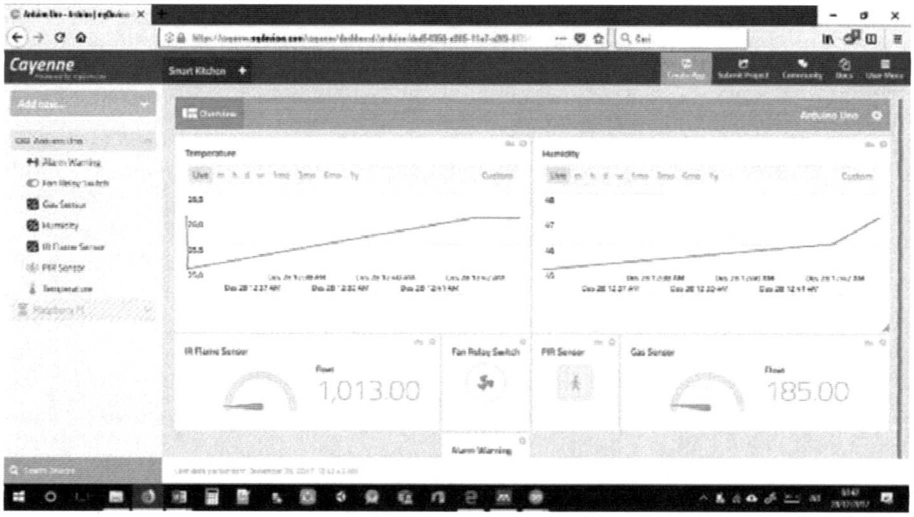

Figure 4. Display of Cayenne Web Dashboard

The dashboard displayed can be customized. The data displayed will also be stored on the server. The stored data can be viewed by opening the chart menu at the top right of each sensor data on a dashboard. The data that received by Cayenne is also displayed on the dashboard in the Cayenne app on a mobile device. The in-app dashboard display can be seen in figure 5.

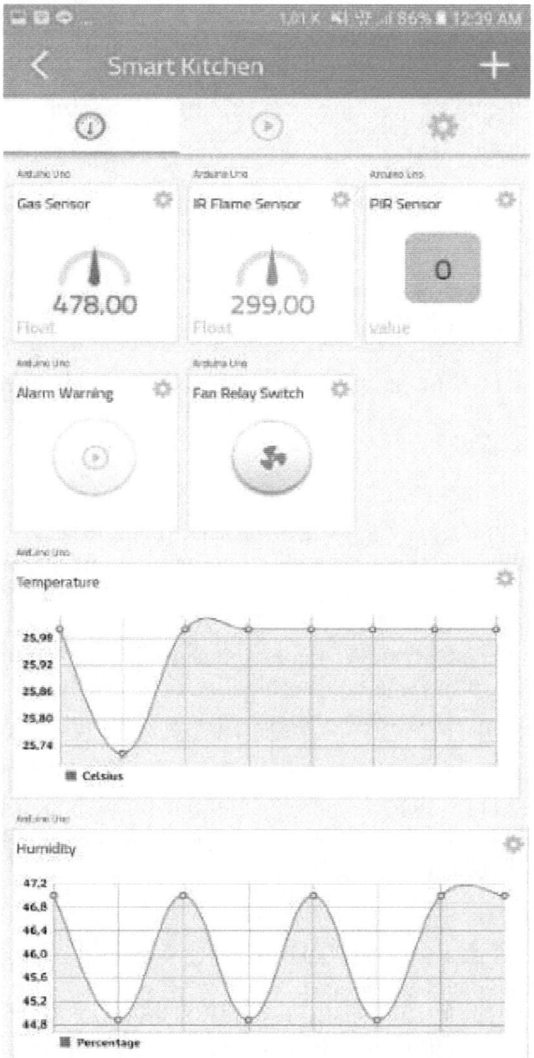

Figure 5. Cayenne application Display

The dashboard displayed in the app can be customized but simpler than Web cayenne dashboard display. The warnings notification in the event of a fire and gas leak also works correctly, as shown in figure 6.

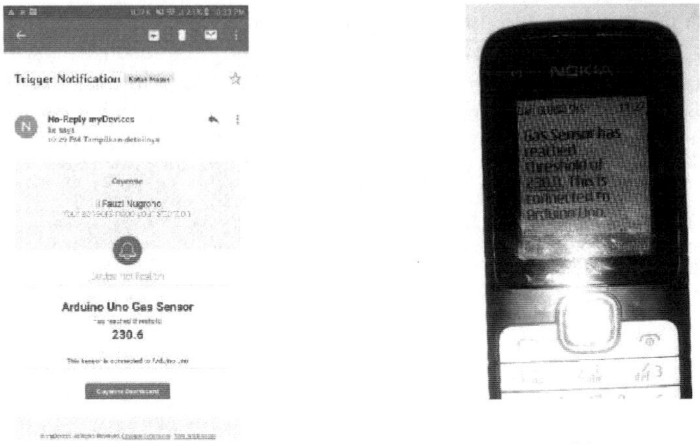

(a) (b)

Figure 6. (a) Email Warning Notification (b) SMS Warning Notification

The system will send warning notification directly to mobile devices via Email and SMS.

4. Conclusion and Future Work

4.1. Conclusion

Based on the design and test of this system, the following conclusions can be taken:

- Based on the test, each of sensors contained in this system works well. After installing, calibration should be done so that the results can be in accordance with the expected.
- All collected data can be displayed on the web and apps. In the delivery of such information, is strongly influenced by the quality of Wi-Fi networks used.
- In simulated fires and gas leaks, the fan can function properly. A warning system can work. Email and SMS can be received directly by the mobile device.

4.2. Future Work

The addition of variant sensors to measure various information about the kitchen is needed. The addition of security systems also needs to be improved. One of the security systems that can be embedded in this system is a fire extinguisher or automatic gas lock.

References

[1] Gusmeroli S, Haller S, Harrison M, Kalaboukas K, Tomasella M, Vermesan O, Vogt H, Wouters K. Vision and challenges for realising the internet of things.

[2] Dohr A, Modre-Opsrian R, Drobics M, Hayn D, Schreier G. The internet of things for ambient assisted living. InInformation Technology: New Generations (ITNG), 2010 Seventh International Conference on 2010 Apr 12 (pp. 804-809). IEEE.

[3] Eisenhauer M, Rosengren P, Antolin P. A development platform for integrating wireless devices and sensors into ambient intelligence systems. InSensor, Mesh and Ad Hoc Communications and Networks Workshops, 2009. SECON Workshops' 09. 6th Annual IEEE Communications Society Conference on 2009 Jun 22 (pp. 1-3). IEEE.

[4] Amaxilatis D, Georgitzikis V. Using Codebender and Arduino in Science and Education. InSystem-Level Design Methodologies for Telecommunication 2014 (pp. 119-134). Springer, Cham.

[5] David N, Chima A, Ugochukwu A, Obinna E. Design of a home automation system using arduino. International Journal of Scientific & Engineering Research. 2015 Jun;6(6):795-801.

[6] Simić M, Stojanović GM, Manjakkal L, Zaraska K. Multi-sensor system for remote

environmental (air and water) quality monitoring. InTelecommunications Forum (TELFOR), 2016 24th 2016 Nov 22 (pp. 1-4). IEEE.

IOP Conf. Series: Materials Science and Engineering **384** (2018) 012008 doi:10.1088/1757-899X/384/1/012008

IoT based Air Quality Monitoring

F N Setiawan* and I Kustiawan

Department of Electrical Engineering Education, Universitas Pendidikan Indonesia, Jl. Dr. Setiabudhi No. 207, Bandung 40154, Indonesia

*finopse@student.upi.edu

Abstract. Air pollution is a mixture of solid particles and gases in the air. Car emissions, chemicals from factories, dust, pollen and mold spores may be suspended as particles. Effect of air pollution has many bad things and the others may cause problems to our health, for instance, asthma, cough, and lung disorders. In addition, the pollutant can cause global warning, acid rain, and disturbing plant growth. Basically, a human cannot determine whether the air is good or not. Hence, it is necessary to have a tool that can measure the air quality. This research is purposed to design an air quality monitoring system by utilizing esp8266 module. As the result, users can monitor the air quality using smartphone connected through ESP8266 Wi-Fi. Therefore the air condition can be monitored every time. Currently, there is so much air pollution cases that actually can be changed if we are aware. In other words we can contribute as part of the solution instead part of the pollution.

1. Introduction

Nowadays the air condition is very polluted. In recent years, car emissions, chemicals from factories, smoke and dust are everywhere. That is the reason why now air condition is very polluted. The effect of air pollution is very bad for our health, especially for place where the air in our body is taken for breathing. In our lungs may cause some diseases, such as asthma, cough, lung disorders[1].

The air pollution cannot be detected by human feelings. The air pollution may contain a lot of dangerous substances, such as LPG gas, smoke, carbon monoxide, methane[2]. Substances in the polluted air are very dangerous. For example, if the carbon monoxide is above 100ppm, it makes human feel dizzy, nauseous, and within minutes they could die.

This research makes human find out which content of the air is polluted. With module nodemcu esp8266, we can monitor the air pollution remotely, because there is a Wi-Fi in nodemcu esp8266. This makes the air condition can be monitored every time.

2. Method

In this research, nodemcu esp8266 is used as the micro controller. This board has Wi-Fi module that acts as the internet connector and information accessing for the air quality[3]. This is the reason why this board is chosen as some of tools for this research, besides the price of this board is very cheap as well. This is the input pin and picture of nodemcu esp8266:

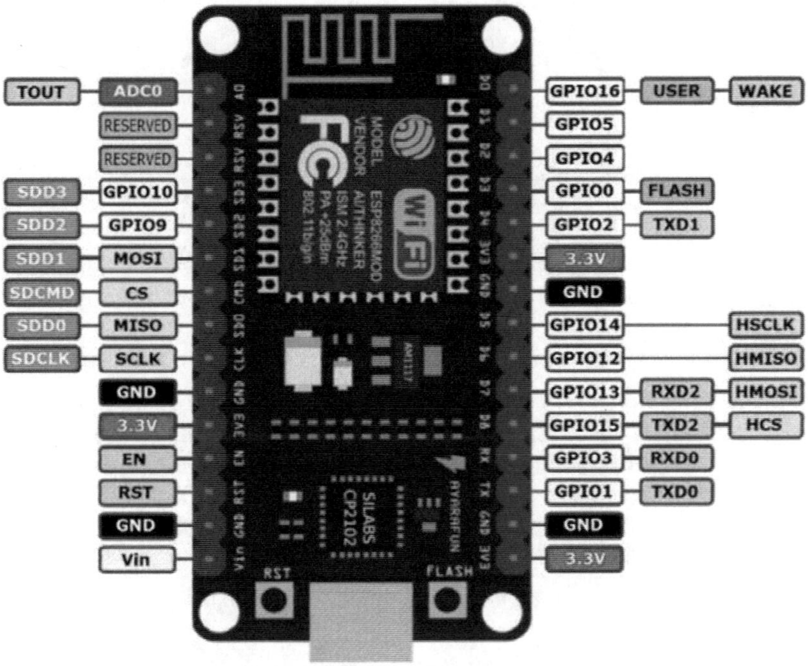

Figure 1. Nodemcu esp8266 pins I/O

In addition, the tools which are needed in this research as in Table 1.

Table 1. Tools of air quality monitoring

1.	Nodemcu ESP8266
2.	MQ2 gas sensor
3.	MQ9 gas sensor
4.	Analog Multiplexer 4051 (CD4051BE)
5.	1.3 inches OLED monitor SSD1306
6.	180ohm resistor x2 and 330 ohm resistor x2
7.	Adapter 5v
8.	ZH03A Laser Dust Sensor

This project is to integrate 3 sensors that detect different kinds of gases and particles with Nodemcu before the data detected is sent to the internet. The 3 sensors are MQ2, MQ9 sensor for carbon monoxide and zh03a particle sensor for PM1.0, PM2.5 and PM10 (dust).

MQ2 sensor module is used for gas leakage detection (home and industry). It is suitable for detecting H2, LPG, CH4, CO, Alcohol, Smoke or Propane. Due to the high sensitivity and fast response time of it. But, in this design, we used MQ2 sensor for smoke situation. This sensor has four outputs, namely, Vcc, ground, D0, and A0. Not all outputs will be connected to nodemcu. However, only D0 that is not connected to nodemcu because this sensor only can choose one output A0 or D0.

Figure 2. MQ2 sensor

The same features with the MQ2 sensor, MQ9 sensor is sensitive and also fast response time. This sensor is suitable for detecting H2, LPG, CH4, CO, Alcohol, Smoke or Propane. The utilization r is for detecting the Carbon Monoxide (CO). The shape of this MQ9 is the same as the sensor MQ2, but it is bigger than MQ2[4]. For the output of this sensor, it is same with MQ2 output.

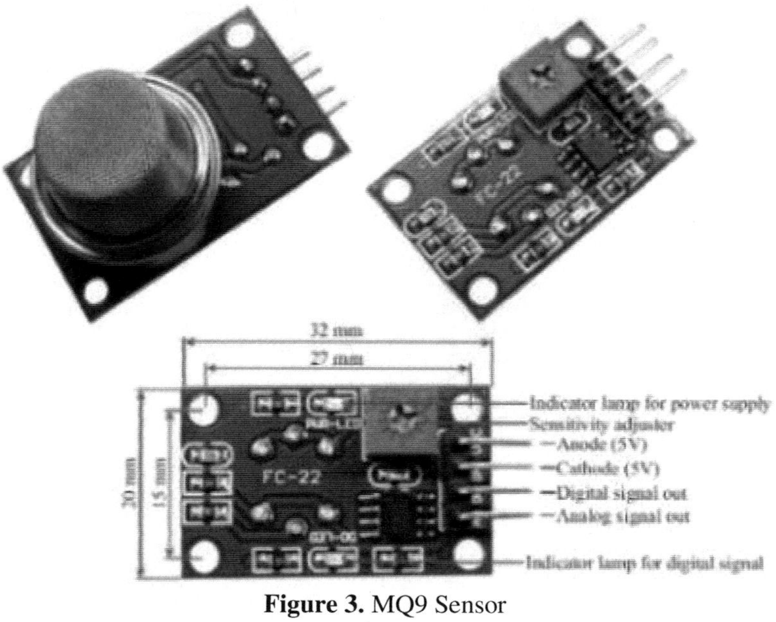

Figure 3. MQ9 Sensor

The last sensor is ZH03A, that is, dust sensor. ZH03A module is a common type, it is a small size sensor that uses laser scattering principle to detect the dust particles in air, with good selectivity and stability[5]. ZH03A has eight pins, but only four pins are needed to be connected to nodemcu. Those are vcc, ground, rx, and tx.

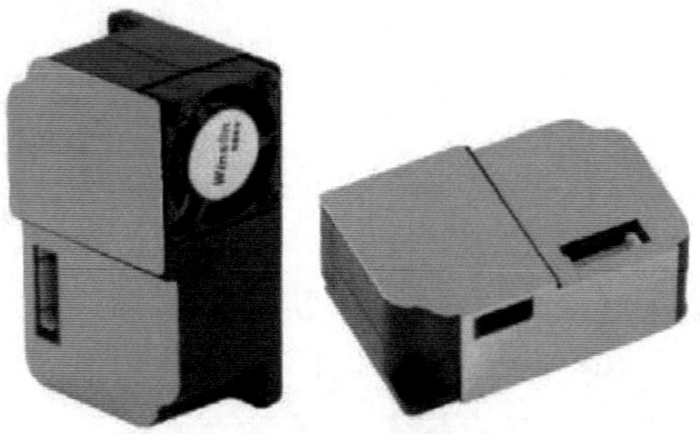

Figure 4. ZH03A dust sensor

The diagram in Figure 5 shows how this research works:

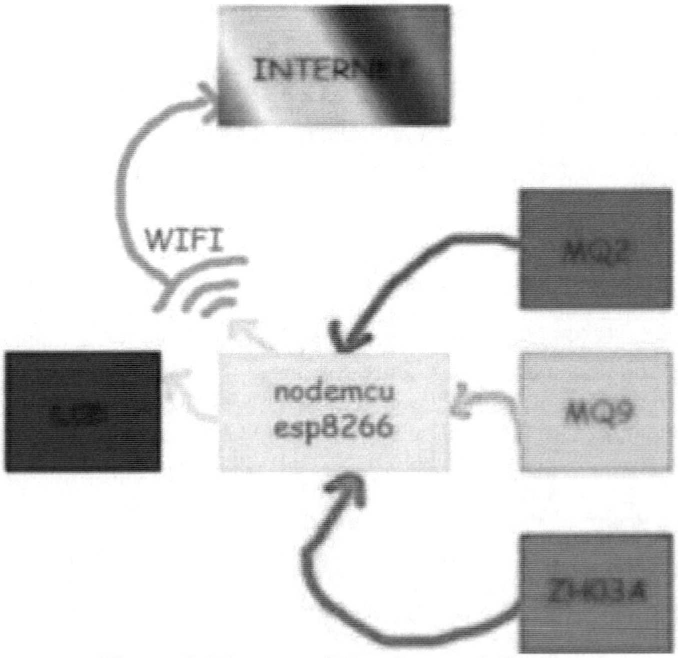

Figure 5. Diagram of How Research Works

Nodemcu as a microcontroller will send the result of sensor readings to LCD and Internet. The internet connection can be served from smartphone tethering or router internet. Thinkspeak is used as media to read the results from nodemcu. As long as nodemcu connects to the Internet, the result readings can be monitored every time.

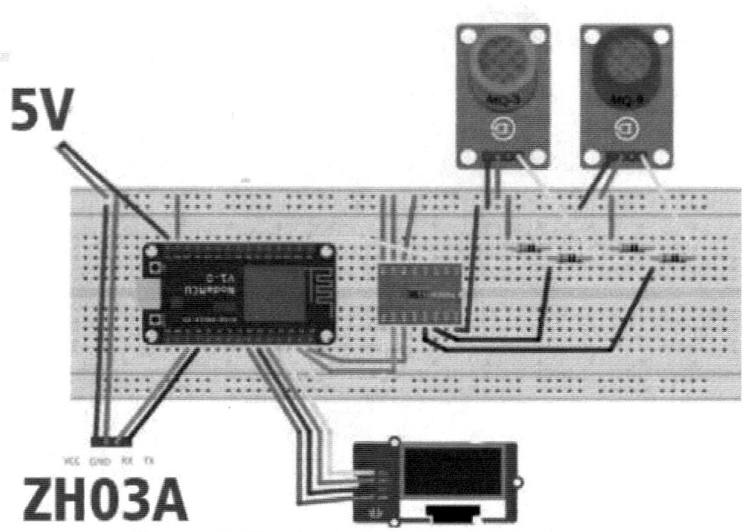

Figure 6. Circuit of air quality monitoring

The picture in figure 6 shows the connection of all different components. One drawback of Nodemcu is it only has one analog I/O pin. In this project, there are 2 sensors, those are MQ2 and MQ9 gas sensor that use an analog output. In this case, a technique called multiplexing can be used. Instead of reading output from 2 sensors at the same time, the data is read one by one. In this project, an analog multiplexer CD4051BE is used. It is a 8-channel multiplexer which can operate with 8 outputs.

Another drawback of Nodemcu is that it only supplies 3.3 V output. While, for other components and three sensors used in this project, they need 5V power supply to operate. Therefore, the adapter 5 V was used to source of all sensors. The analog output of MQ2 and MQ9 are between 0V to 5V whereas the analog pin of nodemcu is only able to read between 0V to 3.3V. This means that nodemcu cannot read the data if MQ2 or MQ9 sensor output is above 3.3V. The data read is not accurate. Therefore, voltage is needed step down.

In this research, voltage divider by two resistors was used. The voltage output is determined by the ratio of the value of two resistors, 180 ohm as R1 and 330 ohm were used as R2 so that Vout is 3.3V. So, Vin is connected to AO pin of MQ gas sensors while Vout is connected to channels of the multiplexer[6].

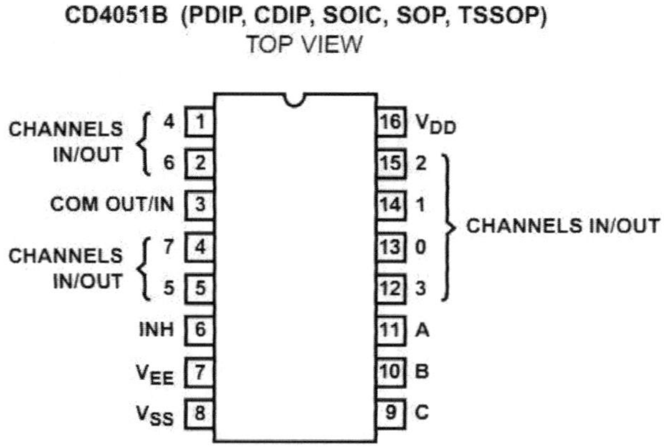

Figure 7. CD4051BE Multiplexer

IOP Conf. Series: Materials Science and Engineering **384** (2018) 012008 doi:10.1088/1757-899X/384/1/012008

Multiplexer connection:

- Vdd (pin 16) to 5V supply
- INH, V_{EE}, V_{SS} (pin 6, 7, 8) to nodemcu GND pin
- Common out/in (pin 3) to nodemcu A0 pin
- A, B, C (pin 11, 10, 9) to nodemcu D0, D1, D2

A, B, C (pin11, 10, 9) are used to select channel. 3-digit binary number is formed in the order of CBA. As we use channel 1 and 2, binary number of 1 and 2 in 3-digit binary number are 001 and 010 respectively. Therefore, when we want output of channel 1, then D0 output 1, D1 and D2 output 0.When we want output of channel 2, D0 output 0, D1 output 1 and D2 output 0.

After all parts are ready to combine, we need PCB to make the board simpler and easier. If PCB is not used, then the flow of current may be disturbed, making all sensors does not work. We use Arduino IDE for software side. But, first nodemcu board must be downloaded first.

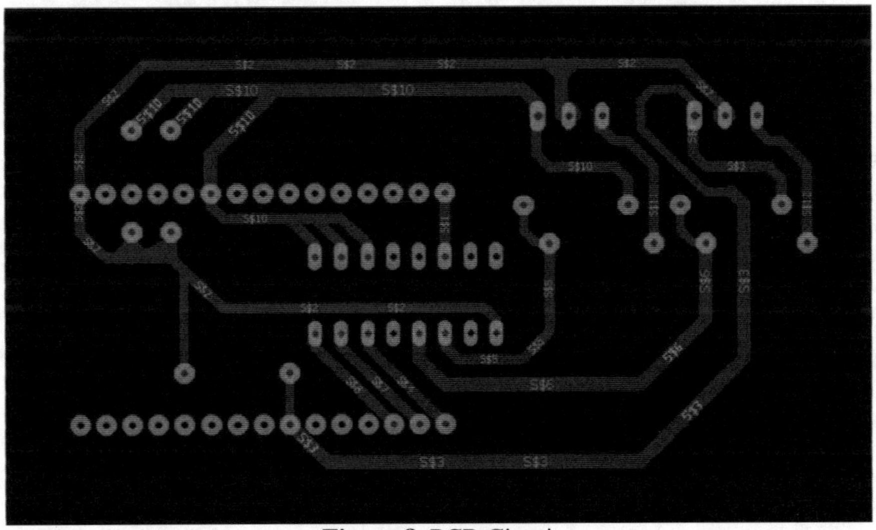

Figure 8. PCB Circuit

When nodemcu connects to internet, nodemcu will send the result of monitoring to thingspeak. ThingSpeak has APIs for collecting data produced by sensors and reading the data from applications[7].

3. Result and Disscusion

Last step after wiring all parts to PCB board is covering all parts with the box. All sensors must be contacted to the air directly. The box needs to drill until the sensors and LCD fit with the box. And then, the all parts seem good looking with the box. The dimension of box is 18 cm x 11 cm x 6 cm. The Figure 9 shows the front and rear view of the box. It can be seen, the box is not too small nor too big as well, so the box is easy to carry anywhere.

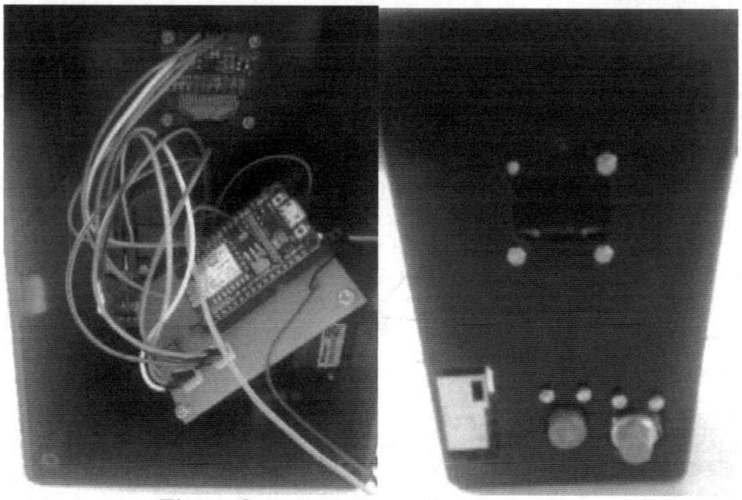

Figure 9. Box of air quality monitoring

Next, we will illustrate the result of the measurement from sensors. The data is taken from the measurement of a cigarette:

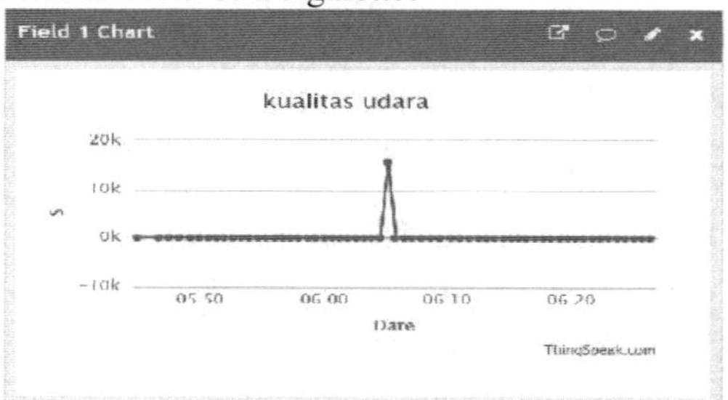

Figure 10. Results of smoke content readings

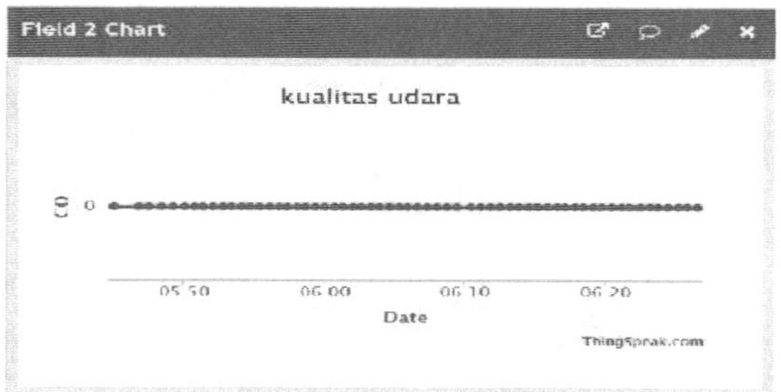

Figure 11. Carbon monoxide reading results

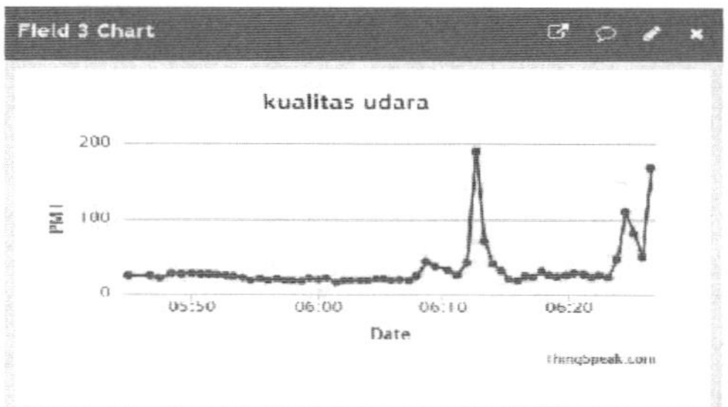

Figure 12. Results of PM1 Reading

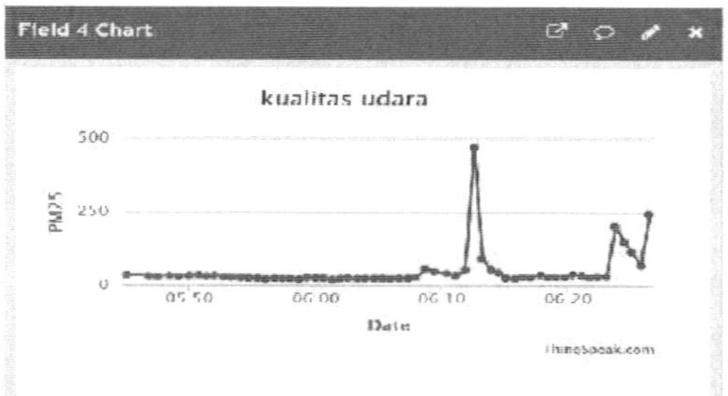

Figure 13. Results of PM2.5 Reading

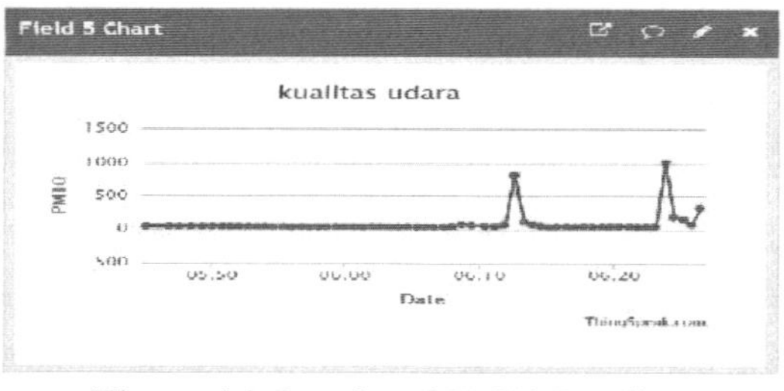

Figure 14. Results of PM10 Reading

The results show that if the red line is upward, the particle level of pollution is higher and dangerous. At level of 0-50 ppm, healthy air falls into the healthy category, while in the range 50 – 100 ppm is categorized medium air (no effect on health). The range of 100-200 ppm, air is not healthy, 200-300 ppm is very unhealthy, while 300 upwards is said to be very dangerous [8]. As shown in figure 10, the smoke point indicates that the smoke level is above 10k, indicating that the ambient air condition is very dangerous.

Carbon monoxide (CO) when inhaled into the lungs will participate in blood circulation and will block the entry of oxygen needed by the body. This can happen because CO gas is toxic metabolism, participating to react metabolically with blood. Like oxygen, CO gas readily reacts with blood (hemoglobin) [9]. Figure 11 shows that no CO content contained in cigarettes, indicating that there is no CO content in the air.

Air pollution problems generally are caused by solid particles of TSP (Total Suspended Particulate or total drift particle) with a maximum diameter of about 45 mm, particles of PM10 (particulate matter) with diameters less than 10 mm and PM2.5 with diameters less than 2.5 mm. These particles are believed by environmental and public health experts to trigger respiratory infections, because solid particles of PM10 and PM2,5 can precipitate in the respiratory tract of the bronchi and alveoli areas, whereas TSP cannot be inhaled into the lungs, but only at upper respiratory tract [10]. Figure 15 shows that the ppm content is almost 500, implying that the air is dangerous and Figure 16 illustrates that the air condition is dangerous because the ppm content is almost 1000.

4. Conclusion

An air quality detector is very important because nowadays air pollution is easy to find. For the air pollution which cannot be easy detected by human, it requires a device as a reader of the air quality. By this research, we can avoid air pollution through monitoring the air quality regularly.

References

[1] A Boubrima, W Bechkit, and H Rivano 2017 *A new WSN deployment approach for air pollution monitoring,* 2017 14th IEEE Annu. Consum. Commun. Netw. Conf. CCNC 2017, pp. 455–460

[2] B Bathiya 2016 *Air Pollution Monitoring Using Wireless Sensor Network* 2016 IEEE International WIE Conference on Electrical and Computer Engineering (WIECON-ECE) 19-21 December 2016

[3] A Škraba, A Koložvari, D Kofjač, R Stojanović, V Stanovov, and E Semenkin 2016 *Streaming pulse data to the cloud with bluetooth le or NODEMCU ESP8266* 2016 5th Mediterr. Conf. Embed. Comput. MECO 2016 - Incl. ECyPS 2016, BIOENG.MED 2016, MECO Student Chall. 2016, pp. 428–431, 2016

[4] K Keshamoni and S Hemanth 2017 *Smart gas level monitoring, booking & gas leakage detector over iot* Proc. - 7th IEEE Int. Adv. Comput. Conf. IACC 2017, pp. 330–332

[5] D Spirjakin, A M Baranov, and V Sleptsov 2015 *Design of smart dust sensor node for combustible gas leakage monitoring* Proc. 2015 Fed. Conf. Comput. Sci. Inf. Syst. FedCSIS 2015, vol. 5, pp. 1279–1283, 2015

[6] H Sohal 2016 *Area and Power Analysis of Adiabatic 2x1Multiplexer Designon 65nm CMOS Technology* ECE

[7] N K Walia, P Kalra, and D Mehrotra 2016 *An IOT by information retrieval approach: Smart lights controlled using WiFi* Proc. 2016 6th Int. Conf. - Cloud Syst. Big Data Eng. Conflu. 2016, pp. 708–712, 2016

[8] M C Budai, Á Lengyel, L Vajda, A Vidács, R Vida, and M Šimek 2015 *Monitoring Health Status of Relatives – ' The Smoke in the C himney ' Approach,* pp. 601–606

[9] M Oprea, M Popescu, and S F Mihalache 2016 A Neural Network Based Model for PM 2 . 5 Air Pollutant Forecasting *International Conference on System Theory, Control and Computing (ICSTCC),* October 13-15 vol. **1**, no. 2, pp. 776–781, 2016.

[10] M J Gonzales and L D Riek 2013 *Co-designing Patient-centered Health Communication Tools*

for Cancer Care 7th International Conference on Pervasive Computing Technologies for Healthcare and Workshops 2013.

IOP Conf. Series: Materials Science and Engineering **384** (2018) 012009 doi:10.1088/1757-899X/384/1/012009

Optimization of Data Communication on Air Control Device Based on Internet of Things with Application of HTTP and MQTT Protocols

F Luthfi*, E A Juanda, and I Kustiawan

Department of Electrical Engineering Education, Universitas Pendidikan Indonesia, Jl. Dr. Setiabudhi No. 207, Bandung 40154, Indonesia

*fityaluthfi@student.upi.edu

Abstract. In this research, we proposed ALDEBARAN (Air Control Device Based on Internet of Things), that is, a device the HTTP protocol which can be applied as a protocol to communicate with the IoT ANTARES (Internet of Things Application and Technology Platform as your Reliable Solution) platform. Wi-Fi connectivity is needed to build a communication between devices and platforms, also to sends data continuously. Because the HTTP has a high overhead protocol resulting in waste of power consumption, so it needs to be optimized. The focus of this research is to optimize data communications on our device to create data communication and power saving efficiently. The method used is implementing the HTTP and MQTT protocols that are combined with several data transmission algorithms. Parameters such as latency and current consumption are compared and analysed to obtain the most optimal combination of protocols and algorithms. The result of this research is that MQTT protocol can be applied to our device and data communication with our scheme becomes more optimal.

1. Introduction

Recently, Internet of Things (IoT) received more attention because of the development of wireless technology [1]. The development of wireless technology allows electronic devices to be interconnected into a network and exchange data. IoT are the objects that are interconnected to one another so as to form a group that can obtain data infrastructure and access data [2].

Every day, IoT is developed by individuals and industries resulting in many web domains being used. And in the coming years will grow rapidly because IoT applications for RFID communication systems, NFC, M2M, and V2V continue to be developed [1]. Industry in Indonesia such as PT. Telekomunikasi Indonesia, Tbk opened a new field of research, that is, IoT platform. The field developed is IoT ANTARES platform (Internet of Things Application and Technology Platform as your Reliable Solution). One of IoT devices which have been developed by the authors and colleagues in internship program was named ALDEBARAN.

In 2013, research of lightweight protocol on smartphones said that MQTT was designed for communication with a very limited resource such as sensors, but the protocol can also be implemented on a smartphone [3]. In 2015, research regarding the evaluation of MQTT protocol in terms of the stability of the mobile communication network said that MQTT performance remained good even though the network is not stable [4]. In 2016, research on comparisons between HTTP and MQTT performance have been compared and declared that MQTT have a better performance than HTTP [5].

The focus of this research is to analyse the data communications on ALDEBARAN by implementing the HTTP and MQTT protocols combined with several data transmit algorithms. The purpose of this research is to optimize data communication, especially to make communication with servers more efficient and make power-efficient data communications on ALDEBARAN devices.

2. Research Method

2.1 Use Case

ALDEBARAN, a device with a room temperature keeper on air conditioner uses an infrared communication. A device can serve controlling the air conditioner via the Internet, making it easy to monitor room temperature in a building or an office which has a lot of air conditioner. The data logger for the temperature of the room is integrated with the IoT ANTARES platform using a Wi-Fi connection. Our device can be seen in Figure 1.

Figure 1. ALDEBARAN Physical Form

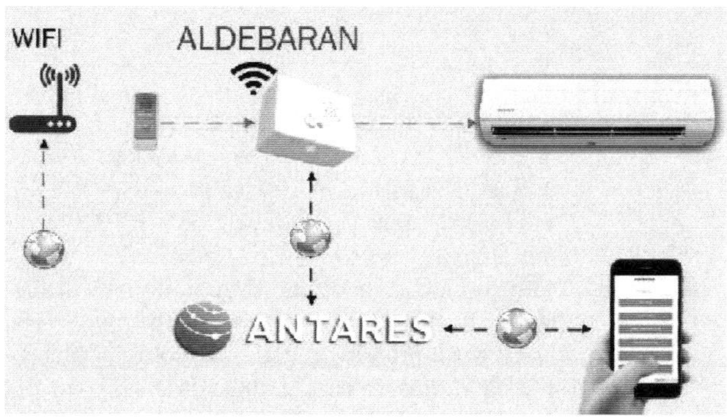

Figure 2. Ilustration of ALDEBARAN Working Principle

The work principle of our device is to replace the conventional air conditioner remote controller into controller remotely via the internet. ALDEBARAN is equipped with LED (Light Emitting Diode) infrared 920 nm as infrared signal transmitter, TSOP34838 infrared receiver as infrared signal receiver, temperature sensor DHT22 as a reader at room temperature, and the microcontroller ESP12-E are packed together with Wi-Fi transceivers so it can be a Wi-Fi client as well as Wi-Fi access point (AP). The device runs with 2 modes, that is, ie running mode and config mode. The working principle of the device can be illustrated in Figure 2.

ALDEBARAN can access an Internet connection from Wi-Fi access point/ router. Communication link is illustrated by a black line so that our device, ANTARES servers, and smartphones can connect via the internet to exchange data. Smartphone data which is sent to the server contain control commands AC, while the data from ALDEBARAN sent to the server contain information of room temperature. Both figures are bridged by IoT ANTARES platform as cloud storage for exchanging data by accessing the API (Application Programming Interface).

2.2 Software Aspect

Algorithms to send and retrieve data that have been designed applied to the pre-existing ALDEBARAN algorithm. The certain algorithm only changes the part of running mode. The algorithm is written in C ++ programming language with Arduino IDE (Integrated Development Environment) as a compiler.

The application of this protocol can be done by adding library on Arduino IDE. The HTTP protocol can use the library <WiFiClient.h> while MQTT protocol can use the library <PubSubClient.h>. Library is used to support all the needs of the protocol implementation.

IoT ANTARES platform is used to store sensor data and device control commands, as well as an ideal platform for local developers, due to free account creation and free API access up to 10,000 access/ day.

2.3 Algorithm Design

In this research, we apply continuous, periodic and adaptive data transmit algorithms to the running mode of ALDEBARAN. This algorithm is divided into several stages, reading data from sensors, connected to the Wi-Fi, establishing a connection with the server, GET data from the server, decision making IR control systems, and POST data to the server. Some stages are made into several design algorithms and will be applied and compared. The purpose of the design of the algorithm is to get the most efficient in terms of power consumption when the algorithm is applied to our device. The designed algorithms can be applied to devices using Wi-Fi connectivity as ESP8266 and other microcontrollers are equipped with Wi-Fi Shield. The algorithm for storing and retrieving data continuously can be seen in Figure 3(a) and periodically can be seen in Figure 3(b).

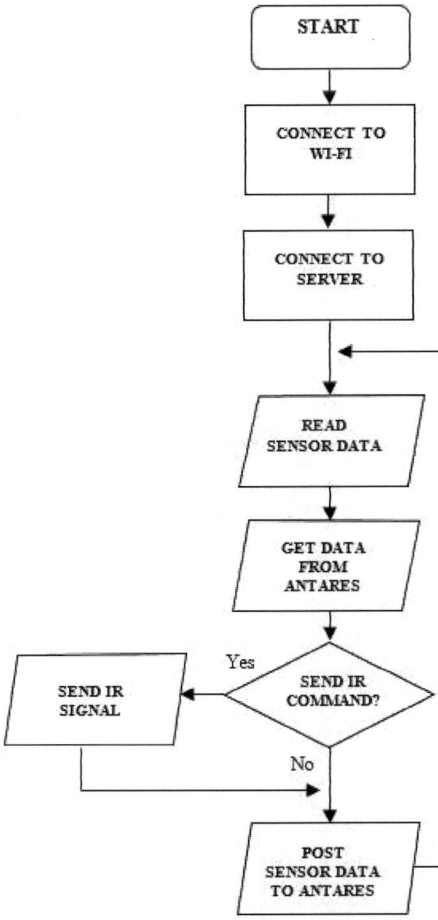

Figure 3(a). Flowchart of Continuous Data Transmit

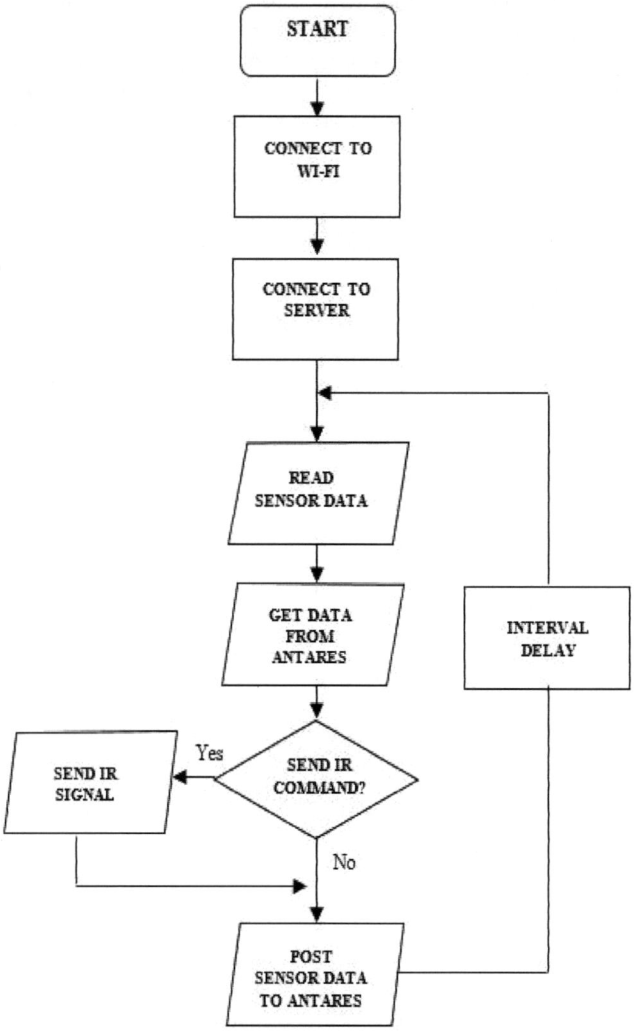

Figure 3(b). Flowchart of Periodic Data Transmit

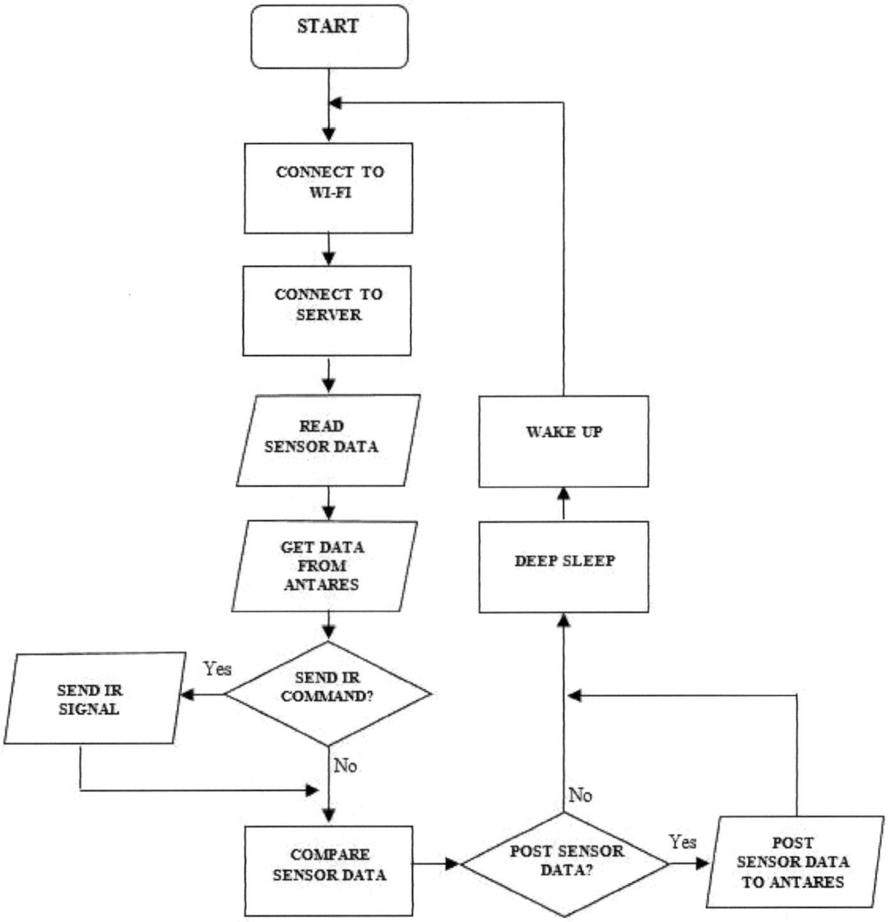

Figure 4. Flowchart of Adaptive Data Transmit

Flow chart on Figure 3 shows several stages including connecting to Wi-Fi, reading sensor data, establishing a connection with the server, retrieving data from the server that contain AC control commands, AC control decision-making functions, and storing sensor data to the server. In step of connection to Wi-Fi, ALDEBARAN tries to enter the Wi-Fi of the selected SSID. Once the device has connected to the Internet, the device will try to connect to the server. In this research, the host used as a server is platform.antares.id with port 8080 for HTTP and 1883 for MQTT. The device reads sensor data in the room such as temperature and humidity values that are accommodated on the variable with float data type. Once connected to the server, then ALDEBARAN retrieves latest data from a server with access to the IoT ANTARES platform API using HTTP requests with the GET method. The data contains an AC control command sent by the smartphone and data received by ALDEBARAN in the form of data containing headers and body with JSON / XML format. The received data is parsed until an AC control command is obtained.

2.3.1. Continuous Data Transmit Algorithm This algorithm describes the stages of storing and retrieving data continuously without delay interval and used for data communication in real time. Our device applies this algorithm to support the air conditioner control functions via infrared communication.

At this step of the decision-making control function, the control command is received then is used as basis for a decision to send an infrared signal. The last step of the algorithm is to store data to the server with access the ANTARES API. The sent data contains value of sensor obtained previously. Furthermore, the program will perform the reading of the sensor data again and repeat the process of storing and retrieving data continuously until the device is turned off.

2.3.2. Periodic Data Transmit Algorithm This algorithm describes the stages of storing and retrieving data periodically with interval delay and used for asynchronous data communication. ALDEBARAN applying this algorithm to support the air conditioner control functions via infrared communication.

The difference in this algorithm is delay interval that can be set when ALDEBARAN are in config mode. As the time delay was performed, the status of Wi-Fi is idle. Furthermore, the program will repeat to perform the reading of the sensor data periodically until the device is turned off.

2.3.3. Periodic Data Transmit Algorithm with Deep Sleep Mode This algorithm describes the stages of storing and retrieving data periodically with interval delay using the RTC in deep sleep mode and is used for asynchronous data communication. Network reliability and overall improvement of energy efficiency can be done using deep sleep [6]. ALDEBARAN applies this algorithm to support the air conditioner control functions via infrared communication.

The difference, delay interval here is the time to perform deep sleep mode. At the time of deep sleep was performed, the Wi-Fi status and CPU are switched off. Once the countdown of the time interval delay has been completed, then the device will wake-up from sleep mode in the state of Wi-Fi is turn off. Devices need to repeat the steps to connect to Wi-Fi and to connect to the internet. Then the program will repeat to perform the reading of the sensor data periodically and repeat the process until the device is turned off.

2.3.4. Adaptive Data Transmit Algorithm This algorithm describes the stages of storing and retrieving data periodically with interval delay using the RTC in deep sleep mode and adaptive phases of data transmission that is used for asynchronous data communication. ALDEBARAN applying this algorithm to support the air conditioner control functions via infrared communication. Storing and retrieving data with adaptive data transmit algorithm can be seen in Figure 4.

Up to the stage decision-making control function, the algorithm is still the same. The next step is to compare the present sensor data with previous data on the last cycle of deep sleep. Sensor data is sent to the server in the previous cycle and used as a threshold decision. Threshold consists of upper and lower threshold determined by summing $\pm$ threshold deviation on the values read previously. The threshold deviation can be set in the system config menu. In this research the adaptive algorithm uses a threshold deviation of 1 degree Celsius which means the threshold range of ± 0.5. The threshold deviation is determined from the standard temperature changes in air conditioner that is every 1 degree Celsius. If the new data is still within the range of the threshold, then the data may not be sent to the server and the device will be directly into deep sleep mode. If the new data from outside the range of the threshold, the threshold is updated and the data is sent to the server and the device may enter the deep sleep mode. As the time of deep sleep was performed, the Wi-Fi status and CPU are switched off. Once the countdown of the time delay interval has been completed, then the device will wake-up from sleep mode in the state of Wi-Fi is turn off. Devices need to repeat the steps to connect to Wi-Fi for accessing the internet. Then the program will repeat to perform the reading of the sensor data periodically and repeat the process until the device is turned off.

The time to perform one iteration cycles depends on the time connected to Wi-Fi, establishing a connection with the server, the reading of sensor data, data retrieval from the server, sending data to the server, and sleep mode. The total amount of time is the system cycle time of this algorithm.

3. Results and Discussion

3.1. Algorithm Performance Comparison

The experiments are done through transmitting data every 20 minutes and the results are compared on any combination of algorithms with protocols. The result of algorithm implementation is system cycle time, then obtained the average current and calculated the battery usage in each combination of algorithm with HTTP and MQTT protocol. Algorithm performance comparison results at intervals of 20 minutes can be seen in Figure 5 and 6.

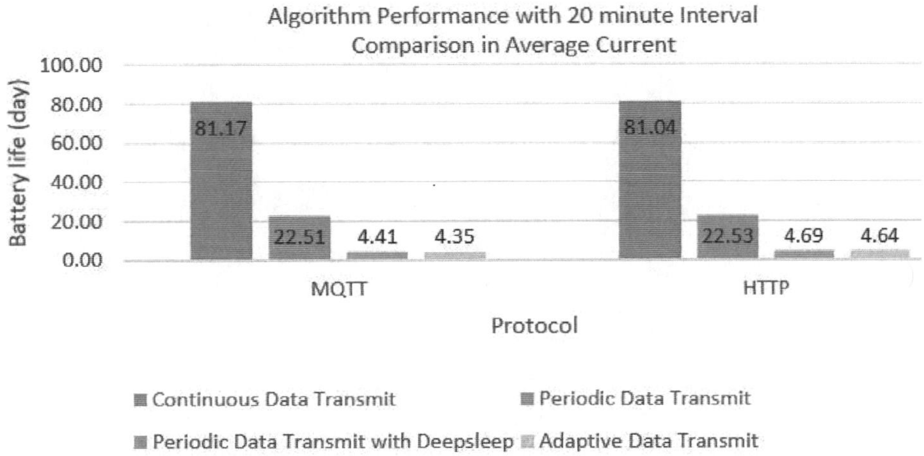

Figure 5. Average Current Comparison

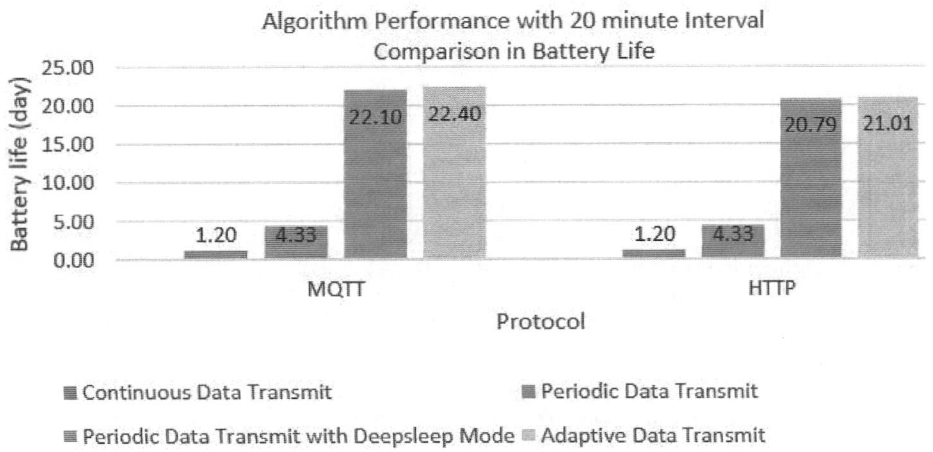

Figure 6. Battery Life Comparison

Based on Figure 5 and 6, it can be seen that the algorithm with the lowest average current is adaptive data transmit algorithm using MQTT protocol with average current of 4.35 mA. Current consumption can be reduced from other data transmit algorithms due to the data on adaptive algorithm should be larger or smaller than the threshold range to send the data. Data sent to the server when the value of the read sensor is changed significantly. Based on the calculation, application of adaptive algorithm with MQTT protocol at 20 minute intervals is able to make ALDEBARAN in active conditions for more than 22 days.

IOP Conf. Series: Materials Science and Engineering **384** (2018) 012009 doi:10.1088/1757-899X/384/1/012009

When compared with the adaptive data transmit algorithm using the HTTP protocol, the average amount of current that is equal to 4.64 mA. The average current of 4.64 mA is resulted in longer battery usage for 21 days. This means that the application of MQTT protocol on adaptive data transmit algorithm at intervals of 20 minutes can make the battery more durable for almost 1 days. The reason is, the implementation adaptive data transmit algorithm is algorithm with the lowest system cycle time in periodic transmission category. System cycle time of each algorithm can be seen in Figure 7.

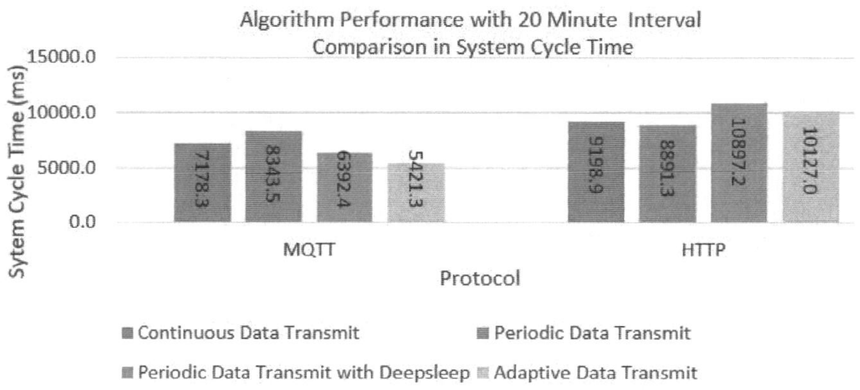

Figure 7. Graph of System Cycle Time Comparison on Each Algorithm

From all the comparison with several parameters, there is a continuous algorithm using the HTTP and MQTT protocol. We can see the comparison of continuous algorithm parameter obtained with a total system cycle time of 7178.3 ms and 9198.9 ms for the MQTT and the HTTP protocols, respectively. Implementation of MQTT protocol can reduce the system cycle time so that when applying to the continuous data transmit algorithm, it will serve better real-time communication.

3.2. Comparison of Latency and Throughput on the HTTP and MQTT Protocol

Next, we discuss on latency for each implementation of the algorithm and protocol. Its value are averaged and compared. The high latency data transmission is resulted in a long delay and low latency value indicates that the data transmitted more quickly to the receiver. Latency on MQTT protocol can be lower because MQTT have a low overhead protocol. MQTT is an application layer protocol on OSI based TCP/ IP model that has a fixed header size of 2 bytes for each data packet. Therefore, MQTT is very suitable for low bandwidth networks [8]. Latency to retrieve the 2 bytes data by using MQTT protocol shows 4.66 times lower than the HTTP. Latency to transmit 5 bytes of temperature data using MQTT protocol shows 5.68 times lower than the HTTP. Latency to send the 5 bytes of humidity using MQTT protocol shows 5.84 times lower than the HTTP. This means that by using MQTT protocol, the efficiency and performance of data transmission time can be improved. Latency comparison results can be seen in Figure 8.

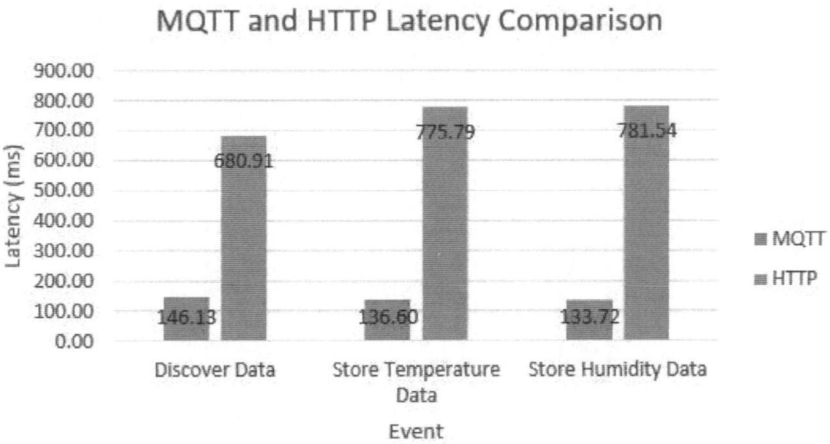

Figure 8. Comparison of Average Latency on the HTTP and MQTT Protocols

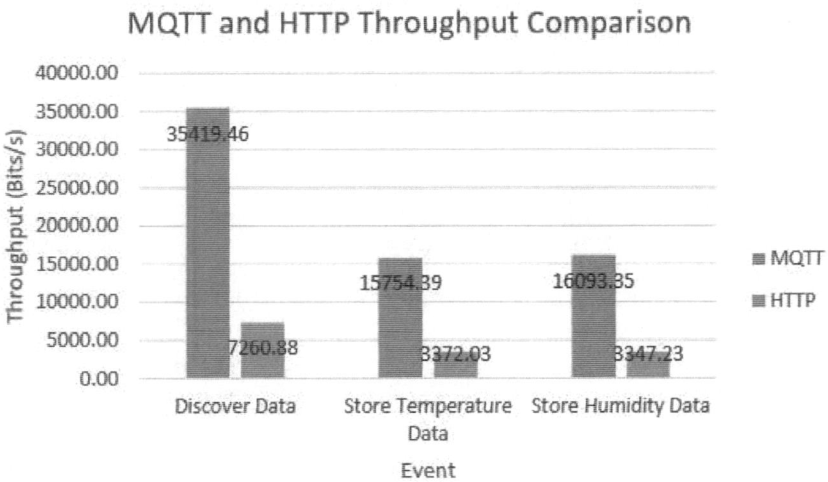

Figure 9. Comparison of Throughput on the HTTP and MQTT Protocols

Large throughput is influenced by the amount of data sent and latency. The lower latency is resulted in the higher throughput, while the higher latency is resulted in lower throughput. Based on previous calculations, MQTT protocol has a low latency as a consequence of higher throughput than the HTTP protocol. The MQTT throughput of retrieving data from the server is 4.88 times higher than HTTP, while sending temperature data to the server is 4.67 times higher than HTTP, and when sending humidity data to the server, the result is 4.81 times higher than HTTP. Comparison of throughput on the HTTP and MQTT protocol can be seen in Figure 9.

4. Conclusion

HTTP and MQTT protocols can be applied to ALDEBARAN devices. Based on our experiments, MQTT protocol characteristics have low latency, low overhead, high throughput and low transmission power compared to HTTP. Each application MQTT algorithm becomes longer battery usage. In comparison, the combination of adaptive data transmit algorithm with MQTT protocol is chosen as the most optimal combination used for non-real time communication on ALDEBARAN devices. This is because of lower power usage, bandwidth efficiency, and increased throughput.

IOP Conf. Series: Materials Science and Engineering **384** (2018) 012009 doi:10.1088/1757-899X/384/1/012009

Another consideration, for real time communication, we can use a combination of Continuous Data Transmit Algorithm with MQTT protocol as the fastest combination for AC control function because using this method can reduce latency and system cycle time, however it may be wasteful in the use of power when using the battery as a voltage source.

References

[1] S. H. Shah and I. Yaqoob, "A survey: Internet of Things (IOT) technologies, applications and challenges," In 2016 IEEE Smart Energy Grid Engineering (SEGE), pp. 381–385.
[2] B. Dorsemaine, J. P. Gaulier, J. P. Wary, N. Kheir, and P. Urien, "Internet of Things: A Definition and Taxonomy," Proc. - NGMAST 2015 9th Int. Conf. Next Gener. Mob. Appl. Serv. Technol., pp. 72–77, 2016.
[3] N. De Caro, W. Colitti, and K. Steenhaut, "Comparison of two lightweight protocols for smartphone-based sensing," In: Proceedings of the 2013 IEEE 20th symposium on communications and vehicular technology in the Benelux (SCVT), Namur, 21 November 2013, pp.1–6. New York.
[4] J. E. Luzuriaga, M. Perez, P. Boronat, J. C. Cano, C. Calafate, and P. Manzoni, "A comparative evaluation of AMQP and MQTT protocols over unstable and mobile networks," 2015 12th Annu. IEEE Consum. Commun. Netw. Conf. CCNC 2015, pp. 931–936, 2015.
[5] T. Yokotani and Y. Sasaki, "Comparison with HTTP and MQTT on required network resources for IoT," 2016 Int. Conf. Control. Electron. Renew. Energy Commun., pp. 1–6, 2016.
[6] H. Lin, "Deep Sleep IEEE 802.11 Enhancement for Energy-Harvesting Machine-to-Machine Communications.pdf," pp. 5231–5236, 2012.
[7] Y. Upadhyay, A. Borole, and D. Dileepan, "MQTT based secured home automation system," 2016 Symp. Colossal Data Anal. Networking, CDAN 2016, 2016.
[8] Chyi-Ren Dow, Syuan Cheng, Shiow-Fen Hwang, "A MQTT-based guide and notification service system," 2016 IEEE 7th Annual Information Technology, Electronics and Mobile Communication Conference (IEMCON) , pp. 1-4 , 2016-10. University of British Columbia.

IOP Conf. Series: Materials Science and Engineering **384** (2018) 012010 doi:10.1088/1757-899X/384/1/012010

Short-term Electrical Load Prediction Using Evolving Neural Network

B Lareno* **and L Swastina**

Department of Informatic Engineering, Indonesia College of Informatics Management and Computer (STMIK), Jalan Pangeran Hidayatullah, Banjarmasin, Indonesia

*blareno@stmik.id

Abstract. A short-term electrical load prediction can use an artificial neural network approach. In this paper, an optimized neural network, namely Evolving Neural Network (ENN) has been developed for short term electric load prediction. ENN uses a genetic algorithm to optimize the weighting of neural networks. After the feedforward algorithm, the process continues with optimization, instead of learning process normally applied to the neural network. The proposed algorithm is implemented in MatLab. Data from April 2010 to April 2011 will be used as training data and data in May 2011 will be used as checking data. To evaluate the performance of Evolving Neural Network, Wavelet Neural Network (WNN) is also involved for comparison. The evaluation is conducted by observing the prediction results. Performance measurements are performed by observing errors that occur. The smaller the error value, the better the accuracy. The experimental result shows that the accuracy performance of ENN is better than WNN.

1. Introduction

After a large deficit of electricity in 2008, as several gas and steam power plants can be realized at the end of 2017 in Kalselteng system, there will be a surplus of 100 MW. The growth rate of electricity demand, from 2009 to 2019, is projected more or less 8.96% per year. The peak load will be 807 MW in 2019 [**1**].

Table 1. Electricity demand forecast of Kalselteng (Barito System)

Desc.	Unit	2009	2010	2011	2012	2013	2014	2015	2016	2017	2018	2019
Demand	GWh	1,740	1,886	2,063	2,257	2,470	2,704	2,960	3,233	3,524	3,835	4,176
-Growth	%	7.09	8.40	9.37	9.40	9.43	9.46	9.48	9.24	8.98	8.84	8.88
Production	GWh	2,070	2,221	2,416	2,629	2,875	3,145	3,440	3,756	4,090	4,448	4,840
Peak Load	MW	372	396	428	462	501	544	591	640	692	747	807
LF	%	63.53	64.01	64.49	64.97	65.46	65.95	66.45	66.94	67.45	67.95	68.46

Table 1 shows the 2008's forecast of demand, production and peak load until 2019. This growth is in line with economic growth of Kalselteng (Kalimantan Selatan and Tengah)[**2**]. By 2019, PLN (state electricity company) will build 1,413 MW of power plants in Kalselteng (see Figure 1), as well as a 150 kV transmission line of 2,606 Kms to achieve the 92 % electrification target in 2019 for the two provinces[**1**].

In 2016, in normal circumstances, the capacity for Kalselteng was 632 MW with 640 MW peak load. However, electrical power could be reduced in November-December because one unit of PLTU (steam power plant) Asam Asam was undergoing routine maintenance so that rota blackouts were needed to overcome power deficit. In addition, damage interruption of the SUTT (high voltage transmission) of 150 kV, could happen anytime. Eg, occurring in the lane between the Gardu Induk (GI/Substation) Barikin and GI Rantau, caused a blackout in almost the entire Kalselteng region. This disturbance made that the power supply from Bangkanai Power Plant (140 MW) could not be distributed. Power loss of 140 MW cause the voltage frequency in the Kalselteng electrical system to drop, causing some large power plants to suddenly stop operating. For recovery, the plants had to be restarted one by one.

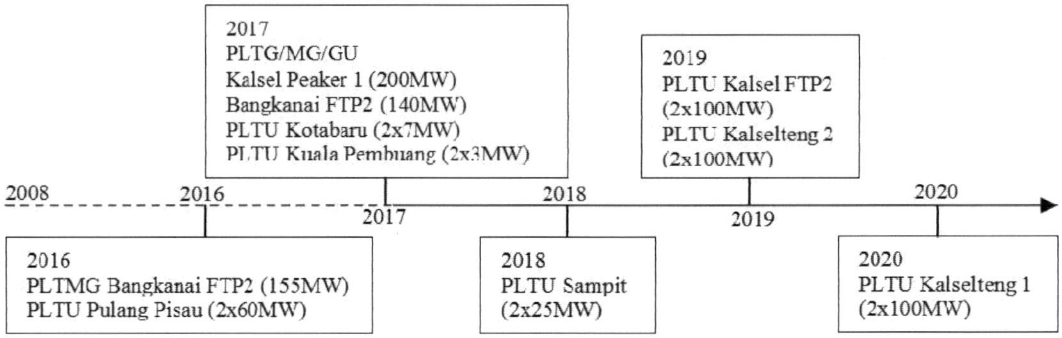

Figure 1. Power plant planning in Kalselteng region

Thus, the fulfillment of dynamic power demanded in critical conditions stated above requires prediction of power consumption to maximize distribution. Electrical power can not be stored in a large quantity so it must be provided when absolutely necessary.

Electrical load data is usually recorded by PLN. It is a time series data, so it can be analyzed and predicted using statistics or soft computing approach. Wang Yong, et al [**3**] proposed a Back Propagation Neural Network (BPNN). They presented an accurate electricity load forecasting algorithm with back propagation neural networks. It contributes to short-term electricity load forecast methodology with a neural network and with weather feature such as max centigrade, min centigrade and weather types. As a result, the electrical load can be predicted more efficiently. But Li Chungui, et al [**4**], who tested the implementation of BPNN in traffic studies, stated that BPNN is not very accurate because it is possibly premature solution. They proposed the processing of genetic algorithms that determine the value of neural network weights. The result, the traffic can be predicted more efficiently. The approach with genetic algorithms has been applied in other areas, such as rainfall prediction and traffic flow. In addition, Yuhui Wang, et al [**5**] declared the Wavelet Neural Network is an excellent tool for predicting rainfall and rainfall-runoff. Gao Guorong and Liu Yanping [**6**] implemented the Wavelet Neural Network (WNN) as an algorithm for predicting the flow of traffic. They claim that WNN is better than BPNN. Lareno [**7**] proposed ENN for short-time traffic flow prediction. The result, ENN is better than WNN, ANFIS and BPNN. Also, Swastina, et al [**8**] proposed ENN for short-time rainfall prediction, with the same result: ENN being better than WNN.

Thus, neural network based algorithm has been known and widely used as an algorithm of time-series data prediction. Therefore, in this paper, an Evolving Neural Network (ENN) algorithm is proposed for predicting short-term electrical load.

2. Research Method

This research uses an experimental research method, which consists of: (1) Data collection and preparation processing, (2) Proposed method, (3) Experimental results and (4) Evaluation and validation of results.

2.1. Data collection and preparation processing

Electrical load data were obtained from PLN Kalselteng. The data required in this study consist of the load from May 1st, 2010 to May 31th, 2011 for the load at 6:00, 10:00, 14:00, 18:00, 19:00, 20:00, 21:00, 22:00, and 23:00 WITA.

The data obtained from the PLN were in the form of data consisting of various parameters, so that they had to be recapitulated first. As a result, data of this process are data with the following attributes: Time and Load (MW). Each row of data consist of data from 9 selected times. Data had been collected for one year and one month, so that there are 3604 data available.

2.2. Proposed method

The algorithm used is the Evolving Neural Network used to predict the electrical load and Wavelet Neural Network (WNN) as compared algorithm. Data from April 2010 to April 2011 will be used as training data and data in May 2011 will be used as checking data. The evaluation is conducted by observing the prediction results. Performance measurements are performed by observing errors that occur through parameters: Root Mean Square Error (RMSE), Mean Absolute Deviation (MAD) and Mean Absolute Percentage Error (MAPE). The smaller the error value, the better the accuracy.

2.2.1. Evolving Neural Network. Evolving Neural Network is a method of weighting between neurons at different layers using the principle of genetic algorithm (GA). Figure 2a, explains the structure of he ENN algorithm [**9**]. Chromosome Representation Encoding ie each gene presents the weights between two neurons in different layers. A chromosome is constructed from a series of genes illustrated in figure 2b. For example, the first gene in the chromosome sequence is W15, the weight that connects neurons 1 and neuron 5. The second gene is W16, the weight of connecting neurons 1 and 6 [**10**]. And so on.

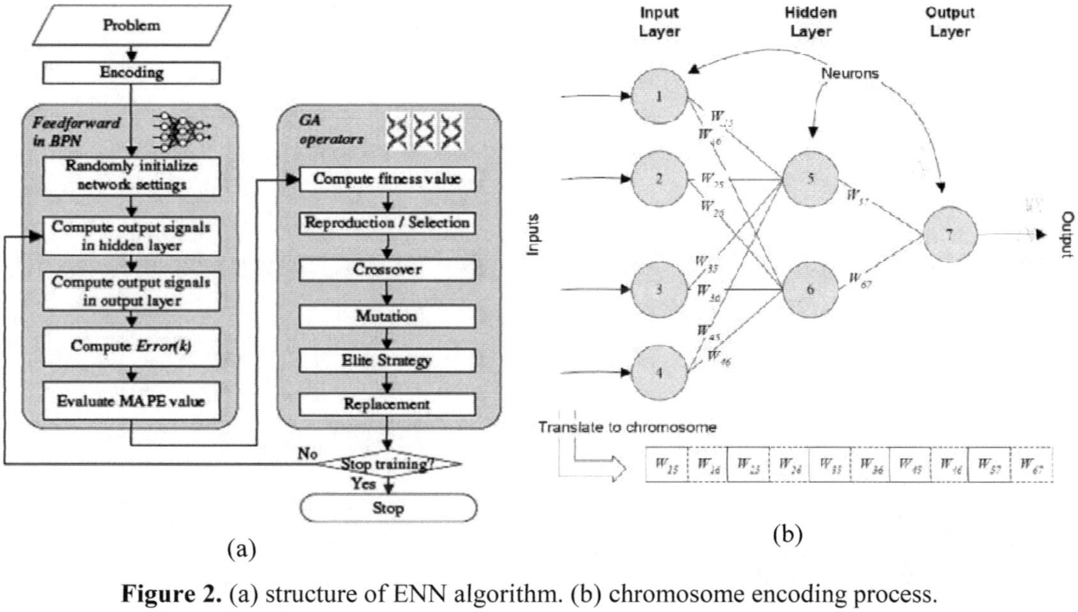

Figure 2. (a) structure of ENN algorithm. (b) chromosome encoding process.

2.2.2. Wavelet Neural Network. Wavelet Neural Network is a neural network based on wavelet transformation. The WNN combines the multiscale analysis ability of the wavelet transform and the classification capability of the artificial neural network by setting the wavelet function as the transfer function of the neural network. The essence of WNN is to find a group of wavelets in the feature space so that the relationship of complex functions contained in the original signal may be accurately expressed **[11]**. Wavelet transformation will be used to correct the weights of relationships between neurons in different layers.

3. Experimental Result and Evaluation

3.1. Structure Test

The experiment to get the best architecture of neural network (without GA or wavelet), indicates structures 9-18-1 and 18-6-1 the lowest RMSE value. Another test performed to get the best architecture for WNN, indicates structures 18-18-1 and 18-32-1.

3.2. ENN

Five important parameters of genetic algorithm according to Suyanto [**10**], are as follows: crossover type and its value, type of mutation and its value and type of selection, arranged using Taguchi experimental design. The Taguchi experimental design [**12**] is used for setting parameters. It's done by calculation of Signal-to-Noise (S/N). The codes and levels for each of the parameters are shown in table 2, as follows:

Table 2. Parameter code and level

Parameter/Code	*Level 1*	*Level 2*	*Level 3*	*Level 4*
Cross over (A)	1-Point	1- Point	2- Point	2- Point
Mutation (B)	1- Point	1- Point	Shift	Shift
Selection (C)	Total	Total	Elite	Elite
Cross over rate (D)	0.2	0.4	0.6	0.8
Mutation rate (E)	0.1	0.3	0.5	0.7

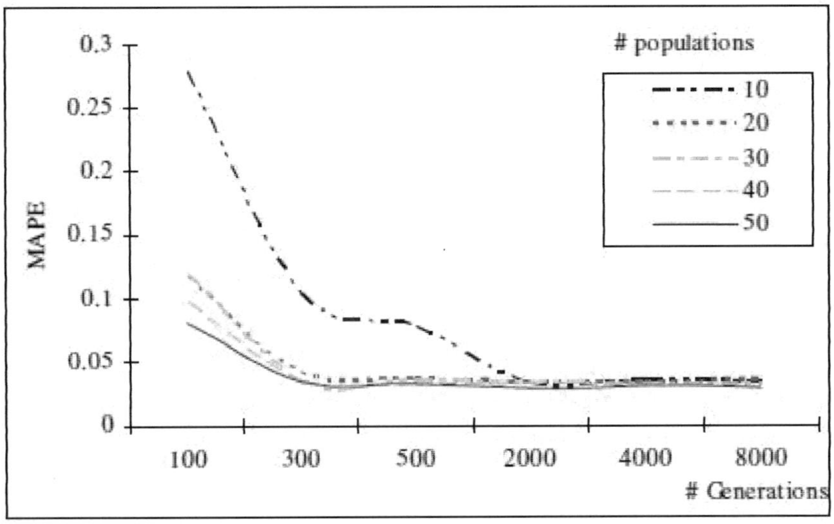

Figure 3. Convergence profile for semi-seasonal time-series data

According to Pei-Chann Chang et al [**9**], a convergence profile for semi-seasonal time-series data suggests that the system can converge after 2,000 generations, even for small population sizes, eg 10. However, for fast and smooth convergence of population sizes are found in populations size of 40 or 50, which converges to a steady state after more or less 300 generations (Figure 3).

Therefore, a population size of 50 will be used as the initial population for the experiment. The electrical load data as input and target, three times experiments of genetic algorithm (without neural network) with 50 populations, then the average S/N ratio calculated at each factor level, the results are shown in table 3 follow:

Table 3. Avarage of S/N rasio

Factor	*A*	*B*	*C*	*D*	*E*
Level 1	13.15	**27.00**	11,76	12.11	13.22
Level 2	10.43	18.66	12.42	15.21	**16.40**
Level 3	**15.28**	9.84	11.69	14.44	12.21
Level 4	13.09	8.77	**18.22**	**16.26**	16.05

From Table 4, the best combination of parameter settings can be found as (A) 3 - (B) 1 - (C) 4 - (D) 4 - (E) 2 (in bold). These codes represent two-point crossover, one point mutation, elite selection, crossover rate = 0.8, and mutation rate = 0.3.

So the structure of NN used is 9-18-1 and 18-6-1. GA operator: two-point crossover, one point mutation, elitist selection, crossover rate = 0.8, and mutation rate = 0.3. For comparation to WNN, ENN tested to 18-18-1 and 18-32-1. The results are shown in Table 4. Figure 4 shows the best result of the ENN for checking data (May 2011).

Table 4. Result of ENN

Architecture NN			ENN (MSE)		
Input	Hidden-1	Out	RMSE	MAD	MAPE
9	**18**	**1**	27.40	21.24	0.1046 (10,46%)
18	**6**	**1**	19.26	11.77	0.0614 (6,14%)
18	**18**	**1**	19.33	11.92	0.0625 (6,25%)
18	**32**	**1**	23.35	15.13	0.0761 (7,61%)
	Avarage		22.34	15.02	0.0762 (7,62%)

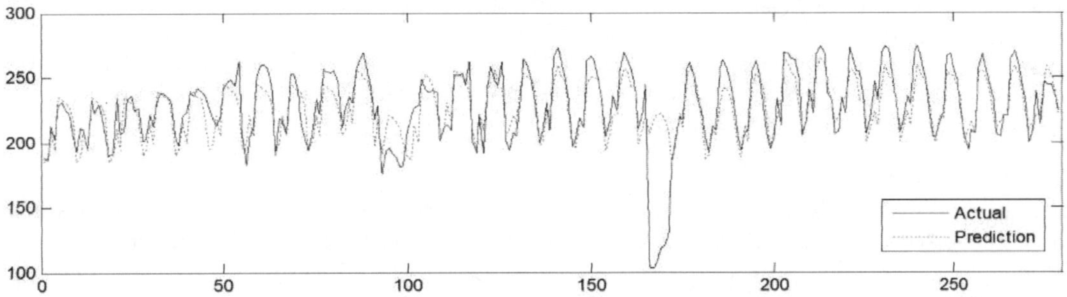

Figure 4. The best result of the ENN (structure 18-6-1)

3.3. WNN

Structure of NN used is 18-18-1 and 18-32-1. For comparation to ENN, WNN tested to 9-18-1 and 18-6-1. The results are shown in Table 5. Figure 5 shows the best result of the WNN.

Table 5. Result of WNN

Architecture NN			WNN (MSE)		
Input	Hidden-1	Out	RMSE	MAD	MAPE
9	**18**	**1**	33.75	26.28	0.1211 (12,11%)
18	**6**	**1**	25.23	17,72	0.0872 (8,72%)
18	**18**	**1**	20.86	13.23	0.0685 (6,85%)
18	**32**	**1**	22.16	14.55	0.0731 (7,31%)
	Avarage		25.50	17.94	0.0875 (8,75%)

3.4. Evaluation

As shown in Table 5 and Table 6, based on MSE, MAD and MAPE, ENN is better than WNN. In Figure 6, smaller error of ENN based on MAPE is 6,14% on 18-6-1 and smaller error of WNN is 6.25% on 18-18-1. Thus, best accuracy performance of ENN based on MAPE is 93,86% on 18-6-1 and best performance of WNN is 93.15% on 18-18-1.

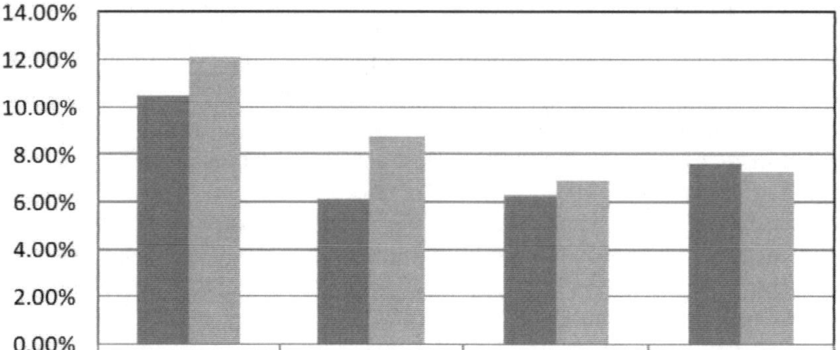

	9-18-1	18-6-1	18-18-1	18-32-1
ENN	10.46%	6.14%	6.25%	7.61%
WNN	12.11%	8.72%	6.85%	7.31%

Figure 5. The best result of the WNN (structure 18-18-1)

4. Conclusion

ENN has been developed for short term electric load prediction and achieved higher precision. Performance measurements were obtained by observing errors that occur through RMSE, MAD and MAPE parameters. Based on RMSE, MAD and MAPE, experimental results show ENN is better than WNN. The best performance of ENN is 93,86% on 18-6-1 and the best performance of WNN is 93.15% on 18-18-1. In general, ENN indicates better accuracy. The Evolving Neural Network (ENN) algorithm approach can be performed to predict short-term electrical load.

Acknowledgments

We wish to thank PLN Kalselteng and Yayasan Bina Ilmu for support this research. For their kind input. We wish to thank Uwe Hummel, Anna Kuswanti and Kinurung Maleh for their kind input.

References

[1] Kementerian ESDM. 2016 *RUPTL PLN 2016-2025*. Jakarta.

[2] Bank Indonesia 2017 *Kajian Ekonomi Regional: Kalimantan Selatan: Triwulan II - 2017*. KPBI Banjarmasin.

[3] Wang Y, Gu D, Xu J and Li J 2009 Back Propagation Neural Network for Short-term Electricity Load Forecasting with Weather Features. *Proc. Int. Conf. on Computational Intelligence and Natural Computing*, pp. 58-61.

[4] Li C, Xu S and Wen X 2010 Traffic Flow forecasting Algorithm Using Simulated Annealing Genetic BP Network. *Proc. Int. Conf. on Measuring Technology and Mechatronics Automation*, pp. 1043-1046.

[5] Wang Y, Jiang Y, Lei X, and HaoW 2010 Rainfall-Runoff Simulation Using Simulated Annealing Wavelet BP Neural Network. *Proc. Int. Conf. on Intelligent Computation Technology and Automation*, pp. 963-967

[6] Gao G and Liu Y 2010 Traffic Flow Forecasting based on PCA and Wavelet Neural Network. *Proc. Int. Conf. of Information Science and Management Engineering*, pp. 158-161.

[7] Lareno B 2014 Analisa Dan Perbandingan Akurasi Model Prediksi Rentet Waktu Arus Lalu Lintas Jangka Pendek. *J. CSRID*, 6(3) ISSN: 2085-1367

[8] Swastina L, Cipta S P, and Lareno B 2014 Penerapan Algoritma Evolving Neural Network Untuk Prediksi Curah Hujan, in *Konferensi Nasional Sistem & Informatika*, Bali, pp. 802-806.
[9] Chang, P C, Wang-Yen W and Tsai C 2005. Evolving Neural Network for Printed Circuit Board Sales Forecasting. *J. Expert Syst. Appl.*, 29(1), 83-92.
[10] Suyanto 2008 *Evolutionary Computing: Komputasi Berbasis Evolusi dan Genetika*. Bandung: Informatika.
[11] Xian-Bin W, Hua Z, and Fa-Yu W 2009 A Wavelet Neural Network for SAR Image Segmentation, *J. Sensors* 9(9) MDPI. ISSN 1424-8220, 7509-7515
[12] Philip J R 1996 *Taguchi Techniques For Quality Engineering: Loss Function, Orthogonal Experiments, Parameters and Tolerance Design*, 2nd ed. New York: Mc Graw-Hill Companies, Inc.

IOP Conf. Series: Materials Science and Engineering **384** (2018) 012011 doi:10.1088/1757-899X/384/1/012011

Compact Portable Industrial Automation Kit for Vocational School and Industrial Training

E S Maarif* and S Suhartinah

Industrial Automation Laboratory, Astra Manufacturing Polytechnic, Jakarta, Indonesia

*eka.samsul@polman.astra.ac.id

Abstract. This Paper presents a compact and portable kit for learning on industrial automation practice for vocational schools and training industry. The kit is compactly designed with reference to basic of standard competences in industrial automation areas. They are PLC programming, PLC input and output wiring, motor control and pneumatic cylinder control. The components used are standard part in industry field. In addition to compact design, the second concept is portable, so the selection of components and layout should be done carefully so that the dimensions of the kit can be maintained at a certain size. This kit has 41 x 23 x 11.8 cm and weight 6.76 kg weight that can be easily carried or to be stored in the cabin of the aircraft during the training trip.

1. Introduction

Industry is always innovating in an effort to provide cost reduction. In this cost conscious environment, industry is looking for employees that can fill numerous roles within their manufacturing facility [1]. On the other hand, many production lines originally operated by current employees have been replaced with automation systems where PLC (Programmable Logic Controller) becomes one of the main controllers. This will also be followed by increasing the need for human resources that can operate PLCs and industrial automation systems. Vocational school students who will later enter industrial field as an automation technician should be equipped with competencies related to PLC and automation. Likewise with the workers who have entered and worked in the industry, their skills about to be improved in order to be able to work with PLC-based automation system.

PLC control programming is a complex cognitive skill that requires hands-on experience to develop proficiency. Successful automation/control engineers must know how to write PLC programs to control and troubleshoot the process being automated [2]. Unfortunately, it is difficult to find suitable devices or training kit to perform hands-on skills aspect during the practical session in executing the syllabus for several courses such as Industrial Automation & Programmable Logic Controller. But, the existing training kits are usually large and difficult to move if the training location is to be carried out in an industrial area.

PLC learning can be done by using industrial standard PLC, microcontroller or just by using personal computer. B.K Rana has worked with C ++ software and for creating a PLC program [3] and PLC with microcontroller and LabView software ever done by K.Bhise that can run program logic well [4]. But PLC learning with this tool is not enough to give enough skill to work in the industry.

Al Mashhanday had designed and implemented PLC for classical control laboratory. He has developed a PLC training kit using LG PLCs equipped with digital IO (input / output), analog IO, seven segment and other simulation devices [5]. So, trainee can learn about programming and wiring but not yet equipped with pneumatic system. Burhan has also built a PLC kit with a variety of IOs as a learning tool. He has been using electric motors and solenoid as output, but does not have a pneumatic cylinder [6]. Training kit with a more complete component with induction motor and Varibale Drive Frequency has also been made by P. Chakraborty, but with a size large enough [7].

This paper describes an effort to design and build a compact portable automation kit teaching basic I/O devices and to evaluate the developed kit and instructional support materials. The kit consists of (1) controller module, (2) plug and play I/O interface, (3) DC motor module, (4) electro-pneumatic module —all integrated in a portable box for easy transport. The controller module contains an industrial programmable logic controller (PLC) and power supply unit.

2. Methodology

Astra Manufacturing Polytechnic has Industrial Automation Laboratory that concern on teaching and development technology in the automation field. In this Lab, students should experience hands-on skills in PLC as it is a major component in engineering courses. Theoretical knowledge and hands-on skills experience in PLC control principles provide more career opportunities besides meeting the emerging of workforce and education need for global industries.

The detailed competence of industrial automation in Indonesia is regulated by the Decree of Ministry of Manpower No. 631 of 2016. Where the outline covers the competence of operating electrical system including relay and electric motor, operates electronics system, operates PLC, and operates pneumatic system. The following is a description of the need for such competence.

2.1. Electrical Device Operation

Electrical competence covers:

- Operate electrical device, this competence deals with the operation of electrical equipment of industrial automation systems according to the operating procedures.
- Operate the electrical system, this competency unit is related to the work of preparation, installation, testing and operation of electrical system for industrial automation system in accordance with the operating procedure of electrical system.
- A assemble equipment and electrical system, this competence unit is related to the implementation of assembly work of equipment and electrical system of industrial automation on network of production system according to work procedure of equipment and electrical system of industrial automation.

To meet the above competency requirements, the kit should have some features of electrical components. The main electrical component is the lamp. The lamp serves as a marker of the condition of an output / indicator. The next electric component is the electric motor, especially the 3 phase electric motor which is the most actuator type in the industry such as for conveyor actuator [8]. Another electric component is a solenoid valve for controlling the direction of the pneumatic cylinder movement. With the solenoid valve student can learn to regulate the valve condition through electrical signal.

2.2. Electronics Device Operation

Electronics competence covers:

- Operate electronic equipment, this unit deals with the operation of any electronic equipment of industrial automation systems in accordance with the manufacturer's operating procedures and in accordance with standard procedures.
- Operates an electronic system, this competency unit deals with preparatory work, set-up of operating parameters, start-up and operation of electronic systems in industrial automation systems in accordance with manufacturer's standard procedures and / or standard procedures.
- Assemble electronic equipment and system, this competency unit is concerned with the execution of assembly work of electronic equipment and industrial automation systems on the network of

production systems in accordance with the procedure of assembling equipment and industrial automation systems.

Electronic devices contained in automation systems include power supply, input devices for PLCs such as push buttons, sensors and relays. This kit should be able to facilitate the Student to get training on how to organize the power supply to meet the power requirements of the control system, connecting the electronic components to the PLC.

2.3. PLC Operation

Competence of PLC field covers:

- Operating PLC, this competency unit is related to PLC operation work including selection of PLC type and peripheral, initialization, program download and testing according to operating work procedure.
- Writing software program, this competency unit is concerned with the design of control program (software) industrial automation system based on the program flow diagram and testing of programming results according to the required specifications and implemented according to programming procedures.

PLC selection hardware is considered on the basis of size, completeness of input and output, ease of getting power supply and ease of programming connection. PLC with programming access via USB is preferred over Serial port because it has high flexibility to all programming devices. Especially nowadays Laptops are more widely used than Personal Computer, while very difficult to find a laptop with serial port.

In software, PLC programming language should be selected the most basic and commonly used in the industry. As per IEC 6-1131 standards, PLC has several languages such as Ladder Diagram (LD), Instruction List (IL), Structure Text (ST) and Sequential Function Chart (SFC) languages [6]. In this kit is limited to Ladder Diagram only. This language is most commonly used by technicians in the image-based installation industry and is logically easy to understand.

2.4. Pneumatic Cylinder Control

Pneumatic competencies covers:

- Operate pneumatic equipment, this competency unit is concerned with the operation of electro-pneumatic equipment and components, including preparatory work, checking, air pressure set-up, performance testing and operation according to operating procedures.
- Operates a pneumatic system, this competency unit is concerned with the operation of the pneumatic / electro-pneumatic system, including preparatory work, checking, air pressure set-up, performance testing and operation of the pneumatic system in accordance with the operating procedures of the pneumatic system.
- Assemble equipment and pneumatic system, this competency unit is concerned with the implementation of assembly work of pneumatic equipment and industrial automation system on the network of production system as per the working procedure of pneumatic equipment and industrial automation system.

The kit must have a pneumatic cylindrical actuator component as one means to train pneumatic cylinder control competence. Student or training participants can learn how to connect a pressurized air hose to the valve, in order to control the movement of the cylinder [9], [10].

3. Design and Fabrication

From the description of competence in chapter II can be drawn a summary that the competencies targeted are operational devices, device assembly or wiring and PLC programming. Therefore, students need to be familiar with functions and general characteristics of hardware devices, to understand how PLC controllers process programs, to be able to understand the control requirements of an Application and write control programs accordingly. Hands-on experience with PLCs and I / O devices is needed to develop these skills. So the automation kits must be able to meet and use with industry needs.

3.1. General Scheme

The kit provides digital input/output that can be used to control the output components. Figure 1 shows the block diagram of the industrial automation kit. It consists of PLC as controller, Input Module, Output Module, 24VDC Power Supply unit and common connection for Input/Output.

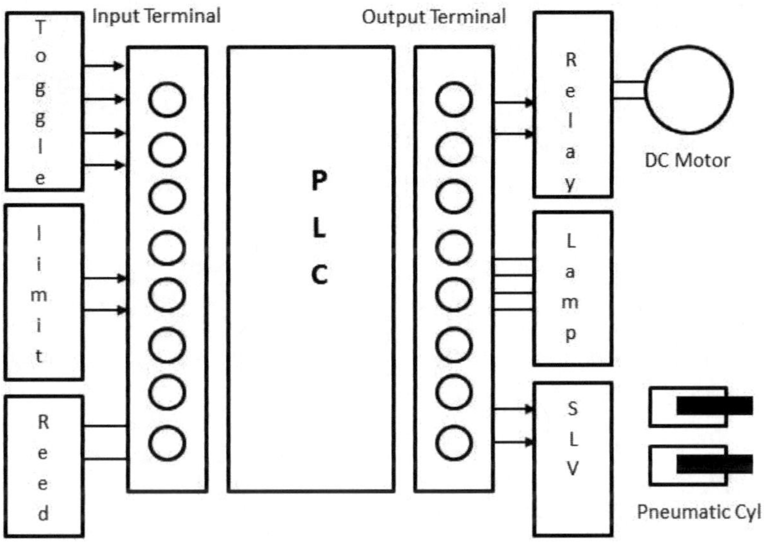

Figure 1. Block Diagram of Industrial Automation Kit

Connection with straight line is permanent wiring and connection with arrow is cable to be wire by student or trainee.

3.2. Portable Concept

Portable concept can be limited by taking the standard size and weight of the kit. To be brought by car transport or shipping in general, no special standards are applied. However, kits containing electrical and electronic components must be carried with special care and caution so as not to damage the working function of the apparatus. The concept of a portable PLC kit has been studied by S.H Hsieh [1] and M.Barret [11] that has the idea of placing control components such as PLCs, push buttons and contactors in one panel. But unfortunately it is not packaged in that is safe and easy to carry.

When traveling by airplane, there is a strict standard on the goods to be brought into the cabin of the aircraft. The size of the suitcase should not exceed 41 x 34 x 17 cm, and with maximum total weight is 7 kg. This standard that is applied as a kit reference can be considered portable. Therefore, the suitcase that is used as the main casing of the kit has size 41 x 23 x 11.8 cm.

Portable concept has given the limit with the size of a rigid suitcase. Thus all the components to be used in the kit should take into account that dimension. Figure 2 and Figure 3 are component layout and kit dimention.

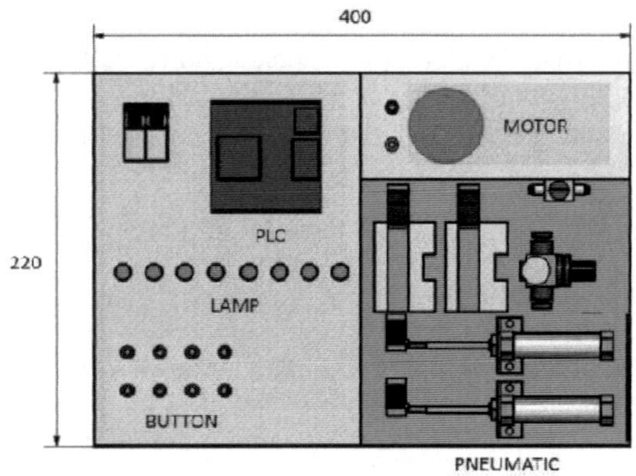

Figure 2. Kit Design Top View

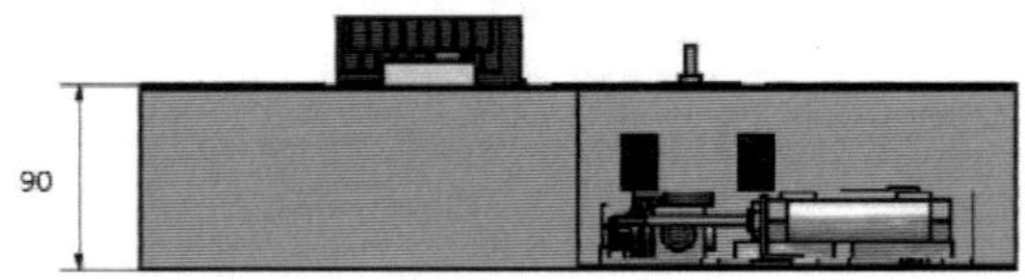

Figure 3. Kit Design Side View

Another requirement to be compiled with portable concept is the weight of the kit. Therefore components selection is important step. Table 1 shows the selected components taking into consideration the available space.

Table 1. Component List

No	Group	Component	Spesification	Quantity
1	Controller	PLC Omron CP1E-N20R	Supply 220V AC 24V DC input (12 bit) Relay Output (8 bit) USB terminal	1 Unit
2	Input	Toggle switch	15 mm	4 pcs
		Reed switch	D-C73	2 pcs
		Limit Switch	-	2 pcs
3	Output	DC lamp	24 V	8 pcs
		Solenoid valve (5/2 single)	SY5120-5D-01	2 pcs
		Relay	Omron MY4	2 pcs
		DC Motor	24 V	1 pcs
4	Terminal	Banana hole	-	28 pcs
5	Supply	Power supply	Omorn 24 V	1 pcs
		Power Socket	240 V 10 A	1 pcs
		Switch	240V 10 A	1 pcs
6	Pneumatic	Pneumatic Cylinder	CDJ2B16-45Z-B	2 pcs
		Pressure gauge	AR4000-04	1 pcs

3.3. Fabrication

The Figure 4 shows the view of the kit that has been designed. In order to keep the box light, it is made from acrylic 3 mm and the main base is acrylic 5 mm.

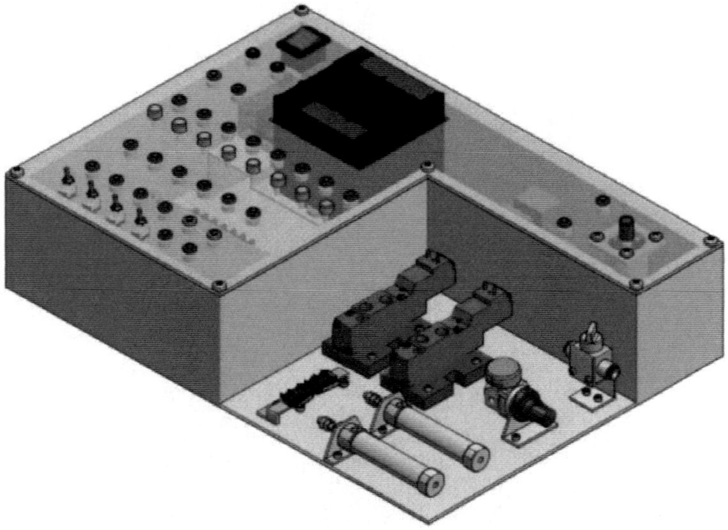

Figure 4. Kit Top View

3.3.1 Output Component and Terminal. The lamp as an input component is selected which has a medium size with a commonly rated 24V, voltage commonly used in the PLC output. But 3 phase electric motor is too big to be placed on kit. Therefore, the motor control simulation can be replaced by setting the direction of DC motor rotation. Clockwise (CW) or Counter clockwise (CCW) adjustments can be done with the help of 2 relays through electrical signals. Each relay replaces of contactor that controls 3 phase electric motor, Figure 6.

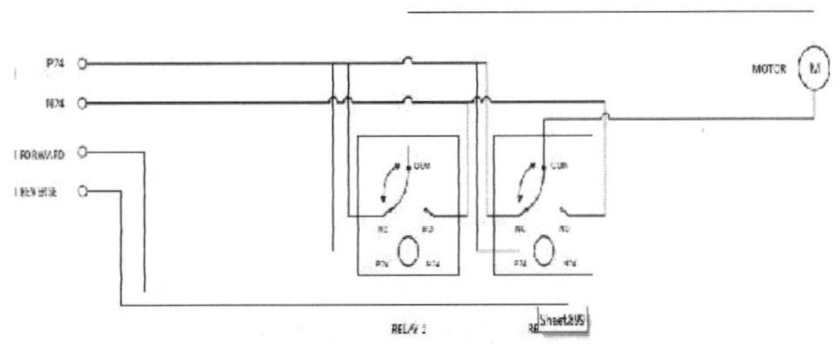

Figure 5. DC Motor Control using Relay

The pneumatic cylinder is selected on a double acting type with a 10 cm stroke dimension controlled by a single solenoid valve 5/2 with spring. They are 2 pneumatic cylinders that controlled by 2 singe solenoid valves.

3.3.2 Input Components. Toggle switches are preferred over push buttons with consideration of smaller size and functional flexibility. The toggle switch can be used as a switch or alternate switch. Other input components are the limit switch and reed switches as the front and rear bounds of the pneumatic cylinder movement.

4. Results and Discussion

A series of manufacturing and testing processes have been undertaken to ensure the condition of the kit is safe to use. The main frame is made using acrylic 3 mm instead of metal. In addition to facilitate the

IOP Conf. Series: Materials Science and Engineering **384** (2018) 012011 doi:10.1088/1757-899X/384/1/012011

process of making and lightweight, acrylic is also useful to avoid the occurrence of short circuit on the main panel. The overall weight of the kit is 6.76 kg, Figure 6. Thus still within the maximum weight standard.

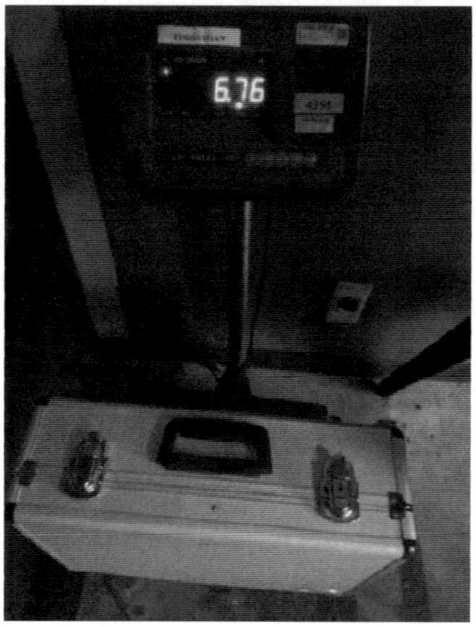

Figure 6. Overall Weight

The Figure 7 shows internal component and Figure 8 is the kit along with the completeness of the power cable, USB cable and connection cable.

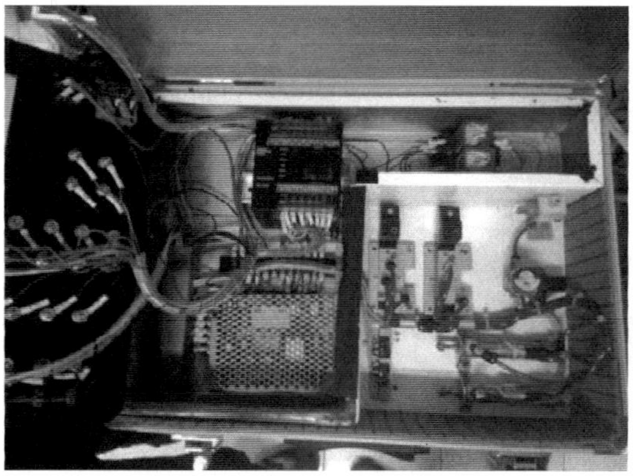

Figure 7. Internal Components

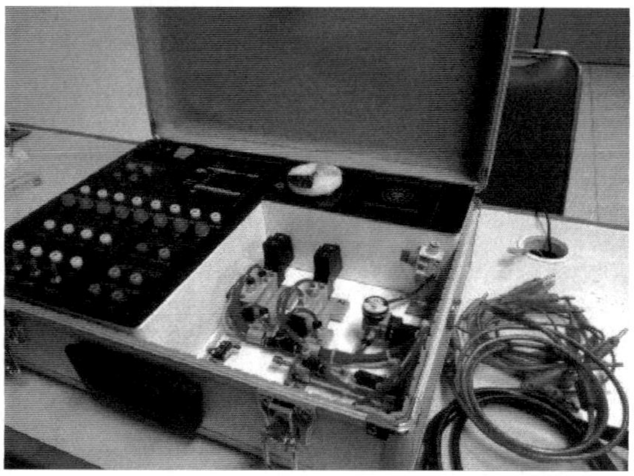

Figure 8. Kit with Cable Connection

5. Conclusion

Through this newly designed, developed and fabricated compact portable kit, training equipment needs and learning in accordance with the target of competence can be met. Trainee can enhance the theoretical knowledge and hands-on skills aspect through ladder diagram design, programming execution, component operational and wiring. The fabrication of compact portable kits has been designed and built with dimension size 41 x 23 x 11.8 cm and weight 6.76 kg, which means it can be considered portable. All the tools needed for teaching and training have been provided by this kit.

References

[1] A R Akparibo, A Appiah, O Fosu-Antwi 2016 Development of a Programmable Logic Controller Training Platform for the Industrial Control of Processes *American Scientific Research Journal for Engineering, Technology, and Sciences (ASRJETS)* Volume **15**, No 1, pp 186-196.

[2] S Hsieh 2015 *Design and Preliminary Evaluation of Portable Kit for Programmable Logic Controller Education* 122nd ASEE Annual Conference & Exposition

[3] B K Rana 2011 Universal Simplest possible PLC using Personal Computer *(IJACSA) International Journal of Advanced Computer Science and Applications*Vol. **2**, No.1

[4] K Bishe and S Amte 2015 Embedded PLC Trainer Kit with Industry Application *International Journal of Engineering Science and Innovative Technology (IJESIT)* Volume **4**, Issue 3

[5] Yousif I. Al Mashhadany 2012 *Design and Implement of a Programmable Logic Controller (PLC) for Classical Control Laboratory* Intelligent Control and Automation, page 44-49.

[6] I Burhan, A A Azman, S Talib 2015 *Multiple Input/Outputs Programmable Logic Controller (PLC) Module for Educational Applications* 2015 Innovation & Commercialization of Medical Electronic Technology Conference (ICMET)" Blue Wave Hotel, Shah Alam, Malaysia, 30 November 2015.

[7] P Chakraborty 2016 *Designing and Assembling of a Programmable Logic Controls (PLC) Laboratory Trainer and Advanced Research Setup* American Society for Engineering Education

[8] Y V Aruna, S Beena 2015 Automatic convey or System with In–Process Sorting Mechanism using PLC and HMI System *Int. Journal of Engineering Research and Application* Vol. **5**, Issue 11, (Part – 3)

[9] K J S Gill, R Kumar, S Kumar 2015 *Designing and Fabrication of Electro-Pneumatic Trainer Kit* International Conference of Advance Research and Innovation (ICARI-2015)

[10] M B Castillo 2015 Low Cost Electro Pneumatic Automation Trainer Kit *International Journal of New Technologies in Science and Engineering,* Vol. **2**, Issue 6

[11] M Barrett 2008 *The Design of a Portable Programmable Logic Controller (PLC) Training System for Use Outside of The Automation Laboratory* International Symposium for Engineering Education, Dublin City University, Ireland

Mobile Video Learning for Improving Programming Competency

G D P Maramis*, V R Palilingan, and M Modeong

Universitas Negeri Manado, Tondano, 95618, Sulawesi Utara, Indonesia

*gmaramis@unima.ac.id

Abstract. Programming competency is a must competency that every student who are studying computer science must have. Mostly of the student are struggling learning this skill especially for those who do not have any computer background. The objective of this research is to design and develop e-learning using videos and tutorial to provide content of programming skill that needed by the student to increase their coding skill. Student can watch this video many times to have a fully understanding about the topic. Research and Development method are used by this research to design the video learning that can be accessed using mobile phone. As the result, the programming skill of the student increased rapidly.

1. Introduction

Nowadays, learning concept using internet technology has been an alternative learning tools for lecturer for making education becomes more interesting, effective and efficiency. This sort of teaching style known as Electronic Learning (e-learning). E-learning it-self is a kind of schooling form that enables students to study anytime and anywhere using internet technology [1]. On the other word, e-learning is a learning method that uses internet [1]. In the scope of education, e-learning is said to be able to convert the conventional learning process into digital, because the learning process using e-learning is an innovation where students can get a material description of the lecturer in visual form with a more dynamic and interactive packaging.

E-learning offers the ability to share material in different types of formats such as video, slideshows, word documents [2], and easy access to lecture assignments. These facilities make e-learning interesting among teachers and students, because students can easily get teaching materials, tutorials, and even tasks with one access. Mobile video learning is one of the e-learning features that is expected to boost student programming skills. Therefore, the existence of video learning can give the students an opportunity to understand and analyze the complexity of the practice of an area of work which will increase student competence in that field [3].

The existence of video learning becomes increasingly popular in the field of education because of its ease in conveying learning and tends to be repetitive, thereby minimizing the possibility for students to be unprofessional to their field. As learning programming that occurred in the Faculty of Engineering UNIMA, lecturers have to repeat the lecture material when there are students who are not familiar with the topics that have passed. This condition can be said to slow down the material that is supposed to go forward. However, the existence of mobile video learning makes it easy for students to repeat past learning by accessing a site via their mobile phone. Brecht [4] supports the existence of learning videos because the existence of this medium can provide a model of evaluation and problem solving to students' understanding, thereby increasing the ability of students to their field of expertise, especially on programming skills.

Another benefit that can be gained by implementing mobile-based learning videos is the space for learning to be broad. Some of the most frequent conditions are student interaction with limited learning. However, with this mobile video learning application enlarge the learning space and student communication to the material being taught. Significantly, the use of mobile video learning is revolutionary because it extends beyond the classroom structural boundaries and communication [5]. In his research, Admiraal & Gielis [3] also stated the existence of video learning can provide five benefits for its users, namely: (a) construction of knowledge, not reproduction; (b) conversations, not receipts; (c) articulation, not repetition; (d) collaboration, not competition; and, (e) reflection, not a prescription.

Mobile video learning is developed with the aim to facilitate students to be able to access course material easily and without being limited by lecture schedule. So that students can learn from anywhere and whenever they want. Furthermore, mobile video learning is intended to improve the skills of programming students majoring in Software Engineering (RPL).

2. Methods

Delivering high-quality and reliable software in modern software industry is a must, therefore, choosing the right method for developing software should be carefully planned. By selected wrong models will cost low-quality and high-cost software products [7]. Extreme programming (XP) is a developing software models. XP is a methodology that emerged as a counter-reaction to the ever increasing focus on processes [6]. XP model is an agile software development methodology founded by Kent Beck in 1996. This kind of methodology is a lightweight method that considered suitable for small and medium sized teams between 3 and 10 engineers [12], small project [8] and used for developing mobile application [9]. Five core values of this method namely simplicity, communication, feedback, courage and respect [10] [11]. Every software development methodology has its own life cycle. Thus, developing mobile video learning employ XP methodology has its life cycle which can be described into five phases, which are exploration phase, planning phase, Iteration to release phase, Production phase, maintenance phase, and finally death phase [8]. These phases are drawn in figure 1.

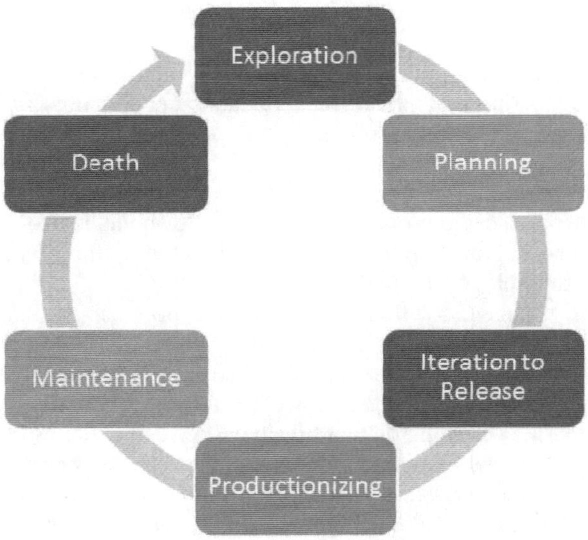

Figure 1. Life cycle of Extreme Programming

2.1. Exploration phase
This phase, is the first phase, in which deals with the requirement, architecture, tools and technologies This phase produced user story cards which is comprises of short name, priority of story and small paragraph without technical details. User story should be detailed enough to help developers have a fully perception of system requirement and time estimation [8].

2.2. Planning phase
Once the exploration phase passed, planning phase should take a place. In this step, priority is set for the user stories, estimation of effort and schedule for each story is decided [12]. Therefore, it is very important to have a full picture about the software that is being developed before move to the next phase.

2.3. Iteration to release phase
Iteration to release phase is the phase taken once the developer has a fully understand about the software requirement. In this stage, some basic development activities like designing, coding, testing and integration takes place. The result of planning phase will break down into number of stages, in which every single stage will take one to four weeks to implement [12]. Finally, the system is ready to be produced.

2.4. Productionizing phase
XP method produces software in small releases. A release is a small part of proposed software that need to implemented in order to fulfill the business process of the project. This phase is about developing software in small releases to check whether the application is ready for production, testing the acceptance rate of the software by the customer, and also testing the performance of the software is taken place. This phase takes an important role as this phase decides whether the system should be revised or not. [12]

2.5. Maintenance phase
Maintenance is a natural process for system software. In this phase, when the functionality is developed, the old system must keep running to ensure that the entirely system will perform well and the new features that has been released are suitable for the old system to satisfied the customer needs [12].

2.6. Death phase
Death phase is the final phase where the customer satisfied with the software and the software is ready to be released and used by the user. As the result, a document paper about five to ten pages is created. This paper describes the future use of the software [8].

3. Results and Discussion
While this e-learning can be accessed properly with the mobile phone, it also can be viewed best through web browser. It is designed and made to be accessed through those kind of technologies so the students can choose which one is suitable for them. When the user first to come to the system, they must sign in into the system before have page that will alow the user to choose subject that he wants to study. If the student has already had an account than he can use the account to access all of the materials offered by the e-learning system. Once the system authenticates the account, student will redirect to the page contains programming subject, in which every subject has a specific title for this lesson. User can choose which title he wants to learn by clicking on that title and the video explains the title will show up to the screen. For convenience, user can download the video so the user will not need the internet if he wants to replay again the video. If the user feels unsatisfied with the video, the user can choose to read from the written tutorial for better explanation. Figure 2 shows the flowchart diagram of this e-learning system.

IOP Conf. Series: Materials Science and Engineering **384** (2018) 012012 doi:10.1088/1757-899X/384/1/012012

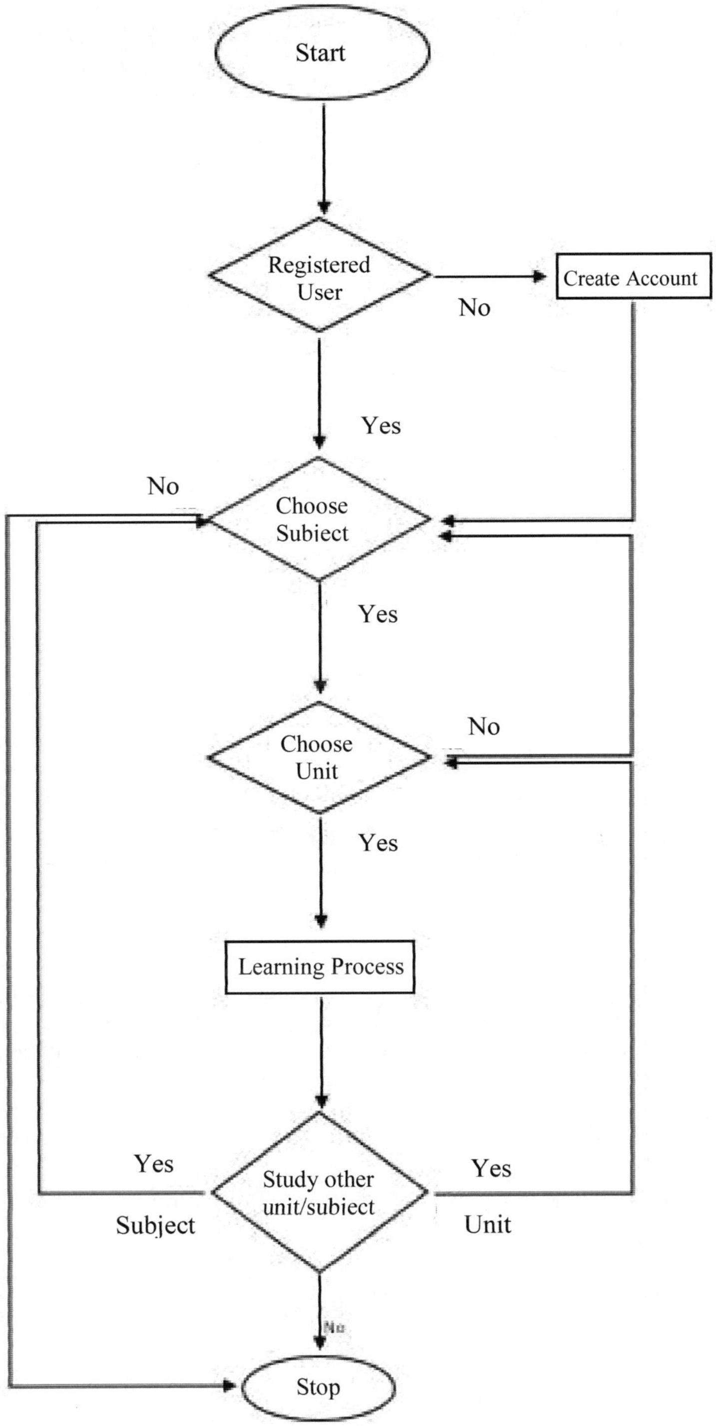

Figure 2. Flowchart diagram

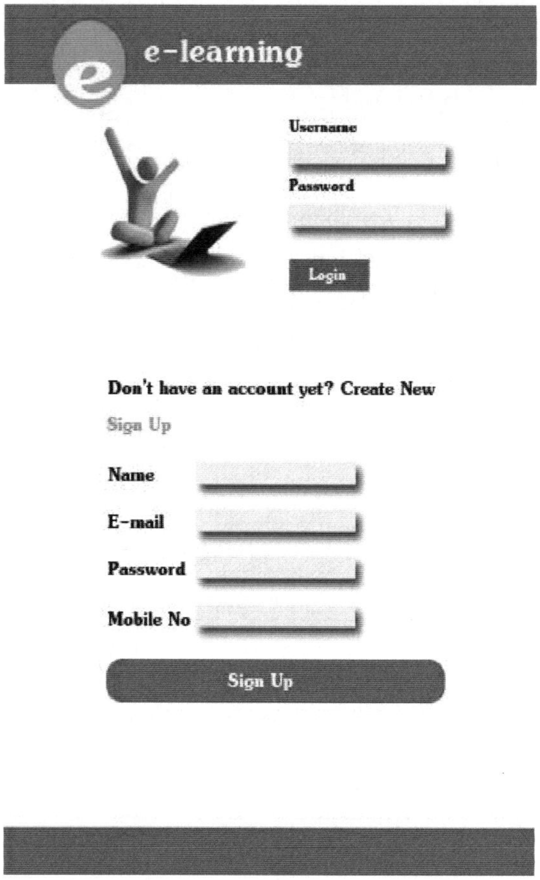

Figure 3. User Interface and Login Screen

As can be seen figure 3 shows the user interface of the e-learning system that will show up for the first time. In this page, user must enter their valid account for accessing the material, otherwise, they must sign up for creating a valid account. The next page will come up is the welcome page. This page show number of programming subjects user may select. Figure 4 shows the welcome page where the student may choose the subject. Every subject contains number of meeting in which every meeting contains different subject that can be taught through video tutorial or written tutorial. This page is shown by the figure 5. As the result, the students can have a better understanding of programming as the material can be accessed many times anywhere and anytime as long as the user has an internet connection to download the materials. The reason why the skill of the student has been increased because the student can watch the video many times until they have a good understanding. When the user seems to struggle with one material, students can ask to the lecturer when they have an off-line class. Therefore, an off-line class also still needed by the students and the lecturer.

Figure 4. Welcome Page

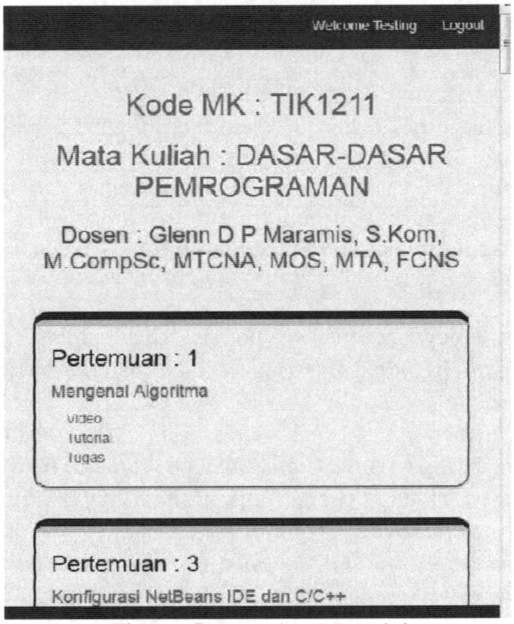

Figure 5. Learning Material

When the user clicked on video link, the video related to the material will shows on the screen. This video is taken and embed in the system. This screenshot video is shown by the figure 5.

Figure 6. Screenshot Video

4. Conclusions

To sum up, mobile video learning has been improved the knowledge and programming skill of the students. When the material can be accessed through the e-learning system, students can learn any time they wants and when they forgot the passed material, the material is still available on the internet and can be accessed anytime and anywhere. For the further research, the e-learning system may be added with chat feature so the user can have a chat with the tutor instantly when the student has a problem.

References

[1] Masrom, M 2007 Technology acceptance model and e-learning. *Article of 12 International Conference on Education.* Sultan Hassanal Bolkiah Institute of Education, Universiti Brunei Darussalam

[2] Epignosis LLC 2014 *E-learning: concepts, trends, applications.* California: Epignosis LLC.

[3] Admiraal, W., Janssen, J., Pijls, M. & Gielis, A 2008 Transfer between Learning and Practice: The Use of Web-based Video in Higher Education in *Proceedings of the Annual Conference on Human Resource Development 2008, Lille, France.*

[4] Brecht, H D 2012. Learning from Online Video Lectures. *Journal of Information Technology Education: Innovations in Practice.* **11**, *p* 227-250.

[5] El-Hussein, M. O. M., & Cronje, J. C. 2010. Defining Mobile Learning in the Higher Education Landscape. *Educational Technology & Society*, **13** (3), *p* 12–21.

[6] Blom, M 2012 Is Scrum and XP suitable for CSE Development? *Elsevier: Procedia Computer Science* **1**, *p* 1511–1517.

[7] Matharu G, Singh H, Mishra A, Upadhyay P 2015 *Empirical Study of Agile Software Development Methodologies: A Comparative Analysis* ACM SIGSOFT Software Engineering Notes, *p* 1-6.

[8] Anwer F, Aftab S, Muhammad Shah S, Waheed U 2017 Comparative Analysis of Two Popular Agile Process Models: Extreme Programming and Scrum. *Internasional Journal of Computer Science and Telecomunications, p* 1-8.

[9] Flora H, Chande S, Wang X 2014 Adopting an Agile Approach for the Development of Mobile Applications *Internasional Journal of Computer Applications, p* 43-50.

[10] Beck K 2000 Extre*me programming explained: embrace change*. Addison-wesley professional.

[11] Juric R 2000 Extreme programming and its development practices *Internasional Conferences Information Technology Interfaces, p.* 97-104.

[12] Kumar G, Bhatia P 2014 *Comparative Analysis of Software Engineering Models from Traditional to Modern Methodologies.*

IOP Conf. Series: Materials Science and Engineering **384** (2018) 012013 doi:10.1088/1757-899X/384/1/012013

Prototype of Arduino Based Parking Rotation System

M Sodiq* and H Hasbullah

Departement Electrical Engineering of Education, Universitas Pendidikan Indonesia, Jl. Dr. Setiabudhi No. 207 Bandung 40154, Indonesia

*muhammad.sodiq@student.upi.edu

Abstract. As car users increase, parking areas can not keep up with car growth. One solution to reduce this problem is to use a rotary parking system. The rotary parking system is an automated parking system that utilizes relatively narrow space using a rotation system. This rotary parking design uses 6 parking shelves that arranged vertically and rotate. The design uses stepper motor as an actuator and some push buttons. The Arduino Uno module is used as the controller that governs all the above devices. The results are done by running the rotary parking system and the system can function properly.

1. Introduction

The land transportation that people use today is a car. In Indonesia alone the total number of cars recorded by KORLANTAS POLRI from 2010 to December 2017 continues to increase to 5.9 million units.

To overcome all the problems mentioned above, we need an efficient parking system which would help to reduce traffic congestion and improve air quality at important locations where traffic rush is more [1].

Increase in the number of cars is not proportional to the available parking area, causing congestion when choosing a vehicle parking lot. From the problem, then created a prototype rotary parking system [2]. The concept of a rotational car park is one of the mechanical car park systems where its controls automatically with drives and cars are stored on vertical rack chains arranged on both sides. The main idea is to piled up six cars in the space usually occupied by two cars [3].

The control using a microcontroller arduino uno module that is able to control various types of electronics related equipment such as servo module, stepper motor, and some push button.

2. Material and method

In table 1 is a list of tools in the prototype design of arduino-based rotation parking system.

Table 1. Material

No	Tools	Specification	Amount
1	Microkontroler	Arduino uno	1 pcs
2	Motor	Nema 17	1 pcs
3	Resistor	3k5 Ohm	5 pcs
4	Resistor	900 Ohm	1 pcs
5	Driver Motor	A4988	1 pcs
6	Push Button	PBS - 11B	5 pcs
7	Led	RGB 2 pin	1 pcs
8	Power Suplay	14 Volt	1 set
9	Framework	Galvanum	2 m
10	Timming Belt	MXL GT-2	1 loop
11	Transmission	Cam Chain	2 loop
12	Pallet	Iron Plat	50 cm^2
13	Gear	Cam Gear	2 pcs
14	Cable	Rainbow Cable	40 pcs
15	Bolt	2mm	30 pcs
16	Bearing	6,8mm	4 pcs

2.1. *Arduino Uno design*

This microcontroller uses the Arduino Uno Module in figure 1 and the specification in table 2 as follows:

Table 2. Specifications Arduino Uno

Microcontroller	ATmega328P
Operating Voltage	5V
Input Voltage (recommended)	7-12V
Input Voltage (limit)	6-20V
Digital I/O Pins	14
PWM Digital I/O Pins	6
Analog Input Pins	6
DC Current per I/O Pin	20 mA
DC Current for 3.3V Pin	50 mA
Flash Memory	32 KB
SRAM	2 KB
EEPROM	1 KB
Clock Speed	16 MHz

Figure. 1 Arduino Uno

To control input push button and output of stepper motor used Arduino Uno module as input processor and give data for stepper motor to rotate. Arduino Uno design is shown in figure 2.

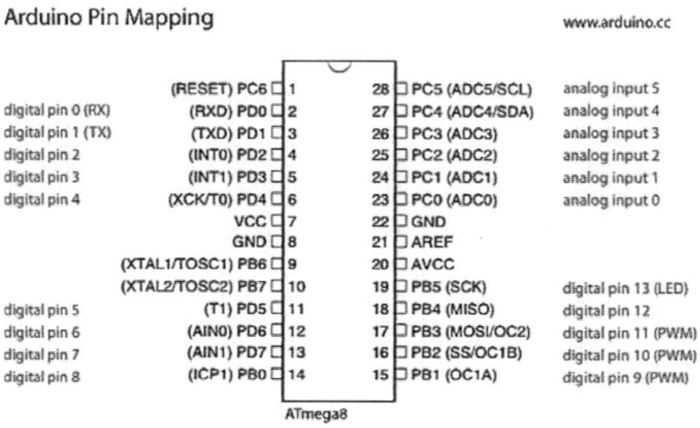

Figure 2. I/O Atmega328P

2.2. Driver Motor Stepper design

The stepper motor control module used is A4988 Stepper Motor Driver module. The function of this driver is to control the direction of rotation and speed of the motor work on step by step according to instructions from Arduino Uno. The image of the A4988 Stepper Motor Driver module is shown in Figure 3.

Figure 3. A4988 Stepper Motor Driver

2.3. Motor Stepper design

The stepper motor used is NEMA 17 which has a voltage input of 12-19 Volts and has a maximum torque of 4.0 kg cm, as shown in Figure 4 (a). This stepper motor has 4 inputs, the axes 1a = negative, 1b = positive, and 2a = negative, 2b = positive, as shown in Figure 4 (b).

IOP Conf. Series: Materials Science and Engineering **384** (2018) 012013 doi:10.1088/1757-899X/384/1/012013

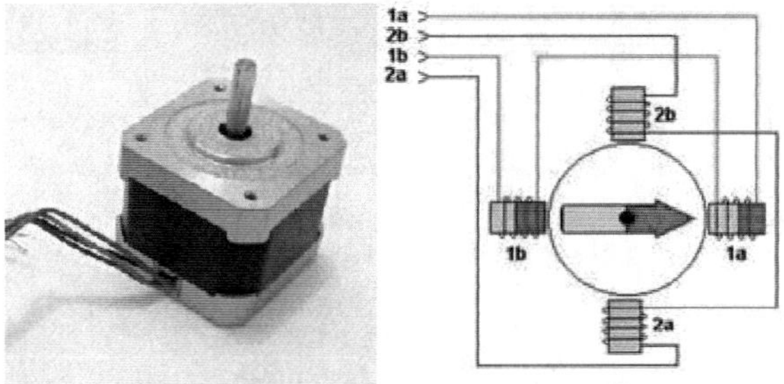

Figure 4. (a) NEMA 17 (b) Conceptual Stepper Motor

2.4. Block Diagram
Includes the system block diagram shown in figure 5,

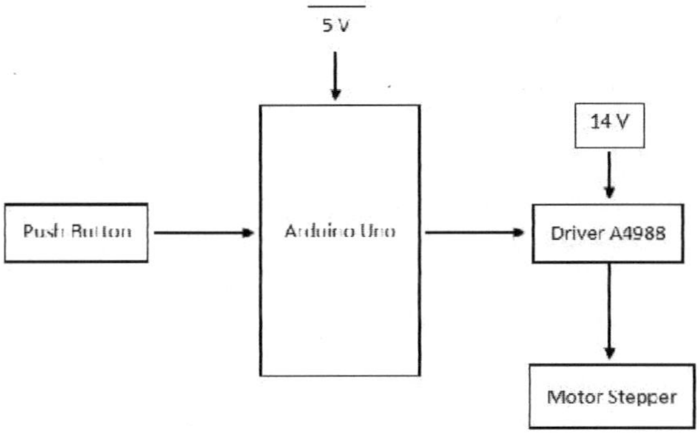

Figure 5. Block diagram

2.5. Design frame
The design of the framework using galvalum materials along 2 m to be cut into a prototype frame rotary parking system according to Figure 6.

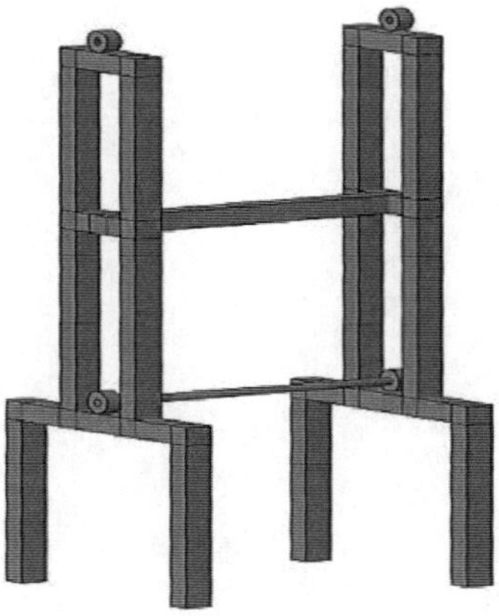

Figure 6. Design Framework

2.6. Tool design
And the design drawing tool shown in Figure 7 [4].

IOP Conf. Series: Materials Science and Engineering **384** (2018) 012013 doi:10.1088/1757-899X/384/1/012013

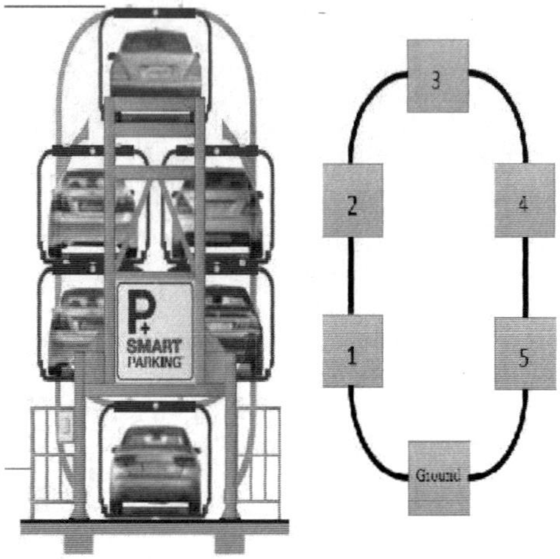

Figure 7. (a) Parking shelf order (b) Parkir Number

The design of rotational parking charging system is designed with the following specifications:

- Model parking rotation system made of 6 parking shelves that can accommodate 6 cars.
- Movement of the rotational parking system model using a stepper motor and using a timming belt to connect to the rotational parking chain [5].
- The rotation parking numbering model can be seen in Figure 7 (b).

2.7. Flowchart

Figure 8 shows the overall flow diagram Prototype of Arduino Based Parking Rotation System.

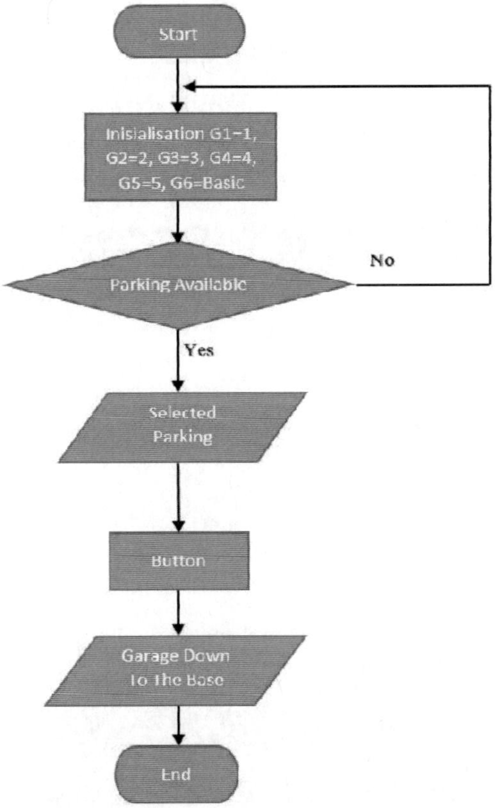

Figure 8. Flowchart

2.8. Design Control Circuits

The control circuit on this tool is shown in Figure 9.

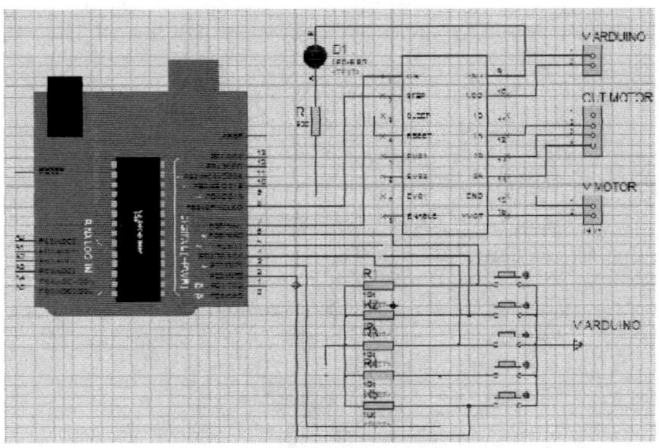

Figure 9. Control Circuits

IOP Conf. Series: Materials Science and Engineering **384** (2018) 012013 doi:10.1088/1757-899X/384/1/012013

3. Result and Discussion

3.1. Arduino IDE Program

Programming used Arduino IDE application, here is the program image for the rotary parking system prototype shown in Figure 10.

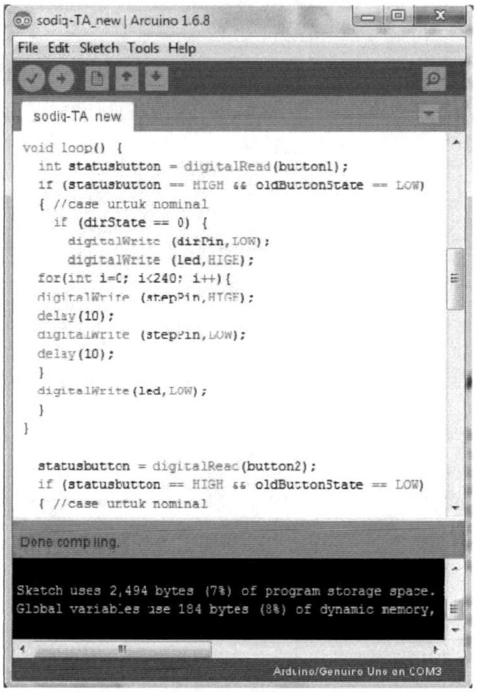

Figure 10. Arduino IDE

The movement of the rotary parking system model is determined by the position of the 1, 2, 3, 4, and 5 toward the position of the basic parking rack on the prototype rotary parking system. The determination of the setpoint is measured by how many step by step movements of the stepper motors to the base parking rack [6].

3.2. Box Panel

Testing each push button as shown in Figure 11.

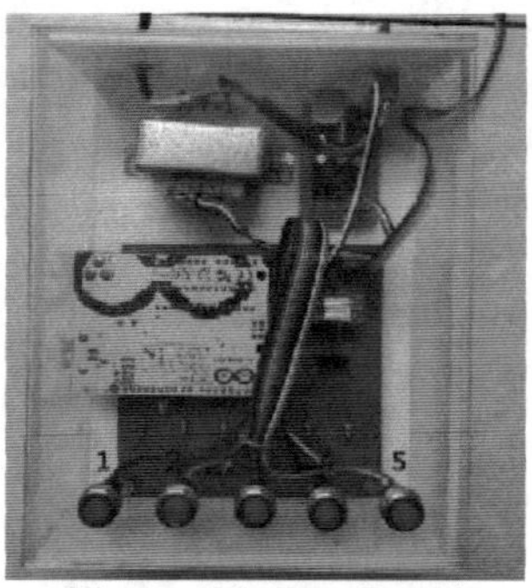

Figure 11. Box Panel

IOP Conf. Series: Materials Science and Engineering **384** (2018) 012013 doi:10.1088/1757-899X/384/1/012013

- Button 1
  Moving to the left by 240 step
- Button 2
  Moving to the left by 477 step
- Button 3
  Moving to the right by 710 step
- Button 4
  Moving to the right by 477 step
- Button 5
  Moving to the right by 240 step

3.3. *Working principles*

The workings of the tool are as follows:

- Using a 5 volt voltage source for arduino uno and 14 volt modules for stepper motor driver.
- The system input comes from several push buttons that will command the program from arduino to the stepper motor driver.
- Each push button has a number corresponding to the laying of the parking shelf in this rotation parking system.
- Push button number 1 is pressed then arduino ordered stepper motor to move perstep and will stop on the basic parking shelf as ordered from arduino.
- The steps if you want to park the car, the first consumer just push the push button according to the empty parking rack, and automatically the empty parking shelves will go down to the basic shelf and the car can be placed directly on the parking shelf. And so next if you want to park the vehicle again.
- The first steps of taking the car is that the consumer only visually looks at the number of the parking lot of his car, and then simply pushes the push button according to the number of car parking shelves to be picked up.

3.4. *Experiment data*

3.4.1. System Error Testing. Real Time Car Parking System using Processing [7]. The test is done by comparing the real time of the test tool with time calculation based on the delay that is controlled by Arduino Uno to find the value of system error. In theory, the error can be calculated by the formula:

$$E = \left[\frac{s \quad x\,ti \quad d}{r \quad ti}\right] x100\%$$

3.4.2. Time and Error Testing. Testing time based on the position of parking shelves to the basic parking shelf, among others:

Table 3. Time Rotary Parking Systems

No. Rack Parking	time(second)	Error (%)
P1	4,5	0,53
P2	9,1	0,52
P3	14	0,50
P4	9,1	0,52
P5	4,5	0,53

4. Conclusion

The conclusions that can be taken in making this research are as follows:

- Use of Arduino Uno as an On-Off controller can be used as a total step by step regulator of stepper motor rotation and parking rack in rotary parking system prototype with step division between 240, 477, 710, 477, 240.

- Time-test results show that the average error rate in the push button experiment 1 is 0,52%, the push button 2 is 0,53%, the push button 3 is 0,50%, the push button 4 is 0,53%, and on the push button 5 of 0,52%. The results of this test indicate that the time taken to generate the response as planned and can be applied to the prototype rotary parking system to achieve the intended position.

References

[1] Chaudhary H. 2017. *Advanced CAR Parking System using Arduino. International Conference on Advanced Computing and Communication Systems (ICACCS -2017),* Jan. 06 – 07, 2017, Coimbatore, INDIA.
[2] Nasir Nursalbiah. 2011. *Rotary Car Park (Pallet Design) Computer Aided Design Analysis Study. IEEE Symposium on Business, Engineering and Industrial Applications (ISBEIA).* Langkawi, Malaysia.
[3] Madhav Bhagdev. 2017. *Vertically Automated Rotary Parking System. IEEE International Conference On Recent Trends In Electronics Information Communication Technology,* May 19-20, 2017, India.
[4] *Advance Parking Solution.* Korea www.dysmart.com (accessed 10 December 2017).
[5] D Y Park 2010 *Vertical Rotary Parking.* http://www.freepatentsonline.com/y2004/0156699.html (accessed on 9 December, 2010).
[6] Sivagnanam G. 2016 *PLC Based Rotary Car Parking System. Coimbatore,* India. 2016.
[7] U. Hartmann. 2006 *Parking Monitor System Based on Magnetic Field Sensor,* Proceedings of the 2006 IEEE Intelligent Transportation Systems Conference (ITSC '06), Toronto, Canada, (2006) 1275-1279.

IOP Conf. Series: Materials Science and Engineering **384** (2018) 012014 doi:10.1088/1757-899X/384/1/012014

Estimation of Elbow Joint Angle Based on Electromyography Using the Sign-Slope Change Feature and Kalman Filtering

T Triwiyanto[1,3*], O Wahyunggoro[1], H A Nugroho[1], and H Herianto[2]

[1]Department of Electrical Engineering & Information Technology, Universitas Gadjah Mada, Jl. Grafika No. 2, Yogyakarta, Indonesia
[2]Department of Mechanical & Industrial Engineering, Universitas Gadjah Mada, Jl. Grafika No. 2, Yogyakarta, Indonesia
[3]Department of Electromedical Engineering, Health Polytechnic of Surabaya, Ministry of Health, Jl. Pucang Jajar Timur No.10, Surabaya, Indonesia

*Triwiyanto123@gmail.com

Abstract. Electromyography (EMG) has widely used in the field of biomedical engineering as a control signal for prosthetic devices and exoskeleton robotic. High accuracy in the prediction of the limb joint is very important to determine the effectiveness of the system. In this study, we propose a new algorithm to improve the elbow joint angle prediction based on electromyography using Sign Slope Change (SSC) feature extraction and Kalman filter (SSC-KF). The EMG signals acquired from the biceps were extracted using SSC feature to get the estimation of the elbow joint angle. The accuracy of the prediction of the elbow joint angle was improved by using Kalman filter. In this study, the SSC-KF algorithm can predict the elbow joint angle with high accuracy. The Pearson's correlation coefficients (mean±S.D.) were 0.95±0.02, 0.96±0.01, and 0.96±0.015 for the motion period of 12 seconds, 8 seconds, and 6 seconds, respectively. Root Mean Square Errors (mean±S.D.) were 10.37°±1.72°, 9.89°±1.11°, and 9.99°±2.2°, for the motion periods of 12 seconds, 8 seconds, and 6 seconds respectively. This SSC-KF algorithm can predict the elbow joint prediction by using a single lead electrode from biceps muscle.

1. Introduction

A good accuracy in the estimation of the limb joint angle for myoelectric control-based devices is very important because it determines the effectiveness of the devices. Recently, a myoelectric control has been used widely in many areas such as a prosthetic device, an assistive or rehabilitative exoskeleton and a teleoperation machine [1]. Previous studies implemented the estimation algorithm by improving the method such as feature extraction, classifier, and pre-processing data. EMG is a signal that is generated when there is an activity in the muscles. The EMG gives much information, including the limb joint angle [2]. The information on the limb joint angle can be found by realizing a feature extraction on the EMG signal. Feature extraction can be realized in the time, frequency or wavelet domain. The time-domain features are still good features to extract the signal because it is less of complexities and there is no transformation stage. Feature extraction is a pre-processing stage in the implementation of the limb joint estimation algorithm. Previous studies [3], [4] have used time domain feature extraction such as Root Mean Square (RMS), Integral of Absolute Value (IAV), Mean Average (MAV), Waveform Length (WL) and Sign Slope Change (SSC) to develop the limb joint angle estimation.

Some previous studies used a pattern recognition (PR) method, such as Artificial Neural Network [4], and Support Vector Machine (SVM) [5], as a classifier to predict the limb joint angle. Even though high accuracy in the estimation of the joint angle is obtained significantly, but the PR method needs a learning step to train the network for every different subject.

The Kalman filter is commonly used as an estimator for a noisy output from the motion sensor, such as an accelerometer or gyroscope sensor. Some previous studies [6], [7] combined Kalman filter with other methods such as SVM, LDA, and Hill-Based to improve the accuracy of the estimation. Improving the accuracy of the elbow joint estimation based on EMG is still a challenging issue. The success of the estimation depends on the selection of the feature extraction. The objective of this study is to develop a new algorithm to predict an EMG-based elbow joint angle by combining zero crossing feature extraction and Kalman filter. Kalman filter is used to improve the estimation by filtering the output of the feature extraction iteratively. By combining the sign slope change feature and Kalman filter (SSC-KF), we have a non-pattern recognition algorithm that can predict the elbow joint angle according to human intention.

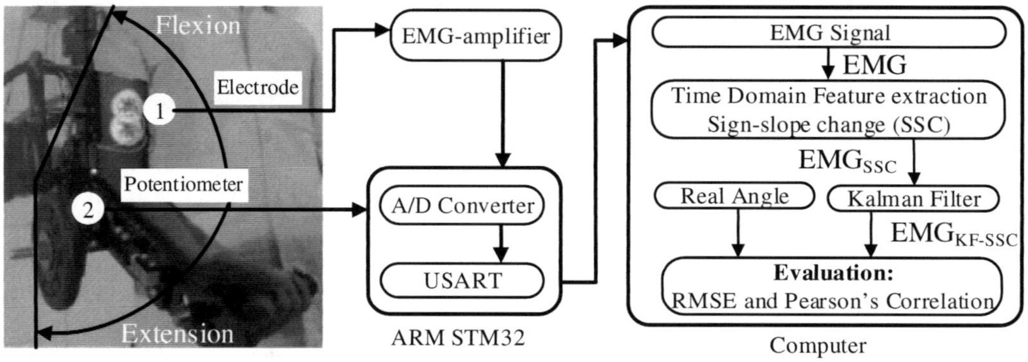

Figure 1. The block diagram of the proposed method.

2. Materials and Methods

2.1. Experimental protocol

Healthy male subjects were selected with no previous injury on the elbow. In this research, the exoskeleton was used to synchronize the movement of the elbow joint. The exoskeleton was constructed by two aluminum beams as shown in the **Figure. 1**. In the exoskeleton joint, we attached a potentiometer to detect the desired elbow joint angle. Prior to the EMG recording process, the subject was trained to move the elbow, flexion and extension direction in the range of 0° to 145° return to 0°. To synchronize the elbow movement, a Windows-based metronome application (TempPerfect version 4.08) was used. In this research, the period of the elbow flexion and extension was adjusted in 12, 8, and 6 seconds period. We collected the EMG signal from biceps muscle using two disposable Ag/AgCl electrodes. The two-channel A/D converters was composed of one channel to collect EMG signal and one channel to collect angle from the elbow joint. The sampling frequency used in the data acquisition process was 1,000 Hz, this is already qualified the Nyquist rules.

2.2. Data Processing

The EMG signal was recorded during the elbow moving in flexion and extension direction. Each segment of the EMG signal was extracted using SSC feature extraction (**Figure 1**). SSC is a feature that detects how many the signals change the slope and exceed the pre-defined threshold value. The featured SSC was written as follows [8]:

$$SSC = \sum_{i=1}^{N-1}[f[(x_i - x_{i-1}) \times (x_i - x_{i+1})]]$$
$$f(x) = \begin{cases} 1, if \rightarrow x \geq threshold \\ 0, otherwise \end{cases} \quad (1)$$

where x_i is the *i*-th EMG signal, *N* is the number of the sample point in each segment and threshold is the level of amplitude that limit the EMG signal. The width of the window sampling is 200 samples as recommended in the previous study [9].

Kalman disseminated a method on the problem of noise by using an adaptive filter approach towards the correction of the error in the system [10] [**Figure 2(a)**]. In this work, Kalman Filter (KF) was applied as a smoothing function on the output of the feature extraction. The Q and R parameter was tuned by experiment which resulted the best performance. Assume the x_k is the result of estimation of Kalman Filter and z_k is the result of the feature extraction. The Kalman filter model [11] is realized in **Algorithm 1 [Figure 2(b)]**. The nomenclature of the Kalman filter variabel was written in **Tabel 1**.

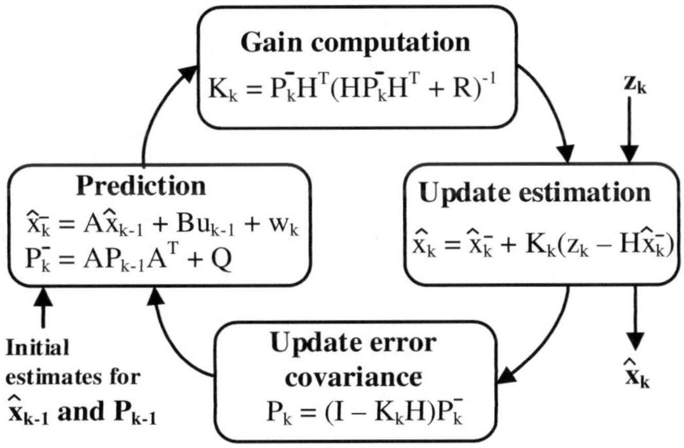

(a)

Algorithm 1: Kalman Filter
Init: $\hat{x}_{k-1} = 0; P_{k-1} = 0;$
Input: $z_k = E$ s
Output: $\hat{x}_k$
1. **For k** → 0 **to** *N* **do**
2. $\hat{x}_k^- = \hat{x}_{k-1}$
3. $P_k^- = P_{k-1} + Q$
4. $K_k = P_k^-(P_k^- + R)^{-1}$
5. $\hat{x}_k = \hat{x}_k^- + K_k(z_k - \hat{x}_k^-)$
6. $P_k = P_k^-(1 - K_k)$
7. **End**

(b)

Figure 2. (a) The block diagram of linear Kalman filter, (b) the algorithm of Kalman filter.

Table 1. The nomenclature of Kalman filter

$\hat{x}_k$	Estimated state	Q	Process noise covariance
$\hat{x}_k^-$	A priori estimated state	R	Measurement noise covariance
$\hat{x}_{k-1}$	Previous estimated state	w_k	Process noise
P_k	Error covariance	v_k	Measurement noise
P_k^-	A priori error covariance	x_i	The *i*-th EMG sample
P_{k-1}	Previous error covariance	N	Number of sample points
K_k	Kalman gain	w_k	Process noise
z_k	Input measurement	A, B, H	Matrix transition
u_k	Input control		

3. Results and Discussion

3.1. Estimating the Elbow Joint Angle

Before the Kalman filtering process, the EMG features were normalized within the range between 0 and 1, as shown in figure 3 (a). Figure 3 (a) shows that the SSC features indicated by little rectangular obviously can follow the movement of the elbow joint angle (indicated by the red color line) but with noise. In this figure 3(a), the Pearson's correlation coefficient and RMSE values were 0.847 and 24.03°, respectively. In order to improve the performance of the estimation, the EMG features extracted using SSC were filtered using Kalman filter as shown in figure 3 (b). After the Kalman filtering process, the performance of the estimation was increased as follows: the Pearson's correlation coefficient and RMSE values were 0.967 and 8.76°, respectively.

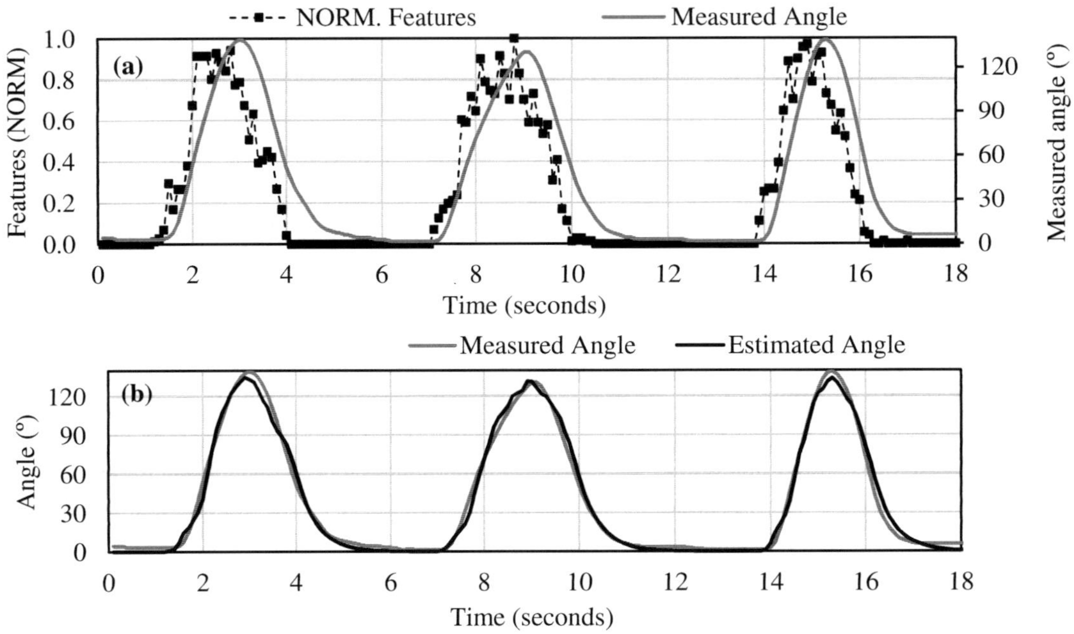

Figure 3. A typical of motion period of 6 seconds. (a) The normalize EMG features, (b) The Kalman filtered EMG features.

In figure 3, the small rectangle indicates as the EMG features, the dashed line shows the estimated elbow-joint angle from EMG features, and the red line indicates the measured elbow-joint angle. To obtain the optimum of the estimation, the Kalman filter parameters consists of the process noise Q and the measurement noise R were tuned to a certain value which resulted in the best performance. After the Kalman filtering process, the predicted angles almost coincide with the estimated angles with small of error [Figure. 3(b)].

3.2. Evaluation of the proposed method

The variation of the performance was examined by calculating the Pearson's correlation coefficient (CC) and RMSE for each cycle of motion during ten continuous cycles. Figures 4(a) and (b) and figures 5(a) and (b) show the variation of the CC and the RMSE before and after Kalman filtering. Before the Kalman filtering process, the mean of CC were 0.85±0.036, 0.88±0.021, and 0.88±0.021 for motion periods of 12 seconds, 8 seconds and 6 seconds, respectively and the mean of RMSE were 25.64±2.47, 23.13±1.96, and 22.15±1.99 for motion periods of 12 seconds, 8 seconds and 6 seconds, respectively.

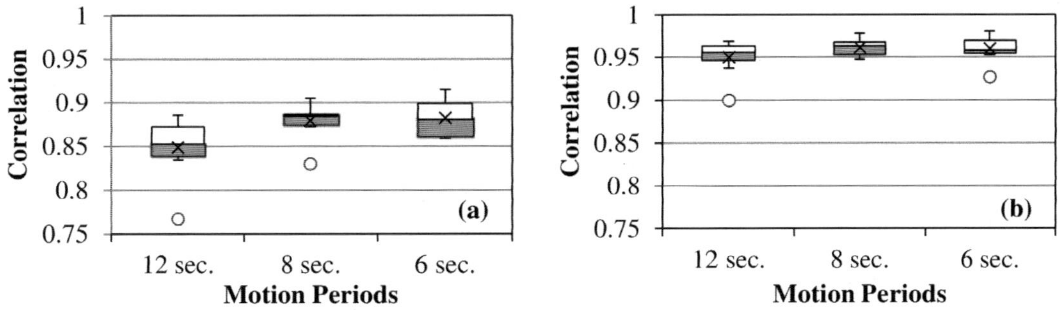

Figure 4. The comparison of the box plot diagram of Pearson's correlation coefficients (a) before and (b) after Kalman filtering process.

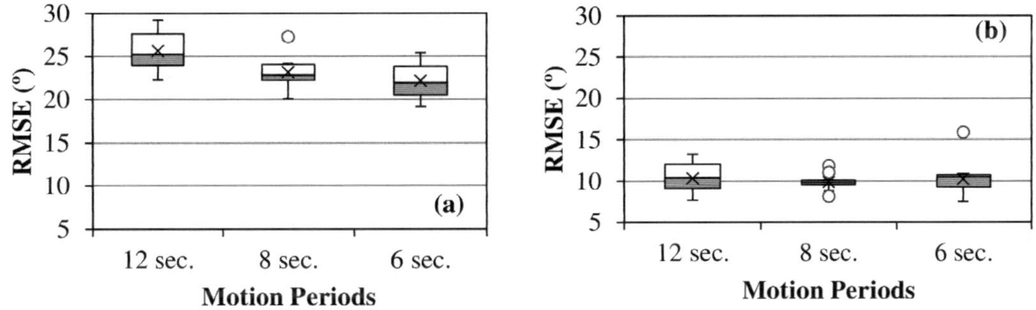

Figure 5. The comparison of the box plot diagram of RMSE values (a) before and (b) after Kalman filtering process.

High accuracy of elbow joint estimation is very important in the development of the exoskeleton robot or prosthetic devices. As shown in Table 2, the mean of RMSE of the estimated angle for 12 seconds, 8 seconds and 6 second periods are 10.39°±0.66°, 9.89°±0.37° and 10.26°±0.82°, respectively. This work was comparable with previous related work. Tang *et al.* [4] reported similar results for three different speeds of motion with mean RMSE of 12.42°, 9.67° and 10.70° for 8 seconds, 4 seconds and 2 seconds period. In their study, the EMG signal collected from biceps, triceps, anconeus and brachialis muscles were extracted in order to obtain the features and to use as the input for the ANN classifier.

Table 2. The summary of the CC and RMSE values after the Kalman filtering process for the motion periods of 12, 8, and 6 seconds.

Descriptive Stat.	CC			RMSE (°)		
	12 sec.	8 sec.	6 sec.	12 sec.	8 sec.	6 sec.
Mean	0.95	0.96	0.96	10.39	9.89	10.26
Standard Deviation	0.021	0.010	0.016	0.60	0.37	0.82
Maximum	0.97	0.98	0.98	13.25	11.85	15.90
Minimum	0.90	0.95	0.93	7.66	8.13	7.48

Table 2 shows that the CC values were higher than 0.9, these indicate that the estimated and measured angle has a high relationship.The consistency of the proposed method was tested for the three different periods namely 12, 8, and 6 seconds. Statistical analysis of single factor ANOVA showed that there was no significant difference of RMSE as well as CC for three different periods of motion (p-value>0.05).

Although the results were very promising for a better performance, there were some factors needs to consider. These factors were the muscle fatigue [12], skin conductance, electrode placement [2] and the number of subjects. Next work, it also needs to explore the Kalman filter parameters such as Q-parameter and R-parameter to optimize the estimation algorithm. Furthermore, this proposed method (SSC-KF) can be applied to any devices based on myoelectric control to get a natural control.

4. Conclusion

This study has demonstrated the effectiveness of the proposed method by using a combination of the Sign Slope Change features and Kalman filtering. The elbow joint angle was estimated using the EMG signal generated from biceps alone. The EMG signals were extracted using the Sign Slope Change and optimized using Kalman filter. Statistical analysis proved that there is no significant difference of RMSE between periods of motion. The performance of the estimation was increased after the Kalman filtering process by 59.45%, 57.23%, and 53.66% for motion period of 12, 8, and 6 seconds, respectively. In next work, this proposed method can be implemented in prosthetics and robotic exoskeleton design.

References

[1] R. a R. C. Gopura, D. S. V Bandara, J. M. P. Gunasekara, and T. S. S. Jayawardane 2013 Recent Trends in EMG-Based Control Methods for Assistive Robots," *Electrodiagnosis New Front. Clin. Res.*

[2] C. J. De Luca, 1997 The Use of Surface Electromyography in Biomechanics, *J. Appl. Biomech.*, vol. **13**, no. 2, pp. 135–163.

[3] K. Kiguchi and Y. Hayashi, "EMG-Based Control of a Lower-Limb Power-Assist Robot," in *Intelligent Assistive Robots, Springer Tracts in Advanced Robotics 106*, Springer, 2015, pp. 371–383.

[4] Z. Tang, K. Zhang, S. Sun, Z. Gao, L. Zhang, and Z. Yang, 2014, An Upper-Limb Power-Assist Exoskeleton Using Proportional Myoelectric Control, *Sensors*, vol. **14**, no. 4, pp. 6677–6694, Apr.

[5] Q. Li, Y. Song, Z. Hou, and B. Zhu, 2013, sEMG based joint angle estimation of lower limbs using LS-SVM," *Lect. Notes Comput. Sci. (including Subser. Lect. Notes Artif. Intell. Lect. Notes Bioinformatics)*, vol. **8226** LNCS, no. PART 1, pp. 292–300.

[6] Z. Li, B. Wang, F. Sun, C. Yang, Q. Xie, and W. Zhang, 2014, SEMG-based joint force control for an upper-limb power-assist exoskeleton robot," *IEEE J. Biomed. Heal. Informatics*, vol. **18**, no. 3, pp. 1043–1050.

[7] Q. Ding, X. Zhao, A. Xiong, and J. Han, 2011, A Novel Motion Estimate Method of Human Joint with EMG-Driven Model," in *2011 5th International Conference on Bioinformatics and Biomedical Engineering*, no. 1, pp. 1–5.

[8] A. Phinyomark, A. Nuidod, P. Phukpattaranont, and C. Limsakul, 2012, *Feature Extraction and Reduction of Wavelet Transform Coefficients for EMG Pattern Classification Feature Extraction and Reduction of Wavelet Transform Coefficients,* vol. **6**, no. JUNE, 2012.

[9] Triwiyanto, O. Wahyunggoro, H. A. Nugroho, and Herianto, 2017, Effect of window length on performance of the elbow-joint angle prediction based on electromyography, *J. Phys. Conf. Ser.*, vol. **853**, p. 12014.

[10] R. E. Kalman, 1960, A New Approach to Linear Filtering and Prediction Problems, *ASME Trans. J. Basic Eng.*, vol. **82D**, pp. 35–45.

[11] P. Y. C. Hwang, 2013, *Introduction to Random Signals and Applied Kalman Filtering with MATLAB Exercises*, Fourth., vol. **53**, no. 9. USA: John Wiley & Sons, Inc.

[12] Triwiyanto, O. Wahyunggoro, H. A. Nugroho, and H. Herianto, 2016, DWT Analysis of sEMG for Muscle Fatigue Assessment of Dynamic Motion Flexion-Extension of Elbow Joint, in *2016 8th International Conference on Information Technology and Electrical Engineering (ICITEE)*, 2016, pp. 1–6.

IOP Conf. Series: Materials Science and Engineering **384** (2018) 012015 doi:10.1088/1757-899X/384/1/012015

Design and Implementation of Productivity Monitoring System in Rubber Injection Molding Machine in Real Time Based on Visual Studio and Android Application

M Rubenstein[1*], D Handaya[1], and W Slamet[2]

[1]Electronics Engineering, Politeknik Gajah Tunggal, 15135 Jatiuwung, Tangerang, Banten, Indonesia.
[2]Department of Research and Development, PT IRC Inoac Indonesia, 15135 Jatiuwung, Tangerang, Banten, Indonesia.

*rubenstein0814@gmail.com

Abstract. The molding machine is used to rubber material cooking process. The one type of molding machine is an injection molding machine. The productivity monitoring process of injection molding machines can be said to still use conventional methods. This is caused by data collection using paper and takes 2 days to be displayed to the computer database. In order to solve this problem, we conduct research by designing real time productivity monitoring system applications on injection molding machines. The monitoring system designed in the form of desktop and mobile applications. Desktop application is used to manage production plans, monitor production and know the results of production data. This test has two stages, the first is to ensure that the counter point in Programmable Logic Controller (PLC) has the same value as desktop and mobile applications and other stages ensure the application can display productivity confarmance indicators. Test conducted show that monitoring systems with dekstop and mobile applications can provide productivity information on injection molding machines in real time, provide indicators of productivity conformance with predefined standards and reduce paper usage. **Keywords:** injection molding machine, productivity, real time monitoring system, paperless.

1. Introduction

Manufacturing industries such as rubber processing currently can not be separated from technology. In the process of processing, expected to produce optimal productivity accompanied by maximum product. One of the machines used is a molding machine for the process of cooking rubber material which is planned to be fully automatic [1]. The process of productivity monitoring conducted on the molding machine still uses conventional methods. The obstacles experienced are relying on the delivery of information by workers. Observations have been made to prove that the processing time of data into daily reports of molding machine production to computers takes 2 days. If the administrator in charge of processing the data can not be present, then the data processing is pending.

The need for fast, precise and accurate information is essential in every aspect of activity in the manufacturing industry. One of the technological innovations in the manufacturing industry is the application of monitoring systems that can be automated remotely by computer-based applications [2] and [3]. The use of information technology is expected to meet all information needs in the manufacturing industry, thereby facilitating the processing, analyzing, and communicating of relevant information in order to efficiently manage resources in the manufacturing industry. The monitoring system ever developed by LIPI is accessible via the web [4]. Monitoring systems can also be applied via smartphones due to the development of significant mobile applications [5]. Currently researchers propose solutions in the form of real time productivity monitoring system on injection molding machine. The system is a desktop and mobile application synchronized with Programmable Logic Controller (PLC) on injection molding machine to increase productivity.

2. Research methods

The research begins by identifying the data used in measuring the productivity of the injection molding machine. Furthermore, programming on PLC is a factor in monitoring productivity. Data generated from PLC are then processed and stored in the database as performed in monitoring system using XAMPP [6]. The type of database used in the study using the local database.

The design of desktop applications is done using Microsoft Visual Studio. Chen once did so on an intelligent vehicle so that it could be proved for the research of the monitoring system being developed [7]. Desktop application created in the form of the main page, the entry page that can be accessed through two types of account that is admin and user, production monitoring page which is the main purpose of this research, the production item page that function to determine the items to be produced, and the production report page as a means of information and printing of production activity data.

Mobile apps created on Android Mobile. This is caused by most developments like [8] and [9] use Android Studio and Android OS-based smartphones. The mobile app loads the login page as a user access security and the main page. The main page contains production monitoring that shows productivity in real time based on desktop application data. The synergy of Visual Studio and Android has been successfully performed for monitoring system [10].

After all system support has been successfully created, the design is connected. The architectural monitoring system is designed as follows.

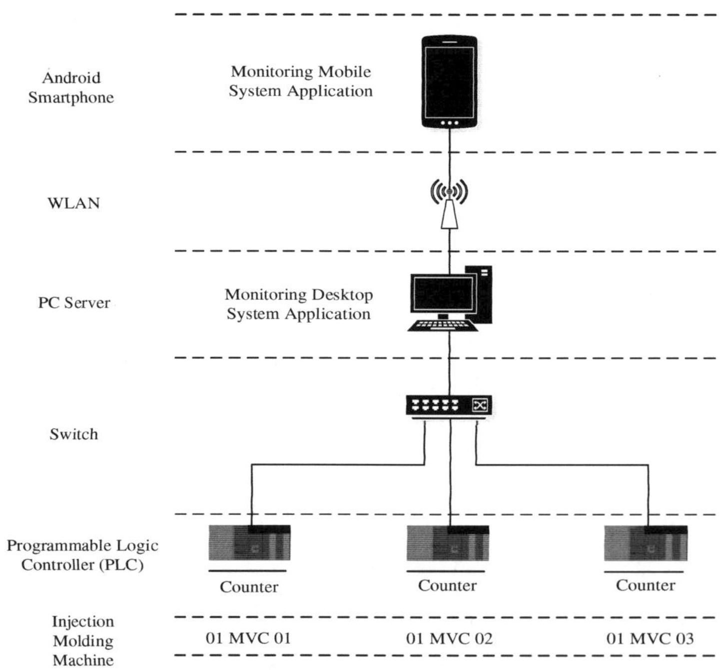

Figure 1. Architectural monitoring system

IOP Conf. Series: Materials Science and Engineering **384** (2018) 012015 doi:10.1088/1757-899X/384/1/012015

Architecture in this research is conducted to know the productivity of injection molding machine through application from 2 different device. The devices used are PC servers and Android smartphones. The application on the PC server is capable of monitoring and entering the data you want to produce and storing it into the database. Applications on Android smartphones can be used to monitor the productivity of injection molding machines when connected in the same network as PC servers.

Testing is done by comparing the counter value of PLC (in the form of shoot) with desktop and mobile applications, both of which have the same value. This test is done at the same time to test the concept of real time monitoring. The next test against the indicator that will appear to know the suitability of productivity standards. The calculation formula included in the application program code is as follows:

$$\text{Standard production value per hour} = \frac{6\ \ m}{s\ \ \ c\ \ ti} \quad (1)$$

$$\text{Actual production value per hour} = \frac{6\ \ m}{a\ \ \ c\ \ ti} \quad (2)$$

The next test displays the results of production activity reports in the desktop application so that the authorized leadership can know the results of production in real time. In addition, it can also find out the data report results each month and every year.

3. Result and discussion

3.1. Real time, monitoring system test

Testing is done by ensuring that exchange rates on PLC, desktop apps, and mobile apps have the same value at the same time. Figure 2 and Figure 3, it can be seen that PLC values, desktop apps, and mobile apps are capable of displaying the same counter value with 8 shooting values at the same time of 15:57.

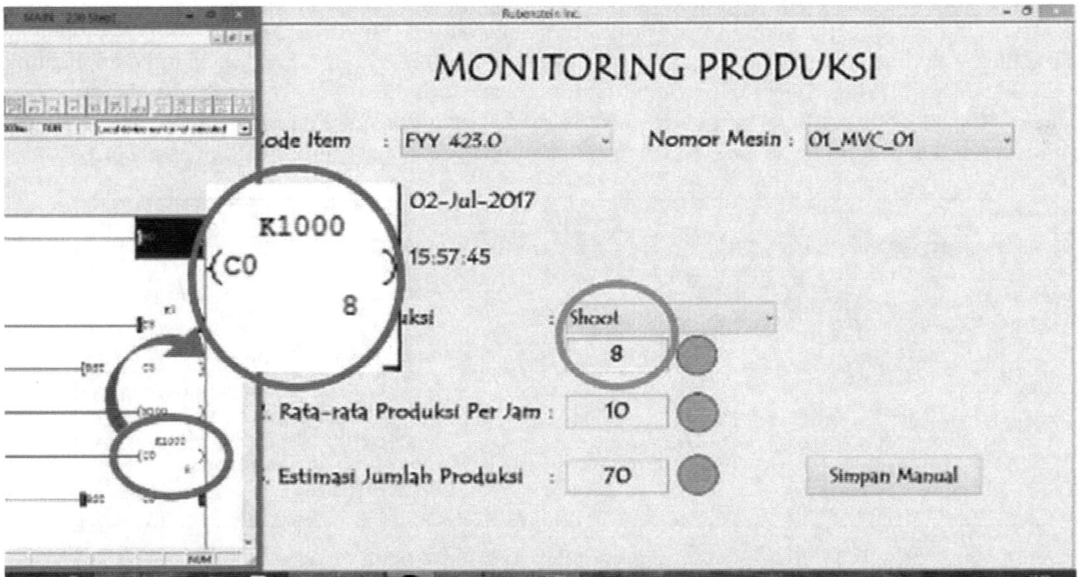

Figure 2. Validity of PLC and desktop application

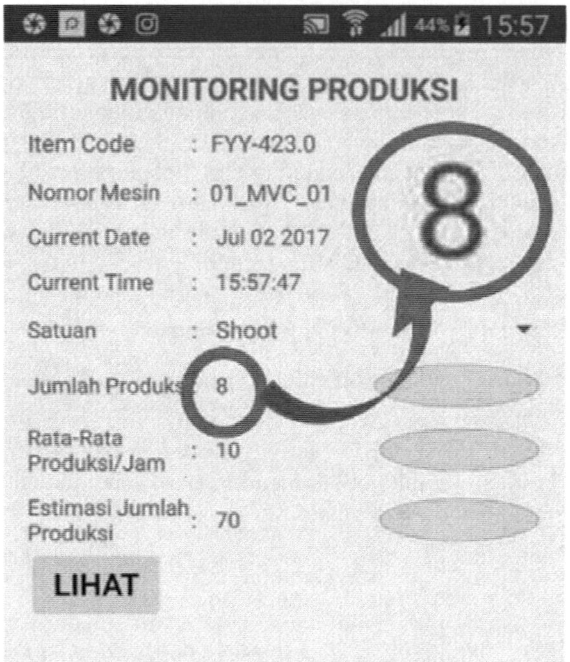

Figure 3. Validity of mobile application

3.2. The standard of suitability productivity

This test is to determine the suitability of productivity standards. When simulated on a PLC counter by giving the cycle time suitable to the specified standard in Figure 4, Figure 5 shows that the application is capable of displaying a green indicator. However, if a certain time cycle exceeds the specified standard, Figure 6 shows that the app is capable of displaying a red indicator. This means that the application is designed to be able to tell when something is not in accordance with the standard set.

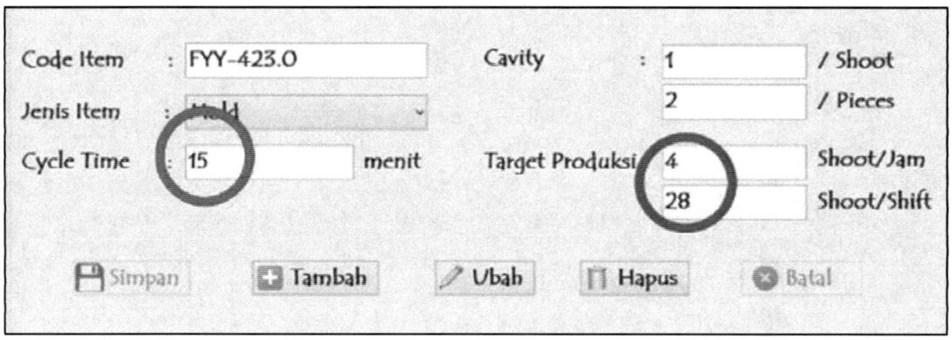

Figure 4. Standard of productivity

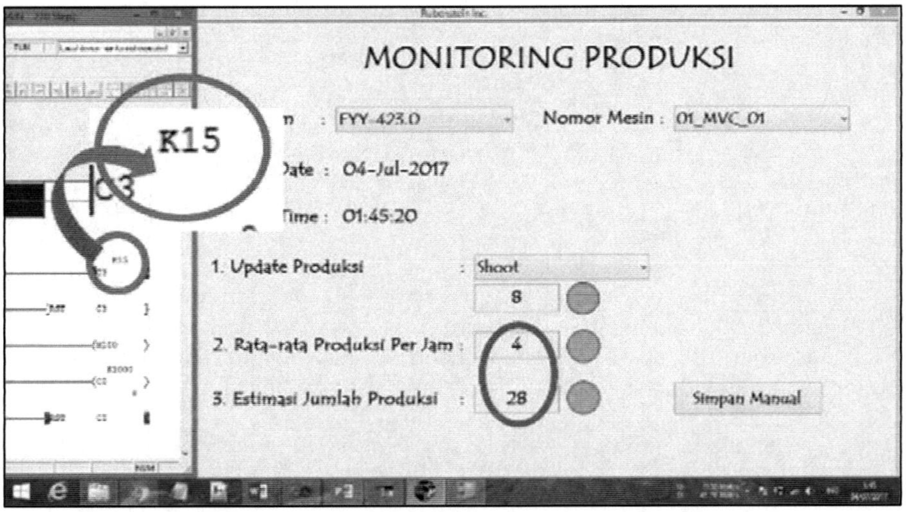

Figure 5. Productivity of counter suitable standard

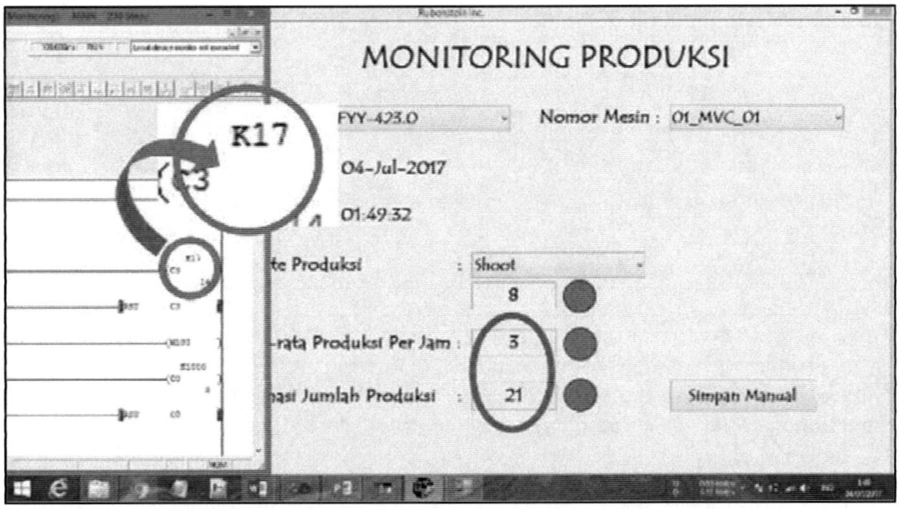

Figure 6. Productivity of counter unsuitable standard

The processing result of the above application can be proven by comparing it with manual calculation. If the standard 15 minutes cycle time is known, then the calculation is as follows:
Counter PLC = actual cycle time = 15 minutes cycle time

$$\text{Actual productivity value per hour} = \frac{60 \text{ minutes}}{\text{actual cycle time}} = \frac{60 \text{ minutes}}{15 \text{ minutes}} = 4 \text{ shoot}$$

$$\text{Standard productivity value per hour} = \frac{60 \text{ minutes}}{\text{standard cycle time}}$$
$$= \frac{60 \text{ minutes}}{15 \text{ minutes}}$$
$$= 4 \text{ shoot}$$

$$\text{Estimation of total production} = \text{Actual productivity value per hour x 7 working hours}$$
$$= 4 \text{ x } 7$$
$$= 28 \text{ shoot}$$

The results of manual calculations show the same value with the desktop application that is the production value per hour which value of 4 shoot and the estimated value of 28 shoot. Here the data of other calculations by using the actual assumption of different cycle time.

Table 1. Test standard of suitability productivity

Max. standard value of cycle time (minute)	Cycle time of PLC counter (minute)	Min. standard value of product in each hour (shoot)	Product actual value in each hour (shoot)	Indicator
15	8	4	7,5	Green
15	10	4	6	Green
15	13	4	4,615384615	Green
15	15	4	4	Green
15	17	4	3,529411765	Red
15	20	4	3	Red
15	22	4	2,727272727	Red
15	25	4	2,4	Red
15	27	4	2,222222222	Red
15	30	4	2	Red

Based on Table 1, the indicator on the monitoring system will show the suitability of productivity by the established production standard. If the actual production value per hour $\geq$ 4 shoot, then the productivity exceeds or equal to the standard set, so the indicator will be green. Whereas if the actual value of production per hour <4 shoot, then the indicator will be red.

3.3. Testing of production activity report

The test strengthens the monitoring system in production activities to be the most important part, especially in the reporting of production activities. As shown in Figure 7, the monitoring system can be used to determine the amount of production activity based on the production report data.

Figure 7. Report of production activity

The report data in the form of monthly and annual reports can be used as an evaluation of company performance. Thus, the digital data report can be hinted at as a reduction in paper usage and increase the productivity of work [11]. In addition, in accordance with [12] that the industry is willing to use information systems.

4. Conclusion

Productivity monitoring systems designed and implemented for rubber injection molding machines can be integrated with Programmable Logic Controller (PLC) as the input of production data on injection molding machine and can provide productivity of injection molding machine production in real time. The applied desktop and mobile applications can provide productivity conformance indicators with predefined color indicator standards. If the actual production value per hour ≥ 4 shoot, then the indicator is green. Applies otherwise, then the indicator will be red. The monitoring system can also report monthly and annual production activities that can reduce paper usage.

References

[1] Shirai, K., Tanaka, H., Banzai, H., & Hosoya, T. 1991 *U.S. Patent No. 5,062,053*. Washington, DC: U.S. Patent and Trademark Office.

[2] Haryanto, Heri, dan Endi Permata. "Sistem Monitoring Proses Produksi pada Mesin Bardi di PT. Tirta Investama (Danone Aqua) Sukabumi Berbasis Web." *Jurnal Ilmiah Setrum (Journal of Electrical Engineering),* Volume **3**:26-34, 2016.

[3] Liang, S. Y., Hecker, R. L., & Landers, R. G. 2002. Machining process monitoring and control: the state–of–the–art. In *ASME 2002 International Mechanical Engineering Congress and Exposition* (pp. 599-610). American Society of Mechanical Engineers.

[4] Gheyb Jhuana Ohara. *Aplikasi Sistem Monitoring Berbasis Web untuk Open Cluster*. Tugas Akhir, Sekolah Tinggi Teknologi Telkom, Bandung, 2005. Tersedia: http://www.komputasi.lipi.go.id/data/1014224400/data/1123986736.pdf

[5] Teng, C. C., & Helps, R. 2010. Mobile application development: Essential new directions for IT. In *Information Technology: New Generations (ITNG), 2010 Seventh International Conference on* (pp. 471-475). IEEE.

[6] Prema Sundaram. "Patient Monitoring System using Android Technology". (International Journal of Computer Science and Mobile Computing), IJCSMC, Vol. **2**, Isuue. 5, May 2013, pg 191-201.

[7] Chen, P., & Liu, S. 2010. Intelligent vehicle monitoring system based on GPS, GSM and GIS. In *Information Engineering (ICIE), 2010 WASE International Conference on* (Vol. **1**, pp. 38-40). IEEE.

[8] Moon, S. W., Kim, Y. J., Myeong, H. J., Kim, C. S., Cha, N. J., & Kim, D. H. 2011, November. Implementation of smartphone environment remote control and monitoring system for Android operating system-based robot platform. In *Ubiquitous Robots and Ambient Intelligence (URAI), 2011 8th International Conference on* (pp. 211-214). IEEE.

[9] Gradl, S., Kugler, P., Lohmüller, C., & Eskofier, B. 2012. Real-time ECG monitoring and arrhythmia detection using Android-based mobile devices. In *Engineering in Medicine and Biology Society (EMBC), 2012 Annual International Conference of the IEEE*(pp. 2452-2455). IEEE.

[10] Lee, J. V., Chuah, Y. D., & Chieng, K. T. 2013. Smart elderly home monitoring system with an android phone. *Int. J. Smart Home*, *7*(3), 17-32.

[11] AMenon, N. M., Lee, B., & Eldenburg, L. 2000. Productivity of information systems in the healthcare industry. *Information Systems Research*, ***11***(1), 83-92.

[12] Chiasson, M. W., & Davidson, E. 2005. Taking industry seriously in information systems research. *Mis Quarterly*, 591-605.

IOP Conf. Series: Materials Science and Engineering **384** (2018) 012016 doi:10.1088/1757-899X/384/1/012016

Design of Process Control System in Water Cooling Processing Interaction Tank using Integral Control and Robust Control in The Tire Industry

D Handaya[1*], P Bakti[2], and Y P Nugraha[3]

[1]Electronic Engineering, Politeknik Gajah Tunggal, Jl. Gatot Subroto KM. 7, Tangerang, Indonesia, 15135

[2]Research Center for Quality System and Testing Technology, Indonesian Institute of Science, Gedung 417, Jl. Kawasan Puspitek, Muncul, Setu, Tangerang Selatan, Indonesia, 15314

[2]Control System and Computer Laboratory, School of Electrical Engineering and Informatics, Bandung Institute of Technology, Jl. Ganesha no. 10 Bandung, Indonesia, 40132

*devi.handaya@poltek-gt.ac.id

Abstract. This laboratory based research build a process control system as water level scanner in interaction tank used for water cooling. The control valve parameter based on water level between tanks are interacted. The control method of tank valve ratio choosed in this reasearch is integral control and robust control. The desired control specification for maximum overshoot is 10%, peak time is 1 second, and error steady state is 0. The design process executed in physic system modelling step and system design criteria. Experiment executed using Matlab Script and Matlab Simulink. The simulation result will be compared to see an effective performance for processing of water cooling interaction tank control method. The design result for integral control can tracking reference accord to system design criteria and robust control can tracking reference without different experience either the system without disturbance or the system with disturbance. Keyword: water cooling tank interaction, process control, integral control, robust control.

1. Introduction

The tire industry has an important role in improving the economic process in Indonesia. This is due to the need for higher tires to support economic processes such as supporting export and import activities. Tires as part of automotive components to deliver the export and import goods. One of the processes in the tire industry, namely the curing process in accordance with the specifications [1]. In the process requires a system to cool the process, which is using water cooling processing. In the water cooling process, it takes a container as a means to prepare or commonly called water cooling process tank. The existing tanks usually only form into one separate tank. In order to maintain water availability in the tank, a control system such as PID Controller [2], [3] is required. The linear control system in this study is intended for the output can track the reference [4]. An integral and robust controller as has been done to solve tracking problems [5], [6]. In this study trials were conducted to control the interacting or paired

IOP Conf. Series: Materials Science and Engineering **384** (2018) 012016 doi:10.1088/1757-899X/384/1/012016

tanks, it has also been done before using the sliding mode control [7], using a linear Model Predictive Control (MPC) [8] and currently researchers use integral and robust control tracking system.

2. Design of Control System

2.1. Physical System Modeling on Interaction Tank

In this research, Matlab simulation was conducted to control the tank connecting valves between the tank and the last tank output valve with the intermediate tank fluid level parameters between the interaction tanks commonly used in the tire industry. The height level of the fluid is translated as tank capacity. In the early stages in design of control system by physical system modeling. The following physical form of the system will be mathematically modeled to be able to determine the transfer function of system [9].

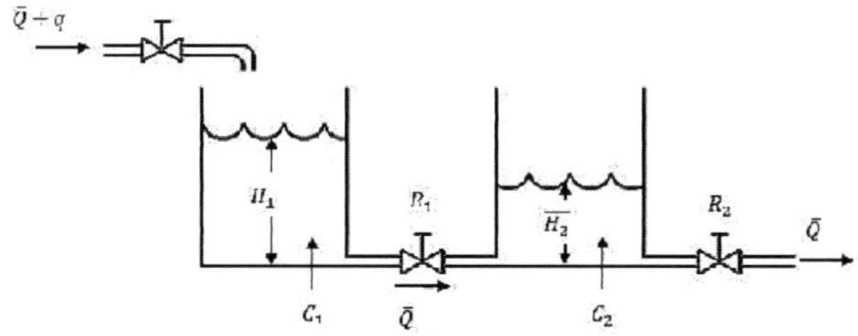

Figure 1. Physical form of fluid capacity system in interaction tank.

where

- $\bar{Q}$: steady state flow rate
- $\overline{H_1}$: steady state of fluid height in tank 1
- $\overline{H_2}$: steady state of fluid height in tank 2
- R : resistance of the fluid capacity system (valve)
- C : capacitancy of the fluid capacity system (capacity of tanks)

This tank system uses two tanks which interact with each other. The two tanks are connected and the flow rate is controlled through the ratio of valve. The system will perform the accumulated value of the comparison between the last tank output and the earliest tank input.

Based on the working principle of the above system, it can be made the block diagram system as follows.

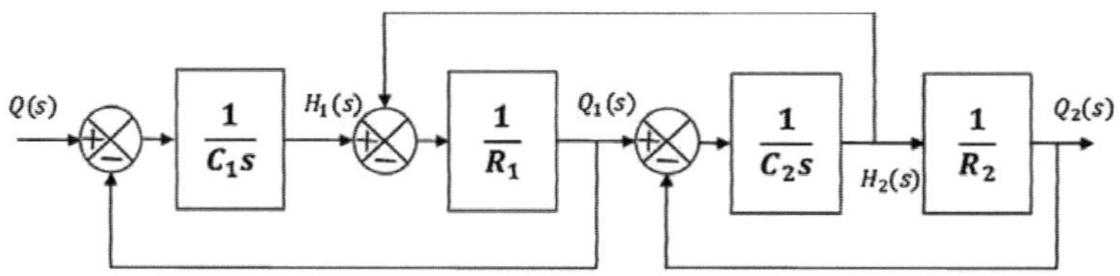

Figure 2. The block diagram of fluid capacity system in interaction tank.

To be able to determine the value of the equation for its transfer function, its component values are assumed as $R_1 = 0.5, C_1 = 3, R_2 = 0.7,$ and $C_2 = 3$. So we get the transfer function system as follows

$$\frac{Q_2(s)}{Q(s)} = \frac{1}{3{,}1\ s^2+5{,}7s+1} \tag{1}$$

2.2. Criteria of System Control Design

Open loop or non-feedback system response is intended to be able to see system performance without feedback by observing performance characteristics such as maximum overshoot, settling time, and error steady state (E_{ss}) [9]. The test results states that the system has a very long time to reach steady state for 30 seconds, and without overshoot so that its performance will be improved by adding controllers. A system to be controlled and observed also needs to be tested by knowing the matrix of controllability and observability [10]. Based on the test of controllability and observability, there is no linear combination so the system can be controlled and observed.

Determination of design system criteria is intended to obtain the system with the desired performance. The system is desired as a reference in determining the reinforcement for the initial system to fit the desired system. In this research, the desired system design criteria as follows maximum Overshoot (M_P) = 10%, Peak Time (T_P) = 1 sec, Tracking reference = 1 and Error Steady State (E_{SS}) = 0.

Based on the design criteria, we can get dumping constants (ζ) is 0.59 and natural frequency constant (ω_n) is 3.89. Based on the characteristic equation, we can get.

$$\alpha_c(s) = s^2 + 4.59s + 15.1321 \tag{2}$$

2.3. Tracking design with Integral Control

To be able get results without steady-state error when the system gets parametric changes, it needs an integral controller [5], [10]. The state space provided from equation (1), the system is represented as.

$$\begin{bmatrix} \dot{x_1} \\ \dot{x_2} \end{bmatrix} = \begin{bmatrix} 0 & 1 \\ -0,317 & -1,81 \end{bmatrix} \begin{bmatrix} x_1 \\ x_2 \end{bmatrix} + \begin{bmatrix} 0 \\ 0,317 \end{bmatrix} u, y = \begin{bmatrix} 1 & 0 \end{bmatrix} x \tag{3}$$

Based on the system equations for integral control [10], we can get.

$$|s \;\; - F + G \;\; | = s^3 + (1,81 + 0,317K_0)s^2 + (0,317 + 0,317K_1)s + 0,317K_i \tag{4}$$

Based on the system equation on integral control, it is necessary to the 3rd order. The order is obtained from 10x ($-\zeta\omega_n$) to obtain the desired pole-pole [-2.2951 + 3.1408i -2.2951-3.1408i -22.9510] and the desired characteristic equation becomes the following.

$$\alpha_c(s) = s^3 + 27,5412s^2 + 120,4818s + 347,2968 \tag{5}$$

So we get the parameter of reinforcement value K = [379.068 81.171] and integral strengthening K_i = 1.095,57. The following block diagram of the system used in Matlab Simulink.

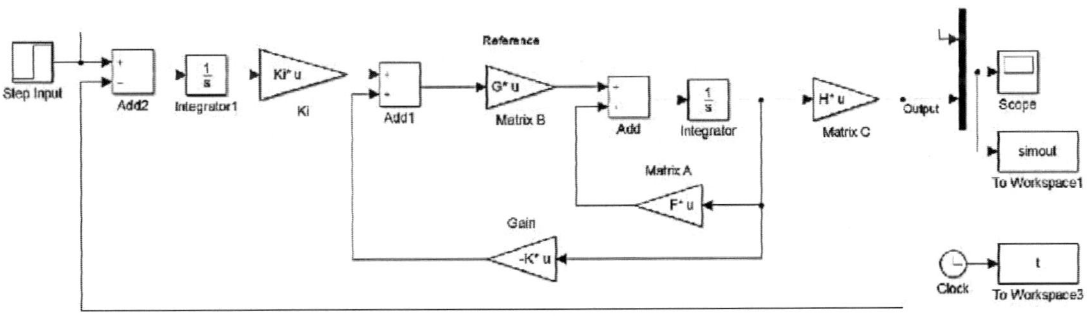

Figure 3. Matlab simulink block diagram system using integral control

2.4. Tracking design with Robust Control

In the robust controller method, it will keep the system tracking the reference and hold it against uninterrupted interference or steady state error equal to zero [6], [10], [11]. If the reference is given a step function with the equation $\frac{1}{s}$, then based on the equation of the reference characteristic and the new system equations in the error space [10] , we can get.

IOP Conf. Series: Materials Science and Engineering **384** (2018) 012016 doi:10.1088/1757-899X/384/1/012016

$$|s - F + G| = s^4 + (2{,}81 + 0{,}317K_0)s^3 + (2{,}127 + 0{,}317K_1 + 0{,}317K_0)s^2 + (0{,}317K_{e0} + 0{,}317 + 0{,}317K_1)s + 0{,}317K_{e1} \quad (6)$$

Based on the system equation on the Robust control [10], then the 3rd and 4th order is [-2.2951 + 3.1408i -2.2951 - 3.1408i -22.9510 -21.9510] and the desired characteristic equation becomes the following.

$$\alpha_c(s) = s^4 + 49{,}5s^3 + 725s^2 + 2992s + 7623{,}5 \quad (7)$$

By equating the desired characteristic equation, we can get the parameter of reinforcement value $K = [2.133{,}069 \quad 147{,}287]$, robust strenghtening $K_{e0} = 7.304{,}42$, and $K_{e1} = 24.048{,}89$. The following block diagram of the system used in Matlab Simulink.

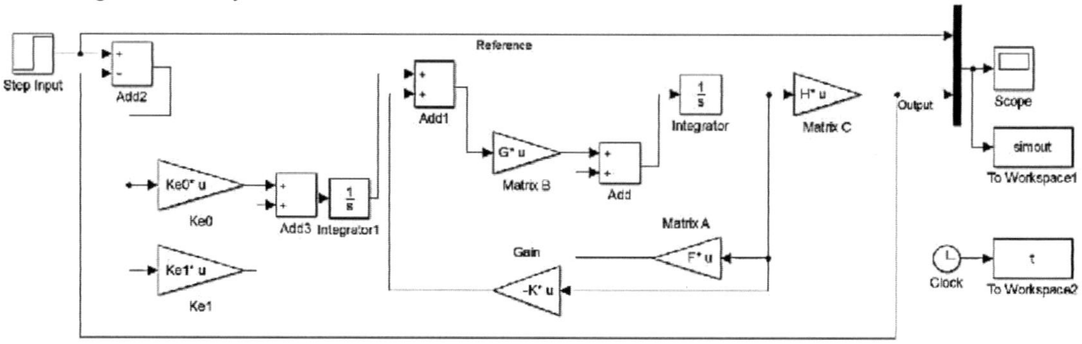

Figure 4. Matlab simulink block diagram system using robust control

3. Results and discussion

3.1. Tracking test of the fluid capacity system in interaction tank with integral control

In testing the fluid capacity system in this interaction tank is made using Matlab software. Researchers use a combination of Matlab Script and Matlab Simulink to find the results. The results obtained from testing the integral control system as follows.

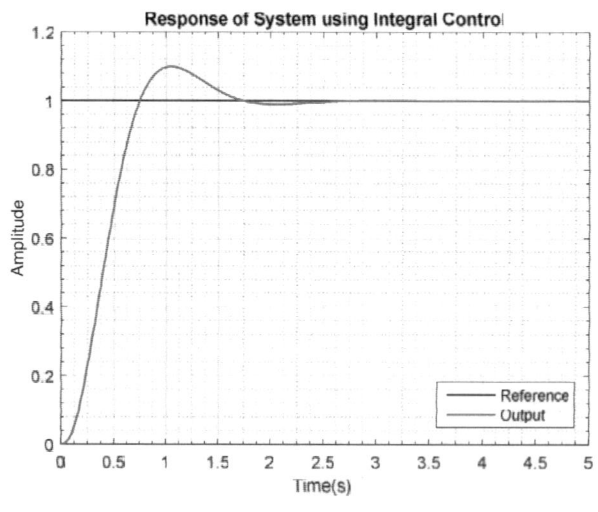

Figure 5. Testing result of the fluid capacity system in interaction tank using integral control.

IOP Conf. Series: Materials Science and Engineering **384** (2018) 012016 doi:10.1088/1757-899X/384/1/012016

Based on the above response graph, it can be seen that the output of the system successfully track the reference or steady state error is 0 with Tracking Reference is 1 in accordance with the design criteria of the desired system it can be said that the output system can track the reference well.

3.2. Tracking test of the fluid capacity system in interaction tank with robust control
In testing the system using combination of Matlab Script and Matlab Simulink to find the results. The results obtained from testing the robust control system as seen in Figure 6. Based on the system response in the Figure 6, with step unit input can be seen that the output can track reference input with Tracking Reference is 1 according to the design criteria of the desired system.

To prove that the robust system against interference will be added another reinforcement as a nuisance. Random noise reinforcement with value K_{d0} is 5 and K_{d1} is 8 is given to system. Based on the Figure 7, the system is still robust or sturdy in maintaining the circumstances to keep track of reference input.

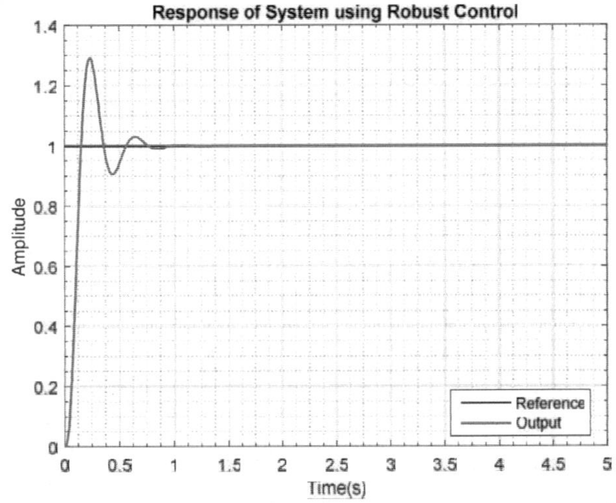

Figure 6. Testing result of the fluid capacity system in interaction tank using robust control.

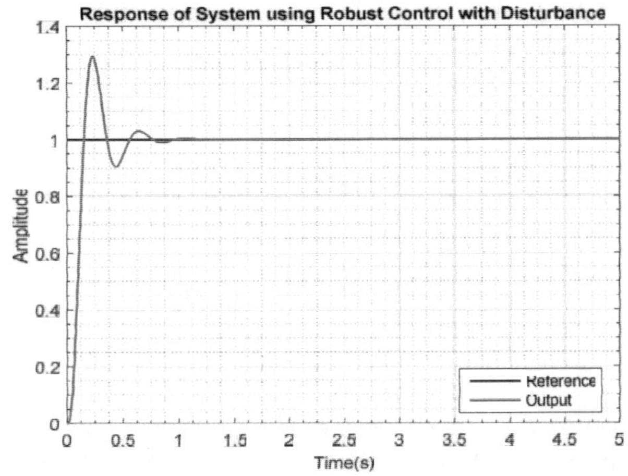

Figure 7. Testing result of the fluid capacity system in interaction tank using robust control with disturbance.

Robustness in maintaining the state of distraction is an advantage in robust control. It is based on the idea of a firm control of bringing the system from the state space to the error space so that the form of fault dynamics can be determined by placing the poles in order to remain stable. Thus, the result of applying robust control to the interaction tank system gets better performance in the form of faster settling time than integral control. The performance of the integral control is in line with its effort in controlling the system by feeding the integral error. However, there is a lack of use of robust control in the form of a little oscillation before steady state but does not affect robust control performance.

4. Conclusion

The simulation test of fluid capacity system in interaction tank in tire industry has been done so that the result of applying integral control has good performance according to the desired system design criteria in the form of maximum 10% overshoot and peak time is 1 second. While the result of robust control implementation has better performance in the form of faster settling time than integral control although having a little oscillation due to higher order of integral control system.

Acknowledgment
The authors would like thank to the Computer Control and System Laboratory, School of Electrical Engineering and Informatics, Bandung Institute of Technology in assisting the provision of facilities for testing using Matlab.

Reference

[1] Formela, K., Wąsowicz, D., Formela, M., Hejna, A., & Haponiuk, J. 2015. Curing characteristics, mechanical and thermal properties of reclaimed ground tire rubber cured with various vulcanizing systems. *Iranian Polymer Journal*, ***24***(4), 289-297.
[2] Sundaravadivu, K., Arun, B., & Saravanan, K. 2011. Design of fractional order PID controller for liquid level control of spherical tank. In *Control System, Computing and Engineering (ICCSCE), 2011 IEEE International Conference on* (pp. 291-295). IEEE.
[3] Tepljakov, A., Petlenkov, E., Belikov, J., & Halas, M. 2013. Design and implementation of fractional-order PID controllers for a fluid tank system. In *American Control Conference (ACC), 2013*(pp. 1777-1782). IEEE.
[4] Raptis, I. A., Valavanis, K. P., & Vachtsevanos, G. J. 2012. Linear tracking control for small-scale unmanned helicopters. *IEEE Transactions on control systems technology*, ***20***(4), 995-1010.
[5] Raffo, G. V., Ortega, M. G., & Rubio, F. R. 2010. An integral predictive/nonlinear H∞ control structure for a quadrotor helicopter. *Automatica*, ***46***(1), 29-39.
[6] Lim, J. S., Park, C., Han, J., & Lee, Y. I. 2014. Robust tracking control of a three-phase DC–AC inverter for UPS applications. *IEEE Transactions on Industrial Electronics*, ***61***(8), 4142-4151.
[7] Abbas, H., Asghar, S., & Qamar, S. 2012. Sliding mode control for coupled-tank liquid level control system. In *Frontiers of Information Technology (FIT), 2012 10th International Conference on* (pp. 325-330). IEEE.
[8] Khalid, M. U., & Kadri, M. B. 2012. Liquid level control of nonlinear coupled tanks system using linear model predictive control. In *Emerging Technologies (ICET), 2012 International Conference on* (pp. 1-5). IEEE.
[9] Katsuhiko, O. (2010). Modern control engineering 5th ed.
[10] Gene F. Franklin., J Davis Powell., Abbas Emami-Naeini. 2010. Feedback Control of Dynamic Systems. 6th ed.
[11] Mobayen, S., & Majd, V. J. 2012. Robust tracking control method based on composite nonlinear feedback technique for linear systems with time-varying uncertain parameters and disturbances. *Nonlinear Dynamics*, ***70***(1), 171-180.

The Effects of Silicon on The Hardness and Wear of Ferritic Grey Cast Iron on Shoe Brake of Train

S Darmo*, L D Setyana, T Tarmono, and N Santoso

Department of Mechanical Engineering, Vocational College, Universitas Gadjah Mada, Indonesia

*suryo.darmo@ugm.ac.id

Abstract. Grey cast iron has been used for a shoe brake of train due to the enough strength and the low cost. The shoe brake made of cast iron sometimes makes a problem in the operation of the train, because it make the wheel of the train is worn-out. The aim of this research is to find the data of cast iron in accordance with the wear rate and hardness properties by addition of various silicon contents. Cast iron returned (scrap) was melted by induction furnace and then poured into the sand mold of the shoe brake. The 75% ferro silicon was added to make the addition of the silicon content to the melted cast iron. Samples (specimen) of grey cast iron was made with five variation of silicon contents. The composition, hardness, and wear rate test was performed to the each sample. Maximum hardness and wear rate was happened at the cast iron of silicon content of 3.3% Si.

1. Introduction

The elements which have a function as graphitiser in grey cast iron are Si, C, Ni, Cu, Ti, Co, Au, Pt, and Al [1,2]. Phosphorous increases fluidity and extends the range of eutectic freezing, thus it increases primary graphitization when the silicon content is high and the phosphorous content is low [3]. Coarse graphite inside ferrite matrix results in low hardness of grey cast iron. Coarse graphite is formed if the silicon and carbon content are high [4]. Ferritic grey cast iron has less wear resistance than pearlitic grey cast iron due to the ferrite structure has less wear resistance than the perlite structure [4]. Ferrite-stabilizers elements consist of Cr, W, Mo, V, Si, Al, Be, Nb, P, Sn, Ti, and Zr, which increase the range of α-phase. Elements which enter only in the ferrite phase are Ni, Si, Al, P, S, and Cu. These elements dissolve easily to form solid solution in ferrite, so they increase the hardness of ferrite [5]. Generally, the addition of silicon content in the cast iron is performed by addition of the ferrosilicon in the liquid of cast iron. Silicon has effect on the increasing of the ferrite and graphite constituents in grey cast irons as well as the change in the shape, size, and graphite distribution. The dissolution of silicon in liquid cast iron contribute to the formation of graphite and is associated with the thermodynamic activity of carbon in silicon-enriched microvolumes of cast iron. Silicon being in a metal solution interacts with carbon monoxide, forming a surface-active monoxide of silicon and graphite. The formation of embryos of graphite is not only due to the supersaturated silicon microvolumes but also by chemical compounds [6]. Aluminium, calcium, zirconium and other impurities that are present in ferrosilicon enhance the graphitizing ability of silicon. The action of FeSi75, Zircinoc, Foundry Grade, and Inogen in cast iron liquid showed that the presence of aluminium, calcium or zirconium contributes to the enhancement of graphitization and improve the distribution of graphite inclusions [7]. In addition, graphitizing effect is also improved by using silicon carbide which does not contain of aluminium and other elements [8]. The main centers for the formation of lamellar graphite is silicate in which it is a thin layer covering the nonmetallic inclusions. The nonmetallic inclusions are of (Mn, X) S, where X is Fe, Al, O, Ca, Si, Sr, Ti, etc [9]. Nucleation of graphite in grey cast irons consists of 3 stages. Stage 1 is the formation of small oxide based sites (usually less than 2·0 μm), stage 2 is the precipitation of complex (Mn,X)S compounds (usually less than 5·0 μm) nucleated by stage 1, and stage 3 is the nucleation of graphite on the sides of the (Mn,X)S compounds [10]. The existence of graphite in the microstructure reduces the strength since it decreases the size of the efficient section and causes notch effect. Sharp edge of flake graphites contributes to stress concentration and crack formation. This character also result in the deterioration of the wear resistance of cast iron. The hardness and wear coefficient of grey cast iron is decreased if silicon content is increased, it is happened on the grey cast iron of silicon content of between 2.04% and 3.33% Si. The highest wear coefficient is 0.45 mm^3/Nm, it is happened on the grey cast iron of 2.04% Si. Lubricating effect of graphite flake influences the wear coefficient of grey cast iron [11].

2. Materials and methods

The wooden pattern which the dimension was the same as the dimension of the shoe brake of train of type T358, was used to make five pair of cope and drag moulds. The mouldings sand were prepared from a mixture of dried fresh silica sand, water, bentonit and dextrin in accordance with BS14 standard [12]. The five mould were labeled as Batch 1, 2, 3, 4, and 5 respectively. The grey cast iron foundry returns of 100 kg were melted in a 500 kg crucible capacity. The melted grey cast iron was poured at 1300°C into moulds. The first batch which represents the control sample was poured into the mould labeled batch 1 without any alloy addition. Thereafter, a known quantity of 75% ferrosilicon granules were added to the remaining molten bath and heated up to 1300°C, stirred manually with a dry wooden stick to facilitate a homogeneous bath and poured into the second mould labeled batch 2. The same process was repeated for the remaining batches of moulds with increasing 75% ferrosilicon granule additions. After pouring, the casting were allowed to solidify and cooled in the mould to room temperature before they were knock out and cleaned up. The cast bars were labeled to ensure traceability. From the cast bars, sample representatives from each bar were cut and machined for microstructural analysis, Brinell hardness and wear test respectively. Composition of cast iron of each batch was determined using composition tester which the model was of metalscan 2500 series. The hardness of the samples was determined using Brinell hardness tester which the model was of HMV-Controlab. Hardness test was carried out under a load of 60 kg and a dwell time of 10 seconds. The microstructural analysis was carried out using metal light microscope which the model was of Olympus. The wear test was carried out using wear tester which the model was of Riken Ogoshi's Universal Wear. Type: OAT-U. Equation (1) was used to calculate the wear rate [11].

$$\text{Wear rate} = (\text{wear volume loss})/(\text{sliding distance x applied load}) \quad (1)$$

Table 1. Chemical composition of experimental grey cast iron (wt%)

SAMPLES	Elemental composition (wt%)														
	Fe	C	Si	Mn	P	S	Cr	Mo	Ni	Al	Cu	Sn	Ti	V	Mg
Batch 1	93.11	3.35	2.42	0.315	0.0771	0.0959	0.283	0.0723	0.0508	<.0070	0.169	<.0200	0.0203	<.0100	0.0016
Batch 2	92.74	3.46	2.94	0.295	0.0231	0.0600	0.161	<.0700	0.0486	<.0070	0.159	0.0205	0.0198	<.0100	0.0018
Batch 3	92.24	3.50	3.33	0.298	0.100	0.0514	0.158	<.0700	0.0465	<.0070	0.159	0.0228	0.0198	<.0100	0.0018
Batch 4	91.87	3.56	3.67	0.295	0.0678	0.0590	0.149	0.0714	0.0485	<.0070	0.159	<.0200	0.0184	<.0100	0.0019
Batch 5	92.01	3.23	3.92	0.296	0.0344	0.0443	0.137	0.0693	0.0492	<.0070	0.163	0.0208	0.0182	<.0100	0.0018

3. Results and Discussion

The results of the compositional of the five batches are presented in Table 1. Micrographs of the grey cast iron samples with the different silicon content are as shown in Figure 1. The differences of the silicon contents cause the variation of wear rates as shown in Figure 2. Wear rate of cast iron is influenced by lubricating effect characteristic of graphite flake [11]. Micrograph of batch 1 (2.42% Si) contains higher pearlitic structure and less graphite structure than batch 2, it causes it's higher wear rate than batch 2 (2.94% Si). The increasing of silicon content make the decreasing of pearlitic structure and increasing of ferritic structure, causes the higher wear rate [4]. This phenomenon is happened on the silicon content of 3.33% (batch 3). The addition of silicon content more than 3.33% (batch 4 and 5) causes the addition of graphite structure and the higher of wear resistance due to the

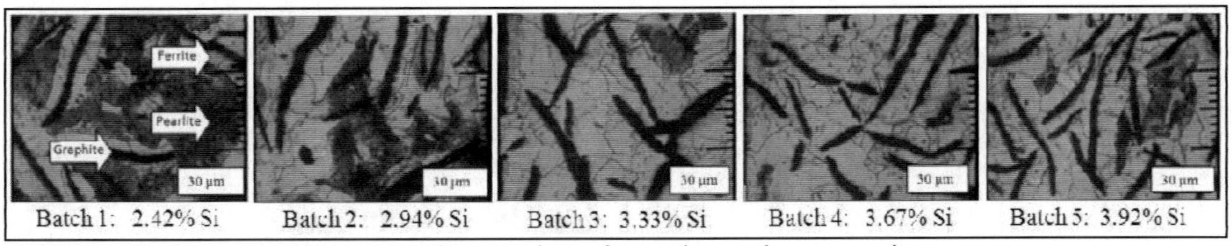

Figure 1. Micrographes of experimental grey cast iron

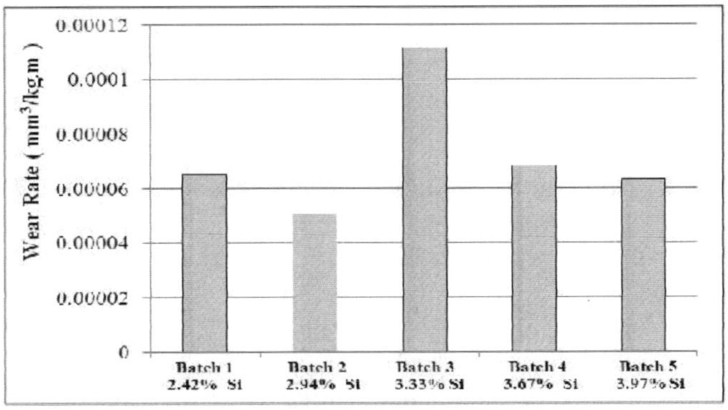

Figure 2. Wear rate of experimental grey cast iron.

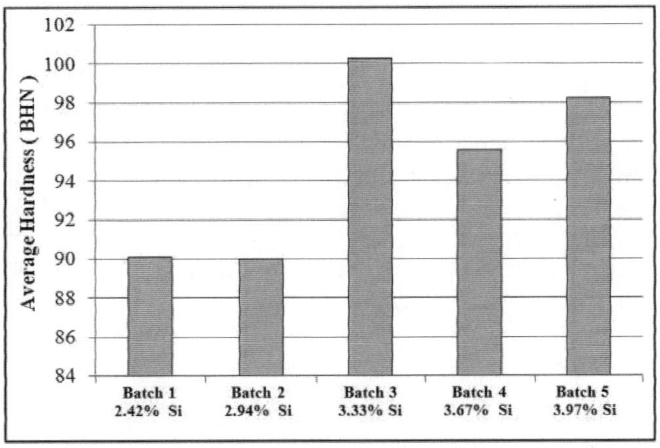

Figure 3. Hardness of experimental grey cast iron.

Self-lubrication of graphite [11]. The hardness of sample of batch 3 (3.33% Si) is higher than the other,it is may be caused by the strengthening of ferrite structure by silicon [5]. The increasing of the silicon content (batch 4, 3.67% Si) results in more graphite structure and decreasing of the hardness of cast iron [4]. Sample with the silicon content of 3.92% Si (batch 5) has higher hardness than the sample with the silicon content of of 3.67% Si (batch 4). The small size of graphite in the batch 5 is the cause of the increasing of the hardness. From the data of the silicon contents, the hardness, and the wear rate of the samples are shown that the variation of the wear rate is not similar to the variation of the hardness, due to of the difference of the kind and size of phase structure and self-lubrication of graphite.

4. Conclusions

This numerical study of the wear rate and hardness of the experimental grey cast iron, in which is used as the candidate of the material of the shoe brake of train. The important points are as follows:

1. Maximum hardness of the grey cast iron is happened on the silicon content of 3.33% Si, this hardness is about of 100.3 BHN.
2. Maximum wear rate of the grey cast iron is also happened on the silicon content of 3.33% Si, that is about of 0.00011 mm^3/kg.m.
3. The variation of the hardness and wear rate is particularly influenced by the kind of phase of the microstructure and the size of the graphite.

Acknowledgments

This research is funded and supported by the Ministry of Research, Technology and Higher Education, Republic of Indonesia. The authors would like to thank to the Mechanical Department of Vocational College and Mechanical Department of Engineering Faculty, Universitas Gadjah Mada for their supports of precise equipments preparations.

References

[1] Angus H T 1978 *Cast Iron Physical and Engineering Properties* 2nd ed. London Butterworths. p 172

[2] Surdia T and Saito S 1999 *Pengetahuan Bahan Teknik* Edisi Keempat Jakarta: PT Paradnya Paramita p 114

[3] Avner S H 1988 *Introduction of Physical Metallurgy* 14th ed. London: McGraw-Hill Company, p 438-439

[4] Surdia T and Chijiiwa K 1991 *Teknik Pengecoran Logam* Edisi Keenam Jakarta: PT Paradnya Paramita p 29-30

[5] Singh V 1999 *Physical Metallurgy* 1st ed. Delhi: Globe Offset Press Chapter 11 p 608-609

[6] Ivanov V, Pirozhkova V, Lunev V 2017 *Eastern-Europian Journal of Enterprise Technologies* p 26-30

[7] Soinski MS, Wawrzyniec A 2010 *Archieves of Foundry Engineering* Vol.10 Issue 2 p 155-158

[8] Stojczew A, Janerka K, Jezierski J, Szajnar J, and Pawlyta M 2014 *Archieves of Foundry Engineering* Vol.14 Issue 3 p 77-82

[9] Riposan I, Chisamera M, Stan S and White D 2009 *China Foundry* Vol. 6 No.4 p 352-357

[10] Riposan I, Chisamera M, Stan S, Hartung C and White D 2010 *Journal of Materials Science and Technology* Volume 26 Issue 12 p 1439-1447

[11] Johnson O, Agunsoye, Talabi S I, Olumuyiwa I A, Afemefuna T O 2013 *Journal of Minerals and Materials Characterization and Engineering* 1 p 61-67

[12] Ademoh N A and Abdullahi AT 2010 *Middle-East Journal of Scientific Research* 5(5) p 402-406

IOP Conf. Series: Materials Science and Engineering **384** (2018) 012018 doi:10.1088/1757-899X/384/1/012018

Wear Prediction of UHMWPE Acetabular Cup against CP Titanium Femoral Head in a Hip Joint Simulator

H Handoko[1*], S Suyitno[1], R Dharmastiti[1], R Magetsari[2]

[1] Department of Mechanical and Industrial Engineering, Faculty of Engineering, Universitas Gadjah Mada, Indonesia
[2] Department of Orthopedics and Traumatology, Sardjito General Hospital and Faculty of Medicine, Universitas Gadjah Mada, Indonesia

[*]handoko.dtm@ugm.ac.id

Abstract. Ultra high molecular weight polyethylene (UHMWPE) is a biomaterial used for the human hip joint prosthesis bearings for decades. The main disadvantage is the wear resistant. The products should be in service in the lifetime of the patients. Tribological assessments by experiments and computations should be made to solve that problem. The aim of this research is to predict the UHMWPE acetabular cup wear volume sliding against commercially pure titanium (cp Ti) femoral head in a hip simulator. The cp Ti has the highest biocompatibility compared with other femoral head metals such as cobalt chrome, titanium alloy and 316L stainless steel. Computation processes use the Abaqus contact mechanics algorithm followed by a custom made Python script for the data extractions and Archard wear model calculations. Models were an assembly of the UHMWPE acetabular cup and 32 mm cp Ti femoral head in a biaxial rocking motion hip simulator. Sliding between the models was a 60 cycles per minute rotation. Calculations conducted for one million cycles cumulative polymer wear volume in five steps of 200,000 cycles. The geometries of the models were updated in every step. Experimental pin on disc tests conducted to obtain the wear factor and coefficient of friction. Those data support the input of computations. Wear volume verification use the experimental data from Röstlund. Results show a numerical error at 31.42 percents.

1. Introduction

Ultra High Molecular Weight Polyethylene (UHMWPE) is a polymer used as bearing materials of the hip joint implants for decades. It is biocompatible and durable for up to 30 years in clinical use [1]. The material is usually paired with other metallic biomaterials known as metal on polymer (MOP) hip implants. The metals are 316L stainless steels, cobalt chromium alloys and titanium alloys. A common problem with the use of MOPs is the wear of polymers. Excessive UHMWPE wear debris would cause osteolysis and inflammation at the tissues around the hip joint implants. In osteolysis, wear debris will trigger the autoimmune of the body to clean it. Macrophages will phagocyte the wear debris. The process absorbs bone minerals. It would release the inflammatory mediators and leads to the bone resorption [2]. Implant loosening, bone breakage and inflammation will follow. The effects are severe pain for the patients and the need for costly revision surgeries. Based on these facts, it is important to study the wear volume of the UHMWPE used as hip bearing materials.

Wear especially in hip implant is a complex phenomenon. The metallic femoral head implants slide along the bearing surface materials when the patient moves their hips. This repeated process combined with the pressure at the implants due to the bodyweight of the patients creates a combined adhesive abrasive wear. Many parameters are involved such as sliding velocity, mechanical properties of biomaterials, lubricant from the body fluid of the patients etc. It is also nearly impossible to accurately measure the volumetric wear in vivo. These problems had been solved with experiments in vitro either by pin on disc tribotesters or hip simulators. A simple Archard wear equation is commonly used to calculate the wear volume. The complexity of the parameters other than contact pressure and sliding distance is represented by a wear factor. The main disadvantage of this method is the time needed to obtain wear data. The tests were usually conducted at a low frequency of one Hertz to mimic the in vivo conditions. Wear data from two to five million cycles would take months to complete with procedures for example as described in [3].

Significant progress on the capabilities of computer hardware and software in the recent decades support the numerical computation methods. Researchers have been using it to study the wear of the hip implants ever since. Maxian et al [4] are the first to use Archard wear law for the hip implant models. The researchers use the finite element method to calculate the contact pressures at incremental sliding distances. Others followed the similar procedures to study the various situations of hip implant models. The most recent published works are the improved solution of hip implant contact problem [5], motion inputs [6], additional device such as hip resurfacing [7] and the type of biomaterial pairs [8]. The outputs of the computation are the wear prediction data. Most of the predicted data are in good agreements or fit well with the experimental data. The method requires lesser time duration to complete. A two days experimental test can be replaced with a two hours numerical computation [9].

Wear predictions on the MOP hip prosthesis biomaterials are dominated by the Co-Cr alloy and UHMWPE pairs. Other femoral implant material rarely studied is the commercially pure titanium (cp Ti). This metal is the most biocompatible material even compared to the titanium alloys. Co-Cr alloys and 316L stainless steels release Ni ions in vivo. This process would lead to inflammation problems for the sensitive patients. The famous Ti-6Al-4V titanium alloys were suspected to release a combined Al and V ions. It would cause diseases such as Alzheimer, neuropathy and ostemomalacia in the long time period of clinical services [10]. Based on those comparisons, attempts can be made to use cp Ti as the femoral implant biomaterial. The aim of this research is to assess the wear volume of UHMWPE acetabular cup paired with cp Ti femoral head. The expectation was that cp Ti could reduce long term clinical effects on the THR (total hip replacement) patients. Numerical investigations predict the wear volume of the UHMWPE cups. The output poses the effectiveness of commonly used Archard wear model to mimic the wear of UHMWPE sliding against cp Ti. It also found out the appropriate pressure of a physiological hip load which contact pressure closes to the in vivo data.

2. Materials and methods

Wear volume prediction in this research relies on the Archard wear law. The law can be derived into a linear wear depth (h) equation as follows:

$$h = k.p.s \tag{1}$$

The equation requires k as a wear factor, p is the contact pressure and s is the sliding distance. Wear factor, k is obtained with the biotribological experiments. Wear volume can be calculated by multiplying the linear wear depth with the contact area at every contact nodes of the models. The calculation of the contact mechanic parameters, i.e. contact pressure and contact area use a commercial finite element software, Abaqus. Its algorithm is mature and used by previous hip modeling and computation studies [11-14]. The hip joint model used was a 46° total excursion biaxial hip simulator model studied experimentally by [15]. Fig 1 shows the model schematically. UHMWPE cup wear volume calculations conducted in repeated steps as shown in Fig 3.

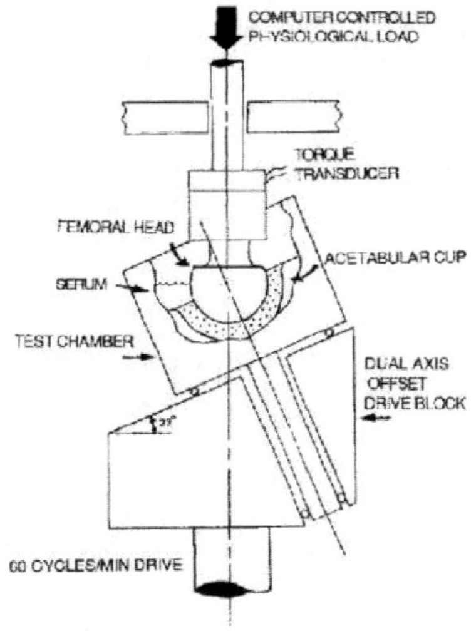

Figure 1. The hip simulator model [15]

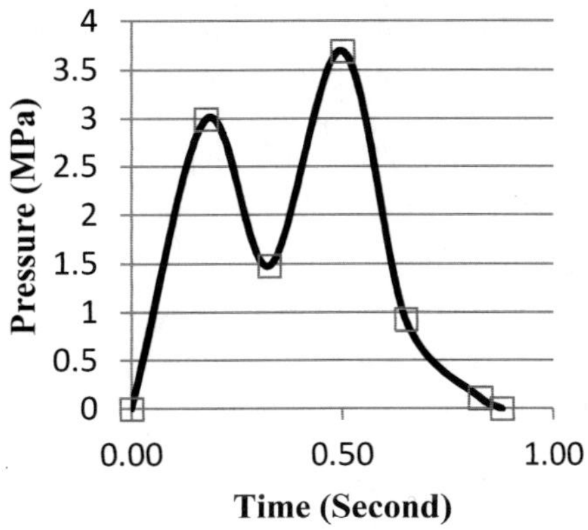

Figure 2. Paul type physiological hip load [15]

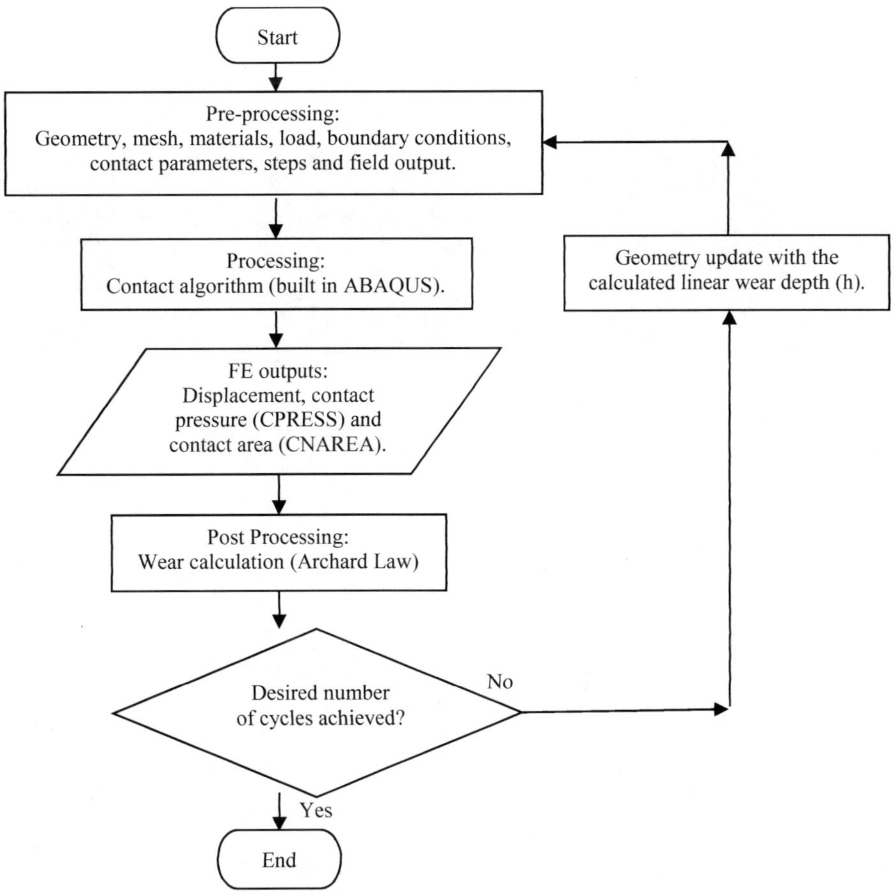

Figure 3. The wear volume calculation steps

The calculations need a coefficient of friction data between the contact materials. It was obtained by experiments with a multi directional pin on disc tribotester. The biomaterials were 50 mm diameter cp Ti discs and 9 mm diameter UHMWPE pins. Contact pressure was set at 3 MPa. Specimens were lubricated with simulated body fluids (SBF). The SBF consists of mainly distilled water with 25 % v/v of bovine serum [16-17]. The temperature of the fluid in the wear test was set between 36 to 38 degrees centigrade to simulate the physiological human body temperature. Polymer weight reductions were measured at every 250,000 cycles.

3. Results and Discussion

The first step to study wear phenomena in the hip joint simulator is to evaluate the contact pressures. This numerical study use the Paul type physiological hip load curve (Fig 2) as experimentally by [15]. The curve has a peak pressure up to 3.5 MPa. It was evaluated with the static three dimensional contact mechanic computations. Contact pressures on the UHMWPE cup was calculated with Abaqus at three physiological pressure variations from 3 MPa close to the curve, 6 MPa and 9 MPa (Fig 4). Results show that the realistic physiological pressure is 3 MPa due to the contact pressure values. High contact pressure (CPRESS) at 42.5 MPa and 40.56 MPa were found from two contact nodes but the other 1869 nodes are below 34 MPa (Fig 5). These numbers were relatively too high compared to the measured in vivo hip joint contacts [18-19]. Other calculations suggest higher values for the peak hip contact pressure of 22.4 MPa at the mid stance of normal walking [20] up to 31.06 MPa [21]. Hence the Paul type physiological hip load curve is realistic, close to the recent results. It can be used for the loads on the hip implant wear prediction.

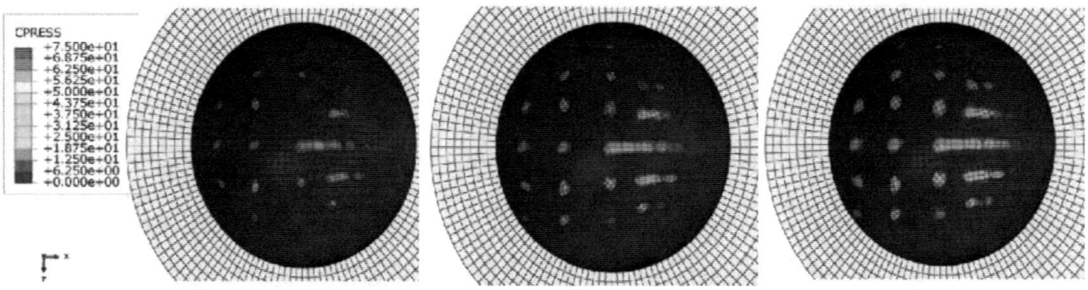

Figure 4. Effect of physiological pressures on the contact pressures at the surface of the UHMWPE cup (denoted as CPRESS by Abaqus), (a) 3 MPa, (b) 6 MPa and (c) 9 MPa

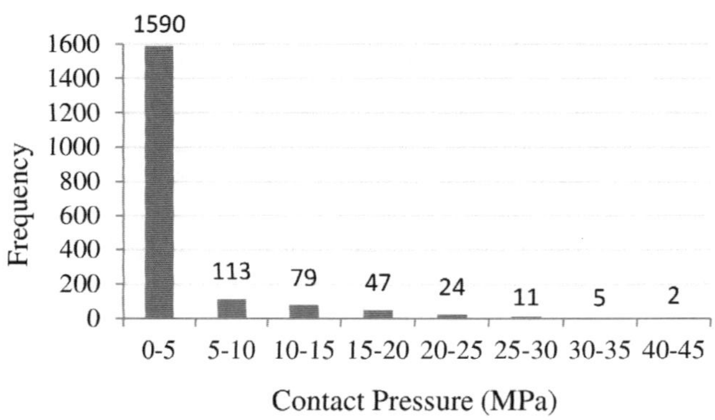

Figure 5. Frequency histogram of contact pressure distributions from the 3 MPa physiological pressure. The acetabular cup model has 1871 contact nodes.

The next step was the contact mechanic computation and wear volume calculations. Experimental wear tests conducted with a three stations multi directional tribotester to support necessary data for the numerical calculations. Results from a million cycles were the coefficient of friction equal to 0.2 and the wear factor, $k = 1.657 \times 10^{-6}$ mm^3 (N.m)$^{-1}$. The coefficient of friction was then fed into the contact mechanics computations. The results were stored in the Abaqus output databases. Contact pressure, contact area and sliding distance data extractions use a custom Python script. Linear wear depth (h) calculations use those data and the wear factor (k). The magnitude of h was used for the wear volume calculations and to update the contact nodes geometries. The result of numerical prediction was a volumetric wear data at the rate of 119.2 mm^3 per million cycles. Experimental data from [15] was 90.7 mm^3 UHMWPE wear per one million cycles. It was a 31.42 % numerical error.

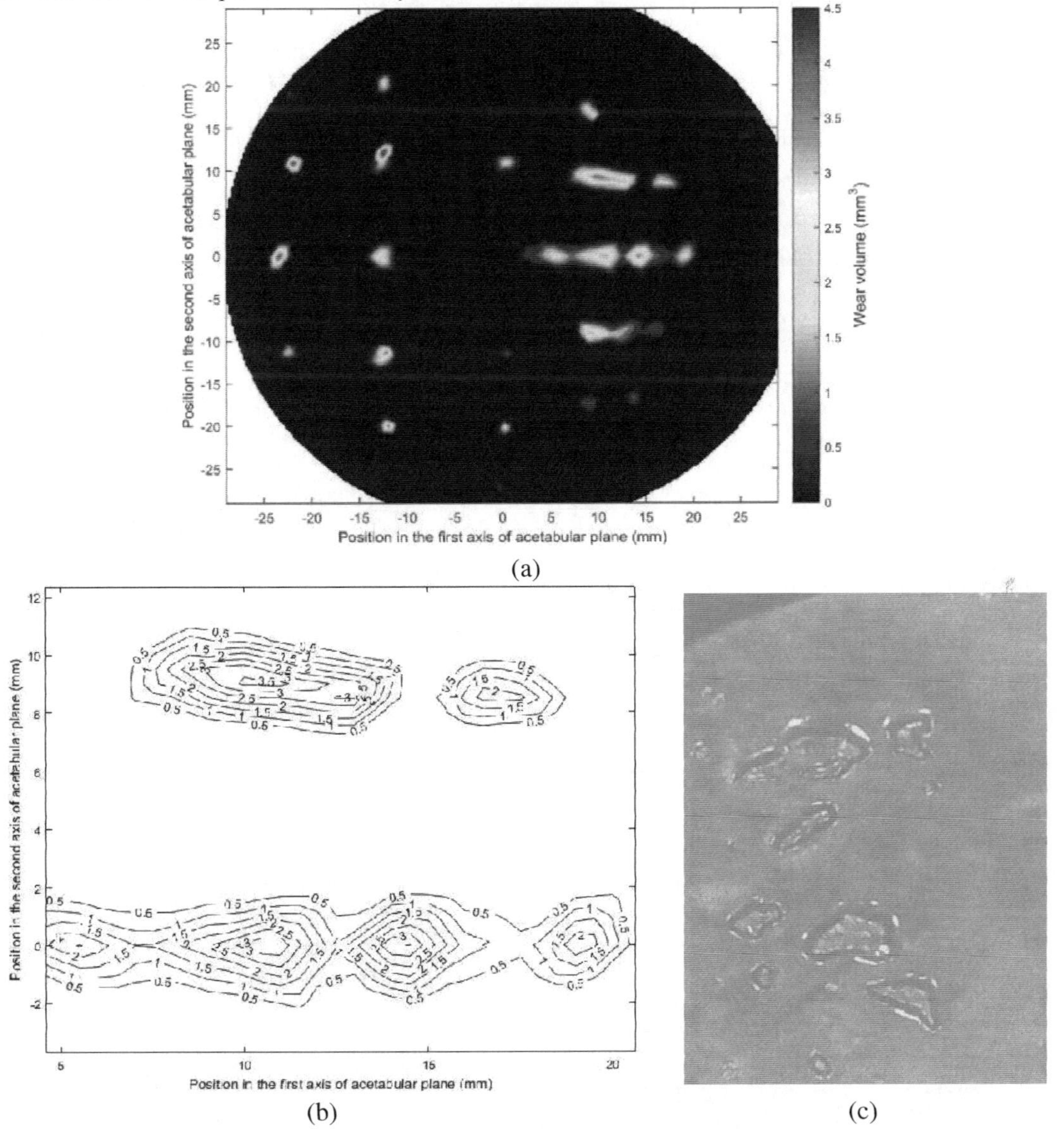

Figure 6. Contour of the acetabular cup wear volume after one million cycles, (a) overall acetabular plane, (b) detailed contour of the large worn area and (c) protuberance found at the surface of UHMWPE pin.

The cause of numerical error in this study is the use of a constant wear factor for the all contact nodes. Different contact pressures at specific contact points could have different wear rates. The points can be treated with different wear factor values. An equation of wear factor as a function of contact pressure is needed and should be obtained experimentally. The equation is available from [23-24] but it was for the UHMWPE and cobalt chromium pairs. The next possible factor affects to the wear calculation errors is the Archard wear model itself. Recent studies began to critically evaluate the use of the law. The polymer deformation remains elastic hence the linear relationship between the load and the contact area might not valid anymore [25]. The quest to modify or even replace Archard wear law for the metal on polymer or the soft on hard biomaterials will support the future numerical wear prediction. Sophisticated future numerical models will reduce the costly and time consuming biotribological experiments.

The contours of the worn acetabular cup (Fig 6a) are similar to the contours of the contact pressure. It is because of the use of Equation (1) relies numerically to the contact pressure values. Local and small pockets of worn surface indicate the potential of protuberance. Experimental pin on disc tests confirm that indication. Protuberance was found at the surface of the UHMWPE pin (Fig. 6c). Surface heterogeneities and micro defects or micro fissures formation lead protuberances in the space between contact surfaces. Elastic properties of the polymer and friction force determine the deformation of protuberances [22]. The repeated deformations due to sliding contact would lead to the plastic deformation and surface strain. From that point, the protuberances turn into wavelets. As the number of wavelets increased, the surface became more defective. Finally the wear becomes catastrophic.

4. Conclusions

This study predicted the wear of UHMWPE cups paired with cp Ti femoral heads in a hip joint simulator model. The results can be summarized as follows:

1. Paul type physiological load applied to the model obtained contact pressures close to the recent calculation results.
2. The numerical wear prediction error is 31.42 % compared to the experimental data.
3. Accurate wear factor equations or wear models should be obtained to increase the accuracy of the wear predictions.

Acknowledgments

This research is funded and supported by the postgraduate scholarship from the Ministry of Research, Technology and Higher Education, Republic of Indonesia. The authors would like to thank to the Faculty of Pharmacy and the Faculty of Veterinary, Universitas Gadjah Mada for their supports of precise equipments and serum preparations.

References

[1] McMinn D J W 2009 *Modern Hip Resurfacing* (London: Springer) p 4
[2] Shanbhag A, Rubash H E and Jacobs J J 2006 *Joint Replacement and Bone Resorption* (New York: Taylor & Francis) p 330
[3] Zdero R 2017 *Experimental Methods in Orthopaedic Biomechanics* (Amsterdam: Elsevier) pp 301-308
[4] Maxian T A, Brown T D, Pedersen D R and Callaghan J J 1996 *Biomechanic* **29** 5
[5] Pakhaliuk V, Polyakov A, Kalinin M and Kramar V 2015 *Procedia Engineering* **100**
[6] Liu F, Fisher J and Jin Z 2013 *Tribology International* **63**
[7] Ali M and Mao K 2014 *Acta of Bioengineering and Biomechanics* **16** 2
[8] Puccio F D and Mattei L 2015 *Tribology International* **83**
[9] Zhao D, Sakoda H, Sawyer W G, Banks S A and Fregly F J 2008 *J. Biomechanic Eng.* **130** 1
[10] Geetha M, Singh A K, Asokamani R and Gogia A K 2009 *Progress in Materials Science* **54**
[11] Gao Y, Jin Z, Wang L and Wang M 2015 *Proc. of the Institution of Mechanical Engineers, Part H* **229** 6 pp 469-74
[12] Hua X, Li J, Wilcox R K, Fisher J and Jones A C 2015 *Proc. of the Institution of Mechanical*

Engineers, Part H **229** 8 pp 570-80
[13] Tudor A, Laurian T and Popescu V M 2013 *Tribology International* **63** pp 158-68
[14] Zhang T, Harrison N M, McDonnell P F, McHugh P E and Leen S B 2013 *Tribology International* **65** 113–27
[15] Röstlund T, Albrektsson B, Albrektsson T and McKellop H 1989 *Biomaterials* **10** pp 176-80
[16] Xiong D, Yang Y and Deng Y 2013 *Surface and Coatings Technology* **228** Supp 1, S442-5
[17] Abdelgaied A, Brockett C L, Liu F, Jennings L M, Fisher J, and Jin Z 2012 *Proc IMechE Part H: J Engineering in Medicine* **227** (1) pp 18-26
[18] Brand R A, Iglic A and Karlj-Iglic 2001 V *Hip International* **11** (3) pp 117-26
[19] Bachtar F, Chen X and Hisada T 2006 *Medical and Biological Engineering and Computing* **44** (8) pp 643-51
[20] Kumagai M, Kim Y H, Inoue N, Genda E, Hua K, Liong B T L, Koo T, Chao E Y S 2003 *Summer Bioengineering Conference*, Florida, p 54
[21] Wang L, Yang W, Peng X, Li D, Dong S, Zhang S, Zhu J, Jin Z 2015 *J Biomechanics* **48** pp 1112-8
[22] Bartenev, G M, Lavrentev, V V 1981 *Friction and Wear of Polymers* (Amsterdam: Elsevier) p 249
[23] Saikko V 2006 *Proc. of the Institution of Mechanical Engineers* **220**, Part H
[24] Kang L, Galvin A L, Fisher J and Jin Z 2009 *Journal of Biomechanics* **42** (7)
[25] Mattei L, Di Puccio F and Ciulli E 2013 *Tribology International* **63**

Design of the Liquid Volume Meter Device Using Ultrasonic Waves Based on Microcontroller

D Zakaria[1,*], T Gunawan[1], Y Somantri[1], A G Abdullah[1], I Widiaty[2], A B D Nandiyanto[3], N Amelia[1], A Juhana[1] and W Arasid[1]

[1]Electrical Engineering Education Department, Universitas Pendidikan Indonesia
[2]Family Welfare Education Department, Universitas Pendidikan Indonesia
[3]Chemistry Education Department, Universitas Pendidikan Indonesia

[*]dikyzak@gmail.com

Abstract. The aim of this study was to design a digital volume meter device using ultrasonic waves based on microcontroller as well as to know the strengths and weaknesses of these device. The method used is an experimental method which consists of a field study and identification of the problem, the study of literature, consultations with specialists, design and manufacturing, testing and analysis of the test results. From the test results in a cylindrical container with radius of 7.5 cm and height of 15 cm, error percentage of water volume measuring is 0.537% and in a square container measured 17.1 x 14.3 x 17.5 cm, error percentage of water volume measuring is 0,486%. From the results, it can be concluded that the volume meter device using ultrasonic waves based on microcontroller is functioning properly to measure the volume of liquid in a cylindrical and square container.

1. Introduction

The volume measuring device in the industry has an important role. One of its functions is to measure the volume of solar used by industrial machinery in a production process. By knowing the exact amount of solar usage, we can know the cost of production appropriately, no less and no more. The use of digital device will provide more precise measurements than analogue device.

This liquid volume meter is a prototype to measure the volume of liquids in a container. The container used is cylindrical and square. When the device is used, the user selects the form and manually inserts the container size using the push button. The user enters the base radius and height of the container if selecting the cylinder menu or entering the length, width and height of the container if selecting the square menu. That is, this volume meter has one program that can be applicable for two types of containers that are cylindrical containers and square. The previous volume meter can only measure volume in one and only cylindrical container [1,2].

2. Method

The method used is an experimental method which consists of a field study and identification of the problem, the study of literature, consultations with specialists, design and manufacturing, testing and analysis of the test results.

3. System designing

3.1. Block diagram

The diagram block of the liquid volume meter using the microcontroller based ultrasonic sensor is shown in Figure 1 as follows:

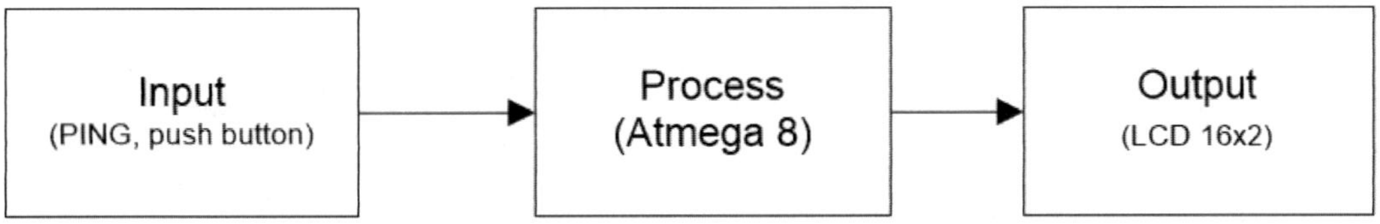

Figure 1. Block diagram system.

Input data from the user are through 4 pieces of push button to select the container form used whether the cylinder or square. The user then inputs the container size data in the form of length, width and height for the square container or the size of the radius and height for the cylindrical container. User inputted data will be stored in EEPROM of the Atmega8 microcontroller [3].

The sensor used is PING Ultrasonic Range Finder Sensor which has a maximum range of 3 meters [4]. Ultrasonic sensor serves to measure the distance from the water surface to the sensor. Sensors detect object distances by emitting ultrasonic waves (40kHz) during transmission time (Tburst = 200μs) and then detecting their reflections. The PING sensor emits ultrasonic waves according to the trigger pulse from the Atmega8 microcontroller as the controller. Then the microcontroller process it in such a way that it gets the height of water in the container to calculate the volume of water. For more details, see figure 2:

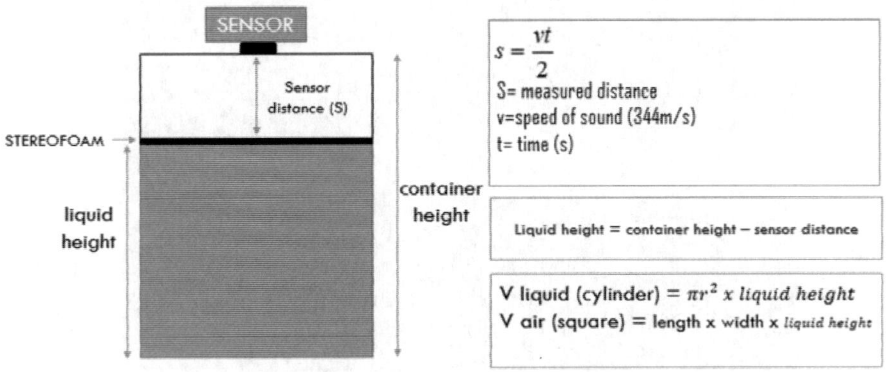

Figure 2. How ultrasonic sensors work.

Process block is the data processing part inputted by the user and data obtained from the ultrasonic sensor. Data received in this part of the process will be processed in such a way as to obtain the volume of the liquid in the container. The main part of this process block is an ATmega8 microcontroller. The output block consists of 16 x 2 LCD. In LCD will appear menu of container type selection, menu to enter container size and menu to display liquid volume in container.

3.2. Hardware designing

Here are a schematic drawing and PCB layout of liquid volume meter using ultrasonic waves based on microcontroller using Cadsoft software EAGLE 7.2.0 (figure 3 and figure 4):

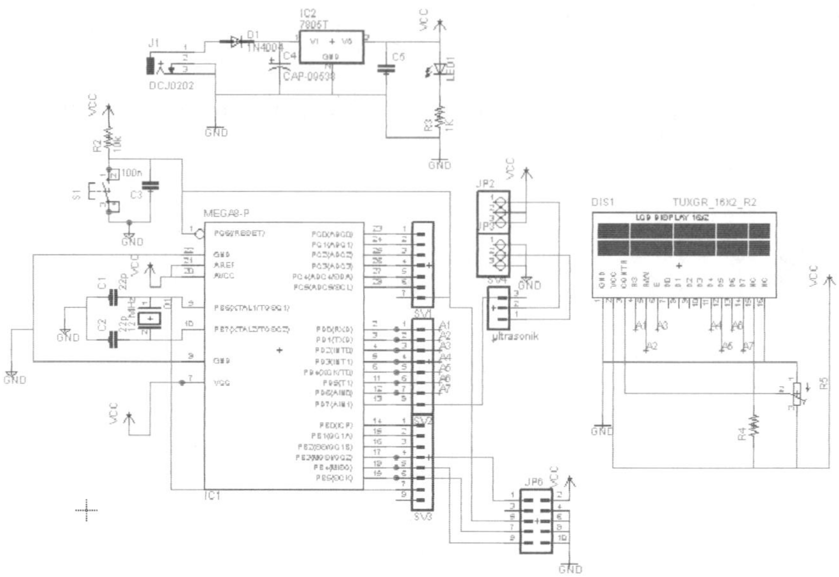

Figure 3. Schematic drawing and PCB layout of liquid volume meter.

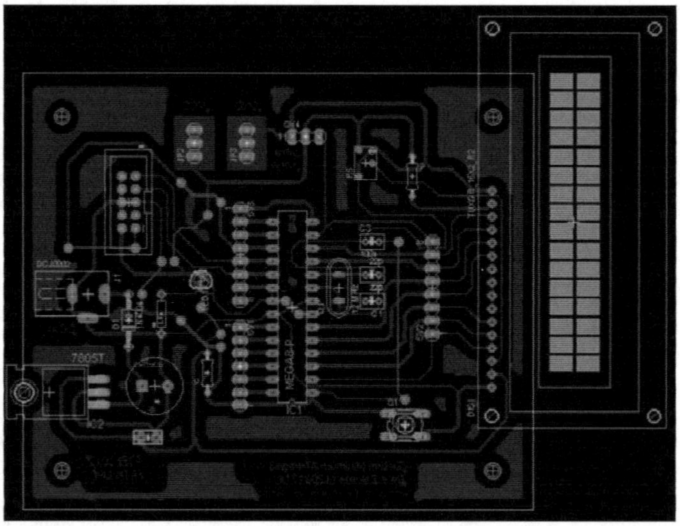

Figure 4. PCB layout.

3.3. Firmware designing

Firmware is a program that serves to control hardware so that hardware can work as desired. Inside the firmware is set the way ultrasonic sensors work to transmit and receive ultrasonic waves, display the menu on the LCD, process the data received from the user input and from the sensor, to get the value of liquid volume then display it in LCD. The completed firmware is then downloaded to ATmega8 using USBasp connector and AVRDUDESS software [4]. Firmware programming uses CodeVisionAVR software which is a C language compiler to Assembly, a language understood by ATmega8 [6]. Here is a figure 5, the flowchart from a liquid volume meter program using ultrasonic waves based on microcontroller:

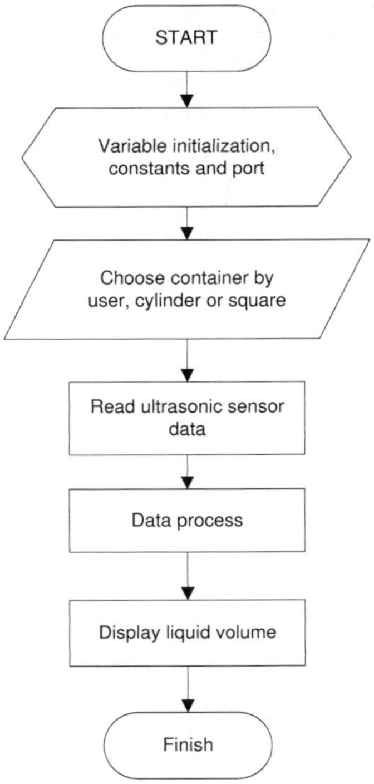

Figure 5. Liquid volume meter program flowchart.

4. Results and discussion

4.1. Device display

Figure 6 is a display of volume meter using ultrasonic waves.

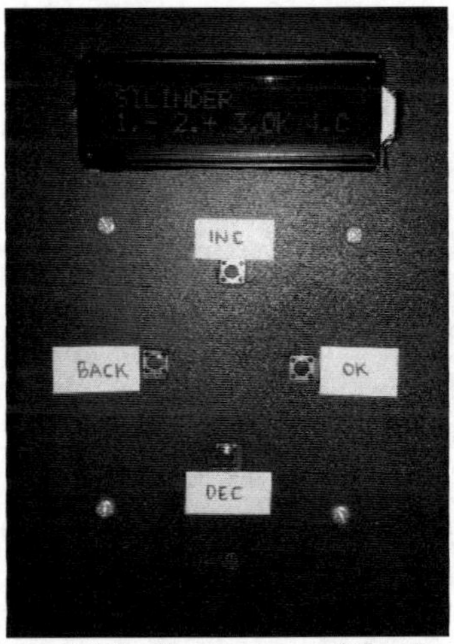

Figure 6. The front view of the device.

Figure 6 shows 4 push buttons to set the selection of container type and to enter the size of the container. Figure 7 is the LCD display to select the shape of the container.

Figure 7. LCD display to select the container shape used.

4.2. Device testing

This device is tested by measuring the volume of the liquid in a cylindrical container with the size of the radius of 7.5 cm and height of 15 cm and in a square container with the size of 17.1 x 14.3 x 17.5 cm. Liquid used is water.

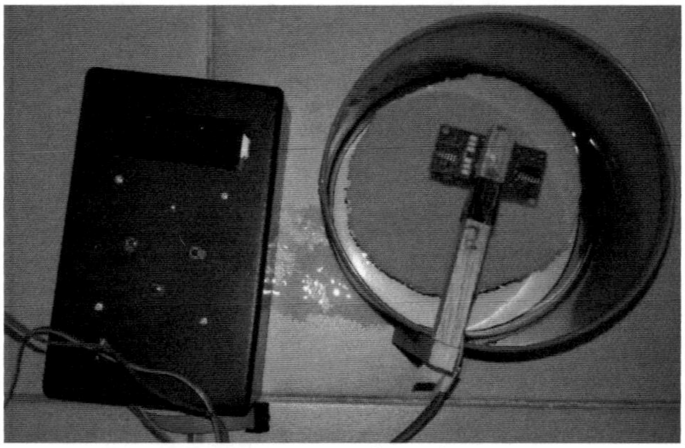

Figure 8. How to use the volume meter.

Figure 8 shows how to install the sensor on the surface of the container. The sensor will emit ultrasonic waves and will shoot into stereo foam with a thickness of 1cm. In the firmware program, the distance indicated by the sensor will be increased by 1cm to estimate the stereo foam thickness. Installation of stereo foam on the water surface due to ultrasonic sensors cannot bounce on water. The sample measurements shown on the LCD in figure 9.

Figure 9. Example of measurement of water volume result.

Figure 9 is the measurement result for the water volume of 600ml and 700ml on a cylindrical container. Here are the results of testing devices that have been done:

Table 1. Device testing on cylindrical container.

No.	Water volume (ml)	Water volume using device (ml) Xn	Error (ml) |e| =|Yn-Xn|	Distance Display (cm)
1	600	597,49	2,51	11,62
2	700	699,42	0,58	11,04
3	800	801,35	1,35	10,46
4	900	903,29	3,29	9,89
5	1000	993,89	6,11	9,37
6	1100	1092,05	7,95	8,82
7	1200	1190,21	9,79	8,26
8	1300	1297,80	2,2	7,65
9	1400	1422,39	22,39	6,95
10	1500	1499,78	0,22	6,51
Total (Σ)	**10500**	**10497,67**	**56,39**	

Table 2. Device testing on square container.

No.	Water volume (ml)	Water volume using device (ml) Xn	Error (ml) |e| =|Yn-Xn|	Distance Display (cm)
1	600	602,23	2,23	15,04
2	700	701,54	1,54	14,63
3	800	798,23	1,77	14,24
4	900	905,38	5,38	13,80
5	1000	1025,59	25,59	13,31
6	1100	1101,38	1,38	13,00
7	1200	1200,69	0,69	12,59
8	1300	1300,00	0	12,18
9	1400	1396,69	3,31	11,79
10	1500	1490,77	9,23	11,40
Total (Σ)	**10500**	**10522,5**	**51,12**	

From Table 1 is obtained:

$$\text{Error percentage} = \frac{\Sigma|e|}{\Sigma Y_n} x\ 100\%$$
$$= \frac{56{,}39}{10500} x\ 100\%$$
$$= 0{,}537\ \%$$

The error percentage of water volume measuring in a cylindrical container is 0.537%. From table 2 is obtained:

$$\text{Error percentage} = \frac{\Sigma|e|}{\Sigma Y_n} x\ 100\%$$
$$= \frac{51{,}12}{10500} x\ 100\%$$
$$= 0{,}486\ \%$$

The error percentage of water volume measuring in a square container is 0,486%.

4.3. Discussion

The percentage error of the device in measuring the volume of water is small enough i.e. 0.537% for cylindrical containers and 0.486% for square containers. From the results of this percentage error can be said that this volume measuring device is good enough in measuring the volume of liquid.

The advantages of this volume meter devices using microcontroller based ultrasonic waves are able to measure the volume of liquids in cylindrical and square containers without change or modify the program. Users can select and enter the size of the container with the available push button. The volume measurements displayed by this device are 2 digits behind the comma so it is more thorough.

The lack of this volume meter using ultrasonic waves based on microcontroller is that there is still a measurement error. In addition, in entering the size of the container, it should push the button long enough to fit the size because it does not use the keypad.

5. Conclusion

This liquid volume meter device using ultrasonic waves based on microcontroller device is able to measure the volume of water well with small error value.

References

[1] Mediaty A U 2011 Pengujian Sensor Ultrasonik PING untuk Pengukuran Level Ketinggian dan Volume Air *Jurnal Ilmah Elektrikal Enjiniring UNHAS* **9** 2 72-77

[2] Wiguna T 2011 *Pengukur Volume Zat Cair Menggunakan Gelombang Ultrasonik Berbasis Mikrokontroler AT89S51* (Semarang: Jurusan Teknik Elektro Universitas Diponegoro)

[3] Winoto A 2008 *Mikrokontroller AVR ATmega8/16/32/8535 dan Pemrogramannya dengan Bahasa C pada WinAVR* (Bandung: Informatika)

[4] Carullo A and Parvis M 2001 An ultrasonic sensor for distance measurement in automotive applications *IEEE Sensors journal* **1** 2 143-147

[5] Fischl T 2011 *USBasp – USB Programmer for Atmel AVR Controllers* Retrieved from http://www.fischl.de/usbasp/

[6] Rahayu E 2014 *Pengenalan CodeVisionAVR* Retrieved from http://www.immersa-lab.com/pengenalan-codevision-avr.htm.

IOP Conf. Series: Materials Science and Engineering **384** (2018) 012020 doi:10.1088/1757-899X/384/1/012020

Design and Simulation System of AHU's Pre-Filter Washing Machine using Microcontroller

A Utami* and D L Hakim

Department of Electrical Engineering Education, Universitas Pendidikan Indonesia, Bandung, Indonesia.

*anisau7@gmail.com

Abstract. Reducing excess water usage on large scale industries is very influential, both in terms of economics and time. An alternative approach proposed to reduce excess water usage to change the process of manual work into machine workmanship. The pre-filter washing machine is a system that can not only reduce the use of excess water discharge but also streamline the work time. However, the pre-filter washers are different from the usual drum-washing machines. The Pre-filter washing machine is designed to follow the cabinets form with 2 vertical stacking racks, so it can accommodate up to 8 Air Handling Unit's (AHU's) Pre-Filter using 2 main valves, 1 water pump, and 1 blower. The washing system designed and simulated by using microcontroller with CIP (Clean In Place) system. This paper expected to save the industrial water usage in terms of Washing AHU's Pre-Filter.

1. Introduction

Water is the source of life [1] because no one can survive without water [2]. With regard the human population in the world, the need of water will increase daily. Among the many clean water sources, around 70% used for the needs of agriculture, 20% used as the needs of the industry and leave only 10% for domestic water needs (including drinking water). There are two main components that form the economic value of water are the energy and capacity [3].

Reduced water sources resulted in the government in some countries make policy to give priority to the needs of domestic water in the water needs of others [4]. As a result, the industry must be able to find other ways to overcome the limitations of the source of the water [5]. One of the other ways to conserve water use. The development of a rapid technology is expected to help in water conservation. By involving the development of technology makes the tendency to live in a practical and economical [6].

Nowadays, the washing machine has become main important part in human life [7]. The washing machine automatically gives freedom for the user to make more working time [8]. It was also used by the industry to save the expenses and the use of water [9]. By adopting from the washing of the pre-filter manually that less effective and efficient from both sides of the use of water, time and expenditures [10]. So this paper's aims to create a pre-filter washing machine. The washing machine pre-filter is one of the washing machines that not only can reduce the use of excessive water intakes can also efficiently working.

IOP Conf. Series: Materials Science and Engineering **384** (2018) 012020 doi:10.1088/1757-899X/384/1/012020

2. Methods

2.1. Block diagram
Setting input and output based on the block diagram:

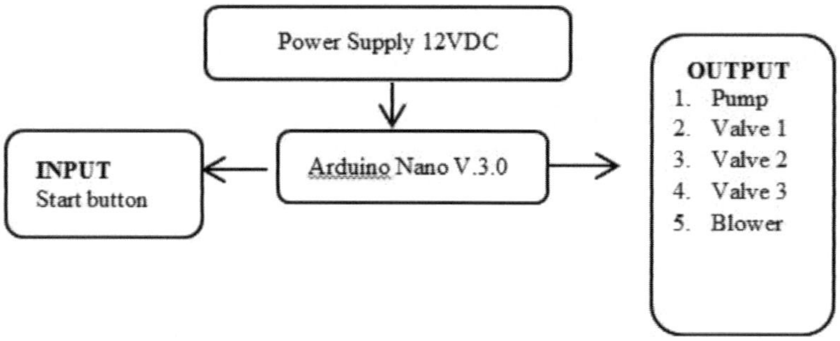

Figure 1. Block diagram.

The block diagram in the figure 1. is describes the overall system [11] structure that the input section consists of a start button, the controller uses Arduino Nano, and the output section consists of a water pump, 2 valves, and blower.

2.2. Hardware design
Air Handling Unit's (AHU's) filter is one part of HVAC (Heating, Ventilation, and Air -conditioning) system that serves to filter controlling, controlling the amount of dust or foreign particles such as the organisms contained in the air before entering the room [12]. The pre-filter washers are designed to be shaped cabinets using a vertical shelf 2 racked shelves as a frame, to accommodate up to 8 Pre-Filters using 2 main valves, 1 water pump, and 1 blower. By using polycarbonate and pipe as a separator between pre-filter. Pipe formed several slits every 2cm in order to spray the water into the dirty pre-filter. By Using an inverted T-shaped iron pipe is accompanied by several slits below so that the wind released by the blower can be spread evenly. And also by making each side into a vacuum for water and air to spread evenly. Washing system is designed and simulated using a microcontroller with CIP (Clean In Place) system. This research is expected to save industry cost in terms of Washing AHU Pre Filter.

2.3. Relay 2 channel

Figure 2. Relay 2-channel.

Relays are switches that are electrically operated [13]. Also, an Electromechanical (Electromechanical) component consisting of 2 main parts with Electromagnet (Coil) and Mechanical Set of Switch Contacts / Switches). By using Electromagnetic Principle by moving contacts Switches with small electric current (low power) could conduct higher voltage electricity. For example, with Relay that uses 5V and 50mA electromagnets capable of moving Armature Relay (which serves as the switch) to conduct 220V 2A electricity.

2.4. Water pump

Figure 3. Water pump.

The Water pump is a mechanical equipment that serves to raise the liquid from the lowlands to the highlands. In principle, the pumps change motor mechanical energy into the fluid flow. The energy received by the fluid will be used to increase the pressure and overcome the resistance in the channel. The pump that selected in the project is Shimizu PS135E Automatic Small Pressure Tank. The water pressure is 1980L/Minutes. It consume 125Watt. The reason that selected the pump are water pressure and the power.

2.5. Water valve inlet

Figure 4. Water valve inlet.

The water valve inlet has main function which to manage water into the washing machine. In addition, the inlet valve acts as a kind of gatekeeper, allowing water to flow to the washing machine while receiving certain electrical signals from the microcontroller. Inside each port there is a special gate. This gate, by default, remains in a closed position. But when you start the wash cycle, the microcontroller sends a signal to the solenoid that controls a particular gate. The solenoid then triggers the gate open so water can enter the engine. The water valve inlet that selected in the project is water valve inlet electromagnetic. It run at 220V AC and 50-60Hz. Water valve inlet is one of the valve type which is usually used in the washing machine.

2.6. *Hair dryer as blower*

Figure 5. Hair dryer rainbow.

The electromechanical devices designed to expel normal or heated air. As usual, it's used to speed up the evaporation of water particles and dry the pre-filter [14]. The blow dryer allows better control by accelerating and controlling the formation of hydrogen bonds while inside each strand. In addition, hair dryer rainbow runs at 220V-50Hz and 850Watt.

2.7. Software design

The design of the software is the most decisive step [15] in the process of making AHU's Pre-Filter Washing Machine based the Arduino Nano. It is a development board (development board) microcontroller based on ATmega328P chip with a very small shape. The function is no different than Arduino Uno. The main difference lies in the absence of DC power jack and the use of Mini-B USB connector. Called as a development board because this board serves as a prototyping arena of microcontroller circuits [9]. The software using Arduino IDE v.1.8.5. software support with the programming language used is the C ++ language with the simplified version with the help of Arduino's libraries. The program has been created and then compiled so that will be obtained file with extension *. This file will be downloaded to Arduino Nano microcontroller.

```
pinMode(valve2, OUTPUT);
pinMode(dryer, OUTPUT);
pinMode(pb_start, INPUT_PULLUP);
digitalWrite(valve1, HIGH);
digitalWrite(valve2, HIGH);
digitalWrite(pompa, HIGH);
digitalWrite(dryer, HIGH);


}
void loop() {
  // put your main code here, to run repeatedly:
int start;
start = digitalRead(pb_start);

if (start == LOW) {
  digitalWrite(pompa, LOW);
  digitalWrite(valve1, LOW);
  delay(7000);
  digitalWrite(valve1, HIGH);
  digitalWrite(valve2, LOW);
  delay(5000);
  digitalWrite(valve2, HIGH);
  digitalWrite(pompa, HIGH);
  digitalWrite(dryer, LOW);
  delay(50000);
  digitalWrite(dryer, HIGH);
}
```

Figure 6. A part of arduino IDE program.

2.8. Flowchart

Program design on AHU's Pre-Filter Washing Machine Based Arduino Nano The following begins with creating a flowchart. The Flowchart will be made in the guidelines create a program for this tool. Flowchart program tool is shown in Figure 7.

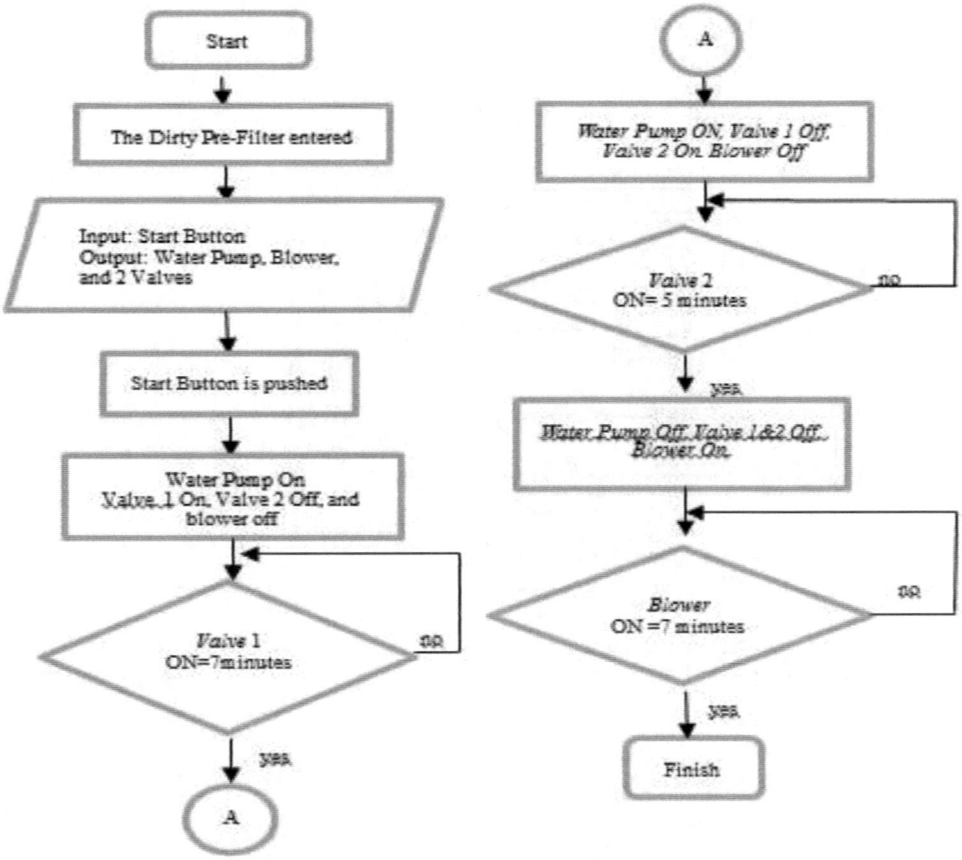

Figure 7. Flowchart.

Based on the flowchart the working principle of AHU's Pre-filter Washing Machine are when the dirty pre-filter entered. When the Start button is pressed then the water pump turning on and valve 1 would open, then water would flow to valve 1 for 7 second. After valve 1 is closed then valve 2 automatically opens and the water would flow for 5 second. After the washing process is complete, the water pump and the valves would off and the blower would light up to dry the pre-filter for 2 minutes. When the blower is ON. When the drying process is complete, it returns to the initial condition where the dirty pre-filters are inserted.

3. Results and discussion

The washing machine could operate for 2 minutes 12 second in a single command from the start button. However, the pre-filter that has been dried on the machine for 2 minutes could not be dried directly. Therefore, it is still necessary for operators to drift filters within a period of at least 1day in a shady place.

In previous, the research used an aquarium pump with a pressure of 300L / H. but the pump does not produce enough water pressure because the water released through the gap does not spray into the pre-filter. Then it replaced with the aquarium water pump with pressurized 800L / H. But, it doesn't work nicely. So with this problem, the pump is replaced with Shimizu SP135E automatic small pressure tank with pressurized 1980L/H. Furthermore, in this study using a hairdryer as a blower, but there are constraints that the wind pressure on Hairdryer is not enough to dry because the wind pressure is less than the solution uses a pipe of iron that is formed into an inverted T letter with a diameter adjusted to the size of a hairdryer. Which the bottom is made for wind distribution channels. In addition, the machine is secured from all sides or vacuum in order to dry the pre-filter well. To make each side into a vacuum, then the bottom of the machine must necessarily be closed. In this study using polycarbonate as its cover. Where the polycarbonate is tilted to the right and made hollow in the section. Then put the hose towards the disposal.

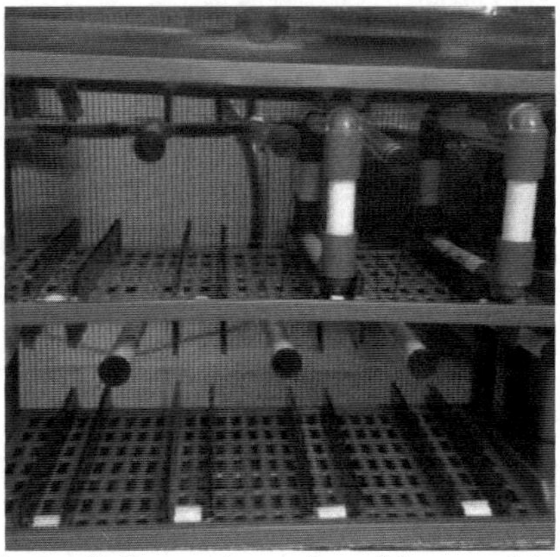

Figure 8. Inner view of washing machine.

Figure 8 is the inner view of washing machine that there are pre-filters separators. The separators using the pipes which is formed several slits every 2cm in order to spray the water into the dirty pre-filter.

Figure 9. Front view of washing machine.

Figure 9 describes the front view of washing machine. The washing machine using polycarbonate as the cover from another side.

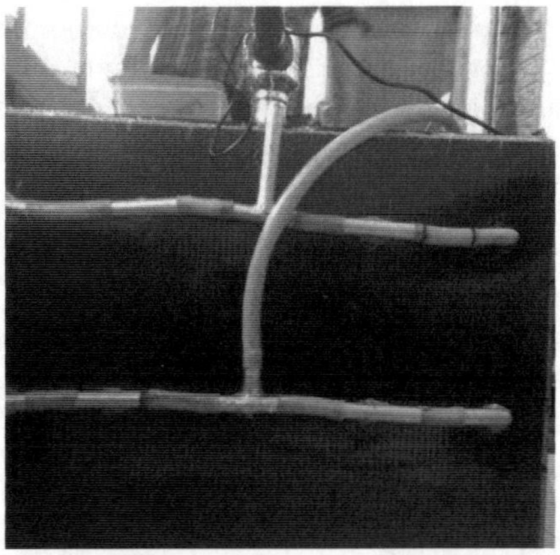

Figure 10. Back view of washing machine.

Figure 10 is the back view of washing machine. On the figure 3.3 there are water stripes using the intervals and its connection.

Figure 11. Valve 1 ON.

Figure 11 is describe when the valve 1 On which showed by spray water from the pipes in the first shelf.

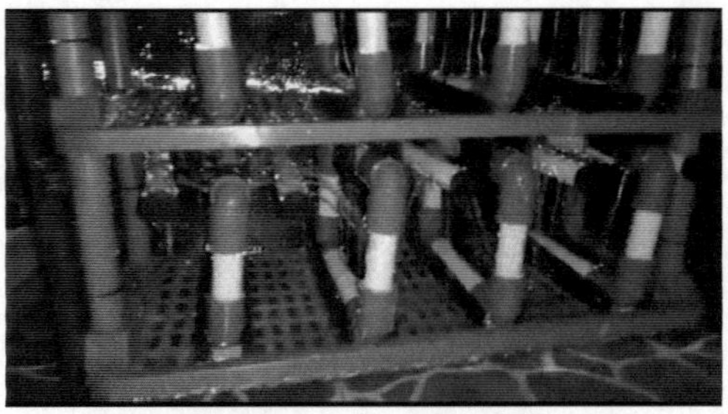

Figure 12. Valve 2 ON.

Figure 12 is describe when the valve 2 On which showed by spray water from the pipes in the second shelf.

4. Conclusions

Based on test results from AHU's Pre-Filter Washing Machine Based Arduino Nano, it could be concluded that AHU's Pre-Filter Washing Machine Based on Arduino Nano, designed from hardware such as Arduino nano, relay, start button, aquarium pump, hair dryer, and water valve inlet. Then the software that is applied to this system is the program built in Arduino programming language. Based on the test already could work well. The washing machine can operate for 2 minutes 12 second in a single command from the start button. AHU's Pre-Filter Washing Machine Based Arduino Nano 100% performance is in accordance with the planned function.

References

[1] US Water Alliance, The Economic Benefits of investing in Water Infrastructure. Value of Water Campaign.

[2] Saseendran S and Nithya V 2016 Automated water usage monitoring system *Int. Conf. Commun. Signal Process. ICCSP 2016* pp. 99–103

[3] Rahmani A 2015 Pengelolaan Air dalam Industri Pangan Pengelolaan Air dalam Industri Pangan *Res. GATE* pp. 0–13.

[4] Gerka A, Bayer F, Eichelberg M, Frenken M, and Hein A 2017 Ambient Water Usage Sensor for the Identification of Daily Activities *Glob. Internet Things Summit* pp. 225–230.

[5] Fraquelli G and Moiso V 2005 Cost Efficiency and Economies of Scale in the Italian Water Industry *Siep, XVII Conf. Finanz. del Sett. pubblico* pp. 15–16.

[6] Allner B Y F A 1932 "Economic Aspects of Water Power" pp. 156–166.

[7] Burhadudin M U, Taufq A J, and Hayat L 2014 "Kendali Mesin Cuci Satu Tabung Menggunakan Mikrokontroler AT89C51" *Simp. Nas. Teknol. Terap.* pp. 1–8.

[8] Ai-zhen W 2012 "The Design of Neural Network Fuzzy Controller in Washing Machine" pp. 136–139.

[9] Du Qiang J G 2015 "The Design of Intelligent Washing Machine Controller Based on FPGA," *Fifth Int. Conf. Instrum. Meas. Comput. Commun. Control*, no. 2/15.

[10] Phenomena S S 2015 "Design of 32-Bit Washing Machine Controller" pp. 220–221.

[11] Paper C et al. 2016 "Development of microcontroller based water pump controller".

[12] Serpong K 2003 "Rekayasa otomatisasi air handling unit (ahu) sistem pendingin" pp. 19–24.

[13] Susilo J 2015 "Aplikasi On Off Otomatis Menggunakan Sensor Ultrasonic".

[14] Ercişli H 2011 "Material Selection Process for Hair Dryer Components".

[15] Gusni M B 2009 "PC Based Washing Machine Controller," Univ. Tek. Malaysa Melaka.

IOP Conf. Series: Materials Science and Engineering **384** (2018) 012021 doi:10.1088/1757-899X/384/1/012021

Noise Properties of the Faraday Effect Measurement Systems

D Handoko[1,*], A Hifzhi[1], L Soedarmawan[1], M Aman[1], I Sugihartono[2], E Handoko[2] and A Sudarmaji[1]

[1]Physics Department, Faculty of Mathematics and Natural Sciences, Universitas Indonesia
[2]Jurusan Fisika, Universitas Negeri Jakarta

*djati.handoko@ui.ac.id

Abstract. Nowadays, nanomagnetic devices have a main role in computer industries. To observe magnetic features of the nanomagnetic device, Magneto-optical effects such as Kerr effect and Faraday effect can be used. We report the fabrication of the Faraday effect measurement setup for characterization of nanomagnetic elements. With achievement of low noise measurement configuration by precision electronic measurements, Faraday effect was observed. The essence of the experiment is to measure a light intensity that is transmitted from the transparent magnetic material. Light intensity is measured by a photodiode detector (Hamamatsu S5870) which produced the laser-induced photocurrent. Using transimpedance amplifier, the signal is converted to the voltage which is measured by the lock in amplifier (Stanford Research SR810). By performing the measurement at different lock-in frequencies and different analyzer angles, we systematically analyzed various random noise sources and their effect to the signal to noise ratio. The interestingly, the noise characteristics of the measured photocurrent exhibit a 1/f noise behavior with a scaling exponent practically 1.

1. Introduction

Spintronics is a multidisciplinary field which has a main role in computer industry nowadays. Mainly, related to speed of data processing and capacity of data storage. Spin term, in this case, according to spin of single electron s which can be detected by its magnetic moment $-g\mu Bs$ (μB is the Bohr magneton and g is the electron g factor) or average spin of a group of electrons, expressed by its magnetization [1]. The main goal of spintronics research is to understand interaction between spin particles with its solid state environment and then to get a useful device from this information.

Kerr effect or Faraday effect is the most famous method to explore the spintronics behavior [2, 3]. The first magneto-optic Kerr effect (MOKE) was founded by John Kerr in 1877. He did it while he was investigating the polarization of light reflected by refined electromagnetic rod. This experiment is a simple experiment but very effective to explore spin behavior in a material. It is presently described full information in the context of macroscopic dielectric theory or microscopic theory.

In this work, we describe very basically Faraday experiment setup. In this setup, we explored a noise which is produced by the instrumentation setup, depend on frequency and power of laser. In the measurement system setup, there are several types of noise like, Johnson noise, Shot noise and 1/f noise [4-8]. It becomes important to know the noise behavior before the experiment setup is completed.

2. Experimental scheme

2.1. Faraday effect noise measurement setup

The Faraday effect noise measurement setup is shown in Fig. 1. In this experiment, we used green laser, 532 nm wavelength, 30 mWatt maximum power, as a light source. This light then is chopped by chopper equipment. The chopper is SR540 by Stanford Research System. Its frequency can be controlled until 4 kHz and has several type of reference mode. In a Faraday effect measurement, light as a measuring signal, should be in polarization condition. In this experiment, we used beam splitter polarizer by Newport as a polarizer. Entering a simple configuration optics device, then the laser will be lain on photodiode sensor. The photodiode sensor is S5870 from Hamamatsu which has 5-10 nA dark current. Current signal from photodiode is converted into voltage using transimpedance amplifier. As a measurement voltage signal tool, we used lock in amplifier SR810 by Stanford Research System. The experiments consisted of measuring the noise as a function of frequency and power of laser.

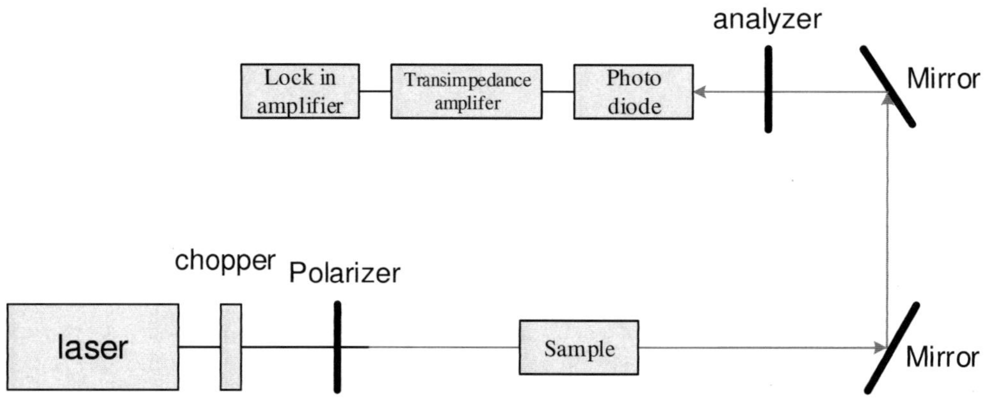

Figure 1. Faraday effect noise measurement setup.

2.2. Transimpedance amplifier

The circuit that we used to measure the photodiode noise current spectral density is shown in Fig. 2. This is feedback negative circuit amplifier that a resistor as a feedback. The photodiode noise current is amplified by this low-noise transimpedance amplifier. The amplifier and setup are considered to ensure that the total noise, measured at the amplifier output, is dominated by the contribution of the photodiode. The reverse bias current of the diode is supplied by the amplifier via the feedback resistor, connected in a negative feedback loop. Cf is a feedback capacitor that the value is same with the total capacitance parallel to the input port of the amplifier caused by stray capacitance and terminal capacitance of photodiode. The noise of the feedback resistor is modelled by thermal noise with voltage spectral density of $V_{noise}(rms) = \frac{1}{2}(4kTR_f\Delta f)$, where k is Boltzmann's constant, T is the absolute temperature, R_f is feedback resistor and Δf is bandwidth in Hz. The amount of noise measured by the lock-in is determined by the measurement bandwidth. The lock-in does not narrow its detection bandwidth until after the phase sensitive detectors. In a lock-in, the equivalent noise bandwidth (ENBW) of the low pass filter (time constant) sets the detection bandwidth.

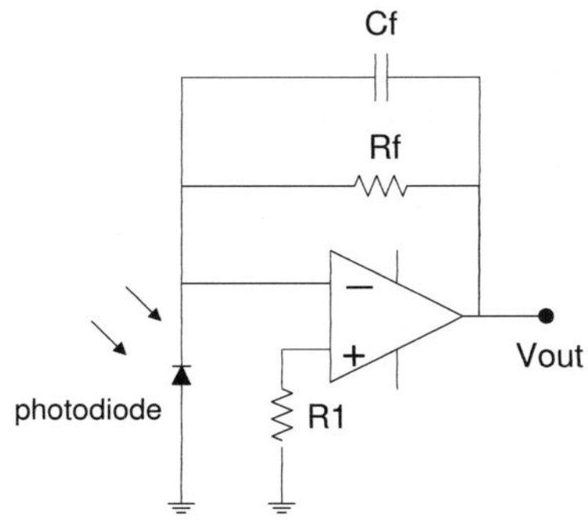

Figure 2. Transimpedance amplifier circuit.

3. Results and discussion

3.1. Frequency effect

The noise from setup instrumentation was measured as a function of frequency. On the log-log scale, the noise behavior of Faraday rotation was divided into 2 part, low and high frequency as illustrated in Fig. 3. For low frequency, the noise response is shown in the first part of Fig. 3, in this term is referred to 45-200 Hz chopper frequency and we used 5f chopper configuration as a frequency reference. At this measurement, the lock in amplifier configures in small frequency reference, such as; synchronous filter slope as of low pass filter and properly time constant. According to this figure, we can see that the noise response decrease as a frequency increase and close to high frequency, it became flat-band and stable.

For high frequency, 400 – 3600 Hz chopper frequency, the noise response is shown in the last part of Fig. 3. At this measurement, the slope of low pass filter of lock in amplifier is 6dB/oct and 3ms time constant. According to this figure, the noise response decrease as a frequency increase and start to be stable at frequency 1000 Hz and absolutely stable over 2400 Hz.

According to this figure, we can see that the noise response has an exponential dependent to frequency. This characteristic is in a good agreement with 1/f noise theory. It makes measurements at low frequencies more difficult.

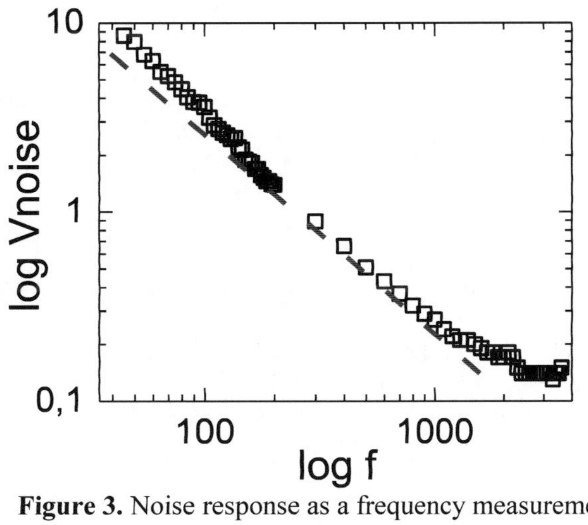

Figure 3. Noise response as a frequency measurement.

3.2. Power of laser effect

We also measured the noise response as a function of power of laser. In several constant frequencies, we demonstrated that noise response depends on power of laser. This behavior is shown in Fig. 4. For a high frequency (1000 – 3000 Hz), this dependency is quiet small. Different behavior appears for a low frequency (100 Hz), noise response increase smoothly at one point as power of laser increase.

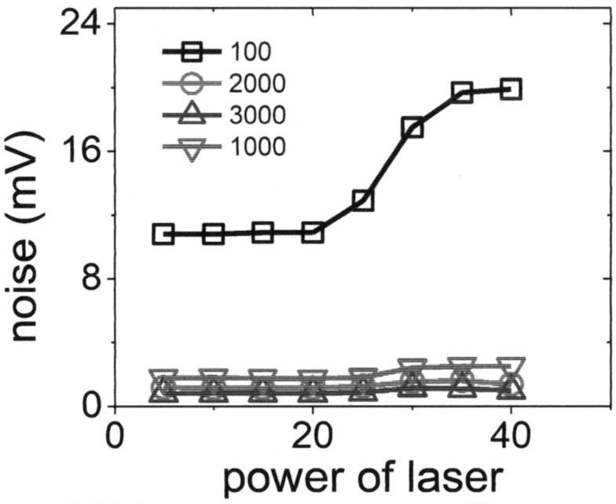

Figure 4. Noise response as a power of laser measurement.

4. Conclusions

The general purpose of this experiment is to setup a Kerr effect experiment. At the first study we find out the noise characteristic of this instrumentation. The noise response for this experiment has two behaviors, stable at high frequencies, more than 1000 Hz and unstable at low frequencies, less than 1000 Hz. For this setup, the maximum signal to noise ratio is about 1.572×10^8. For stability and easily measurement, frequencies chopper should be more than 2400 Hz.

Acknowledgements

The author would like to thank PITTA 2017 no 645/UN2.R3.1/HKP.05.00/2017 which support this work.

References

[1] Abragam A 1961 *The Principles of Nuclear Magnetism* (Oxford University Press, London).
[2] Sethna J P, Dahmen K A and Myers C R 2001 *Nature* **410**, 242.
[3] Faulkner C C, Cooke M D, Allwood D A, Doroth´ee Petit, Atkinson D and Cowburn R P 2004 *J. Appl. Phys.* **95**, 6717.
[4] Su Y K, Juang F S, Chang S M, Chiang C D and Cherng Y T 1999 *IEEE Journal of Quantum Electronics* **35**, 5.
[5] Milotti 2001 *1/f noise: a pedagogical review*. Invited talk to E-GLEA-2, Buenos Aires, Sept. 10-14.
[6] Lee J W, Jeong J R, Kim D H, Ahn J S, Kim J and Shin S C 2000 *Rev. Sci. Instr.* **71**, 3801.
[7] Qiu Q and Bader S D 2000 *Rev. Sci. Instr.* **71**, 1243.
[8] Libbrecht K G, Black E D and Hirata C M 2003 Am. J. Phys. **71**, 11.

IOP Conf. Series: Materials Science and Engineering **384** (2018) 012022 doi:10.1088/1757-899X/384/1/012022

Electrodeposited Ni-TiAlN Composite Coating on Tungsten Carbide: Effect of Surfactant Concentration on Physical and Mechanical Properties

E Budi[1,2,*], Y Ermadiana[1], I Sugihartono[1], D Syafindra[1] and M Fajrin[3]

[1] Physics Program, Faculty of Mathematics and Science, Universitas Negeri Jakarta, Jl. Rawamangun Muka I, Jakarta 13220, Indonesia
[2] Physics Education Program, Faculty of Mathematics and Science, Universitas Negeri Jakarta, Jl. Rawamangun Muka I, Jakarta 13220, Indonesia
[3] Material Physics Laboratory, Faculty of Mathematics and Science, Universitas Negeri Jakarta, Jl. Rawamangun Muka I, Jakarta 13220, Indonesia

*esmarbudi@unj.ac.id

Abstract. The effects of concentration of *Sodium Dodecyl Sulfate* ($C_{12}H_{25}NaSO_4$) as surfactant on structure and hardness of electrodeposited Ni-TiAlN composite coating has been investigated. The composites were deposited at fixed current of 3 mA, 12 V for 15 minutes. The electrolyte consists of 0.49 mol/l H_3BO_3, 0.17 M $NiCl_26H_2O$, 0.38 M $NiSO_46H_2O$, 2 gr/lit AlN and 2 gr/lit TiN. The concentration of *Sodium Dodecyl Sulfate (SDS)* was varied into 0.2 , 0.4 and 0.6 gr/lit. The electrodes used were Pt wire as counter electrode, silver (Ag) wire as reference electrode and tungsten carbide (WC) rod as working electrode. The surface morphology and composition of composites were characterized by using Scanning Electron Microscopy (SEM) dan Energy Dispersive Spectroscopy (EDS), respectively. The crystal structure of comosites was characterized by using X-Ray Diffraction (XRD) and the hardness was tested by using hardness Vickers tester. The results showed that a fewer aglomeration was formed on the composite surface for SDS concentration of 0.4 gr/lit. However, the increase of SDS concentration lead to the decrease of crystal size and increase the hardness of composite.

1. Introduction

Hard nitride particle-reinforced nickel (Ni) based-composite coating deposited by electrode position has been widely developed due high hardness, wear and corrosion resistance [1, 2]. The improvement is contributed by incorporation of hard particle that is uniformly distributed as a second phase of composite in nickel metal matrix thus it possess high hardness and good wear and corrosion resistant [3]. Nitride particles such as TiN, AlN and TiAlN are common ceramic material and have high hardness and wear resistance. Therefore these particles are widely used as reinforcement particle in nickel matrix composite coating [4-6].

The mechanical properties (e.g. hardness) of electrodeposited composite coating is influenced by electrode position parameters through controlling physical properties (e.g. morphology and structure) of composite. The electrode position parameters that influences the composite properties include electrode position current, particle and electrolyte concentration, surfactant and etc. [1, 7].

The control of physical properties of electrodeposited composite can be done by adding surfactant into electrolyte. Since the hard particle compound tend to agglomerate within the electrolyte thus it is needed to be dispersed by using surfactant [8]. Sodium Dedocyl Sulfate (SDS) is an anionic surfactant that can reduce surface tension of particle [9]. The objective of this experiment is to investigate the effect of surfactant concentration on physical and mechanical properties of Ni-TiAlN composite coating.

2. Experiment

The composites were deposited at fixed current of 3 mA, 12 V for 15 minutes. The electrolyte consists of 0.49 mol/l H_3BO_3, 0.17 M $NiCl_26H_2O$, 0.38 M $NiSO_46H_2O$, 2 gr/lit AlN and 2 gr/lit TiN. The concentration of *Sodium Dodecyl Sulfate (SDS)* was varied into 0.2 , 0.4 and 0.6 gr/lit. The electrodes used were Pt wire as counter electrode, silver (Ag) wire as reference electrode and tungsten carbide (WC) rod as working electrode. The surface morphology and composition of composites were characterized by using Scanning Electron Microscopy (SEM) dan Energy Dispersive Spectroscopy (EDS) of Jeol, JED-2300, respectively. The crystal structure of comosites was characterized by using X-Ray Diffraction (XRD) of Panalytical Empyrean Philips with Cu-Kα radiation source (λ =1.54 Å) and the hardness was tested by using micro hardness Vickers tester HV-1000.

3. Results and discussion

3.1. Surface morphology and composition

The experimental result on SEM image of surface morphology of Ni-TiAlN composite coating at various SDS surfactant concentration is shown in Figure 1. In general, the greater the SDS surfactant concentration, the greater inhibited TiN and AlN aggregation in the electrolyte. The aggregation appears on the composite surface at SDS concentration of 0.2 gr/lit and it disappears at SDS concentration of 0.4 gr/lit. However, the aggregation appears again at further increasing SDS concentration up to 0.6 gr/lit.

(a)

Figure 1. Surface morphology Ni-TiAlN composite coating deposited at SDS surfactant concentration of (a) 0.2 gr/lit.

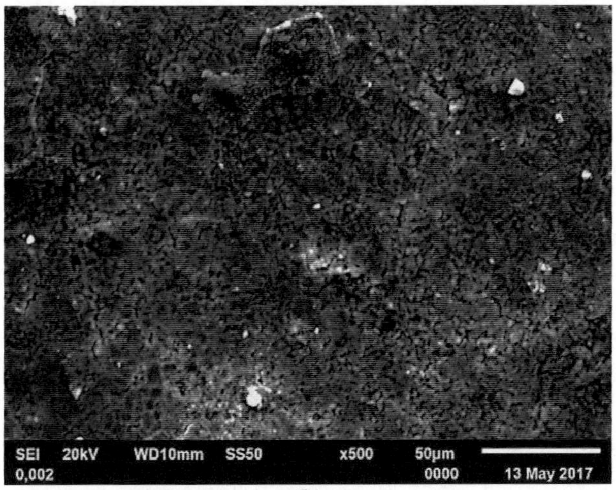

(b)

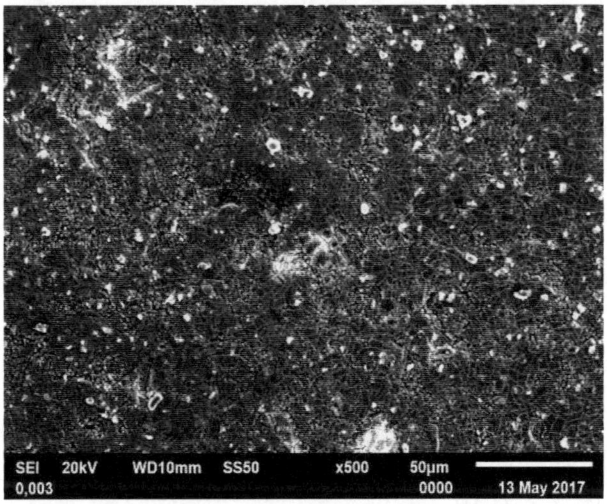

(c)

Figure 1 (Cont.). Surface morphology Ni-TiAlN composite coating deposited at SDS surfactant concentration of (b) 0.4 gr/lit and (c) 0.6 gr/lit.

The EDS composition of Ni-TiAlN composite coatings at various SDS concentration are shown in Figure 2. It shows that metal element particle of Al tend to increase while Ti element tend to decrease as SDS concentration is increased. It seem that the SDS surfactant is more reactive to the electronegative elements (Al) than Electropositive element (Ti). As a result more Al is deposited in the composite while Ti is loosely absorbed and lead to the Ni grain growth.

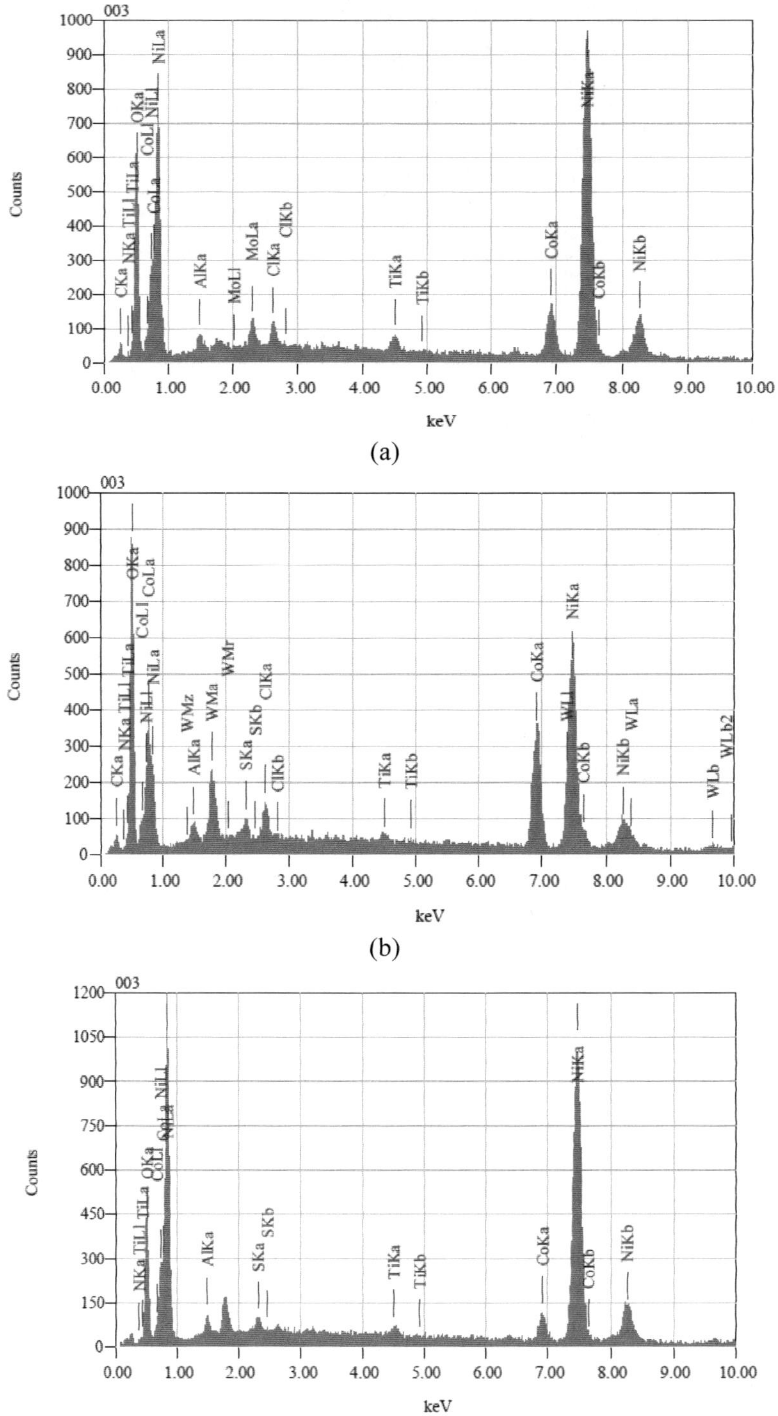

(a) (b) (c)

Figure 2. EDS analysis of Ni-TiAlN composite coatings at SDS concentration of (a) 0.2 gr/lit, (b) 0.4 gr/lit, (c) 0.6 gr/lit.

One of the major function of surfactant is to provide the stability of the colloid and disperse the particle in the electrolyte [9]. Without surfactant, the particles may agglomerate due to polar and ionic surface forces. The surfactant can control adatom and kinetically delay the particle segregation due to cluster attachment onto the surface. The reduce agglomeration is due to the diminished strength of interfacial Coulomb interactions that causes the minimization of local dipole moments. It is known that the increase of particles concentration may accelerate the effect of agglomeration and subsequent gravitational sedimentation [10]. In this study, it seem that the sedimentation effect appear at further increasing in surfactant concentration up to 0.6 gr/lit.

3.2. Crystal structure
XRD spectrum of Ni-TiAlN composite coating in varying SDS concentration is shown in Figure 3.

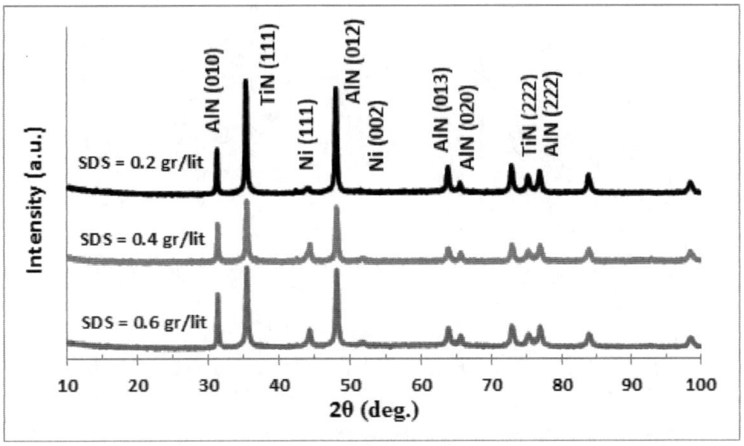

Figure 3. XRD spectrum of Ni-TiAlN composite coating deposited at SDS concentration of 0.2 gr/lit, 0.4 gr/lit and 0.6 gr/lit.

From the calculation result, it is shown that, generally, the crystal size of composite is about 4 nm and it tend to decrease as increasing in SDS concentration.

Ni grains are shown by the diffraction peaks at 2-theta of 44.33^0 and 51.72^0 which correspond to the (111) and (002) crystal plane, respectively. TiN grains are displayed by the peaks at 35.51^0 and 75.63^0 which correspond to the (111) and (222), respectively. AlN grains are identified by peaks at 31.23^0, 47.90^0, 63.97^0 , 65.57^0 , 76.82^0 and 83.46 which correspond to (010), (012), (013), (020), (222) and (014), respectively. The intensity of Ni (111) and Ni (222) peaks show a tendency to increase as the SDS concentration is increased. Meanwhile, the intensity of TiN (111) and AlN (012) peaks tend to decrease as the SDS concentration is increased.

The surfactant changes the (hkl) surface energy of particles thus it influences the crystal structure [9]. The increase of SDS concentration may increase the cathodic reactivity of the surfactant thus the nitride particles are more easily embedded on the substrate [11]. Increase of embedded particles on the substrate improve the nucleation site number and perturbs the crystal growth of Ni-metal matrix and consequently reduces the crystallite size [12] as can be seen in XRD spectrum result.

3.3. Hardness
The hardness test result is shown in Table 1. It suggests that the increases of SDS concentration enhances the composite hardness. The reason may be contradictely different by other explanation [13]. The increasing composite hardness is due to the interaction between nickel matrix with reinforced nitride particles (TiN and AlN) which improves upon the SDS surfactant addition as the surfactant increases the tension between matrix and reinforced particles. The SDS surfactant may promote the ionization of

TiN (and AlN) particles and facilitate their adsorption and deposition on the substrate [7]. Thus it may cause the increase of hardness.

Table 1. Experimental result of Vickers hardness test of Ni-TiAlN composite coating at various SDS concentration.

No	SDS concentration (gr/lit)	HVN (kg/mm^2)
1	0.2	2355.07
2	0.4	2415.40
3	0.6	2520.20

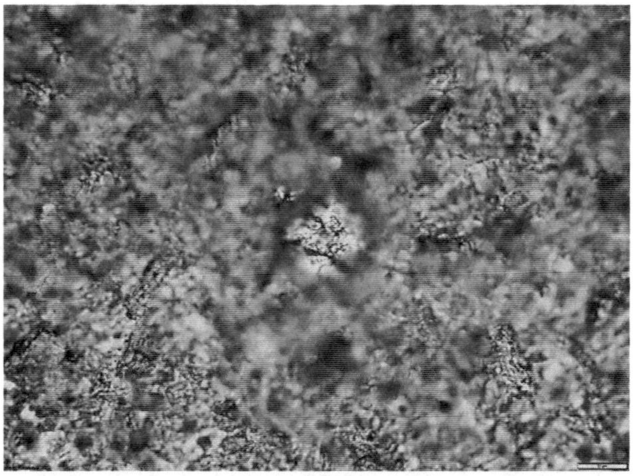

(a)

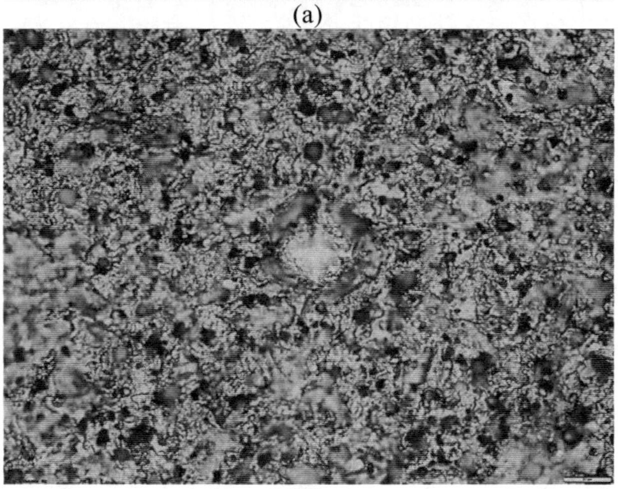

(b)

Figure 4. Vickers hardness indentation on surface of Ni-TiAlN composite coating deposited at SDS concentration of 0.2 gr/lit, 0.4 gr/lit and 0.6 gr/lit.

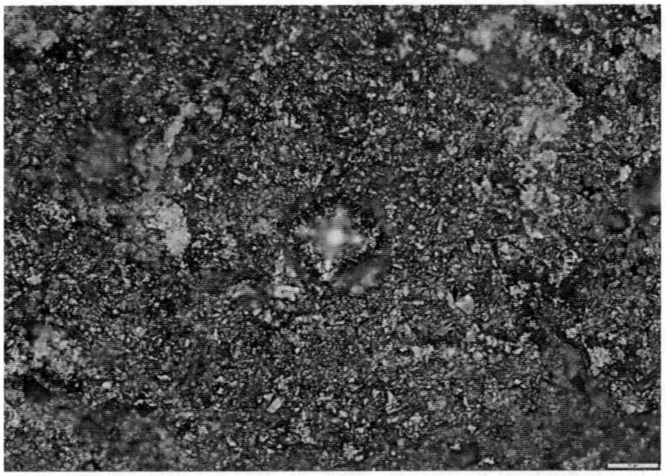

(c)

Figure 4 (Cont.). Vickers hardness indentation on surface of Ni-TiAlN composite coating deposited at SDS concentration of 0.2 gr/lit, 0.4 gr/lit and 0.6 gr/lit.

4. Conclusions

The Ni-TiAlN composite coatings were successfully prepared by electrodeposition with varying *Sodium Dodecyl Sulfate* (SDS) surfactant. The SEM/EDS results indicate the incrase of aggregation as the increase of SDS concentration. The Al element is more reactive with SDS surfactant and as a result Al is more absorbed by substrate than Ti. The XRD and hardness test results indicate the reduction of crystallite size of composite lead the improve of its hardness. Thus it is concluded that the electrodeposition at high SDS concentration is the best parameters for the preparation of Ni-TiAlN composite coating is the higher.

Acknowledgments

The authors gratefully acknowledge research facilities support from Laboratorium of Material Physics, Physics and Physics Education Programs, Faculty of Mathematics and Science, Universitas Negeri Jakarta. This research was financially supported by BLU FMIPA UNJ with contract No. 24/SPK/Penelitian/6.FMIPA/2017 and DIKTI-Ministry of Education with contract No. 7/SP2H/DRPM/LPPM-UNJ/IV/2017.

References

[1] Jiang M, Ma C, Xia F F and Zhang Y 2016 Application of artificial neural networks to predict the hardness of Ni–TiN nanocoatings fabricated by pulse electrodeposition *Surface & Coatings Technology* **286** p. 191–196.

[2] Xia F F, Liu C, Wang F, Wu M H, Wang J D, Fua H L and Wang J X 2010 Preparation and characterization of Nano Ni–TiN coatings deposited by ultrasonic electrodeposition *Journal of Alloys and Compounds* **490** p. 431–435.

[3] Shi, Lei, Sun C, Gao P, Zhou F and Liu W 2006 Mechanical properties and wear and corrosion resistance of electrodeposited Ni–Co/SiC nanocomposite coating *Applied Surface Science* **252** p. 3591–3599.

[4] Xia F F, Yue W, Wang J, Liu C, Wang F and Lia Y 2015 Synthesis of Ni–TiN composite nanocoatings by magnetic pulse current deposition *Ceramics International* **41** p. 11445–11448.

[5] Xia F F, Xu H, Liu C, Wang J, Ding J and Ma C 2013 Microstructures of Ni–AlN composite coatings prepared by pulse electrodeposition technology *Applied Surface Science* **271** p. 7–11.

[6] Budi E, Muarief, Khasanah M, Permatasari A L, Sugihartono I, Fahdiran R and Budi A S 2017 Crystal Structure and Corrosion of Electrodeposited Ni-TiAlN Composite Coatings *J. Tech. Soc. Sci.* **1**(2).

[7] Wu M, Jia W and Lv P 2017 Electrodepositing Ni-TiN nanocomposite layers with applying action of ultrasonic waves *Procedia Engineering* **174** p. 717 – 723.

[8] Chen L, Wang L, Zeng Z and Zhang J 2006 Effect of surfactant on the electrodeposition and wear resistance of Ni–Al2O3 composite coatings *Materials Science and Engineering* **A 434** p. 319–325.

[9] Heinz H, Pramanik C, Heinz O, Ding Y, Mishra R, Marchon D, Flatt R J, Lopis I E, Llop J, Moya S and Ziolo R F 2017 Nanoparticle decoration with surfactants: Molecular interactions, assembly, and applications *Surface Science Reports* **72** p. 1–58.

[10] ZHU Xu-bei, CAI Chao, ZHENG Guo-qu, ZHANG Zhao and LI Jin-feng 2011 Electrodeposition and corrosion behavior of nanostructured Ni-TiN composite films /Trans *Nonferrous Met. Soc. China* **21** p. 2216–2224.

[11] Sudagar J, Lian J and Sha W 2013 Electroless nickel, alloy, composite and nano coatings – A critical review *Journal of Alloys and Compounds* **571** p. 183–204.

[12] Ramesh Bapu G N K R and Jayakrishnan S 2012 Development and characterization of electro deposited Nickel–Titanium Carbo Nitride (TiCN) metal matrix nanocomposite deposit *Surface & Coatings Technology* **206** p. 2330–2336.

[13] Nordina M N, Asari N, Mahaidin A A, Shad K M and Aziz N M A N A 2016 Immobilization of Bromocresol Purple in Inorganic-Organic Sol-Gel Thin Film with Presence of anionic and non-ionic surfactants *Procedia Chemistry* **19** p. 275 – 282.

International Symposium on Materials and Electrical Engineering (ISMEE) 2017 IOP Publishing
IOP Conf. Series: Materials Science and Engineering **384** (2018) 012023 doi:10.1088/1757-899X/384/1/012023

Nominal of Money and Colour Detector for the Blind People

F Andika and J Kustija*

Department of Electrical Engineering Education, Universitas Pendidikan Indonesia
Jl. Dr. Setiabudhi No. 207 Bandung 40154, Indonesia

*j.kustija@upi.edu

Abstract. Blind people often face a detection to identify the colour object. One of important of inability from the blind is to identify a nominal of money. It is very likely occurrence of undesirable things, such us as the possibility of exchanged money, or even act of fraud. It is necessary to design an equipment that allows for detecting a colour and a nominal of money. This tool is designed using TCS3200, RGB LED, LDR as the sensors, Arduino Nano as the controller and speaker. The sensors will detect the colour of the money, then sensors will transforms the colour into analog data and Arduino Nano will process this analog data into a digital data. Arduino Nano will convert the data into the form of audio as the output result. The speaker will release the audio according to the money or colour which detected, the audio output will be identified by the blind people.

1. Introduction

Blind people often face a detention to identify the colour of an object. One of important of inability from the blind is to identify a nominal of money. So far the blind identify a nominal of money by making a fold or arrange their money with help by the other people, but it has a weakness, that if blind forget with the fold or honesty from people who make a transaction with them giving a right nominal or not [1]. From this case, it is necessary to design an equipment that allows for detecting a colour and nominal of money to support the blind. This paper describes the design of nominal of money and colour detector using a sensor and microcontroller.

2. Methods

2.1. List of materials

Table 1 is List of Materials (LoM) required in this research.

Table 1. List of materials.

Materials	Specification	Quantity
Arduino	Arduino Nano	1
Colour Sensor	TCS3200	1
RBG LED	Common cathode	1
Resistor	LDR	1
TP4056	5 V, 1A	1
Amplifier	PAM8403	1

Table 1. Cont.

Materials	Specification	Quantity
Switch	2 pin	1
Baterry	Li-ion 18650 4,2 V, 1200 mAh rechargable	1
Mp3 module	DFPlayer mini	1
Acrylic	2mm, size A4	1
Micro SD Card	4 Gb	1

2.2. Block diagram

The block diagram of this tools is represented by figure 1.

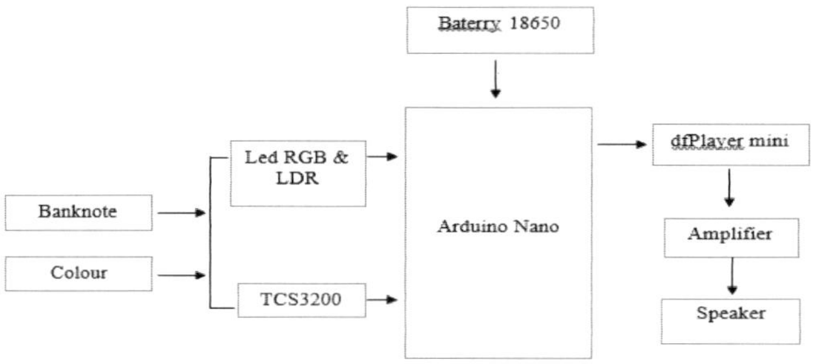

Figure 1. Block diagram.

2.3. Flowchart

The flowchart of this tools is represented by figure 2.

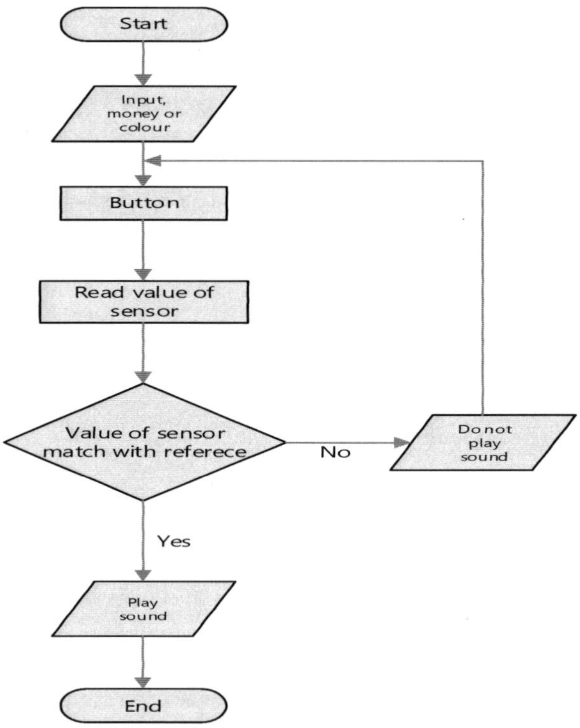

Figure 2. Flowchart.

IOP Conf. Series: Materials Science and Engineering **384** (2018) 012023 doi:10.1088/1757-899X/384/1/012023

2.4. Circuit diagram

Figure 3 show the circuit diagram of this tool.

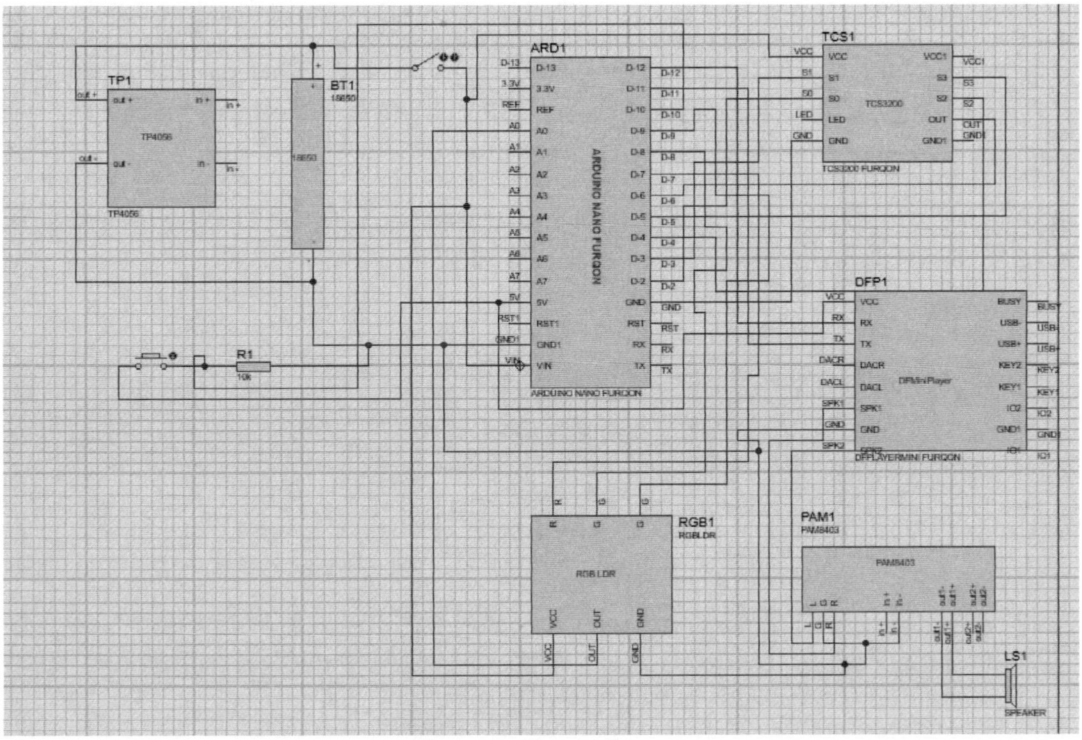

Figure 3. Circuit diagram.

2.5. Arduino nano

The controller of this tools is using Arduino Nano, the specification of Arduino Nano is represented by Table 2.

Table 2. Arduino nano specification.

Spesification	Component
Microcontroller	ATMega328
Architecture	AVR
Operating Voltage	5 V
Flash Memory	32 KB
SRAM	2 KB
Clock Speed	16 Mhz
Analog I/O	6
Digital I/O	14
DC Current Per I/O	40 mA
Input Voltage	7-12
Weight	7 g

Arduino Nano could be easily programmed by using Arduino IDE with another electronic device such as a computer by using USB port [2, 3].

IOP Conf. Series: Materials Science and Engineering **384** (2018) 012023 doi:10.1088/1757-899X/384/1/012023

2.6. Colour sensor

Colour sensor is component that could identify the colour of the object to be electrical value. In this research use TCS3200. It has two main component, there are photodiode and frequency converter [4, 5]. Photodiode will release a current which value is proportional to the colour detected, this current will be converted into a signal which frequency is proportional to its current. TCS3200 has 4 filter mode, red filter, clear, green filter and blue filter [6]. TCS3200 identify the colour and give serial output in RGB value to the Arduino Nano [7]. To set filter mode is by use S2 pin dan S3 pin as the Table 3 [6]. Figure 4 is the pin of TCS3200.

Table 3. Filter mode TCS3200.

S2	S3	Photodiode
L	L	Red
L	H	Blue
H	L	Clear
H	H	Green

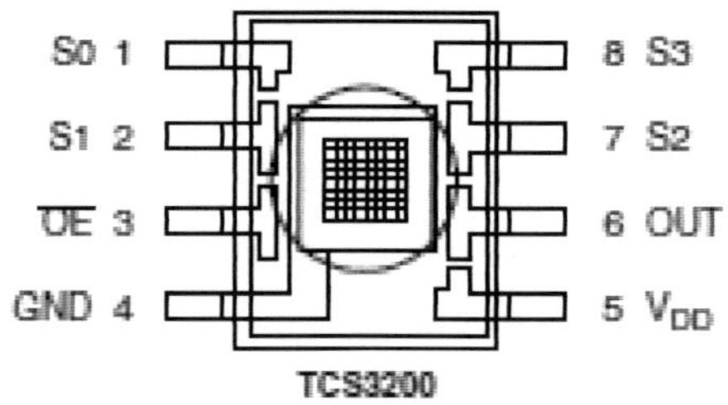

Figure 4. TCS3200.

2.7. RGB led and LDR (Light Dependent Resistor)

Both of this component could be used as RGB sensor to detect the colour of an object. At the first the RGB led will emit red, green and blue alternately to the object that will be detected. The reflected colour will affect the value of resistance of the LDR and expressed in the RGB output. Figure 5 is the Schematic of sensor RGB from LDR.

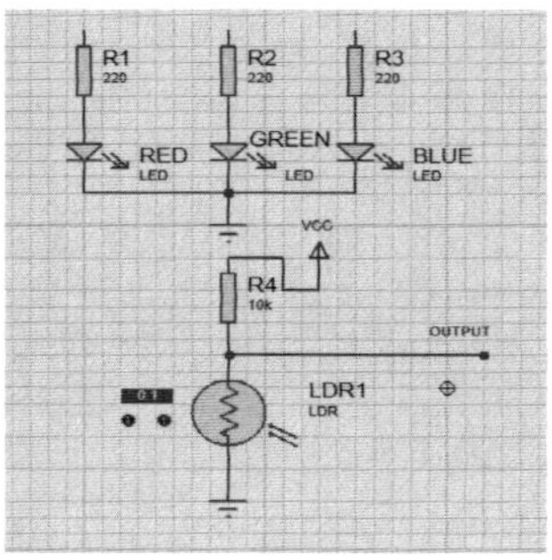

Figure 5. Schematic of RGB led and LDR.

2.8. TP4056

TP4056 is a module that used to recharge the battery. This module has 2 led indicator, which is red and blue. Red led indicates battery charging is still going, while blue led indicates the battery is full. When the battery is fully charged, the system will cut off, so the current does not flow to the battery. TP4056 work with USB and adapter [8].

2.9. DFPlayer mini
DFPlayer mini is a module that used for the music player. This module could work alone (standalone) or work with another microcontroller with serial communication RX/TX [9].

2.10. PAM8403
PAM8403 is an amplifier that can be used for a sound amplifier, the power of this amplifier is 3 watts with 5 volt DC source. The magnitude of this amplifier can be adjusted with potentiometer contained in this module [10].

2.11. Design
This equipment is using acrylic 16 cm x 5 cm x 7 cm for the case, all materials will be inserted in this case. The case looks like as the follows. Figure 6 is the design and component position of this tools.

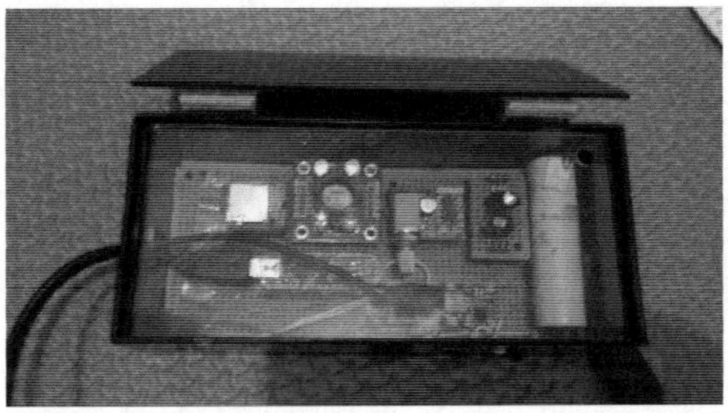

Figure 6. Design and component position.

3. Results and discussion

3.1. Sensor recitations and data mapping
To read the sensor recitations is by using serial communication in Serial Monitor from Arduino IDE. Sensors recitation is done for 4 times to get ADC value for each position. The output from TCS3200 are R, G, B meanwhile the output of LDR are RL, GL, and BL. There are 4 positions of money, it has 4 possibilities positions. This tool is tested with Rupiah 2016 emissions. The left table is the ADC of the money or colour, and the right table is the mapping data of ADC. Mapping of ADC is by making an interval of data (See in Table 4 until Table 19).

Table 4. ADC of Rp.1000.

Position	R	G	B	RL	GL	BL
1st	58	59	51	23	31	64
2nd	63	68	56	18	23	64
3rd	56	63	60	18	26	53
4th	61	66	59	22	29	62

Table 5. Mapping ADC of Rp.1000.

1st Position	2nd Position	3rd Position	4th Position
56≤ R ≤60	61≤ R ≤65	54≤ R ≤59	59≤ R ≤63
57≤G ≤61	66≤ G ≤71	61≤ G ≤66	64≤ G ≤68
49≤ B ≤54	58≤ B ≤63	58≤ B ≤63	57≤ B ≤62
21≤RL≤26	16≤ RL ≤22	16≤RL ≤21	20≤ RL ≤26
31≤GL≤37	21≤ GL ≤27	24≤GL ≤30	27≤ GL ≤32
62≤BL≤68	48≤ BL ≤53	51≤ BL≤58	60≤ BL ≤67

Table 6. ADC of Rp.2000.

Position	R	G	B	RL	GL	BL
1st	63	70	61	20	27	61
2nd	60	65	57	20	26	51
3rd	58	64	57	25	31	58
4th	66	68	57	23	30	62

Table 7. Mapping ADC of Rp.2000.

1st Position	2nd Position	3rd Position	4th Position
61≤ R ≤66	58≤ R ≤62	56≤ R ≤60	64≤ R ≤68
68≤ G ≤72	63≤ G ≤68	62≤ G ≤667	66≤ G ≤71
59≤ B ≤64	55≤ B ≤60	55≤ B ≤59	55≤ B ≤60
18≤RL ≤25	18≤ RL ≤24	22≤ RL ≤27	22≤RL ≤27
25≤ GL≤31	24≤ GL ≤30	29≤ GL≤33	28≤GL ≤33
59≤BL ≤65	49≤ BL ≤56	56≤ BL ≤62	60≤BL ≤66

Table 8. ADC of Rp.5000.

Position	R	G	B	RL	GL	BL
1st	58	60	60	17	24	57
2nd	56	57	52	15	25	55
3rd	50	64	60	19	28	61
4th	59	70	62	21	31	61

Table 9. Mapping ADC of Rp.5000.

1st Position	2nd Position	3rd Position	4th Position
56≤ R ≤60	54≤ R ≤58	48≤ R ≤59	57≤ R ≤62
58≤ G ≤68	55≤ G ≤65	62≤ G ≤67	68≤ G ≤72
58≤ B ≤62	50≤ B ≤60	58≤ B ≤62	60≤ B ≤65
15≤ RL ≤21	13≤ RL ≤18	17≤ RL ≤22	19≤RL≤24
22≤GL ≤27	23≤ GL ≤28	26≤ GL ≤32	29≤RL≤34
55≤ BL ≤61	53≤ BL ≤60	59≤ BL ≤65	60≤BL≤66

Table 10. ADC of Rp.10.000.

Position	R	G	B	RL	GL	BL
1st	63	71	56	20	32	59
2nd	59	67	52	18	26	45
3rd	57	60	47	18	27	48
4th	62	69	54	24	31	53

Table 11. Mapping ADC of Rp.10.000.

1st Position	2nd Position	3rd Position	4th Position
61≤ R ≤66	57≤ R ≤62	55≤ R ≤60	60≤ R ≤65
69≤ G ≤73	65≤ G ≤70	58≤ G ≤63	67≤ G ≤72
54≤ B ≤59	50≤ B ≤55	45≤ B ≤49	52≤ B ≤54
18≤ RL ≤25	16≤ RL ≤21	16≤RL≤23	22≤RL ≤26
30≤ GL ≤35	24≤ GL ≤30	25≤GL≤30	29≤GL ≤34
57≤ BL ≤64	43≤ BL ≤49	46≤BL≤51	51≤BL ≤58

Table 12. ADC of Rp.20.000.

Position	R	G	B	RL	GL	BL
1st	60	59	51	18	22	47
2nd	52	52	48	20	23	48
3rd	52	61	59	18	22	48

Table 13. Mapping ADC of Rp.20.000.

1st Position	2nd Position	3rd Position	4th Position
58≤ R ≤63	50≤ R ≤54	50≤ R ≤60	63≤ R ≤67
57≤ G ≤62	50≤ G ≤56	59≤ G ≤63	62≤ G ≤66
49≤ B ≤53	46≤ B ≤50	57≤ B ≤60	54≤ B ≤58
16≤ RL ≤21	18≤ RL ≤22	16≤ RL ≤22	21≤RL≤28
20≤GL ≤25	21≤ GL ≤26	20≤ GL ≤26	27≤GL≤32
45≤ BL ≤51	46≤ BL ≤52	46≤ BL ≤52	65≤BL ≤72

Table 14. ADC of Rp.50.000.

Position	R	G	B	RL	GL	BL
1st	63	62	50	24	28	58
2nd	60	60	51	20	23	42
3rd	63	60	52	20	26	48
4th	71	68	52	26	30	60

Table 15. Mapping ADC of Rp.50.000.

1st Position	2nd Position	3rd Position	4th Position
61≤ R ≤65	58≤ R ≤63	61≤ R ≤65	69≤ R ≤74
60≤ G ≤64	58≤ G ≤63	58≤ G ≤66	66≤ G ≤71
40≤ B ≤53	49≤ B ≤53	50≤ B ≤55	50≤ B ≤56
22≤ RL ≤27	18≤ RL ≤23	18≤ RL ≤24	24≤RL≤29
26≤GL ≤32	21≤ GL ≤27	24≤ GL ≤30	28≤GL≤34
56≤ BL ≤61	40≤ BL ≤46	44≤ BL ≤50	58≤BL≤64

Table 16. ADC of Rp.100.000.

Position	R	G	B	RL	GL	BL
1st	56	68	56	18	28	53
2nd	51	64	52	15	25	46
3rd	57	67	55	15	27	48
4th	54	67	54	19	32	62

Table 17. Mapping ADC of Rp.100.000.

1st Position	2nd Position	3rd Position	4th Position
54≤ R ≤58	49≤ R ≤54	55≤ R ≤59	52≤ R ≤57
66≤ G ≤71	62≤ G ≤66	65≤ G ≤70	65≤ G ≤72
54≤ B≤58	50≤ B ≤56	53≤ B ≤58	52≤ B ≤56
16≤ RL ≤22	13≤ RL ≤17	13≤ RL ≤19	17≤RL≤22
26≤ GL≤31	23≤ GL ≤28	25≤ GL ≤30	30≤GL≤36
51≤ BL ≤57	44≤ BL ≤49	46≤ BL ≤52	60≤BL≤67

Table 18. ADC of colour.

Colour	R	G	B
Red	45	90	70
Yellow	48	57	57
Green	71	53	62
Light green	50	47	47
Blue	90	73	50
Dark blue	100	93	64
Orange	33	79	6
Pink	32	78	54
Black	103	109	87
White	43	46	36

Table 19. Mapping ADC of colour.

Colour	RL	GL	BL
Red	43≤ RL≤47	89≤GL≤93	69≤ BL≤73
Yellow	46≤ RL≤50	55≤GL≤59	65≤ BL≤70
Green	69≤ RL≤75	50≤GL≤55	60≤ BL≤66
Light green	48≤ RL≤53	45≤GL≤49	45≤ BL≤50
Blue	88≤ RL≤93	71≤GL≤76	48≤ BL≤52
Dark blue	98≤ RL≤103	90≤GL≤100	60≤ BL≤68
Orange	30≤ RL≤40	77≤GL≤80	61≤ BL≤72
Pink	30≤ RL≤35	75≤GL≤84	52≤ BL≤58
Black	102≤ RL≤107	108≤GL≤113	85≤ BL≤93
White	40≤ RL≤50	43≤GL≤54	33≤ BL≤45

3.2. *Trial*

This tool is tested by using some different sample of rupiah 2016 emission and some colour, this trial is 4 times for each colour or money. Table 20 is the result of the sensing colour trial, and Table 21 show the result of trial sensing of nominal.

Table 20. Result of colour.

Colour	Result
Red	Red
	Red
	Red
	Red
Pink	Pink
	Pink
	Pink
	Pink
Green	Green
	Green
	Green
	Green
Light green	Light green
	Light green
	Light green
	Light green
Blue	Blue
	Blue
	Blue
	Blue

Table 20. Cont.

Colour	Result
Dark blue	Dark blue
	Dark blue
	Dark blue
	Dark blue
Orange	Orange
	Orange
	Orange
	Orange
White	White
	White
	White
	White
Yellow	Yellow
	Yellow
	Yellow
	Yellow
Black	Black
	Black
	Black

Table 21. Result of nominal detector.

Money	Result
Rp.1000	Undetected(error)
	Rp.1000
	Rp.1000
	Rp.1000
Rp.2000	Rp.1000 (Error)
	Rp.2000
	Rp.1000 (Error)
	Rp.2000
Rp.5000	Rp.5000
	Rp.5000
	Rp.5000
	Rp.5000
Rp.10.000	Rp.10.000
	Rp.10.000
	Rp.10.000
	Rp.10.000
Rp.20.000	Rp.20.000
	Rp.20.000
	Rp.20.000
	Rp.20.000
Rp.50.000	Rp.50.000
	Rp.50.000
	Rp.50.000
	Undetected(error)
Rp.100.000	Rp.100.000
	Rp.100.000
	Rp.100.000
	Rp.100.000

The result of the trial for the colour is always true for 4 time trial, but the tool has a some error to detecting a nominal of money, it is because the data ADC of money with different nominal is almost same. Sensing of nominal money depends on the physical condition of money.

$$error = \frac{total\ error}{total\ trial}\ x\ 100\%$$

$$error = \frac{4}{28}\ x\ 100\%$$

$$error = 14{,}3\%$$

4. Conclusions

Based on the test that has been done can be taken several conclusions:

- The design of colour and nominal of money detector works well but has some error.
- This tool has an error if the nominal value of money included in the box is the 2004 emission money, this is because the ADC value of the nominal of money 2004 has the same ADC value at different nominal on emissions 2016 banknotes.
- Sensing of money depends on the physical condition of the money to be detected.

References

[1] Siwindarto I P, Probadi D A and Rif'an M 2014 "Alat Deteksi Nominal Uang Kertas Untuk Penyandang Tuna Netra" *Jurnal Mahasiswa TEUB* **2** (1).
[2] Sudhan R H, Kumar M G, Prakash A U, Devi S A N U R and Sathiya P 2015 "Arduino ATMEGA-328 microcontroller" *International Journal of Innovative in Electrical, Electronics, Instrumentation and Control Engineering* **3** (4) pp. 27-29.
[3] Arduino LLC 2015 "Arduino Nano".
[4] Joy A 2014 "Object Sorting Robotic Arm Based on Colour Sensing" *Int. J. Adv. Res. Electr.* **3** (3) pp. 7741–7746.
[5] S Shirgave, A Salunkhe, K Shirgave and U S Y 2017 "Color Sorting Robot" *Ijarcce* **6** (3), pp. 403–405.
[6] F B Diagram 2004 "Programmable Color Light-To-Frequency Converter Texas Advanced Optoelectronic Solutions Inc . Programmable".
[7] T G Gaikar 2016 "Object Sorting using Color Sensor and Arduino" pp. 483–486.
[8] N T P A Corp 2015 "TP4056 1A Standalone Linear Li-lon Battery Charger with Thermal Regulation in SOP-8".
[9] P Map and W Mode 2014 "DFPlayer Mini SKU : DFR0299".
[10] PAM 2007 "Pam8403".

Traffic Light Monitoring System based on NodeMCU using Internet of Things

G S Miratunnisa and A H S Budi*

Department of Electrical Engineering Education, Faculty of Technology and Vocational Education, Universitas Pendidikan Indonesia. Jalan Dr. Setiabudhi No.229 Bandung 40154

*agusheri@aol.com

Abstract. The increasing numbers of the private car cause the road traffic congestion. That is becoming important problems in the big cities. Another solution to reduce the traffic jam is developing intelligent transportation system such as intelligent traffic light system. However, the solution of such problems is one of the mandatory concern. In recent years, a smart city is the hottest topic for the efficiency of traffic light using the Internet of Things. Based on that issue, this research proposed traffic light monitoring system that can monitor and display a real-time traffic congestion through smartphones. The system utilized NodeMCU (ESP 8266 12E) equipped by IR obstacle sensor to indicate the road traffic congestion. To connect the internet, the system uses ESP8266 12E wifi in the chip of NodeMCU. It will send a notification to smartphones user. With this research, it is expected to improve road traffic tremendously such as to predict traffic congestion and the system can give some efficient routes for a user.

1. Introduction

Nowadays, Indonesia become urban country because every year the number of private vehicles always increases about six million units per year, 10-15 % comes from the cars. But the infrastructure to support that so bad. That is becoming important problems, because according from TomTom navigation devices Indonesia has the fourth worst evening rush hour traffic in the world cited by CNN.

Some countries in the world have the same problem [1], such as Japan, USA, and Europe countries. To reduce that problem their reformed the road system and improved the traffic management method using intelligent transportation system [2, 3]. The development of ITS has been a popular research for the traffic monitoring using The Internet of Things [1, 3]. IoT concept constructs some object around us become smarter for connection to each other such as roadside sensor systems for supporting road infrastructure systems[4], so that can communicate with each other for gather everyday information about the traffic environment with the minimum human intervention[4]. From that issues, this research was design an intelligent transportation system using internet of Things platform. The system can monitor a real-time condition the traffic congestion on road [5, 6], use some roadside sensor system for dynamic management of traffic signals. And to avoid the traffic jam, the user can able to find out optional routes to go to destination.

IOP Conf. Series: Materials Science and Engineering **384** (2018) 012024 doi:10.1088/1757-899X/384/1/012024

2. Methods

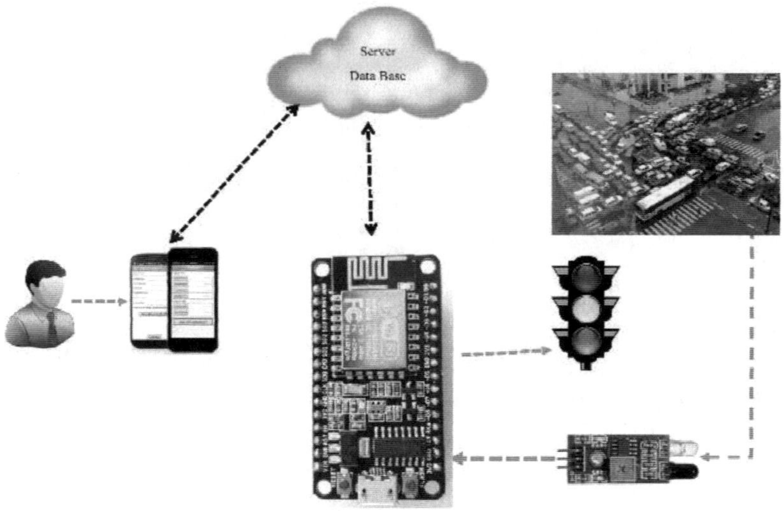

Figure 1. Architecture of the system.

The figure 1 approximately overall architecture of the traffic light monitoring system. The system divided into two module such as hardware module and software module. In hardware module, utilized the Android Phone/Device, NodeMCU (ESP8266 12E), IR obstacle Sensors (FC-51), Resistors, LED, etc. And in software used the MIT App Inventor for an android application, a web server for saving the database [7, 8].

In this system, sensors will send information to a NodeMCU [9]. Then NodeMCU will receive that information and transfers to the android device via the internet. Android device will send all the collected information to the server over the web server and store it in the database [8].

According to the traffic signals from the sensors will be managed dynamically to avoid a traffic jam. When a user needs a real-time traffic data or condition the routes, the user can log in to the Android application, the application will request to the web server for the user requirement [7] and the server will get results from requirement and user will see the traffic data and traffic condition.

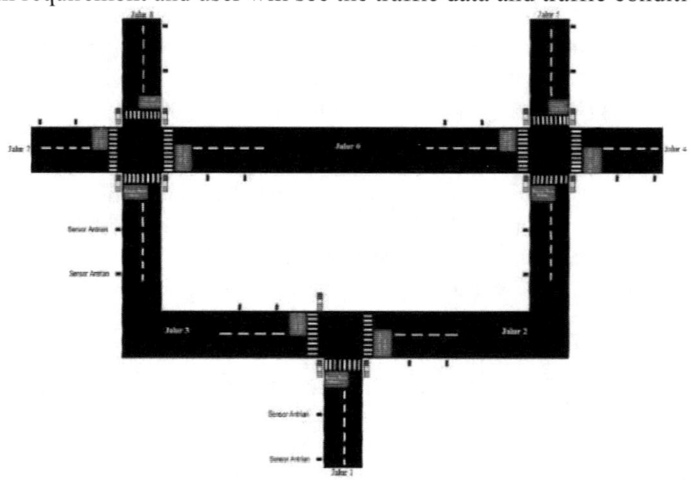

Figure 2. Design of the system.

Figure 2 is a miniature of 2 crossroads and 1 T junction in an urban area with the maximum speed of the vehicle is 50 km / h with a road width of 7 meters and 3 meters, with a road length of 25 kilometers. Which equipped 3 sensors at a certain distance to the traffic signals. The first sensor indicates the route is normal, the second sensor indicates the route is jammed, and the third sensor indicates total traffic jam.

2.1. Hardware
The hardware for this system using some components includes: Phone/Device, NodeMCU (ESP8266 12E), IR obstacle Sensors, Resistors, LED etc.

2.1.1. NodeMCU (ESP8266 12E). ESP8266 is a microcontroller designed for Expressive Systems. ESP8266 is a solution for Wi-Fi networks from existing Micro Controllers to Wi-Fi and is also capable of running standalone applications. Connection with PC using the micro USB cable and there are 17 GPOI, with a consumed current of 10uA ~ 170mA and RAM of 32K + 80K [10]. It is designed for wireless location-aware devices, wireless positioning system signals, industrial wireless control, etc. ESP8266 is used to process and transfer information to the web server so that the smartphone can access the information.

2.1.2. IR obstacle sensors. It detected object range of 10-30 cm. This sensor can be used for most indoor applications where no important ambient light is present [11]. The IR sensor will used to sense the road traffic congestion.

2.1.3. Step down module (MB-V2-102). It is a voltage-reducing module and current stabilizer, the input voltage for the module is 6.5V-9.0V DC from datasheet [12]. While the output voltage of 3.3V Dc and 5 V Dc with current 700mA. The module can use a USB or jack from the adapter.

2.1.4. LED. A light-emitting diode (LED) is a semiconductor device that emits visible light when an electric current passes through it. In this system, LED is a miniature of traffic light system in the real life.

2.1.5. Resistors. It is a passive two-terminal electrical component. It implements electrical resistance as a circuit element and acts to reduce current flow. Also, act to lower voltage levels within circuits, at the same time.

2.2. Software
Software module for developing a system using various software tools which includes: Arduino IDE, MIT App Inventor, and Firebase.

2.2.1. Arduino IDE. Arduino is programmed with a C/C++. Then Arduino IDE software is written in Java based on the processing project from writing and uploading the code to the Atmega chip in Arduino. It is compatible software to program with some module such as ESP.

2.2.2. Firebase. It is middleware between the hardware module and the end user. Firebase is platform for internet of things for web server with open big data from sensors. Which is MQTT platform (Message Queuing Telemetry Transport), it is a protocol that runs on top of the TCP / IP stack and has a small data packet size with a low overhead (minimum 2 bytes) so that it affects the consumption of the power supply is also quite small.

2.2.3. MIT app inventor. The first one the application will receive all the sensor values and transmit/ upload that sensor values to the server another app will use to monitor the traffic condition. The software is interfacing from android smartphone user with a web server, so user able to see the condition of the road from a smartphone.

IOP Conf. Series: Materials Science and Engineering **384** (2018) 012024 doi:10.1088/1757-899X/384/1/012024

3. Results and discussion

In this section discussion about the design overall of the system. This screenshots of the Android application are as follows:

Figure 3. Android application.

Figure 4. Arduino IDE program.

IOP Conf. Series: Materials Science and Engineering **384** (2018) 012024 doi:10.1088/1757-899X/384/1/012024

The system was adapted from the Android application as shown in figure 3 was developed using App inventor software to connect with Wifi module, web server and display given sensor values which are collected from the hardware module and finally upload this values on the web server. This developed android app will active on any android device and it will always connect to the hardware module. Figure 4 showed the program from Arduino IDE to write in the microcontroller NodeMCU, it is important to program for this research because of all program from the system in here.

In the Android application showed some button for user requirement, the application provided some option for monitoring traffic light in the first crossroads or second crossroads or T junction in some condition, also user can select another button for monitoring traffic density. The application can show the condition of the route, the user chooses the location and destination in this system then application show how condition in the route. If the route is jam the system will give some optional routes.

In figure 5 showing the hardware system for monitoring traffic light. This prototype has several that is the mismatch between the goals with the results achieved. These constraints include:

- In the planning of the prototype using NodeMCU ESP8266 but I/O from the microcontroller is less than other microcontroller but NodeMCU stabil on connectivity.
- The output voltage from NodeMCU is 3V then the sensors need 5 volts to work and has a large current consumption, which that requires a separate adapter. Then NodeMCU did not have a voltage stabilizer so ESP always does the reset, to reduce it needed a voltage stabilizer using a step down the module.
- NodeMCU support with another web server, so that can be modified the web server or MQTT broker. And NodeMCU can programme with LUA. And for the Android application can use AndroidApp, its support for NodeMCU.
- The sensor can not implement in the real life because of many disadvantages, readability of the sensor so small from 0-30 cm. The sensor can not detect glass objects.

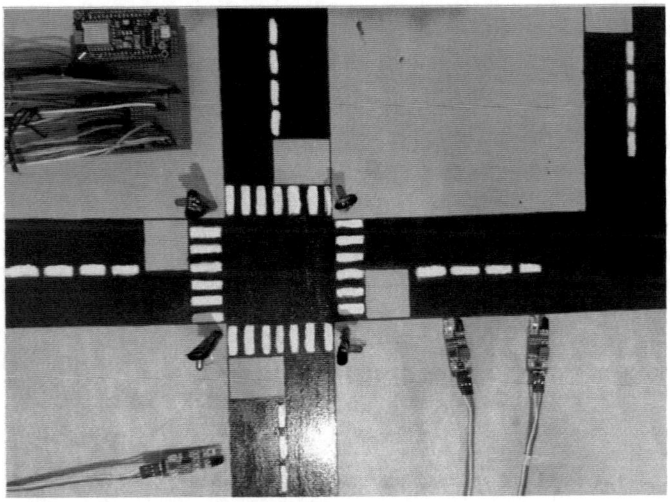

Figure 5. Hardware system.

4. Conclusions

The design of monitoring traffic light system on this research is not too much developing as a whole. Need future research about ITS. The monitoring traffic congestion system was presented traffic connection routes with IoT to develop a real-time software for managing traffic conditions. This system used a web server and an Android client application for reporting and displaying traffic condition. The system for administrators includes a web application, and for the interface of the client application, it is easy and safe to use while driving. The focus on this research about efficiency and traffic density using ITS and IoT technologies.

References

[1] Wang Y and Qi H 2012 "Research of Intelligent Transportation System based on the Internet of Things frame" *Wirel. Eng. Technol.* **3** (3) pp. 160–166.

[2] Xu X, Gao X, Zhao X, Xu Z and Chang H 2016 "A novel algorithm for urban traffic congestion detection based on GPS data compression" *Int. Conf. Serv. Oper. Logist. Informatics* pp. 107–112.

[3] Wu N and Fang H 2017 "A Novel Traffic Light Recognition Method for Traffic Monitoring Systems" *2nd Asia-Pasific Conference on Intelligent Robot System (ACIRS)* pp. 141–145.

[4] Chepuru A and Pradesh A 2015 "a Survey on Iot Applications for Intelligent Transport Systems" (8) pp. 116–127.

[5] Pyykonen P, Laitinen J, Viitanen J, Eloranta P and Korhonen T 2013 "IoT for intelligent traffic system" *Proc. - 2013 IEEE 9th Int. Conf. Intell. Comput. Commun.* pp. 175–179.

[6] Sherly J and D.Somasundareswari, "Internet of Things Based Smart Transportation Systems," *Int. Res. J. Eng. Technol.* **2** (7) pp. 1207–1210.

[7] Nor, Mohd R F A, Zaman F H K and Mubdi S 2017 "Smart Traffic Light for Congestion Monitoring using LoRaWAN" *IEEE 8th Control and System Graduate Research Colloquium (ICSGRC)* pp. 4–5.

[8] Choosri N 2015 "IoT-RFID Testbed for Supporting Traffic Light Control" *Int. J. Inf. Electron. Eng.* **5** (2) pp. 102–106.

[9] Perumalla B K and Babu M S 2016 "An Intelligent Traffic and Vehicle Monitoring System using Internet of Things Architecture" *International Journal of Science and Research* **5** (11) pp. 2013–2016.

[10] U M V "ESP8266 Handson Technology" pp. 1–22.

[11] De Nunzio G and Marsella G "FC-51 : IR Infrared Obstacle Detection Sensor Module 2 - 30cm" pp. 0–8.

[12] P. Specifications, "Breadboard Power Supply Specifications," no. Dc, pp. 1–2, 2014.

IOP Conf. Series: Materials Science and Engineering **384** (2018) 012025 doi:10.1088/1757-899X/384/1/012025

Design of Battery Charging Control System (Fast Charging) on Electric Vehicle based on Atmega32 Microcontroller

H A Hidayah and M Somantri*

Department of Electrical Engineering Education, Universitas Pendidikan Indonesia, Jl. Dr. Setiabudhi 207, Bandung 40154, Indonesia.

*msomantri@upi.edu

Abstract. Electric Vehicles are vehicles that do not release toxic gases that can reduce environmental pollution caused by combustion. The electric vehicle requires a lithium ion battery as a replacement energy source of fuel oil so it needs charging a battery that can speed up charging safely. This research uses fast charging method through high current method and adjusts the current capacity needed by the battery in accordance with the result of circuit specification. This research uses rectifier circuit, buck converter, Atmega microcontroller. The results of this study are expected to increase current capacity and speed up the process of charging the battery.

1. Introduction

The development of automotive technology is growing rapidly around the world [1]. Automotive technology generally uses an internal combustion system or ICE (Internal Combustion Engine) as its prime mover, but the impact of this combustion system causes air pollution or exhaust fumes that disrupt human breath and damage the environment. Along with the development era of many scientists from various countries and manufacturers engaged in the field of automotive that has applied electric propulsion technology that is electric motors used for electric vehicles. On the continent of Europe already almost all transportation vehicles use electricity. Not less behind in Asia, especially Japan and China, the country has long ago developed this friendly environmental transportation technology and the State has been able to create the main driving force that is Electric Motors, Indonesia is also now - re-develop electric cars, one of them namely the electric car named Selo type of electric vehicle is a vehicle intended for high-speed.

Electric vehicles require lithium-ion batteries to store energy sources to drive electric motors. Lithium-ion batteries are rechargeable batteries and can be used many times. These batteries consist of positive electrodes containing $LiCoO2$ and negative electrodes made of carbon graphite (C6) and separator made of a thin layer of plastic that can be passed by ions. This type of battery is strongly recommended for storage of energy used electric vehicles, because Lithium is a highly reactive element, meaning that bias stores more energy in bonds atoms. Lithium-ion battery loses only 5% of its content every month meaning this battery is durable and very suitable for energy source of electric vehicle [2],[3].

In order to develop an electric car, the need for fast and safe battery charging, therefore in this study examined about the design of charging control system (fast charging) on electric vehicle based on Atmega32 microcontroller. The design of this battery charging design has two charging conditions

IOP Conf. Series: Materials Science and Engineering **384** (2018) 012025 doi:10.1088/1757-899X/384/1/012025

that are at the time of normal charging condition and at the time of full charging condition by using high current (fast charging) [4]. This method can be achieved in a simple way without using complex mathematical calculations, by using a MOSFET control based on Atmega32 microcontroller [5]. Microcontroller will control high voltage and current through PWM buck converter when charging electric car battery.

2. Methods

2.1. Block diagram design of fast charging circuit

Figure 1 illustrates our design related to fast battery charging system. As shown in Figure 1, we utilized microcontroller, driver MOSFET, buck converter DC to DC, and voltage detector and current sensors for supporting the efficiency of our proposed design.

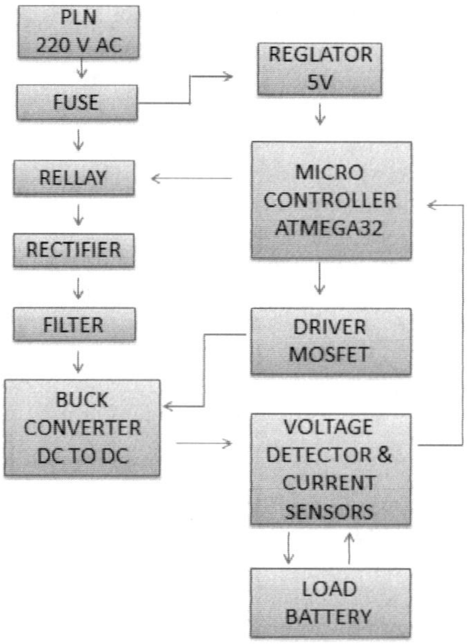

Figure 1. Design diagram of fast battery charging system.

2.2. Design of buck converter circuit

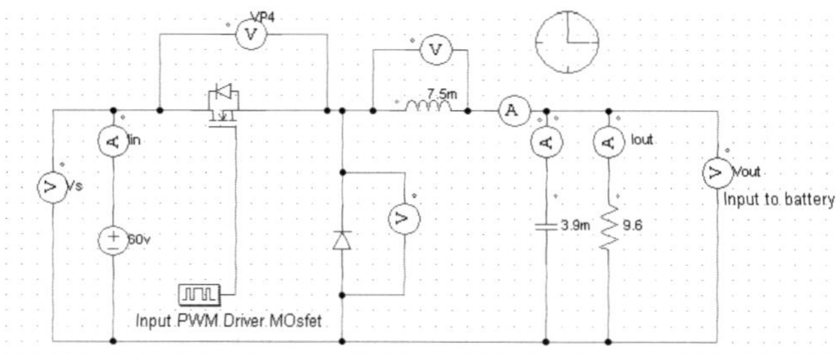

Figure 2. Schematic buck converter circuit.

The buck converter circuit on Figure 2 has functions to convert DC voltage into DC voltage which is variable. The working principle of Buck-Converter is by using a continuous switch (ON-OFF) known as PWM (Pulse Width Modulation) and Duty Cycle in controlling the speed of the switch's work.

Calculation duty cycle:

$$D = \frac{t_{on}}{t_{on} + t_{off}} = \frac{t_{on}}{T}$$
$$t_{on} = DT$$
$$t_{off} = (1 - D)T$$
$$L = \frac{V_S - V_R}{\Delta I} t_{on}$$

Calculation of inductor value:

$$L = \frac{V_S - D\,V_S}{\Delta I} t_{on} \leftrightarrow L = \frac{V_S\,(1 - D)\,D}{\Delta I\ f}$$

$$\Delta V = V_{\max} - V_{\max}\,(e^{-\frac{1}{RC}t_{off}})$$
$$= V_{\max}\,(1 - e^{-\frac{1}{RC}t_{off}})$$
$$= V_{\max}\left(\frac{t_{off}}{RC}\right)$$
$$\Delta V = V_{\max}\left(\frac{1 - DT}{RC}\right)$$
$$C = V_{max}\,\frac{1-D}{f\,R\,\Delta V} \qquad ;\ V_{\max} = V_R + \frac{\Delta V}{2}$$
$$C = \left(D\,V_S + \frac{\Delta V}{2}\right)\frac{1 - D}{f\,R\,\Delta V}$$

Calculation of capasitor value

$$\Delta V = V_{max} - V_{min}$$
$$V_{\max} = V_R + \frac{\Delta V}{2} \quad ...(1) \quad \text{Capacitor voltage when charging}$$
$$V_{\min} = V_R - \frac{\Delta V}{2} \quad ...(2) \quad \text{Capacitor voltage when discharging}$$
$$\Delta V = V_{\max} - V_{\max}\left(e^{-\frac{1}{RC}t_{off}}\right) \qquad ;\ V_{min} = V_{\max}\left(e^{-\frac{1}{RC}t_{off}}\right)$$
$$= V_{\max}\,(1 - e^{-\frac{1}{RC}t_{off}})$$
$$= V_{\max}\left(\frac{t_{off}}{RC}\right)$$
$$\Delta V = V_{\max}\left(\frac{1 - DT}{RC}\right)$$
$$C = V_{max}\,\frac{1 - D}{f\,R\,\Delta V}$$
$$C = \left(D\,V_S + \frac{\Delta V}{2}\right)\frac{1 - D}{f\,R\,\Delta V}$$

2.3. Designing MOSFET driver circuit

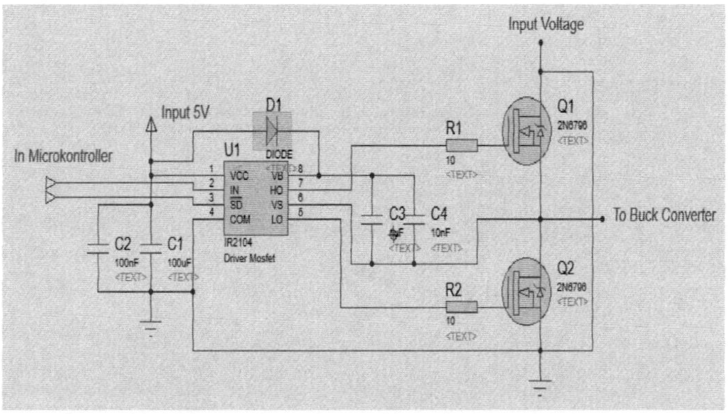

Figure 3. Schematic mosfet driver circuit.

This circuit depicted in Figure 3 serves as a liaison between the microcontroller with a buck converter and as a separator between power circuits with control circuit. The goal is to avoid damage to the microcontroller in case of voltage behind the buck converter.

2.4. Designing a series of relay modules

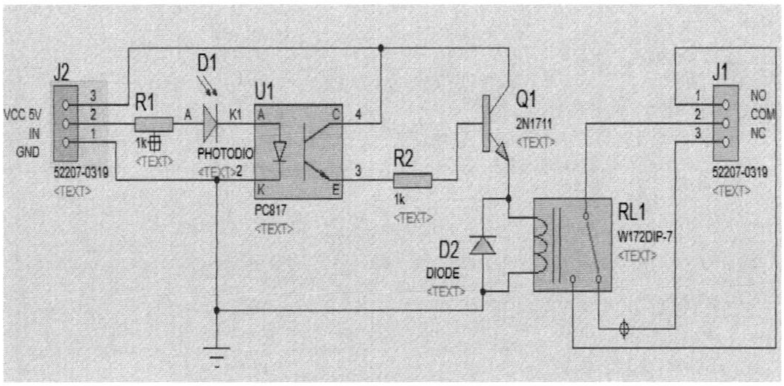

Figure 4. Schematic of relay modules.

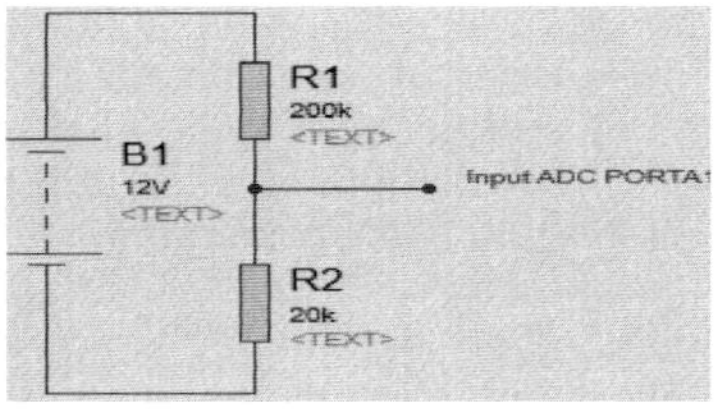

Figure 5. Schematic of voltage sensors circuit.

The relay circuit in Figure 4 serves to disconnect and connect the voltage from the voltage source to the battery fast charging circuit set by the microcontroller whose input adjusts from the voltage sensor and current sensor.

2.5. Design of voltage detector

This voltage divider circuit in Figure 5 works for voltage divider or voltage sensor circuit, because microcontroller system can not accept voltage more than 5 volts and the circuit can measure DC voltage of 48V – 54V to 0V – 5V.

2.6. Design of current sensor circuit

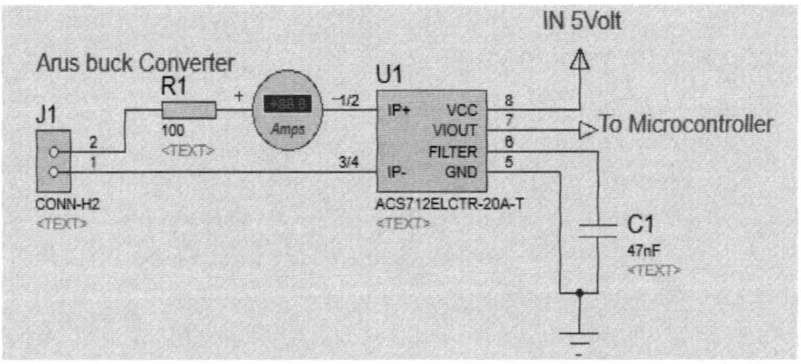

Figure 6. Schematic of current sensor circuit.

This series of current sensors serves as a safety and monitoring current that flows on the battery charging circuit.

2.7. System design of Atmega32A microcontroller

The design of the minimum system set of atmega32A functions as the brain of all battery charging devices. Microcontroller Atmega 32A consists of PORTA, PORTB, PORTC and PORTD, each PORT different function. And the schematic of microcontroller Atmega32 circuit is shown in Figure 7.

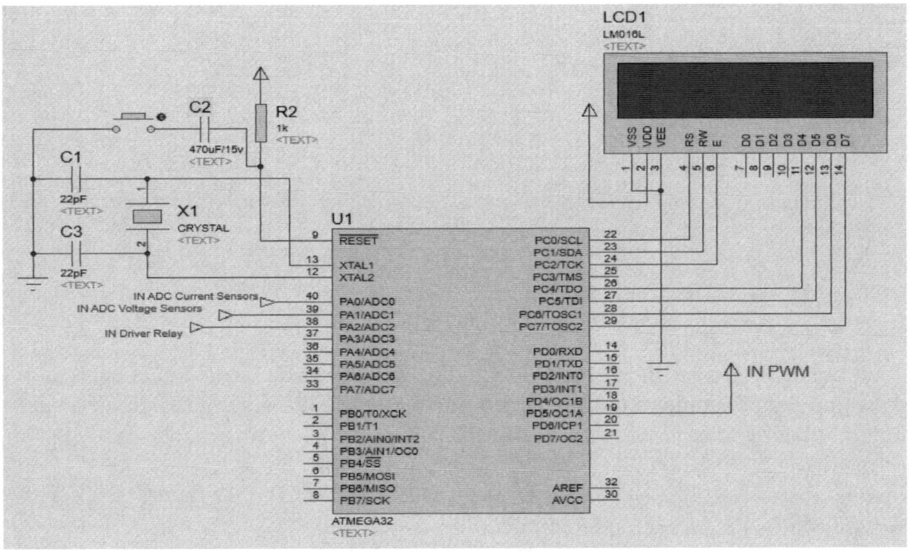

Figure 7. Schematic of microcontroller Atmega32 circuit.

3. Results and Discussion

Here is a quick battery charging device test that uses voltage changes through the buck converter set by frequency changes using the mosfet component. Current sensors and voltage sensors in the input to the system of this circuit device are to determine the condition of the battery. And the battery capacity is in full or empty and also the sensor as a safety at the time of charging the battery take place. All devices are in control using Atmega32 microcontroller. The real of our product design is shown in Figure 8.

Figure 8. Fast charging device.

Testing device tool is done by testing all parts of each series of simulation results and measurements of the circuit. Our experiments show the results in Figure 9 and Figure 10 which include:

3.1. Rectifier and filter circuits

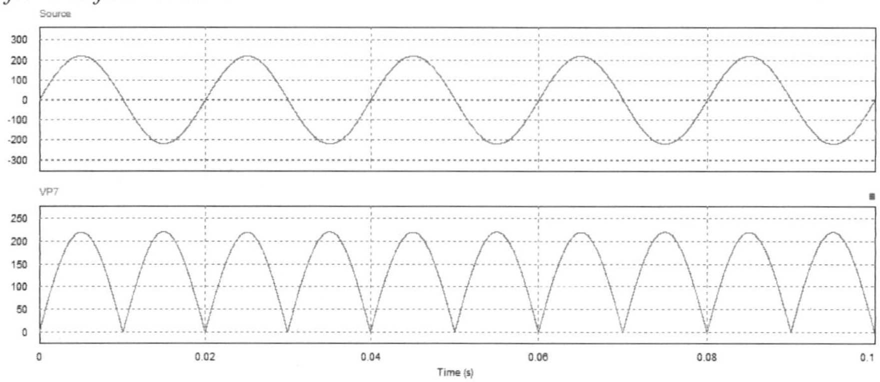

Figure 9. Wave simulation results.

This circuit uses a 2A bridge diode component, 220 AC input voltage and is converted to 220 DC. Figure 9 shows the input simulation of power line, while Figure 9 presents the signals after filtering. The output of regulator circuit is shown in Figure 10. The parameters for simulation are listed in Table 1.

3.2. Regulator circuit

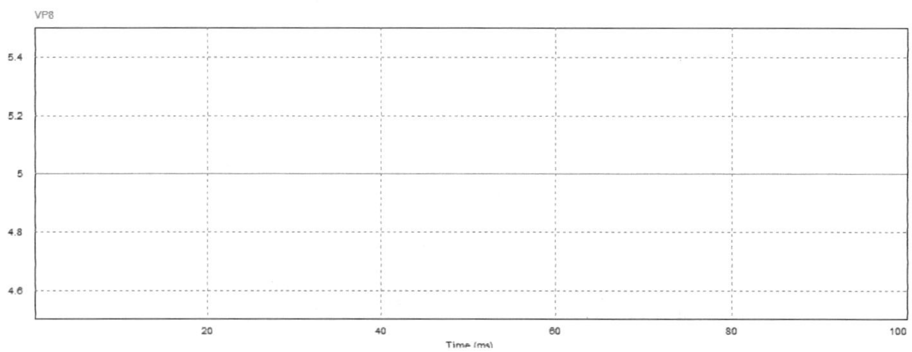

Figure 10. Output voltage waves.

3.3. Buck converter

Table 1. Main Circuit Parameter.

Parameter name	Symbol	Value
Input voltage	Vin	220 V
Output voltage	Vout	54 V
indutor	L	7.5mH
Capacitor	C	7.9uF
Freqency	F	50000
resistance	R	9.6 Ohm

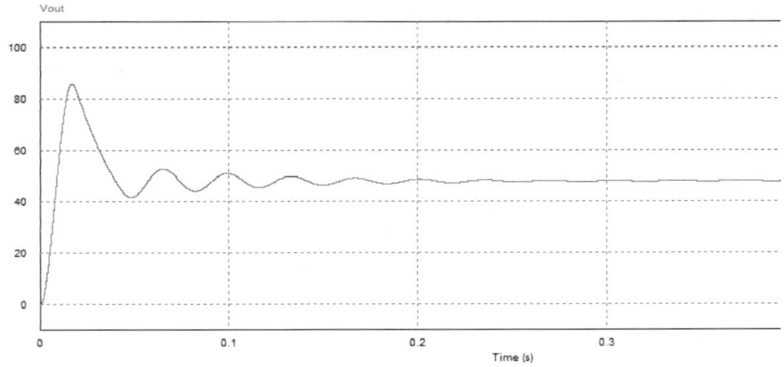

Figure 11. Voltage output buck converter.

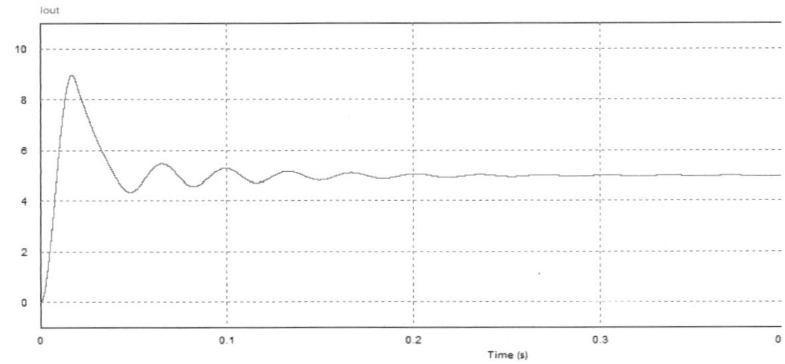

Figure 12. Current Output buck converter.

The above wave simulation is the result of the buck converter circuit with the value of the component that has been specified (Figure 11 and Figure 12).

3.4. Microcontroller Atmega32A

Table 2. Port-specific functions.

PORT	Function
VCC	IN + DC 5V
GND	IN – DC 5V
PORT A0	IN Current Sensors
PORT A1	IN Voltage Sensors
PORT B	Dwonloader Program
PORT C	Output LCD
PORT D0	Output Module Relay
PORT D1	Output High Driver Mosfet
PORT D2	Output Low Driver Mosfet

For testing microcontroller, systems need to unite all the sensor modules and all the driver modules that include: current sensor module, voltage sensor, MOSFET driver and relay driver. Relay works when receiving input I / O from sensors and in process by microcontroller using C language program using CV AVR software. Table 2 lists the functions of ports in microcontroller.

Figure 13 and 14 show the relation of voltage and current over time. While, using lithium-ion battery with 10 Ah and 5 Ah charger is compared as listed in Table 3.

Table 3. Battery charge measurement using 5A charging device.

Type Battery	Output Voltage	Output Current	Long battery charge time
Lithium –	46 V	4.88 A	0 Minute
Ion	47 V	4.88 A	12,5 Minute
10 Ah	48 V	4.88 A	25 Minute
	49 V	4.88 A	37,5 Minute
Charger	50 V	4.88 A	50 Minute
5A	51 V	4.88 A	62,5 Minute
	52 V	4.88 A	75 Minute
	53 V	4.88 A	87,5 Minute
	54 V	0.78 A	100 Minute
	54.28 V	0.66 A	112,5 Minute
	54.28 V	0.59 A	125 Minute
	54.28 V	0.51 A	137,25 Minute
	Times		2 Hour 12.5 Minute

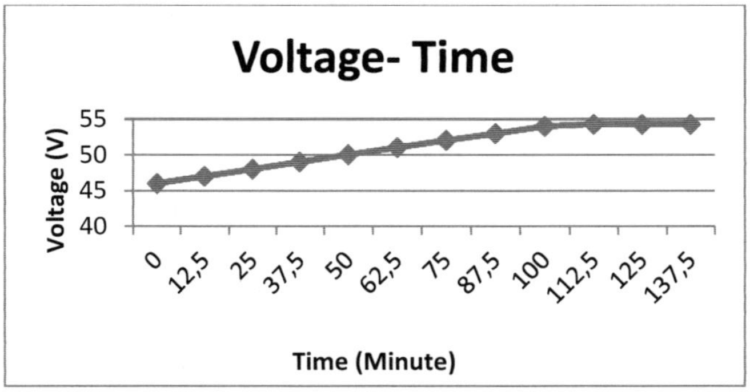

Figure 13. Chart of Voltage – Time.

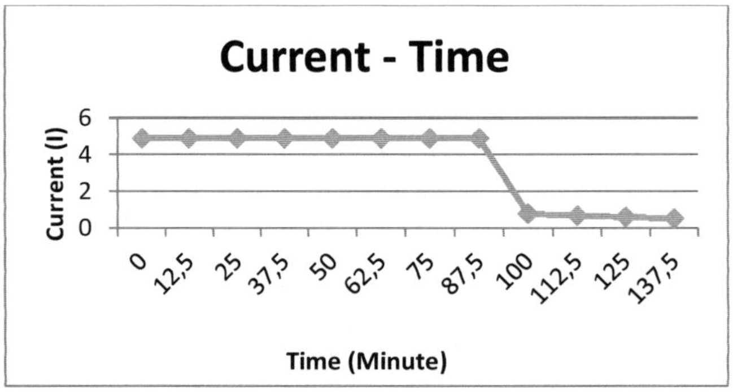

Figure 14. Chart of Current – Time.

4. Conclusions

The conclusion of the design is that the device is not maximal functioning properly because there is one circuit device that does not work properly, namely the power circuit. This circuit consists of various circuit-shaped circuit modules from the start of the sensor and driver circuit modules. Each circuit module and sensor system is tested one by one for the device until work properly. However, at the time of testing or simulation on the device, there are several circuit systems that do not match with the simulation in the software. Like MOSFET in buck converter circuit system is very susceptible to heat because of fast on-off switch for voltage. On this project, the voltages and currents are 48-54V and 4.88 - 0.51 A respectively of buck converter with a battery charging time spend around 2.5 hours from a blank status until the battery is full.

References

[1] Friedrich H and Schumann S 2001 Research for a "new age of magnesium" in the automotive industry *Journal of Materials Processing Technology* **117** (3) 276-281
[2] Lu L, Han X, Li J, Hua J and Ouyang M 2013 A review on the key issues for lithium-ion battery management in electric vehicles *Journal of power sources* **226** 272-288
[3] Thackeray M M, Wolverton C and Isaacs E D 2012 Electrical energy storage for transportation—approaching the limits of, and going beyond, lithium-ion batteries *Energy & Environmental Science* **5** (7) 7854-7863
[4] Zhang P, Sadler C M, Lyon S A and Martonosi M 2004 Hardware design experiences in ZebraNet *Proceedings of the 2nd international conference on Embedded networked sensor systems* (pp. 227-238) ACM
[5] Dewangan A K, Chakraborty N, Shukla S and Yadu V 2012 PWM based automatic closed loop speed control of DC motor *International journal of engineering trends and technology* **3** (2) pp.110-112

IOP Conf. Series: Materials Science and Engineering **384** (2018) 012026 doi:10.1088/1757-899X/384/1/012026

Prototype of Pico Hydro – Solar Photovoltaic Hybrid System

I W Ratnata*, S Sumarto and W S Saputra

Faculty of Technology and Vocational Education, Universitas Pendidikan Indonesia, Bandung, Indonesia

*wratnata@gmail.com

Abstract. In this research will be developed a prototype of Hybrid Power Plants (HPP) which integrates pico-hydro power and Solar Photovoltaic (SPV) system. This HPP prototype will be developed in two stages; the first stage or prototype-1 has a capacity 100 - 200 watts (W) and the second stage or prototype-2 with a capacity of 500-1000 watts. Prototype-1 was used as a study and evaluation object for the development of prototype-2 in hopes of producing better output. From the development of prototype-1 it is known that pico-hydro generates a maximum power 86 W (from a capacity 100 W) and the SPV produces a maximum power 30 W in sunny weather and 23 W in cloudy weather (from capacity 100 W). In this prototype-1, the electric energy of pico-hydro in the form of alternating current electric energy, after being converted into direct current electric energy, then combined with the output of the SPV (in the form of direct current electric energy) and stored in the battery (accumulator). At this combination, the input voltage to the battery is in the range 12-13 V and the current 2-3 A.

1. Introduction

According to 2015 Electricity Statistics [1], up to the end of 2015 the installed capacity of power plants in Indonesia reaches 55,528.10 MW consisting of PLN power plants amounting to 38,314.23 MW and Non-PLN of 17,213.87 MW; compared to 2014 of 53,065.50 MW, the installed capacity of power plants increased by 2,462.60 MW or 4.64%. The installed capacity of national power plants based on the type of power plants in 2015 can be seen in Figure 1.

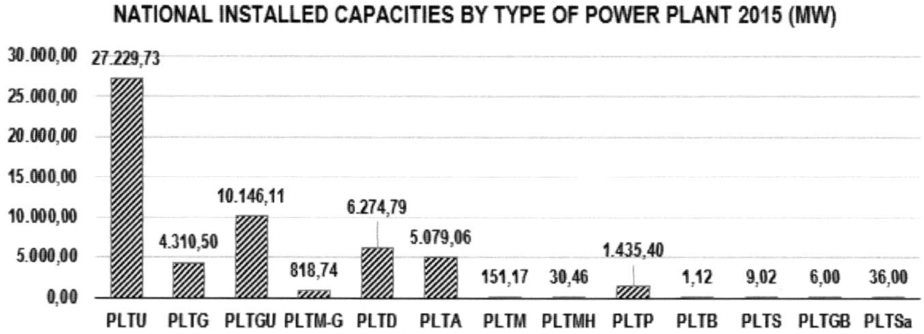

Figure 1. National installed capacities by type of power plant 2015 (MW).

Based on the data in Figure 1, it can be seen that the dependence of power plants in Indonesia on fossil fuels (such as coal, petroleum, and natural gas) is enormous. By 2015, the installed capacity of fossil fuel-based power plants is 48,779.87 MW or 87.85% of total installed capacity. In fact, as is known, fossil fuels are so limited that at some point they will run out, and their effects can contaminate and even damage the environment. Based on the above, it is necessary to think about alternative energy supply that has character:

- Can reduce dependence on the use of fossil energy, especially coal and oil;
- Can provide electrical energy in a regional local scale;
- Able to exploit the potential of local / local energy resources, as well as
- Environmentally friendly, in terms of production processes and disposal of its products do not damage the surrounding environment.

An electrical energy supply system that can meet the above criteria is an energy conversion system that utilizes new and renewable energy (NRE) resources [2]. Renewable energy resources such as geothermal energy, water, bioenergy, solar, wind, and currents/ocean waves; while those including new energy include liquefied coal, coal methane gas, depressed coal, nuclear, and hydrogen. Up to 2015, the installed capacity of new NRE-based power plants is about 6,748.23 MW or 12.15% of total installed capacity [1].

To meet the electricity needs of the people of Indonesia from Sabang to Merauke, the government has committed to realize the supply of electricity amounting to 35000 Megawatts (MW) within 5 years (2014-2019). In that time period, the government together with PLN and the private sector will build 109 plants; each comprising 35 projects by PLN with a total capacity of 10,681 MW and 74 projects by private/Independent Power Producer (IPP) with a total capacity of 25,904 MW. In 2015, PLN will sign a power plant contract of 10,000 MW as the first phase of a total of 35,000 MW. In this 35,000 MW program, the share of NRE-based power generation is 25% or 8,750 MW, with details: 1.751 MW geothermal (20%), hydro 2,438 MW (28%), biofuels 1,156 MW (13%), and solar, wind, sea and other NRE 3,405 MW (39%). The required budget is USD 29.8 billion or approximately Rp.402 trillion.

In relation with new and renewable energy source, just behind the Electricity Engineering Laboratory, the Department of Electrical Engineering Education, Indonesia University of Education, there is a small river in which according to the observation, potency the flow of water can be utilized for power plants with the power output of tens to hundreds of watts. Based on this, in this research will be conducted a study and trial to develop a power plant by utilizing potency of the river water flow, in the form of pico-hydro power plant. Furthermore, this power plant will be combined with solar power plant or Solar Photovoltaic (SPV) which will also be developed at the site to form a prototype of Hybrid Power Plants (HPP).

2. Literature review

The hydro power plant is a power plant that uses water energy as its original energy source. The mechanical energy contained in the water energy is used to rotate the turbine; this rotating turbine is connected to the generator to produce electrical energy [3],[4].

Solar power plant is a power plant that converts solar energy into electrical energy. Power plants can be done in two ways, namely directly using photovoltaic and indirectly with the concentration of solar energy. Photovoltaics directly converts light energy into electricity using the photoelectric effect. The concentration of solar energy using a lens or mirror system is combined with a tracking system to focus solar energy to a point to move the engine of heat [5].

Hybrid Power Plants (HPP) is a power plant system that combines several types of power plants, in general between fuel-based power plants and renewable energy-based power plants. Generally, power generating systems that are widely used for HPP are diesel generators, solar photovoltaic (SPV), microhydro, and power plant. HPP is a solution to overcome the fuel crisis and the lack of electricity in remote areas, small islands and also in urban areas [6]. The illustration of hybrid photovoltaic-wind energy system is shown in Figure 2.

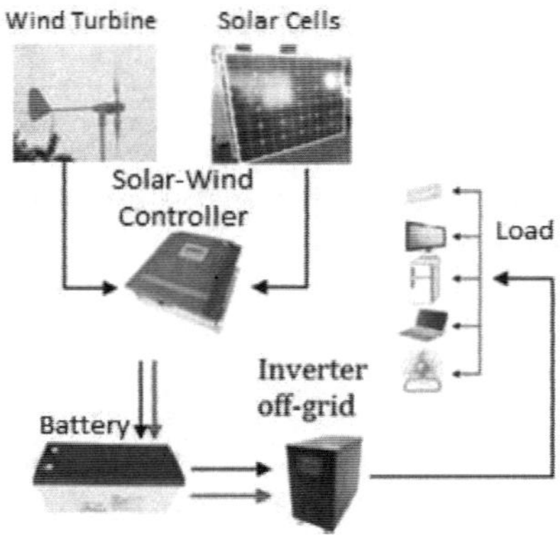

Figure 2. Hybrid photovoltaic-wind energy system.

3. Research Methods

This research is a descriptive study that discusses the provision of renewable electric energy through the manufacture of prototype of Pico Hydro-Solar Photovoltaic hybrid power plant, at the Electrical Engineering Laboratory, Indonesia University of Education. This research was conducted through making of prototype of Pico Hydro-Solar Photovoltaic power plant by utilizing small river water flow around the laboratory. Solar Photovoltaic is installed on the roof of the laboratory building, where the existing area is sufficient for the installation of photovoltaic electrical panels.

The research related to the manufacture of prototype of PLTPH-PLTS hybrid power plant is done with the following steps:

- Literature review, potential analysis and feasibility study of pico-hydro power plant;
- Literature review, potential analysis and feasibility study of solar power plant;
- Collection of relevant data and information;
- Making lay out of hybrid power plant: Pico-Hydro – Solar Photovoltaic;
- Civil construction for turbine open flume 100 watts and 600 watts;
- Procurement of 100 watt and 600 watt electricity generators and their control systems;
- Pico-hydro Power Plant trial;
- Civil construction for the installation of solar cell;
- Procurement of solar cells, control systems and batteries;
- Solar Power plant trial;
- Analysis of energy forecasts to be generated from HPP system;
- Supply transformer and AC to DC converter (rectifier), as well as DC to AC converter (inverter);
- Installation of HPP: Pico-Hydro – Solar Photovoltaic;
- Testing and Commissioning test;
- Preparation of reports.

4. Results and discussion

Currently it has been made prototype Hybrid Power Plants (HPP) which combines Pico-Hydro Power Plants and Solar Photovoltaic (SPV) as diagramed in Figure 3. This prototype is a prototype-1, which uses generators and solar panels with a small capacity. The performance of Prototype-1 afterward measured and evaluated as input to develop prototype-2 with greater capacity.

In this prototype-1, pico-hydro power plant is built using an open flume turbine with a capacity 100 W, while solar photovoltaic uses solar cells with a capacity 2 x 50 WP (watt peak). Currently, pico-hydro power plant produces a maximum power of 86 W, while the solar photovoltaic generate a maximum power of 30 W in sunny weather and 23 W in cloudy weather.

IOP Conf. Series: Materials Science and Engineering **384** (2018) 012026 doi:10.1088/1757-899X/384/1/012026

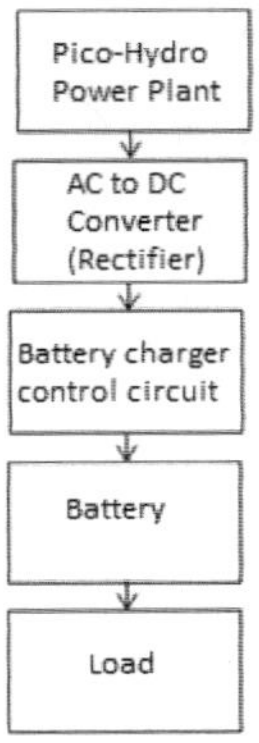

Figure 3. Block diagram of pico-hydro power plant.

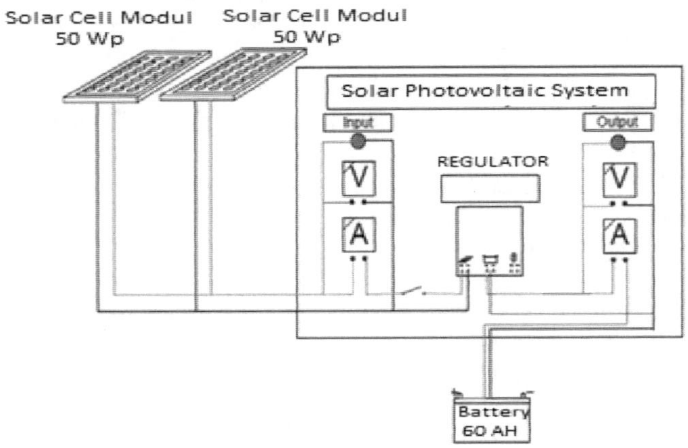

Figure 4. Wiring diagram of solar power plant.

Electrical energy sourced from pico-hydro, once set through Auto Load Control (ALC) then rectified using rectifier, and then stored in battery (accumulator). Electrical energy derived from the solar photovoltaic, firstly stabilized through the regulator, directly stored in the accumulator. For use of loads requiring an alternating current (ac) power source, the electrical energy stored in the accumulator is first converted through the inverter; while for direct current (dc) electric load can directly use electrical energy from accumulator. Figure 4 shows the wiring diagram of solar power plant in this study.

5. Conclusions

- Utilization of small river water flows behind the Electrical Engineering Laboratory, Indonesia University of Education using open flume turbine TC-60 with a capacity 100 W produces electricity with a maximum power of 86 W. Muddy water and a lot of waste is an obstacle in the utilization of river water.
- The muddy water causes turbine rotation decreases and effected to generator rotation as well and can causes decreasing of the power produced by generator become roughly 66 W; even impact the waste or rubbish disruption could result in the generator not operating at all.
- Utilization of solar energy into electrical energy using solar cells 2 x 50 WP in this study has not yet been optimal. The total maximum power generated is 30 W in sunny weather and only 23 W in cloudy weather.

- In this study, the electrical energy generated from the utilization of water flow around electrical engineering lab in the form of pico-hydro power and the utilization of solar energy in the form of Solar Photovoltaic (SPV) System is integrated into a Hybrid Power Plant (HPP). After going through several settings and conversions, the energy from both sources is stored in the battery. At this combination, the input voltage to the battery is in the range of 12-13 V and the current of 2-3 A.

References

[1] Directorate General Electricity, Ministry of Energy and Mineral Resources of Republic of Indonesia 2016, Electricity Statistics 2015.
[2] Hong T, Koo C and Kwak T 2013 Framework for the implementation of a new renewable energy system in an educational facility *Applied energy* **103** pp 539-551
[3] Paish O 2002. Small hydro power: technology and current status *Renewable and sustainable energy reviews* **6** (6) pp. 537-556
[4] Hadjipaschalis I, Poullikkas A and Efthimiou V 2009 Overview of current and future energy storage technologies for electric power applications *Renewable and sustainable energy reviews* **13** (6-7) pp.1513-1522
[5] Pavlović T M, Radonjić I S, Milosavljević D D and Pantić L S 2012 A review of concentrating solar power plants in the world and their potential use in Serbia *Renewable and Sustainable Energy Reviews* **16** (6) pp.3891-3902
[6] Arsie I, Marano V, Nappi G and Rizzo G 2005 A model of a hybrid power plant with wind turbines and compressed air energy storage *ASME 2005 Power Conference* (pp. 987-1000) American Society of Mechanical Engineers

Low Cost Handheld Digital Oscilloscope

I Fushshilat[1,*] and D Barmana[2]

[1]Department of Electrical Engineering Education, Universitas Pendidikan Indonesia, Bandung, Indonesia
[2]Departement of Mechanical Engineering Education, Universitas Pendidikan Indonesia, Bandung, Indonesia

*imanf@upi.edu

Abstract. This paper discusses the design and development of handheld digital oscilloscope at low cost using embedded microcontroller system. Oscilloscope is an important instrument that must exist in every electrical and electronic laboratory, but the oscilloscope price is relatively expensive so the purpose of this paper will be describe of research to create oscilloscope using a low cost digital system. Unlike general oscilloscopes that have large dimensions, the oscilloscopes planned and built in this research will have smaller dimensions with smaller packages, so the oscilloscope will be easy to held and mobile, easy to carry anywhere, and easy to use. Performance will be measured from capability of accuracy measurement parameter and efficiency will be describe about cost and other parameter specification are obtained from comparation with other existing oscilloscope. This Research is using experimental method, outcome from research is prototype of low cost handheld digital Oscilloscope.

1. Introduction

Oscilloscope is a basic needs at engineering laboratory which is used on telecommunication, electrical, electronic analog, or digital fields. Oscilloscope is different from any measurement instruments, because oscilloscope is able to show visualization from electrical signal form which it can make user could analysis further and get deeper comprehension. Digital oscilloscope is better than analog oscilloscope (CRT) seen from the shape, efficiency capability, and storage capability [1].

There are various kinds of standalone oscilloscope product from any brand which has a high price. As an example, the price of Fluke digital oscilloscope is around 1000-1500 US $ [2]. Another product like Digilent Analog Discovery Kit and myDAQ which has been made by National [3, 4]. The instruments are fulfill the student needs, but it need a computer to show the output and data processing. In another research before, couple electrical or electronic measurement instruments are using computer help as data processing and also showing output data which has been made [5]. Digital oscilloscope in virtual form using Visual Basic (VB) software which is installed on computer has been successfully created [6].

Digital oscilloscope also can make by using Field Programmable Gate Arrays (FPGA) which is a compact electronic board. Inside the FPGA there is a Microprocessor which has been equipped with various Input/Output (I/O) and programmable. FPGA has embedded system form that could be more practical for user because it is standalone. After it has been programmed using computer for special purpose, FPGA is already working well alone without any computer, but for the FPGA board price is expensive than other programmable embedded system. The research about designing oscilloscope using

FPGA has been done. Where Terasic DE0 used as a hardware and Quartus II as a software. For the result, this design could measure and display the data with chart form from any signal which produced by Function Generator. And it can display the monitor output screen with 640x480 resolution [7].

This research is focused on discussing the design and the construction of portable oscilloscope which has a small size and can be grab with hand. This oscilloscope has many advantages such as practical, easy to used, easy to carried anywhere, and it is using low power than others oscilloscope.

The oscilloscope is designed by using an embedded system namely Arduino. Arduino is an open-source electronics platform based on easy-to-use hardware and software. It's intended for anyone making interactive projects [8]. Arduino can be developed to become the main brain from various electronic device according with the purposes of the program and the design made by the user. The advantages of Arduino from any others platform embedded system is more cheaper from the price aspect, easy to used, and there are many literature, program code, hardware or project which is shared free on internet by others Arduino user. The Arduino user usually using or develop with Arduino on their project. Beside of that, Arduino is open source and also can build the Arduino hardware by ourselves.

2. Methods

The method used in this research is experiment. It started with case study, formulation of the problem including solution design, finding the literature about the problem to find the solution, designing hardware and software and then examination.

2.1. Block diagram system

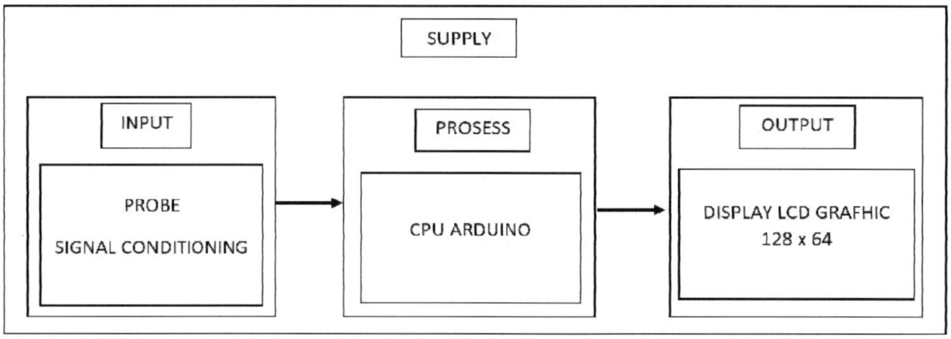

Figure 1. Block diagram.

Based on Figure 1, there are three part of system block from Handheld Digital Oscilloscope sequence. First, input system block has probe part which used to receive input signal then the signal will forwarded to Arduino processing unit part. The probe is consist signal conditioning circuit which also used as voltage scalers in hardware. After that is the process system block, in this block the data from input signal will be processed to become digital data and it will be ready to display on output block. Arduino have a role as "brain" to processing the data. And the last is output block. In this block has LCD graphic 128x64 as a digital data display that will show several parameters like peak to peak voltage (Vpp), Average Voltage (Vavg), Root mean square voltage (Vrms), Frequency and also shown the form of input signal wave. All the component from every system block is working with DC voltage that supplied by Battery or DC Power Supply.

2.2. *Hardware*

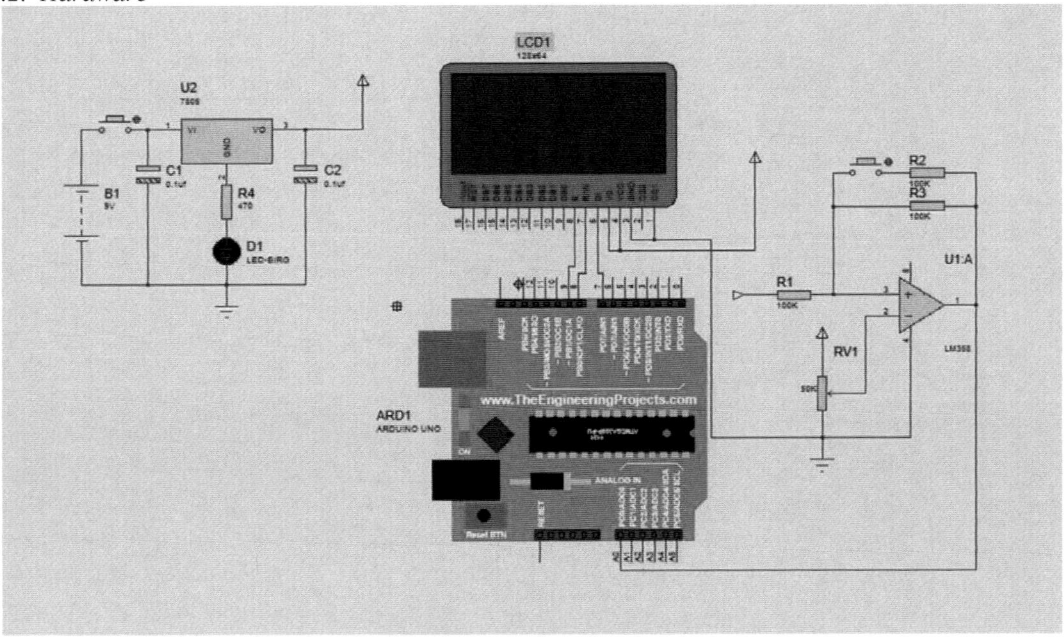

Figure 2. Hardware circuit.

Based on Figure 2, the hardware design begins with a signal conditioning circuit formed from the OPAMP IC LM358 which is configured as an inverting amplifier and it can be set in the great strengthening of 1x and ½ times. This circuit will receive signals from the input device in the form of electrical waves and then adjusted to the calibration voltage on a non-inverting input. Whose magnitude can be set through a potensio that is mounted on a non-inverting input to produced voltage. The result of voltage corresponds to the signal conditioner output will be inputted to the Arduino as a data processor. The potensio on non-inverting input LM358 will served as a horizontal line position calibrator on screen of the Oscilloscope LCD display.

The functions of Arduino as a data processor, the signal sent from the signal conditioner and it will be read by the analogue input of the Arduino. Using the ADC feature of the signal in the form of analog data is converted into digital data. And then the digital data will be processed and formed values from the parameters to be displayed on LCD Graphic with 128x64 pixel resolution. The LCD will displayed voltage, frequency, and wave form from input signal.

2.3. *Algorithm*

Generally the workflow of the device which created in this research is shown as seen on the flowchart on Figure 3.

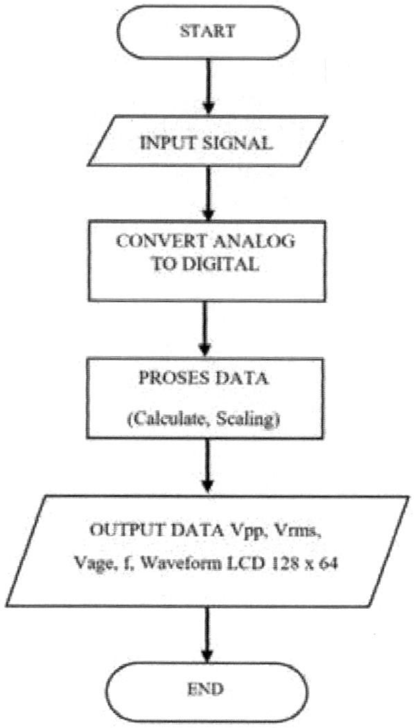

Figure 3. Algorithm flowchart.

The analog signal from the signal conditioner is converted to the digital data by ADC feature on Arduino then the digital data will be processed to get parameter value of voltage and frequency which it will be displayed. In addition, it will be processed of scaling manipulation to make the data appear on the LCD graphic. The data will show the signal waveform image or character data for information of voltage and frequency.

3. Results and discussion

Low Cost Handheld Digital Oscilloscope has been tested to know about accuration rate of voltage and frequency parameter and also the capability to display waveform from signal which has been inputted. The oscilloscope performance was tested in a way to give a big voltage signal and the frequency will came out with various from accurate signal source to oscilloscope input. And then it will displaying the result of signal reader, involve the reading of value pertinence from voltage and frequency parameter and also the perfection of waveform which is displayed on oscilloscope screen.

The signal source for the testing is from GATTEN Function waveform generator with ATF05C type. The test done with giving 3 kind of signal waveform, there are sinus, square, and triangle waveform. With voltage value is 1 Vpp and 2 Vpp, and for the frequency value it was variative. It started with 1-90 Hz with measurement interval is in every 10 Hz, if 100-900 Hz the interval measurement is in every 100 Hz and for 1-7 KHz the interval measurement is in every 1 KHz. The result of oscilloscope performance seen on table 1.

Table 1. Performance measurement.

No.	f	Voltage Accuracy		Frequency	Waveform Signal		
		1 Vpp	2 Vpp	Accuracy	Sine	Square	Triangle
1	0- 90 Hz	96.18%	94.36%	70.46%	Appear	Appear	Appear
2	100 - 900 Hz	95.77%	94.44%	79.98%	Appear	Appear	Appear
3	1 KHz - 3 KHz	94%	95.66%	92.94%	Appear	Appear	Appear
4	4 KHz - 6 KHz	94%	96%	90%	Disappear	Disappear	Disappear
	Average Sub Total	94.99%	95.12%	83.23%			
	Total Average Accuracy	**91.11%**					

As seen at Table 1 the oscilloscope capability for displayed signal waveform is around 10 Hz until 3 KHz, meanwhile for the voltage and frequency value measurement until 6 KHz with accuration mean value is 94.99% for 1 Vpp voltage measurement, and 95.12% for 2 Vpp, and 83.23% for frequency measurement until 6 KHz with the total Average Accuracy is 91,11%.

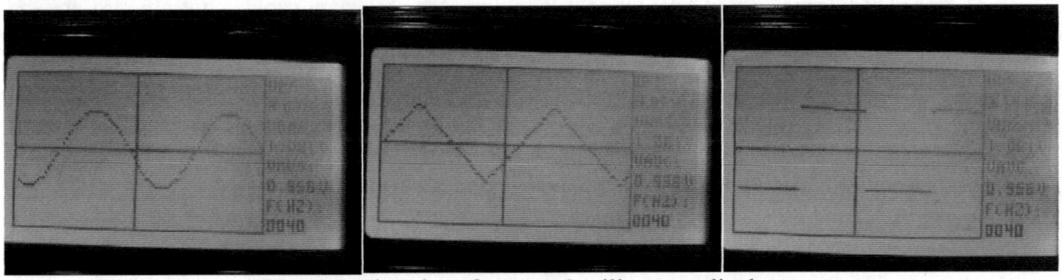

Figure 4. Signal perform on Oscilloscope display.

Beside did test to the performance of the oscilloscope tool that has been made, it is also done with comparative specification and price of couple oscilloscope variants which is show on table 2.

Table 2. Comparation of specification with existing product.

No.	Specification	Gwistek 1072	ATTEN ATH 01	Low-cost Handhold Digital Oscilloscope
1	Power Supply	220V AC	220V AC/ 12V DC	9V DC
2	Range Voltage Input	50 vpp	50 Vpp	20 Vpp
3	Bandwidth	70 MHz	10Hz - 50 MHz	10 Hz - 3 KHz
4	Dimension	30 cm x 11,5 cm x 13,7	20 cm x 13,5 cm x 5,2 cm	14,5 cm x 9,5 cm x 5 cm
5	Cost	8.500.000	Rp. 4.500.000	Rp. 350.000

In Table 2 appears that the advantages of the oscilloscope that has been made are on the price aspect and size dimensions.

4. Conclusions

This paper presents results of research about design and development Laboratory Kit, which is quite important for electrical and electronic laboratory, that is the oscilloscope with low price with tiny dimensional size, so it is easy to grip with hand (Handheld) and also it is easy to carry everywhere. Besides that, another advantage of this oscilloscope is the low power consumption, even it can be run by using batteries 9V which has been sold in the market. So, it can be concluded from the advantages the oscilloscope is named Low Cost Handheld Digital Oscilloscope.

There are many limitation from input frequency segment, however this oscilloscope is ready to use for research or electrical/electronic practical needed which is no need high frequency measurement, as an example is Electrical or Basic Electronics. The efforts to fixing input frequency rate (bandwidth) and the resolution could be done for increase oscilloscope performance.

References

[1] Wagh A, Dave Z, Singh G, Dange V, Tambe A and Gengaje S 2014 "A low cost portable oscilloscope for educational platforms using a Programmable System on Chip" *In Advances in Communication and Computing Technologies (ICACACT), 2014 International Conference on* (pp. 1-4), IEEE Publisher.

[2] Product Catalog Flukewebsite 2016 "Fluke" [Online] Available on http://en-us.fluke.com/products/portable-oscilloscopes/.

[3] Digilant Discoverkit website 2016 [Online] Available on http://www.digilentinc.com/Products/.

[4] National Instruments website 2016 "National Instrumen" [Online] Available on http://www.ni.com/mydaq/.

[5] Zheng Y, Guan X, Wang Y and Li W 2012 "Design for portable virtual instrument with USB interface" *International Conference on Informatics, Electronics & Vision*, IEEE Publisher.

[6] Guili L and Quancun K 2013 "Design of virtual oscilloscope based on GPIB interface and SCPI" *In Electronic Measurement & Instruments (ICEMI), 2013 IEEE 11th International Conference on* (Vol. 1, pp. 294-298), IEEE Publisher.

[7] Başa B and İskefiyeli M 2015 "Realization of Digital Oscilloscope with FPGA for Education" *Procedia-Social and Behavioral Sciences* **174** pp. 814-820.

[8] Introduction Arduino [Online] Available on http://www.arduino.cc.

IOP Conf. Series: Materials Science and Engineering **384** (2018) 012028 doi:10.1088/1757-899X/384/1/012028

Miniature Applications PLC for Traffic Light and Intelligent Meticulous

I R Putri* and Y Somantri

Department of Electrical Engineering Education, Universitas Pendidikan Indonesia, Jl. Dr. Setiabudhi No. 207, Bandung 40154, Indonesia

*intanrp@student.upi.edu

Abstract. Education is an important role in facing the era of globalization. Human resources are the key to competitiveness in the era of globalization later, therefore the education system must also be upgraded in order to produce human resources that are ready and able to compete for the sake of survival in the future. This paper discusses how to create media that can be useful as a teaching material in the world of education. The goal achieved in this research is to design the build-up of an educational media in Vocational High School (SMK) using Omron CP1E PLC. The method used is Research and Development, with steps including problem analysis, data collection, product design, and product testing. The results obtained that the learning media is in accordance with the planned design system with satisfactory results, based on trial application program.

1. Introduction

The Era of globalization like now many things to be prepared one of them is the quality of education [1].Vocational High School (SMK) is one form of formal education units are required to be able to follow the development of technology so as to produce graduates who are competent psychomotor, cognitive, and effective [2].

Students of SMK especially those who take Electrical Engineering Study Program are expected to have skill in using and program Programmable Logic Control (PLC) with better. Therefore, the skills of graduates in the field of PLC needs to be improved, so that graduates are able to compete and ready to work in the industrial world [3]. Based on the results of the observations, control systems used in the industry are also still using the PLC system.

PLC is a special form of microprocessor-based controller that uses a programmable memory to store instruction and to implement function such as logic, sequencing, timing, counting and arithmetic in order to control machines and processes and are designed to be operated by engineering with perhaps a limited knowledge of computer and computing languages[4].PLC was created to replace automation systems that use hundreds or thousands of relays, PLC is designed as a control tool that has many input and output lines, operating ability at higher temperatures, immunity to electrical noise, and resistance to vibration. Programs are stored in a memory that is not easily lost [5, 6]. To help students to improve understanding of the PLC need an interesting media. Through this media, students are expected to be more active in following the learning [7, 8].

Teaching aid is a means of channeling messages or learning information to be conveyed by the source message to the target or recipient of the message. In this case, the source message is the teacher and the

IOP Conf. Series: Materials Science and Engineering **384** (2018) 012028 doi:10.1088/1757-899X/384/1/012028

recipient of the message is a student. Limited media used in the class is thought to be one of the causes of the weak understanding of student learning [8, 9]. So in facilitating the problems that exist then this paper discusses how to design, create and test instructional media on applications of PLC usage. Where the planning is done in the form of applications from a traffic light and intelligent meticulously using PLC control system that can be used as a medium in the learning process.

2. Methods

In general the application of this instructional media is the same as the working system of the traffic light and the use of intelligent meticulously as we see in general, but is created as a miniature simulator. The stages of the method used can be seen in Figure 1.

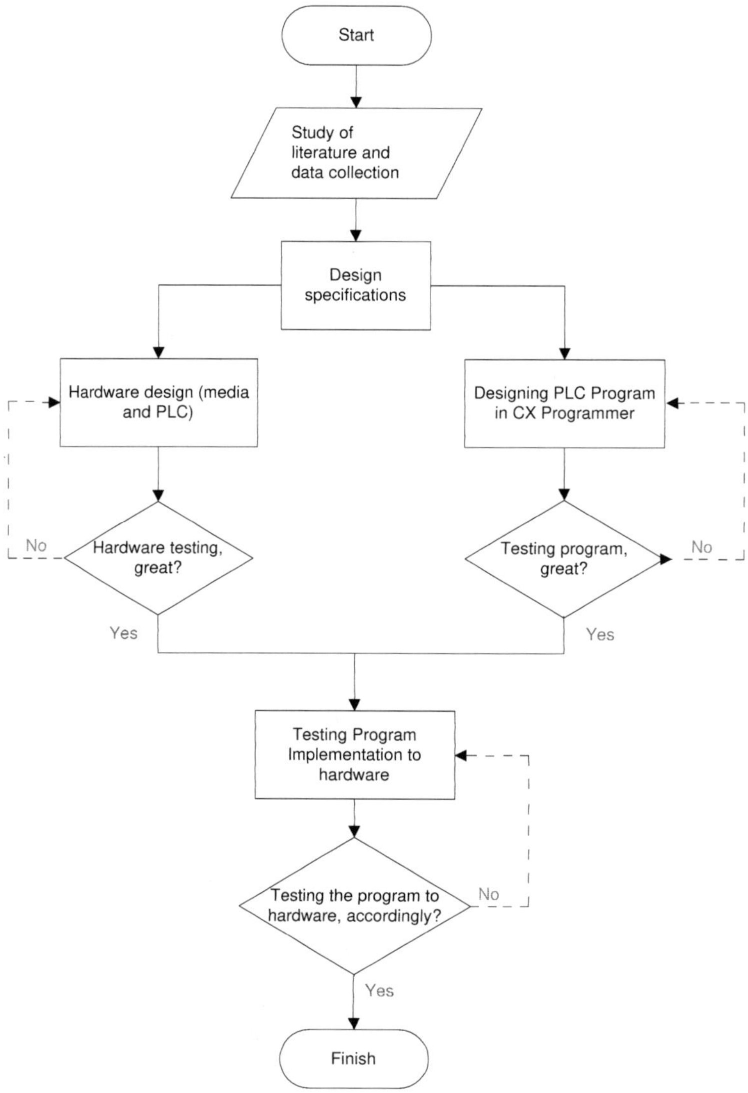

Figure 1. Flowchart research method.

Table 1. Equipment and materials used.

No	Name of equipment	Explanation
1.	PLC	CP1E NA-20-DR-A
2.	*Personal Computer*	Acer
3.	*Acrylic*	Thick 2 mm
4.	*Plat Alumunium*	Thick 1 mm
5.	*Push Button*	6 pieces
6.	*Lamp Indicator*	5 pieces
7.	*Pilot Lamp*	12 pieces

This research is focused on the design, assembly, and testing. The first stage is the planning stage, which is done by doing the circuit design process hardware using Microsoft Visio. The design of the hardware circuit that is done is the design of a series of traffic light media and intelligent meticulous. Figure 2 shows the design of the series of traffic light media while the Figure 3 is a series of media intelligent meticulously. It can be seen from Figure 2, the result of wiring design of traffic light media has four intersections.

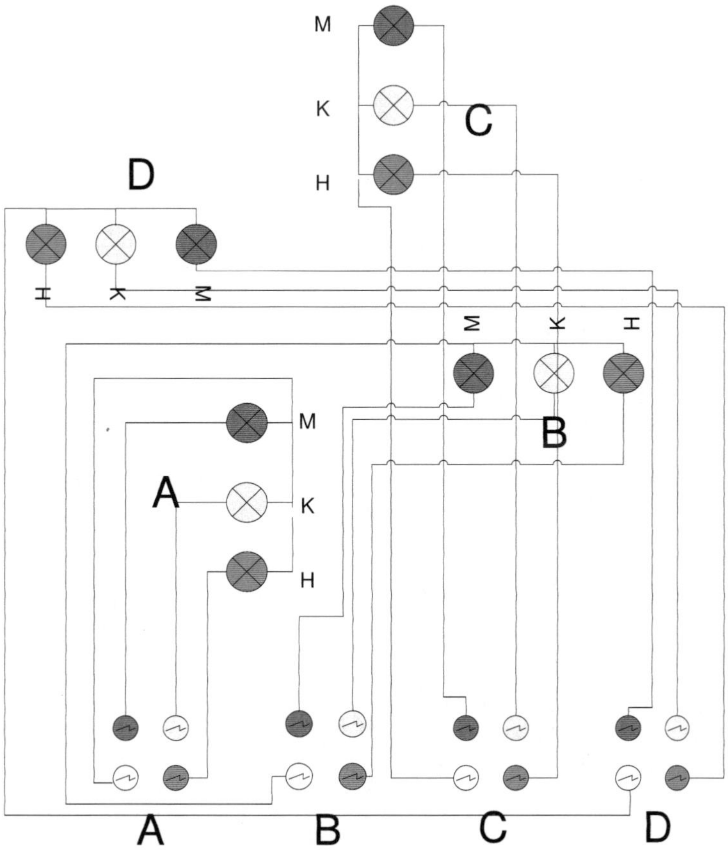

Figure 2. The circuit of traffic light traffic wiring.

IOP Conf. Series: Materials Science and Engineering **384** (2018) 012028 doi:10.1088/1757-899X/384/1/012028

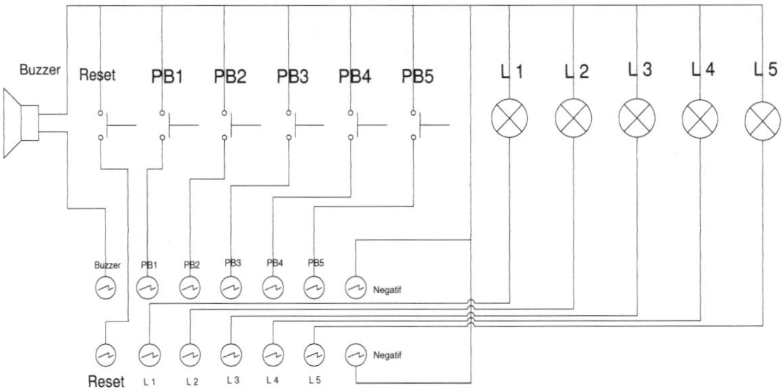

Figure 3. The circuit wiring of intelligent meticulous media.

Figure 3 is the result of wiring design intelligent meticulous media with five group intelligent meticulous light. The second stage is the design stage of program making, This PLC programming using Ladder language. Ladder diagram is one type of programming process control using PLC.

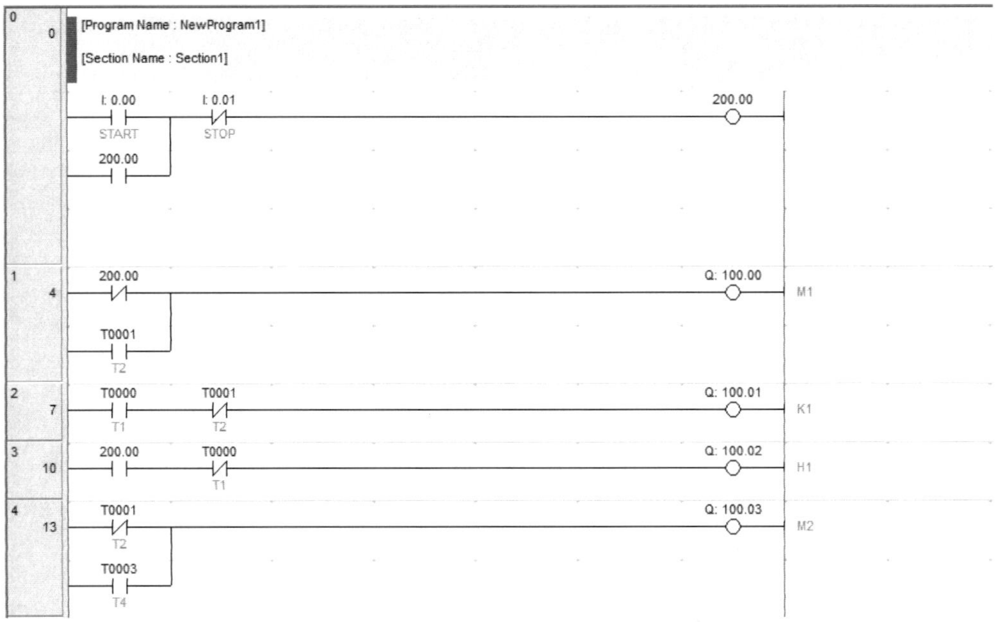

Figure 4. Design of ladder media traffic light diagram.

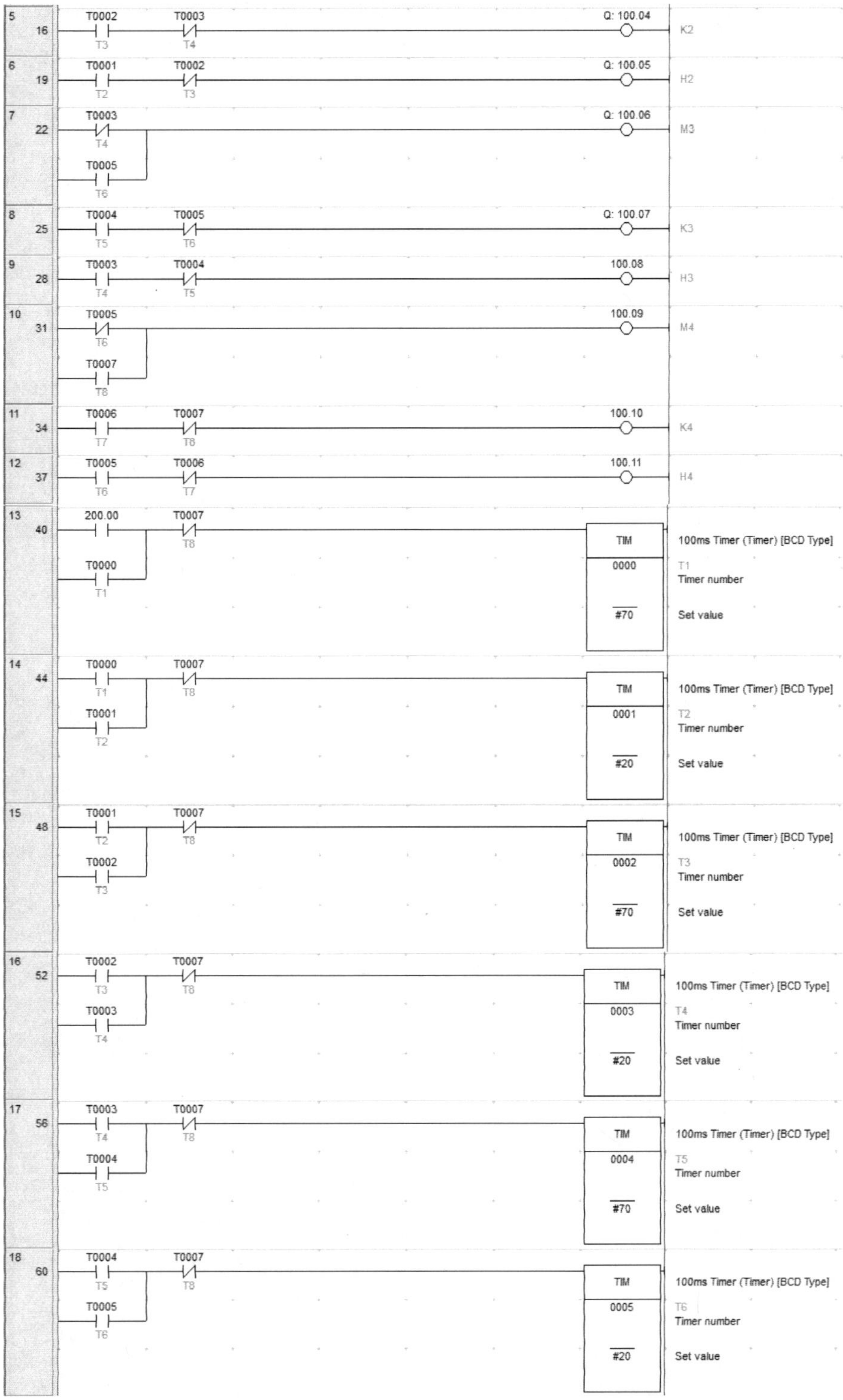

Figure 4. Cont.

IOP Conf. Series: Materials Science and Engineering **384** (2018) 012028 doi:10.1088/1757-899X/384/1/012028

Figure 4. Cont.

Figure 5 illustrates the traffic light program that has outputs in the form of lights from each intersection, and input as a reset.

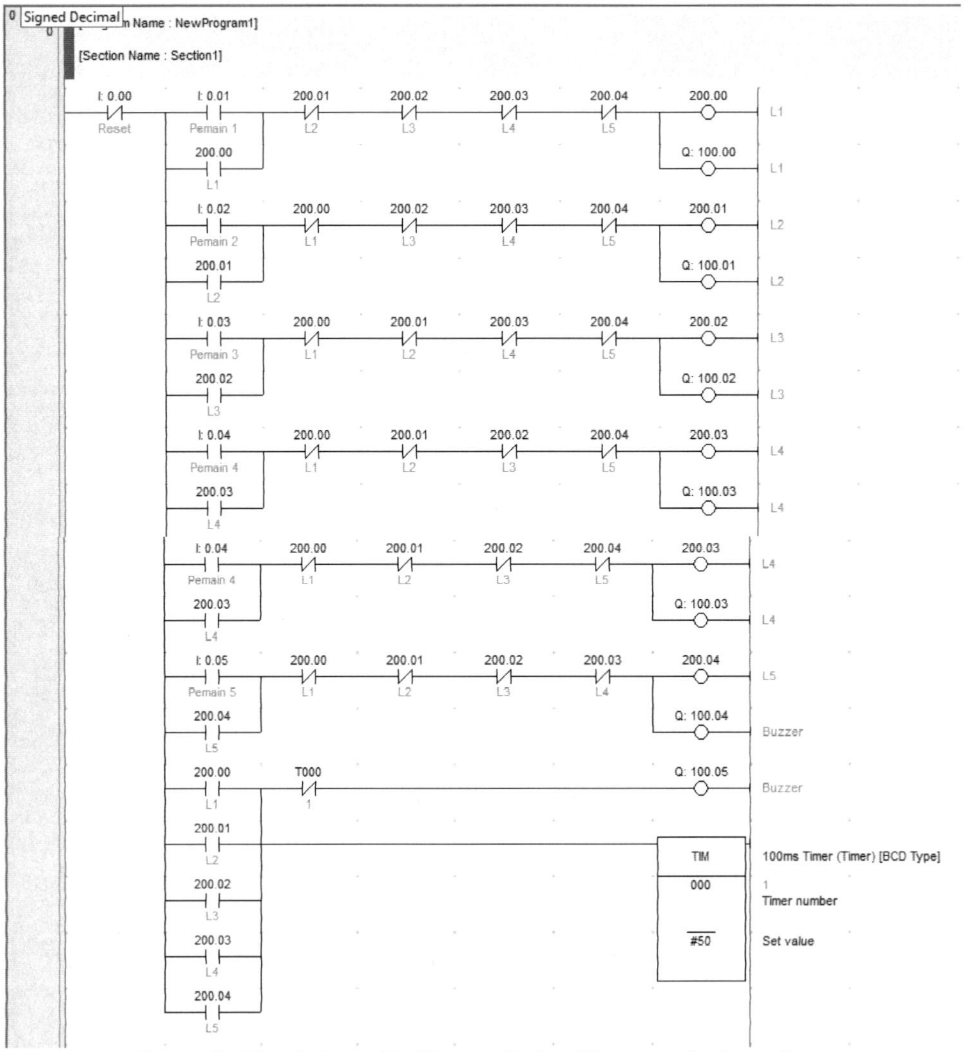

Figure 5. The design of ladder media intelligent meticulous diagram.

Figure 5 shows the result of the design of program intelligent meticulous, the output of this program in the form of five lights and one buzzer and input of six push buttons. The third stage is to test the performance of learning media whether it is in accordance with the program and control system.

3. Results and discussion

The media is created with the concept of mobile. There are 2 boxes, the first box is a media box application of the use of PLC, while the second box is a PLC box. Why are two made to be used in other applications or using other types of PLC. There is a banana slot that functions as a control system assembly of a media application. Where if the program has been completed then the next step is to assemble the control system using jumper cable. If the control system set made in accordance with the program then the media should run.

Programming on the application of PLC traffic light and intelligent metric is using ladder language which ladder this diagram is one way to program the control process. Programming using ladder language programming language that uses a line diagram that displays the interconnection between input devices and output devices. The use of input output is actually in accordance with the program created by the user, and the control system to be assembled on the application media also adjusts to the input output used in the ladder program. At the time of the program design problems that occur is difficult to adjust the timer for each intersection of the traffic light itself. But overall the program created has been successfully completed and runs in accordance with the work of the media application.

The results of this study are obtained from testing every part of the media that has been made. The test is performed in accordance with the media usage procedure. Hardware testing is done by performing multiple tests on each component used. After the test has been done got results that all components contained on the media work well.

Software testing is done by testing to test the program on the PLC that has been made whether in accordance with the design. From the results of experiments, it can be taken the conclusion that the ladder program runs according to the initial plan of the program.

At this stage of testing, the stages are performed that assemble the control system on the ladder program using a jumper and test whether the program and control system run according to the program command or not. However, the problems often times occur when the media is not running because there are errors in assembling a control system and wiring.

4. Conclusions

We can conclude that the design of miniature PLC application for a traffic light and intelligent meticulous match with the planned system. The results obtained are the design of the hardware has been in accordance with the initial planning. Then, the design of ladder diagram program follows the working principles of each PLC application. And finally, the whole system of applications for the PLC is already running well based on trials that have been done on a regular basis.

References

[1] Dinni N F 2013 "*Simulator Traffic Light Sebagai Media Pembelajaran Mata Pelajaran Perakitan dan Pengoperasian Sistem Kendali di SMKN 2 Yogyakarta*" (Doctoral dissertation, Fakultas Teknik).

[2] Mahnun N 2012 "Media Pembelajaran (Kajian terhadap Langkah-langkah Pemilihan Media dan Implementasinya dalam Pembelajaran)" *ANIDA'* **37** (1) pp. 27-34.

[3] Suprianto B 2015 "Pengembangan Media Pembelajaran Trainer PLC untuk Aplikasi Kontrol Otomatis Pintu Air Sungai pada Mata Pelajaran Teknik Kontrol Terprogram di SMK Negeri 3 Surabaya" *Jurnal Pendidikan Teknik Elektro* **4** (1).

[4] Bolton W 2015 *Instrumentation and control systems* (Newnes).

[5] Harini B W, Setiyani T P A and Prabowo P S 2016 "Peningkatan Ketrampilan Programmable Logic Controller (PLC) pada Guru SMKN di Purworejo, Jawa Tengah" *Kaji Tindak: Jurnal Pemberdayaan Masyarakat* **3**(2) pp. 182-190.

[6] Barrett M 2008 "The design of a portable programmable logic controller (PLC) training system for use outside of the automation laboratory" *In International Symposium for Engineering Education* (pp. 1-5).
[7] Pranowo L D and Lion D 2008 "Prototipe Lift Barang 4 Lantai menggunakan Kendali PLC" *Media Teknika* **8** pp. 27-36.
[8] Kozma R B 1991 "Learning with media" *Review of educational research* **61**(2) pp. 179-211.
[9] Saputra F K 2015 "Pengembangan Media Pembelajaran Interaktif Pengenalan Komponen Elektronika Berbasis Adobe Director di SMK" *Jurnal Pendidikan Teknik Elektro*, **5**(3).

IOP Conf. Series: Materials Science and Engineering **384** (2018) 012029 doi:10.1088/1757-899X/384/1/012029

Bluetooth Low Energy (BLE) Based Power Window System

J Purba* and D Wahyudin

Department of Electrical Engineering Education, Universitas Pendidikan Indonesia, Bandung 40154, Indonesia

*jonathanprb31@student.upi.edu

Abstract. In driving the safety factor, comfort, and convenience become things to note. This study discusses the control of window power using Bluetooth low energy (BLE) and applications on Android. Many of us experience anxiety and discomfort when we forget to close our window or enter our car in the hot air. IOT system can be applied to solve this problem by controlling the window power. This research makes it easy for us either to open or to close car windows from around 10 to 30 meters separated from the vehicle through Android smartphone. With this approach, the user can open or close the car window.

1. Introduction

The development of increasingly sophisticated communication technology. Currently with Bluetooth feature we can communicate between gadgets. By utilizing this feature we can use as a controller tool. No exception in terms of setting the Power Window on the car.

At this time Power Window control is still by pressing the button by the rider. This manual way sometimes creates problems when the car keys lag inside. We have to force open the keyhole or retrieve the backup key.

Much research has been done on controlling the power windows, both national and international. As research by Meydianto from Yogyakarta State University. Then Aditya et al from Narotama University Surabaya and also Binglin Li and Quan Tao from University of Technology, China. From several previous studies many have managed to control this power window. These studies mostly use Arduino as a microcontroller. In this project ESP32 is used as a microcontroller as well as data communication using Bluetooth.

With the current technology is very possible for us to make use of Power Window automation, so it can be remotely controlled by using Android. So if something happens, like a key lags in us there is no need for confusion to solve the problem.

Based on the problems and review of the above research, this research makes an easy and fast power window controller tool used. This tool uses Android / Smartphone, ESP32 as a Bluetooth microcontroller and communication.

2. Methods

2.1. System power window

Power window system is a series of electrical body that serves to open and close the door glass by using a switch, where the power window switch is mounted on the inside of the door. When the power window switch is pressed it will rotate the motor power window and this rotation will be changed by the regulator into up and down motion to close or open the door glass. How the power window works. The lifting mechanism (regulator power window) is the most important component of the power window system. A small electric motor attached to the regulator by using the gear ratio that provides enough rotary power to lift the car window glass, while keeping the window pane capable of rising / falling smoothly.

2.2. Component

The components used in making this project are as follows:

ESP32 is a single 2.4 GHz Wi-Fi chip and Bluetooth combo chip designed with ultra low-power TSMC 40 nm technology. It is designed and optimized for best power performance, performance, robustness, flexibility, features and RF reliability, for a variety of applications, and different power profiles. ESP32 is designed for mobile applications, wearable electronics, and the Internet of Things (IoT). It has many advanced low-power chip features, including good clock resolution gating, power mode, and dynamic power scaling [1].

Figure 1. ESP32.

Motor power window is a motor drive regulator rotates clockwise or reverse direction moves the window regulator to be converted into up and down motion. The type of motor used in the power window system is a DC motor. Electric motors use electrical energy and magnetic energy to produce mechanical energy. Operation of the motor depends on the interaction of two magnetic fields. It is simply said that electric motors work on the principle that two magnetic fields can be made interacting to produce motion. The purpose of the motor is to produce a force that drives (torque) [2].

Figure 2. Motor power window.

Power Supply is an electronic component that has a function as a supplier of electric current by first changing the voltage from AC to DC. So the electric current of PLN which is Alternating Current (AC) goes into power supply, then the voltage is converted to Direct Current (DC) and then flows to other components that require.

Figure 3. Power supply.

A relay is an electronic component in the form of a switch or an electrically operated switch. Relays are also commonly referred to as electromechanical components consisting of two main parts namely the coil or the electromagnet and the switch or mechanical. Relay components use the electromagnetic principle as the driving contact of the switch, so by using a small electric current or low power, can deliver a current that has a higher voltage. Here is a picture and also a symbol of the relay component [3].

Figure 4. Relay 5V, 2chanel.

Smartphones are high-powered mobile phones, with functions that almost resemble a computer. Android is a Linux-based operating system designed for touch screen mobile devices such as smart phones and tablet computers [4].

Figure 5. Smartphone/Android.

2.3. Block diagram

Figure 6 shows the block diagram of the power window control. From the figure can be seen section and function below.

- Android / Smartphone serves as a communication tool between users and microcontrollers.
- ESP32 serves as a given data processing microcontroller.
- Relay serves as a voltage control switch.
- DC motor serves as a driver of the power window system.
- Battery serves as a power supply.

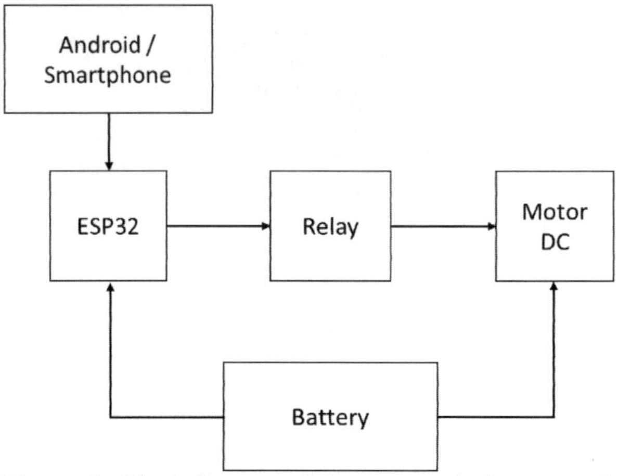

Figure 6. Block diagram of the power window control.

2.4. Flowchart

Figure 7 shows the flowchart for the process of reading the data up to the process of moving the power window. The process begins by reading data from the keystrokes that are on the remote control transmitter. If there is a keystrokes, then the data will be sent via transmitter. The data will be received by the receiver, then processed by ESP32. After the process is complete, the dc power window motor will rotate according to the data received from the keystrokes. The dc power window motor will stop when there is a keystroke once again or the switch is touched by the mechanical window [5].

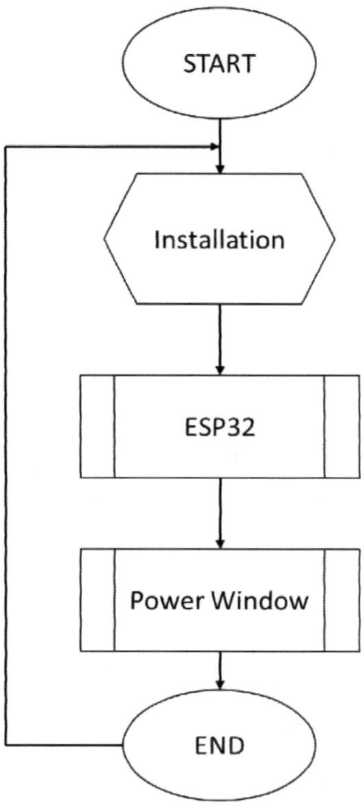

Figure 7. Flowchart.

3. Results and discussion

In testing the DC Power Window motor control circuit, it is done by giving ESP32 program to play DC motor. The ideal results that are omitted are listed in Table 1 [5].

Table 1. Reference table.

Relay 1	Relay 2	Motor Output
0	0	0
0	1	Clockwise
1	0	Anti-Clockwise
1	1	0

In fact, the results of the work for this study were obtained in Table 2.

Table 2. Table of research results.

Relay 1	Relay 2	Motor Output
0	0	0
0	1	Clockwise
1	0	Anti-Clockwise
1	1	0

Temporary test result is DC motor can be rotated according to given input. When relay 1 is given HIGH motor will rotate clockwise, relay 2 is given HIGH motor will rotate counter clockwise.

4. Conclusions

After going through system design and testing, the result of power window automation system with remote control using ESP32 can be summarized as follows, the whole tool is running well as desired. The motor has also rotated according to the input given. To develop this tool in the future, will be made some improvements to the system, display on the application and others to create maximum results for its users.

References

[1] Espressif Systems 2016 ESP32 Datasheet.
[2] Miu N S C a K N 2015 A Study of Time Window Selection for Electric Power Distribution System Analysis *IEEE*.
[3] Relay-Kontaktor-Sensor-Industri.
[4] Open Handset Alliance, "Android Overview," [Online]. Available: https://www.openhandsetalliance.com/android_overview.html.
[5] Aditya Dwi Aryanto A Z F S W 2016 Otomatisasi Power Window Dengan Remote Control Menggunakan Arduino *NARODROID*.

IOP Conf. Series: Materials Science and Engineering **384** (2018) 012030 doi:10.1088/1757-899X/384/1/012030

Development of K500 Signal Processing System using Raspberry PI in the 3-electrodes Electrochemical Sensor

K A M Akbar[1,*], R V Manurung[2] and B Mulyanti[1]

[1]Department of Electrical Engineering Education, Universitas Pendidikan Indonesia Bandung 40154, Indonesia
[2]Indonesia Institute of Sciences, Jl. Sangkuriang, Bandung 40135, Indonesia

*akbarkarim@student.upi.edu

Abstract. The electrochemical sensor exploits current or voltage on the electrodes when the chemical reaction occurs between the sensitive part of the electrode and the heavy metal ion detected. The output of current or voltage changes corresponds to the amount of ion concentration of heavy metal in water. This research aims to develop signal processing system using raspberry PI as a supporting electronic scheme of electrochemical sensor to detect and analize heavy metal pollutant. The system contains of potentiostat, signal conditioning and data acquisition schemes. This research covers the activities of the design and development of Raspberry Pi based supporting electronic scheme, PCB and the electronic components, as well as the software program for the measurement of signal input from the electrochemical sensor. The performance of K500 system developed in the research is a little bit lower than that of the system published by another researcher. Despite its performance, this system costs only $45 cheaper than previous system that costs $100. For full version, i.e. including Raspberry Pi as a server, the K500 system with the dimension of 110x90x70 mm costs only $70, cheaper than the other system using laptop computer as a server that costs $60 without its server.

1. Introduction

Heavy metal is hazardous pollutant which contaminates human being health as a result of accumulation in environment because it can't be degradated. In normal concentration heavy metal is not too dangerous in water and sedimentation, some heavy metal are required for metabolism of human being in chemical process. But too dangerous if the content is too high and in a long time it can harm in water ecosystem (river, sea, etc.) [1]. Therefore the concentration of heavy metal in water has to be monitored. One of the techniques for heavy metal monitoring is the system of solid electrodes technology integrated into small sensor system based on thick film technology (TFT) [2]. This TFT based sensor system offers low fabrication cost [3]. However, the cost for signal processing is too expensive and the dimension is too big. Meanwhile, Raspberry Pi can be used as a server and microcontroller programing which also offers low component cost and low dimension [4]. Therefore, in this research, we developed the K500, the signal processing system using Raspberry Pi integrated into heavy metal monitoring system. The supporting circuit required to support the performance of 3-electrodes sensor devices is potentiostat. Potentiostat is an instrument to measure potential difference between working and reference electrodes in the electrochemical cell. The working principal of potentiostat is by giving electrical current from working electrode cell to counter electrode cell passing the reference electrode and measuring current difference in the cell [5].

2. Hardware description

The working principal of basic circuit can be seen in Figure 1.

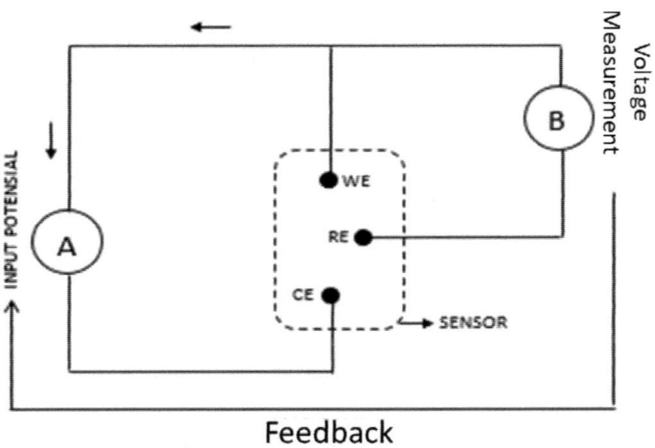

Figure 1. The sensor diagram scheme with 3 electrodes connected between input bias circuits with potential measurement. The controlled variables are the cell potential and the cell current.

The illustration of the diagram of potentiostat working block in this system can be seen in Figure 2.

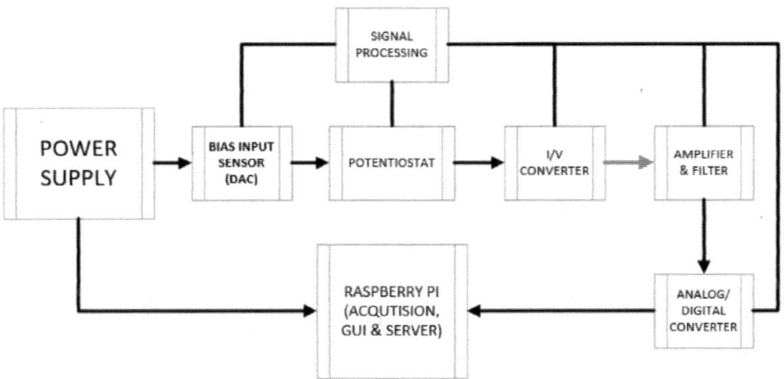

Figure 2. Working diagram block of potentiostat with DAC (Digital/Analog Converter) gives controlled potential input conversion to SENSOR through CONTROL-AMP and the current measurement will be converted by ADC (Analog/Digital Converter) circuit for data acquisition.

2.1. Bias sensor input

The bias sensor input uses IC TLC082N Op-Amp which functions as amplifier with high precision, low bias current, low offset voltage which can be controlled according to the need using IC LM385 as a reference voltage connected with Trimmer Potensio (Figure 3)[6,7].

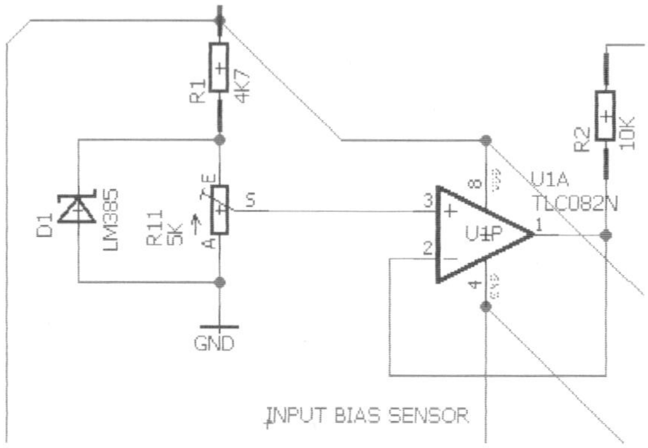

Figure 3. Bias sensor input with IC TLC082N and IC LM385.

2.2. *I/V converter*

The output of the sensor used is electrical current or called as amperometric sensor. In order to be able to be process by Raspberry Pi the current should be converted into voltage by using IC CA3140AE where the Op-Amp will process the input to be the voltage for output (Figure 4) [8].

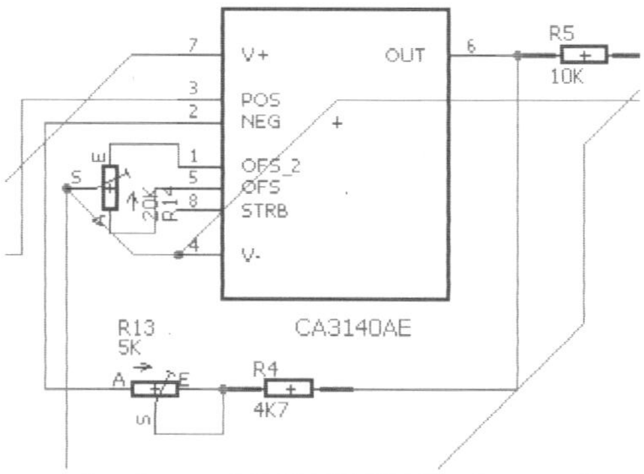

Figure 4. The I/V Converter circuit uses IC CA3140AE where voltage offset can be controlled according to the need with the help of potentio trimmer.

2.3. *Amplifier and filter*

The current from thick film technology (TFT) sensor of this 3 electrodes is very small, nano scale [2], so the amplifier circuit is required and placed after I/V Converter circuit by using IC TLC081N which has capacity almost the same as IC TLC082N. The difference is only in built-in IC TLC081N [6]. Because the output of IC TLC081N is rough enough so addition of Capacitor 0.01F is required as a filter (Figure 5).

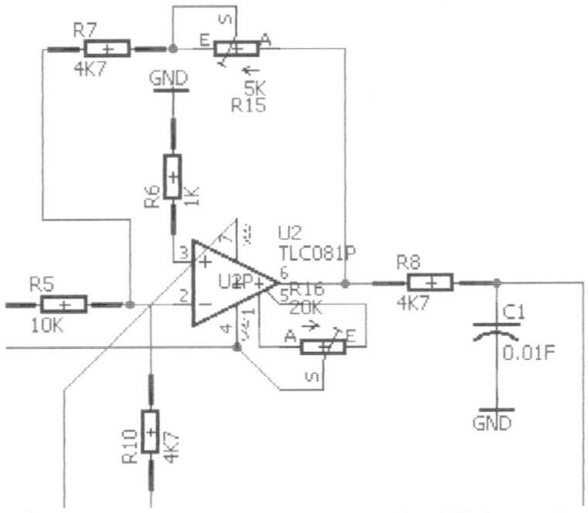

Figure 5. The amplifier circuit with IC TLC081P and addition of capacitor 0.01F as a filter.

2.4. *A/D converter*

Raspberry Pi only receive input of voltage signal in the form of digital, so the Analog to Digital Converter circuit required by using IC MCP3008 with 10 bit-resolution and technology of Low Power CMOS (Figure 6) [9].

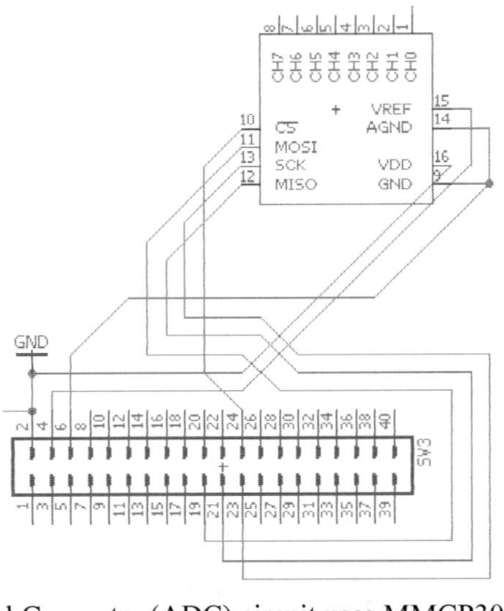

Figure 6. Analog to Digital Converter (ADC) circuit uses MMCP3008 where input channel 0 is obtained from output of amplifier circuit, ADC output is directly connected to GPIO Raspberry Pi 40pin.

2.5. *Power supply*

The signal processor in this research is made portable using battery 9V with capacity >1.000 mAH for 1 hour operation, because Raspberry Pi consumes ±16 mAH/minute. The voltage required of the signal processor is devided in to two parts that is 9V for potentiostat and 5V for Raspberry Pi. The potentiostat input is based on maximum input voltage every IC [5, 7, 8] and the Raspberry Pi input is based on datasheet [10]. Therefore IC LM7805 is required as voltage regulator 5V [11] and in Figure 7 is found ICL7660 as simple voltage multiplication (Vout=(-)nVin) [11, 12].

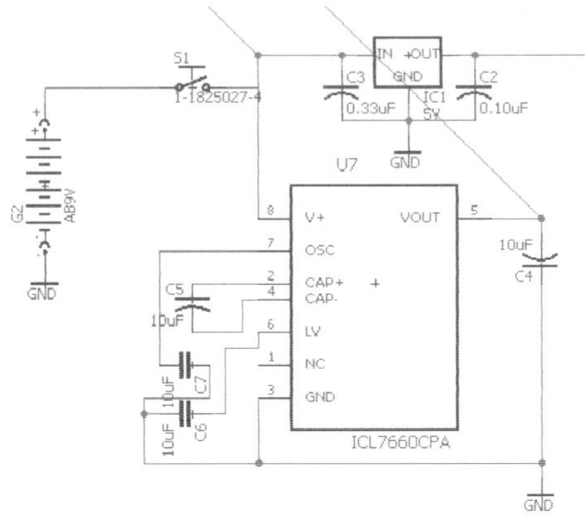

Figure 7. Power supply circuit with supply battery 9V (G2) and multiplication of voltage with ICL7660 and voltage regulation 5V using IC LM7805 (IC1).

2.6. *Raspberry Pi*

The Raspberry Pi used in this research is Raspberry Pi 3 Model B which has more working process speed with processor 1.2GHz Cortex-A53 compared with previous version using processor 900MHz Cortex-A7. Raspberry Pi 3 is even more efficient in power consumption because in idle condition Raspberry Pi 3 is more efficient 22% than Raspberry Pi 2. The advantages which must required is the built-in wifi feature in Raspberry Pi 3 to facilitate the monitoring process of this research [10].

3. Production operation

The production operation of the hardware in this research covers:

3.1. *Circuit simulation*

For the circuit simulation before the production of PCB layout, the researcher applies NI Software.

3.2. *The schematic design and PCB board*

The schematic desain and PCB layout applies the help of CadSoft Eagle Ver.7 software, meanwhile the 3D product model applies Google Sketch Up 2017 which can be seen in Figure 8 & 9.

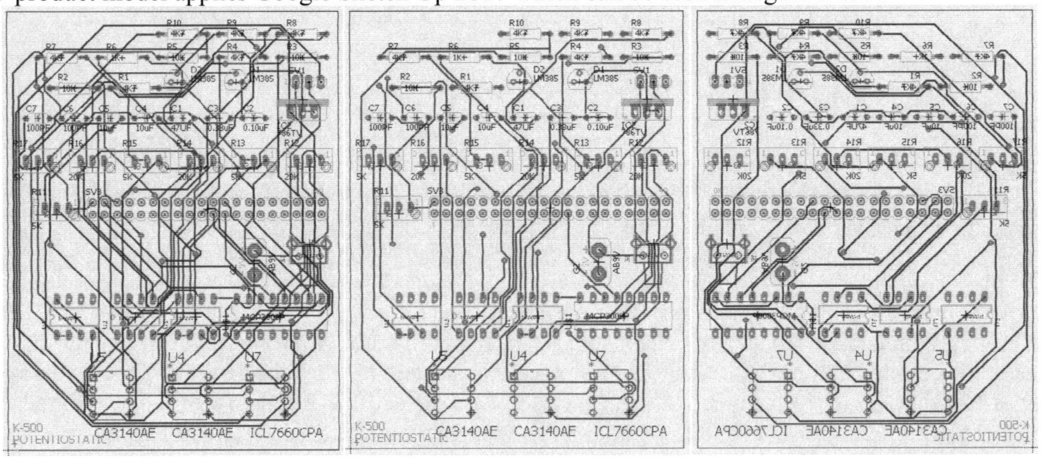

(a) (b) (c)

Figure 8. Complete schematic design of PCB (a), Front schematic design of PCB (b), Back schematic design of PCB (c).

IOP Conf. Series: Materials Science and Engineering **384** (2018) 012030 doi:10.1088/1757-899X/384/1/012030

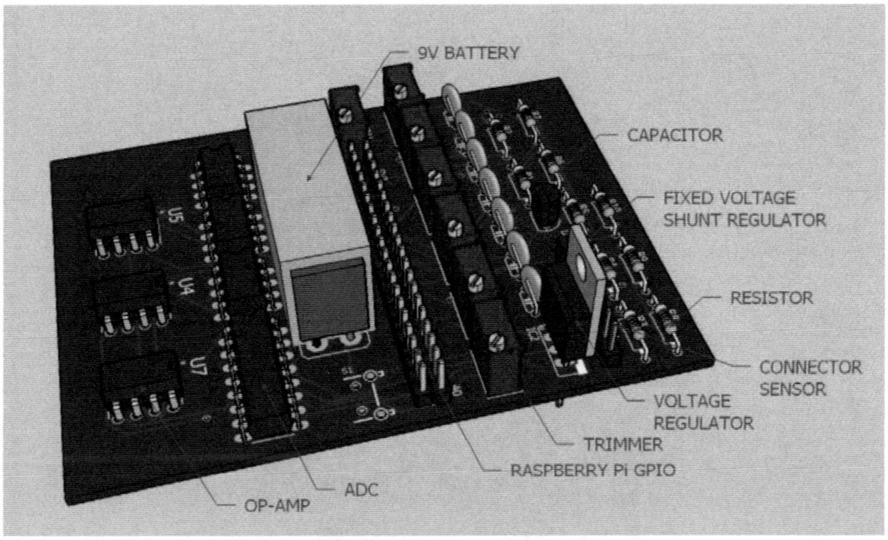

Figure 9. 3D model PC.

3.3. Communication
The communication required from potentiostat to Raspberry Pi applies Phyton programming which can be accessed online in https://learn.adafruit.com/raspberry-pi-analog-to-digital-converters/mcp3008GUI.

3.4. GUI
Graphical User Interface (GUI) which researcher produces has simple interface with Phyton3 base. As for the assisting software for data acquisition such as Numpy and Scipy, for real-time plot applies PyQtGraph and for interface applies PyQt which can be accessed on their official website for tutorial [13-16].

4. Examination and results
For the physical form of potentiostat hardware can be seen in Figure 10 and for examination the device already design and made in this research should be compared with the existing device and internationally recognized. The researcher compares his research result with research result of T. Dobbelaere in HardwareX about “USB-controlled potentiostat/galvanostat for thin-film battery characterization” in the characteristic of cyclic voltammetry (Figure 12). Therefore the measurement technic has to be the same using “dummy cell” as replacing electrochemical sensor, where the dummy cell consist of resistor 1000KΩ and capacitor 1000μF which is set serial and connected from port WE to port RE/CE (Figure 11). The above measurement is based on 1st orde differential equation:

$$I = \frac{U_R}{R} = C\frac{dU_C}{dt}, U = U_R + U_C \tag{1}$$

It usefull to verify potentiostat measurement by comparing the potential value to the result expected.

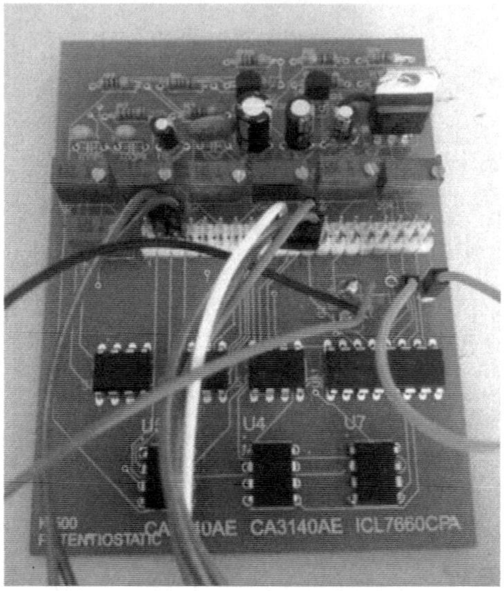

Figure 10. Potentiostat hardware.

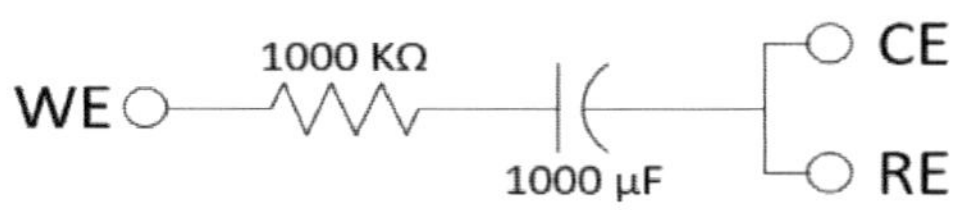

Figure 11. Dummy cell schematic.

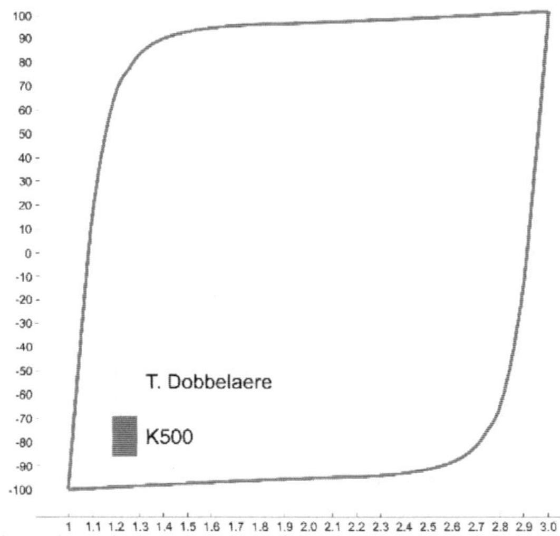

Figure 12. The result of cyclic voltammetry graphic, the result which is not too far can be considered that the device made has good performance.

Although the performance of K500 is a bit under the performance of T. Dobbelaere it can be concluded that k500 has a good performance, economically k500 is cheaper where the production cost of K500 is under 45$ compared with T. Dobbelaere which costs almost 100$ [17]. From point of full version (including server) this research is even more excellent because the role of Raspberry Pi as a server with total cost about 70$ and with the dimension 110x90x70 mm including server, compared to Dstat, Emstat and Cheapstat which still uses laptop as a server and costs over 60$ excluding server [18].

5. Conclusions

The performance of K500 system developed in the research is a little bit lower than that of the system published by another researcher. Despite its performance, this system costs only $45 cheaper than previous system that costs $100. For full version, i.e. including Raspberry Pi as a server, the K500 system with the dimension of 110x90x70 mm costs only $70, cheaper than the other system using laptop computer as a server that costs $60 without its server.

As the design is made open source and not many newest precision components are available in local market, there should be a special modification to process electrochemical signal sensor with new method. The limitation in this journal is that not all the result and the process can be presented due to the number of pages stipulated by the publisher, so for more information please contact the researcher through enclosed email above. New experiment in python programming is really needed to maximize calculation, GUI, and also performance of the device used in this research.

Acknowledgements
The researchers appreciate LIPI (Indonesia Institute of Sciences) for fully support by providing research fund and laboratory.

References

[1] Jenkins S H 1971 Advances in environmental sciences and technology *Atmos. Environ.* **5** 79
[2] Prasek J and Adamek M 2004 Development of new thick film sensor for heavy metals detection *Proc. IEEE Sensors* 749–52
[3] Wiranto G, Idayanti N, Tayubi Y R and Batara D S 2016 Design and Fabrication of Low Cost Thick Film pH Sensor using Silver Chlorinated Reference Electrodes with Integrated Temperature Sensor *MATEC Web Conf.* **1** 1–5
[4] Deshmukh A D 2016 A Low Cost Environment Monitoring System Using Raspberry Pi and Arduino with Zigbee *Int. Conf. Inven. Comput. Technol.* 1–6
[5] Electrode W, Electrode R, Electrode W, Electrode W, Electrode W and Electrode W Potentiostat Fundamentals 1–3
[6] Information D 2015 TL08xx JFET-Input Operational Amplifiers 82
[7] Description G and Diagrams C 2000 LM185 / LM285 / LM385 Adjustable Micropower Voltage Reference Adjustable Micropower Voltage References 1–12
[8] Instrumentation P, Amplifiers H, Low O and Systems S V 2005 Ca3140, ca3140a 1–24
[9] Inl L S B 2007 Characteristics Technology 1–34
[10] Pi R 2016 DATASHEET Raspberry Pi Compute Module (CM1) Raspberry Pi Compute Module 3 (CM3) Raspberry Pi Compute Module 3 Lite (CM3L) 3 0–21
[11] Information O 2003 Series positive-voltage regulators *Volt. Regul. Funct.* 134–53
[12] Components P and Range V 2010 Icl7660a, icl7660a 1–11
[13] "NumPy — NumPy." Online Available: http://www.numpy.org/
[14] "SciPy.org — SciPy.org." Online Available: https://www.scipy.org/
[15] "PyQtGraph - Scientific Graphics and GUI Library for Python." Online Available: http://pyqtgraph.org/
[16] "Riverbank | Software | PyQt | What is PyQt?" Online Available: https://www.riverbankcomputing.com/software/pyqt/intro
[17] Dobbelaere T 2017 A USB-controlled potentiostat/galvanostat for thin-film battery characterization *HardwareX* **2** 34–49
[18] Dryden M D M and Wheeler A R 2015 DStat: A versatile, open-source potentiostat for electroanalysis and integration *PLoS One* **10** 1–17

IOP Conf. Series: Materials Science and Engineering **384** (2018) 012031 doi:10.1088/1757-899X/384/1/012031

Trainer of Traffic Light System Based on PLC

K Kurniati* and Y Mulyadi

Department of Electrical Engineering Education, Universitas Pendidikan Indonesia, Jl. Dr. Setiabudhi No. 207, Bandung 40154, Indonesia

*karlinkurniati@student.upi.edu

Abstract. Currently traffic congestion becomes one of the problems especially in big cities. Therefore, the traffic system needs to be applied in order to avoid severe traffic jams and traffic accidents. This research discusses about traffic system at an intersection that is simulated on the trainer that made in such a way that resembles the street intersection of five in general with based PLC Omron CQM1H-CPU51 series. This traffic system simulation works using a timer that has been set in such a way. On the trainer, there are five paths in which each path is installed green, yellow and red lights and 2 buttons to start and end the program. The green light is on for 25 seconds and the yellow light is on for 5 seconds while the red light keeps on when the green light at another intersection lights up.

1. Introduction

Currently, the traffic system becomes one of the most important things, especially in densely populated urban areas, the more people who have vehicles and roads that remain and many intersections will cause serious congestion. Therefore traffic control system needs to be designed time calculation certain and controlled data can be monitored via computer [1-3].

The task of the traffic system is to arrange a certain time at each lane meeting to pass the car to avoid accidents and accumulation of vehicles in several lanes and also to give time for pedestrians to cross. Traffic light instructions are marked with green, orange and red lights [3-5].

In this research using programmable logic control (PLC) for setting traffic lights of 15 lamps for the intersection of five rows each still 3 lights per line with different colors i.e. green, orange and red as a hint for the vehicle.

2. Methods

In this study using PLC Series CP1L-L20 which has 8 inputs and 16 inputs. In this system used 2 inputs and 15 lamp output because of the amount of output, less expansion unit used to increase the amount of output that can be used. Lights are placed on miniature roads connected to PLC using block terminals [1].

Using source 220 VAC and 24 VDC resulting from internal PLC. The system works when the green button is pressed, the system will turn on continuously and reset when the red button is pressed. Figure 1 illustrates the block diagram system.

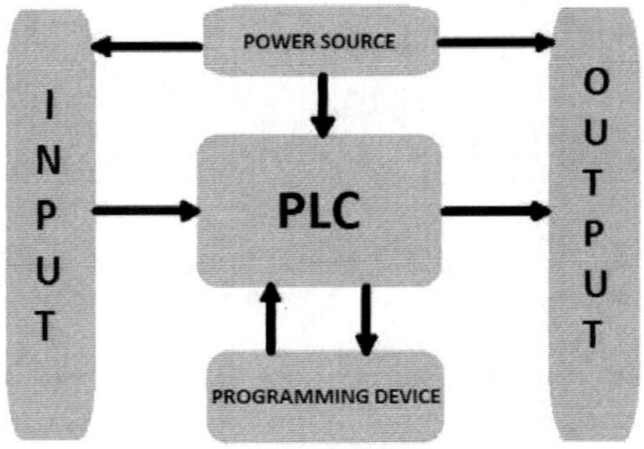

Figure 1. Block diagram system.

This project uses ladder programming language diagram and in the program by software CX programmer 9.6. In this software, the program of traffic light system is designed by utilizing the facilities available in the software such as timers.

This program is designed by an internal timer on the PLC. The timing of each lamp can be seen in figure 2 with the overall time in one cycle up to 60 seconds.

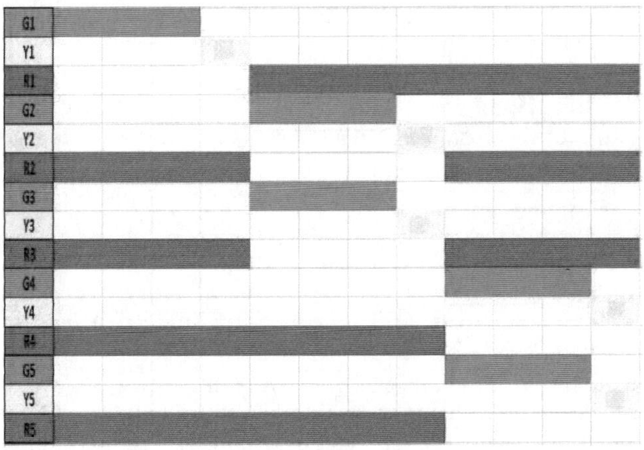

Figure 2. Time diagram.

The component used is a component that has a functional equation with the actual tool. The components used have dimensions and small size simulations can be made with a size that is not too large. In this project a traffic light indicator using a 24VDC lamp pilot placed on a board that has been designed to resemble a five-junction road.

3. Results and discussion

After several experiments, there are several conditions that occur in the process of this traffic light system. At initial condition, all lights are off, after the start button at press system on condition 1 as seen in figure 4 after 25 seconds switch on condition 2 for 5 seconds before switch to condition 3 for 25 seconds and switch to condition 4 with time for 5 seconds as well as on condition 5 lasted for 25 seconds and condition 6 for 5 seconds before return to initial condition. The system will not stop and reset it before the stop button is pressed.

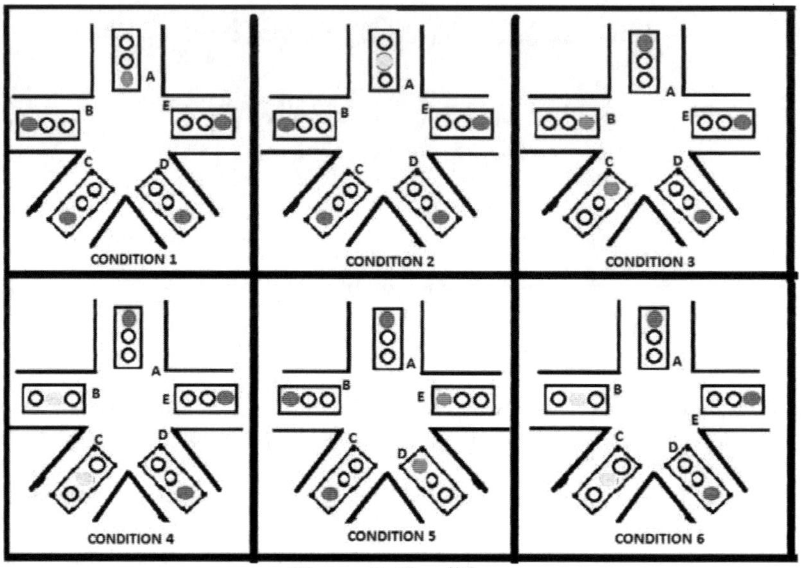

Figure 3. Condition.

4. Conclusions

This research makes the traffic system program with 5 intersections that are displayed directly through the simulation and miniature of the crossroads. The Traffic system program runs continuously before the program is stopped by pressing the stop button.

References

[1] Liu Y and Chen X F 2009 "Design of traffic lights controlling system based on PLC and configuration technology" *1st Int. Conf. Multimed. Inf. Netw. Secur.* **2** pp. 561–563.

[2] Jaksic N 2002 "Improving Self-Efficacy in Engineering Students using PLC Based Traffic Light Experiments" *Age* **7** (1).

[3] Qi L, Zhou M and Luan W "An emergency traffic light strategy to prevent traffic congestion" *ICNSC 2016 - 13th IEEE Int. Conf. Networking, Sens. Control* pp. 5–10.

[4] Abdulhussein M M 2017 "Traffic Light (Vehicles And Pedestrians) Control System Using PLC" **2** (3) pp. 41–44.

[5] Khattak M A 2011 "PLC based intelligent traffic control system" *International Journal of Electrical & Computer Sciences IJECS-IJENS* **11**(6).

Prototype of Water Level Control System

K Karwati* and J Kustija

Department of Electrical Engineering Education, Universitas Pendidikan Indonesia, Jl. Dr. Setiabudhi No. 207, Bandung 40154, Indonesia

*karwati_cdeon@student.upi.edu

Abstract. Global climate change and unpredictable rain circle cause uncertainty of water availability. Therefore, the technology which can increase water distribution efficiency is needed. This paper describes prototype water level control system that has an important role in providing convenience in the drainage system. If usually the water gate at the dam is operated manually, this prototype is simulation open and close automatically water gate based on rainfall levels using Arduino. To automatically control open and close the water gate, we use Ultrasonic sensors as input values. The water gate will be driven using a Servo Motor. The output of the water level status is displayed in the form of LCD display and is also followed by sound output for people with visual impairments (blind). The design results of this prototype are expected to be one of the good contributions the drainage system. With this automatic sluice of, the course will minimalize the risk from flood or another risk.

1. Introduction

Global climate and unpredictable rain circle may cause uncertainty of water availability [1]. Recently, a very important problem from all over the world is the management of water resources [2]. Water is commonly used in households, agriculture, and industry [2, 3]. Therefore, we need technology which can increase water distribution efficiency.

In practice are known that many types of water level control system that can be done [2-6]. Such system includes providing flood prediction, environmental protection, water discharge, power plant system, providing water control in industry, simple water level control in the home [7].

The most efficiency of water distribution system to every aspect can be enhanced through dam automation. For that application can be done by designing open and close automatic water gate at the dam for a drainage system. The drainage system efficiency is not more than 40% if manually operated, through some automation, the drainage system efficiency can be enhanced to 50% [5].

This paper design prototype water level control system by open and close automatically the water gate use ultrasonic sensor. Open and close automatically the water gate becomes one of the means of service in an effort to control the distribution of more efficient drainage system to every aspect so as to minimalize the risk of the flood [6].

2. Methods

2.1. Software design

In this paper, the ultrasonic sensor reads the water level in the aquarium as input to the Arduino. As output, the servo would move the door aquarium, and water level conditions would be visible on LCD and led, as illustrated in the following schematic capture by Proteus (Figure 1).

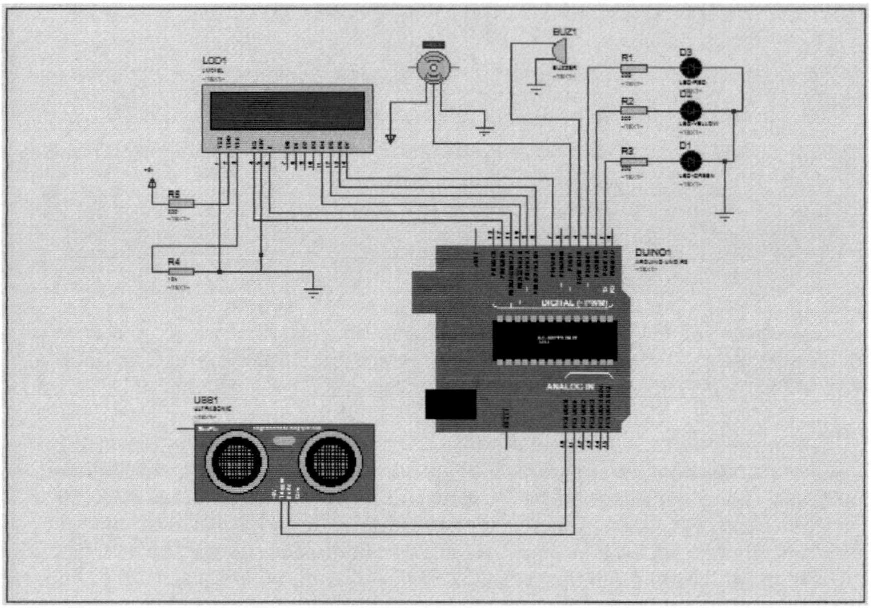

Figure 1. Schematic capture.

The open and close automatically water gate operation is illustrated in block diagram system (Figure 2). The diagram contains several parts, that is, water level detected by the ultrasonic sensor, input data from ultrasonic sensor to Arduino, Arduino set servo motion, the servo for moving the water gate (the door aquarium), and the ultrasonic sensor for making feedback to Arduino. If the water level is low so the door would close.

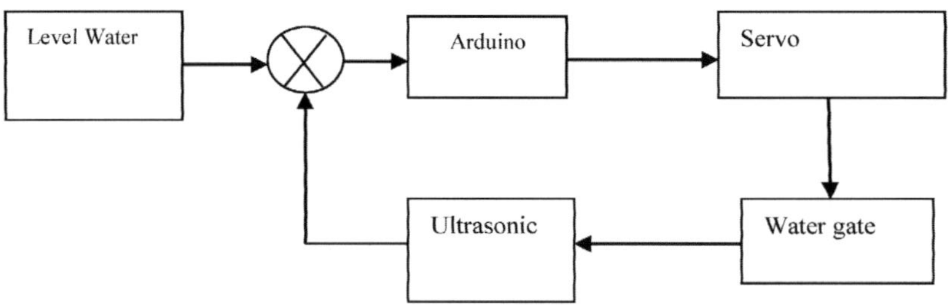

Figure 2. Block diagram system.

The flowchart of system is as follows (Figure 3).

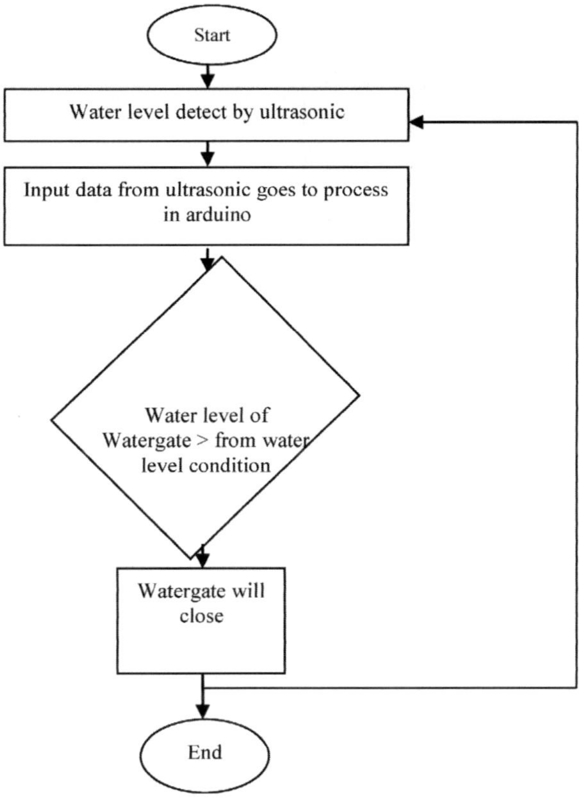

Figure 3. Flowchart system.

The paper programs is set to 3 water level conditions, that is "AMAN" at water level ≤ 5 cm from ultrasonic sensor, "SIAGA" at 5 cm ≥ dist ≤ 10 cm from ultrasonic sensor, and "AWAS" at ≥ 10 cm from ultrasonic sensor. Designing software programs using Arduino software with C programming language.

2.2. Hardware design
Plant uses a glass aquarium with a door hole on the front. Aquarium measuring 50 x 30 x 20 cm. The door opening is 5 x 5 cm in size. The door is used using acrylic material with a size of 7 x 7 cm.

The working principle of the ultrasonic sensor is the transmitter sending ultrasonic waves, then measuring the time required until the arrival of the reflection of the object. Ultrasonic sensor used is SRF-05. The time difference when transmitted to capture returns is used as a reference calculation of how far the distance between the sensors with the object that reflects the ultrasonic waves [8]. Ultrasonic Sensor Devantech SRF-05 with the following specifications:

- Works on a 5 volt DC voltage
- Current load of 30 mA - 50 mA
- Generate a wave frequency of 40 KHz
- The range of detectable distance of 3 cm - 400 cm
- Requires a minimum input trigger of 10 uS

It can be used in two choices of input trigger mode and echo output mounted on different pins or input trigger and echo output mounted in one pin the same [9].

The ultrasonic sensor SRF-05 has the following views (Figure 4).

Figure 4. Ultrasonic SRF-05.

The SRF-05 ultrasonic schematic can be seen as follows (Figure 5).

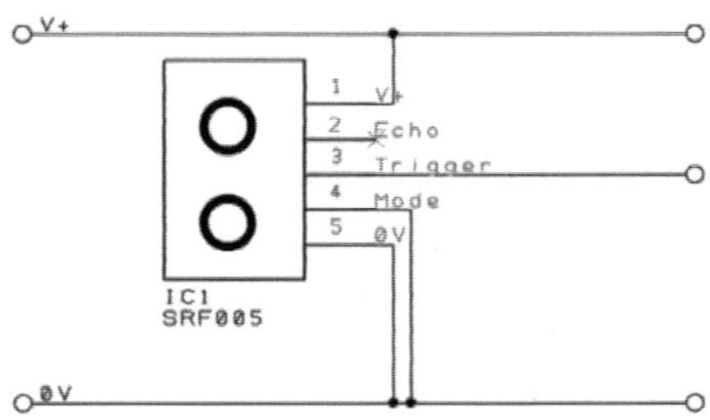

Figure 5. Schematic ultrasonic sensor.

2.2.1. Servo motor. In general there are 2 types of servo motors. Namely servo standard and continuous servo. Standard servo motors are only capable of moving in a semi-circle or 180 degrees, while continuous-type servo motors are capable of rotating by 360 degrees. The servo motor used is a continuous-type servo motor. Continuous types can rotate up to 360 degrees. Servo motors have the following display (Figure 6).

Figure 6. Servo motor 360 degree.

The schematic of the servo motor can be seen as follows (Figure 7).

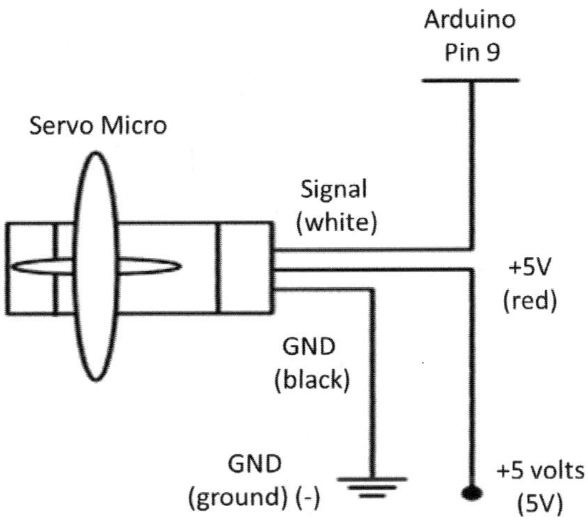

Figure 7. Schematic servo motor 360 degree.

2.2.2. Arduino Nano. Arduino Nano is a microcontroller based on Atmega 328P. Arduino Nano has more or less the same function as Arduino Duemilanove, but with different parts. Arduino Nano has a DC power cable, and works with Mini-USB cable. Arduino Nano has 22 digital pin I / O (consisting of 6 pins PWM output), 8 pin analog I / O, and reset button [12]. Parts of the Arduino Nano have the following appearance (Figure 8).

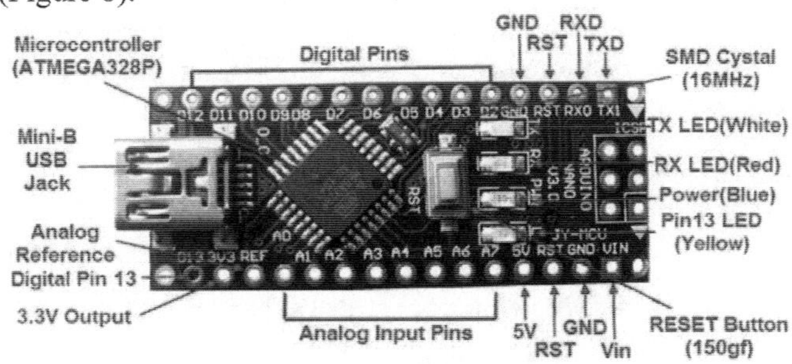

Figure 8. Arduino nano.

The schematic of Arduino Nano can be seen as follows (Figure 9).

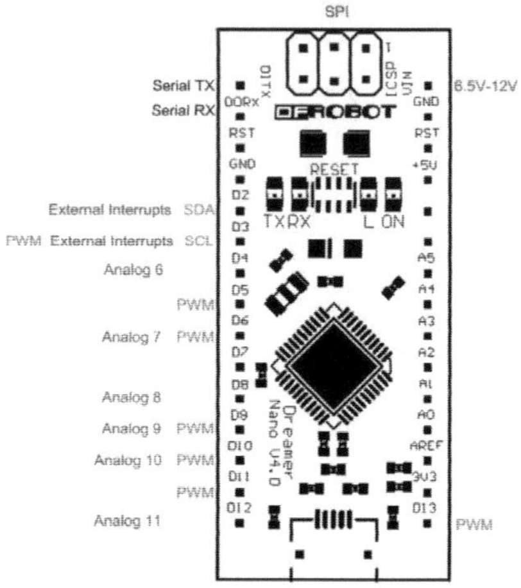

Figure 9. Schematic of arduino nano.

2.2.3. LCD 16 x 2. LCD (Liquid Crystal Display0) is a type of display media that uses liquid crystals as the main viewer LCD is already used in various fields such as electronic devices such as television, calculator, or computer screen LCD used dot matrix with character number 2 x 16. The LCD is very functional as a viewer which will later be used to shown working status of the tool. The features presented in this LCD are:

- Consists of 16 characters and 2 lines.
- Has 192 characters stored.
- There is a programmable generator character.
- Can be addressed in 4-bit and 8-bit mode.
- Equipped with back light [2].

The schematic of the 16 x 2 LCD has the following display (Figure 10).

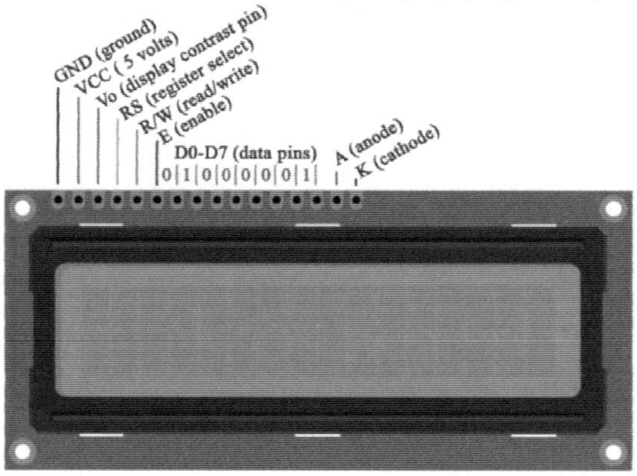

Figure 10. Schematic of LCD 16 x 2.

3. Results and discussion

The SRF-05 ultrasonic sensor is used as the sensor that detects the water level of the plant in 3 states, which is "AMAN" status at water level with distance of $\geq$ 10 cm from ultrasonic sensor, "SIAGA" at water level 5 cm distance $\geq$ distance $\leq$ 10 cm of the ultrasonic sensor, and "AWAS" at the water level with a distance of $\leq$ 5 cm from the ultrasonic sensor.

The results of prototype testing can be seen in Table 1.

IOP Conf. Series: Materials Science and Engineering **384** (2018) 012032 doi:10.1088/1757-899X/384/1/012032

Table 1. Testing result.

No	Distance	Led			Buzzer	Servo
		Green	Yellow	Red		
1.	15,9	On	OFF	Off	Off	Off
2.	12,5	On	Off	Off	Off	Off
3.	8	Off	On	Off	On	On
4.	7,15	Off	On	Off	On	On
5.	2	Off	Off	On	On	On

Based on the test results performed, it can be seen that the prototype succeeded in accordance with the desired. The components used can work properly as desired. Test results in some conditions can be seen as follows.

Figure 11. Test results on "AMAN" status.

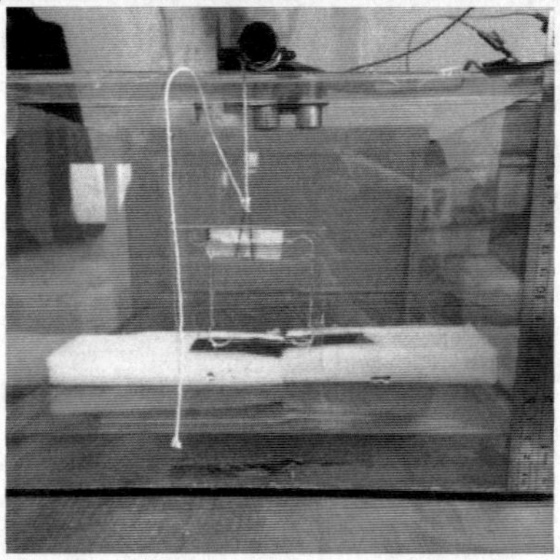

Figure 12. Test results on "AMAN" status.

The test results of Fig. 11. and Fig. 12 produce "AMAN" status with live green LED and the servo motor remains silent.

IOP Conf. Series: Materials Science and Engineering **384** (2018) 012032 doi:10.1088/1757-899X/384/1/012032

Figure 13. Test results on "SIAGA" status.

Figure 14. Test results on "SIAGA" status.

The test results of Fig. 13 and Fig. 14 produce "SIAGA" status with live yellow LED and rotating servo motor.

Figure 15. Test results on "AWAS" status.

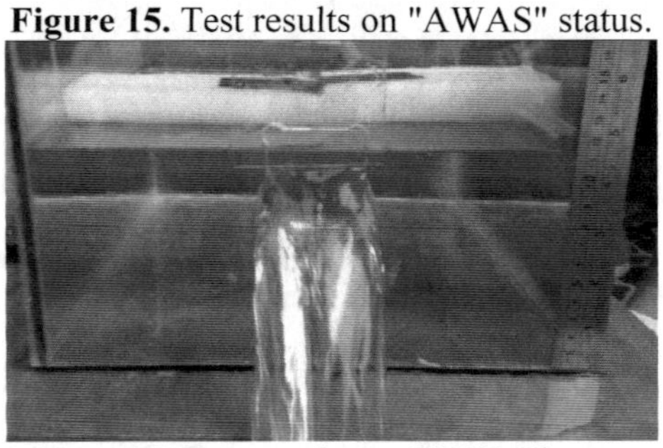

Figure 16. Test results on "AWAS" status.

The test results of Fig. 15. and Fig. 16. resulted in the status "AWAS" with the red LED on and the rotating servo motor. Based on the test results, ultrasonic sensors can function properly. Testing of ultrasonic sensors can be seen in Table 2 as follows.

Table 2. Ultrasonic sensor test result.

NO	DIST (CM)	SENSOR READING (CM)	ERROR (%)	AVERAGE ERROR (%)
1.	2	1.67	1.6	
2.	3	2.89	3.6	
3.	4	3.80	5	
4.	5	4.90	2	
5.	7	6.84	2.2	
6.	8	7.82	2.25	2.15
7.	9	8.86	1.5	
8.	10	9.86	1.4	
9.	14	13.78	1.5	
10.	17	16.9	0.5	

Table 2 is a comparison test of ultrasonic water distance measurement. Obtained by the comparison result of ultrasonic sensor distance measurement with water got an error of the difference of reading. The following error is obtained by the measurement:

$$\text{Error\%} = \left(\frac{sample\ distance - sensing}{sample\ distance}\right) x\ 100\ \%$$

Based the data of sensor sensing test results showed that the sensor has a precision measurement level with an average error of 2.15%. This project is experiencing some constraints that are not in accordance with the desired process in the process of working in a 360 degree servo motor system. Servo motors are understood to be adjustable rotation using the degree of degree, not working as desired. 360 degree servo motors are the same as continuous motors that can only rotate clockwise and clockwise, so there is no time to stop.

Based on these constraints, this project tries to use 180 degree servo motor. However, because the rotation is only 180 degrees, less appropriate to open the door on this prototype. Constraints on this can be solved by adding TIP components, diodes, and resistors in a 360-degree servo motor circuit. By using TIP, the servo motor can be set to spin and stop.

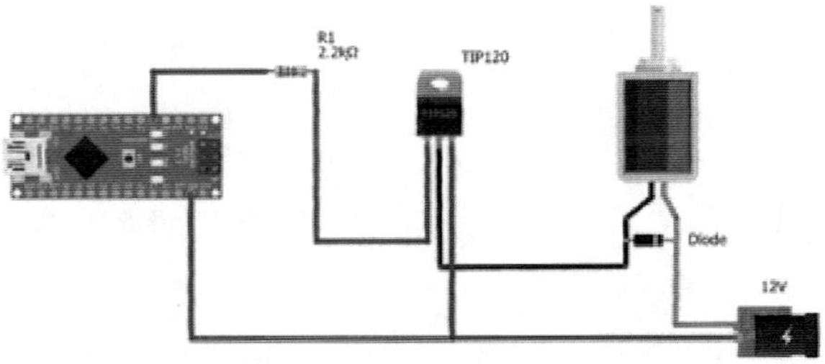

Figure 17. TIP 120 circuit with servo 360 degrees.

4. Conclusions

Our design is in conformity with the intended purpose. The prototype of water level control system works well in distributing water in accordance with the water level, so it can be one of the contributors in the drainage system especially in the dam area to the surrounding environment. Based on the test results, the prototype can function well. From the sensor side, ultrasonic sensor is a good distance detection sensor with error 2.15% based on test results. Constraints in the use of 360 degree servo motors can be handled using the TIP 120, so the 360 servo motor can function as desired.

References

[1] Kulkarni N K and Shete V V 2015 "Hybrid neuro-fuzzy approach for flood prediction and Dam gate control" *Int. Conf. Inf. Soc. i-Society 2014* pp. 213–218.
[2] Illes C, Popa G N and Filip I 2013 "Water Level Control System Using PLC and Wireless Sensors" *IEEE 9th Int. Conf. Comput. Cybern.*, pp. 195–199.
[3] Pratilastiarso J, Tridianto E, Elvian G P H, Budi U E, Vera N and Ika C A 2017 "Simulation water level system with feedback feedfoward control" *Proc. - 2016 Int. Electron. Symp. IES 2016* pp. 42–47.
[4] Jeswin C J, Marimuthu B and Chithra K 2017 "Ultrasonic Water Level Indicator and Controller Using Avr Microcontroller" *In Information Communication and Embedded Systems (ICICES), 2017 International Conference on* (pp. 1-6). IEEE.
[5] Bhat S P and Hirekhan S R 2016 "Automation of water discharge process at canals" *Proc. 2015 Int. Conf. Appl. Theor. Comput. Commun. Technol. iCATccT 2015* pp. 609–613.
[6] Litrico X, Belaud G and Fromion V 2007 "Stability analysis of automatic water level control gates in open-channels" *2007 46th IEEE Conf. Decis. Control* pp. 1591–1596.
[7] Hy Cao N D 2016 "Design of water tank level PID control based on kingview" pp. 2–5.

IOP Conf. Series: Materials Science and Engineering **384** (2018) 012033 doi:10.1088/1757-899X/384/1/012033

The Development of Smart Parking System based on NodeMCU 1.0 using the Internet of Things

L Anjari and A H S Budi*

Department of Electrical Engineering Education, Universitas Pendidikan Indonesia, Jl. Dr. Setiabudhi 207, Bandung 40154, Indonesia

*agusheri@aol.com

Abstract. Recently the number of vehicle ownership has been increasing, hence it needs a large parking area. The shortage is how to look for the empty parking space in the large parking area and also the amount of illegal parking car that does not follow the rules of the parking space unit. In this paper, we proposed the smart parking system using the Internet of Things (IoT) that can be part of a solution for the parking problem. This system help in organizing the parking lot and helps the driver to reach their parking spots easily as they known which space is vacant. The parking space can be detected using an Infrared sensor that connects to the ESP12-E (NodeMCU) module that was programmed through Arduino IDE. Users can access parking space information using a smartphone via an application. Especially for users who have been registered before, they have a code for login the app as the requirement for security system and user parking convenience. The system can work with the purpose of the research appropriately.

1. Introduction

In recent years, the Internet of Things has been applied in many ways. The smart parking system is one part of the technology of the Internet of Things [1]. The concept of the Internet of Things starts from a device that can be traced, controlled, or monitored over the internet [2]. One of the systems of smart parking is to know the condition of parking lot via the internet. This is related to parking problems which one of them is the difficulty of knowing the condition of vacant space in the wide parking lot so that the driver spends his time just to find a parking place and tend to more difficult along with the increasing number of vehicle ownership [2-4]. Problems related to parking can be solved if the driver can be informed beforehand about the availability of parking space around the desired destination [2]. As the result, the concept of the Internet of Things applies to the smart parking system.

Various approaches and research have been done to overcome parking problems. Since the early 1970s, smart parking has been implemented throughout Europe, the UK, and Japan. The initial system is displayed in the driver's parking information such as availability status and/or the amount of space available [5]. More complex smart parking incorporates more advanced technology to serve customers with different needs [6]. A recent survey conducted by the International Parking Institute [6] reflects an increasing number of innovative ideas related to parking systems. Currently, there are certain parking systems that are able to provide real-time information about available parking spaces. Such systems require efficient sensors to be placed in parking lots to monitor parking spaces and rapid data processing units to gain practical insights of data collected from various sources [2].

IOP Conf. Series: Materials Science and Engineering **384** (2018) 012033 doi:10.1088/1757-899X/384/1/012033

According to these data, this study was to design a prototype of parking monitoring system as well as to develop parking system with reference to the pre-existing system [7]. In this system, we used NodeMCU as the microcontroller which was supported by Wi-Fi system. The proposed smart parking system was implemented using a mobile app that connects to the cloud. This system helped users to know the availability of parking space in real time.

2. Methods

The project of this smart parking system was built by some system such as hardware architecture, software architecture and system.

2.1. Hardware architecture

The main of the hardware architecture in this project was built by 3 component, as followed:

2.1.1. NodeMCU (ESP-12). The NodeMCU (Node MicroController Unit) in figure 1 is an open source software and hardware development environment that is built around a very inexpensive System-on-a-Chip (SoC) called the ESP8266. The ESP8266, designed and manufactured by Espressif Systems, contains all crucial elements of the modern computer: CPU, RAM, networking (wifi), and even a modern operating system and SDK. NodeMCU in this prototype used a microcontroller module with a Wi-Fi system which was programmed using Arduino IDE. The version of NodeMCU was using NodeMCU v.1 board or commonly also called ESP-12E. Because the I / O is limited, then in this project used 2 NodeMCU to comply the goal of the project.

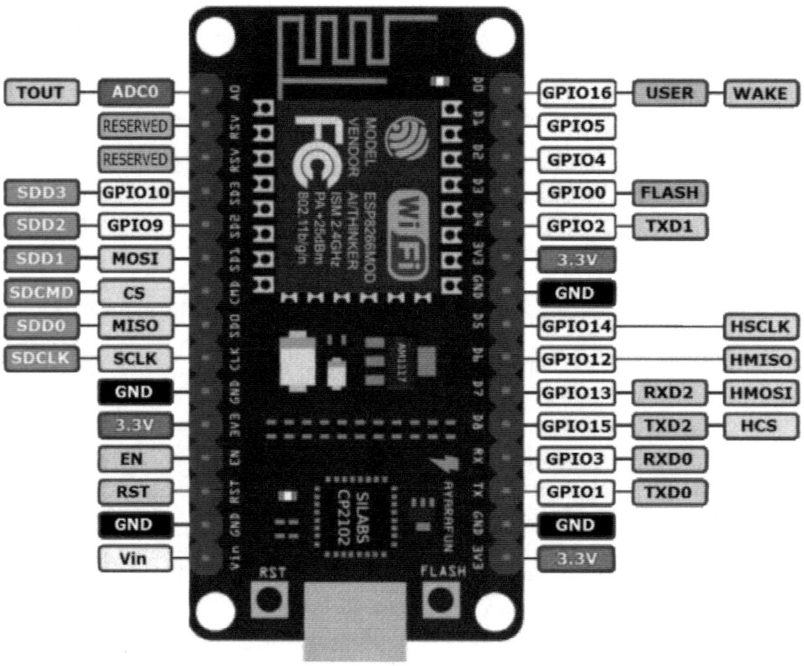

Figure 1. NodeMCU 1.0.

2.1.2. Infra-red obstacle. IR Obstacle is used for detecting the parking space, the way it works is when there is an obstacle that blocks the sensor, the LED will be active as a sign or information that the parking space is already filled by someone.

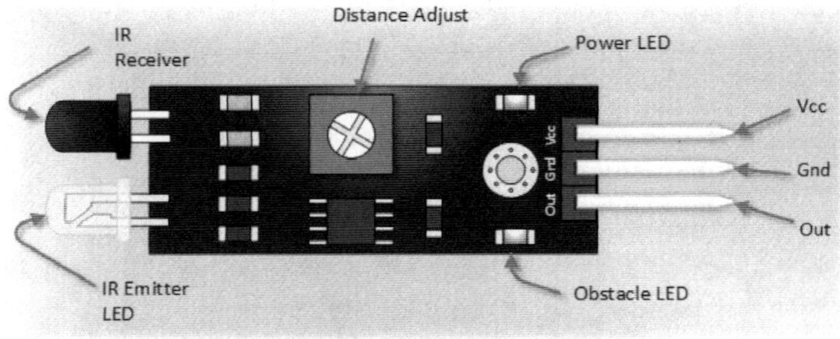

Figure 2. Infrared obstacle sensor.

In this prototype, 2 Sensors Infrared in figure 2 used to detect the car with the logical AND for showing on the app. The use of AND logic in sensor placement is intended to be more accurate parking space detection and also to avoid possible misuse of sensors in parking spaces.

2.1.3. Power supply adapter. The adapter is a circuit to convert AC voltage into DC. Power Supply Adapter is an adapter that can convert a large AC power voltage into a small DC voltage egg. From 220v AC voltage to 6v, 9v, or 12v DC voltage. The adapter used in this prototype is using 5V / 1.5A adapter in order to support the power consumption requirement for all components.

2.2. *Software architecture*

Software architecture is built by several systems, which is programmed by Arduino and sent into the cloud and then the app synchronizes with the real-time database on the cloud.

2.2.1. Arduino IDE. The Arduino Integrated Development Environment (IDE) is a software for write and upload the program into the NodeMCU for communicating between microcontroller and Cloud or a system.

2.2.2. Firebase. Firebase is a mobile and web application development platform developed by Firebase, Inc. in 2011, then acquired by Google in 2014. Firebase provides a real-time database and backend as a service. The service provides application developers with an API that allows application data to be synchronized across clients and stored on Firebase's cloud. The database is also accessible through a REST API. The REST API uses the Server-Sent Events protocol, which is an API for creating HTTP connections for receiving push notifications from a server. Firebase in this project is used to store the data of prototype. Firebase here acts as a cloud. Cloud Firestore is a NoSQL document database that lets to easily store, sync, and query data for your mobile and web apps - at a global scale.

2.2.3. MIT app inventor. MIT App Inventor is an open source web application. It allows users to create an Android application. The app inventor replaces the complex language of text-based coding into a visual drag and drops building blocks. The app inventor can be used even by inexperienced users to create the basic and fully functional app. The MIT App Inventor was provided by Google, and now maintained by the Massachusetts Institute of Technology (MIT) and the inventors are Mark Friedman and MIT Professor Hal Abelson. MIT app inventor in this project is used for display the real-time database on the firebase. The real-time database is the data of the sensor of smart parking project.

2.3. System architecture

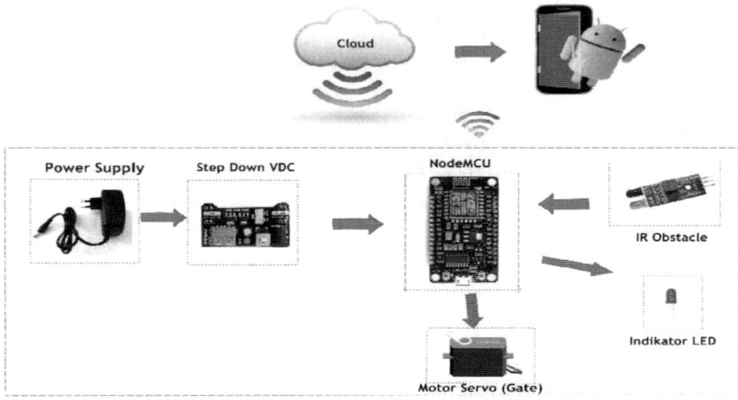

Figure 3. Architecture of smart parking system.

The architecture of smart parking system as shown in figure 3 has depicted many components. The IR Obstacle sensors are connected to the microcontroller NodeMCU. NodeMCU chip comprises of a self-contained SOC (system on a chip) with integrated TCP/IP protocol stack that allows any microcontroller to access a Wi-Fi network. NodeMCU acts like an intermediate between the sensors and cloud. Cloud acts as a database to store all the records related to parking areas and users that have access to the system. The NodeMCU then transmits this data to the Firebase through Arduino IDE. The mobile application acts as an interface for the users to interact with the system on Cloud Firebase.

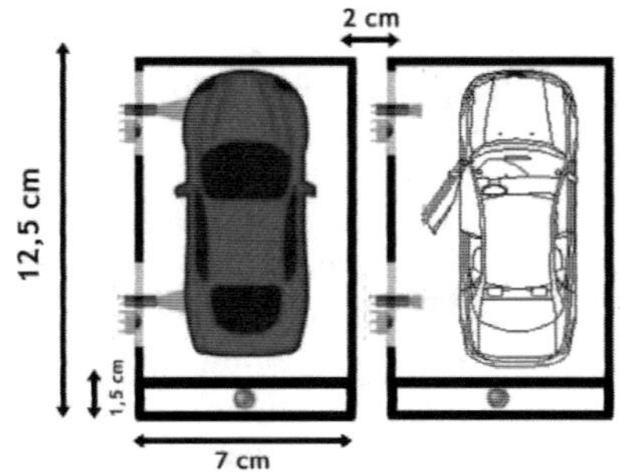

Figure 4. Architecture of prototype for parking space dimension.

A parking space as shown in figure 4 is a location that is designated for parking. Space many are delineated by road surface markings. The car fits inside the space, either by parallel parking, perpendicular parking or angled parking. The basic dimensions based on the Parking Planning and Operation Manual are dependent on door openings, vehicle type. The width of the door open will affect passenger comfort in and out of the vehicle. On the prototype of smart parking system is used perpendicular parking because it's commonly used in car parking lots.

3. Results and discussion

The result of this project, one of which is creating a smart parking app that is connected to the prototype that has been created. This app consists of several screens such as depicted in figure 5, and 6. The first screen in figure 5 is for the registered user login, the second screen in figure 6 for the menu, and the next screen in figure 7 is to display the contents of the menu one which is the parking lots menu of the prototype made.

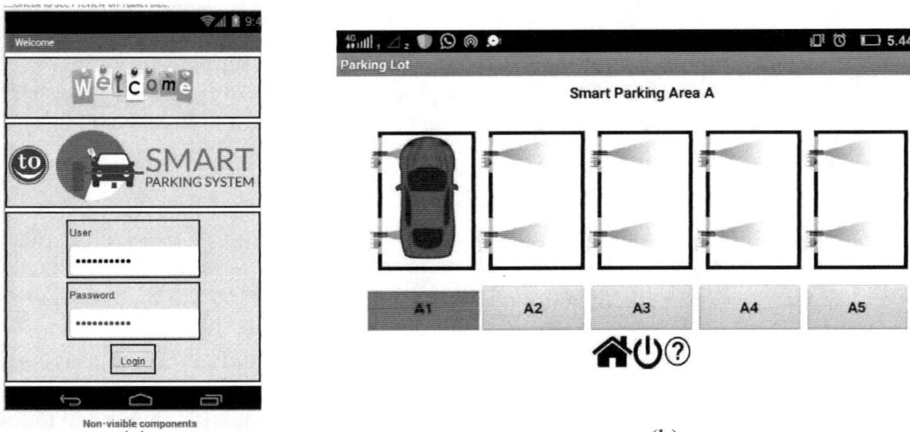

(a) (b)

Figure 5. (a) The registered user login, (b) Display the contents of the menu.

Based on the prototype as shown in figure 10, the space parking A1-A5 will be detected filled or detected there is a car if the IR Obstacle sensor obstructed by the car. LED will be active as a sign or information that the parking space is already filled by someone. This system using a Wi-Fi connection to communicate between ESP8266 and Cloud Firebase. The data is sent from the sensor and synchronized with the data in the application. And then, the application will display the data parking lot in real-time.

The application is synchronized with a real-time database on firebase that acts as a Cloud as shown in figure 6. This is done to test the concept of the Internet of Things (IoT) that is when the data can be accessed by everyone anywhere and anytime via an internet connection. This project has several constraints that are the mismatch between the goals with the results achieved. These constraints include:

- In the planning of this project, the microcontroller that used is ESP32 because it has more I/O than NodeMCU. Because of the limitations of science that are understood and IoT platforms that support ESP32 is more complicated than ESP8266 there is a human error where the system created is not in accordance with the planning.
- NodeMCU has only 10 digital I/O, so the system is not built according to the planning that there are 2 sensors in each parking space.

Figure 6. Real-time database on firebase platform.

4. Conclusions

Smart Parking System on this project has been successfully made in accordance with the purpose of the smart parking system. One of the purposes is to know the condition of the parking lot through the internet that is connected with the application in real-time. This application can allow parking users to search for empty space in the parking lot whenever and wherever using applications based on Internet of Things (IoT).

References

[1] Lee C, Han Y, Jeon S, Seo D and Jung I 2016 "Smart parking system for Internet of Things" *IEEE Int. Conf. Consum. Electron.* pp. 263–264.

[2] Khanna A and Anand R 2016 "IoT based smart parking system" *Int. Conf. Internet Things Appl.* pp. 266–270.

[3] Basavaraju S R 2015 "Automatic Smart Parking System using Internet of Things (IoT)" *Int. J. Sci. Res. Publ.* **5** (12) pp. 629–632.

[4] Ibrahim F et al. 2016 "Smart Parking System Based on Embedded System and Sensor Network" *Int. J. Comput. Appl.* **140** (12) pp. 975–8887.

[5] Chinrungrueng J, Sunantachaikul U and Triamlumlerd S 2007 "Smart Parking: An Application of Optical Wireless Sensor Network" *Int. Symp. Appl. Internet Work.* pp. 66–66.

[6] Rodier C J, Shaheen S and Kemmerer C 2008 "Smart Parking Management Field Test: A Bay Area Rapid Transit (BART) District Parking Demonstration Final Report" *Transp. Res. Rec.* **2038** (1) pp. 62–68.

[7] Septriyaningrum I A, Nugrahadi D T and Ridwan I 2016 "Perancangan Dan Pengembangan Prototype Sistem Parkir" *J. Ilmu Komput.* **3** (2) pp. 146–155.

IOP Conf. Series: Materials Science and Engineering **384** (2018) 012034 doi:10.1088/1757-899X/384/1/012034

System of Water Quality Monitoring and Feeding on Freshwater Fish Cultivation

M D Ramadhona* and D L Hakim

Department of Electrical Engineering Education, Universitas Pendidikan Indonesia, Jl. Dr. Setiabudhi No. 207, Bandung 40154, Indonesia

*mdr_dhona@student.upi.edu

Abstract. Monitoring system has been utilized in many ways. The monitoring system is very useful and helps to alleviate human activities. Starting from industry, livestock, fishery and plantation. For example is fishery monitoring system. This system has been widely used to monitor the quality of water used. Currently fish cultivation is in great demand because of the promising income, especially on freshwater fish. However, many cultivators have problems because of the many fish are dead and the crops that are not maximal. One of the factors that affect these is weather changes that result in an unstable water state. In order to get the expected results, factors that must be considered are water content, temperature and water salt levels that affect the quality of fish metabolism. The standard of water content in freshwater fish cultivation reaches 4 - 9 pH and the temperature must be maintained at 25,5 °C- 32,7 °C and the salt content is 105μS - 120μS. For the many factors that need to be considered then a system that can monitor the state of water at any time is required. This monitoring system uses Arduino uno as micro controller, equipped with ph sensor, temperature sensor, salinity sensor, LCD, and buzzer. The readable data will appear on the LCD and the buzzer will sound when one of the moisture levels is not at a standard number. Equipped with automated fish feeding system remotely to support this monitoring system.

1. Introduction

Technological advancements are now growing rapidly. This is evidenced by the many technological innovations. Technological innovations ranging from simple to sophisticated. Technological developments have long existed since ancient times and evolved to this day until it creates systems, that is techniques that can help human life.

Technology has been widely used by humans in helping everyday life. One of them is monitoring system. Monitoring system has been widely used by freshwater fish cultivators [1]. This system is used in improving the production output [2]. Cultivators have many problems, such as income which is get is not maximum and many fish that are dead [3]. Factors that cause one of these cases are temperature, acidity (pH), uncontrolled water salt content by cultivators due to weather changes [4]. Thus, a monitoring system that can monitor these factors at any time with accurate data accuracy is required.

From the results of sensors monitoring placed on the pool or water, it will get the data according to circumstances. The data will be sent to the Microcontroller IC, which can determine the survival values of freshwater fish. Data will be displayed on the LCD so it can be viewed at any time. If the value of any of the factors is not at the standard number then the warning buzzer will sound.

IOP Conf. Series: Materials Science and Engineering **384** (2018) 012034 doi:10.1088/1757-899X/384/1/012034

From the automatic monitoring system, the authors hope to help the cultivators in improving knowledge and technology so that it can help to improve the results of fish cultivation as a livelihood [5].

2. Methods

In general, this monitoring system works by reading the measurement numbers of the sensor, then buzzer will sound when the reading of numbers is not at a standard number. The readable data will appear on the LCD and feeding can be done remotely. In assisting the completion of this study, it requires equipment and materials. The equipment and materials used is in the table 1 as follow.

Table 1. Equipment and materials used in the assembly of monitoring and feeding system.

No	Equipment and Materials	Information
1	Arduino UNO	CH340/ATmega328PA
2	*Personal Computer*	Lenovo AMD A8
3	*Adapter DC*	*5 V*
4	*LCD*	16x2 *character*
5	Sensor Ph	SEN0161
6	Temperature Sensor	DS18B20
7	Salinity Sensors	Conductivity/TDS
8	Motor Servo	SG90
9	*Box*	1 piece
10	*Push Button*	2 pieces
11	*Switch on/off*	1 piece
12	*Potentiometer* 5k	1 piece
13	Resistors 470kΩ	1 piece
14	Resistors 100kΩ	2 pieces

The first stage is the design stage schematic circuit using Fritzing software while the second stage is schematic design on PCB board using Eagle software then print the design result on PCB. The next stage is assembling the tool using tools and materials according to the design. The third stage is the design stage of the program using Arduino software. The fourth stage is the making of mechanical design using Inkscape software as well as the design of the tool on the mechanical design. The last stage is the testing tool. Testing tools is done ranging from testing temperature sensors, salinity sensors and pH sensors. For feeding system, testing tools is done by open and close testing of motor servo and motor delay.

2.1. Hardware design

The monitoring system consists of DC 5v source, sensor, servo, buzzer and LCD, also Arduino Uno microcontroller. The function of the sources above is to activate the entire system of measuring instruments and indicators. Block diagram monitoring system and feeding system can be seen in figure 1.

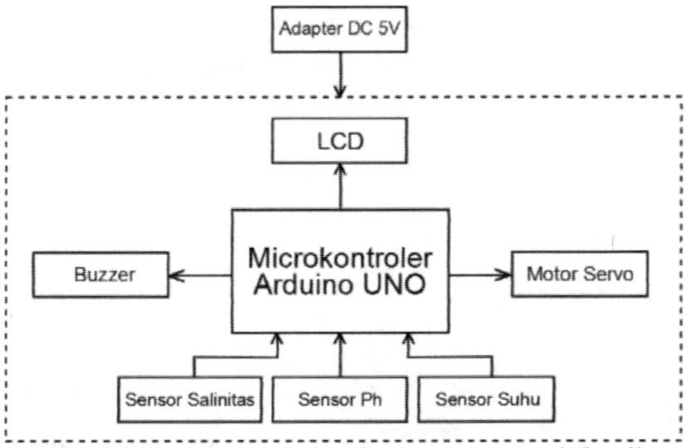

Figure 1. Block diagram monitoring system and feeding system.

From figure 1, it can be seen that the initial phase of DC 5v source activates the entire system while the PH Sensor is for measuring acidity of water. The sensors used in this monitoring system are glass electrodes that act as pH sensors connected to analog signals 0 arduino [1], vcc 5v and GND. Salinity sensors is used to measure water salinity. This sensor is one of the chemical sensors designed based on the electrical properties of water. The salinity sensor consists of two electrodes dipped in water. The data signal of the sensor is connected to analog data 1 arduino, vcc 5v and GND. Temperature sensor is DS18B20. The data obtained from the sensor will be sent to Arduino. Arduino will receive and manage data then displayed on the LCD. The buzzer will sound when one of the measurement values is not at the number specified in the Arduino program. Feeding system is done by pressing push button.

Figure 2. Mechanical design system monitoring and feeding system.

The black box in Figure 2 contains a series of monitoring system support with dimensions of 18 x 11 x 6.5 cm. This box consists of 2 push buttons, namely for the sensor menu button and the feeding system button.1 LCD is as an indicator to display the data measurement results. The back is a buzzer and potentiometer that has function to adjust the LCD brightness. The bottom box is as a fish feed container which contains a servo motor that open the lid of the box.

2.2. Software design

Software functions is to provide instructions and run the programs. Instruction is done to take the information obtained from the reading sensors Ph, salinity and temperature. Then it is processed on micro controller with program language C. Arduino Compiler.

Fritzing schematic design serves as the design of electronic equipment after the manufacture of prototypes using the actual equipment. The prototype is made above the breadboard board so that if there is an error it is easy to fix. It is also connected to Arduino to give commands on the tools used. The schematic design on fritzing software is shown in figure 3.

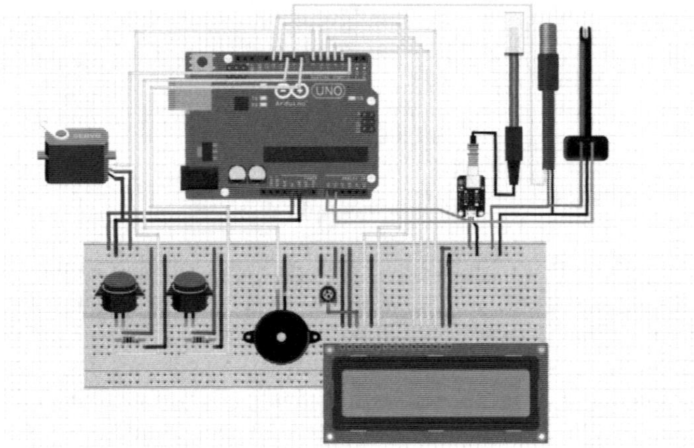

Figure 3. Schematic design on fritzing.

Schematic design of the circuit on PCB using eagle-7.7.0 software and then it is printed. The circuit design will be used instead of the breadboard board which will then be assembled on the box. The following (Figure 4) is the circuit design on the PCB board on eagle-7.7.0 software.

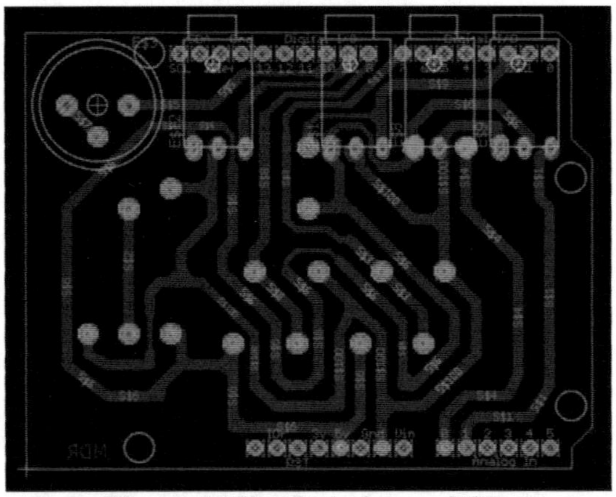

Figure 4. Circuit design on PCB.

3. Results and discussion

The results obtained from the experiment and through identifying data that includes the effects of weather changes and time. This monitoring system can measure the pH, temperature and salinity. The pH level in water can measure on a scale between pH 0 and pH 14, for temperatures it can measure on a scale of -10o C to 100o C, whereas for salinity the pH level can measure 0 to 1000μS salt content.

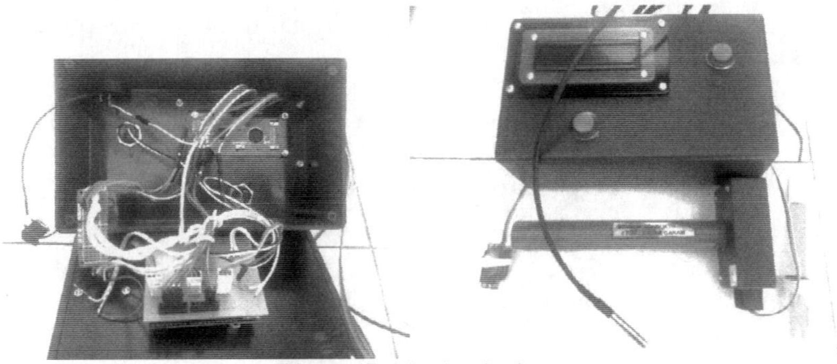

Figure 5. Mechanical system.

The experiment was conducted in freshwater fish cultivation ponds. The data is then processed and conducted data comparison with research ever undertaken by previous researchers. Data retrieval is taken on average and generates data and is presented in tables and graphs.

Data are presented in tables and graphs. Then determine the rate of the measurement error. The error rate can be determined using the relative error formula is.

$$\text{Error} = \frac{\text{measurement data} - \text{accuare data}}{\text{accurate data}}\ 100\%$$

The value of measurement data was obtained from the experimental results of the tool. The data in the form of tables and graphs from the results of pH sensor experiments on freshwater fish ponds are as follow:

Table 2. Results of pH.

	PH		
NO	**Sensor pH (n)**	**Research Data**	**Error (%)**
1.	7.4	7,3	1.3%
2.	7.5	7.45	0.6%
3.	7.5	7.4	1.3%
4.	7.4	7.5	1.3%
5.	7.5	7.15	4.8%
Error Average			**1.8%**

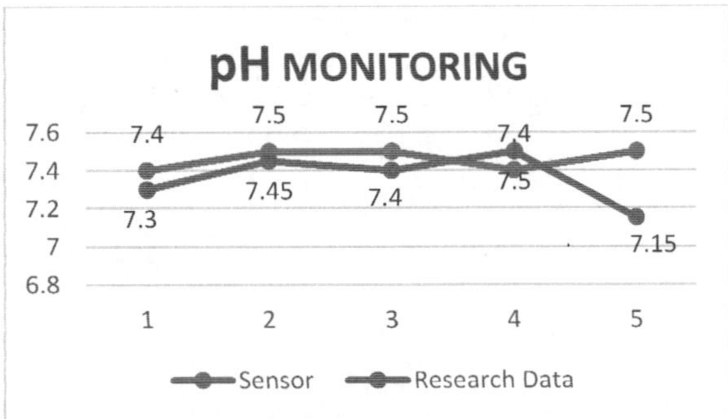

Figure 6. Graph of pH.

Data in the form of tables and graphs from the experimental results of temperature sensors in freshwater fish ponds is as follow.

Table 3. Result of temperature.

NO	TEMPERATURE		
	Temperature Sensor	**Research Data**	**Error (%)**
1.	25.3°C	24.81 °C	1.9%
2.	26.2°C	26.5 °C	1.1%
3.	26.3° C	26.62 °C	1.1%
4.	25.2°C	26.75 °C	5.6%
5	26.2°C	27 °C	2.9%
Error Average			**2.5%**

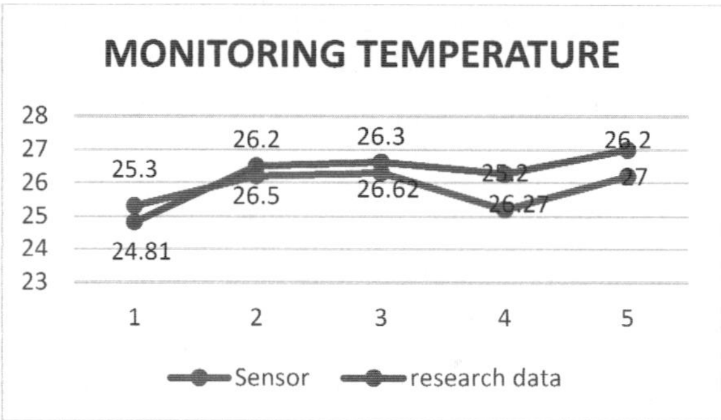

Figure 7. Graph of temperature sensor.

Data in the form of tables and graphs from the experimental results of temperature sensors in freshwater fish ponds are in Table 4.

Table 4. Results of salinity.

NO	SALINITY		Error (%)
	Salinity Sensor	**Research Data**	
1.	106μS	107μS	0.9%
2.	107μS	108μS	0.9%
3.	108μS	110μS	1.8%
4.	107μS	109μS	1.8%
5.	107μS	109μS	1.8%
Error Average			**1.4%**

IOP Conf. Series: Materials Science and Engineering **384** (2018) 012034 doi:10.1088/1757-899X/384/1/012034

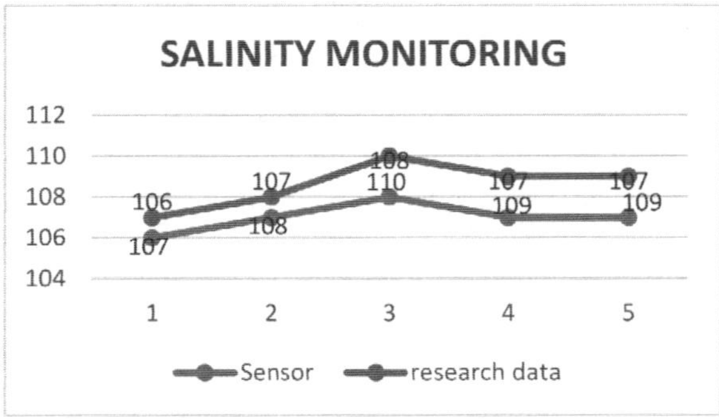

Figure 8. Graph of the salinity.

The following error rate data (Figure 9) from the three sensors in the form of experiment result graph.

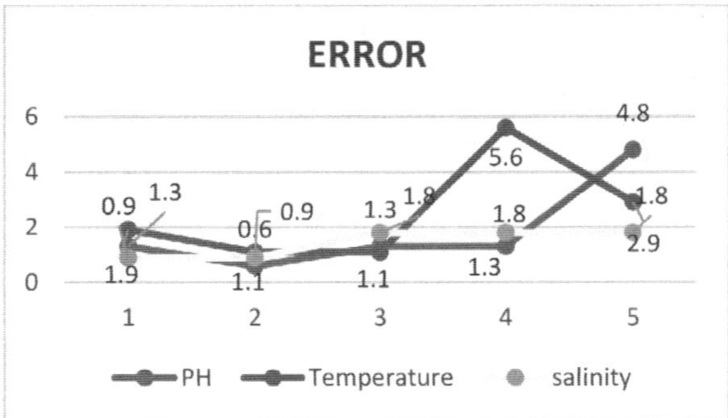

Figure 9. Graph of error rate.

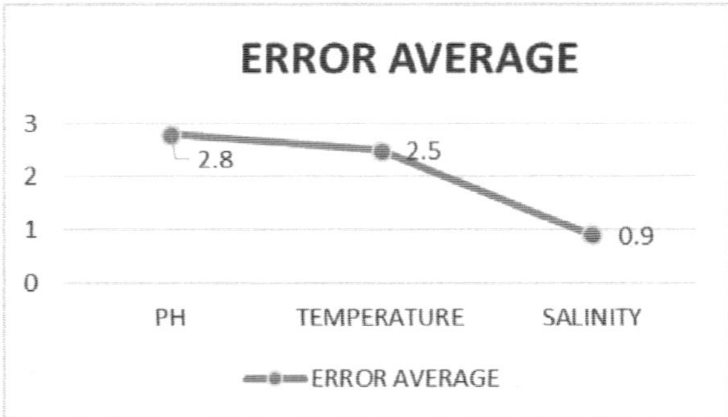

Figure 10. Graph of error average.

From figure 10, the rate of error monitoring system is fairly small. The average measurement error compared with research data is said to be quite accurate. The rate of monitoring system is below 3% and can be applied to freshwater fish cultivation.

4. Conclusions

In this research the monitoring system works by using Arduino UNO ATmega328PA micro controller as controller. Arduino is programmed using C language to run instructions. The monitoring system is arranged in several sections ranging from circuit box, ph sensors, salinity sensors, temperature and box feeding system sensors. This system makes it easier to monitor fresh water content in freshwater fish cultivation with a small error rate in every weather change comes. This monitoring system also has 3 sensors or measuring devices at once and also equipped with a remote feeding system. This study is expected to meet and increase production, income and reduce the rate of fish death in freshwater fish cultivation.

References

[1] Shidiq M and Rahardjo P M 2008 "Pengukur Suhu dan pH Air Tambak Terintegrasi dengan Data Logger" *Eeccis* **II** (1) pp. 22–25.
[2] Chen J H, Sung W T and Lin G Y 2016 "Automated Monitoring System for the Fish Farm Aquaculture Environment" *Proc. - 2015 IEEE Int. Conf. Syst. Man, Cybern.* pp. 1161–1166.
[3] Randhawa S, Sandha S S and Srivastava B 2017 "A Multi-sensor Process for In-Situ Monitoring of Water Pollution in Rivers or Lakes for High-Resolution Quantitative and Qualitative Water Quality Data" *Proc. - 19th IEEE Int. Conf. Comput. Sci. Eng. 14th IEEE Int. Conf. Embed. Ubiquitous Comput. 15th Int. Symp. Distrib. Comput. Appl. to Business, Engi* pp. 122–129.
[4] Qiu J 2014 "The polution of Heavy Metals in Sediments from Beng river" pp. 62–65.
[5] Garcia M, Sendra S, Lloret G and Lloret J 2011 "Monitoring and control sensor system for fish feeding in marine fish farms" *IET Commun.* **5** (12) pp. 1682–1690.

IOP Conf. Series: Materials Science and Engineering **384** (2018) 012035 doi:10.1088/1757-899X/384/1/012035

Analysis of 20 kV Power Distribution System Reliability using the Section Technique

M C W Lestari* and Y Mulyadi

Department of Electrical Engineering Education, Universitas Pendidikan Indonesia, Jl. Dr. Setiabudhi No. 207 Bandung 40154, Indonesia

*madecitrawidyal@student.upi.edu

Abstract. Electrical power is one of the basic human needs in life. The development of electrical power, as time progresses, requires power systems to have a better level of reliability. This research aims to know the level of 20 kV power distribution system reliability at feeder. The feeder analyse is a PKW feeder from PLN Area Bandung Utara The level of reliability is obtained from index SAIDI (System Average Interruption Duration Index) and index SAIFI (System Average Interruption Frequency Index). This research uses the method of Section Technique. From the addition on each section's index, the value of SAIDI is 3.112934347 hour/year and SAIFI is 1.1152 times/year.

1. Introduction

Power distribution system has a role on well distribution of electricity. Electricity energy is very important necessity for human activity. Usually, reliability of distribution power can be defined as ability of a system to give enough supply of power with satisfying quality [1]. Therefore, power distribution system should have a good reliability so human need of electricity power won't be disturbed.

The level of reliability can be disturbed by some obstacles like natural ecosystem, human activity, old instruments, even animals around the power distribution feeder [2]. If power distribution feeder often get a problem, the quality of power distribution will worsen [3]. There are known indexes to use for knowing the reliability of power distribution such as SAIDI (System Average Interruption Duration Index) and SAIFI (System Average Interruption Frequency Index).

There are some factor that should be known and counted before analyzing the level of reliability, which is frequency of failure and duration of the failure. In this research, author is using Section Technique method for process data which is obtained and produced a reliability level. This method is similar to FMEA (Failure Mode and Effect Analysis) method which counts reliability of a system based on effects of failure with the system and how it will affect the load point. Consequences from individual failure of an instrument will systematically identified with analysis on what happened when problem occurs [4]. But on the method used by author that the calculation will be separated on some sections. After that it will be compared with SPLN 59:1985.

2. Reliability on power distribution

2.1. Definition and basic theory of reliability
Reliability is defined as probability of an instrument of system to be able to function as it supposed to be, during some time and under some work conditions. So, reliability on distribution system is probability of distribution system for be able to work functionally under certain time range and under certain work conditions [5].

2.2. Reliability on distribution of power
Level of distribution system's reliability can be measured from how far the power distribution continuously occurred to consumers without power outage. Along with time, there are growing numbers of power load which can be seen on growing numbers of industrial site, economic growth, also new housing that demanding a high reliability level over time [5].

2.3. *Calculation on reliability index*

2.3.1. Average failure rate (λi). Average failure rate is frequency of failure on each component that will affect load point [4].

$$\lambda s = \sum \lambda i \quad (1)$$

Where λi is Failure Rate on load point i.

2.3.2. Average annual outage time (Us). Average annual outage time is duration on power outage that happens on each component that will affect load point [4].

$$Us = \sum \lambda i.ri \quad (2)$$

Where λi is Failure rate on load point i and ri is Repair time or Switching.

2.3.3. System Average Interruption Duration Index (SAIDI). SAIDI is index that used to calculate level of reliability where the index counts the median on outage duration that inflicted on customer in certain period of time [6].

$$SAIDI = \frac{\sum Ui.Ni}{\sum N} \quad (3)$$

Where:
Ni = Number of consumer on load point
Ui = Duration of instrument failure on load point
N = Number of consumer on feeder

2.3.4. System Average Interruption Frequency Index (SAIFI). SAIFI is index that used to calculate level of reliability where the index counts the median on outage frequency that inflicted on customer in certain period of time [6].

$$SAIFI = \frac{\sum \lambda i.Ni}{\sum N} \quad (4)$$

Where:
Ni = Number of consumer on load point
λi = frequency of failure rate on load point
N = Number of consumer on feeder

2.3.5. SPLN 59:1985. Standard that used for Failure rate and repair time on power distribution system instrument 20kV is SPLN 59:1985, listed at Table 1.

Table 1. SPLN 59:1985.

Component	λ (Failure Rate)	ri (Repair Time) (hour)	ri (Switching Time) (hour)
Distribution Trafo	0,005/unit/year	10	0,15
Circuit Breaker	0,004/unit/year	10	0,15
Recloser/sectionalizer	0,003/unit/year	10	0,15
Overhead line	0,2/km/year	3	0,15
Underground Cable	0,07/km/year	10	0,15

3. Research methods

Research on Reliability of power distribution system 20 kV on PLN Area Bandung Utara PKW feeder using descriptive analysis method, which is research method that uses collective data directly and analyze to make a conclusion that can be used as base for consideration on good power distribution, and result of maintenance can be reference on fixing system performance on power distribution to consumer.

Analysis used is using section technique method which is a method that evaluate reliability by dividing system on smaller section to minimalize error on shorter time. This method on evaluating reliability is based on how failure of an instrument affects system operation. Effects or consequences from individual instrument's failure will systematically identified with analysis on what happened if error happened. Then each of instrument failure will be analyzed on each load point using approach from bottom to top that considered on one failure at a time [1]. Figure 1 shows the SAIDI and SAIFI calculation steps using Section Technique Method.

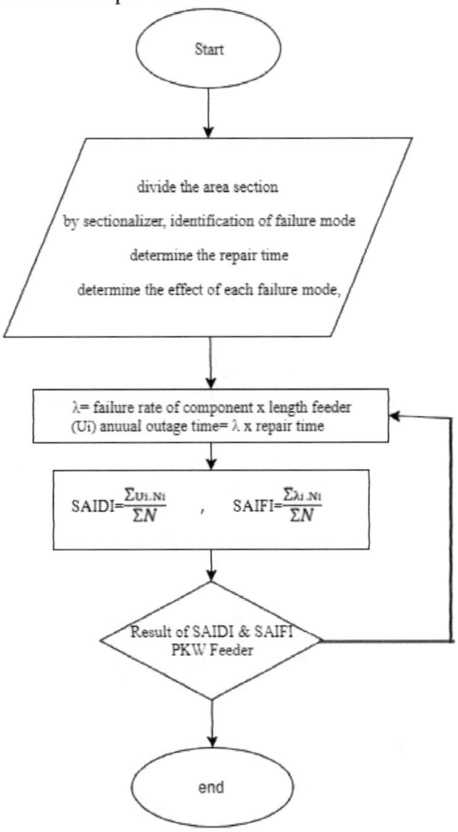

Figure 1. Steps of section technique method.

IOP Conf. Series: Materials Science and Engineering **384** (2018) 012035 doi:10.1088/1757-899X/384/1/012035

4. Results and discussion

4.1. *Single line diagram of feeder*

On this Reliability level analysis the data used from one feeder from Bandung Utara that is PKW Feeder. This feeder supplied from Dago Bengkok Electric Power Substation. To analyses component at PKW feeder needs a single line feeder diagram as shown in Figure 2.

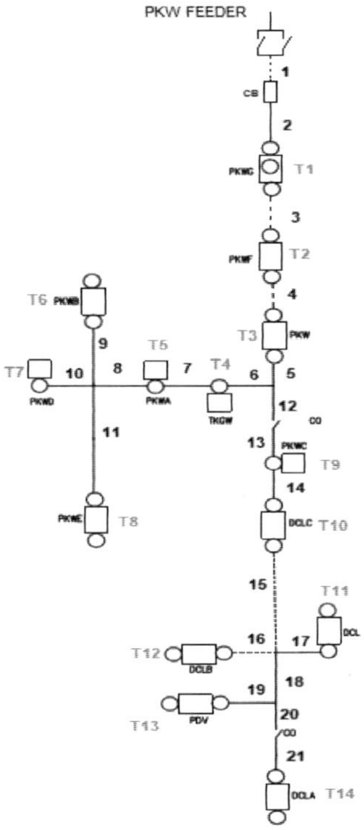

Figure 2. Single line diagram PKW feeder.

PKW feeder has 14 load points in form of distribution transformator and 2 sectionalizer. With 1.796 consumer, total length of feeder is 6,482 kms. This PKW feeder will be devided on 2 section.

Total of consumer data on each analyzed feeder load point can be seen on Table 2 and Table 3.

Table 2. Total consumer on load point PKW feeder section 1.

Load Point	Power Substation Distribution	Total Consumer
1	PKWG	1
2	PKWF	330
3	PKW	330
4	TKGW	53
5	PKWA	54
6	PKWB	36
7	PKWD	67
8	PKWE	27

Table 3. Total consumer on load point PKW feeder section 2.

No	Load Point	Power Substation Distribution	Total Consumer
1	9	PKWC	28
2	10	DCLC	1
3	11	DCL	550
4	12	DCLB	120
5	13	PDV	26
6	14	DCLA	173

PKW feeder is a feeder that doesn't have a lot of line. Total length of 6,482 kms feeder is divided into 21 lines. Data on length of each line on PKW feeder can be seen on Table 4.

Table 4. Channel length.

Channel	Length (Kms)	
	Overhead line	Underground cable
Line 1	0.57	
Line 2		0.375
Line 3	0.155	
Line 4	0.153	
Line 5		0.09
Line 6		0.095
Line 7		0.127
Line 8		0.253
Line 9		0.398
Line 10		0.253
Line 11		0.52
Line 12		0.214
Line 13		0.223
Line 14		0.253
Line 15	0.48	
Line 16	0.132	
Line 17		0.389
Line 18		0.41
Line 19		0.314
Line 20		0.285
Line 21		0.793
Total	**1.49**	**4.992**

4.2. Reliability system index analysis using section method technique
On reliability system index analysis in PKW feeder is divided on two section. This calculation is using failure rate and recovery time on 20kV distribution network standard based on SPLN 59:1985[1]. Calculation of reliability on each section is explained on below.

4.2.1. Section 1. To know failure rate and recovery rate on system can be seen on failure mode list. Failure mode list for section 1 can be seen on Table 5.

Table 5. Failure mode list section 1.

Instrument Data		Load Point Affected by Repair Time	Load Point Affected by Switching Time
Number of failure	**Instrument**		
1	CB	LP 1 – LP 14	-
2	T1	LP 1	-
3	T2	LP 2	-
4	T3	LP 3	-
5	T4	LP 4	-
6	T5	LP 5	-
7	T6	LP 6	-
8	T7	LP 7	-
9	T8	LP 8	-
10	L1	LP 1 – LP 14	-
11	L2	LP 1 – LP 14	-
12	L3	LP 1 – LP 14	-
13	L4	LP 1 – LP 14	-
14	L5	LP 1 – LP 14	-
15	L6	LP 1 – LP 14	-
16	L7	LP 1 – LP 14	-
17	L8	LP 1 – LP 14	-
18	L9	LP 1 – LP 14	-
19	L10	LP 1 – LP 14	-
20	L11	LP 1 – LP 14	-
21	L12	LP 1 – LP 14	-

Next to count frequency on failure and duration on failure in each load point. One sample case took from LP1. LP1 score gained from adding number of instrument failure by multiply failure rate of overhead line/underground cable with length of the line according to (1) formula. U LP1 score can also be gained by multiplying number of λ with repair time which can affect LP1 according to (2) formula. These can be seen on Table 6.

Table 6. Calculation of λ and U LP 1 section 1.

Component	Failure rate of component	Length Fedder	λ (fault/yr)	Repair time (hr)	Switching time (hr)	U (hr/yr)
CB	0.004		0.004	10		0.04
T1	0.005		0.005	10	0.15	0.05
L1	0.07	0.57	0.0399	10	0.15	0.399
L2	0.2	0.375	0.075	3	0.15	0.225
L3	0.07	0.155	0.01085	10	0.15	0.1085
L4	0.07	0.153	0.01071	10	0.15	0.1071
L5	0.2	0.09	0.018	3	0.15	0.054
L6	0.2	0.095	0.019	3	0.15	0.057
L7	0.2	0.127	0.0254	3	0.15	0.0762
L8	0.2	0.253	0.0506	3	0.15	0.1518
L9	0.2	0.398	0.0796	3	0.15	0.2388
L10	0.2	0.253	0.0506	3	0.15	0.1518
L11	0.2	0.52	0.104	3	0.15	0.312
L12	0.2	0.214	0.0428	3	0.15	0.1284
	TOTAL		0.53546	TOTAL		2.0996

On load point 1, condition occurred at all the instrument only repair time condition, which means no instrument on switching time condition. Therefore to seek value of U on table 6 only repair time condition only. After getting the number of λ and U on each load point, next is to count the value of SAIDI and SAIFI on section 1. Counting the value of SAIDI is using formula (3) and to count the value of SAIFI is using formula (4). The result of calculation of SAIDI and SAIFI value is can be seen on Table 7.

Table 7. SAIDI and SAIFI score section 1.

Load point	λLp	Ulp	Total of customer	SAIDI	SAIFI
1	0.53546	2.0996	1	0.001169	0.0002981
2	0.53546	2.0996	330	0.385784	0.0983863
3	0.53546	2.0996	330	0.385784	0.0983863
4	0.53546	2.0996	53	0.0619592	0.0158014
5	0.53546	2.0996	54	0.0631283	0.0160996
6	0.53546	2.0996	36	0.0420855	0.0107331
7	0.53546	2.0996	67	0.0783258	0.0199754
8	0.53546	2.0996	27	0.0315641	0.0080498
9	0.53046	2.0496	28	0.0319537	0.00827
10	0.53046	2.0496	1	0.0011412	0.0002954
11	0.53046	2.0496	550	0.6276615	0.162446
12	0.53046	2.0496	120	0.1369443	0.0354428
13	0.53046	2.0496	26	0.0296713	0.0076793
14	0.53046	2.0496	173	0.1974281	0.0510966
	TOTAL		**1796**	**2.0746**	**0.53296**

From calculation result, the scores of SAIDI and SAIFI on section 1 are respectively 2.0746 hour/year and 0.5326 times/year.

4.2.2. Section 2. List of failure mode on section 2 can be seen on Table 8.

Table 8. Mode failure list section 2.

Instrument Data		Load Point Affected by Repair Time	Load Point Affected by Switching Time
Number of failures	Instrument		
1	T9	LP 9	-
2	T10	LP 10	-
3	T11	LP 11	-
4	T12	LP 12	-
5	T13	LP 13	-
6	T14	LP 14	-
7	S1	LP 9 – LP 14	LP1-LP8
8	S2	LP 14	LP1-LP13
9	L13	LP 9 – LP 14	LP1-LP8
10	L14	LP 9 – LP 14	LP1-LP8
11	L15	LP 9 – LP 14	LP1-LP8
12	L16	LP 9 – LP 14	LP1-LP8

IOP Conf. Series: Materials Science and Engineering **384** (2018) 012035 doi:10.1088/1757-899X/384/1/012035

Table 8. Cont.

Instrument Data		Load Point Affected by Repair Time	Load Point Affected by Switching Time
Number of failures	Instrument		
13	L17	LP 9 – LP 14	LP1-LP8
14	L18	LP 9 – LP 14	LP1-LP8
15	L19	LP 9 – LP 14	LP1-LP8
16	L20	LP 9 – LP 14	LP1-LP8
17	L21	LP 14	LP1-LP13

Next is to calculate failure frequency value and duration of failure on each load point. This can be seen on Table 9.

Table 9. λ and U LP 1 calculation section 1.

Component	Failure rate of component	Length feeder	λ (fault/yr)	Repair time (hr)	Switching time (hr)	U (hr/yr)
S1	0.003		0.003		0.15	0.00045
S2	0.003		0.003	10	0.15	0.00045
L13	0.2	0.223	0.0446	3	0.15	0.00669
L14	0.2	0.253	0.0506	10	0.15	0.00759
L15	0.07	0.48	0.0336	10	0.15	0.00504
L16	0.07	0.132	0.00924	3	0.15	0.001386
L17	0.2	0.389	0.0778	3	0.15	0.01167
L18	0.2	0.41	0.082	3	0.15	0.0123
L19	0.2	0.314	0.0628	3	0.15	0.00942
L20	0.2	0.285	0.057	3	0.15	0.00855
L21	0.2	0.793	0.1586	3	0.15	0.02379
	TOTAL		**0.58224**	**TOTAL**		**0.087336**

After the value of λ and U on each load point, next is to calculate value of SAIDI and SAIFI on section 2. The result of SAIDI and SAIFI calculation on section 2 can be seen on Table 10.

Table 10. SAIDI and SAIFI score section 2.

Load point	λLp	Ulp	Total of customer	SAIDI	SAIFI
1	0.58224	0.087336	1	4.86281E-05	0.000324187
2	0.58224	0.087336	330	0.016047261	0.106981737
3	0.58224	0.087336	330	0.016047261	0.106981737
4	0.58224	0.087336	53	0.002577287	0.017181915
5	0.58224	0.087336	54	0.002625915	0.017506102
6	0.58224	0.087336	36	0.00175061	0.011670735
7	0.58224	0.087336	67	0.00325808	0.021720535
8	0.58224	0.087336	27	0.001312958	0.008753051
9	0.58224	1.89656	28	0.029567751	0.009077238
10	0.58224	1.89656	1	0.001055991	0.000324187
11	0.58224	1.89656	550	0.5807951	0.178302895
12	0.58224	1.89656	120	0.126718931	0.03890245
13	0.58224	1.89656	26	0.027455768	0.008428864
14	0.58224	2.37812	173	0.229072806	0.056084365
TOTAL			**1796**	**1.038334347**	**0.58224**

From the calculation above, the value on SAIDI and SAIFI are 1.038334347 hour/year and 0.58224 times/year respectively. After gained value on reliability index for each section, the reliability index can be gained by adding each value from each section as can be seen on Table 11.

Table 11. Analysed feeder reliability index.

Section	SAIDI	SAIFI
1	2.0746	0.53296
2	1.038334347	0.58224
TOTAL	**3.112934347**	**1.1152**

From the addition on each section's index, the value of SAIDI and SAIFI are 3.112934347 hour/year and 1.1152 times/year respectively. This result will be maximum limit on reliability.

SAIDI and SAIFI value can be compared with other SAIDI and SAIFI value from PT. PLN rayon Bandung Utara on year 2017, where SAIDI value is 0.0567 hour/year and SAIFI value is 0.0875 times/year. It can be seen from SAIDI and SAIFI value on PKW feeder is reliable because it has value below the maximum reliability limit.

5. Conclusions

From the result of calculation on this research, can be concluded that good value for PKW feeder's reliability index are 3.112934347 hour/year for SAIDI and 1.1152 times/year for SAIFI. After being compared to PT. PLN Rayon Bandung Utara on 2017 it can be concluded that PKW feeder is reliable.

References

[1] Wicaksono H P and Hernanda I G N S 2012 "Analisis Keandalan Sistem Distribusi Menggunakan Program Analisis Kelistrikan Transien dan Metode Section Technique" *J. Tek. Its* **1** (1) pp. 153–158.

[2] Ramdan M and Herawan F 2016 "Analisis Kehandalan Kinerja Penyulang 20 kV Di PT . PLN (Persero) Distribusi Jawa Barat Area Cirebon Rayon Ciledug" no. 1301475.

[3] Erhaneli E 2016 "Evaluasi Keandalan Sistem Distribusi Tenaga Listrik Berdasarkan Indeks Keandalan Saidi Dan Saifi Pada PT . PLN (Persero) Rayon Bagan Batu Tahun 2015" *Jurnal Teknik Elektro-ITP* **5** (2).

[4] Santoso R 2016 "Evaluasi Tingkat Keandalan Jaringan Distribusi 20 kV Pada Gardu Induk Bangkinang Dengan Menggunakan Metode FMEA (Failure Mode Effect Analysis)" *Jurnal Online Mahasiswa Fakultas Teknik Universitas Riau* **3** (2) pp. 1-7.

[5] Fatoni A 2017 "Analisa Keandalan Sistem Distribusi 20 kV PT. PLN Rayon Lumajang dengan Metode FMEA (Failure Modes and Effects Analysis)" *Jurnal Teknik ITS* **5** (2) pp. 462–467.

[6] Fauziah F, Penangsang O and Soeprijanto A 2012 "Studi Perbaikan Keandalan Jaringan Distribusi Primer Dengan Pemasangan Gardu Induk Sisipan Di Kabupaten Enrekang Sulawesi Selatan" *Jurnal Teknik ITS* **1** (1) pp. B119-B124.

IOP Conf. Series: Materials Science and Engineering **384** (2018) 012036 doi:10.1088/1757-899X/384/1/012036

A Framework of Counseling System for Student Guardianship using Case Based Reasoning (CBR) Inference

M Muslim[1,2*], A Alwi[1,2] and E Erika[1]

[1] Muhammadiyah University, Ponorogo, Indonesia
[2] Gadjah Mada University, Yogyakarta, Indonesia

*munirah.mt@gmail.com

Abstract. Case Based Reasoning (CBR) based systems are systems that use case-based inference methods, in this case identical to previous experiences used to predict solutions to some new problems. In this study, the framework of the CBR-based counseling system was developed, which is expected to support and represent the guardian's lecturers to face students in counseling, where students can receive solutions of all kinds of problems encountered in this study. Within this framework, counseling data will be processed into the basis of knowledge, whether it be a problem or a solution into the counseling system. Using a CBR based algorithm, the data will be training data to be able to provide a solution to the next incoming problem.

1. Introduction

The definition of CBR was created by Riesbeck and Schank [1]: "A case-based reasoned solves problems by using or adapting solutions to old problems." while the methodology for CBR was describes by Peter Checkland [2] as a "set of principles which guide action". What are they? The first of these is a desire by the problem solver to solve a problem by explicitly trying to reuse a solution from a similar past problem.

Thus, a CBR must retrieve cases from a case-library and in some way assess the similarity of cases in the library to the current problem description. Second, a CBR system should attempt to reuse the solution suggested by a retrieved case, either with or without revision. Finally, a CBR system should seek to increase its knowledge by retaining new cases. The subsequent sections will show how four different applications use this set of principles, defined as CBR, to solve real-world problems [3]. CBR is a problem-solving methodology that aims at reusing previously solved and memorized problem situations, called cases. A case is a concrete problem-solving experience. One of the main assets of CBR is its eagerness to learn.

Learning in CBR can be as simple as memorizing a new case or can entail refining the memory organization or meta-learning schemes [4]. CBR working cycle can be described best in terms of four processing stages (R4 model) as shown in Figure 1 [5]: (a) RETRIEVE the most similar case(s). (b) REUSE the case(s) to attempt to solve the current problem. (c) REVISE the proposed solution if necessary. (d) RETAIN the new solution as a part of a new case. Other paragraphs are indented.

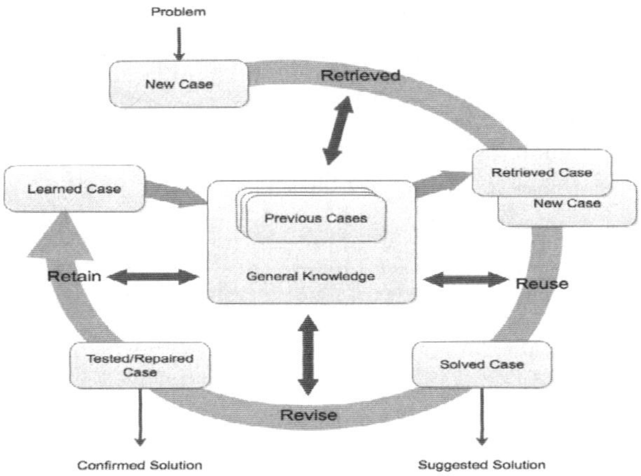

Figure 1. CBR life cycle.

From model R4, then developed model R5 by Gavin and Zhaohao [6]. They propose an additional feature by adding the REPARTITION feature.

2. Discussion

Currently, there have been many papers discussing the implementation of CBR in various fields, such as medicine [7][8][9], education [10], and many research and technology fields. For this paper, we will discuss how to build a counseling system using CBR method. Student counseling is usually done by students who want to share with lecturers about the various academic problems they by meet face to face on campus. It was conducted between the lecturers who were appointed as guardian lecturers to the students under their guardianship. Things students usually consult are formal matters of lectures or about family issues that lie behind the academic difficulties they face.

Official matters such as consultations on what courses are appropriate for them to take in the next semester, how much credit they can take, about the final project, the fees to be paid, the procedures administrative or academic to pass and so forth. Things that are informal or semi-official e.g. previously problematic courses should be faced in what way, about the conflict between the student and his lecturer, about his financial difficulties in financing the study, how to study which is effective, how the strategy of dealing with a lecture.

2.1. Concept of CBR algorithm for counseling expert system

From the concepts that have been made by Munirah and Aslan [8], has constructed and described algorithms for counseling systems based on CBR. The Expert System based CBR implies that the knowledge base of the expert system is a set of cases, cases taken from CBR, and the inferencing machine is the CBR method. In the implementation, the case in question is a card that has a general structure as follows:

- Section Questions asked by students (next called as Head)
- Section Questions asked by the expert system and answered by the student (next called as Body)
- Section Solutions that answer student questions (next called as Tail).

Furthermore, related to how the CBR process in the basis of cards can be explained by the following steps:

Step 1. All the cards in the system form clusters called clusters of cards.
Step 2. Each card may only join one cluster.

Step 3. Each card in the same cluster will have the same set of parameters, the contents or answers of those parameters may be different.
Step 4. Each card in the same cluster has different or equal solutions.
Step 5. Each cluster has a cluster keyword.
Step 6. Cluster keyword = the combined number of keywords from card heads joined in the cluster.
Step 7. If there is a new card, the new card selects the closest cluster as its cluster.
Step 8. The proximity of a cluster with a new card, measured by a metric formulated as follows:

$$\Delta(X, A) = \frac{1}{\Delta key\_word(X, A) + 1} \tag{1}$$

Where: Δkey_word(X, A) = the number of cluster X keywords that are the same as the A card keyword

Step 9. The formation of a new cluster is determined by setting a threshold of distance between any clusters with a new incoming card. The assignment of a threshold is set based on assumptions, which can be corrected as the system goes where users can estimate better accuracy.

- Example of threshold setting:
  A new card A forms a new cluster, if and only if for all X cluster on the system is filled with d (X, A)> 1/5.
- Example of threshold use:
  Suppose in the system there are two clusters, namely cluster X and cluster Y, then there is a new card A that enters the system. And threshold is set equal to 1/5. For example:
  Δkey_word (X, A) = 4
  Δkey_word (Y, A) = 3
  Then the cluster proximity is obtained:
  d (X, A) = 1/4
  d (Y, A) = 1/3
  Then the result is:
  d (X, A)> 1/5 and d (Y, A)> 1/5
  Means, card A forms a new cluster different from the cluster X and cluster Y, call it for example cluster Z, where A ∈ Z.

Step 10. When a new card, entered into a cluster, then automatically the new card has the same parameters with all cards in the cluster. Then the new card is returned to the user to manually fill those parameters.
Step 11. When a new card has been filled in its parameters, the new card is returned to its original cluster, to find a solution that matches the card.
Step 12. The determination of the card solution is done by creating a metric that measures the distance between the new card and the other cards in the same cluster.
Step 13. Metrics that measure between cards in the same cluster are defined as follows:
Suppose that card B and card C are in cluster X, and there is a new card A that is returned to cluster X after parameter A is filled by the student.
The similarity between card A to card B is expressed by a metric formulated as follows:

$$J(A, B) = \frac{1}{\Delta parameter\ of\ (A, B)} \tag{2}$$

Where: $\Delta parameter\ of\ A, B$ = the number of answers with the same parameters between A and B

For Example:
$\Delta parameter\ of\ (A, B)$ = 5 (there are 5 answers with the same parameters between A and B), then the similarity is: J (A, B) = 1/5, while $\Delta parameter\ of\ (A, C)$ = 6 (there are 6 answers to the same parameters between A and C), so the similarity is: J (A, C) = 1/6

Conclusion:
A is closer to C then to B, so solution A = solution C

Step 14. Related to how a new solution is formed, a threshold is created for the J (A, B) metric that determines when a new solution is formed. Suppose the threshold is set to 1/5, this means:
If for every Y card is any in the X cluster, and the new card A, where J (A, Y) <1/5, then there is no solution matching one of the cards in cluster X, so card A forms a new solution. New solutions are provided by an expert or expert related to the field in question on card A, then stored (adapted) back into cluster X for next use.

2.2. Framework for the counseling expert system
Based on the concept that has been described above, all the work steps in the above algorithm are depicted in the framework scheme as shown in Figure 2.

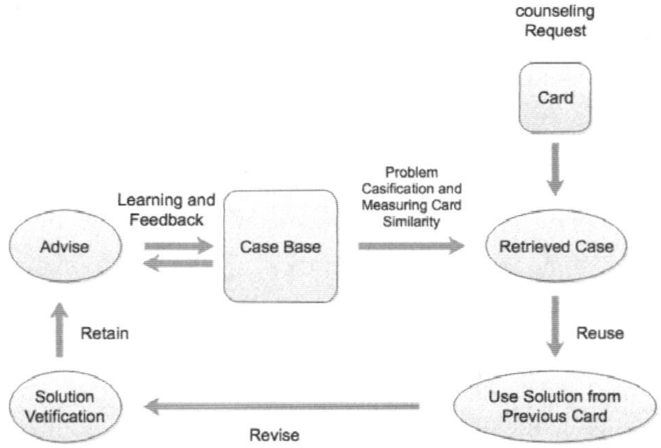

Figure 2. The framework for the counseling expert system.

3. Conclusions
From the results of this study, can be concluded:
- Construction of counseling expert system can be a solution to facilitate student counseling system and make it easier for guardian lecturers to handle more students in their guardianship even though they cannot meet face to face.
- CBR inference method is implemented by formulating metrics based on matching words on student questions, and it is hoped that the use of these metrics into the system can flexibly provide solutions to student problems.
- This scheme is a framework of counseling system based on an expert system with CBR inference method.

References
[1] Riesbeck C K, Schank R 1989 *Inside Case-based Reasoning* (Erlbaum, Northvale, NJ).
[2] Checkland P, Scholes J 1990 *Soft Systems Methodology in Action* (Wiley, NY).
[3] Watson I 1999 Case-based reasoning is a methodology not a technology **12** p 303–308.
[4] El-sappagh S H and Elmogy M 2015 Case Based Reasoning: Case Representation Methodologies **6**(11) p 192–208.
[5] Aamodt A and Plaza e 1994 CBR: foundational issues, methodological variations and system approaches *AI Communications* **7**(1) p 39-59.
[6] Gavin F and Zhaohao S 2003 R5 model for case-based reasoning *Knowledge- Based Systems* **16**(1) p 59–65.

[7] Aribowo A S 2010 Pengembangan Sistem Cerdas Menggunakan Penalaran Berbasis Kasus (Case Based Reasoning) Untuk Diagnosa Penyakit Akibat Virus Eksantema *Jurnal Telematika* **7**(1).

[8] Octaviani F dkk. 2011 Implementasi Case Based Reasoning Untuk Sistem Diagnosis Penyakit Anjing *Jurnal Informatika* **7**(2).

[9] Romadhan A N 2013 *Implementasi Case-Based Reasoning Untuk Pendukung Dokter Jaga Dalam Mendiagnosa Penyakit Pada RSU PKU Muhammadiyah Delanggu* (STMIK AMIKOM, Yogyakarta).

[10] Munirah M and Aslan A 2016 Konsep Skema Sistem Cerdas Buatan Konseling Mahasiswa Berbasis Case Based Reasoning (CBR) *Seminar Nasional Telekomunikasi dan Informatika (SELISIK)*, Universitas Pasundan Bandung, ISSN 2527-5488.

IOP Conf. Series: Materials Science and Engineering **384** (2018) 012037 doi:10.1088/1757-899X/384/1/012037

Fabrication and Characterization of Thick Film Ceramics $La_{0,9}Ca_{0,1}FeO_3$ for Ethanol Gas Sensor using Extraction of Fe_2O_3 from Yarosite Mineral

N I Ariyani[1], D G Syarif[2] and E Suhendi[1,*]

[1] Program Studi Fisika, Universitas Pendidikan Indonesia, Jl. Dr. Setiabudhi 229 Bandung 40154, Indonesia
[2] PSTNT BATAN, Jl. Tamansari 71 Bandung 40132, Indonesia

*endis@upi.edu

Abstract. Fabrication of $La_{0,9}Ca_{0,1}FeO_3$ thick film ceramics using Fe_2O_3 powder extracted from yarosite mineral as ethanol gas sensor has been successfully performed. Fe_2O_3 powder extracted from yarosite mineral as the basic material in this research can increase the added value yarosite mineral. Fe_2O_3 powder and 10% mol of CaO dissolved in HCl were mixed with $LaCl_3.7H_2O$ powder dissolved in aquades.The solution of Fe_2O_3, $LaCl_3.7H_2O$ and CaO mixed and then precipitate using NH_4OH. The $La_{0,9}Ca_{0,1}FeO_3$ precipitate was calcined at temperature 800^{0}C for 2 hours to produce $La_{0,9}Ca_{0,1}FeO_3$ powder. $La_{0,9}Ca_{0,1}FeO_3$ powder was crushed and mixed with Organic Vehicle to produce a $La_{0,9}Ca_{0,1}FeO_3$ paste. Using the screen printing technique, the $La_{0,9}Ca_{0,1}FeO_3$ paste is coated on the alumina substrate and then fired at 600^{0}C for 2 hours to produce the thick film ceramic $La_{0,9}Ca_{0,1}FeO_3$. Based on XRD and SEM characterization data, the thick film ceramics $La_{0,9}Ca_{0,1}FeO_3$ were made even though some of Fe_2O_3 did not react and the grain size is almost uniform, and there are many pores. Measurement of electrical characteristics shows a good response to the presence of ethanol gas, has high electrical sensitivity value and low optimum working temperature that is in the range 290^{0}C - 295^{0}C.

1. Introduction

A gas sensor is an electronic device that can generate electrical signals a response to chemical interactions with gases [1]. The ethanol gas sensor plays a role in detecting the presence of ethanol gas in food safety testing [2], testing of ethanol in room, in food and the human body via an alcohol test at the mouth (for drivers) [2-3]. Metal Oxide Semiconductor (MOS) is a capable material used to make gas sensors because it has many advantages, such as cheaper price, fast response time and can detect various types of gas [4]. Currently, many researchers employed p-type semiconductors as gas sensors, one is thick film ceramics of $LaFeO_3$. The thick film ceramics of $LaFeO_3$ has good response to the presence of ethanol gas [2, 5-6].

In this research, the thick film ceramic of $LaFeO_3$ was added with 10% mole Ca^{2+} to make a thick film ceramic of $La_{0,9}Ca_{0,1}FeO_3$ and the Fe_2O_3 used was that extracted from yarosite mineral. The addition of Ca^{2+} ions in $LaFeO_3$ can increase the conductivity value as well as the sensitivity of the gas sensor

[7]. Purity of the Fe_2O_3 is 91,30% [3]. Yarosite mineral contains materials such as SiO_2 and TiO_2 other than Fe_2O_3 [8, 9].

2. Material and method

Thick film ceramic of $La_{0,9}Ca_{0,1}FeO_3$ was synthesized from Fe_2O_3 powder extracted from yarosite mineral, $LaCl_3.7H_2O$ powder and CaO powder using co-precipitation method. The compounds contained in the Fe_2O_3 powder extracted from yarosite minerals are shown in table1 [3]. The amount of Fe_2O_3, $LaCl_3.7H_2O$ and CaO powder used in the fabrication of $La_{0,9}Ca_{0,1}FeO_3$ is 0.8244 gram, 2.3963 gram, and 0.0967 gram respectively. Fe_2O_3 and CaO powder were dissolved in 10 M HCl solvent, and $LaCl_3.7H_2O$ powder dissolved an aquades solvent. The solutions of Fe_2O_3, CaO and $LaCl_3.7H_2O$ were mixed, then added a NH_4OH to produce the $La_{0,9}Ca_{0,1}FeO_3$ precipitate. The $La_{0,9}Ca_{0,1}FeO_3$ precipitated was dried at 100^0C and then calcined at 800^0C for 2 hours to produce $La_{0,9}Ca_{0,1}FeO_3$ powder. The $La_{0,9}Ca_{0,1}FeO_3$ powder was crushed to produce a nanoparticle-sized powder. The crushed $La_{0,9}Ca_{0,1}FeO_3$ powder mixed with Organic Vehicle (OV) with a ratio between $La_{0,9}Ca_{0,1}FeO_3$ and Organic Vehicle (OV) powder is 70%: 30% to yield the $La_{0,9}Ca_{0,1}FeO_3$ paste. Paste $La_{0,9}Ca_{0,1}FeO_3$ coated on alumina substrate (Al_2O_3) using screen printing technique, then fired at 600^0C for 2 hours to produce the thick film ceramic $La_{0,9}Ca_{0,1}FeO_3$.

Table 1. Content of compounds in yarosite mineral.

No	Compound	% Weight
1	Fe_2O_3	91.30
2	Al_2O_3	3.30
3	SiO_2	2.05
4	TiO_2	3.02
5	CaO	0.16
6	MnO	0.17

The crystal structure and morphology of the thick film ceramic $La_{0,9}Ca_{0,1}FeO_3$ can be seen from the analysis result of XRD (X-Ray Diffraction, X-ray XRP PRO series, $\lambda = 1{,}540598\ \dot{A}$, PAN) and SEM characterization (Scanning Electron Microscope, JEOL JSM – 6360 LA). The electrical resistance of the thick film ceramics of $La_{0,9}Ca_{0,1}FeO_3$ was measured in a chamber without and with ethanol gas of 50 ppm, 100 ppm, and 200 ppm. The Measurement was conducted at Material Laboratory of PSTNT – BATAN, Bandung. The sensitivity value is calculated using the equation (1):

$$S = \frac{R_g - R_o}{R_o} \quad (1)$$

with, S is the sensitivity, R_g and R_o are respectively the resistance with ethanol gas and without ethanol gas.

3. Result and discussion

3.1. Characterization results X-Ray Diffraction (XRD) ceramic thick film $La_{0,9}Ca_{0,1}FeO_3$
Pattern result XRD characterization thick film ceramic of $La_{0,9}Ca_{0,1}FeO_3$ is shown in figure1. The resulting pattern XRD thick film ceramics of $La_{0,9}Ca_{0,1}FeO_3$ has been made, matched with the $LaFeO_3$ database (JCPDS No. 37 - 1493) and CaO (JCPDS No. 37 - 1497).

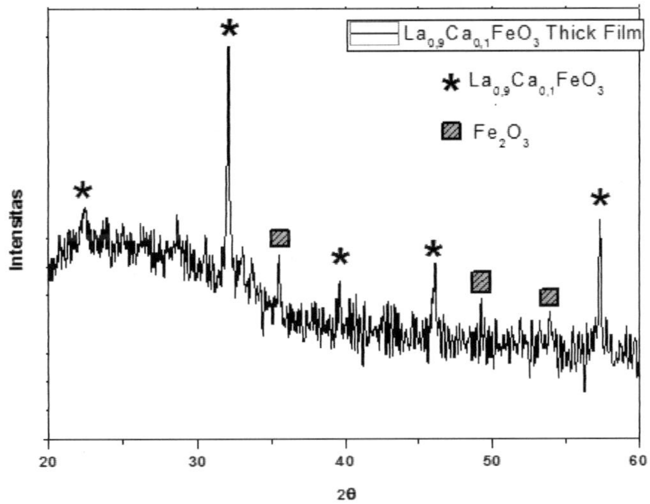

Figure 1. Pattern result XRD characterization of ceramic thick film $La_{0,9}Ca_{0,1}FeO_3$.

The results show that the maximum peak data matchs with $LaFeO_3$ database (JCPDS No. 37 - 1493) and unmatched to the CaO database (JCPDS No. 37 - 1497). It shows that CaO dissolved to form $La_{0,9}Ca_{0,1}FeO_3$. The presence of Fe_2O_3 peaks indicated that Fe_2O_3 did not react well to form $LaFeO_3$. This situation occurs because the fired time is short and is due to the effect of Ca^{2+} addition in $LaFeO_3$. Figure.1 shows several peaks that undefined or not fit with the $LaFeO_3$ database (JCPDS No. 37 - 1493) and CaO (JCPDS No. 37 - 1497) i.e peak with 2θ in the range 270^{0}C - 280^{0}C. These peak may be from reacted material of on purity (table 1). The raw material contains not only the Fe_2O_3 compound but also other compounds such as SiO_2, Al_2O_3, K_2O, and Na_2O which may affect grain growth.Thick film ceramics of $La_{0,9}Ca_{0,1}FeO_3$ that has been made has an orthorhombic crystal structure. Calculation of crystallite size using Debye Scherrer equation shows that the average crystallite size of thick film ceramic $La_{0,9}Ca_{0,1}FeO_3$ was fabricated is 38,7625 nm.

3.2. Characterization Scanning Electron Microscope (SEM) of thick film ceramics $La_{0,9}Ca_{0,1}FeO_3$
Characterization SEM of thick film ceramics $La_{0,9}Ca_{0,1}FeO_3$ is shown in figure 2. From the figure it can be seen that the micro structure of the thick film ceramics of $La_{0,9}Ca_{0,1}FeO_3$ has almost uniform grain size, and contains many pores. The grain size of the thick film ceramics $La_{0,9}Ca_{0,1}FeO_3$ from SEM characterization data obtained by calculating some sample grain size, it is found that the average grain size of the thick film ceramic $La_{0,9}Ca_{0,1}FeO_3$ is 0.1785 μm (178,5 nm).

Figure 2. Morphologyof thick film ceramics $La_{0,9}Ca_{0,1}FeO_3$.

3.3. Electrical characterization of thick film ceramic $La_{0,9}Ca_{0,1}FeO_3$

Electrical characterization was showed by profile resistance function temperature of the thick film ceramics of $La_{0,9}Ca_{0,1}FeO_3$. Response gas sensor indicated by the sensitivity value and the optimum working temperature. Figure.3 (a) showed the graph resistance function temperature of the thick film ceramics of $La_{0,9}Ca_{0,1}FeO_3$ with varying ethanol content. The graph showed that the much ethanol content, the value of resistance at the same temperature is increased. Substitution of Ca^{2+} in La^{3+} ion has produce oxygen vacancy, this mechanism based on the KrÖger Vink equation (2).

$$2CaO \rightarrow 2Ca_{(La)}^{-1} + 2O_{(o)} + V_o^{+2} \qquad (2)$$

The substitution of Ca^{2+} in La^{3+} ion may inhibited grain growth. $LaFeO_3$ added Ca^{2+} ion has smaller grain size compared to $LaFeO_3$ without Ca^{2+} ion added. Smaller grain size can decrease electrical resistance and reduced interconnection between the grains. This situation has result the electron conduction path becomes smaller. So, the thick film ceramics of $La_{1-x}Ca_xFeO_3$ based gas sensor has a lower working temperature compared to $LaFeO_3$ based gas sensor. Figure 3 (b) showed the sensitivity function temperature graph of thick film ceramics of $La_{0,9}Ca_{0,1}FeO_3$ based gas sensor. The effect of ethanol content as shown in figure 3 (b) may increase the sensitivity value.

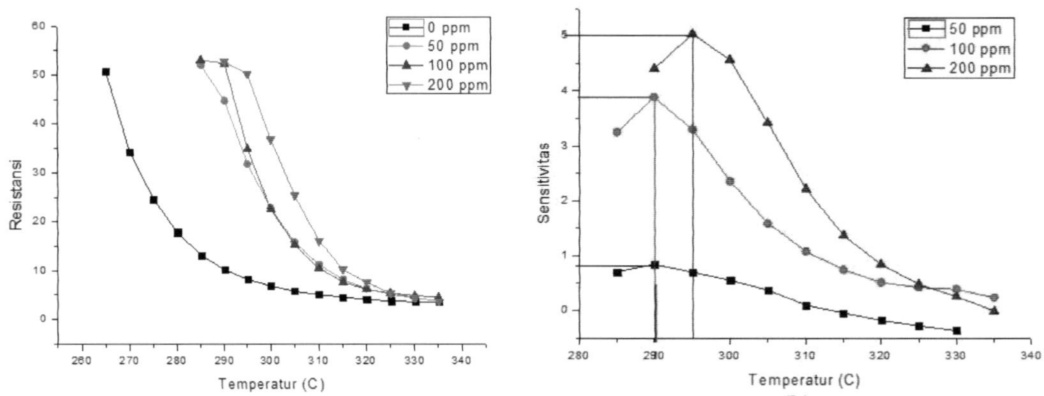

(a) (b)

Figure 3. (a) The graph resistance function temperature and (b) The graph sensitivity function temperature of thick film ceramics $La_{0,9}Ca_{0,1}FeO_3$.

4. Conclusions

The powder of $La_{0,9}Ca_{0,1}FeO_3$ from yarosite minerals, $LaCl_3.7H_2O$ and CaO has been successfully synthesized using a co-precipitation method. Thick film ceramics of $La_{0,9}Ca_{0,1}FeO_3$for ethanol gas sensor has been successfully made by screen printing techniques. The $La_{0,9}Ca_{0,1}FeO_3$ thick films have orthorhombic crystal structure, with average crystallite size calculated using the Debye Scherrer equation of 38.7625 nm. The Thick film ceramic of $La_{0,9}Ca_{0,1}FeO_3$ has average grain size of 0.1785 μm. The thick film ceramics of $La_{0,9}Ca_{0,1}FeO_3$ contain many pores. The ethanol gas sensor made from Fe_2O_3 extracted from yarosite mineral has working temperature in range 290^0 - 295^0C.

References

[1] Yadav A K, Singh R K and Singh P 2016 Fabrication of Lanthanum Ferrite Based Liquefied Petroleum Gas Sensor *Sensor and Actuators* **B 229** p. 25-30.

[2] Haron W, Wisitsoraat A and Wongnawa S 2017 Nanostructured Perovskite Oxide-$LaMeO_3$ (M = Al, Co, Fe) Prepared by Co-Precipitation Method and Their Ethanol-Sensing Characteristics *Ceramics International Another reference*.

[3] Suhendi E, Novia H, Syarif D G and Prajitno D H 2010 Studi Awal Pembuatan Keramik Film Tebal (Thick Film) Berbasis Fe_2O_3 dari Bahan Dasar Lokal untuk Sensor Gas Alkohol *Prosiding Seminar Nasional Penelitian, Pendidikan dan Penerapan MIPA, Fakultas MIPA, Universitas Negeri Yogyakarta*.

[4] Xiao H, Xue C, Song P, Li J and Wang Q 2015 Preparation of Porous $LaFeO_3$ Microspheres and Their Gas-Sensing Property *Applied Surface Science* – 29743.

[5] Suhendi E, Witra, Hasanah L and Syarif D G 2017 Characteristics of a Thick Film Ethanol Gas Sensor Made of Mechanically Treated $LaFeO_3$ Powder *AIP Conference Proceedings* **1848**, 050008.

[6] Fan H T, dkk. 2011 Preparation of $LaFeO_3$ Nanofibers by Electrospinning for Gas Sensors with Fast Response and Recovery *Nanotechnology* 22(11).

[7] Kong L B and Shen Y S 1996 Gas Sensing Property and Mechanism of $Ca_xLa_{1-x}FeO_3$ Ceramics *Sensor and Actuators* **B 30** p. 217-221.

[8] Suhendi E, Novia H and Syarif D G 2013 Karakteristik Listrik Keramik Film Fe_2O_3 dengan Variasi Ketebalan yang dibuat dari Mineral Lokal di Atmosfir udara dan Atmosfir Alkohol **7** (1) ISSN : 1979 – 8911.

[9] Syarif D G, Guntur D S and Yamin M 2005 Pembuatan Keramik Termistor NTC Berbahan Dasar Yarosit dan Evaluasi Karakteristiknya *Prosiding Seminar Nasional Sains dan Teknik Nuklir, P3TkN-Batan, Bandung.*

IOP Conf. Series: Materials Science and Engineering **384** (2018) 012038 doi:10.1088/1757-899X/384/1/012038

Home Scale Yogurt Incubator Control System using Microcontroller ATmega328

N A Emelda* and B Mulyanti

Department of Electrical Engineering Education, Universitas Pendidikan Indonesia
Jl. Dr. Setiabudhi No.229, Bandung 40154, Indonesia

*nisrina_athirah@student.upi.edu

Abstract. The process of producing yogurt has been known since long time ago, but usually it carried out traditionally without using control technology. This paper aims to produce easy use of home-scale yogurt incubator system using Microcontroller ATmega328. The incubator was modified from two-liter volume of rice cooker controlled by ATmega328, where the temperature is measured by the DS18B20 temperature sensor. ATmega 328P also acts as a time-controller during the process of producing yogurt. The yogurt incubator system resulted in this research works well according to the system designed, including pasteurization, cooling inoculation, and incubation processes. The system can work automatically and very easy to use. Compare with conventional system, this system even works faster and able to produce better quality yogurt.

1. Introduction

Many scientists have confirmed the importance of dairy products because milk contains essential nutrients for body health [1]. One of the products that can be processed from milk is yogurt. The process of making yogurt utilizes the bacteria Streptococcus thermophilus and Lactobacillus bulgaricus [2] which can improve the balance of human intestine [1]. The fermentation process in yogurt should be controlled to produce good quality yogurt [3]. The process of producing yogurt in incubator including pasteurization, inoculation cooling, and incubation processes [4]. The incubator is a temperature controlled heater [5].

The incubator used in the process of producing yogurt has been widely available in the market and the price is quite affordable [6]. But the incubator has a disadvantages, such as not automatic as well as requires a long time of about 8-10 hours. Meanwhile, in the market has also available automatic incubator, however the price is quite expensive.

From the existing problems, in this paper, we propose the system of yogurt incubator by modifying the rice cooker and based on microcontroller ATmega328. The advantages of the system developed are works automatically and low price as well as easy to use. By using a used rice cooker for the incubator it can help government program for reducing electronic waste, which increase in line with technology developments.

2. Methods

2.1. Research design
The design process of this tool through several stages, shown in Figure 1 flowchart follows:

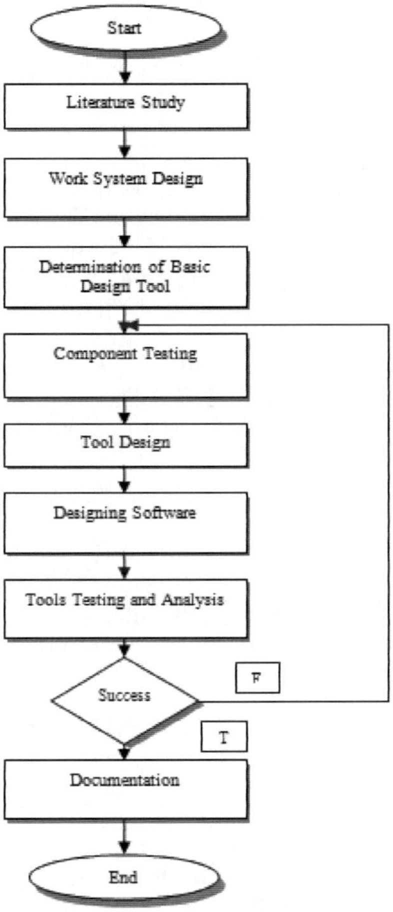

Figure 1. Research design flowchart.

2.1.1. Literature study. At this stage is done the collection of theory and information about the final project. Theories and information obtained are drawn from various sources such as books, journals, and scientific articles on the internet. Theories and information collected learn about yogurt, ATmega328 microcontroller, DS18B20 sensor, and others related to the final project.

2.1.2. Work system design. Designing work system tools from start to end process.

2.1.3. Determining of basic design tool. Determining the desired design and collecting materials and tools to be used.

2.1.4. Component testing. Checking the components used to run well or not, so when the process of assembling the tool running smoothly.

2.1.5. Tool design. Make tools with tools and materials that have been prepared.

2.1.6. Designing software. The final stage of the research by photographing or video of the tool has been successful.

2.1.7. Tools testing and analysis. Testing after the tool has been completed. To see if the tool created is successful or not.

2.1.8. Documentation. The final stage of the research by photographing or video of the tool has been successful.

2.2. Design system

The design of this incubator is controlled by ATmega328 which is connected to the DS20B18 temperature sensor. The temperature sensor will measure the temperature during the yogurt making process. The temperature and timing results will be displayed on the LCD. This tool is focused on the temperature sensor because all components work according to a predetermined temperature.

Block diagram design system can be described in the figure 2.

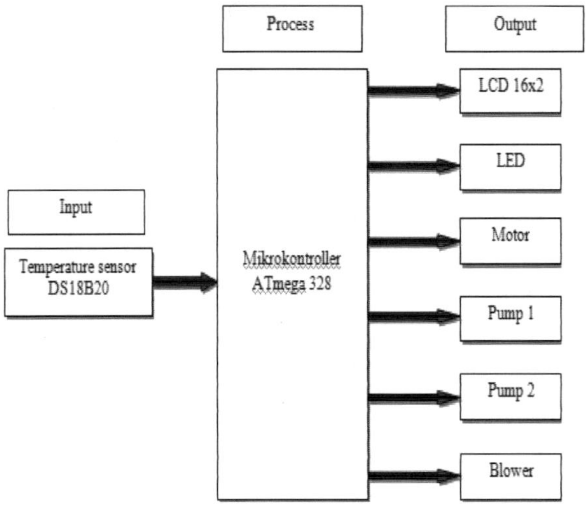

Figure 2. Block diagram design system.

2.2.1. Temperature sensor DS18B20. DS18B20 is a 1-Wire digital temperature sensor from IC Maxim. Display degrees in Celsius with precision 9 to 12 bits, from -55 to 125 (+/- 0.5). This temperature sensor has 3 feet that are, ground, data, and Vcc. In this design serves to measure the temperature of the milk. Serves as a measurement of milk temperature during the process of making yogurt.

2.2.2. Microcontroller ATmega328. Microcontroller ATmega328 is an AVR 8 bit microcontroller output. ATmega328 has 3 main PORT that is PORTB, PORTC, and PORTD with total input/output pin of 23 feet. This microcontroller output from Atmel having RISC architecture (Reduce Instruction Set Computer). It has a 14 pin digital I / O pin of which is PWM (Pulse Width Modulation). In this design, the microcontroller serves as the control that governs all movement of the tool [6].

2.2.3. LCD. LCD (Liquid Crystal Display) is one of the important display techniques to use for applications [7]. Usually to display information as well as an image. Here serves to display the amount of temperature and the length of time the process of making yogurt.

2.2.4. Motor. In theory, a DC motor is an electric motor that requires a direct current voltage supply or DC current (Direct Current) on the field coil to be converted into mechanical energy. The field coil on

IOP Conf. Series: Materials Science and Engineering **384** (2018) 012038 doi:10.1088/1757-899X/384/1/012038

the motor is called the stator, and the anchor coil is called the rotor. In this design, it is evacuated to move the stirrer.

2.2.5. LED. LED (Light Emitting Diode) is an electronic component that can emit monochromatic light when given the forward voltage made of semiconductor material. LEDs perform less power and have more efficient power [8]. LEDs have a variety of colors and their appearance can be changed in the form of flip-flops. In this design LED functions as a sign of the system has been active.

2.2.6. Pump 1. The pump is a machine to drive fluid. The pump moves the fluid from the low pressure to the higher pressure place, to overcome this pressure difference, energy is required. In this design to drain the bacteria needed in making yogurt.

2.2.7. Pump 2. The pump is a machine to drive fluid. The pump moves the fluid from the low pressure to the higher pressure place, to overcome this pressure difference it takes energy. In this design to drain the anti-yeast needed in the manufacture of yogurt.

2.2.8. Blower. A blower is a machine or apparatus used to raise or enlarge the air or gas pressure to be drained in a particular chamber as well as the suction or vacuum of air or certain gases. Helps speed up the process of cooling milk before it is given bacteria.

Work system of incubator can be described in the figure 3.

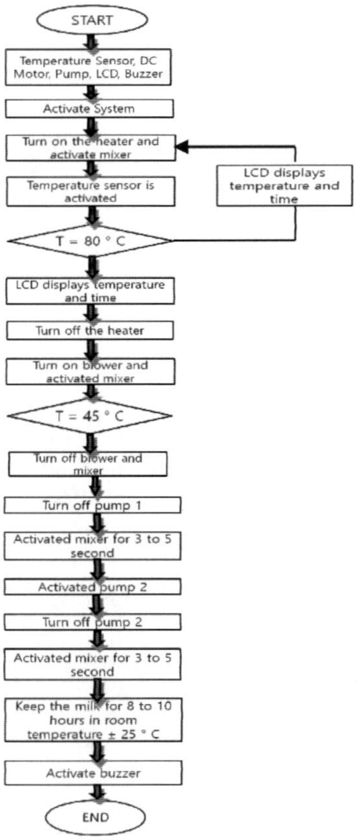

Figure 3. Work system flowchart.

IOP Conf. Series: Materials Science and Engineering **384** (2018) 012038 doi:10.1088/1757-899X/384/1/012038

This incubator work system starts from pasteurization process for ± 30 minutes reaching 80 ° C, by activating heater and motor. Then cooling the temperature inoculation ± 25 minutes to reach 45 ° C temperature that activates the blower and motor. After the temperature has reached 45 ° C given the seeds of bacteria and anti-yeast flowed by pumps 1 and 2 assisted stirring briefly by the motor. Last incubation process takes 8-10 hours with temperature below 45 ° C and above 25 ° C.

2.3. Designing software (program)
Software used in the design of this tool is Arduino IDE. Programming using C ++ language. Program from software uploaded to ATmega328 via Arduino. The data goes into the microcontroller IC which will be the system's work regulator of the tool [9]. Designing software can be described in the figure 4.

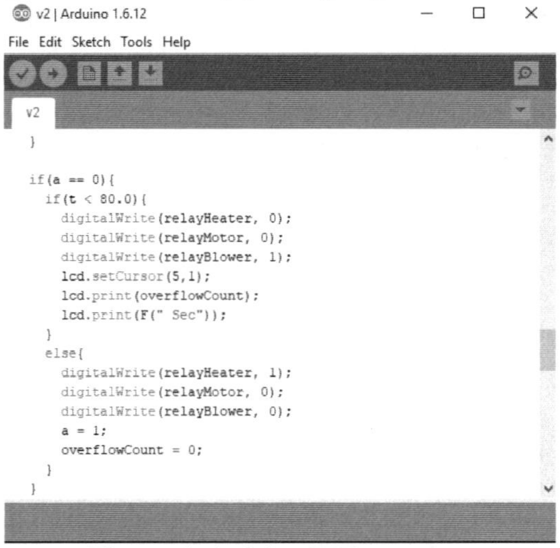

Figure 4. Arduino IDE program.

Arduino IDE consists of 3 parts:
- Editor
- Compiler
- Uploader

3. Results and discussion
From the design of the yogurt incubator and the test results of the tool works well and in accordance with the work system that has been designed. The process of this tool through 3 stages of the process of pasteurization, cooling inoculation, and incubation. Calculation of time and temperature can be told in table 1:

Table 1. Temperature and time result.

Process	**1 Liter**		**2 Liter**	
	Temp	**Time (s)**	**Temp**	**Time (s)**
Pasteurization	80 °C	907	80 °C	1478
Cooling Inoculation	45 °C	753	45 °C	1400
Incubation	25 °C < 45 °C	28800	25 °C < 45 °C	28800

In the above table shows the results of comparative analysis of the time of manufacture of yogurt 1 liter and 2 liters. It can be seen that the less milk volume is used, the more rapid the time. First, there is the pasteurization process, the process by which the initial temperature of milk is 26.87 ° C for the 1 liter and 28.25 ° C for which 2 liters is heated using a heater to a temperature of 80 ° C. In this process takes 907 seconds and 1478 seconds. Comparison of temperature rise of milk in process of making yogurt 1 liter and 2 liters can be seen in figure 5 and 6:

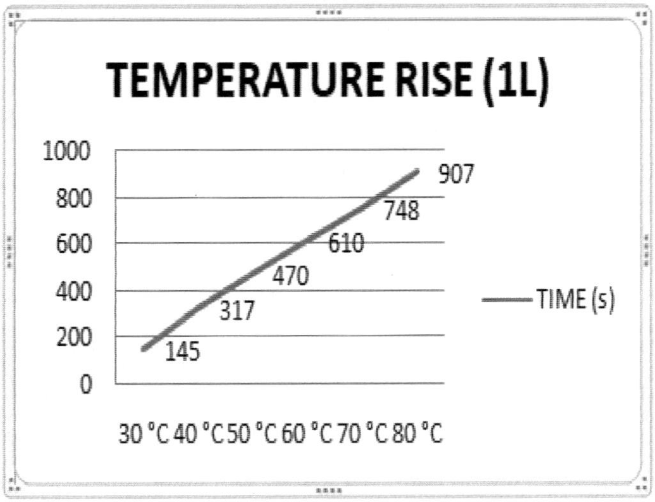

Figure 5. Temperature rise in 1L volume.

A temperature rise of 1 liter of milk from 30 ° C to 80 ° C takes 907 seconds.

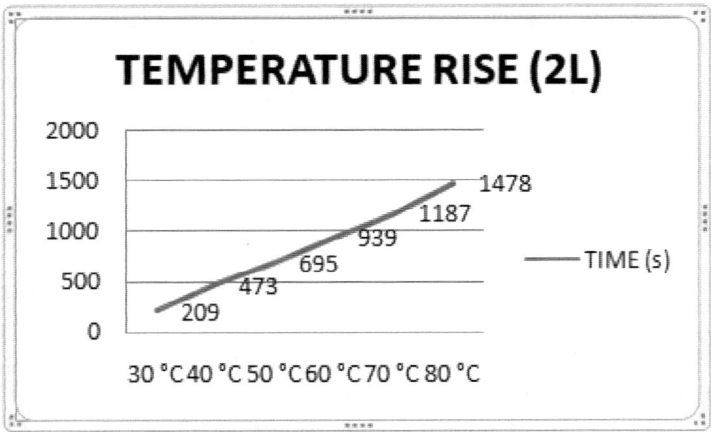

Figure 6. Temperature rise in 2L volume.

A temperature rise of 2 liter of milk from 30 ° C to 80 ° C takes 1478 seconds.

Second, there is an inoculation cooling process that is, the temperature drop from 80 ° C to 45 ° C to put bacterial seeds and anti-yeast. At this temperature reduction process takes 753 seconds and 1400 seconds. Comparison of temperature rise of milk in process of making yogurt 1 liter and 2 liters can be seen in figure 7 and 8:

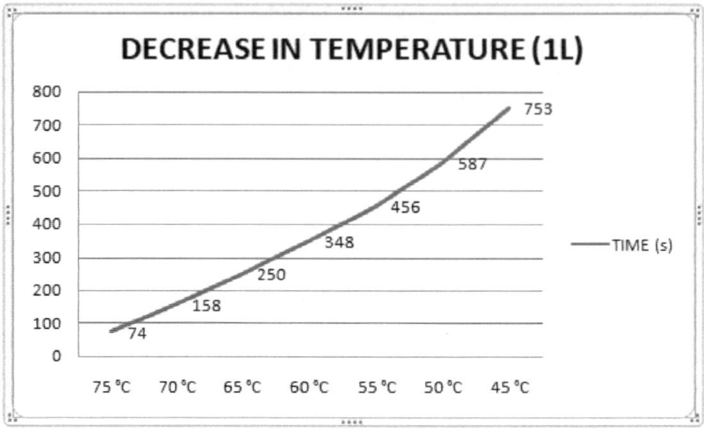

Figure 7. Decrease in temperature in 1L volume.

A decrease in temperature of 1 liter of milk from 80 ° C to 45 ° C takes 587 second.

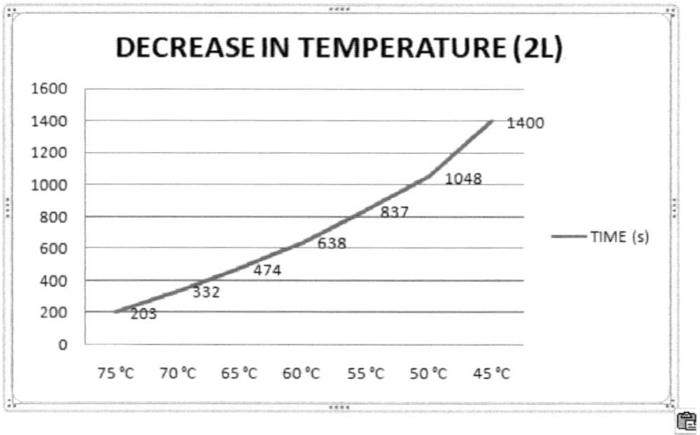

Figure 8. Decrease in temperature in 2L volume.

A decrease in temperature of 2 liter of milk from 80 ° C to 45 ° C takes 1400 second.

The third is an incubation process in which the milk that has been given bacteria and seeds is silenced for 8 hours with temperatures below 45 ° C and above 25 ° C. After all the process is done yogurt. Yogurt results can be seen from the following figure 9:

Figure 9. Yogurt results.

This incubator has a deficiency in the pasteurization process, especially as the volume of milk increases. The greater the volume, the longer the pasteurization process. The heating of milk during the pasteurization process with incubators or manually is almost the same time. The time required in the process of making yogurt manually about 30 minutes. In this experiment as much as 2L takes ± 25 minutes, if the volume increases again then the time can be increased again or beyond the time of making yogurt manually. The heater settings follow the working system of the rice cooker, so the heat increase cannot be controlled. The process will be even better if the heater can be controlled so the results of the process become faster.

4. Conclusions

The yogurt incubator system resulted in this research works well according to the system designed, including pasteurization, cooling inoculation, and incubation processes. The system can work automatically and very easy to use. Compare with conventional system, this system even works faster and able to produce better quality yogurt. The system still has the disadvantage, especially during the pasteurization process. The heat should be controlled so that the warming of milk can be faster.

References

[1] Liu H, Xu Y, Wang Y and Wei Y 2011 "Notice of Retraction Investigation of Production of Sanitarian Yogurt Drinking" *In 5th International Conference on Bioinformatics and Biomedical Engineering* (pp. 1-4) IEEE.

[2] Rizka V 2015 "Sistem Pengendali Suhu Otomatis Pada Inkubator Fermentasi Yoghurt Berbasis Mikrokontroler Dengan Metode Logika Fuzzy" (Doctoral dissertation, Universitas Andalas).

[3] Yılmaz C and Gökmen V 2017 "Formation of tyramine in yoghurt during fermentation–Interaction between yoghurt starter bacteria and Lactobacillus plantarum" *Food Research International* **9** (7) pp. 288-295.

[4] Fakharudin A S, Embong A, Hamid R A, Hamza M A M, Ajid K A, Ali N M, ... and Zain W S W M 2007 "Optimisation of fermentation process using data mining techniques for small-medium industry" *In International Conference on Intelligent and Advanced Systems* (pp. 273-275) IEEE.

[5] Rajan D K, Verho J, Kreutzer J, Välimäki H, Ihalainen H, Lekkala J, ... and Miettinen S 2017 "Monitoring pH, temperature and humidity in long-term stem cell culture in CO 2 incubator" *In 2017 IEEE International Symposium on Medical Measurements and Applications* (pp. 470-474) IEEE.

[6] "Electric yogurt maker Important safeguards Save these instructions."

[7] Huang Y L, Chen F C and Chien S Y 2016 "Algorithm and Architecture Design of Multi-Rate Frame Rate Up-Conversion for Ultra-HD LCD Systems" *IEEE Transactions on Circuits and Systems for Video Technology*.

[8] Sukanya S, Gayathri K and Thenmozhi R 2017 "A power quality improved isolated bridgeless sepic converter for LED lamp" *In 2017 International Conference on Innovations in Green Energy and Healthcare Technologies* (pp. 1-7) IEEE.

[9] Stusek M, Pokorny J, Masek P, Hajny J and Hosek J "A Non-invasive Electricity Measurement within the Smart Grid Landscape: Arduino-based Visualization Platform for IoT"s.

IOP Conf. Series: Materials Science and Engineering **384** (2018) 012039 doi:10.1088/1757-899X/384/1/012039

Power Transmission and Distribution Monitoring using Internet of Things (IoT) for Smart Grid

N A Hidayatullah[1,*], A C Kurniawan[2] and A Kalam[3]

[1]Electrical Engineering Department, State Polytechnic of Madiun, Indonesia
[2]Electrical Engineering Department, State University of Malang, Indonesia
[3]College of Engineering and Science, Victoria University, Melbourne, Australia

*asyik@pnm.ac.id

Abstract. In the last decade, much of the attention is made towards introducing the smart systems and appliances to meet the requirement of the century and make life comfortable. During the same period, electric power sector also made the necessary innovation to compensate the demand of today's electric supply and to make use of electric resources effectively by introducing "Smart-Grid". The smart grid is a part of transformation and reformation in the power industry sectors. The smart grid is a future modern power system that utilizes internet of thing to monitor, control and create various intelligent communications in the electrical system. In this paper, the author first presents the fundamental architecture of Internet of Things (IoT). They then outline the key technologies of IoT for Smart Grid. Finally, the ideas of applied communication and framework of transmission and distribution monitoring based IoT for Smart Grid is discussed. The deployment of IoT technologies in Power System infrastructures would accelerate the smart grid development and enhance the electricity delivery services becoming more robust, attractive, responsive and communicative.

1. Introduction

Electric power transmission and distribution is an integral part of electric power system. It mainly focuses on delivering the electric power to the end user. Most of them on a radial topology with generation located in a remote area far from the end user. The transmission and distribution system mostly fails to provide real-time data monitoring to the service provider, operator and end user. Smart grid system allows the utilities to control the electric usage. It is the intelligent process by which supplier can increase the overall efficiency, stability and make the electric power transmission and distribution in more appropriate way. In recent years, researchers are making efforts to work in order to make electrical power system more intelligent by means of power transmission and distribution monitoring using the Internet of Things (IoT) for the smart grid. IoT is a communication network with the objective for connecting, exchanging and communicating all kind of sensing information devices with the internet which use radio frequency identification (RFID), global positioning system (GPS), infrared sensors and laser scanner through secure gateway protocol. By using IoT one can achieve intelligent monitoring, managing, tracking, positioning, identification and tracking some particular system [1].

Smart grid is a new technology paradigm for electricity supply to answer the afore-mentioned challenges. The term Smart Grid is widely used with different meanings and diverse definitions. Basically, the definition of a smart grid is the integration of information technology (IT), IoT, intelligent devices and advanced control method with the existing power system networks to enable power generation, transmission and distribution making it more robust, attractive, responsive and communicative. The main advantages of smart grid implementation in terms of utility benefits include lower maintenance and operational cost; increased large-scale renewable energy utilization; reduced outage and perturbation; minimal power loss; enabled energy management system; and real-time power flow monitoring. Correspondingly, the advantages of smart grid implementation in terms of customer benefit include enabled electricity consumption monitoring and customer energy management systems.

One of the key successes of the smart grid is all-in-one integration and communication between power system infrastructures as hardware and information sensing, processing, analyzing and controlling it as intelligent systems [2]. To integrate these systems needs an emerging information one processing and data acquisition method, which is known as Internet of Things (IoT) technology. IoT is the right response to address the all-in-one integration and communication in the smart grid. Nowadays, most of the current studies in Indonesia only emphasize the broad theoretical concept of a smart grid, and there are very few detailed studies on power transmission and distribution monitoring system using IoT. This paper zooms in the technical aspect of IoT for transmission and distribution monitoring that could be implemented in Indonesia. Implementing IoT in the current existing power grid is an important approach to speed up the smart grid development.

In this paper, Section 2 describes the fundamental architecture of IoT technology. The key technology of IoT for Smart grid is discussed in Section 3. The ideas of applied communication and framework of transmission and distribution monitoring based on IoT in smart grid is proposed in Section 4. Finally, Section 5 concludes the paper.

2. Internet of Things (IoT) architecture in smart grid

In recent years, IoT gaining significant interest from researchers due to its potential for information processing and data exchange in various aspects of our everyday life. In order to form complete smart grid system, it needs various IoT elements as shown in figure 1. The IoT system architecture consists of four layers, including the application layer, cloud management layer, network layer, and device layer as described in Figure 1 [3].

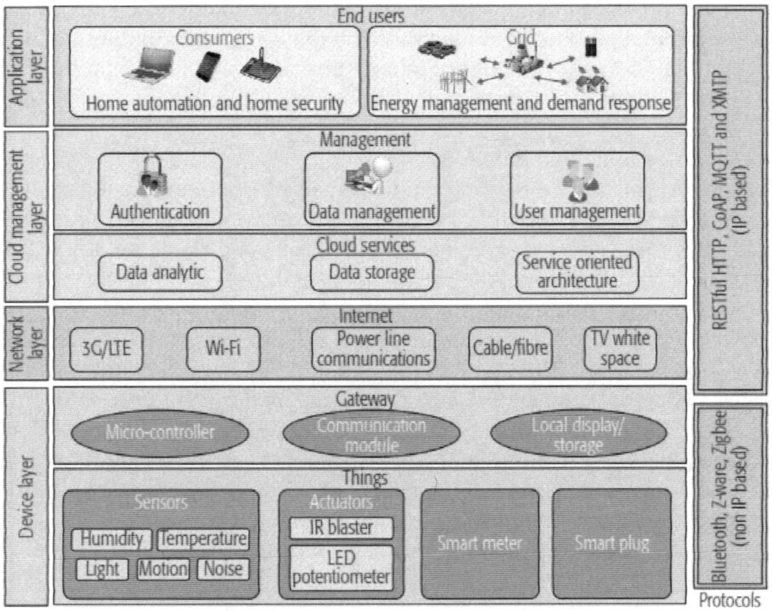

Figure 1. The system architecture of IoT [3].

The top layer is the "application layer", which is responsible for collecting, processing and calculating information from end users. The application layer provides services to consumers through broad of intelligent application infrastructures such as real-time energy management, demand response management, visual monitoring platform and other basic services solution. The application layer also enables the end user to interact with smart grid through the user interface and responds to the information provided by the application devices and software features.

The second layer is "cloud management layer". The cloud service layer is cloud storage in which digital data is managed, maintained, processed and backed-up remotely. This cloud layer is available and accessible for the user to retrieve the data over internet services anytime. On the other hand, the management layer is an application where the digital database is managed. It enables users to create, control and manage their authentication and account. The key point of the cloud management layer is data storage, data management, data processing and computing.

The third layer is "network layer". The network layer integrates all kinds of devices between the application layer and device layer within the smart grid. The network layer consists of communication networks such as the Internet, fiber optic cable, power line communication, wireless broadband and 3G/4G mobile communication network.

The fourth layer is "device layer" which contains things sub-layer and gateway sub-layer. The things layer is responsible for collecting raw information which is acquired from sensing equipment, smart meter, smart plug-in devices and actuators. Meanwhile, the gateway sub-layer connects all the things layer devices and mediates multiple accesses to the things layer through a microcontroller, communication module and local storage. The main tasks of device layer are gathering, identifying, processing and analyzing all necessary information data using a variety of smart equipment and transmitting the information to the cloud management layer.

3. Key technologies of IoT for smart grid

In order to utilize IoT for Smart grid, there are several key technologies which must be used in transmission and distribution monitoring system. It can be described as follows:

3.1. Smart Sensor Technology (SST)

The smart sensor can be considered as one of the most important devices of IoT for the smart grid. Smart sensor is a device that informs the control system about certain parameters and what is actually occurring in the physical object that is monitored. Smart Sensor provides raw data for information processing, analyzing and feedback. Currently, several numbers of advanced sensors technology have been applied in various fields. The major objectives are to provide a technical solution, achieve a high level of precision, enhance system quality and reliability [4].

In transmission and distribution monitoring system, smart sensors can be used at multiple places and spread over the grids such as on the transmission tower, transmission and distribution lines, distribution transformer and sub-stations. They are two types of smart sensors that can be utilized. The first type is used to measure weather condition around the assets such as wind direction sensors, temperature sensors, wind speed sensors, humidity sensors and rain sensors [5]. These sensors are responsible for natural disaster mitigation and prevention that is not available on the current grid. The other sensor type is used for asset and power quality monitoring such as GPS, strain sensors, accelerometer sensor for vibration and line monitoring, conductor temperature sensor and magnetic field sensor [5]. These sensors are mainly responsible for monitoring states of the towers, lines, and sub-stations. For power quality monitoring the sensors including voltage sensor, a current sensor, temperature and fault sensor. By installing these sensors in the distribution line, the quality of distribution line can be monitored in real time basis. In addition, the sensor also can be used to monitor or identify particular information in substation transformers such as temperature, primary voltage, and secondary voltage.

3.2. Information and Communication Technology (ICT)

To enable the IoT on smart grid, information and communication technologies are required in order to create end-to-end coordination and perform bi-directional communication for all IoT layer elements. According to data transmission coverage area, ICT can be classified into two types: short-range communication and wide range communication technologies [6]. The short range ICTs including Bluetooth, Ultra Wide Band (UWB) and Zigbee. Zigbee is the latest wireless communication technology based on IEEE 802.15.4 standard and was developed for transmission line monitoring [5]. The wide range of ICTs includes General Packet Radio Service (GPRS), IP based Internet/Wireless Sensor Network, 2G/3G/4G mobile communication network, satellite communication network and Fiber Optic Cable. In the smart grid system, ICT is important and essential to assist wireless sensor network in delivering and forwarding data.

4. System design

In designing the transmission and distribution based IoT one can consider using Wireless Sensor Network (WSN) and General Packet Radio Services (GPRS). A WSN is a wireless network formed by large numbers of sensor nodes where each node equipped with sensors to monitor physical conditions. GPRS is a GSM-based wireless communication packet network that provides wide-range IP protocol connection from point to point node. GPRS is a high-speed data processing, for which allows multiple packet data transmission, extensive signal coverage, fast installation feature and low maintenance cost [7]. GPRS technology seems out of date if compared to the 3G/4G technologies. However it's still reliable and good alternative for data processing and communication. Moreover, several transmission and distribution networks in Indonesia are located in a remote area where the 3G/4G network is not available.

Figure 2 illustrates the typical system architecture of IoT for the power transmission and distribution monitoring system. On the left side, one can deploy sensors node where physical events are measured. By using single-hop or multi-hop, the data measured by individual sensors are collected and aggregated which subsequently is transmitted through GPRS network to the database or server for further analysis, making decisions and taking appropriate actions. The server part of cloud management layer in which all the data is stored and it is a place for data retrieval.

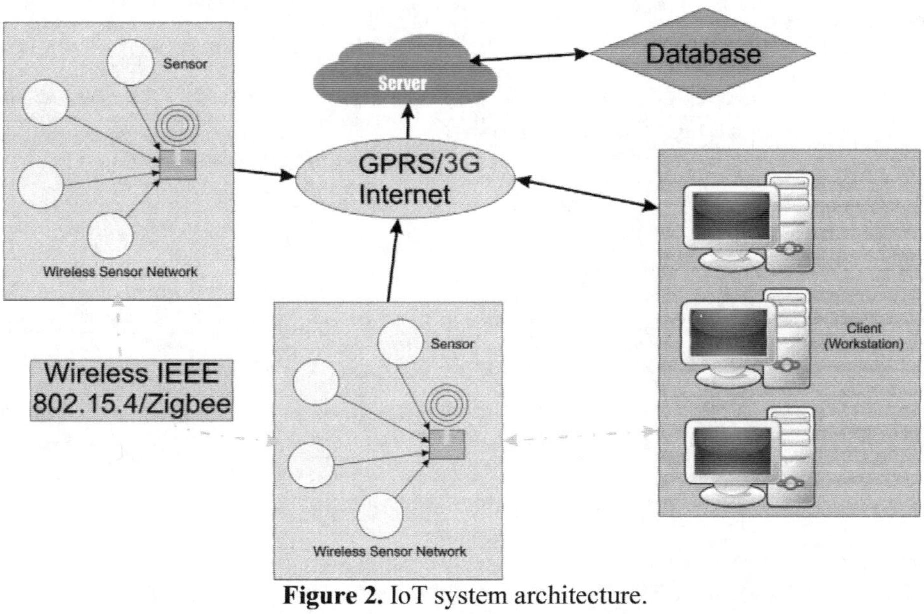

Figure 2. IoT system architecture.

To transmit data from the sensors, the system uses a combination of GPRS technology and Zigbee module. In this framework, one can adopt a two-layer wireless network to give more freedom and flexibility. The first level uses IEEE 802.15.4 or Zigbee wireless which is low cost, low energy consumption and has a communication range between 100 up to 1000 meters in each node [8]. With regard to Zigbee networking capability, mesh topology is proposed in this system. The main benefit of the mesh topology is that every single node is able to communicate with any other node directly within the coverage area range. Furthermore, it increases network reliability and sustains wireless connection if one of the nodes has a fail, loss signal or disabled. The second level uses GPRS module to assist the network in sending or forwarding data. The GPRS technology can improve network performance for real-time data as well as it is more efficient and robust [5]. GPRS module also supports TCP or IP protocol that enables to send data directly to the server using particular node's address. By using the two-layer wireless sensor networks, collecting and forwarding data to server or cloud management layer become more reliable.

4.1. Transmission hardware design

Figure 3 depicts deployment of the equipment in the transmission system based on IoT. It consists of four parts, namely wireless sensor network, data transmitter, the server (cloud management) and workstation. The WSN is capable of sensing the physical world in the transmission line and towers such as temperature, humidity, conductor temperature, tension sensor, wind velocity and leakage current etc. Sensors collect sensing data and deliver them through data transmitter IEEE 802.15.4 / Zigbee module between node to node or node to node server. Then the merged data is transmitted to GPRS to achieve long-distance and high-speed information transmission between WSN and server. Through GPRS communication module the server collects, processes, analyzes and stores all sensing data. The cloud management is connected to the internet and has a specified IP address to receive and forward information data to the workstation. The workstation is a real-time monitoring center that monitors transmission parameter obtained from the WSN. It is connected with the internet enables access from a remote area by the operating personnel.

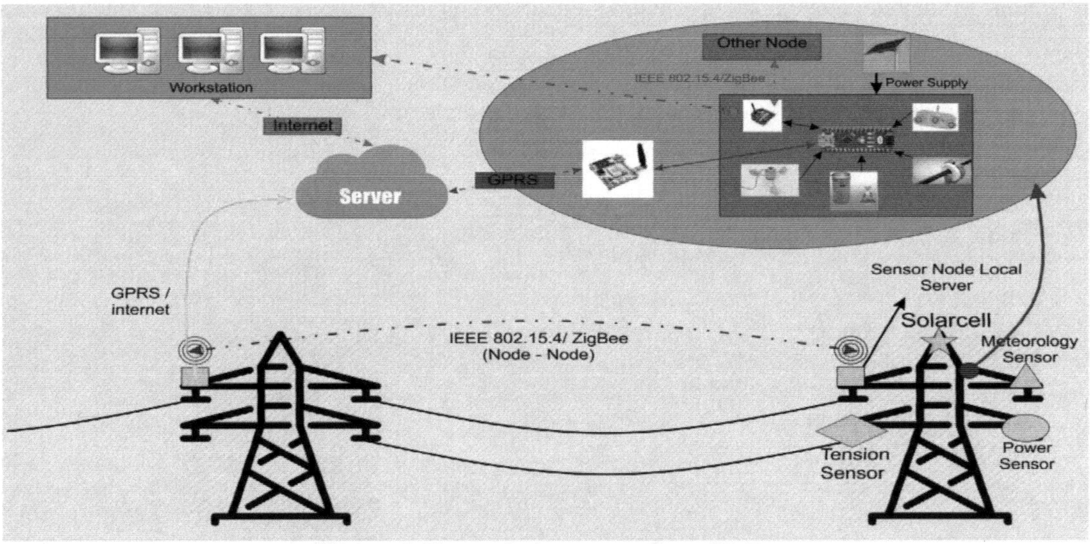

Figure 3. Transmission system based IoT.

4.2. Distribution hardware design

Figure 4 depicts the deployment schematic of the equipment in the distribution system based IoT. The WSN device, network model, communication topology and data processing are similar to the transmission system. However, monitoring status of sub-station is of the main concern in the distribution system. Therefore, sub-station should be well equipped with a smart sensor and smart meter, through which important parameter such as current and voltage profile, transformer temperature, oil level, load profile, fault and outage can be monitored. Monitoring on transmission and distribution system based IoT is a powerful tool to increase system reliability, realizes early warning and real-time monitoring through the cloud information management. The status of power transmission and distribution system can be displayed visually on PC, laptop and smartphone.

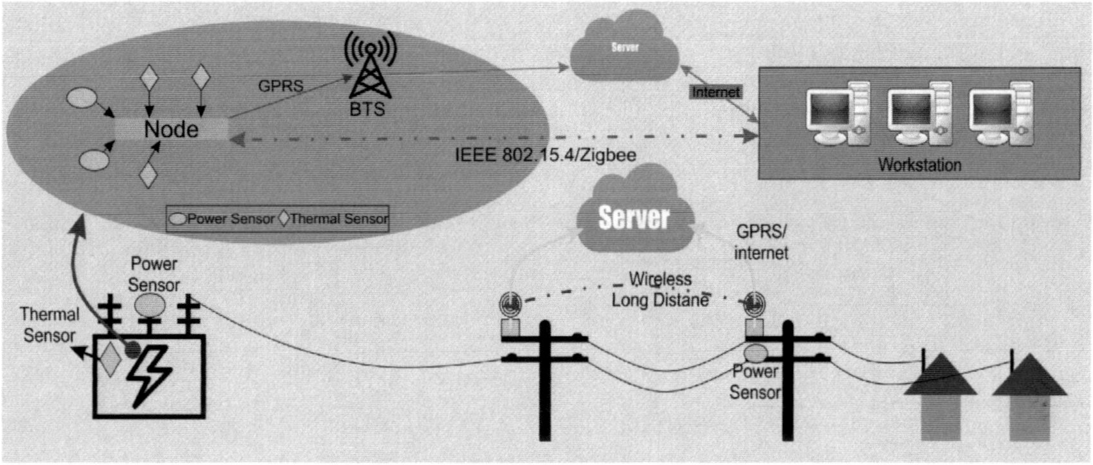

Figure 4. Distribution system based IoT.

4.3. Software design

In general, the software of this system is composed of WSN software and IoT software. Both of them are customized software developed and installed in the sensor nodes, coordinator nodes, embedded operating system, applications and server. Figure 5 shows flowcharts of the software.

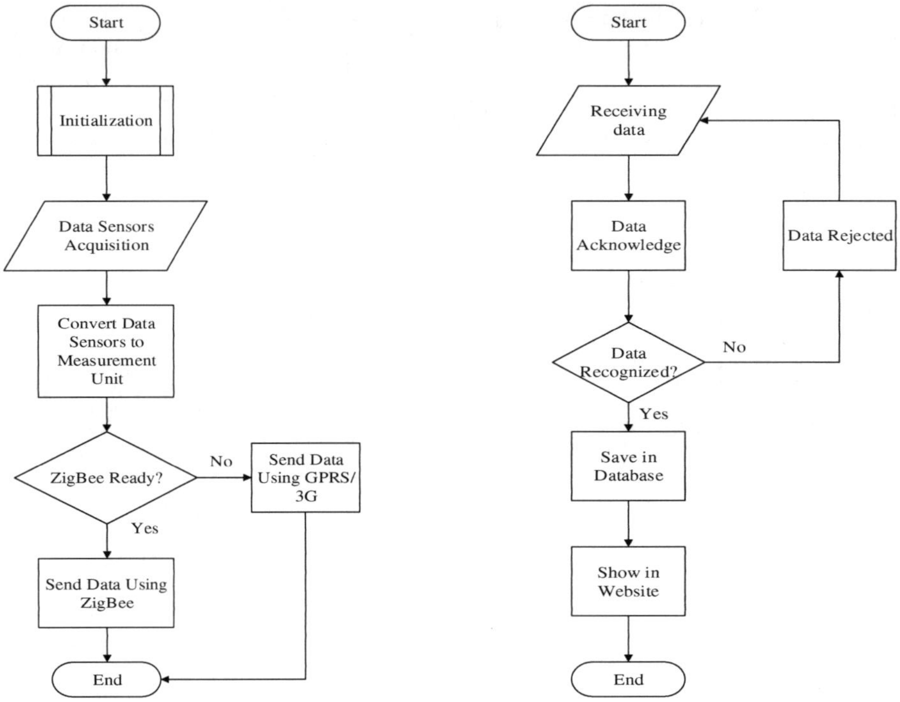

(a) (b)

Figure 5. (a) Flowchart of WSN, (b) Flowchart of IoT.

The WSN software is mainly responsible for system initialization, sensors data acquisition, and data processing from analog to digital. Then the data are sent to the server through Zigbee network or GPRS. On the other hand, the IoT software is management software that connected to the internet and has a particular IP address. This software is responsible for data verification, data reception, data processing and data storage. All of these data are stored in the database and accessible through a web server.

5. Conclusions

This paper introduces fundamental Internet of Things architecture for power transmission and distribution monitoring system. The system and its hardware design are analyzed and discussed. In designing WSN based IoT one must consider the sensor requirement as well as system architecture suitable for specific local power transmission and distribution system. Therefore for archipelagos and developing countries like Indonesia, a combination of wireless transmission with Zigbee and GPRS is suggested to meet the basic requirement in terms of communication data, real-time and long distance monitoring system. Through IoT technology, the power transmission and distribution parameters can be visually displayed on PC and even smartphone. In future work, the authors will focus on developing the design to experiment devices for power distribution system, which could be beneficial to Indonesia.

Acknowledgements

This work was supported by State Polytechnic of Madiun under DIPA No 042.01.2.400873/2017.

References

[1] Daniele M, Sabrina S, Francesco D P and Imrich C 2012 Internet of things: Vision, applications and research challenges *J.Ad Hoc Networks* **10** p 1497-1516.

[2] Xinghuo Y and Yusheng X 2016 Smart Grids: A Cyber-Physical Systems Perspective *IEEE Proc* **104** p 1058-1070.

[3] Sanjana K V. *et al* 2016 System design of the internet of things for residential smart grid *IEEE Wireless Communication* p 90-98.
[4] Olfa K and Hans-Rolf T 2004 Sensor Technology and Future Trend *IEEE Transaction on Instrumentation and Measurement* **53**(6) p 1497-1501.
[5] Junru Lin. *et al* 2014 Monitoring Power Transmission Lines using a Wireless Sensor Network *Wireless Communication and Mobile Computing* (John Wiley & Sons, Ltd).
[6] Qinghai Ou. *et al* 2012 Application of Internet of Things in Smart Grid Power Transmission *3rd FTRA International Conference on Mobile, Ubiquitous, and Intelligent Computing* p 96-100.
[7] Abel A B, Gehm J O, Quevedo M L, da Silva Martins and R C Beltrame 2014 A low-cost GPRS based communication system for a smart transformer *11th IEEE/IAS International Conference on Industry Applications, Juiz de Fora* p 1-6.
[8] Made A S, Farhad S, Sumedha R and Arindam G 2015 ZigBee-Based Communication System for Data Transfer Within Future Microgrids *IEEE Transactions On Smart Grid* **6**(5) p 2343-2355.

IOP Conf. Series: Materials Science and Engineering **384** (2018) 012040 doi:10.1088/1757-899X/384/1/012040

Reduced Graphene Oxide/Polyaniline Nanocomposite as Efficient Counter Electrode for Dye Sensitized Solar Cells

R Rahmawati[1,3], V Suendo[1*] and R Hidayat[2]

[1]Department of Chemistry, Inorganic and Physical Chemistry Research Division, FMIPA, Institut Teknologi Bandung
[2]Department of Physics, Magnetic and Photonic Physics Research Division, FMIPA, Institut Teknologi Bandung
[3]Chemistry Education Study Program, FTK, UIN Sunan Gunung Djati Bandung

*vsuendo@chem.itb.ac.id

Abstract. Reduced graphene oxide (rGO) was synthesized using sonication-assisted oxidation of graphite followed by reduction with ascorbic acid as environmentally friendly reductor. The polyaniline was prepared via chemical polimerization of aniline in 0°C. The rGO and polyaniline were blend in three mass compotitions of 1:3, 1:1, and 4:1 to produced rGO/polyaniline nanocomposites. The rGO/polyaniline nanocomposites were characterized using Raman spectroscopy, Fourier transform infrared spectroscopy and LCR meter. The rGO sheets as the support materials could provide more active sites for the nucleation of polyaniline, where polyaniline acts as conducting support. The rGO/polyaniline nanocomposites was deposited on synthetic graphite substrate laminated the glass substrate and applied as the counter electrode for efficient dye sensitized solar cells (DSSC). The rGO/polyaniline nanocomposite is an efficient counter electrode because of Pt/TCO free. The usage of rGO/polyaniline nanocomposites film as counter electrode in DSSC was demonstrated. The best performance was obtained for rGO/Polyaniline ratio of 4:1. Here, the optimum short circuit current density (Jsc), the open circuit current (Voc), the fill factor (FF), and the overall conversion efficiency under AM 1.5, 100 mW cm^{-2} illumination are 7.853 mA/cm^2, 0.64 Volt, 47.2% and 2.64%, respectively.

1. Introduction

Dye sensitized solar cells (DSSC) have received much attention due to their low cost, easy fabrication, environmental friendliness, and respectable energy conversion efficiency [1-3]. DSSC consist of several main components n-type semiconductor electrode (photo anode), a sensitizer (dye), a redox couple electrolyte, and a counter electrode (CE) [1]. CE is one of the most important component in DSSC. The role of CE is to transfer electrons from the external circuit back to redox electrolyte which is the mediators for regenerating the sensitizer (dye) after the electron injection and to catalyze the reduction of I_3^- to I^- [1, 4]. A CE material in DSSC should possess a high catalytic activity and stability toward electrolyte. Platinum (Pt) is the most preferred material for CE in DSSC, due to its excellent physical and chemical properties, such as a high electrical and thermal conductivity, catalytic activity toward I_3^-, and corrosion resistance to iodide ions present in redox electrolyte [3-5]. However, because Pt is precious metal, it is high cost and prevented it from being widely used in commercial DSSC. Recently, Pt-free materials with lower cost and good stability have been investigated as counter electrode for DSSC, mainly based on combination carbon materials and conducting polymers [1, 4-6].

Graphene, one of the allotropes of abundantly carbon is one of the most promising materials for application in solar cells. Graphene is possessing high conductivity, great mechanical strength, high carrier mobility, high specific area, and high optical transparency [1-2, 4-6]. However, graphene is not perfect candidate for a counter electrode in DSSC, because it has limited number of active sites for I_3^-/I^- electro catalysis. Polyaniline is a conducting polymer and has high catalytic activity which has been employed as the counter electrode for DSSC. The combination of graphene and polyaniline are believed as an efficient route [1-6]. The graphene /polyaniline nanocomposite was synthesized using insitu polymerization and deposited on fluorine-doped tin oxide glass has conversion energy efficiency reached 6.09% which is comparable to 6.88% of the cell with Pt counter electrode [2].

In this paper, reduced graphene oxide (rGO) was synthesized using sonication-assisted oxidation of graphite followed by reduction with ascorbic acid as environmentally friendly reductor [7, 8]. The polyaniline was prepared via chemical polymerization of aniline in 0°C [9]. The rGO and polyaniline were blend to produced rGO/polyaniline nanocomposites. The rGO/polyaniline nanocomposites was deposited on synthetic graphite substrate laminated the glass substrate and applied as counter electrode. The counter electrode was the Pt and TCO free, it was low cost, environmental friendly, and efficient.

2. Experimental

2.1. Synthesis of reduced graphene oxide (rGO)

1 g graphite flakes was added to 50 mL concentrated sulphuric acid (97%, Merck) while stirring in ice bath. 3 g potassium permanganate (>99%, sigma Aldrich) was gradually added by maintaining temperature under 10°C. Then, the suspension stirred at room temperature for 25 min followed by 5 min sonication. After repeating the stirring-sonication process for 12 times, the reaction was quenched by addition of 200 mL distilled water. An extra 2h ultrasonic treatment was carried out. After adjusting pH 6 by addition of 1 M sodium hydroxide (>98%, Merck) solution, the suspension was further sonicated for 1h, 10 g of L-ascorbic acid (>99%, sigma Aldrich) was dissolved in 100 mL distilled water and slowly added to the exfoliated graphite oxide suspension at room temperature. The reaction was performed at 95°C for 1h. The resultant black presipitates were washed with 1 M hydrochloric acid (37%, merck) and distiled water to pH 7. Finally, the filtrate was dried to obtain rGO powder.

2.2. Synthesis of polyaniline

Emeraldine salt polyaniline (PANI-ES) was synthesized by chemical oxidation, conventional emulsion polymerization. In two different flask, 9,8 g Ammonium peroxydisulfate (APS) (sigma Aldrich) was dissolved in 100 mL distilled water, in another flask 3.72 g aniline (sigma Aldrich) was dissolved in 100 mL hydrochloric acid (>37%, Merck) 1M. And then, both of flasks was kept out in ethilenglycol bath at temperature 0°C for 1h. APS solution was mixed with aniline solution and kept out at temperature 0°C for 24h. The green presipitates were simply filtered by cellulose filter paper and further were washed with aseton and 1 M hydrochloric acid solution.

2.3. Assembly of rGO/polyaniline nanocomposite film

The rGO and polyaniline were blend in three mass compotitions of 1:3, 1:1, and 4:1 to produced rGO/polyaniline nanocomposites. The rGO/polyaniline nanocomposites was deposited on synthetic graphite substrate laminated the glass substrate using screen printing technique, followed by sintering in air at 150°C for 15 min and applied as the counter electrode for efficient dye sensitized solar cells (DSSC).

2.4. Assembly of DSSC

A TiO_2 photo anode film were prepared by coating the TiO_2 colloid on ITO glass substrate using screen printing technique, followed by sintering in 500°C for 1h. Subsequently, the TiO_2 film was soaked in a 0.5 mM N719 (purchased from solaronic, Poland) ethanol solution for 24h. The DSSC was fabricated by sandwiching liquid electrolyte between a dye sensitized ITO-TiO_2 photo anode and graphene/polyaniline film counter electrode.

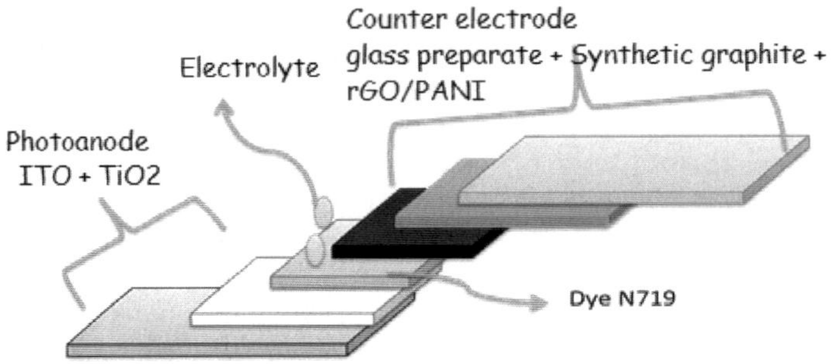

Figure 1. Design of DSSC using rGO/PANI nanocomposite CE.

2.5. *Other characterizations*

Raman spectroscopy measurements were taken using Bruker with a 532 nm laser. FTIR spectra were collected using a Bruker tensor instrument. The solar simulator measurements were taken using ORIEL S013A 100mW/cm^2 dan Khitley 2400.

3. Results and discussion

Raman spectroscopy is a non-destructive technique in use to study structural change and disorder of graphite, GO, and rGO during the oxidation and reduction steps [7, 8]. As shown in Fig. 2a, graphite has a prominent peak at 1566 cm^{-1} (G band) which is due to the first order scattering of E_{2g} mode and two weaker peaks located at 1340 cm^{-1} and 2687 cm^{-1} (D and 2D bands, respectively) originated from the second order double resonant process between non-equivalent K point in Brillouin zone of graphene. Structural defects induced by attachment of OFGs to the basal plane of graphene during the oxidation of graphite significantly intensified the D band and up-shifted the G band to 1585 cm^{-1} in the Raman spectrum of GO. The blue shift of the G band has been attributed to the activation and merging of the Raman –inactive D' band with the G band, the exfoliation of graphite to a single graphene sheet, and the resonation of isolated double bonds at higher frequencies than the G band of graphite [7]. On the other hand, the red shift of G band in Raman spectrum of rGO to 1579 cm^{-1} was observed that can ascribed to the reduction of GO. The relative intensity of the D and G band (I_D/I_G), which is an estimation of the disorder level in graphene, exhibits a non monotonic behavior. The 2D band is found at around 2697 cm^{-1} which is in reference with the reported result for the single layer graphene within the experimental accuracy [8].

The FTIR spectrum of graphite, GO and rGO (Fig. 2b) showed various peaks corresponding to different oxygen functionalities verifying successful oxidation of graphite into GO, and reduction of GO to rGO. The broad signal around 3400 cm^{-1} was due to O-H stretching vibrations and adsorbed water molecules. The peak near 1700 cm^{-1} is attributed to various skeletal vibrations of C=C aromatic unoxidized graphitic domain. Other prominent signals in the GO's spectrum which have been attenuated in rGO's spectrum are 1399 cm^{-1}, and 1071 cm^{-1} originating from O-H deformation, and the C-O stretching respectively [7, 8].

In Fig. 2c shows the main bands of PANI Raman spectra situated at 1621, 1624, 1491 dan 1505 cm^{-1} may be attributed to the following: vibration of C-C semiquinonoid ring, and vibration of benzenoid ring. Fig 2d. Shows FTIR spectra of PANI. The main bands situated at 1575, 1490, 1301 and 1138 cm-1 may be attributed to the following vibrations: stretching of C=C in quinoid ring, stretching vibration of C=C benzenoid ring, stretching of C-N secondary aromatic amine, stretching of C=N, respectively [10].

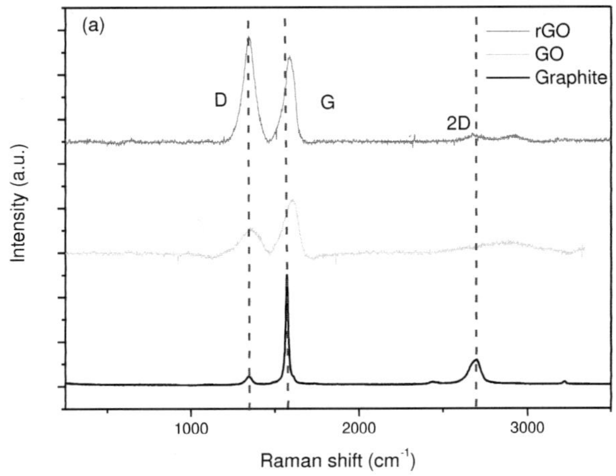

(a)

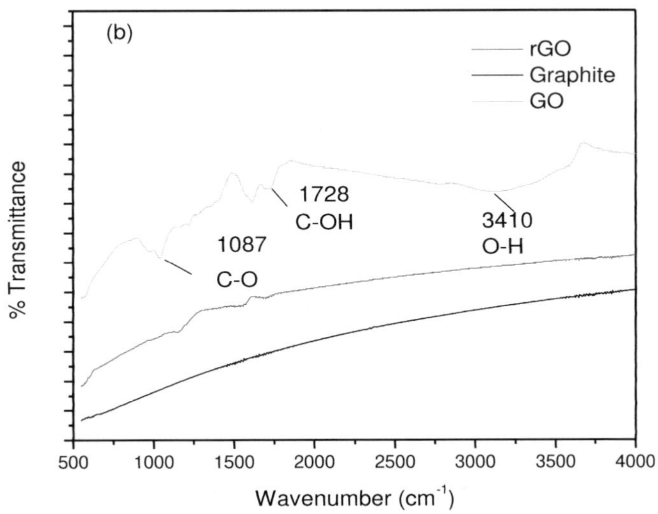

(b)

Figure 2. Characterization of graphite, GO, rGO powder, and polyaniline emeraldine salt (a) Normalized Raman spectra of graphite, GO, and rGO. (b) FT-IR spectra of graphite, GO, and rGO.

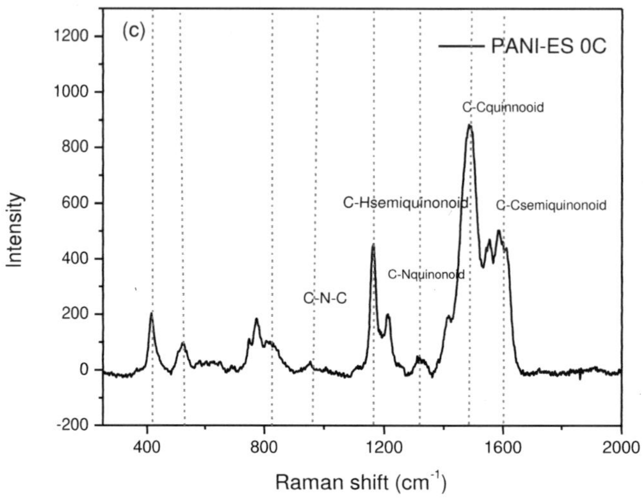

(c)

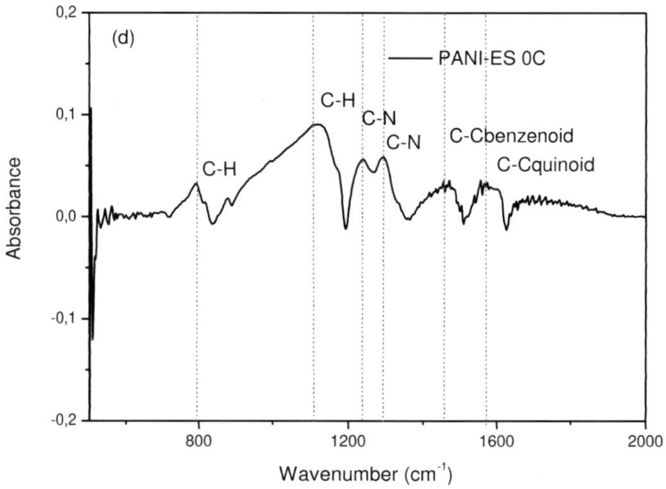

(d)

Figure 2 (Cont.). Characterization of graphite, GO, rGO powder, and polyaniline emeraldine salt (c) Normalized Raman spectra of PANI. (d) FT-IR spectra of PANI.

Graphene is a good electron acceptor. On the other hand, aniline is good electron donor [11]. When polyaniline are blended into graphene sheets, polyaniline can immediately absorp onto the surface of the graphene sheet due to the electrostatic attraction. And then the structure of PANI covering nanosheets forms slowly [12]. During polimerization, graphene sheets acting as the support material can supply a large number of active sites for nucleation of PANI. where polyaniline acts as conducting support [12, 13]. Fig. 3 shows characterization of rGO/PANI nanocomposites film using Raman spectroscopy (Fig. 3a) and FTIR spectroscopy (Fig. 3b). The Raman spectras of rGO/PANI nanocomposites film 4:1, 1:1, and 1:3 (Fig. 3a) exhibit similar signals to that rGO, the D-band and G-band still in same positions. But, the G-band of rGO/PANI nanocomposites film of 4:1, 1:1, and 1:3 have the shoulder. The broader shoulder showed by rGO/PANI nanocomposites film 1:3, apparently as aresult of influence coming from PANI. Therefore, Raman spectras results confirm the formation of PANI nanocomposites coating on the surface of graphene sheet. Fig. 3b shows FTIR spectras of rGO/PANI nanocomposites film 4:1, 1:1, and 1:3. FTIR spectras of rGO/PANI nanocomposites film1:1, and 1:3 have similar signals to the PANI, but the rGO/PANI nanocomposites film 4:1 has similar signal to the rGO. The spectra of FTIR depends on the composition of the composites.

Fig. 4 shows rGO/PANI nanocomposites conductivity. In Fig. 4a, rGO has typically as semiconductor. Fig. 4b shows that by introducing graphene into PANI matrix, the conductivity increases with increasing graphene content. The highest conductivity achieved by rGO/PANI nanocomposite of 4:1. Based on the graph (Fig. 4b), the rGO/PANI nanocomposites have typically as conductor, conductivity decreases with increasing frequency. Homogeneous PANI coating on the surface of graphene sheets increases both conductivity, and thus is preferred for the enhancement of the electrocatalytic performance as a counter electrode for DSSC [2].

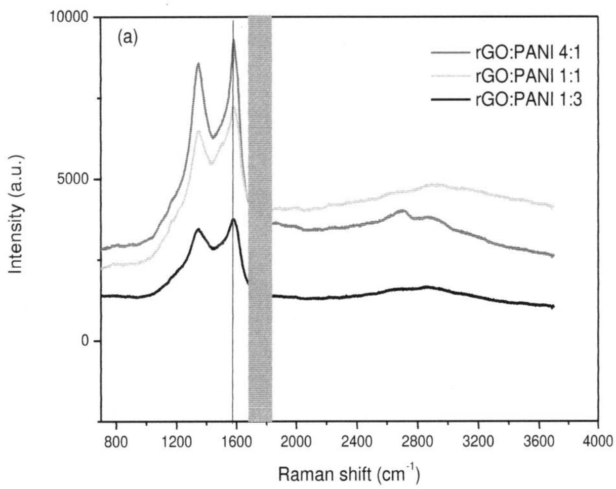

(a)

Figure 3. Characterization of rGO/PANI nanocomposites film (a) Raman spectra of rGO/PANI nanocomposites film of 4:1, 1:1, and 1:3.

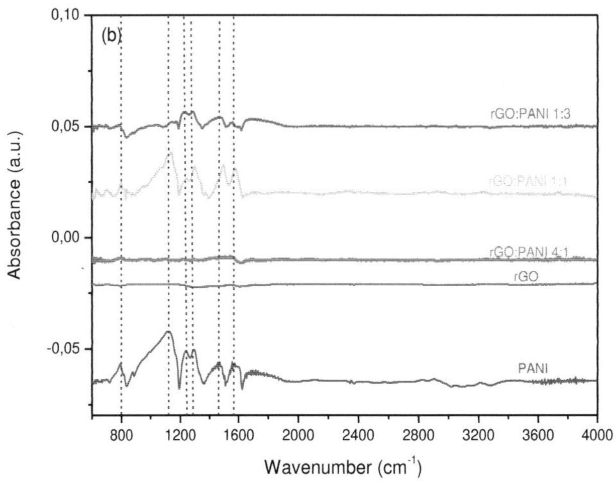

(b)

Figure 3 (Cont.). Characterization of rGO/PANI nanocomposites film (b) Normalized FT-IR spectra of rGO/PANI nanocomposites film of 4:1, 1:1, and 1:3.

The photocurrent-voltage curve of DSSC with rGO/PANI nanocomposites are compared in Fig. 5. The best performance of rGO/PANI nanocomposites counter electrode achieved by rGO/PANI of 4:1 with photoconversion efficiency 2.65%, mainly due to highest fill factor 0.472. The PCE increased with the increase in mass composition of rGO. The derived photovoltaic parameters, such as the open-circuit voltage (V_{OC}), the short circuit current density (J_{SC}), the fill factor (FF) and the photoconversion efficiency (PCE) are summarized in Table. 1. Thus, the as-prepared rGO/PANI nanocomposites is good candidate as the counter electrode in DSSC.

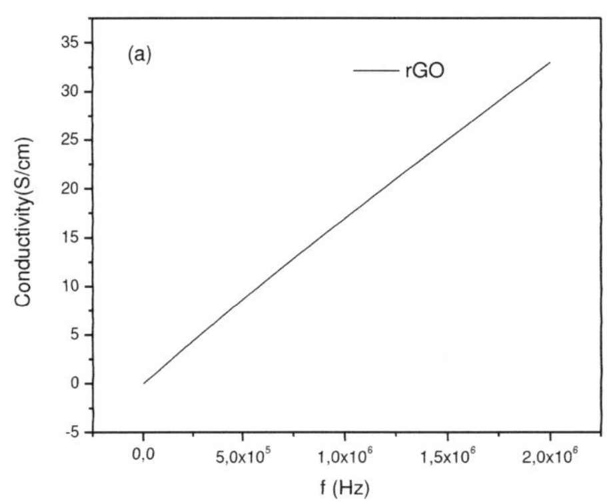

(a)

Figure 4. rGO/PANI nanocomposites conductivity (a) conductivity of rGO.

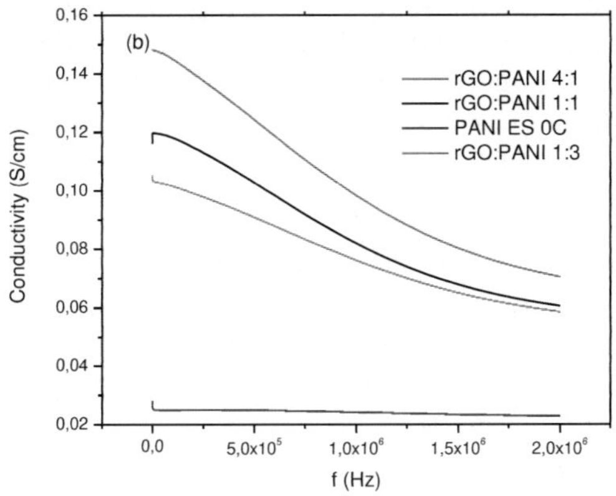

(b)

Figure 4 (Cont.). rGO/PANI nanocomposites conductivity (b) conductivity of rGO/PANI 4:1, 1:1, and 1:3.

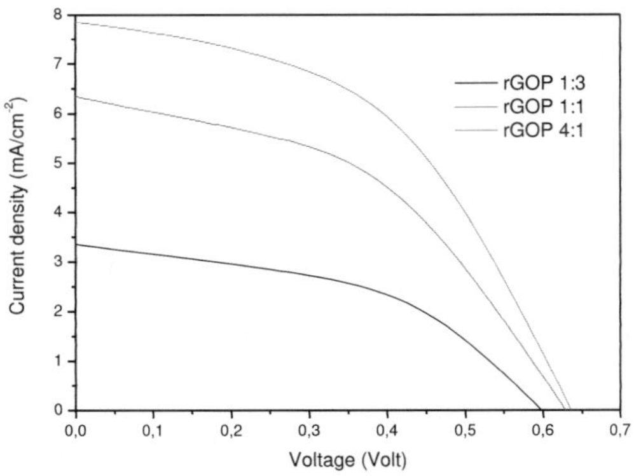

Figure 5. I-V curve of DSSC based on rGO/polyaniline nanocomposites counter electrode.

Table 1. photovoltaic parameters of DSSC with rGO/PANI CEs.

Films	PCE (%)	FF	Voc (Volt)	Jsc (mA/cm^{-2})
rGOPANI1:3	1.04	0.463	0.6	3.359
rGOPANI1:1	2.01	0.446	0.637	6.351
rGOPANI4:1	2.64	0.472	0.64	7.853

4. Conclusions

The best performance was obtained for rGO/Polyaniline ratio of 4:1. Here, the optimum short circuit current density (Jsc), the open circuit current (Voc), the fill factor (FF), and the overall conversion efficiency under AM 1.5, 100 mW cm^{-2} illumination are 7.853 mA/cm^2, 0.64 Volt, 47.2% and 2.64%, respectively.

References

[1] Shalini S, Balasundaraprabhu R, Prasanna S, Mallick T A, and Senthilarasu S 2015 *Renewable*

and Sustainable Energy Reviews **51** p 1306-1325.
[2] Singh E and Nahwa H S 2015 *Science of advanced Materials* **7** p 1863-1912.
[3] Janani M, Srikrishnarka P, Nair S V and Nair A S 2015 *J. Mater. Chem. A* **3** p 17914–17938.
[4] Hardin B E, Snaith H J and McGehee M 2012 *Nature Photonics* **6** p. 162-169.
[5] Konwer S, Gogoi J P, Kalita A, Dolui S K 2011 *J. Mater Sci: Mater Electron* **22** p. 1154–1161.
[6] Baniasadi H, Ramazani A, Mashayeksakhan S and Ghaderine F 2014 *Synthetic Metals* **196** p. 199-205.
[7] Wang G, Zhuo S and Xing 2012 *Materials Letters* **69** p. 27-29.
[8] Konwer S 2016 *J.Mater Sci: Mater Electron* **27** p. 4139-4146.
[9] Harish C, SreeHarsha V S, Santhos C, Ramachandran R, Saranya M, Mudallar T V, Govardhan K and Grace N A 2012 *Advanced Science, Engineering and Medicine* **4** p. 1-9.
[10] Goswami S, Maiti U N, Maiti S, Nandy S, Mitra M K and ChattoPadhyay K K 2011 *ScienceDirect:Carbon* **49** p. 2245-2252.
[11] Sun H, Luo Y, Zhang Y, Li D, Yu Z, Li K and Meng Q 2010 *J. Phys. Chem. C* ***114*** p. 11673–11679.
[12] Lee J W, Ramasamy E, Lee D Y and Song J S 2009 *ACS.Applied Materials and Inter* **4** p. 114-120.
[13] He B, Tang Q, Wang M, Chen H and Yuan S 2014 *ACS Applied Materials and Interfaces* **6** p. *8230-8236.*

IOP Conf. Series: Materials Science and Engineering **384** (2018) 012041 doi:10.1088/1757-899X/384/1/012041

Designing Energy and Power Monitoring System on Solar Power Plant Using Raspberry Pi

R H P Putra*, D Wahyudin and T Sucita

Department of Electrical Engineering Education, Universitas Pendidikan Indonesia, Jalan Dr. Setiabudhi No. 207, Bandung, Indonesia 40154

*regy.roy.hidayat@student.upi.edu

Abstract. The application of Information Technology (IT) has been growing rapidly recently. IT utilized to monitor flowing power and stored electrical energy which is produced by solar cell. This project explains how to build and design interface system. Electrical energy need to be monitored in order to keep energy stored. Single Board Computer (SBC), microcontroller, sensors, and transceivers are used in logging electrical power for this project. These nodes are connected wirelessly. Raspberry Pi 2 B, Arduino Nano, NRF24L01, Voltage sensor and ACS712 sensor are required to run this project. IDLE, Apache, BIND, hostapd, udhcpd, and highcharts are programs used in Raspberry Pi 2 B. The values of power are displayed into graphical chart and log energy table are able to be accessed by using computer or smartphone. By connecting to local network via Wi-Fi and go to http://solmon.home to see measurement results.

1. Introduction

The current development of technology produce a tremendous number of electronic devices with its functionality. Electronic device for controlling and automating is one of the most developed technology. Electronic device is being utilized in every aspect, such as in house as smart home [1]. Raspberry Pi has been used for multipurpose monitoring such for surveillance [2]. By using Raspberry Pi as a server solar power plant can be monitored both power and stored energy.

An independent home of producing electrical energy by installing solar panel. Using a solar panel, there is no any sound noise as mechanical electrical energy generator [3]. This energy production could be used for lighting or home appliance. Before energy are being used, it should flow first into battery. Monitoring of flowing energy is needed in order to check whether the energy is available or not.

Microcontroller and single board computer (SBC) are already used for load monitoring on previous works [4-7]. However there is some lacks, as the system is unable to be accessed in flexible range in the reason of limitation by using wire for the media connection. The others consideration, cost to build the systems are pricey, rare or unavailability components to find, and complexity for user to use the system.

Our goal is to design a system for monitoring power and flowing energy from the solar panel through battery and monitoring a simulated load by using a Raspberry Pi 2 B as main component or server to display data and two Arduino Nano as data fetcher for power and energy [8]. The communication among server, data fetcher and user are wirelessly.

IOP Conf. Series: Materials Science and Engineering **384** (2018) 012041 doi:10.1088/1757-899X/384/1/012041

2. Methods

There are bunch of module communication to build, the communication established from Arduino Nano to Raspberry is using NRF24L01, is a RF module that only communicate between those two devices. NRF24L01 is used since its effective and efficient for this proposed system. While for connection between Raspberry Pi to user's device is using Wi-Fi connection. TP-Link TL725WN is attached to Raspberry Pi since Raspberry Pi 2 B does not have built-in Wi-Fi. Thus this module is chosen for its compatible with Raspberry Pi 2 B to work as hotspot and it is more familiar in several online forums.

The objects is to monitor solar cell and a simulated load. We used solar cell system at High Voltage Laboratory Universitas Pendidikan Indonesia. The simulated load is a single resistor with a DC supplied.

For the designing software, we split into two parts that is server side and node side. This begins from designing Arduino Nano script as node side to make sensors are working well. Compiled the script using Arduino IDE from PC. The calibration occurs in this progress. Voltage sensor should read voltage in parallel circuit and current sensor read current as in series circuit. Calibrating current sensor need a special sampling. Current sensor we are using is ACS712 which manufactured by Allegro MicroSystems. It performs based on Hall effect means converting DC or AC current into proportional voltage output within 0-5 volt range. Magnetic field is applied for this reading current sensor. This sensor should be programmed by using sampling data to get smooth value of reading current. Since Arduino Nano has an analogue pin input, it is just a common reading analogue pin. Then, configure networking of NRF24L01. Arduino IDE using C as a programming language. ACS712 and voltage sensor need to be calibrated in order to get good measurement.

Once sensor management is done, we move on into server side it is Raspberry Pi. The server using Python programming language to run the script. This device works as an intermediate to user's device. User device can be personal computer or smartphone as long it has web browser to access and connected to Wi-Fi from Raspberry Pi. The whole activities occur as shown in Figure 1.

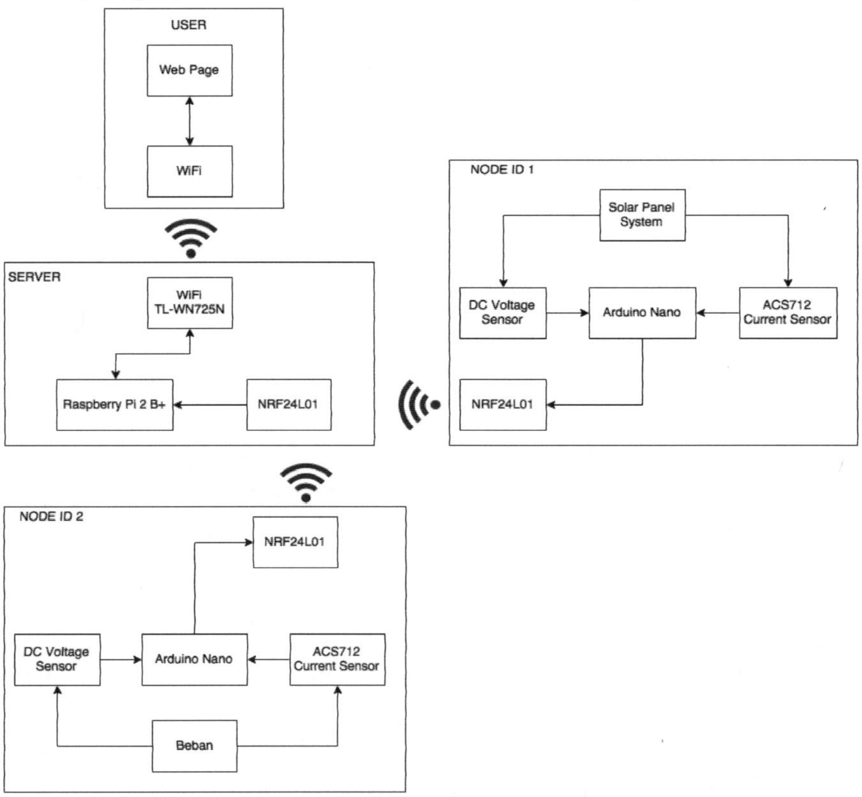

Figure 1. Block diagram of system.

Programs is not packed so need to be configured themselves independently. One program could impact to other programs. For example if hotspot program is not working, user could not access data. The programs become a whole system to perform at the same time. The programs are installed on the server are driver on Raspberry Pi to communicate between Raspberry Pi and TL-725N, Raspberry Pi and NRF24L01; Apache2 for web server, MySQL and PHPmyAdmin for logging and configuring database, hotspot program which consist of udhcpd for giving IP address automatically to user's device, hostapd for making wireless access point from server; and the last is DNS server for converting IP address into name domain.

To identify energy flow from solar panel, we log to database of the value which is sent by node side into server. After logging, the web server program performs a reading database and convert the value into a graphical display. In order to make easier to troubleshoot, figure 2 shows the flowchart of this designed system.

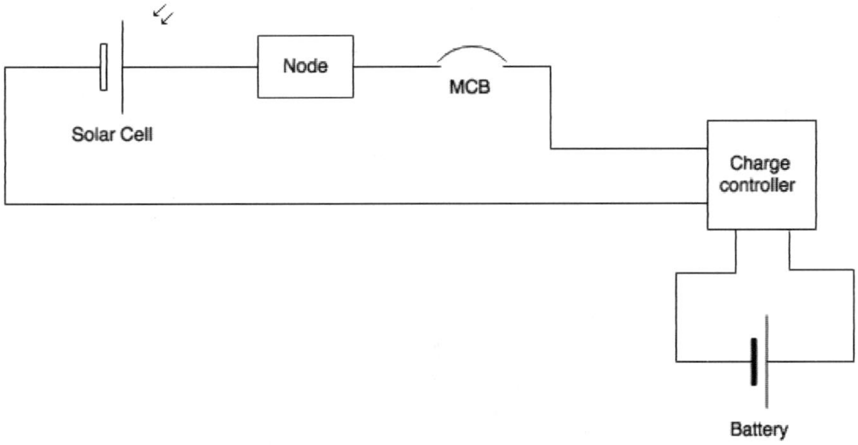

Figure 2. Node on solar cell circuit.

Figure 2 shows that Node is placed between solar cell and charge controller. The placement of node for real is shown in figure 4.

3. Results and discussion

According to experiment from designing this system at High Voltage Laboratory FPTK UPI, we obtained the results:

- A prototype of system logging power solar panel.
- Designs and configuration programs to make this system.
- The output which can be accessed by user in a local network.

The devices are connected each other can communicate as well as designed.

Figure 3. Raspberry Pi as a server is package in an acrylic with Raspberry Pi logo.

Raspberry Pi is bundled into a single acrylic box and only has one input power source from USB cable. Since Raspberry using wireless communication both to nodes and user, thus transceiver is inside the box. This device is the central unit of the system [9].

Figure 4. Arduino Nano as node in an acrylic box with Arduino Logo.

Two Arduinos as nodes is shown in Figure 4. Each has same two power input by using USB source cable. For reading media, using four cables connector. Two cables for reading voltage positive side and negative side; two others cable are used for reading current.

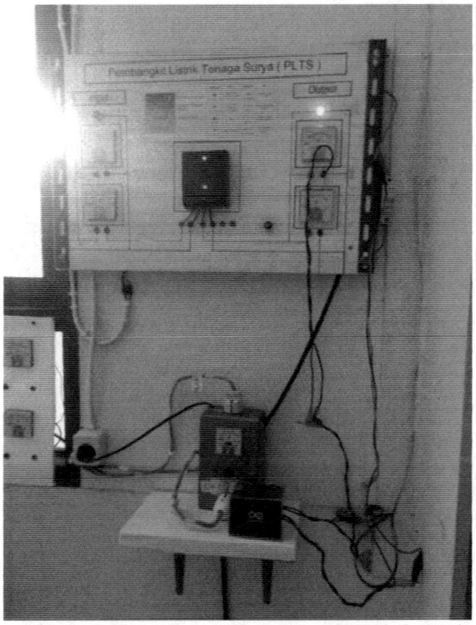

Figure 5. Integrated node in solar cell system in High Voltage Laboratory.

This calibrating sensor is needed to make sure the sensors are in standard condition. By comparing the value result from 10 multi meters to a sensor voltage. This calibrating using an object which is rheostat as load and DC power supply. Set to 100Ω and 12V as the object. Calibrating current sensor apply the same object as calibrating voltage sensor. Table 1 shows the measurement.

Table 1. Voltage measurements.

Voltmeter	Value (V)
1	10.22
2	10.17
3	10.21
4	10.22
5	10.24
6	10.25
7	10.26
Sensor	10.14

Table 2. Current measurements.

Ampere meter	Value (A)
1	2.64
2	2.60
Sensor	2.68

These calibrating result obtain a low error. By using descriptive statistic, we obtain error on Voltage is 0.005% and current is 0.02%. This has fulfilled for some applications which requires error 0.1%-1% [10].

Once user access to network, the user's device will obtained automatic IP address and just typing on address bar the IP address 10.1.30.1 or domain name http://solmon.home the server will redirect user into a web page which contains graphical information. Also there is an option to access log which allow user to view the last value. It could be chosen per date. Figure 4 shows the result of web page accessed by personal computer. Figure 5 shows the result of web page accessed by mobile phone. Figure 6 shows the option to view the log.

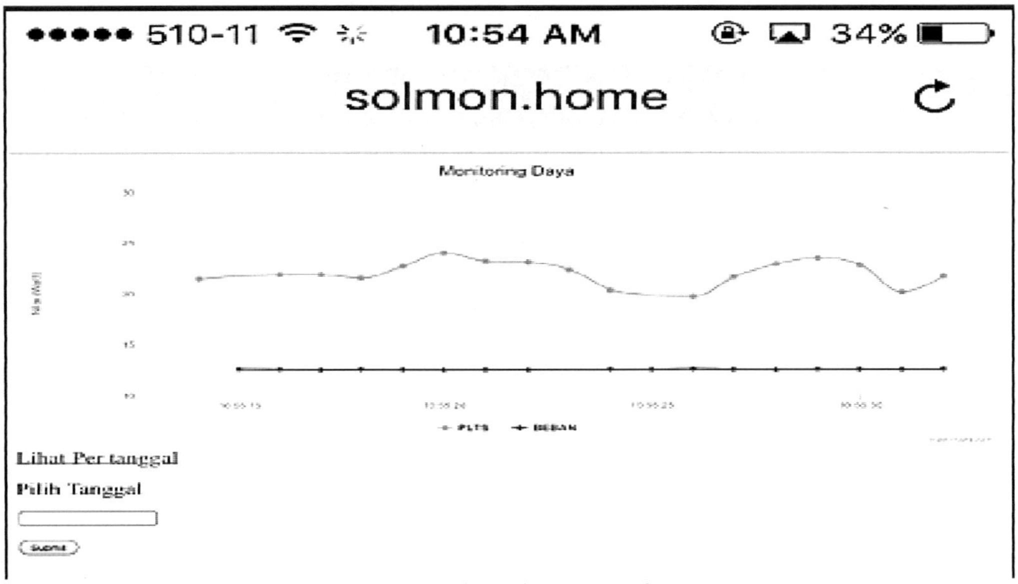

Figure 6. Main web page on phone.

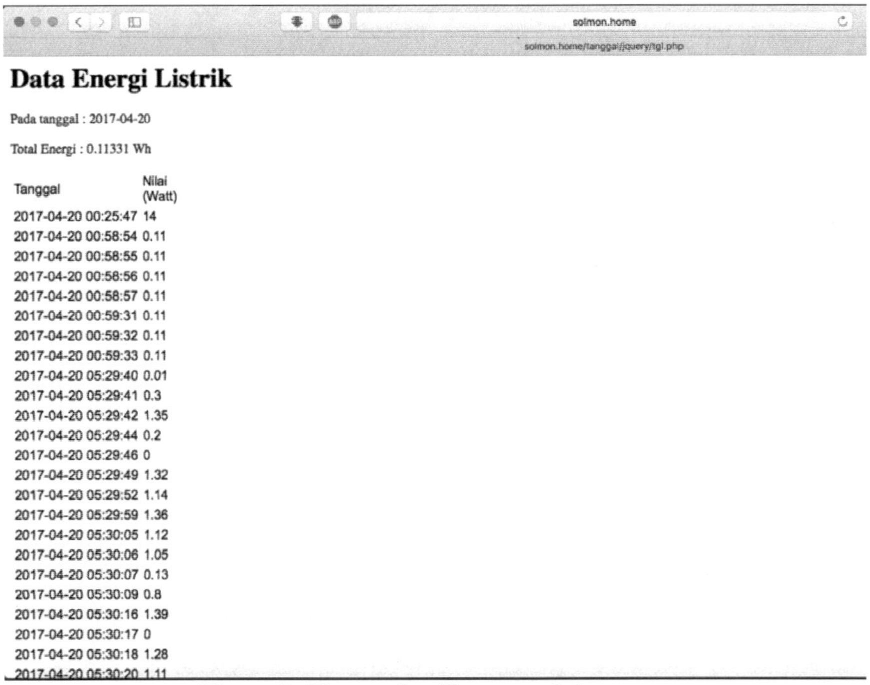
solmon.home

solmon.home/tanggal/jquery/tgl.php

Data Energi Listrik

Pada tanggal : 2017-04-20

Total Energi : 0.11331 Wh

Tanggal	Nilai (Watt)
2017-04-20 00:25:47	14
2017-04-20 00:58:54	0.11
2017-04-20 00:58:55	0.11
2017-04-20 00:58:56	0.11
2017-04-20 00:58:57	0.11
2017-04-20 00:59:31	0.11
2017-04-20 00:59:32	0.11
2017-04-20 00:59:33	0.11
2017-04-20 05:29:40	0.01
2017-04-20 05:29:41	0.3
2017-04-20 05:29:42	1.35
2017-04-20 05:29:44	0.2
2017-04-20 05:29:46	0
2017-04-20 05:29:49	1.32
2017-04-20 05:29:52	1.14
2017-04-20 05:29:59	1.36
2017-04-20 05:30:05	1.12
2017-04-20 05:30:06	1.05
2017-04-20 05:30:07	0.13
2017-04-20 05:30:09	0.8
2017-04-20 05:30:16	1.39
2017-04-20 05:30:17	0
2017-04-20 05:30:18	1.28
2017-04-20 05:30:20	1.11

Figure 7. The result of log on web browser.

Raspberry Pi 2 B is chosen as a server in this system hence it has a parts to support this system works such a GPIO ports to connect NRF24L01. This server performs like a PC yet only using low voltage than a PC [11, 12]. This device has a wide community and support. There are other sort of this device like Banana Pi, Intel Galileo, BeagleBone, Orange Pi, C.H.I.P, etc. [13]. The Raspberry Pi is a device which can purchased and available in many robotic store.

The Arduino Nano is node side which is the compact, effective and efficient circuit to support this system. There are many sort of this circuit board such as Arduino UNO, MKR1000, etc. This circuit can be replaced by these circuit board for later works and development.

This system can monitor how much energy stored from solar cell. At the database server, energy values shows in a table as shown on figure 7. User could monitor by choosing date to show energy values.

4. Conclusions

This system logging for power and energy from solar panel and simulated load are built by node. Both are connected to server to show values from sensors in graphical and stored to database in server. The sensors must be calibrating first, especially for current sensor. Hence, sensors are affected by noise and sensitivity, also this sensor using hall effect which convert current into magnetic field and the output become a proportional 0-5 volt. The server installed some programs which is not much configuration. The user could use a PC or mobile phone to access server via Wi-Fi networks.

Acknowledgements
This paper is based on Tugas Akhir Teknik Elektro which is guided by Didin Wahyudin, Ph.D and Dr. Tasma Sucita.

References

[1] Augusto J C and Nugent C D 2006 *Designing Smart Homes The Role of Artificial Intelligence* (New York: Springer-Verlag Berlin Heidelberg).
[2] Jain A, Basantwani S, Kazi O and Bang Y 2017 "Smart Surveillance Monitoring System" pp. 269–273.
[3] Palz W 2011 *The Emergence of Electricity from the Sun Power for the World* (Singapore: Pan Stanford Publishing Pte, Ltd.).
[4] Firnandi I, Kustija J and Trisno B 2016 "Rancang Bangun Sistem Kontrol Beban dan Akuisisi Data Berbasis Web dengan menggunakan Single Board Computer" **14**(1).
[5] Sulistyowati R, Febriantoro D D, Elektro J T, and Industri F T 2012 "Perancangan Prototype Sistem Kontrol dan Monitoring Pembatas Daya Listrik Berbasis".
[6] Setiono A and Suharto 2009 "Prototipe Aplikasi KWh Meter Digital Menggunakan Mikrokontroler ATMEGA8535 untuk Ruang Lingkup Kamar" **26** pp. 32–39.
[7] van Staden P and Kotze B 2017 "Wireless Node Energy Monitor using common Development Platforms" pp. 1514–1519.
[8] Melgar E R, Diez C C and Jaworski P 2012 "Arduino and Kinect Projects Design, Build, Blow Their Minds".
[9] Guleci M and Orhun M 2017 "Android Based Wi-Fi Controlled Robot using Raspberry Pi" pp. 5–9.
[10] Moghe R, Iyer A R, Member S, Lambert F C, Member S and Divan D M 2014 "A Low-Cost Wireless Voltage Sensor for Monitoring MV / HV Utility Assets" 5(4) pp. 2002–2009.
[11] Pardeshi V, Sagar S, Murmurwar S and Hage P 2017 "Health Monitoring Systems using IoT and Raspberry Pi - A Review" pp. 134–137.
[12] Sunehra D and Ramakrishna P 2016 "Web Based Patient Health Monitoring System using Raspberry Pi" pp. 568–574.
[13] Saari M, Hyrynsalmi S and bin Baharudin A M "Survey of Prototyping Solutions Utilizing Raspberry Pi" pp. 991–994.

IOP Conf. Series: Materials Science and Engineering **384** (2018) 012042 doi:10.1088/1757-899X/384/1/012042

Authentication Security in Radio Frequency Identification with IDEA Algorithm

H Nurdiyanto[1], R Rahim[2,*], R Hidayat[3], P Harliana[4], G Gunawan[5], H A Adam[5], Y Sonatha[3] and M Azmi[3]

[1]Department of Informatics Engineering, STMIK Dharma Wacana, Metro Lampung, Indonesia
[2]School of Computer and Communication Engineering, Universiti Malaysia Perlis, Malaysia
[3]Department Information Technology, Politeknik Negeri Padang, Padang, Indonesia
[4]Department of Informatics Engineering, Sekolah Tinggi Teknik Harapan, Medan, Indonesia
[5]Department of Computer Engineering and Informatics, Politeknik Negeri Medan, Medan, Indonesia

*usurobbi85@zoho.com

Abstract. Data security on Radio Frequency Identification (RFID) cards for parking security applications need to be applied for more protection especially information on RFID cards, as well as to improve security on RFID access authentication and verification using cryptography algorithm, one of the algorithms that can be used for authentication and verification in RFID communication process is International Data Encryption Algorithm (IDEA) algorithm, the results obtained that the data security on RFID can work well.

1. Introduction
Radio Frequency Identification (RFID) [1] is an object identification method by utilizing radio waves using RFID Reader and RFID Transponder (RFID tag) where each RFID Tag has a different Identification number [1, 2], RFID has widely used in various activities such as attendance, passport, grocery store [3], E-Toll and also for parking at malls or companies.

RFID is not a perfect technology and certainly has a weakness [4] and even the information contained in RFID tags can be taken by irresponsible [5], it could happen because most RFID Tag data stored in text form that is easy to read when RFID Tag is scanned [5], there are many ways can use to avoiding the scanning process on RFID Tags such as metal or aluminum foil [5] but that also does not guarantee data on RFID Tags cannot be taken, another solution is to secure data available on RFID Tag by using cryptography [6-8] techniques.

IDEA [9-11] is one of block cipher cryptography algorithm that operates on 64 bit plaintext block and 128 bit key length [9, 12], one of IDEA algorithm difference with other algorithm is IDEA algorithm using confusion and diffusion technique for encryption process so the result do not have exact pattern, for data security experiments on RFID using IDEA algorithm require some tools such as ARDUINO UNO R3 and Micro Controller ATMega328 [13] whereas for RFID Tags using Mifare S50 RFID card as well as IDEA algorithm implement on RFID Tag, the implementation of IDEA algorithm in RFID doesn't guarantee data cannot be taken, but it can definitely make it difficult for irresponsible parties to reading the data [14].

2. Material and method

RFID Tag data security can be applying in any cases such as attendance, E-Toll, Identification Card, Parking, etc., and for this experiment, the application of security with IDEA algorithm on RFID Tag is made for consumer data self-service parking, implementation of IDEA and RFID algorithm in the picture below:

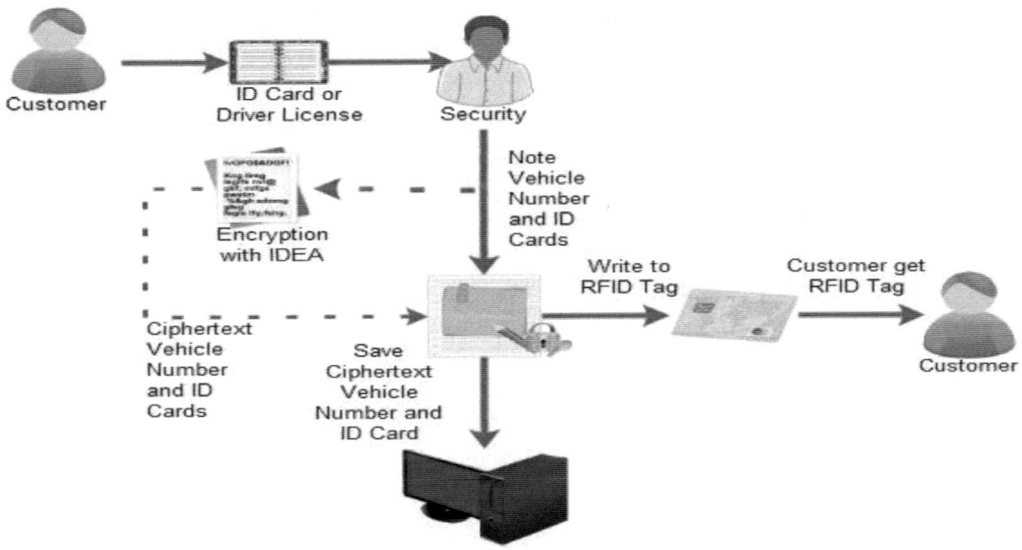

Figure 1. Unique code (cipher text) writing to RFID tag.

Figure 1 shows the process of writing data into RFID Tags where the data in the form of ID card and Vehicle Plates are converted into cipher text form by using IDEA algorithm before written to RFID Tag, and then stored in database, for process of reading data contained in RFID Tag can be seen in Figure 2 of the following:

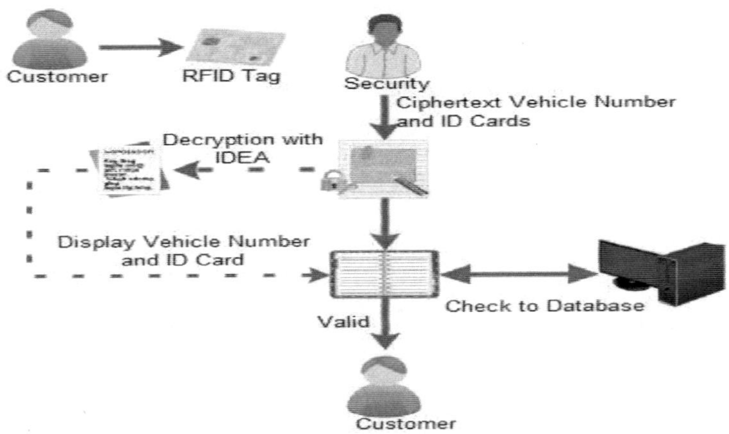

Figure 2. Read and decryption unique code (cipher text) in RFID tag.

Figures one and two above are the flow of the process writing and reading data contained in RFID where information in the form of cipher text encrypted by using IDEA algorithm, for this experiment the use of RFID module looks like in Figure 3 below:

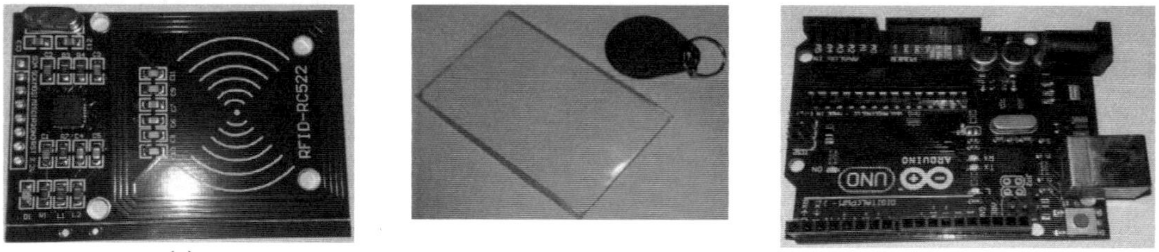

(a) (b) (c)

Figure 3. RFID module MFRC522 (a), Mifare S50 *tag* RFID (b), Arduino uno R3 (c).

RFID Module MFRC522 is used as a unique code reader stored in tag cards using Mifare S50 RFID Card while Arduino Uno R3 is used as a connecting board between RFID reader with the host computer as well as RFID reader with Mifare S50 card (RFID tag) and also for writing unique code into RFID tag.

Data security on RFID Tags by using IDEA algorithm has some function such as:

- Key generation
  The formation process begins by dividing 128-bit keys into eight pieces of 16-bit sub-key [9, 10], see Figure 4 below:

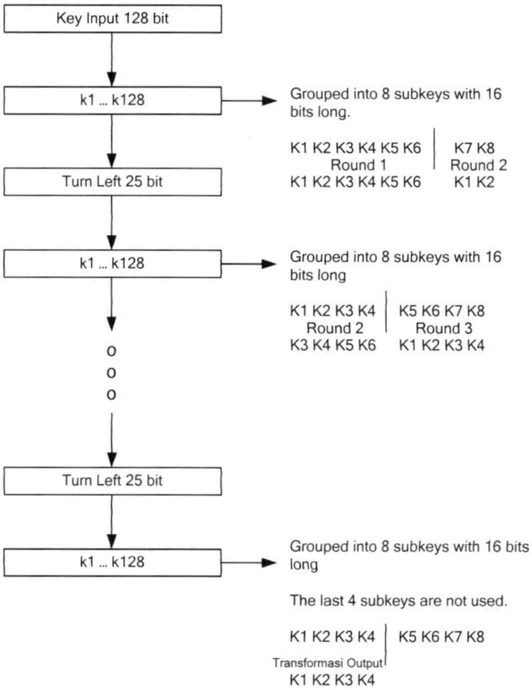

Figure 4. Key generation.

- Encryption and decryption
  Split 64-bit plaintext into four sub-blocks with 16 bits long, i.e., X1, X2, X3, X4 as the first-phase iteration, and there are eight iterations, at each iteration, four sub-blocks are XOR-aligned, added, multiplied with six 16-bit sub-key, for detail see Figure 5 below:

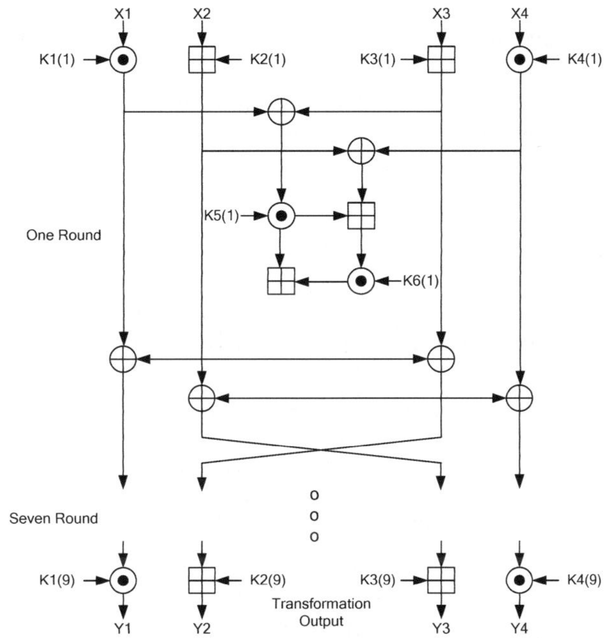

Figure 5. Key encryption and decryption process.

3. Results and discussion

Experiments Data security in RFID tags are made step by step so the writing, reading and encryption process can be known step by step, and for RFID modules that have been assembled could see in Figure 6 below:

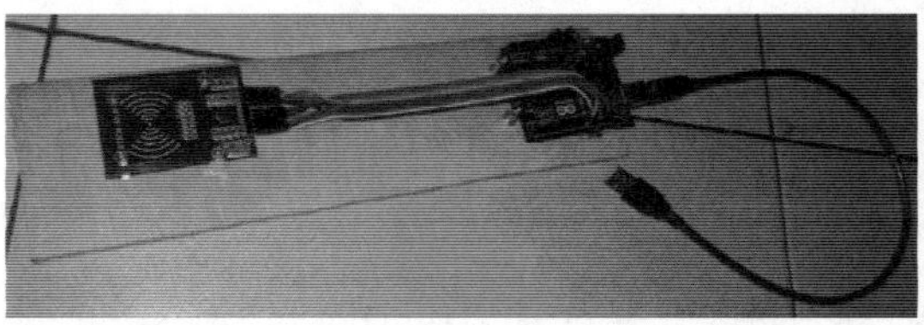

(a)

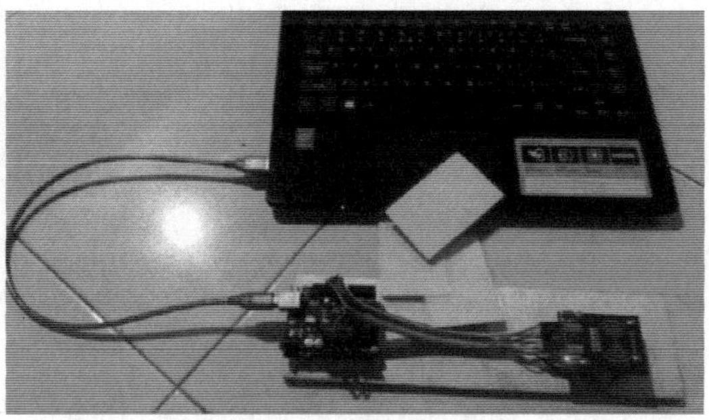

(b)

Figure 6. Arduino uno assembled results with RFID reader (a), RFID reader connection with computer (b).

For example, data security performed on the number of vehicle plates that enter into the parking area, for testing the encryption with IDEA algorithm as in the process below:

Plaintext = BK4087AE
Key = robbirahimkeeren

The encryption process with the IDEA algorithm consists of 8 rounds with each round having a relatively long process, for example, iteration one and two as below:

FIRST ROUND ITERATION

01) L#1 = (X1 * K1) mod (2^16 + 1) = 424B * 726F mod (2^16 + 1) = 06E3
02) L#2 = (X2 + K2) mod 2^16 = 3430 + 6262 mod 2^16 = 9692

SECOND ROUND ITERATION

01) L#1 = (X1 * K1) mod (2^16 + 1) = 314E * 6572 mod (2^16 + 1) = A733
02) L#2 = (X2 + K2) mod 2^16 = 9604 + 656E mod 2^16 = FB72

The round process is performed as much as eight rounds to get the output transformation as below:

TRANSFORMATION

01) Y1 = (X1 * K1) mod (2^16 + 1) = FDF0 * 989A mod (2^16 + 1) = AB02
02) Y2 = (X2 + K2) mod 2^16 = 0CF6 + 5C98 mod 2^16 = 698E
03) Y3 = (X3 + K3) mod 2^16 = 9A99 + 5A1A mod 2^16 = F4B3
04) Y4 = (X4 * K4) mod (2^16 + 1) = 861F * 5B5A mod (2^16 + 1) = FC0B

Ciphertext:
Y1 = AB02 = «
Y2 = 698E = iŽ
Y3 = F4B3 = ô³
Y4 = FC0B = ü
Cipher text = «iŽô³ü

The cipher text writes into RFID Tag and then if the irresponsible party takes cipher text in RFID Tag then the party take a long time to decipher, data security on RFID Tags does not guarantee the data is safe from irresponsible parties but can make it difficult to read it.

4. Conclusions

The IDEA cryptographic algorithm can be used well to secure data on RFID Tags so that RFID data is increasingly difficult to know by irresponsible parties, then the use of ARDUINO, MFRC522 and Mifare S50 modules of RFID tags as a tool to test the use of writing and reading on RFID can be well used, for the next development of security is not only limited to the data on RFID Tag but also can do the encryption on the frequency of electromagnetic waves that exist in RFID.

References

[1] Want R 2006 An introduction to RFID technology *IEEE Pervasive Computing* **5**(1) p. 25–33.
[2] Nambiar A N 2009 RFID technology: a review of its applications *Proc. World Congr. Eng. Comput. Sci.* **2** p. 20–22.
[3] Garrido Azevedo S and Carvalho H 2012 Contribution of RFID technology to better management of fashion supply chains *Int. J. Retail Distrib. Manag.* **40**(2) p. 128–156.
[4] Kaur M, Sandhu M, Mohan N and Sandhu P S RFID Technology Principles, Advantages, Limitations & Its Applications *Int. J. Comput. Electr. Eng.* 3(1) p. 1793–8163.

[5] Randall A 2017 *RFID Can Be Hacked: Here's How, & What You Can Do To Stay Safe* [Online] Available: http://www.makeuseof.com/tag/rfid-hacked-stay-safe/. [Accessed: 03-Sep-2017].

[6] Hariyanto E and Rahim R 2016 Arnold's Cat Map Algorithm in Digital Image Encryption *Int. J. Sci. Res.* **5**(10) p. 1363–1365.

[7] Rahim R and Ikhwan A 2016 Cryptography Technique with Modular Multiplication Block Cipher and Playfair Cipher *Int. J. Sci. Res. Sci. Technol.* **2**(6) p. 71–78.

[8] Rahim R and Ikhwan A 2016 Study of Three Pass Protocol on Data Security *Int. J. Sci. Res.* **5**(11) p. 102–104.

[9] Almasri O and Jani H M 2013 Introducing an Encryption Algorithm based on IDEA **2**(9) p. 334–339.

[10] Ludwig G A and Alegre P 2007 International Data Encryption Algorithm (IDEA) 8(1) p. 6–11.

[11] Singh H P, Verma S and Mishra S 2013 Secure-International Data Encryption Algorithm *Int. J. Adv. Res. Electr. Electron. Instrum. Eng.* **2**(2) p. 780–792.

[12] Chen J, Xue D and Lai X 2008 An analysis of international data encryption algorithm(IDEA) security against differential cryptanalysis *Wuhan Univ. J. Nat. Sci.* 13(6) p. 697–701.

[13] Ansari A N, Navada A, Agarwal S, Patil S and Sonkamble B A 2011 Automation of attendance system using RFID, biometrics, GSM modem with .Net framework *2011 International Conference on Multimedia Technology, ICMT 2011* p. 2976–2979.

[14] Boyle R J, Challa C D and Clements J A 2017 Valuing Information Security: A Look at the Influence of User Engagement on Information Security Strength *J. Inf. Priv. Secur.* **13**(3) p. 137–156.

IOP Conf. Series: Materials Science and Engineering **384** (2018) 012043 doi:10.1088/1757-899X/384/1/012043

Adaptive Content Server for Optimizing Battery Life in Mobile Devices

M Suryawinata[1]* and A G Abdullah[2]

[1]Departemen Informatika, Universitas Muhammadiyah Sidoarjo, Jl. Majapahit 666b, Sidoarjo, 61215, Jawa Timur, Indonesia
[2]Departemen Elektro, Universitas Pendidikan Indonesia, Jl. Dr. Setiabudi no 229, Bandung 40154, Jawa Barat, Indonesia

*suryawinata@umsida.ac.id

Abstract. The latest growth of mobile technology has resulted in more users accessing network services via mobile devices, such as tablets or smartphones. Personalized look and content becomes more important. Mobile devices have different properties than desktop computers. Mobile devices do not have a large screen, convenient input-output devices, and have limited computing capabilities. Content adaptation is a common definition used for adaptation or customization of content transferred from the server via the HTTP protocol to the client. The research method is to adjust the content which will be transferred to the user based on the device resources i.e. the battery on the client side so that higher service quality will be achieved. In this study, researchers will conduct the development of content adaptation methods that consider the condition of battery resources on the devices. Battery resource estimation is done to determine the current condition of existing battery in accessing device. The appropriate content determination mechanism will be based on the value generated from the estimated resource of the battery on the device that has been done.

1. Introduction

Mobile devices are now a very popular device and many users are doing daily activities through this device. By the end of 2013 the number of mobile devices has grown to 7 billion devices with 77% of them smartphone [1]. Smartphones and tablets dominate the largest amount of traffic by 88% of total internet traffic [1]. The website was originally designed specifically for desktop computer use. The website has a size (resolution) of 800 x 600 pixels at first, and continues to grow with the development of desktop monitors [2]. This is because in a large screen, a web designer can include a lot of important information in one page. Another reason is that desktop computers come with convenient input devices and have great computing capabilities [2], so images and videos can be displayed via websites for desktop

Lately, many users access the website via mobile devices, such as tablets or smartphones. Personalized look and content becomes more important. Mobile devices have different properties than desktop computers. The mobile device does not have a large screen, convenient input-output device, and has limited computing capabilities [2] [3]. If the website for a desktop computer is instantly displayed on a mobile device, then the information that appears on the screen of the mobile device will be difficult to read. Therefore a different website appearance is required for the level of legibility of information to be higher. Web browsers are becoming one of the most important apps in any mobile device. Almost all smartphones have web browser software. However, this default web browser does not provide content adaptation or support features [4]. The average web server that exists only provides two service schemes based on its accessory devices. When a user accesses a web server from a desktop device, it will be given a page that is usually displayed on a desktop device. Meanwhile, if users access through mobile devices such as smartphones or tablets, it will be given a page that is adjusted for the average mobile device.

The content customization mechanisms commonly performed by service providers are still less than optimal. With the diversity of mobile device resources and specifications and conditions, it needs a content adjustment mechanism that can support the diversity of the device in terms of its resources.

From the above problems, the authors have conducted research to provide solutions by presenting the results of research on the mechanism of content adjustment based on information resources on mobile devices. In addition the authors will also present the concept for the development of a further model of content adjustment.

The authors' contribution to this research is the proposal to develop a content adjustment mechanism that takes into account the conditions of the dynamic battery resources on the device so that the batteries on the mobile device can last longer. The server will provide adaptive content that will be transferred according to realtime condition of accessing mobile device.

2. Related Research

Research related to research conducted by the author has been done, but most of the research that has been done only provides an adaptation mechanism that only provides the concept and minimize the amount of data sent to the user. In other words, the content will be kept to a minimum so that quality can be achieved.

Research conducted by Tian, et al [5] has a primary focus on measuring mobile website performance. The background of the research conducted by Tian is the use of XML in the web service resulting in the size of the web the greater, therefore it is necessary to use compression on web services, especially on mobile clients that tend to have poor connectivity and high communication costs. Compression is one way to overcome the problem of message size in the web service. This study also shows that compression is very helpful to clients who have poor connectivity and limitations of the device used.

Compression process also has some limitations of CPU time on the client device used for compression and decompression. The compression process also decreases CPU performance on the server due to the additional time used to compress and decompress the content. The method used by Tian is by compressing the XML response data in the web service. This study shows that on both ends of computing, both servers, and clients with poor connectivity, show that compression provides benefits through an approach made using XML compression.

Meanwhile, research from Guirguis et al [6] has a major focus on content adaptation on the device. Mobile website users experience a common problem that most web content is designed for desktop computer users who have a large screen size and have a fast internet connection. On mobile devices, the screen size is limited and has a slower Internet connection. Therefore Guirguis, et al consider to deal with the restrictions that exist in this mobile device.

Mobile devices have limited computing capabilities, smaller memory, smaller screen size, small bandwidth, small resources, and short durability [2] [3] [6]. Guirguis et al proposed the framework for web content and resources adaptation in mobile device (WRAMD). WRAMD can be used by webmaster as a tool to develop mobile website and desktop website in one process. WRAMD can facilitate the production of content on the website, both mobile and desktop and will make the website become lighter. The purpose of this study is to provide a good website view, minimize the use of bandwidth for access to mobile devices and provide a specific look at each mobile device.

Content adaptation is an important technique for mobile devices. The existing content adaptation system has been developed for specific purposes, Jiang proposes an extensible content adaptation system called Xadaptor [7]. Jiang uses a rule-based approach to facilitate an extensible and systematic content adaptation system. This integrates adaptation mechanisms for some content types and groups these content into the rule base. Rules will be organized based on information from individual clients.

Jiang classifies objects on HTML pages into structures, content and pointer objects. Content adaptation techniques that already exist, are generally focused on the content and do not consider adaptation techniques for structured HTML documents. Xadaptor uses an object structure on an HTML page for an adapted content adaptation system.

With the increasing diversity of content access on devices and network technology, as well as user preferences, content delivery has evolved into a ubiquitous service issue where production, delivery and access to content form new challenges for content providers, network operators and users. Content adaptation has been recognized as one of the important aspects of ubiquitous content delivery. But this is still a big challenge not only because of the complexity and diversity of technology, but also the implications of system development that impact content providers, network operators, users, and third-party service providers. Ding developed a framework called the Adaptation Management Framework (AMF) [8]. AMF targets the distribution of content evenly through the content delivery chain.

3. Experimental Method

The system to be created, is illustrated in Figure 1 below.

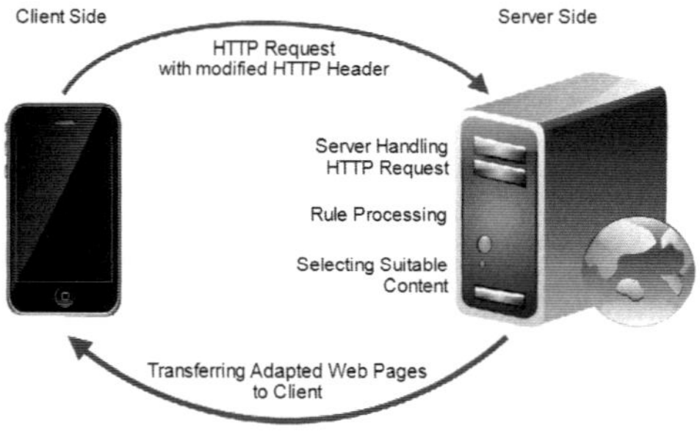

Figure 1. Proposed Research Model.

The algorithm begins by performing resource detection on client devices such as battery levels. After some of these parameters are known the value will be made the process of determining adaptation strategy. This process of determining adaptation strategies will determine the appropriate adaptation scenarios for clients.

When HTTP Request takes place, the browser on the mobile device will send information in the modified HTTP header to the server shown in Figure 2. This modified HTTP header difference with the default HTTP Header is just one line that shows the battery levels in the accessing device.

```
Get request from 202.67.46.26 on 22-08-2017 07:54:22 with
header detail:
  --> Battery-level = [82.0]
  --> Connection = [keep-alive]
  --> Host = [36.85.90.238:8080]
  --> User-agent = [Wee Browser]
  --> Accept-language = [en-US]
  --> X-requested-with = [com.example.adaptivewebbrowser]
--> Accept=
[text/html,application/xhtml+xml,application/xml;q=0.9,image/we
bp,*/*;q=0.8]
```

Figure 2. Modified HTTP Header.

Battery resource detection in accessing device is done using BatteryManager class. The batteryManager.EXTRA_SCALE method is used to determine the battery scale of the mobile device. Of the two variables generated by the method, the realtime battery condition will be compared to the scale on the mobile device so that it will know the remaining value of battery power in percentage form. Flowchart of the process for adapting the content in the server side is shown in Figure 3.

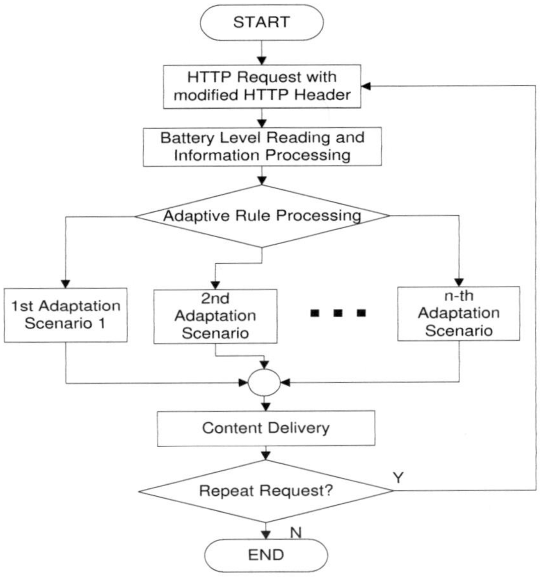

Figure 3. The process of adaptive content in server side.

4. Results and Discussion

The test was conducted to find out whether the research optimization of battery resources has been achieved. The following is the testbed used in this study:

Web Server

Hardware Specifications:

- Processor: Intel Core2Duo T6400 2.40 GHz
- Memory: 2 GB DDR3
- Hard drive: 250 GB SATA
- NIC: TP Link Gigabit Ethernet
- OS: Ubuntu Server 14.04

Mobile Device

Hardware Specifications:

- Processor: Intel Atom dual core 1.2 GHz
- Memory: 2GB
- Storage: 8 GB NAND Flash Memory
- WiFi: 802.11.b / g / n
- Network: 2G & 3G
- OS: Android 4.3 Jelly Bean
- Browser: Android Java Browser

Trials based on residual battery power on mobile devices are performed to measure the conformity of content sent by the server by considering the remaining battery power on mobile devices. If the battery is still high, it will deliver good quality content. Conversely, if the battery is low, it will send content that will not drain the battery quickly.

Table 1. Rule Based on Battery Power Level.

No	Battery Level	Content	Quality	Size
1	91% - 100%	Content 1	High	Biggest
2	81% - 90%	Content 2	High	|
3	71% - 80%	Content 3	Middle-High	|
4	61% - 70%	Content 4	Middle-High	|
5	51% - 60%	Content 5	Middle	|
6	41% - 50%	Content 6	Middle	|
7	31% - 40%	Content 7	Middle-Low	|
8	21% - 30%	Content 8	Middle-Low	|
9	11% - 20%	Content 9	Low	|
10	0% - 10%	Content 10	Low	Smallest

The test results based on the remaining battery power are shown to find out how far the browser is tailoring the service in the form of content provided to users based on the remaining battery power on the mobile device. Figure 4 Shows the remaining battery power available on a mobile device of 82% or nearly full when HTTP requests take place.

```
Get request from 202.67.46.26 on 08-08-2017 07:54:22 with
header detail:
   --> Battery-level = [82.0]
   --> Connection = [keep-alive]
   --> Host = [36.85.90.238:8080]
   --> User-agent = [Wee Browser]
   --> Accept-language = [en-US]
   --> X-requested-with = [com.example.adaptivewebbrowser]
--> Accept=
[text/html,application/xhtml+xml,application/xml;q=0.9,imag
e/webp,*/*;q=0.8]
```

Figure 4. HTTP Header Scenario.

The test results based on the remaining battery power are also performed for dynamic conditions. For example in some other test scenario, it is shown that the server access test is performed under the condition of the battery resources in the accessing device is below 82% when the HTTP request takes place. In all of scenarios tested, the modified HTTP header that sent from the mobile device always sent the battery information accurately. The server response in accordance to the received HTTP header that shown in Figure 5.

```
Get request from 36.82.97.110 on 09-08-2017 17:06:42 with
header detail:
  --> Accept-encoding = [gzip,deflate]
  --> Accept = [image/webp,*/*;q=0.8]
  --> Connection = [keep-alive]
  --> Referer = [http://36.85.90.238:8080/index.html]
  --> Host = [36.85.90.238:8080]
  --> User-agent = [Mozilla/5.0 (Linux; Android 4.4.2;
ASUS_T00F Build/KVT49L) AppleWebKit/537.36 (KHTML, like
Gecko) Version/4.0 Chrome/30.0.0.0 Mobile Safari/537.36]
  --> Accept-language = [en-US]
  --> X-requested-with = [com.example.adaptivewebbrowser]
  --> Cache-control = [max-age=0]
C:/Wee/MiniWebServer/Perpus7.jpg
```

Figure 5. Server Response.

The experiment was done in repetitively until the smartphone shuts down itself. Table 2. shows the results has been adapted to the remaining battery power conditions on mobile devices. It shows the content conformity to the rules that have been set before. The conformity test result shows that the conformity of the content is 100%.

Table 2. Results on Battery Power Level and Content.

No.	Battery Level	Content	Rule Conformity
1	100	Content 1	Conform
2	96	Content 1	Conform
3	92	Content 1	Conform
4	88	Content 2	Conform
5	84	Content 2	Conform
6	80	Content 3	Conform
7	76	Content 3	Conform
8	72	Content 3	Conform
9	68	Content 4	Conform
10	64	Content 4	Conform
11	60	Content 5	Conform
12	56	Content 5	Conform
13	52	Content 5	Conform
14	48	Content 6	Conform
15	44	Content 6	Conform
16	40	Content 7	Conform
17	36	Content 7	Conform
18	32	Content 7	Conform
19	28	Content 8	Conform
20	24	Content 8	Conform
21	20	Content 9	Conform
22	16	Content 9	Conform
23	12	Content 9	Conform
24	8	Content 10	Conform
25	4	Content 10	Conform

From the results of tests conducted found that the study of adaptive browsers on resource information on the accessory device has been successfully done in accordance with the rules designed. This will provide a new alternative for web app developers to adapt content according to resource conditions on accessory devices. If this is applied, there will be an increase in battery life in the access device.

This result can also be viewed in Figure 6. This figure, shows the difference between the non adaptive content with the adaptive ones. The trial results of network usage on WiFi networks show that from 157 times the request experiments show that the average network usage in smaller adaptive browsers is 406.65 kB compared to non-adaptive browsers that show an average network usage rate of 7350 kB. This is because the higher the quality of content received, the greater the network traffic. In an adaptive browser, the content quality level is determined on the resources of the accessing device. Content received by accessory devices has a more optimal quality of resource conditions. The positive impact of this adaptive browser usage of network usage on accessory devices is getting lower.

Figure 6 also shows that in the conducted stress test, the adaptive server can give the device more than 500 times request-reply whereas the non adaptive server can access just 157 times. This is shows that it is a great deal for adaptive system to be implemented in mobile environment.

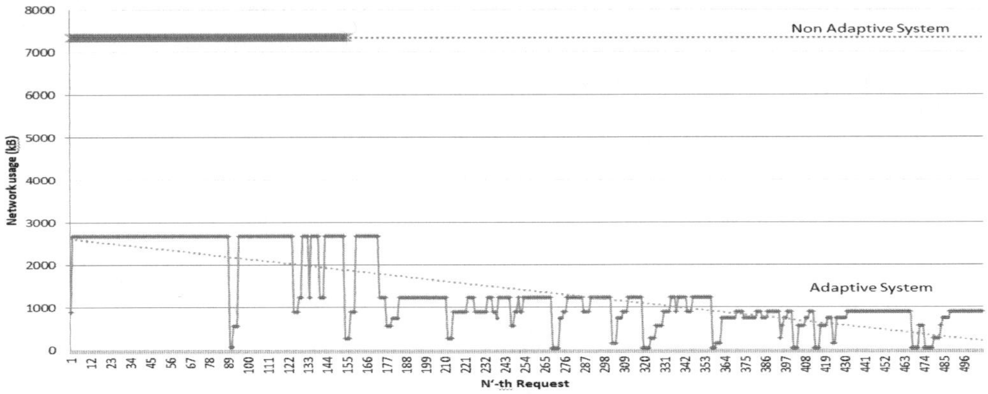

Figure 6. Network Usage Test Result.

5. Conclusions

In today's modern era, mobile technology and devices have become a primary need to access information. More than 80% of internet users use mobile devices. Limitations of the battery power on mobile devices also become one of the constraints of users to access information in a long time. In this research battery consumption in mobile device can be prolonged up to 300%. With this research, expected battery power will last longer because the information obtained can be adjusted to the condition of the battery resources in the device. The next development suggestion is to insert other information on the HTTP header that can support the quality of content delivery, both in terms of response time, processing speed, memory load and so on.

Acknowledgements

We acknowledged Universitas Muhammadiyah Sidoarjo and Universitas Pendidikan Indonesia for facilitating this research

References

[1] Cisco 2018 *Cisco Visual Networking Index: Global Mobile Data Traffic Forecast Update, 2013–2018* [Online] available at: http://www.cisco.com/c/en/us/solutions/collateral/service-provider/visual-networking-index-vni/white_paper_c11-520862.html.

[2] B'Far R 2004 *Mobile Computing Principles: Designing and Developing Mobile Applications with UML and XML* (Cambridge: Cambridge)

[3] Ojala O 2005 *Service Oriented Architecture in Mobile Devices: Protocols and Tools* (Espoo: Helsinki University of Technology)

[4] Wang H, Kong J, Guo Y and Chen X 2013 Mobile Web Browser Optimizations in the Cloud Era: A Survey *IEEE 7th International Symposium on Service Oriented System Engineering (SOSE)* 527 – 536

[5] Tian M, Gramm A, Ritter H and Schiller J H 2007 Adaptive QoS for Mobile Web Services through Cross-layer Communication *Computer Section* **40** 2 59-63

[6] Guirguis S K and Hassan M A 2010 A Smart Framework For Web Content And Resources Adaptation In Mobile Devices *The 12th International Conference on Advanced Communication Technology (ICACT)* **1** 487-492

[7] Jiang H, Tong G, Wei H, I-Ling Y and Bastani F 2007 A Flexible Content Adaptation System Using a Rule-Based Approach *Knowledge and Data Engineering, IEEE Transactions* **19** 1 127 140

[8] Jie D and Ning L 2011 A Distributed Adaptation Management Framework in Content Delivery Networks *Wireless Communications, Networking and Mobile Computing (WiCOM), 2011 7th International Conference* **1** 4 23-25

IOP Conf. Series: Materials Science and Engineering **384** (2018) 012044 doi:10.1088/1757-899X/384/1/012044

Design and Simulation of Washing Machine using Fuzzy Logic Controller (FLC)

N Wulandari* and A G Abdullah

Department of Electrical Engineering Education, Universitas Pendidikan Indonesia, Jl. Dr. Setiabudhi No. 207, Bandung 40154, Indonesia

*wulandarinur67@student.upi.edu

Abstract. This paper aims to design and simulate Fuzzy Logic control system in washing machine. Washing machine is one of the main machines in the household. Therefore it is necessary to control the washing machine system to facilitate the user, reduce the time, electricity consumption, and water in washing water. Fuzzy Logic is an efficient approach in complex control systems compared to conventional approaches. In this paper, Fuzzy Logic control is used to determine the value of washing speed based on 4 different inputs and the stability of motor speed of washing machine based on 3 inputs. The stability setting of the motor speed of the washing machine is controlled to have a constant speed despite the load changes. In the washing speed control system, the output is the value of washing speed. Based on the value of the output, the input value is a reference speed that will be processed on the motor stability control system. The output generated on this system is a control voltage for the inverter, which in case of input changes in the other two inputs will not affect the output value. In paper in, the washing machine control system is designed and simulated using Fuzzy logic toolbox in Matlab.

1. Introduction

One of the main machines in the household is the washing machine. In the early development of washing machine technology all motor control operations are set manually by operating timer [1]. Timer is used to determine the time the machine will work. With the ever-evolving technology, the use of microcontroller is being applied in the washing process based on predetermined time [2], [3]. In addition, the technology of using sensors in washing machines started to be applied [2], [3].

Currently, the washing machine control system has started using Fuzzy Logic as its controller [4]-[8]. The use of Fuzzy Logic controls is easier to implement than conventional controls such as PID [4], [5], and [9]. The weakness of the PID control is difficult to determine the constant value of P, I, and D in order to obtain the desired result [10]. The use of Fuzzy Logic controls is also applied to the control of the induction motor [10] - [13] commonly used in washing machines especially the 1 phase induction motor. The speed of an induction motor is affected by changing the load given. In addition, the speed control performance of the washing machine is also influenced by the effects of mechanical and electrical subsystems [14].

This paper aims to design and simulate a washing machine control system using Fuzzy Logic. The controlled parameters are washing speed based on the weight input of clothing, the type of defilement, the level of dirtiness, and the sensitivity of clothing. What distinguishes this research from others is that the output value of the speed control system washes into the motor stability control system. The value becomes input as reference value of motor speed. In this system there will be 2 other input that is error and derror. The output value of motor stability control system will not affect the value of error and derror given. Where the system output is the control voltage to be supplied to the inverter to run the motor.

Fuzzy Logic controls are smart system controls. Fuzzy Logic controllers have the ability to solve problems on complex systems that are not owned by conventional controllers [10]. The basic structure of the Fuzzy Logic system is shown in Figure 1.

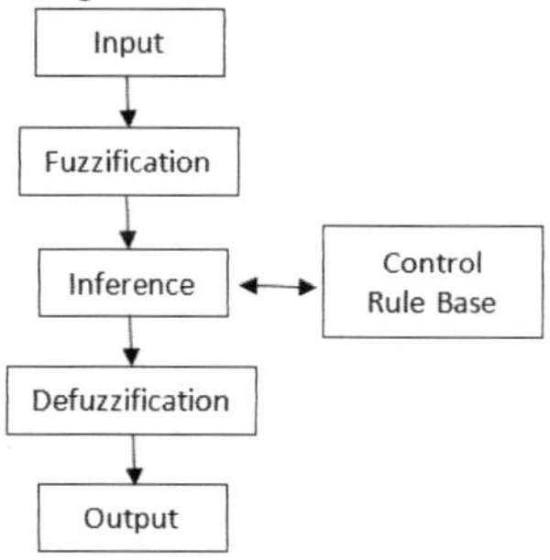

Figure 1. Fuzzy Logic system structure.

In the fuzzification process, there is a conversion of crisp input values into fuzzy values through mapping techniques of input variables into membership functions and truth values. Membership function is a curve showing the mapping of data input points into membership values that have intervals between 0 and 1 [15]. Membership function sets the function to declare membership of a value. There are several forms of membership functions such as triangular, trapezoidal. Each rule on fuzzy will relate to another fuzzy relation. Generally fuzzy rules are expressed in the form of IF-THEN logic. Evaluate fuzzy rules to generate the output of each rule. The input of the defuzzification process is a fuzzy set obtained from the composition of fuzzy rules, while the resulting output is a number in the fuzzy set domain. So if given a set fuzzy within a certain range, then it should be taken a certain crisp value as output.

2. Method

In this research is to make the control system simulation and make the design of Fuzzy Logic system using Fuzzy Logic Controller Simulink on software MATLAB. Figure 2 shows the proposed system design.

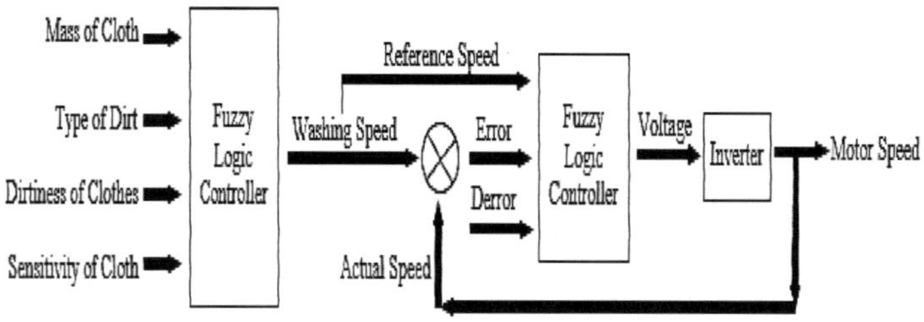

Figure 2. Proposed System.

In this design have 2 different systems. The first fuzzy logic control system is used in determining the washing speed with 4 inputs. These 4 inputs can be provided manually or automatically by using the corresponding sensor. After the first system produces the output the system then performs a process whereby the output on the first system is defined as the reference speed at the second system input. The second system runs the process of regulating motor speed stability with 3 inputs. Two other inputs include error is the difference between the reference speed with the actual speed and the current error difference derror with the previous error.

The first step in using Fuzzy Logic system is to determine the input and output. In this model the desired output is obviously the washing speed. But for more details can be show in Figure 3. This membership function is based on the analysis of several related journals [4] - [6].

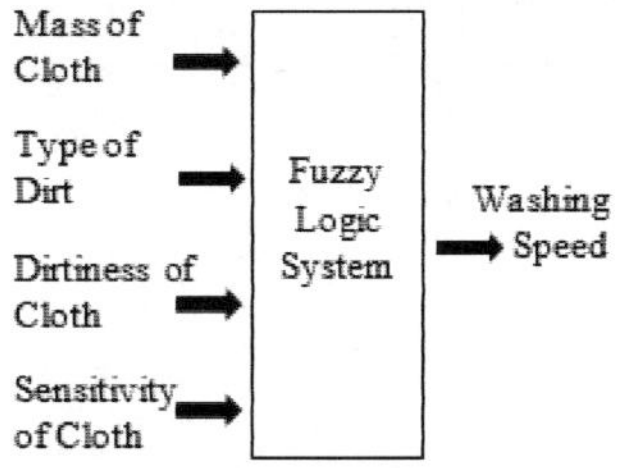

Figure 3. Fuzzy Logic Control for washing machine.

Each linguistic input and output is set into the membership function. The membership function used is the triangular membership function. It can be seen in figures 4 and 5 show the membership functions of input and output and also lower and upper limits.

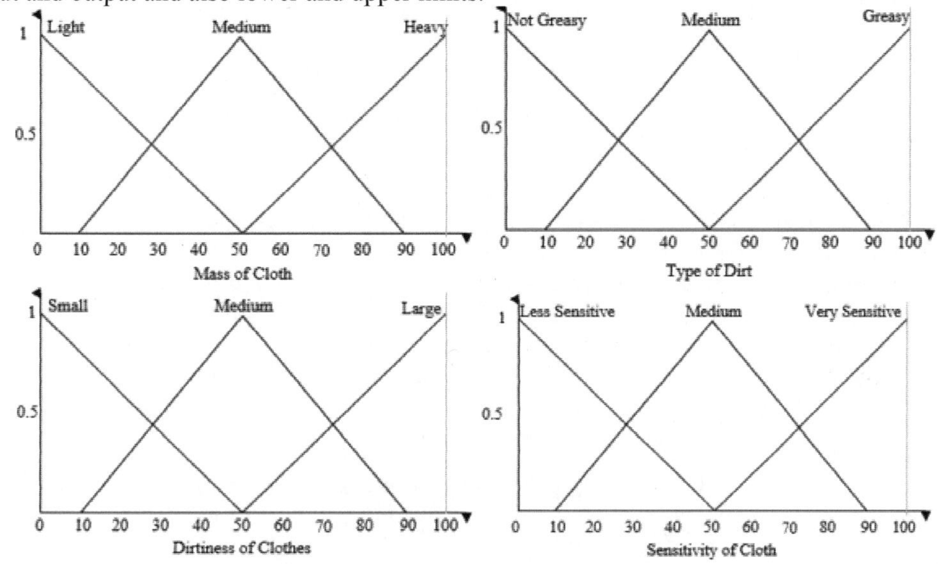

Figure 4. Input membership function.

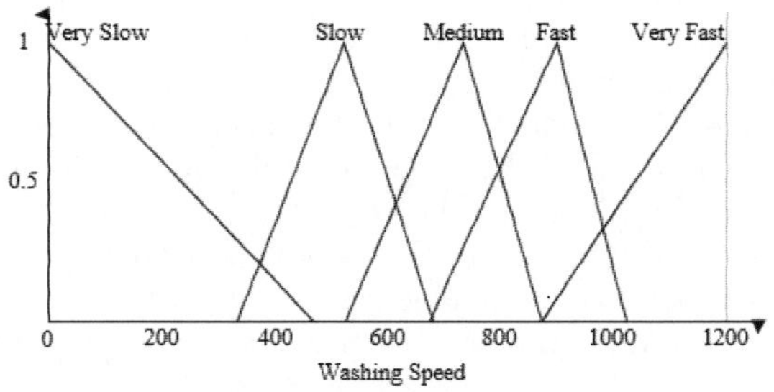

Figure 5. Output membership function.

IOP Conf. Series: Materials Science and Engineering **384** (2018) 012044 doi:10.1088/1757-899X/384/1/012044

After determining the membership function and upper and lower limits for the input and output parameters, the total IF-THEN rule base required is 81 rules. Rules can be seen in table 1. The rule setting obtained results from the analysis of several related journals [4] - [6].

Table 1. If-Then Rule base required is 81 rules.

No	MoC	ToD	DoC	SoC	WS
1	Lt	NG	S	VS	VS
2	Lt	NG	S	M	S
3	Lt	NG	S	LS	M
4	Lt	NG	M	VS	S
5	Lt	NG	M	M	M
6	Lt	NG	M	LS	F
7	Lt	NG	Le	VS	M
8	Lt	NG	Le	M	F
9	Lt	NG	Le	LS	VF
10	Lt	M	S	VS	S
...	...	...	...	...	...
81	H	G	Le	LS	VF

Notes: Lt=Light, M=Medium, H=Heavy, NG=Not Greasy, G=Greasy, S=Small, Le=Large, LS=Less Sensitive, VS=Very Sensitive, VS=Very Slow, S=Slow, F=Fast, VF=Very Fast.

The inference system used is the Mamdani Method. In this method the implication function used is Min that is in implies a rule takes its minimum value.

As for the method of defuzzification used the method of Center of Gravity (CoG) or another name centroid. In this method the crisp output value is obtained by enchanting the center point (z *) fuzzy area. The output on this system will be forwarded to the next system.

The input and output for Fuzzy Logic Modeling of Induction Motor Speed can be seen in figure 6.

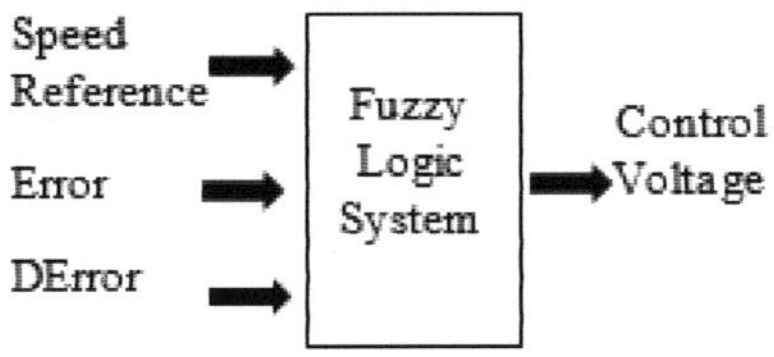

Figure 6. Fuzzy Logic Control for induction motor speed.

In this model the input parameters are reference speed, error, and derror value. The reference speed is obtained from the previous system output. Error is the difference between set point speed and actual speed. Derror is the current error increment with the previous error. While output is a voltage (volts) that will be given to the inverter as a control voltage to run the motor. Membership functions used are triangle and trapezium.

The membership function used can be seen in figures 7 and 8. And the rules required as many as 125 rules shown in table 2. Rules are based on several journals [11] - [13].

The method used is Mamdani Method with Min implication function and defuzzification method using CoG.

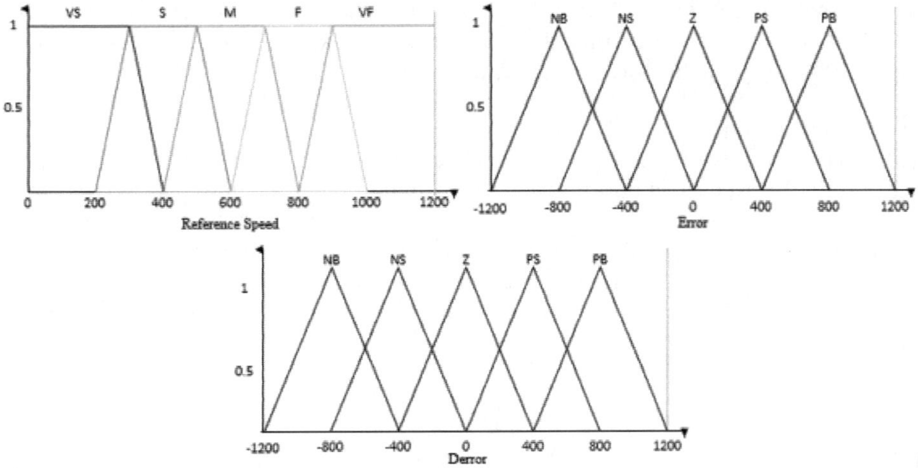

Figure 7. Input membership function used are triangle and trapezium.

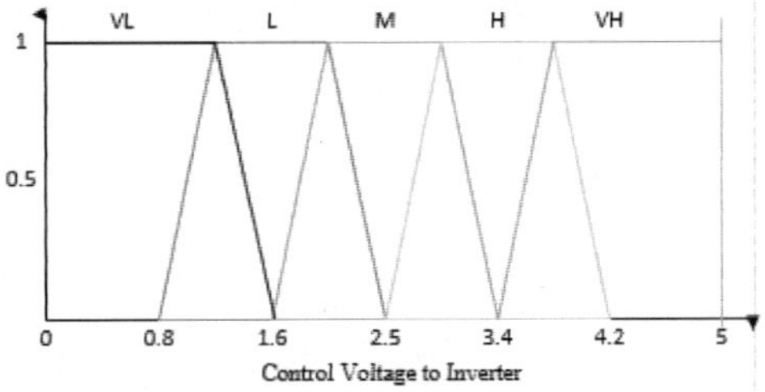

Figure 8. Output membership function used are triangle and trapezium.

Table 2. If-Then Rules.

No	RS	E	DE	VC
1	VS	NB	NB	VL
2	VS	NB	NS	VL
3	VS	NB	Z	VL
4	VS	NB	PS	VL
5	VS	NB	PB	VL
6	VS	NS	NB	VL
7	VS	NS	NS	VL
8	VS	NS	Z	VL
9	VS	NS	PS	VL
10	VS	NS	PB	VL
...	...	....	...	...
125	VF	PB	PB	VH

Notes: VS=Very Slow, S=Slow, M=Medium, F=Fast, VF=Very Fast, NB=Negative Big, NS=Negative Small, Z=Zero, PS=Positive Small, PB=Positive Big, VL=Very Low, L=Low, H=High, VH=Very High.

From a combination of 3 inputs it can be concluded that the rule at VS reference speed will output VL, S yields output L, M produces output M, F produces output H, and VF produces output VH. Although there is a change in the input error and derror.

3. Results and Discussion

Using Fuzzy Logic controls can result in different washing rates according to different inputs on the weight of clothing, the type of impurity, the level of impurities, and the sensitivity of clothing. For example, the case on the first system will be given 40 inputs on the weight of the clothing, 20 on the type of clothing defilements, 20 on the level of clothing defilements, and 30 on the clothing sensitivity. The data is influenced by 8 rules, after doing the implications on the rule is taken minimum value. Once we get the implication function of each rule in use Max method to do the composition between all rules. So, get the output area can be seen in Figure 9.

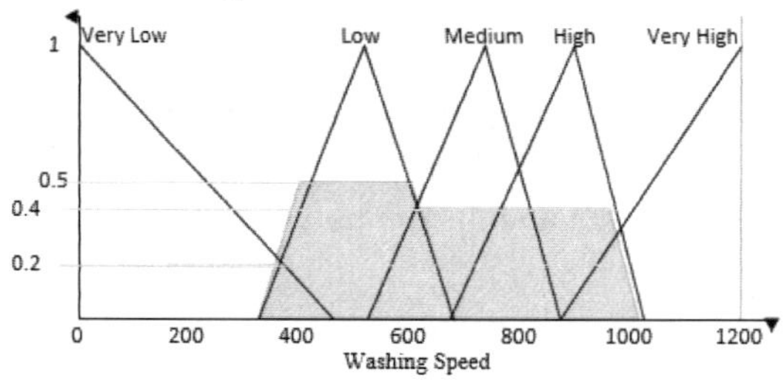

Figure 9. Graph of the Composition of the Rules (The data is influenced by 8 rules).

From picture 9 done defuzzification with method of centroid hence generated output value equal to 600 rpm.

While in the second system to stabilize the speed of the induction motor, the output voltage value to be given to the inverter will remain the same according to the reference value of the speed. Where previously been done to initialize for motor speed with control voltage. Despite the change in input error and derror values. For example, given the reference value of the speed of the first system output value of 600 rpm, error = 100 and derror = 100. The data is influenced by 4 rules, after doing the implications of the rule is taken the minimum value. Once we get the implication function of each rule in use Max method to do the composition between all rules. So get the output area can be seen in Figure 10.

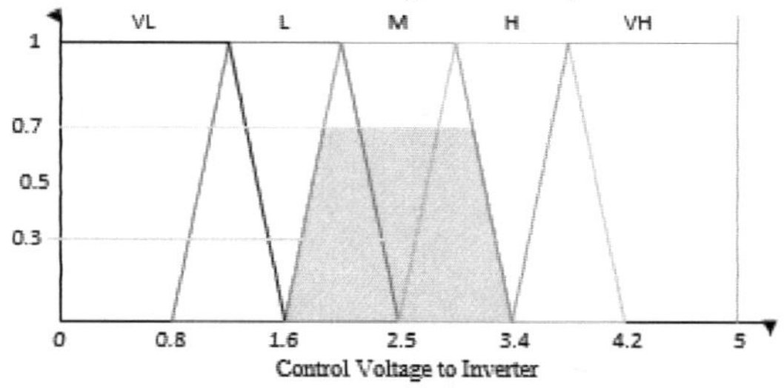

Figure 10. Graph of the Composition of the Rules (The data is influenced by 4 rules).

It can be seen in Figure 11 shows a surface response relationship between input and output on the case of washing speed. Figure 12 shows the response surface of the relationship between the input and output on the stability control system of induction motor speed.

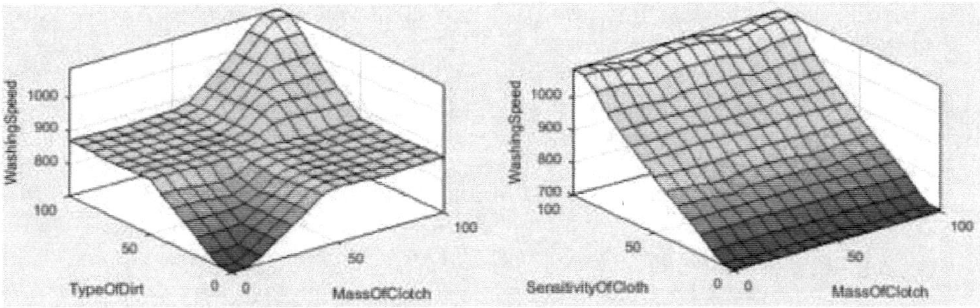

Figure 11. Surface response relationship between input and output on the case of washing speed.

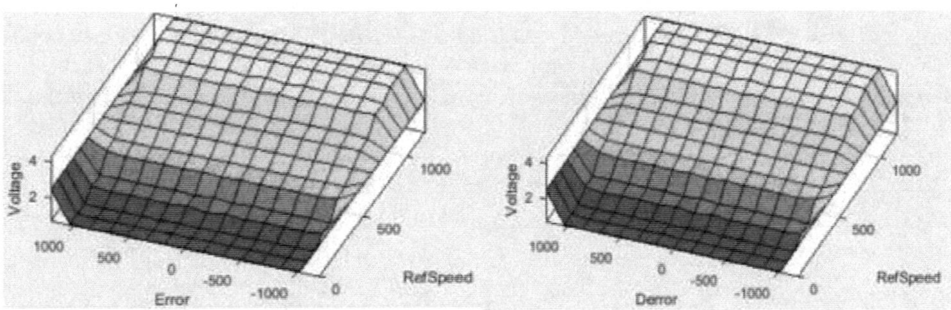

Figure 12. Surface response relationship between input and output on the stability control system of induction motor speed.

4. Conclusions

In this paper, developing fuzzy logic for washing machines can help improve automation systems in washing machines and reduce electricity, water and time consumption. Based on the weight of clothing detected by the weight sensor, while other input parameters are given externally as well as with related sensors. The parameters can determine the appropriate washing speed. The results of the washing speed can enter the next control system to stabilize the motor speed despite any changes in load. Looking at the surface graph obtained from the combination illustrates the desired operating achievement. Using a fuzzy logic system is easier to implement into complex systems than conventional controls.

References

[1] Russell W 1942 Automatic Control for Washing Machines February, pp 89–92

[2] Perwita C D, Susanto A, and Soedjatmiko 2010 Mikrokontroler ATMega 16 untuk Pengaturan Mesin Cuci *Jurnal Penelitian Teknik Elektro* **3** 285–289

[3] Nando F., A S R and Raharjo Y 2012 *Design Model of Automation Washer for Two Tubes Aperture (Twin Tube Top Loader) Microcontroller Based ATMEGA32* pp 112–122

[4] Hatagar S, and Halase S V 2015 Three Input – One Output Fuzzy logic control of Washing Machine *Int. J. Sci. Res. Eng. Technol.* **4** 1 2278–882

[5] Hatagar S S and Halase S S V 2015 Simulating Matlab Rules in Fuzzy Controller Based Washing Machine Simulating Matlab Rules in Fuzzy Controller Based Washing Machine *International Journal of Emerging Trends in Science and Technology* **2** (07) 2796–2802

[6] Demetgul M, Ulkir O and Waqar T, 2014 Washing Machine Using Fuzzy Logic *Autom. Control Intell. Syst.* **2** 3 27

[7] Spasojevic B 2004 Fuzzy optical sensor for washing machine *7th Semin. Neural Netw. Appl. Electr. Eng. 2004. NEUREL 2004* 237–242

[8] Handaya D, Handaya D, Prima Y and Auditia I 2016 Simulasi Mesin Cuci Industri Tekstil

Berbasis Kendali Fuzzy dan Interface LabVIEW 2014 *Prosiding Seminar Nasional Fisika (E-Journal)* **6** SNF2017-CIP

[9] Wahjono E 2016 Pengaturan kecepatan motor induksi sebagai penggerak mobil listrik dengan kontroler fuzzy logic berbasis direct torque control. *Jurnal Ilmiah Mikrotek* **1** (3) 136-144

[10] Ardana I W R and Sutawinaya I P 2017 Pemodelan Sistem Kontroler Logika Fuzzy pada Pengaturan Kecepatan Motor Induksi Menggunakan Perangkat Lunak Matlab/Simulink *Matrix: Jurnal Manajemen Teknologi dan Informatika* **7** (1) 1-6

[11] Devi K, Singh R, Gautam S and Nagaria D 2015 Speed control of induction motor using fuzzy logic approach. *International Journal* **3** (4) 694-697

[12] Firmansyah 2003 Penerapan Logika Fuzzy pada pengendali PID Kecepatan Motor Induksi sebagai Penggerak Conveyor *J. Swateknologi* **2** 10 17–26

[13] Neema D D, Patel R N and Thoke A S 2011. Speed control of induction motor using fuzzy rule base *International Journal of Computer Applications* **33** (5) 21-29

[14] Bascetta L, Rocco P, Zanchettin A M and Magnani G 2012 Mechatronics Velocity control of a washing machine : A mechatronic approach *Mechatronics* **22** 6 778–787

[15] Barrett J D 1997 A First Course in Fuzzy Logic *Journal of the American Statistical Association* **92** (439) 1217-1219

IOP Conf. Series: Materials Science and Engineering **384** (2018) 012045 doi:10.1088/1757-899X/384/1/012045

Simulation of Working Age Effect on the Geothermal Power Plant Generators Efficiency

R A Aziz* and T Sucita

Department of Electrical Engineering Education, Universitas Pendidikan Indonesia, Jl. Dr. Setiabudhi No. 207 Bandung 40154, Indonesia

*rizalabdulaziz04@student.upi.edu

Abstract. The efficiency of the generator engine is very influential on the performance of the entire steam generating system, the efficiency is proportional to the performance of the generator engine. Also Efficiency depends on the working age of the machine, therefore an analysis of the efficiency of the generator engines based on A Certain period of time is needed. One way to find out the feasibility of a generator is by calculating the efficiency value of the generator now (actual) Compared to the efficiency value of the generator at the first time the generator is operated (commissioning). Manual in analyzing efficiency calculations are quite complicated because they are concerned with steam (thermodynamic). To facilitate the calculation of the required mathematical analysis software support. This research is focused on making simulation for efficiency value based on Certain time period According to the operational age of the generator using MatLab software with the aim of more variation result and can be presented in the form of graph is easy to understood. This result can also be made as a prediction of the feasibility of a steam generator engines. This research was conducted PLTP Kamojang, West Java, Indonesia.

1. Introduction

The energy consumption in the world is now a little over 400 EJ per year. Available energy resources in the world are large, and the energy shortage is not expected in the foreseen future. On the other hand, most (86%) of the energy used in the world at present is coming from finite energy resources, whereas renewable energy sources are more suitable for sustainable development. The highest share of the use of renewable energy resources is in Iceland, where renewable energy comprises approximately 70% of the primary energy resources and approximately 30% is derived from fossil fuels. This unique position has been achieved by an extensive and advanced use of geothermal energy [1].

In Indonesia is very rich in geothermal energy sources, according to data sources obtained from Indonesia investment on 21 December 2017 where Indonesia is ranked third in generating energy sourced from geothermal with the amount of power generated by 1.197 Mwe, About 40% of geothermal energy reserves the world lies beneath the land of Indonesia, then the country is estimated to have the largest geothermal energy reserves in the world and therefore has a high potential for renewable energy sources. However, most of this potential has not been used. Currently, Indonesia uses only 4-5% of its geothermal capacity [2].

One of the sources of geothermal powered electricity generation in Indonesia is in the Garut area of the plant Kamojang geothermal power, which is located at 40 kilometers southeast of the city of

Bandung and 24 km northwest of the town of Garut, with a height of 1500 meters above sea level. Having air temperature peak at 15°C-25°C with an average rainfall of 2,885 mm annually [3].

Generators and turbines become a major component generation tool, where the turbine and generator function serves as a tool to convert, which serves as a driving turbine first, that converts potential energy of steam into energy kinetic and then converted into mechanical energy in the form of rotation of the turbine shaft. While the generator is a tool that serves to change the mechanical energy into electrical energy through the electric field generated due to the intersection of the lines of magnetic force lines are then cause electric lines of force GGL, Based on the main component of the generation system there is certainly an efficiency value generated by a generator The efficiency value is determined from the value of Pin and Pout generated and the basis for declaring a system is said to be efficient or not to the load [4].

Therefore, this research is done in the form of efficiency analysis of Kamojang Geothermal Power Plant (PLTP). With reference to the comparison between the first times the data generator is operated in commissioning and on the actual data. This research is a form of development from previous research where the calculation of efficiency of geothermal power generator which refers to the law of thermodynamics which become the reference in determining Thus the authors try to simulate the calculations and analyze the efficiency of a system of geothermal power plant by accumulating an overall efficiency value to obtain a comparison impairment generator efficiency by using the GUI in Matlab. Data comparison will be found to efficiency values when first generator is operated and when the actual data [5].

2. Methods

This research was done at the company PT Indonesia Power UPJ Kamojang geothermal power. Figure 1 shows the flowchart groove in the execution of the research, the method described in the settlement of a simulation analysis of generator.

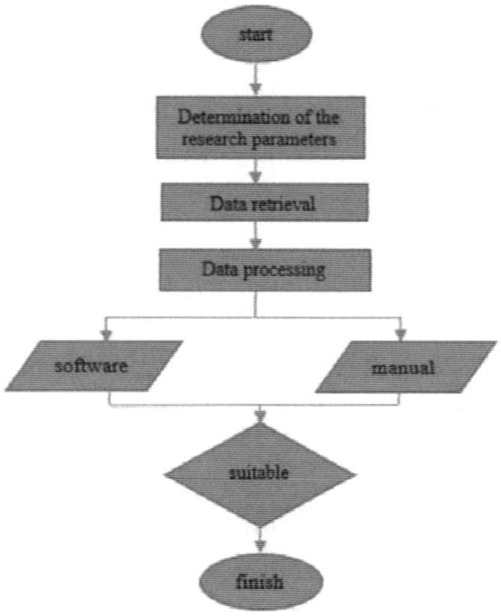

Figure 1. The research flow shows the research process starting from the determination of parameters to the data processing.

2.1. Determination of the research parameters

In this research, parameters used are turbine input and output vapor temperature, steam mass flow rate, to calculate the incoming power of the generator, and generator output power to calculate efficiency value in the generator.

The steam cycle used in this discussion is the Separated Steam Cycle, wherein geothermal power plant (PLTP) Kamojang comes out of the wellhead as a mixture of two-phase fluid (vapor phase and liquid phase). Then the first process of separation on the fluid. This is possible by passing the fluid into the separator, so that the vapor phase will be separated from the liquid phase. The vapor fraction generated from the separator is then used in the calculation of turbine power. Therefore, this energy conversion system is called the Separation Cycle of Separation. [6] Figure 2 shows the Schematic diagram of Steam Cycle Results of Separation [6].

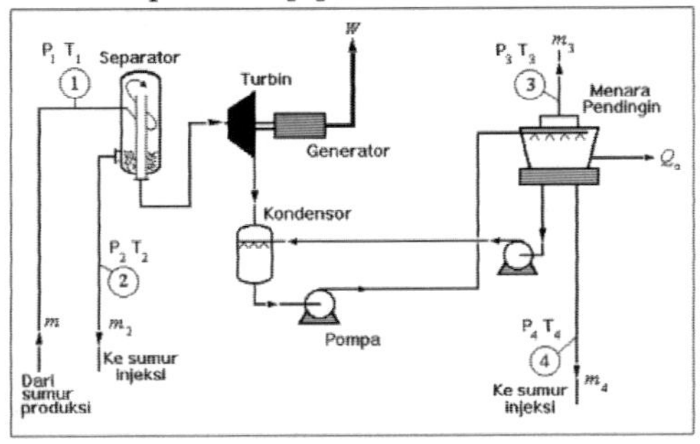

Figure 2. Schematic diagram of Steam Cycle Results of Separation.

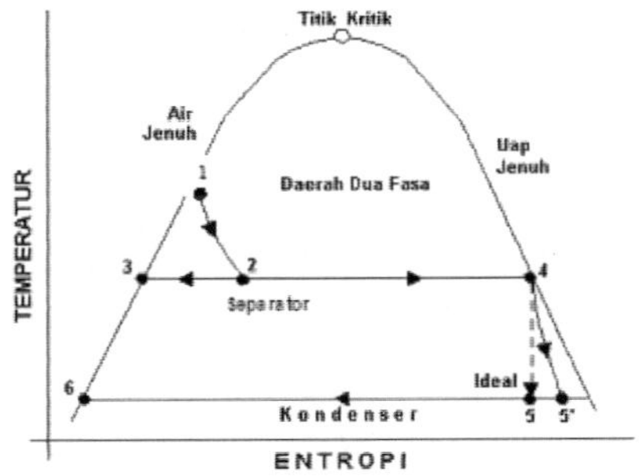

Figure 3. T - S Diagram for Steam Conversion System Results of Separation.

Figure 3 shows the steam cycle that occurs through the T-S chart. At point 1 geothermal fluid is a mixture of two phases. Before entering the fluid turbine undergo isenthalpic process from point 1 to point 2. At the well head known mass flow rate of fluid vapor fraction (steam quality at well head). At point 2 the fluid enters the separator, the vapor fraction enters the separator, while the water fraction is discarded. At the pressure and temperature of this turbine input is known enthalpy and entropy fluid from the steam table. Entropy at point 4 and point 5 (turbine input and output) is considered to be the same (the process occurring within an isentropic turbine). The calculation of turbine power in this system is similar to that in the Single Evaporation Cycle, the difference lies in determining the initial conditions of the fluid. At point 1 the fluid is a mixture of two phases (liquid phase and vapor phase), so the fluid enthalpy equals the sum of the enthalpy of the two phases. Furthermore, the turbine power determination procedure is the same as the calculation procedure in the Single Evaporation Cycle.

In this journal, the researcher conducts research in order to find efficiency value in Kamojang 2 PLTP generator unit, where the power generated by generator unit 2 is 55MW. Generating electrical energy in a geothermal power plant involves two energy conversions, converting steam heat energy into kinetic energy in a turbine and using a rotary generator to convert energy Data retrieval mechanical turbine into electrical energy [7]. There are several criteria in determining efficiency value of a generator which in discussion of this paper is the vapor temperature, and flow rate of vapor that enter into generator where this become main parameter in determining power coming in generator. In this discussion also thermodynamic table into a reference where in determining the power into the generator through some calculation enthalpy and entropy incoming steam temperature [8].

Data collection was performed at the thermal power plants of the earth where the data taken is the data commissioning which the data were first generator operated with regard to steam and speed of the flow of steam into a turbine which is converted into a form of mechanical energy into and become the power to drive the generator, Where data is obtained from the commissioning of the generator unit 2 PLTP Kamojang, 1987.

Besides commissioning of data required, the actual data was needed to determine the percentage decrease in the efficiency of the generator as a result of the influence of the working age generator actual data is the data obtained in June 2017. In the table 1 and table 2 below the commissioning of data and actual data obtained.

2.2. Data processing

Data processing is done by using the formula and based on thermodynamic table with the result of data that has been in can with the following flow:

2.2.1. Determine enthalpy and entropy of the incoming and outgoing steam turbines on the thermodynamic table.

Table 1. The enthalpy table and the entropy of a temperature based on the thermodynamic table.

Enthalpy	Entropy
hf =	sf =
hfg =	sfg =
hg =	sg =

A temperature has a value of enthalpy and entropy where the value is already on the Table 1 thermodynamic so, we can simply look at the tables.

2.2.2. Finding quality steam dryness fraction (X).

$$x = \frac{s_{fg1}}{s_{fg2}} \quad (1)$$

Information

x = Quality steam dryness fraction
s_{fg1} = Value of entropy evap when vapors enter
s_{fg2} = Value of entropy evap when the steam out

Evap entropy values can be seen in the table so we stayed split thermodynamic entropy values entered by the entropy evap out

2.2.3. Finding value on exit saturated steam turbine.

$$hg3=hf2+x\times hfg2 \quad (2)$$

Information

hg3 = Value of saturated vapor as it exits the turbine
x = Quality steam dryness fraction
hf2 = The value of the enthalpy of liquid when the steam out
hfg2 = The value of the enthalpy evap when the steam out

2.2.4. Searching for businesses that do steam turbine (W).

$$w=\Delta h=h\_g1-h\_g3 \quad (3)$$

Information
w =Δh = Efforts turbine
hg1 = the value of the enthalpy of vapor when the steam out
hg3 = Value of saturated vapor as it exits the turbine

2.2.5. Looking for steam turbine power (P).

$$Pturbin = m \times w \tag{4}$$

Information
Pturbin = Power the steam turbine
M = Mass flow rate of steam
w = Efforts turbine

2.2.6. Calculating unit 2 generator efficiency (based on data commissioning & actual data) [9].

$$\eta = Pout / Pin \times 100\% \tag{5}$$

Information
η = efficiency generator
Pout = power out on the generator
Pin = power in on the generator

Data processing is done by manual calculation and calculation using software, where software used is MatLab R2015b software using Graphical Using Interface method on its calculation.

2.3. Graphical using interface
GUI is a graphical display on one or more windows that contain controls, components that are called, which allows users to interactive tasks. Users do not have to write a script or typing commands on the command line to complete the task. Unlike the encoding program to complete the task, the user does not need to understand the details of how the task is done. GUI components can include menus, toolbars, push buttons, radio buttons, list boxes, and slider-just to name a few. GUI created using MATLAB® tool can also do all kinds of computing, read and write data files, communicate with other GUI, and displays the data as a table or as a plot [10].

In software design simulation of GUI calculations in matlab R2015b, we need to know first [11]. In calculating simulation design using Graphical Using Interface there are some tools used in displaying the programs that we want to show, tools used in this final project work described in the Figure 4.

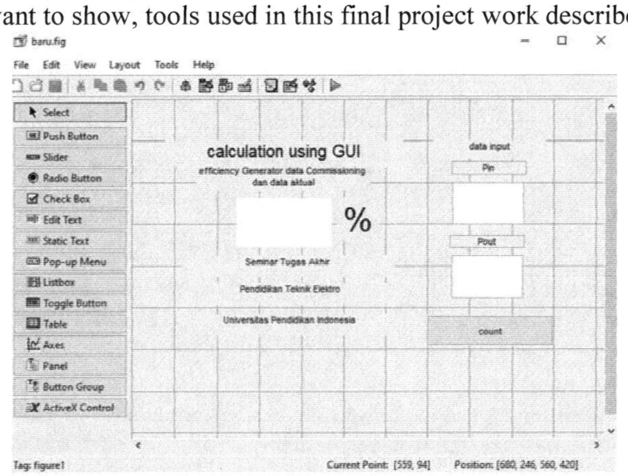

Figure 4. Schematic display of GUI program.

In this project the tools used are push button, statistic text, and edit text, where in the previous explanation explains that in this GUI software does not require scrip command in finishing task because automatically format of command has been input on script page but, in the use of tools there are tags and strings that need to be set to call the command when one of the tools want to be done or operated, and the operation function we want to do, its example as in the push button tool we tag and the string we want to write. And the variables we will input need to be in the settings will be operated like what, in the case of generator efficiency calculation generator equations call the pushbutton tag so that when pushbutton push button on the input into the string Pin and Pout can operate according to the input command. Schematic setting tool on the GUI is shown in Figure 5.

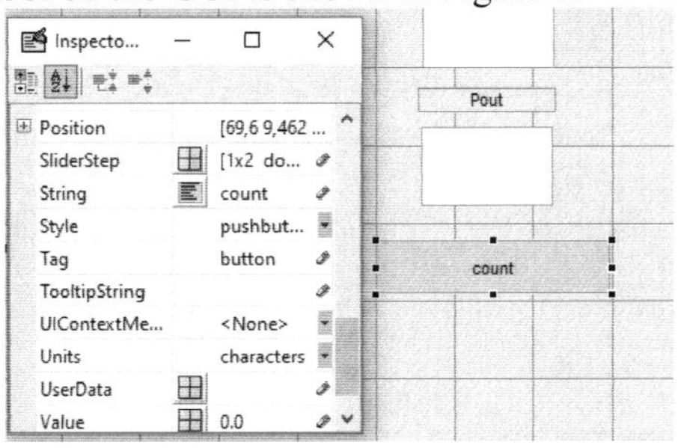

Figure 5. Schematic setting tool on the GUI.

The tag in question is the identity tool in the GUI program, functions as a tag to be included into the program call. After the design that we want to have finished we do, we live operate simulation by clicking the play button on the top of the GUI display so that it will appear graphical display that want to show with various graphics functions that want to be displayed, in this study graphic display wanted is as shown in Figure 6.

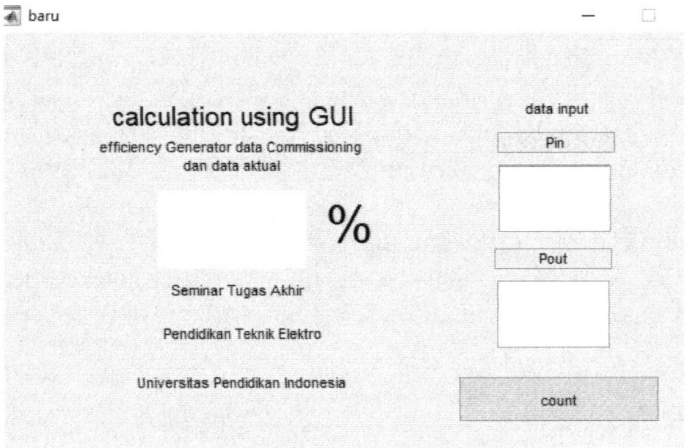

Figure 6. GUI simulation display.

3. Results and discusions

Analysis is done by taking the data commissioning and actual data in the can from the data retrieval results in PLTP kamojang on 11 June 2017, where the data commissioning and actual data in can be described in the table 2 below:

3.1. Commissioning data [12]

After the calculation of data in Table 2 then the efficiency value of each hour can be determined by using the calculation model that has been described based on thermodynamic calculation efficiency value of generator unit 2.

Table 2. Data commissioning on August 7, 1987.

hour	Generator power (MW)	Steam Temperature in (° C)	Steam Temperature out (° C)	Steam Mass Flow rate (tons / hour)
17.00	52.4	165	64	410
18.00	55	165	56	410
19.00	55	165	56	410
20.00	55	166	56	410
22.00	54.9	165	56	410
24.00	54.8	165	56	410

Table 3. Manual calculation results P input.

Hour	Turbine power (kJ/s)	Generator Efficiency (%)
17:00	102.339,15	51
18:00	112.731,44295	49
19:00	112.731,44295	49
20:00	112.849,89895	49
22:00	112.731,44295	49
24:00	112.731,44295	49
Average		49.3

Table 3 is Manual calculation results P input. After the next manual calculation we compare it by using the simulation calculation by input power value and generator output power in simulation that has been made, the simulation result shown in Figure 7 where the efficiency value shown by matlab simulation with GUI method, efficiency value in show by simulation is the result of data on the calculation at 17:00

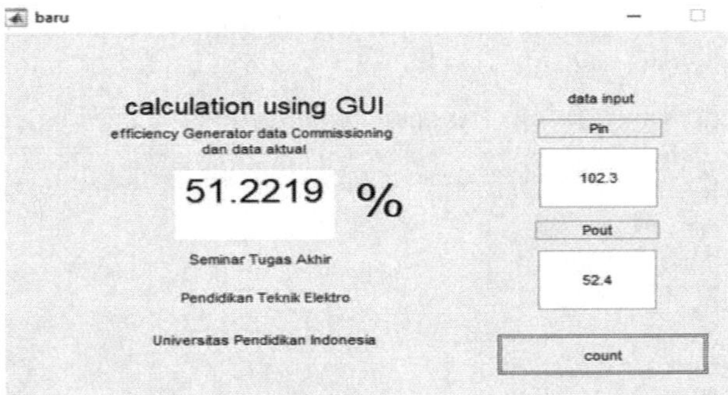

Figure 7. GUI simulation results based on commissioning data.

In addition shown by the GUI method shown also the graph of efficiency values shown by Figure 8 where there is a decrease in efficiency over time, the decrease is caused by a decrease in output power output of the generator due to the outflow vapor process that changes the value of the vapor enthalpy on the thermodynamic table, seen efficiency value at 17:00 hours decreased at 18:00.

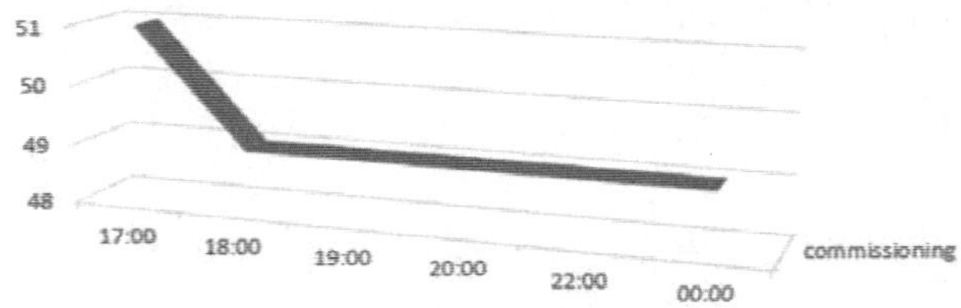

Figure 8. Graph of generator efficiency based on commissioning data.

3.2. Actual data [13]

Table 4. Actual data of PLTP Kamojang on 11 June 2017.

Hour	Generator power (MW)	Steam Temperature in (° C)	Steam Temperature out (° C)	Steam Mass Flow rate (tons / hour)
00.00	56,1	164,4	49	423
01.00	56,1	164,5	49	425
02.00	56,2	164,5	49	425
03.00	56,3	164,5	48,5	425
04.00	56,4	164,5	48,5	418
05.00	56,1	164,3	48,5	414
06.00	56	164,5	48,5	416
07.00	56	164,5	48,5	416
08.00	55,8	164,6	48,5	423
09.00	55,6	165	48,5	419
10.00	55,5	164,3	48,5	422
11.00	54,9	164,6	48,5	418
12.00	54,8	164,4	49	422
13.00	54,9	165	49	421
14.00	54,7	165	49	419
15.00	54,6	164,6	49,5	420
16.00	54,8	165	48	414
17.00	54,7	164,8	48	415
18.00	54,8	164,4	48,5	419
19.00	55	164,7	48,3	413
20.00	55,2	164,6	49	421
21.00	55,4	164,6	49	421
22.00	55,4	164,6	50	424
23.00	55,6	164,2	49,5	424
24.00	55,6	164,4	49,5	428

Table 4 is actual data of PLTP Kamojang on 11 June 2017. After calculation of data in Table 5 then the efficiency value every hour can be determined by using the calculation model that has been in paparka based on thermodynamic calculation efficiency value of generator unit 2.

Table 5. Manual calculation results P input and generator efficiency based on actual data.

Hour	Turbine Power (kJ/s)	Generator Efficiency (%)
00.00	123,97	45
01.00	124,556	45
02.00	124,556	45
03.00	124,556	45
04.00	122,505	46
05.00	121,332	46
06.00	121,918	46
07.00	121,918	46
08.00	124,556	44
09.00	122,848	45
10.00	123,677	45
11.00	122,505	45
12.00	123,677	44
13.00	123,434	44
14.00	122,848	44
15.00	123,091	44
16.00	121,382	45
17.00	121,625	45
18.00	122,798	44
19.00	121,039	45
20.00	123,384	45
21.00	123,384	45
22.00	121,193	46
23.00	124,263	46
24.00	125,435	44
	Average	44,96

After the next manual calculation we compare using the simulation calculation by input power value and generator output power in simulation that has been made, the simulation result shown in Figure 9 below where the value of efficiency shown by matlab simulation with GUI method, efficiency value in show by simulation is the result of data on the calculation at 13:00.

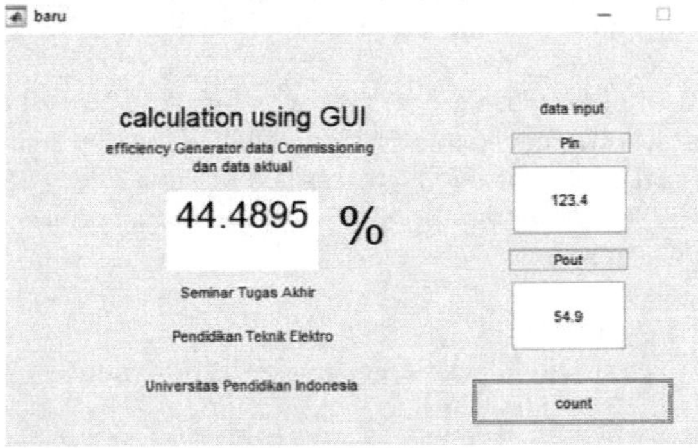

Figure 9. GUI simulation results based on actual data.

In addition shown by the GUI method shown also the graph of efficiency values shown by Figure 10 where there is fluctuation rise and fall of efficiency over time, the decrease is caused by a decrease in output power output from the generator due to the outflow vapor process that changes the value of vapor enthalpy on tables of thermodynamics, seen efficiency value at 17:00 hours decreased at 18:00. And fluctuations gradually rose by 19:00 hours and the peak of highest efficiency was at 22:00 on the chart with efficiency rating reaching 46%.

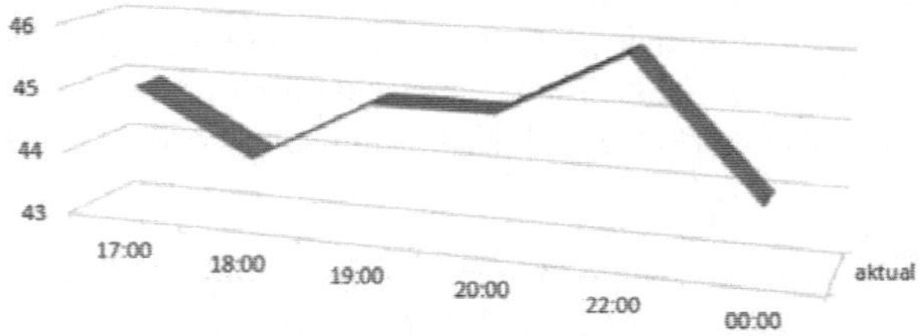

Figure 10. Graph of generator efficiency based on actual data.

If compared to the data commissioning chart with actual data then Figure 11 shows the difference graph where the visible difference can be caused by the influence of working age on the generator where the value of efficiency in the commissioning data has decreased up to the last actual data obtained.

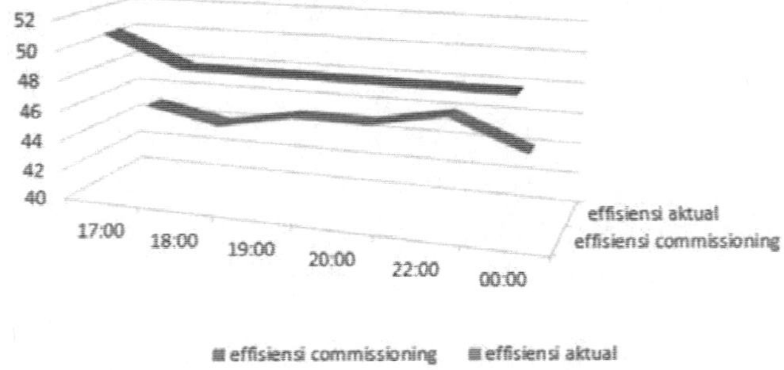

Figure 11. Graph of comparison between efficiency value based on commissioning data and actual data.

4. Conclusions

Based on the research that has been done there is a decrease in the number of efficiency presentation of generator unit 2 PLTP kamojang where the decline occurs within the working age of the generator for approximately 30 years, where the impairment of efficiency is obtained from the comparison of 49.3% commissioning average data value, average efficiency of actual data is 44.96%, which decreased efficiency by 4.34%.

The difference in the difference between the efficiency of the generator when first generated with the actual data, has a fairly large difference, this is also affected by the age of work and quality of maintenance, where the normal generator engine to replace the unit every 35 years, but it can be said maintenance at PT Indonesia Power good with duration 30 years working age decrease efficiency of generator only decrease 4.34% only.

References

[1] Hammons T J 2003 Geothermal power generation worldwide. *In Power Tech Conference Proceedings, 2003 IEEE Bologna* **1** 8

[2] Indonesia Investments 2017 *Energi Panas Bumi Indonesia - Geothermal* [Online] Available at: https://www.indonesia-investments.com/id/bisnis/komoditas/energi-panas-bumi/item268? [Accessed: 29-Dec-2017].

[3] Ruliandi, D. (2015, July). Geothermal power plant system performance prediction using artificial neural networks. *In Technologies for Sustainability (SusTech) 2015 IEEE Conference* 216-223

[4] Williamson K H, Gunderson R P, Hamblin G M, Gallup D L and Kitz K 2001 Geothermal power technology *Proc. IEEE* **89** 12 1783–1792

[5] Hawley R and Marlow B a 1984 Efficiency of steam turbine generators for central power stations *Electron Power* **30** 1 23

[6] Saptadji N 1979 *Sekilas Tentang Panas Bumi* (Bandung: ITB)

[7] Astra I M, Sugihartono I and Catherine L 2010. Hasil Perhitungan Efisiensi Termal Pltgu Dan Peluangnya Sebagai Penyumbang Pemanasan Udara (Studi Pada Pltgu Priok Dengan Pola Operasi 2-2-1 Menggunakan Metode Newton-Raphson) *Jurnal Meteorologi dan Geofisika* **11** (1)

[8] Cengel Y A and Boles M A 2002. Thermodynamics: an engineering approach *Sea* 1000 8862

[9] rizasulistiati A K 2010 *termodinamika* (yogyakarta: graha ilmu)

[10] Malinowski J, McCormick J and Dunn K 2004 Advances in construction techniques of AC induction motors: Preparation for super-premium efficiency levels. *IEEE Transactions on Industry Applications* **40** (6) 1665-1670

[11] MathWorks Inc. 2015 *Creating Graphical User Interfaces* MATLAB User Guid. p 502

[12] Perusahaan Umum Listrik Negara dan Kamojang Geothermal Power Station Units 2 (2/2) 1988 *Commissioning Test Report Mechanical* (Bandung: Perpustakaan P.T Indonesia Power)

[13] Indonesia Power 2017 *Laporan beban unit 2 P.T. Indonesia Power UBP PLTP Kamojang tanggal 28 Oktober 2013* (Bandung: P.T. Indonesia Power UBP PLTP Kamojang)

Analysis of Additional Bentonit and Salt for Improving the Resistance of Electrode Rods

R Pratama* and W S Saputra

Department of Electrical Engineering Education, Universitas Pendidikan Indonesia, Jl. Dr. Setiabudhi No. 207, Bandung 40154, Indonesia

*Reffky14@student.upi.edu

Abstract. Grounding is one way to assure the safety of the hazards and the damage caused by the incurring currents. A grounding system, according to the General Electrical Installation Requirements 2000: 11, is a connecting point of an electrical circuit or a carrier not part of an electrical circuit with an earthing earth. If there is any isolation damage to a tensioned electrical installation, the danger of touch tension can be avoided as the current continues to flow into the ground through the earthing or grounding system. This aims to analyze or know the value of soil type resistance, which later can be analyzed and used as a reference for planning an earthing. In addition, this discussion aims to determine the effectiveness of earthing system improvements using additives substance. To perform the data collection, it is done by direct measurement method in the field by using earth tester and other supporting tools, and recording the result of measurement using earth tester.

1. Introduction

Grounding is the most important factor for improving power system safety and electrical appliance, soil conditions influence for the value of resistance in the electrical grounding process [1]. Good grounding accordance with applicable standards which cannot be exceed 5 Ω. The factors affecting grounding resistance are soil type, soil salinity, soil temperature, and soil moisture. Soil moisture can be created and maintained by the addition of additive substances that are absorbing/ adsorption for liquids and gases. The sample of additives including gypsum, charcoal powder, salt, zeolite, bentonite [1].

There is a significant difference between the ground without addition the additive substances in the form of salt with the addition of salt. The grounding system with the addition the addictive substances in the form of bentonite resulted a decrease in the value of earth resistance. Both types of the addictive substances can decrease the value of earth resistance [2].

In this research will discuss about the difference the value of earth resistance between the ground which added bentonit, salt and not given additive substances. This research will be made to the clay soil.

2. Method

2.1. Research method

The method used in the research "Analysis of additional bentonite and salt for improving the resistance of electrode rods" is a method with a qualitative approach, where the method used is a case study. Technique of collecting data in the form of data from measurement using earth tester with method "I".

2.2. Research proses

In the research process, will be explained by the flowchart in figure 1:

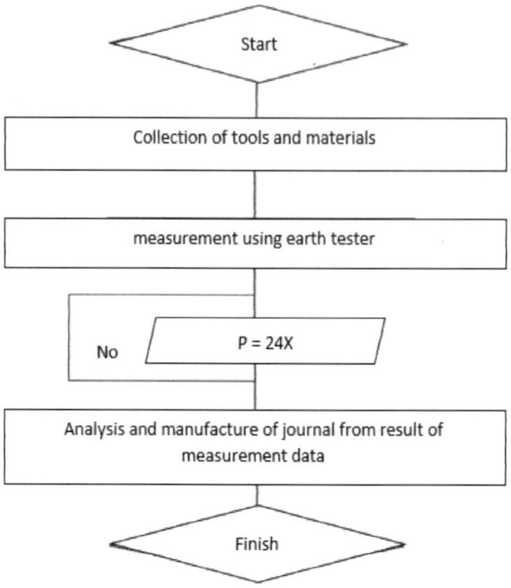

Figure 1. Research Process.

2.3. Selection of the tools and materials used
For the first step, selection of tools and materials will be needed to help get the data of soil resistance measurement value:

2.3.1. Electrode Rod. Electrode rod or often called a grounding rod serves to drain the charge of electric current from the conductor cable to the grounding ground implanted in the ground.[3] The rod electrode used has a diameter of 1.5 cm and a length of 1.80 m can see in figure 2.

Figure 2. Electrode Rod.

2.3.2. Bentonit. Bentonite is a type of clay that mostly contains montmorillonite with minerals such as quartz, calcite, dolomite, feldspars, and other minerals [4]. Bentonite has properties can absorb water and hold water on the structure, consisting of tetrahedral layer and octahedral layer then the interlayer layer where the absorption of water occurs in the interlayer layer. In this interlayer layer there are water molecules and cations [5]. The sample of Bentonit can see in figure 3.

Figure 3. Sample Bentonit.

2.3.3. Salt. In chemistry, salt is an ionic compound consisting of positive ions (cations) and negative ions (anions), thus forming a neutral compound (without charge). Salt is formed from the reaction of acids and bases. There are many kinds of salt, among others, neutral salts, alkaline salts, acid salts [3,6]. Sample of salt can see in figure 4.

Figure 4. Sample Salt.

2.3.4. Earth Tester. The grounding measurement device used is Digital Earth Resistance Tester KYORITSU Model 4105A as shown in Figure 5. This tool serves to display the measured earth prison value with the following technical specifications [7,8]:

- Measurement range for ground resistance is 0 - 20 Ω, 0 - 200 Ω, 0 -2000 Ω
- The terminal number 3 pieces are E, P and C.

Figure 5. Earth Tester Kyoritsu model 4105A.

2.4. Making holes for measurement

At this step, the hole making has a function to store bentonite, salt and the installation of rod electrodes, then below there is an explanation of the holes made as follows figure 6:

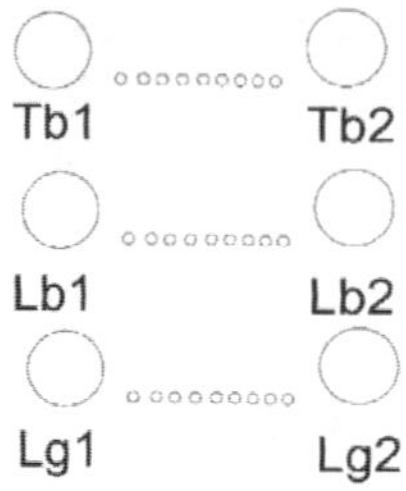

Figure 6. Hole for Measurements.

Explanation:
TB1 = Holes not filled with additive substance with a depth of 125 cm
TB2 = Hole not filled with additive with a depth 90 cm
LB1 = Holes containing 5kg bentonit with a depth of 125 cm
LB2 = Holes containing 5kg bentonite with a depth 90 cm
LG1 = Holes containing 5kg salt with a depth of 125 cm
LG2 = Holes containing 5kg salt with a depth of 90 cm

2.5. Measurements using earth tester

In the use of earth tester tool in this research there are several ways as follows:

- Prepare the measurement tool.
- Plug the main electrode in the hole for measurement, then plug the number 1 and 2 auxiliary electrode. Distance 5 meter for auxiliary electrode number 1 and 10 meter for auxiliary electrode number 2 from the main electrode rod.
- By reason in this measurement using method "I". So, the position of the main electrode with the 1 and 2 auxiliary electrodes shall be equal.
- Attach the red and yellow cables to the auxiliary electrode and green wire on the main electrode, Note that if there is dirt on clam wire or dirt on the electrode, must be cleaned first because that dirt can affect the measurement result.
- Then turn on the measurement limit button at 200 Ω because this research is done on clay type soil that having 100Ω res resistance value.
- After that push "PRESS TO TEST" button. If there are not many points and the measuring lights do not blink then there is no error on the cable or the tool. Then turn the "PRESS TO TEST" button to the right to hold the measurement results to help with the documentation process.
- At the Last, record the measurements that appear on the earth tester and repeat the measurements up to 24x on the next day with the same hour.

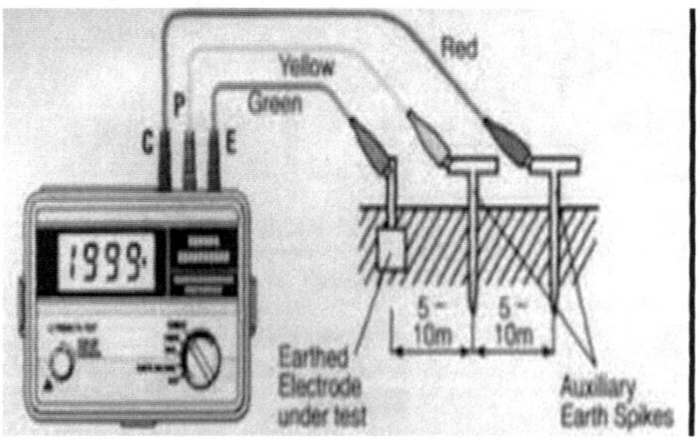

Figure 7. Measurement Using Method "I".

3. Results and Discussion

3.1. Measurements result without added additive substance

In this measurement, no additive substance used. Because, the researcher want to knows how much soil resistance value before applying method how to improve the earth system by adding additive substance. Below is the result of the calculation without use of additive substance:

Table 1. Measurement result without added additive substance.

Measurement	Time	Earth Resistance (Ω)	
		Depth 80 cm	Depth 110 cm
	12.00	17.5	15.6
I	16.00	17.5	15.6
	20.00	17.5	15.6
	12.00	17.5	15.6
II	16.00	17.5	15.6
	20.00	17.5	15.6
	12.00	17.2	15.1
III	16.00	17.2	15.1
	20.00	17.2	15.1
	12.00	17.5	15.6
IV	16.00	17.3	15.3
	20.00	17.3	15.3
	12.00	17.3	15.3
V	16.00	17.3	15.3
	20.00	17.3	15.1
	12.00	17.3	15.1
VI	16.00	17.3	15.1
	20.00	17.3	15.3
	12.00	17.3	15.3
VII	16.00	17.3	15.1
	20.00	17.3	15.1
	12.00	17.3	15.3
VIII	16.00	17.3	15.3
	20.00	17.3	15.3
Average		17.345833	15.320833

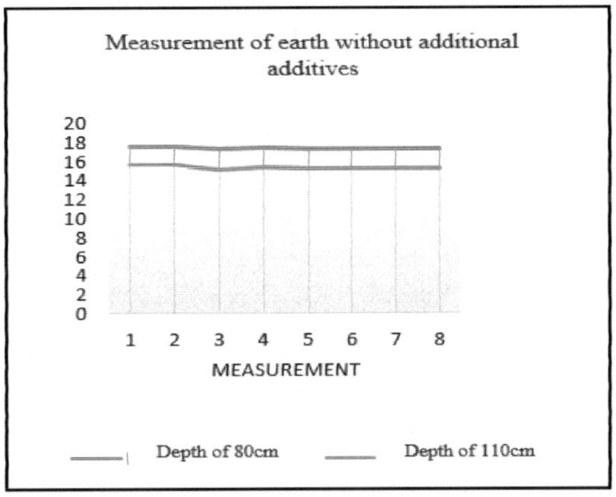

Figure 8. Graphic measurement result without added additive substance.

Can be seen from the table 1 and figure 8, the value of soil resistance in the hole that is not added using salt or bentonit results at a depth of 80 cm has an overall average value of 17.345833 while at a depth of 110 cm has an average value of 15.320833 overall. The value of the grounding measurements without additive additions is bigger than the value of the PUIL standard which requires under 5 Ω. Figure 9 is measurement result without added additive substance with earth tester.

Figure 9. Measurement result without added additive substance.

3.2. Measurements result with added bentonit

Table 2. Measurement result with added bentonit.

Measurement	Time	Earth Resistance (Ω)	
		Depth 80 cm	Depth110 cm
	12.00	6.6	4.8
I	16.00	6.6	4.8
	20.00	6.6	4.8
	12.00	6.5	4.8
II	16.00	6.5	4.8
	20.00	6.6	4.8
	12.00	6.4	4.6
III	16.00	6.4	4.6
	20.00	6.4	4.6
	12.00	6.4	4.6
IV	16.00	6.5	4.6
	20.00	6.5	4.6
	12.00	6.4	4.6
V	16.00	6.4	4.6
	20.00	6.5	4.6
	12.00	6.4	4.6
VI	16.00	6.4	4.6
	20.00	6.4	4.6
	12.00	6.4	4.6
VII	16.00	6.4	4.6
	20.00	6.4	4.6
	12.00	6.4	4.6
VIII	16.00	6.4	4.6
	20.00	6.4	4.6
Average		6.454166	4.841666

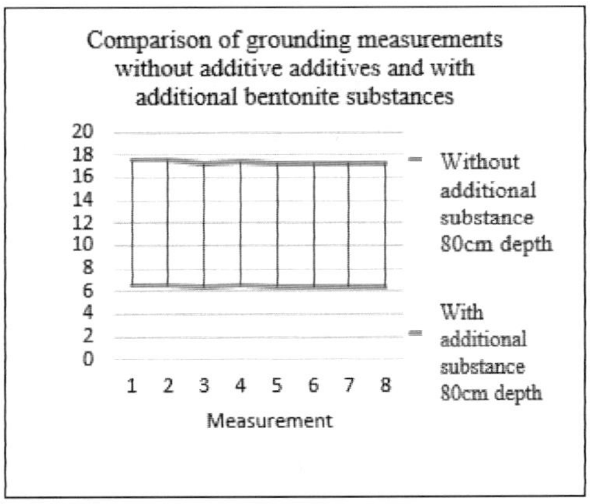

Figure 10. Comparison measurement results between without added additive substance and added bentonit at a depth 80 cm.

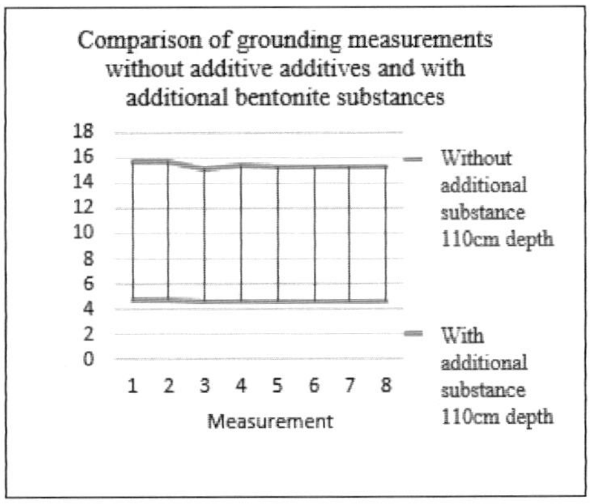

Figure 11. Comparison measurement results between without added additive substance and added bentonit at a depth 110 cm.

Can be seen from the table 2, figure 10 and figure 11, in contrast to the holes that are not added bentonit substances. In the non-additive holes and additive holes added to the depths of 80 cm and 110 cm there is a difference, wherein the bentonite added hole improves the resistance value of the soil. At a depth of 80 cm the hole added with bentonite has an average earth value of 6.454166 while the hole with added bentonite at a depth of 110 cm has an average value of 4.841666. Although holes with added bentonite can improve the resistance value of grounding resistance, but still have not reached the expectations of researchers who hope to achieve 2 or 3 Ω. Figure 12 is measurement with erath tester.

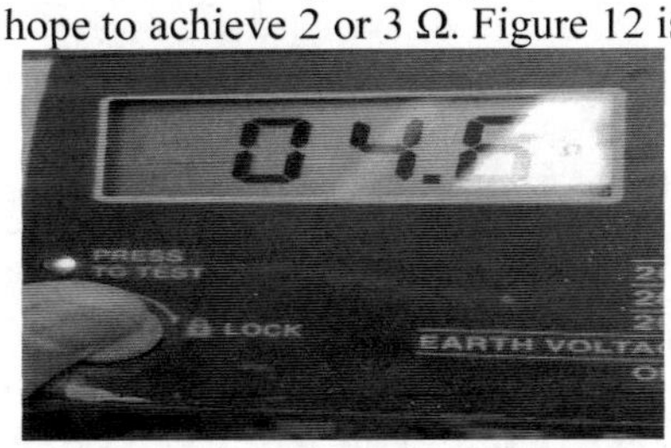

Figure 12. Measurement result with added bentonite.

3.3. *Measurements result with added salt*

Table 3. Measurement result with added salt.

Measurement	Time	Earth Resistance (Ω)	
		Depth 80 cm	Depth 110 cm
I	12.00	9.3	8.8
	16.00	9.3	8.8
	20.00	9.3	8.8
II	12.00	9.3	8.8
	16.00	9.3	8.8
	20.00	9.3	8.8
III	12.00	9.2	8.7
	16.00	9.2	8.7
	20.00	9.2	8.7
IV	12.00	9.2	8.7
	16.00	9.3	8.8
	20.00	9.3	8.8
V	12.00	9.2	8.7
	16.00	9.2	8.7
	20.00	9.3	8.8
VI	12.00	9.2	8.7
	16.00	9.2	8.7
	20.00	9.2	8.7
VII	12.00	9.2	8.7
	16.00	9.2	8.7
	20.00	9.2	8.7
VIII	12.00	9.2	8.7
	16.00	9.2	8.7
	20.00	9.2	8.7
Average		9.23	8.73

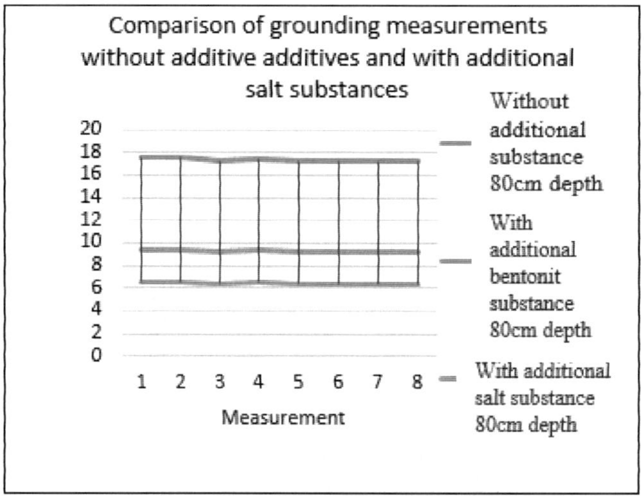

Figure 13. Comparison measurement results between without added additive substance, added bentonit, and added salt at a depth 80 cm.

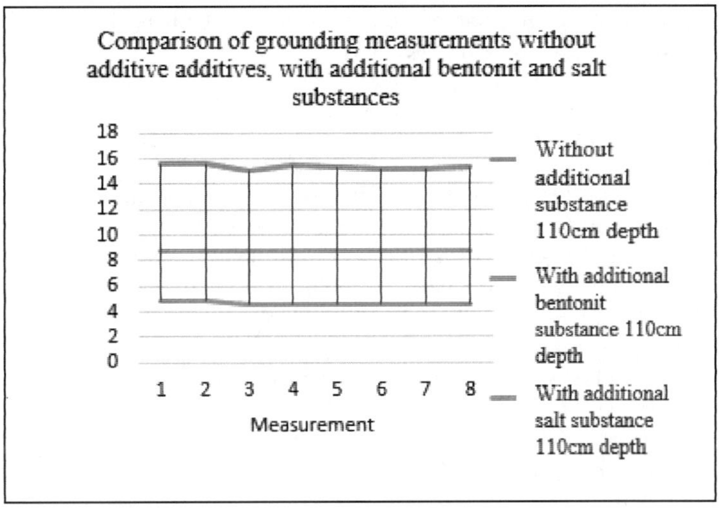

Figure 14. Comparison measurement results between without added additive substance, added bentonit, and added salt at a depth 110 cm.

Based on the measurements listed on the table 3, figure 13 and figure 14, the salt-added holes at depths of 80 and 110 cm can also decrease the resistance value of the soil such as holes with additional bentonite, but the result is still no better with the hole with additional bentonite which at a depth of 110 holes cm can improve the value of ground resistance up to 4.8 Ω. At a depth of 80 cm a hole added with salt can improve the soil resistance value by an average value of 9.23 and in a hole with a depth of 110 cm the average value is 8.73. It can be said that the hole with additional salt has not reached the value of ground resistance according to the rules of PUIL under 5Ω. Figure 15 is Measurement result with added salt.

Figure 15. Measurement result with added salt.

4. Conclusions

Based on the research "Analysis of Additional Bentonite and Salt for Improving the Resistance of Electrode Rods" which has been done, can be said as proving research on how to improve the earthing system. This study adds additives substance to improve the value of soil resistance. From the calculation results, it shows the difference of soil resistance value from the rod electrode directly implanted to the ground and the rod electrode which first added the additive substance in the hole. Besides added with additives substance, depth and soil type can also affect the value of grounding resistance.

Disadvantages possessed of the results from this study are not yet able to find out what effect on electrode rod which used in the next few years. Then, what is the effect on the surrounding environment.

This research is still not perfect, so it needs refinement in order to add ways to improve the value of soil resistance is better, simpler and lower cost. Some suggest that can be developed include the addition

IOP Conf. Series: Materials Science and Engineering **384** (2018) 012046 doi:10.1088/1757-899X/384/1/012046

of hole depth for measurement and also use of the other materials such as charcoal or gypsum, because bentonite price is too expensive for some people.

References

[1] Abidin Z 2017 Karakteristik Batang Pentanahan Sistem Arang-Garam (Sigarang) Sebagai Upaya Perbaikan Sistem Pentanahan *Jurnal ECOTIPE* **4** (1) 11-16

[2] Simamora J, Martin Y and Gusmedi H 2016 Pengaruh Penambahan Asam Sulfat (H2SO4) pada Bentonit untuk Penurunan Nilai Tahanan Pentanahan *Electrician* **10** (1) 36-43

[3] Janardana I G N 2005 Perbedaan Penambahan Garam Dengan Penambahan Bentonit Terhadap Nilai Tahanan Pentanahan Pada Sistem Pentanahan *Jurnal Teknologi Elektro* **4** (1)

[4] Janardana I 2005 Pengaruh Umur Pada Beberapa Volume Zat Aditif Bentonit Terhadap Nilai Tahanan Pentanahan *Teknol. elektro* **4** 2 1–6

[5] Andini D, Martin Y and Gusmedi H 2016. Perbaikan Tahanan Pentanahan dengan Menggunakan Bentonit Teraktivasi. *Electrician* **10** (1) 50-59

[6] Abidin Z and Ghufron A 2005 Analisa Perbaikan Sistem Pentanahan Instalasi Listrik di Tanah Kapur dan Padas Menggunakan Metode Sigarang (Sistem Grounding Arang dan Garam) *J. Tek. Elektro* 1–5

[7] Meliopoulos S 2010 Testing and Evaluation of Grounding Systems : The Revision of the IEEE Std 81 and Bonding of is Grounding and Reliable Power System 1–46

[8] Petrache E 2014 Methods for Improving Ground Resistance/Transient Ground Impedance of Transmission Structures 1–25

Teacher Performance Assesment Application using Naive Bayes Classifier Method

Y Findawati[1*], C Taurusta[1], I Widiaty[2] and A B D Nandiyanto[3]

[1]Jurusan Informatika, Universitas Muhammadiyah Sidoarjo, Jl. Raya Gelam No. 250, Candi, Sidoarjo, Jawa Timur, Indonesia
[2]Family Welfare Education Department, Universitas Pendidikan Indonesia
[3]Chemistry Education Department, Universitas Pendidikan Indonesia

*yulianfindawati@umsida.ac.id

Abstract. In this study to determine the classification of teacher performance feasibility is by using naive bayes Classifier method. The input data comes from teacher data SMPN 1 Jabon Sidoarjo. The teacher data sample will be used as training data and data testing. The Naive Bayes classifier is a simple probability classifier that applies Bayes's theorem.In this study, the feasibility assessment of teacher performance there are 8 criteria there are Service Orientation, Integrity, Commitment, Discipline, Cooperation, Teaching, Certification and Diploma. The output classification is the feasibility of teacher performance. The system is developed using under the programming language PHP and MySQL. Data used is 46 data which divided in two data that is training data as much 36 data and 10 data as test data. Tests on sample data with real data indicate the suitability of 90%.

1. Introduction

Teachers are an important element in the quality of education. To improve the quality of education required qualified and professional teachers. It is therefore necessary to assess the feasibility of teacher performance as an effort to improve the quality of education.

Teachers are an important element in school education. Teachers as professionals have a very important in achieving the educational vision of creating smart and competitive Indonesian people [3]. To be able to participate in the provision of quality education, then a teacher must meet the qualifications as stated in Law RI Number. 20/2003 on National Education System. Law Number 14/2005 on Teachers and Lecturers, and Government Regulation Number 19/2005 on National Education Standards stating that teachers are professional educators.

With the development of junior high school students from year to year continues to increase of course also encourages the development and improvement of educators needed. Improving the quality of educators is needed in the development of quality learning and teaching students. In achieving the quality of good education is strongly influenced by the performance of teachers in performing their duties so that teacher performance becomes an important demands to achieve educational success. The act of assessing teacher performance is one way of improving the quality of education.

In the study Sandi Fajar Rodiyansyah and Edi Winarko in 2013 with title Post Classification of Twitter Traffic Congestion Bandung Using Naive Bayesian Classification of the test results, the application shows that the smallest accuracy value of 78% produced on the test with a sample of 100 and produce a high accuracy value 91.60% in the test with 13106 samples [7].

In the study Sri Kusuma Dewi in 2009 with the title of classification of nutritional status using naive bayes classification resulted that in this research, the NBC algorithm will be applied for measuring the human nutrient status by using anthropometry data as input system. The result of this research shows that NBC can solve this problem adequately. Research results shows total performance of this system as 93.2% [2].

Assessment of teacher performance feasibility there are 8 criteria that is Service Orientation, Integrity, Commitment, Discipline, Cooperation, Teaching Hours, Certification and Diploma. The performance appraisal will use the naive bayes classifier method which is a simple probabilistic based prediction technique with a strong independence assumption [5]. The case study of the feasibility assessment of the performance of this teacher is from SMP Negeri Jabon Sidoarjo addressed at Jalan Dukuhsari 01 Jabon village / District Jabon, District / Sidoarjo, East Java Province. The system is developed using the programming language PHP and MySQL. Data used as many as 46 data that is training data as much 36 data and 10 data as test data. The formulation of the problem based on the above background is How to develop teacher performance appraisal apps using the web based Naive Bayes Classifier method.

2. Experimental Method

Data mining is the mining or discovery of new information by searching for patterns or rules of some very large data [1]. Data mining is also referred to as a series of processes to explore the added value of knowledge that has been unknown manually from a data set [6]. Data mining, often referred to as knowledge discovery in database (KDD). KDD is an activity that includes the collection, use of data, history to find regularities, patterns or relationships in large data sets [8].

Naive Bayes Classifier is one of the algorithms in data mining techniques that apply bayes theory in classification. Bayes's decision theorem is a fundamental statistical approach to pattern recognition. Naive Bayes is based on the simplifying assumption that attribute values are conditionally independent when given an output value. In other words, given the value of output, the probability of observing collectively is the product of the individual probability. By entering equation 1 to equation 2 we will find the approach used in NBC.

$$P(a_1, a_2, a_3, \ldots, a_n \mid V_j) = \prod_i P(a_i \mid v_j) \quad (1)$$

$$V_{NB} = \arg\max_{vj \in V} P(V_j) \prod P(a_i \mid v_j) \quad (2)$$

V_{NB} : the output value of naive bayes classification results

$P(a_i|v_j)$: the ratio between nc / n, where nc is the amount of training data for v = vj and a = ai; and n is the total possible output.

3. Results and Discussion

In this research will be used Data training and data testing. Data will be used as mining and testing process, in the form of parent data sample and teacher data SMPN 1 Jabon. This data has attributes of Service Orientation, Integrity, Commitment, Discipline, Cooperation, Teaching, Certification and Diploma. The research stages from finding literature studies about naive bayes and teacher performance assessments, after which identify problems and collect teacher performance data. Then performed data cleaning by eliminating noise and inconsistent and irrelevant data which continued the selection of data used which then combined into data with the appropriate format and select the data set to be used for research. Then proceeded to analyze the case study using the naive bayes method.

The test is done on 37 data. Position = teacher; service orientation = good enough; integrity = good enough; commitment = good; discipline = good; cooperation = good; hours of teaching = excellent; certification = yes; employment status = pns, then how are predictions for teacher performance feasibility?

The calculation steps are as follows:

P (Performance = Yes) = 0.75
P (Performance = no) = 0.25

Complete P (X | H):
p (orientation = good enough | Feasibility = Yes) = 5 = 0.1851
P (Orientation = good enough | Feasibility = No) = 6 = 0.6667

Complete P (X | H):
p (integrity = good enough | Feasibility = Yes) = 2 = 0.07407
P (integrity = good enough | Feasibility = No) = 3 = 0.33333

Complete P (X | H):
p (Commitment = good | Feasibility = Yes) = 17 = 0.6296
P (Commitment = good | Feasibility = No) = 5 = 0.55556

Complete P (X | H):
p (Discipline = good | Feasibility = Yes) = 9 = 0.33333
P (Discipline = good | Feasibility = No) = 2 = 0.22222

Complete P (X | H):
p (Cooperation = good | Feasibility = Yes) = 19 = 0.70374
P (Cooperation = good | Feasibility = No) = 9 = 1

Complete P (X | H):
p (teaching hours = Very good | Feasibility = Yes) = 19 = 0.70
P (teaching hours = Very good | Feasibility = No) = 9 = 1

Complete P (X | H):
p (Certificate = yes | Feasibility = Yes) = 27 = 1
P (Certificate = yes | Feasibility = No) = 8 = 0.8889

Complete P (X | H):
p (Status = pns | Feasibility = Yes) = 27 = 0.931
P (Status = pns | Feasibility = No) = 9 = 1

p (X | Feasibility = Yes) = 0,0015
p (X | Feasibility = No) = 0,0216

p (X | Feasibility = Yes) P (Feasibility = Yes) = 0.0011
p (X | Feasibility = No) P (Feasibility = No) = 0.0054

So the prediction of feasibility from teacher with criteria service orientation = good enough, integrity = good enough, commitment = good, discipline = good, cooperation = good, hours of teaching = excellent, certification = yes, employment status = pns is not feasible.

Implementation of the application can be seen in the following forms. Figure 1 shows Login Form. Login administrator is a process to access data mining application determines the best teacher performance by entering User Admin username and password.

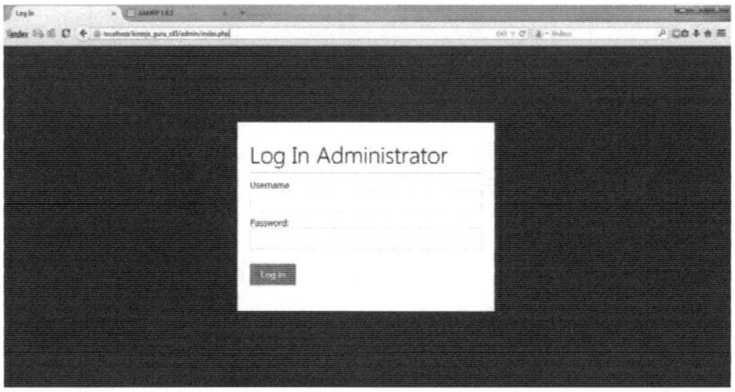

Figure 1. Login Form.

Figure 2 shows Form Manage Teacher data. In this form the user can input teacher data as desired. The system will be automatically input as training data. The system is also directly inputted with the decision.

Figure 2. Form Manage teacher data.

Figure 3 shows Homepage thats contains the menus that will run on the data mining process from the teacher data input to the teacher performance feasibility calculation.

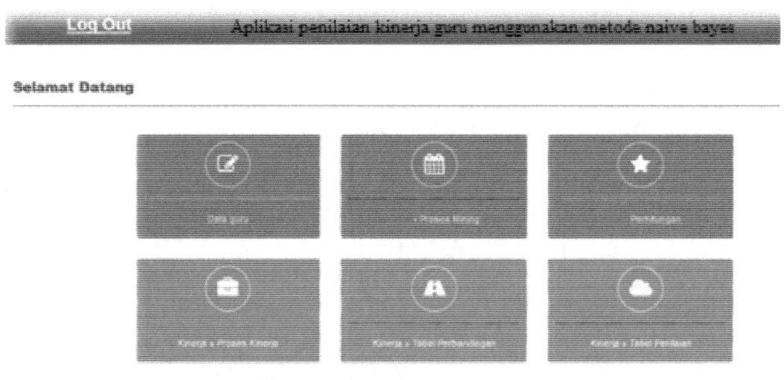

Figure 3. Homepage.

Figure 4 shows Calculation form. On this calculation menu page, the system automatically calculates the existing training data input and the system also calculated in accordance with the naive Bayes algorithm.

NO	NAMA	JABATAN	ORIENTASI PELAYANAN	INTEGRITAS	KOMITMEN	DISIPLIN	KERJASAMA	JAM MENGAJAR	SERTIFIKASI	STATUS KEPEGAWAIAN	KELAYAKAN
1	Hadi Utomo, S.Pd	Guru	Cukup Baik	Cukup Baik	Sangat Baik	Baik	Baik	Sangat Baik	Belum	PNS	YA
2	Machammad Fau	Guru	sangat Baik	Baik	Baik	Cukup Baik	Baik	Sangat Baik	Belum	PNS	YA
3	Hasril Dwi Susilo,	Guru	Baik	Sangat Baik	Baik	Cukup Baik	Baik	Sangat Baik	Belum	PNS	YA
4	Dwi Indatiningrum	Guru	sangat Baik	Baik	Baik	Cukup Baik	Baik	Sangat Baik	Belum	PNS	YA
5	Surya Indah Purn	Guru	Cukup Baik	Cukup Baik	Cukup Baik	Baik	Sangat Baik	Baik	Belum	PNS	YA
6	Ismakhil Makhfud	Guru	Cukup Baik	Cukup Baik	Cukup Baik	Baik	Baik	Sangat Baik	Belum	PNS	YA
7	Tia Rizki Siswanto	Guru	Baik	Sangat Baik	Baik	Cukup Baik	Baik	Baik	Belum	PNS	YA
8	Dwi Puspitasari, S	Guru	sangat Baik	Sangat Baik	Baik	Cukup Baik	Sangat Baik	Sangat Baik	Belum	PNS	YA
9	Andye Susanto, S.	Guru	Cukup baik	Baik	Baik	Cukup Baik	Baik	Sangat Baik	Belum	GTT	TIDAK
10	Nurul Khofifah, S.	Guru	Cukup baik	Sangat Baik	Sangat Baik	Cukup Baik	Baik	Baik	Belum	GTT	TIDAK

Figure 4. Calculation form.

To know the performance of the application, then testing the system. And the result can be seen in table 1.

Table 1. testing results.

Data	Results		Conformity
	Real Data	*NBC Results*	
1	feasible	feasible	suitable
2	feasible	feasible	suitable
3	Not feasible	Not feasible	suitable
4	feasible	feasible	suitable
5	feasible	feasible	suitable
6	feasible	feasible	suitable
7	feasible	feasible	suitable
8	feasible	Not feasible	Not Suitable
9	feasible	feasible	suitable
10	feasible	feasible	suitable

Based on the above test results then obtained the data test data with the data rill is 90%.

4. Conclusions

Based on the results of research and discussion that has been done, it can be concluded that the method of naive bayes classifier can be used as an assessment of teacher performance based on the criteria of Service Orientation, Integrity, Commitment, Discipline, Cooperation, Teaching, Certification and Diploma. Based on the test results then obtained the data test data with the data rill is 90%

References

[1] Davies and Paul B 2004 *Database Sistem Third Edition* (New York: Palgrave macmillan).

[2] Kusumadewi S 2009 *Klasifikasi Status Gizi Menggunakan Algoritma Naive Bayes Classification* (Yogyakarta: Universitas Islam Indonesia)

[3] Law RI No. 20 2003 *on National Education System*

[4] Law No. 14 2005 *on Teachers and Lecturers, and Government Regulation No. 19/2005 on National Education Standards*

[5] Prasetyo E 2012 *Data Mining Konsep dan Aplikasi Menggunakan Matlab* (Yogyakarta: Andi Offset)

[6] Pramudiono I 2003 *Pengantar Data Mining: Menambang Permata Pengetahuan di Gunung Data*

[7] Rodiyansyah F S and Winarko E 2013 Klasifikasi Posting Twitter Kemacetan Lalu Lintas Kota Bandung Menggunakan Naive Bayes Classification *IJCSS* **7** 13-22

[8] Santoso B 2007 *Data Mining Teknik Pemanfaatan Data Untuk Keperluan Bisnis* (Yogyakarta:Graha Ilmu)

Parametric Study of an Ultra-wideband 2-18 GHz Microstrip to Coplanar Stripline Transition

R R Anandhitto, T Hariyadi and Mukhidin

Department of Electrical Engineering Education, Universitas Pendidikan Indonesia, Jl. Dr. Setiabudhi No. 207, Bandung, Indonesia

*reynatto@student.upi.edu

Abstract. Microstrip line has become the most utilized transmission line for microwave circuits, but it has unbalanced characteristic, while coplanar stripline is a popular balanced transmission line for designing balanced circuits. Most of the ultrawideband antenna is a balanced antenna type such as dipole, bow tie, or two arms antenna. Therefore, we need a transition line to feed a balanced antenna because the port is usually an unbalanced structure like an sma connector. In this paper, a microstrip to coplanar stripline transition design is proposed. The goal is to design a compact and low-cost transition line that works at 2-18 GHz band to feed a balanced antenna. To achieve the goal, we propose a transition line with FR4 as the substrate. FR4 chosen as the substrate because it is available at the local market and has a low-price tag. The transition line has been constructed and simulated in CST Studio Suite to find the most appropriate design.

1. Introduction

A transition structure or balanced-to-unbalanced (balun) is a device that used to convert signals between an unbalanced circuit structure and a balanced circuit structure. Directly connecting an unbalanced structure like an sma connector to a balanced antenna can affect the radiation. Therefore, we need a transition line to avoid the unwanted radiation. A transition line can be made by connecting two transmission line with different characteristic into one structure. Microstrip (MS) line has become the most utilized transmission line for microwave circuits, but it has unbalanced characteristic, while coplanar stripline (CPS) is a popular balanced transmission line for designing balanced circuits.

Various method of designing MS-CPS transition have been reported. A transition with ground shaped taper using RT/Duroid 5880 as the substrate have been reported in [1]. In [2], a transition with radial stub on the top layer using RT/Duroid 5870 as the substrate have been reported. While transition with microstrip line equal to quarter wavelength using RO4003 as the substrate have been reported in [3] and a transition with three via holes on the CPS using Rogers 4003C as the substrate have been reported in [4]. There is no transition design using FR-4 as the substrate even though it has cheaper fabrication cost compared to other substrate as mentioned before. In this paper, a microstrip to coplanar stripline transition using FR-4 substrate is proposed. FR-4 that used on our design has dielectric constant of 4.4 and thickness of 1.6 mm. The transition targeted to work at 2-18 GHz band to feed a balanced antenna. The impedance of microstrip (unbalanced) adjusted at 50 Ω and the impedance of coplanar stripline (balanced) adjusted at 150 Ω.

2. Transition design

Our transition line design refer to the design of microstrip to coplanar stripline transition with ground shaped taper proposed by Kim, et al [1] as shown in Figure 1.

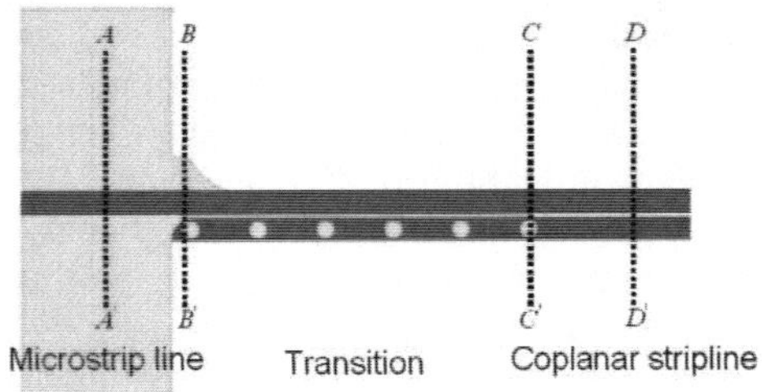

Figure 1. Refered transition structure [1].

The structure has three parts, there are a microstrip section, a transition section and a coplanar stripline section. Microstrip transmission line has a formula to determine the width of the strip. Using the microstrip line formula [5] below we can calculate the width for microstrip with 50 Ω impedance.

$for \frac{W}{d} < 2$

$$\frac{W}{d} = \frac{8e^{A}}{e^{2A} - 2} \tag{1}$$

$for \frac{W}{d} > 2$

$$\frac{W}{d} = \frac{2}{\pi}\left[B - 1 - \ln(2B - 1) + \frac{\epsilon_r - 1}{2\,\epsilon_r}\left\{\ln(B - 1) + 0.39 - \frac{0.61}{\epsilon_r}\right\}\right] \tag{2}$$

where

W = width of microstrip

d = substrate thickness

$$A = \frac{Z_0}{60}\left(\frac{\epsilon_r + 1}{2}\right)^{1/2} + \frac{\epsilon_r - 1}{\epsilon_r + 1}\left(0.23 + \frac{0.11}{\epsilon_r}\right) \tag{3}$$

$$B = \frac{377\pi}{2\,Z_0(\epsilon_r)^{1/2}} \tag{4}$$

Using the formula (2) we can determine the width to make a 50 Ω microstrip line on FR-4 substrate, the calculated result is 2.96 mm. After simulated with CST Studio Suite, the 2.96 mm width has higher impedance value. The closest width for 50 Ω impedance is 2.9 mm. Therefore, we use 2.9 mm for the microstrip line width.

Microstrip is a ground backed transmission line, while coplanar stripline composed by two symmetric stripline without ground. Because the width of microstrip has been fixed at 2.9 mm, to make a coplanar stripline we have to add another separated strip with a determined gap.

There is a synthetic formula [6] to calculate the width or the gap of coplanar stripline. Because we already have the width dimension so we have to calculate the gap part. The formula (5) and (6) shows how to calculate the gap between two strips on coplanar stripline.

$$S = W\,.\,G \tag{5}$$

where

$$G = 21.385\left\{\left(\frac{\eta_0}{Z_0(\epsilon_r + 0.86)^{1/2}}\right)\left[1 + \exp\left(\frac{0{,}1\,W\,(\epsilon_r - 1{,}52)}{H\,\epsilon_r}\right)^{0{,}68}\right]\right\}^{-3{,}753} \tag{6}$$

IOP Conf. Series: Materials Science and Engineering **384** (2018) 012048 doi:10.1088/1757-899X/384/1/012048

Using the formula (5) and (6), to get 150 Ω coplanar stripline impedance, we have to set 6.8 mm for the gap between two strips with 2.9 mm width. After simulated, the impedance value of the coplanar stripline is 166.6 Ω. Multiple simulation has been done to make a coplanar stripline with 150 Ω impedance and the result is 3.5 mm gap has the closest impedance value that is 149.5 Ω.

In order to optimally match the characteristic impedances between the microstrip line (50 Ω) and the coplanar stripline (150 Ω), a taper consist of 'ground-shaped micro-coplanar stripline with vias' [1] is used. The taper design shown in Figure 2.

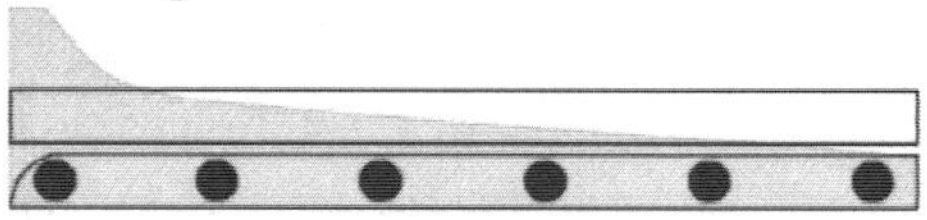

Figure 2. Taper design of the transition structure.

3. Simulation results and discussion

Our proposed transition line design shown in Figure 3 have to meet the following specifications,

- Range frequency: 2-18 GHz
- Return loss: < - 10 dB
- Coplanar stripline (balanced) impedance: 150 Ω
- Microstrip (unbalanced) impedance: 50 Ω

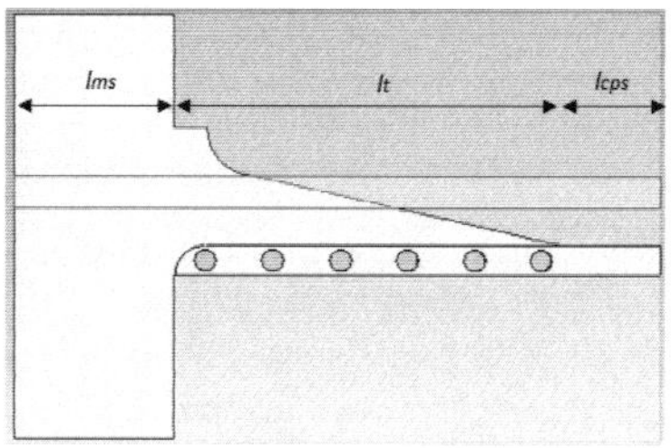

Figure 3. Layout of the initial proposed MS-CPS transition structure.

The simulation results from the initial design are shown in Figure 4. S-Parameters shows return loss and insertion loss from both transmission line ports. S11 is the return loss of the microstrip line, S21 is the microstrip line insertion loss, S12 is the insertion loss of the coplanar stripline, S22 is the return loss of the coplanar stripline.

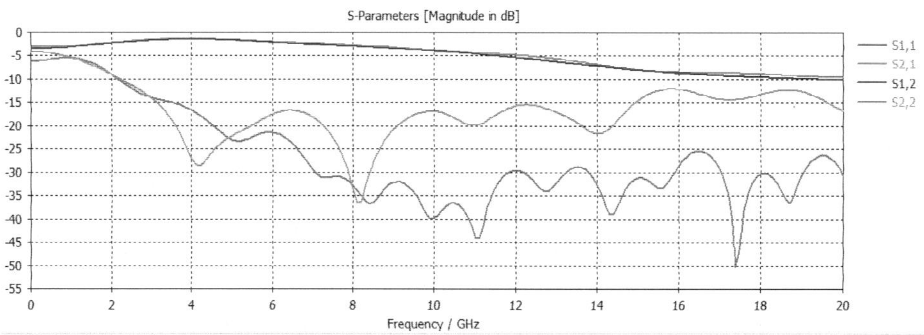

Figure 4. S-Parameters of initial design.

It can be observed from Figure 5 that the simulation results of return loss from the initial design still more than -10 dB at 2 GHz for both microstrip and coplanar stripline. This result means that the transition structure has not complied with the specification.

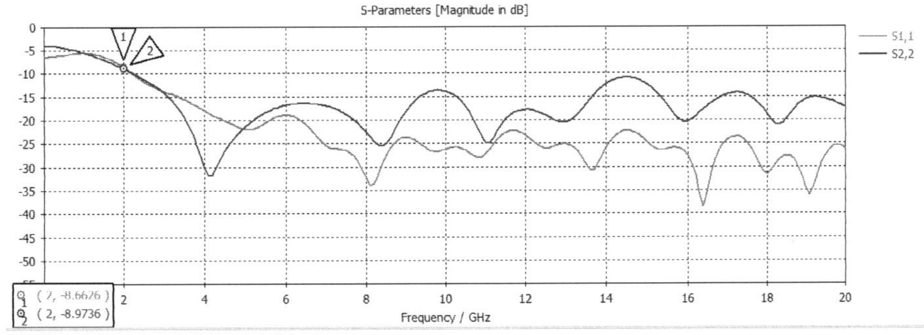

Figure 5. Return loss (S11 and S22) of initial design.

Figure 6 shows the insertion loss of the simulated initial design. The results from initial design have insertion loss almost reaching -10 dB at 18GHz.

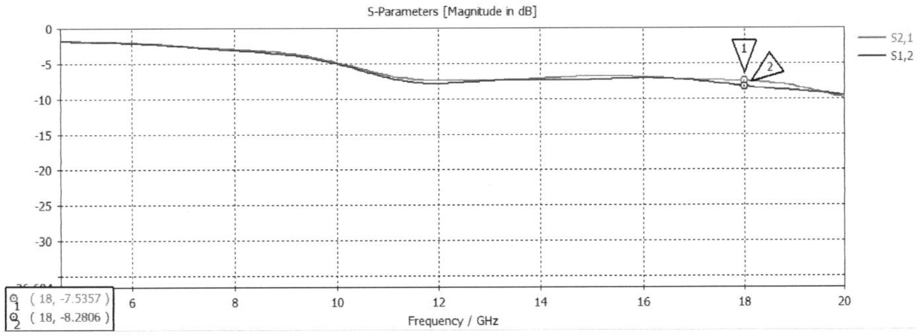

Figure 6. Insertion loss (S21 and S12) of initial design.

The microstrip impedance from initial design is observed at 49.56 Ω, because it is close to the 50 Ω target impedance, there is no need to change the width of microstrip line. While the impedance of coplanar stripline is observed at 166.6 Ω, it is necessary to change the distance between two strips (gap) of the strip in coplanar stripline to achieve the 150 Ω of impedance.

To find the ideal distance or gap close to achieve coplanar stripline 150 Ω of impedance, other simulations have been done by changing the distance between two strips (gap). The simulation results of changing the gap from 1-4 mm are listed in Table 1.

Table 1. Coplanar Stripline impedance with various gap.

Coplanar Stripline	
Gap between Strips	Impedance
1 mm	107.6 Ω
2 mm	129 Ω
3 mm	142.4 Ω
4 mm	151.4 Ω

As listed on the Table 1, we can see the closest impedance value to 150 Ω is within 3 - 4 mm of gap. When the design simulated using the gap value of 3.1 - 4 mm between the two strips, the closest impedance value to 150 Ω achieved at 3.5 mm with 149.5 Ω impedance value.

After found the most ideal gap for the coplanar stripline, other simulations have been done to find the effect of the via hole diameter. Figure 7 shows the microstrip line return loss (S11) results obtained from a parameter sweep simulation using different via hole diameter from 1 – 2 mm.

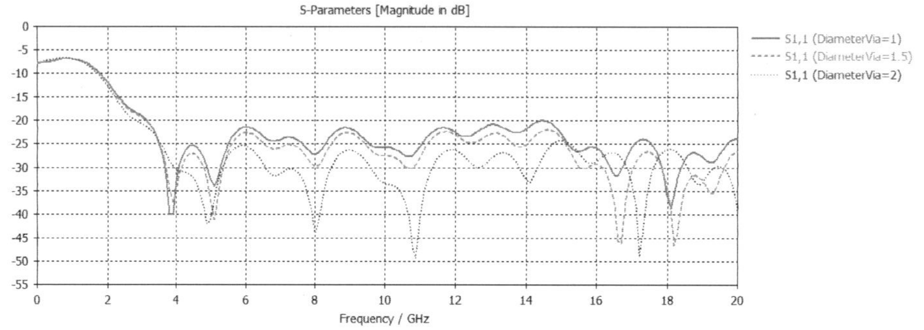

Figure 7. Microstrip return loss (S11) with different via hole diameter.

Figure 8 shows the coplanar stripline return loss (S22) results obtained from a parameter sweep simulation using different via hole diameter from 1 – 2 mm.

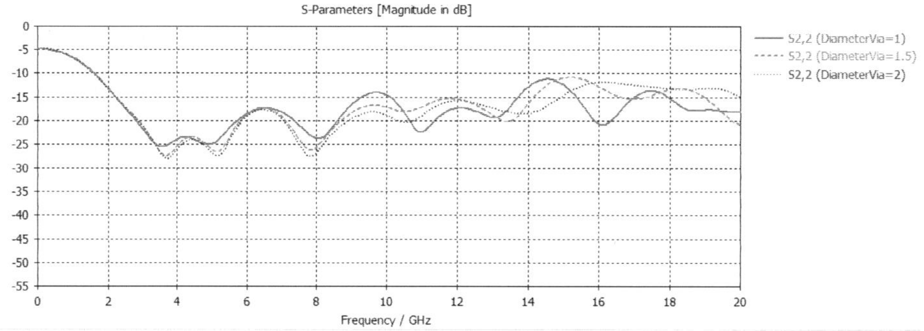

Figure 8. Coplanar stripline return loss (S22) with different via hole diameter.

It can be observed that the smallest magnitude of the microstrip return loss (S11) and coplanar stripline return loss (S22) are achieved by using 2 mm via hole diameter compared to the 1 and 1.5 mm via hole diameter. Therefore, the design of transition structure use 2 mm via hole diameter. Simulation using via diameter larger than 2 mm were not performed because it is too close to the edge of the strip.

Since there are no formula to calculate the ideal transmission line length, multiple simulations have been done by changing the length of the two transmission lines. Simulations are performed comparing the two different lines with different length. Figure 9 shows the simulation result that have most significant differences of insertion loss (S21) when the microstrip line adjusted at 20 mm with different coplanar stripline length from 10, 20 and 30 mm.

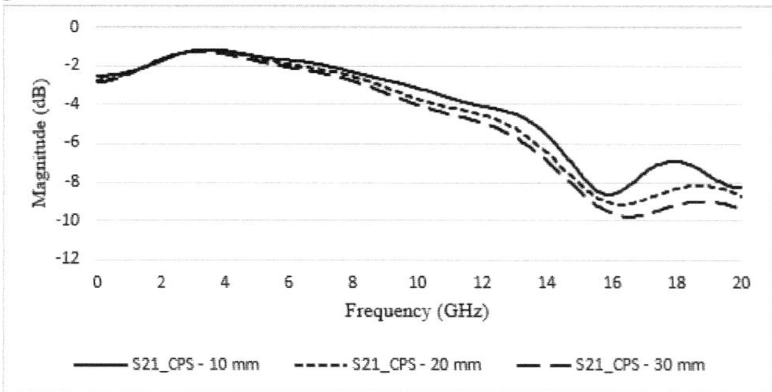

Figure 9. S21 result using different coplanar stripline length.

Figure 10 shows the simulation result that have most significant differences (S11) when the coplanar stripline adjusted at 20 mm with different microstrip length from 10, 20 and 30 mm.

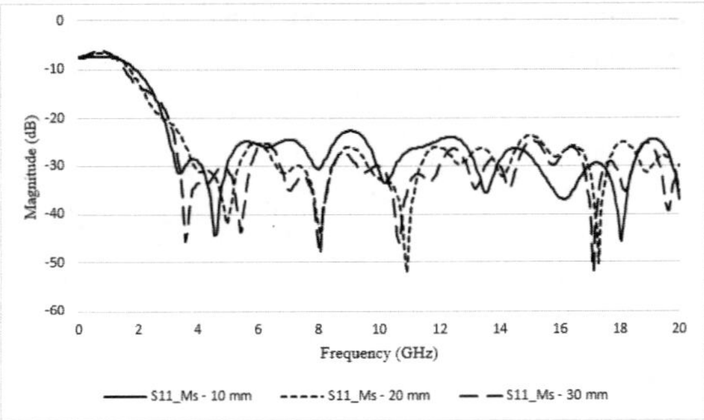

Figure 10. S11 result using different microstrip length.

Simulation results shows coplanar stripline length affects the insertion loss. At the longest CPS length simulated (30 mm), the insertion loss was observed has a value -10 dB at 18 GHz, at the shortest CPS length (10 mm) insertion loss observed has a value -7 dB at 18 GHz. The impact of the microstrip line length is observed affects the lowest frequency. When the design using the shortest microstrip length (10 mm) the simulation results shows a return loss value of -10 dB higher than 2 GHz frequency, at the longest microstrip length (30 mm) the return loss value of -10 dB is observed at 1.5 GHz frequency. Therefore, to create a design with the lowest insertion loss, the coplanar stripline has to be made as short as possible. To create a design with lowest frequency, the microstrip line has to be made as long as possible.

After observing the simulation results, both of the transmission lines length is adjusted. The length of the microstrip line is adjusted at 15 mm with the length of the coplanar stripline adjusted at 9.2 mm to obtain the total length of the structure is 60 mm. The details of the optimized structure length described in Table 2.

Table 2. Optimized structure length.

Optimized Structure Length	
Microstrip line length	15 mm
Transition length	35.8 mm
Coplanar stripline length	9.2 mm
Total length	60 mm

Another simulation has been done by adding via holes to see the effect of the number of via holes on S-Parameter. There is no change in the transition structure beside adding via holes. Adding more via holes is done because the previous simulation shows via diameter affecting the simulation results. Therefore, the effect of adding via hole needs to be observed. Using the structure dimension described in Table 2, the simulation begins with the gap between two strips set to 3.5mm. Figure 11 shows microstrip line return loss (S11) results obtained from a parameter sweep simulation using different number of via holes.

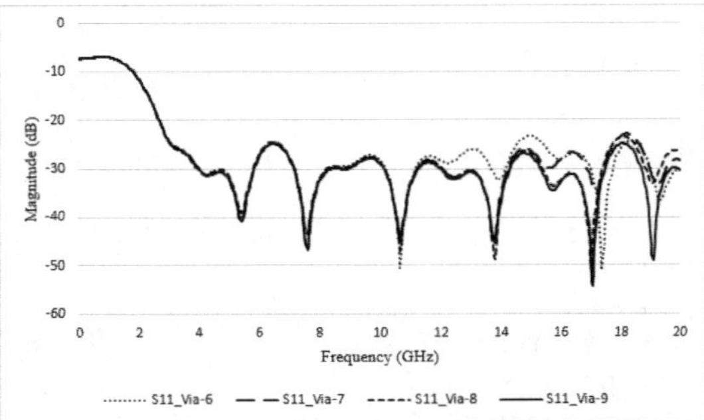

Figure 11. S11 with different number of via holes.

Figure 12 shows coplanar stripline return loss (S22) results obtained from a parameter sweep simulation using different total via holes.

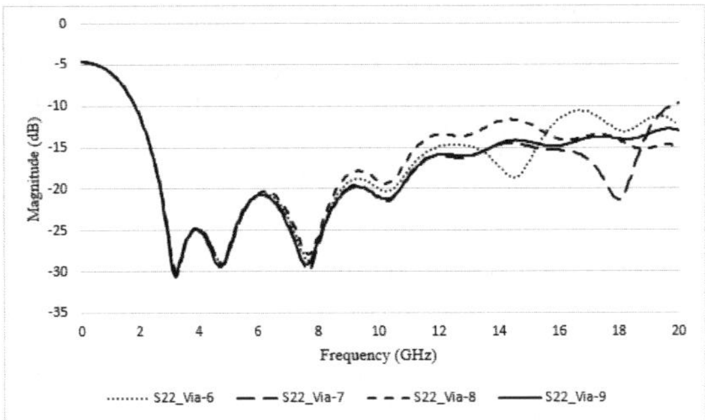

Figure 12. S22 with different number of via holes.

Figure 13 shows insertion loss (S21 and S12) results obtained from a parameter sweep simulation using different total via holes.

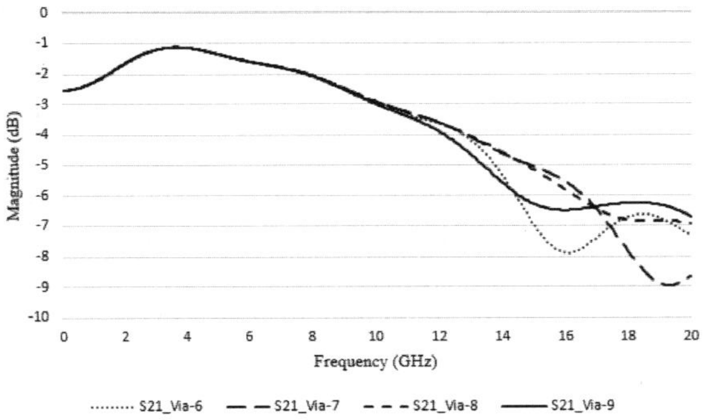

Figure 13. Insertion loss with different number of via holes.

From the simulation results, it can be observed that the most optimal transition line performance achieved when the design using 9 via holes. Return loss of both lines (S11 and S22) has less than -10 dB value at 2-18 GHz frequency with smallest insertion loss value (S21 and S12). The optimal design shown in Figure 14. The dimensions of the structure are described in Table 3. Simulated optimal design is described in the next section.

(a) (b)

Figure 14. (a) Top view of the optimal design. (b) Bottom view of the optimal design.

Table 3. Detail dimension of the optimal design.

Detail Dimension of The Optimal Design	
Microstrip line length	15 mm
Coplanar stripline length	9.2 mm
Transition structure length	35.8 mm
Gap between two strips	3.5 mm
Strip width (MS, CPS, Transition)	2.9 mm
Via hole diameter	2 mm

3.1. S-Parameters of the optimal design
The insertion loss (S21 and S12) observed in Figure 15 is -7 dB at 18 GHz with the return loss of both lines (S11 and S22) less than -10 dB at 2 – 18 GHz. This result shows that the transition line has achieve the range frequency specification at 2-18 GHz with the smallest insertion loss compared to the initial design.

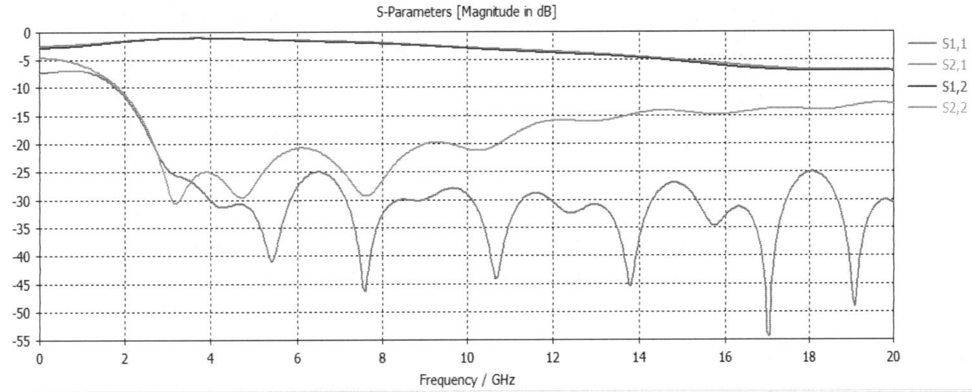

Figure 15. S-Parameters of the optimal design.

3.2. Microstrip (Unbalanced) Impedance
The optimal design simulation result shown in Figure 16 has microstrip line impedance value of 49.56 Ω. Since it is close to the impedance target of 50 Ω, the microstrip impedance specification considered has been achieved.

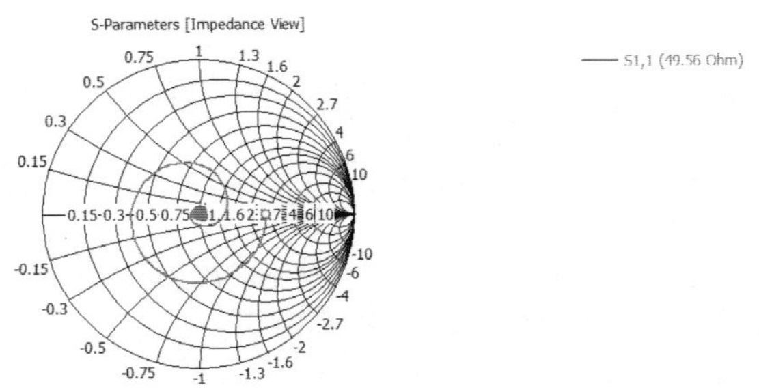

Figure 16. Microstrip line impedance of the optimal design.

3.3. Coplanar Stripline (Balanced) Impedance
The optimal design simulation result shown in Figure 17 has coplanar stripline impedance value of 149.5 Ω. Since it is close to the impedance target of 150 Ω, the coplanar stripline impedance specification considered has been achieved.

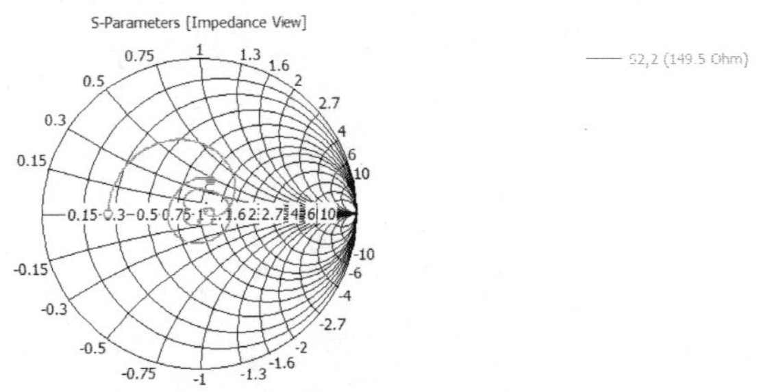

Figure 17. Coplanar stripline impedance of the optimal design.

IOP Conf. Series: Materials Science and Engineering **384** (2018) 012048 doi:10.1088/1757-899X/384/1/012048

Overall, the optimal design achieved all the specifications of a balanced-unbalanced transition line. The optimal simulation results are described in Table 4.

Table 4. Simulation Results

Specification	Target	Achievement
Frequency range	2 – 18 GHz	2 – 18 GHz
Return loss	< -10 dB	< -10 dB
Coplanar Stripline impedance	150 Ω	149.5 Ω
Microstrip impedance	50 Ω	49.56 Ω

Based on the simulation results, the insertion loss (S21 and S12) has a big value (-7 dB at 18 GHz). This is caused by FR4 characteristic that has a typical loss tangent in the region of 0.02, which is too high for many circuit applications [7].

4. Conclusions

A transition between microstip and coplanar stripline using FR-4 has been designed. The transition structure has three parts, there are a microstrip section, a transition section and a coplanar stripline section. Simulation result shown the return loss (S11) has less than -10 dB value from 2-18 GHz. The impedance of the microstrip is obtained at 49.5 Ω and 149.5 Ω for the coplanar stripline. The best result is achieved by optimizing the transition line structure. Several parameters have been changed like strip gap of coplanar stripline, via holes diameter, transmission lines length and adding more via holes to achieve all the determined spesifications.

References

[1] Kim Y, Woo D, Kim K W and Cho Y 2007 A New Ultra-wideband Microstrip-to-CPS Transition *IEEE MTT-S Int. Microw. Symp.* 1563–1566

[2] Suh Y and Chang K 2001 A Wideband Coplanar Stripline to Microstrip *IEEE Microw. Wirel. COMPONENTS Lett.* **11** 1 28–29

[3] Mandal M K and Chen Z N 2013 Compact Ultra-Wideband Microstrip-to-Coplanar Stripline Transitions *IEEE MTT-S Int. Microw. RF Conf.* 1 8–11

[4] Jiang Q, Domier C and Luhmann N C 2012 A Wideband Low Loss Planar Microstrip-to-CPS Balun *Proc. APMC 2012, Kaohsiung, Taiwan.* 1205–1207

[5] Pozar D M 2011 *Microwave Engineering* (Hoboken, NJ: JohnWiley & Sons, Inc,)

[6] Chan S W 2005 New and Very Simple Synthesis Formulas for Coplanar Stripline," *Microw. Opt. Technol. Lett.* **44** 2 199–202

[7] Aguilar J R, Beadle M, Thompson P T and Shelley M W 1998 The Microwave And RF Characteristics Of FR4 Substrates *IEE Colloqium Low Cost Antenna Technol. - IET Conf.* 0–5

IOP Conf. Series: Materials Science and Engineering **384** (2018) 012049 doi:10.1088/1757-899X/384/1/012049

Directional 2x2 MIMO Microstrip Antenna for LTE 1.8 GHz Application

R P Megantara* and T Hariyadi

Department of Electrical Engineering Education, Universitas Pendidikan Indonesia, Jl. Dr. Setiabudhi No.207, Bandung, Indonesia

*rizkisev@student.upi.edu

Abstract. LTE is a standard of wireless communication access by using GSM/EDGE and UMTS/HSPA technology for high-speed data transfer rates. This technology can download and upload with the speed up to 300 Mbps and 75 Mbps respectively. LTE technology using MIMO system can increase the data rate and capacity on sending and receiving simultaneously on a radio frequency channel so that the antenna which is used for LTE band-3 works at 1.8 GHz also using MIMO structure. In this paper, the authors will propose a MIMO 2x2 antenna with directional radiation patterns that works at LTE band-3 using two types of patches, which are the parasitic and main patch. In this simulation, we use FR-4 as the substrate. This substrate has a dielectric constant of 4.5, and the thickness is 1.6 mm. The method that is used in this research is design and simulation. The calculation, design and simulation process on this research the authors use the CST Studio Suite Software. The target results from this research are the antennas works at 1.8 GHz LTE band; the gain is more than 6 dBi; diversity polarization is vertical and horizontal, coupling between two antennas is less than -20 dB, and the antenna bandwidth is 200 MHz from 1.7 to 1.9 GHz for VSWR less than 2 with directional radiation patterns.

1. Introduction

High demand of data access and communication service in the digital world can be seen from the increase of cell phone and internet usage. Based on the fact above, there is a requirement of a concept that can manage the demand LTE is one of the breakthrough that can solve the problem. LTE is a 4G technology where the Internet Protocol as the transmission medium to transmit Digital Broadband Packages with capacity of 100-300 Mbps [1].

In order to make the technology, antenna is one of the component that is required for the signal transmission with required bandwidth capacity. With this research, the antenna will be designed with the synthesis of other existing concept. Based on the previous study, feeding microstip slot is one of the antenna construction methods. However, the result of that method has the radiation pattern that spread to all directions [2]. Therefore, the technique used to solve the problem is to make conventional patch so that the radiation pattern will spread to only one direction and the usage of two parasitic patches placed between the main patch to increase the bandwidth capacity [3]. In this paper we propose a directional 2x2 MIMO antenna with polarization diversity using similar method in [3] but for LTE 1.8 GHz application.

2. Antenna Design

In this segment, the antenna construction with CST Studio Suite will be described. Standard bandwidth capacity for LTE band-3 is 1.71 GHz-1.88 GHz with the total bandwidth is 170 MHz [4]. To increase the bandwidth, 2x2 MIMO antenna with two patches design of main and parasitic patch are used [3]. The antenna material used is FR-4 with 4.3 permittivity. The antenna system contains feeding line microstrip with 2x2 MIMO microstrip antenna with rectangular patch. The length of the main patch antenna dimension is around ½ λ on FR-4 substrate with λ can be found from equation (1) $\lambda = \frac{c}{f\sqrt{\varepsilon_r}}$ [5].

$$\lambda = \frac{c}{f\sqrt{\varepsilon_r}} \tag{1}$$

Where,
λ = Wavelength (m)
c = Speed of light (3×10^8 m/s)
f = Center frequency (Hz)
ε_r = Substrate permittivity
With using analytical line impedance in CST Studio Suite, the width of feeding line is 3 mm for 50 Ω of impedance.

2.1. Two-element arrays design

For the design of horizontal polarization antenna, two parasitic patches that has smaller dimension compared with main path and coupling placement placed close to the end of the main patch. For the vertical polarization antenna design, two parasitic patches with more or less the same dimension with main patch and coupling placed exactly in fuse$x = 0$.

For the two antenna designs, parasitic patch was placed exactly in the middle and the distance between parasitic and main patch was the same. This is done to avoid the clash of signal from one another. The dimension length in the main patch is ½ λ in FR-4 in 1.8GHz frequency with 40mm result, but with optimization of dimension length construction, 38mm result was acquired.

2.2. Microstrip feeding system

System send the signal from the transmitter to element arrays and sent it back from element arrays to the receiver. The microstrip feeding line strip is based on the quarter-wave or ½ λ in FR-4. This part make it possible to find a good match in 50 Ω transmitter/receiver with array elements within the vertical and horizontal polarization with the same microstrip feeding line concept [3].

The dimension of Microstrip feeding line can be seen in Table 1. The antenna structure that designed with two FR-4 substrate layers with the same formation structure. Meanwhile, the core of antenna structure is the ground that located within two substrates in which the ground is passed by feeder that connected with the ground that has wider diameter than the feeder. The patch position of antenna transmitter is located above the substrate with the height of 3.2 mm and the feeding line in the lowest layer.

Table 1. Microstrip feeding system.

Parameter	Value (mm)
Substrate Height	1.6
Transformer Width	5.5
Transformer Length	20
50 Ω Line Width	3

2.3. MIMO 2x2 system

Multiple input multiple output (MIMO) communication system is a reliable new technology that provides higher data transmitted via radio frequency. System arrays on the antenna can be used on MIMO technology with the transmitter and receiver of an antenna contained in one device [6]. The 2x2 MIMO antenna design on FR-4 substrate produce a good isolation on reflection coefficient. The final dimension of 2x2 MIMO antenna is shown by Table 2 and 3. The geometry of 2x2 MIMO antenna can be seen in Figures 1 to 4.

Table 2. The dimension of antenna with vertical polarization.

Parameter	Value (mm)	Parameter	Value (mm)	Parameter	Value (mm)
Lf	3	Ls	80	Ppb	37.15
Lg	80	Ls2	80	Ppk	35.5
Lgr	7	Lsl	5.5	Ps	210
Lpb	42	Lslt	81.5	Ps2	210
Lpb	40	Pf	18.3	Pslt	3
Lpt	3	Pg	210	Tg	0.315

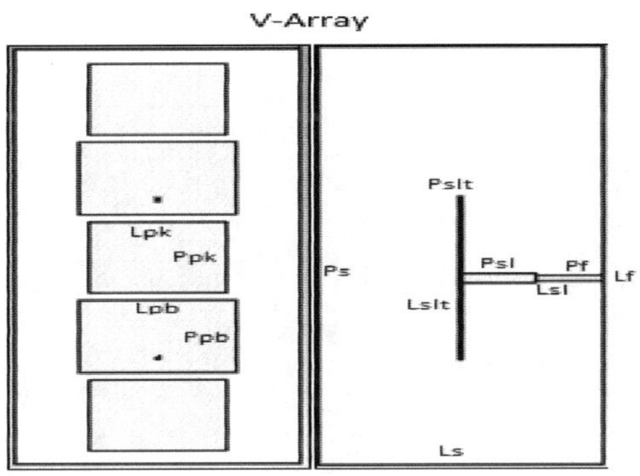

Figure 1. The geometry of antenna with vertical polarization (front and back view).

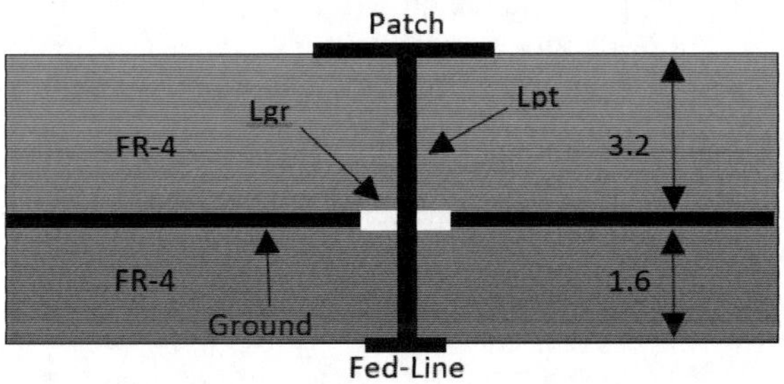

Figure 2. The geometry of antenna with vertical polarization (side view).

Since the stack between the substrate and the ground layer has the same length and width then Ps = Ps2 = Pg and Ls = Ls2 = Lg. For vertical polarization antenna, the value of the parameter affecting to the operating frequency is Ppb. Using equation (1), we obtain the length of Ppb that is 40 mm but with the optimization process Ppb length is obtained 37.15 mm.

Table 3. The dimension of antenna with horizontal polarization.

Parameter	Value (mm)	Parameter	Value (mm)	Parameter	Value (mm)
Lf	3	Ls	80	Ppb	40
Lg	80	Ls2	80	Ppk	23
Lgr	5	Lsl	5.5	Ps	210
Lpb	37	Lslt	84.5	Ps2	210
Lpb	36.3	Pf	8.5	Pslt	3
Lpt	3	Pg	210	Tg	0.315

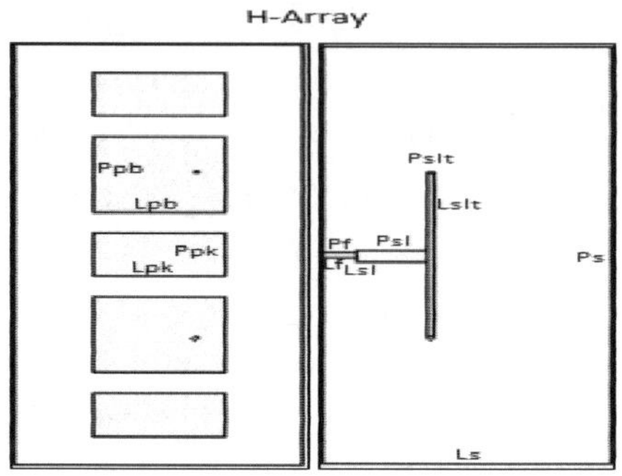

Figure 3. The geometry of antenna with vertical polarization (front and back view).

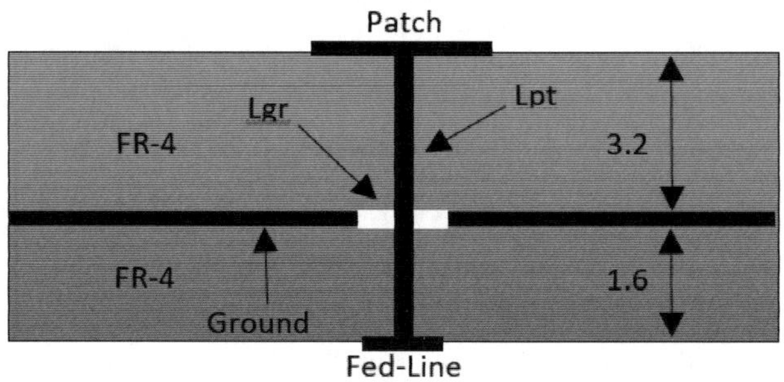

Figure 4. The geometry of antenna with vertical polarization (side view).

Since the stack between the substrate and the ground layer has the same length and width then Ps = Ps2 = Pg and Ls = Ls2 = Lg. For horizontal polarization antenna, the value of the parameter affecting to the operating frequency is Lpb. Using equation (1), we obtain the length of Lpb that is 40 mm but with the optimization process Lpb length is obtained 37 mm.

The antenna is designed using an electromagnetic coupling technique where there are two substrate layers where the substrate layer on top must be thicker than the substrate layer underneath and between the two substrate layers there is ground [7]. This technique allows more optimal antenna bandwidth.

3. Results and Discussion

The parameters that most affect the operating frequency of the antenna are Lpb and Ppb. These parameters lead to the achievement of wideband with the expected bandwidth that affects to S-parameters. Four parameters are interrelated because wideband is generated by 2 operating frequencies either the main patch or the parasitic patch. Figure 5 explain how the effect of the Ppb parameter on the S-parameters of antenna with vertical polarization. While Figure 6 explain how the effect of the Lpb parameter on the S-parameters of antenna with horizontal polarization.

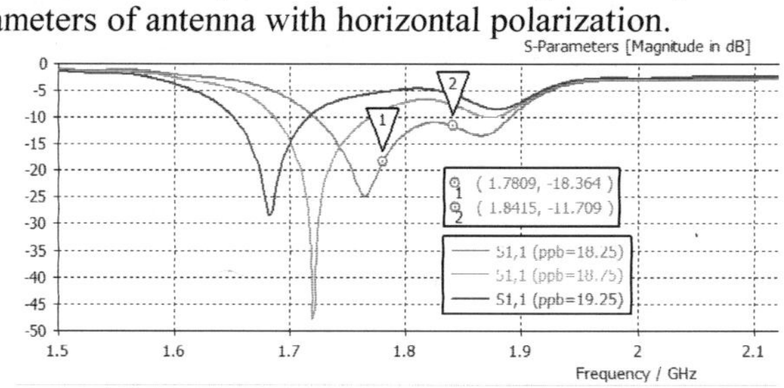

Figure 5. Effect of Ppb length to S-parameters of antenna with vertical polarization.

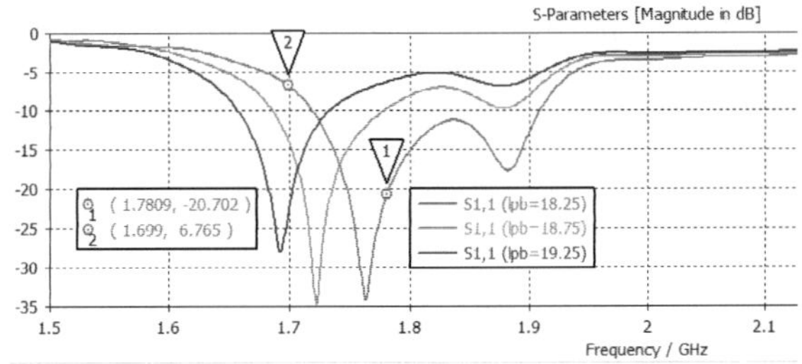

Figure 6. Effect of Lpb length to S-parameters of antenna with horizontal polarization.

From the experiment results Ppb has an effect on return loss of antenna with vertical polarization while Lpb has an effect on return loss of antenna with horizontal polarization. If the value of the two antenna parameters gets larger than the antenna is not suitable, then the optimization process needs to be done. The optimization result of both parameters is 18.25 mm for Ppb and Lpb dimension for good return loss result.

After we obtain the optimal design for both antenna with vertical and horizontal polarization, we design the 2x2 MIMO antenna. Figures 7 show the final design of 2x2 MIMO antenna. The distance between antennas is defined as ½ λ where the distance affects the return loss and insertion loss on 2x2 MIMO antenna.

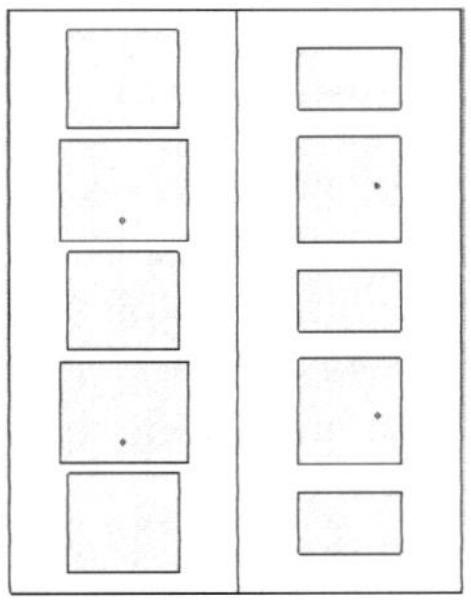

Figure 7. The geometry of 2x2 MIMO antenna.

The simulation goal of 2x2 MIMO antenna is to obtain the return loss as good as each antenna when they are not in MIMO configuration and insertion loss is less than -20 dB [8]. The simulation results show that the bandwidth of the antenna with the horizontal polarization is 1.72 – 1.92 GHz (200 MHz), meanwhile the antenna with vertical polarization is 1.73 – 1.90 GHz (170 MHz) for return loss less than -10 dB. The S-parameters result of 2x2 MIMO antenna can be seen in Figure 8.

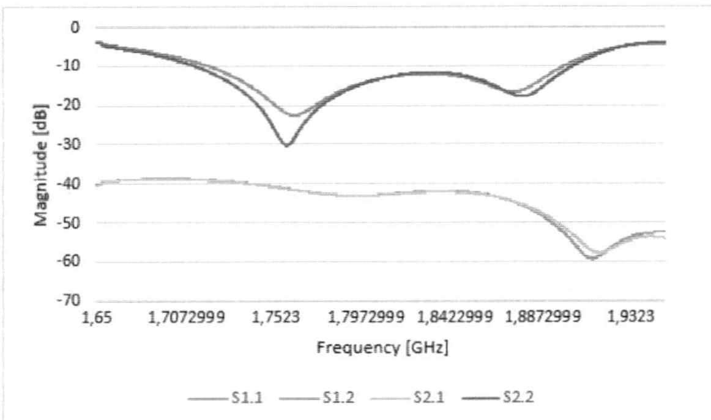

Figure 8. S-parameters result of 2x2 MIMO antenna.

IOP Conf. Series: Materials Science and Engineering **384** (2018) 012049 doi:10.1088/1757-899X/384/1/012049

Meanwhile, from the antenna radiation pattern, eventually the result has achieved the required directional radiation pattern with different of antenna gain on each frequency caused by the matching of main and parasitic patch. Tables 4 and 5 show the antenna gain and VSWR was caused by antenna matching.

Table 4. The simulation result of 2x2 MIMO antenna gain.

Frequency (GHz)	Horizontal polarization (dBi)	Vertical polarization (dBi)
1.73	5.718	5.107
1.77	6.418	5.87
1.8	6.373	6.088
1.84	6.206	6.242
1.88	5.062	3.953

Table 5. The simulation result of 2x2 MIMO antenna VSWR.

Frequency (GHz)	Horizontal polarization	Vertical polarization
1.73	1.65	1.8
1.77	1.08	1
1.8	1.3	1.4
1.84	1.7	1.7
1.88	1.4	1.3

The location of the via hole determines the polarization of the transmitted antenna. The location of the via-hole on the antenna with vertical polarization is at $x = 0$ and the location of the via-hole on the is antenna with horizontal polarization is at $x = -10$ with each distance between via-holes is λ. Figure 9 show if the vertical antenna polarization and Figure 10 show if the antenna polarization is horizontal. This is shown by the difference of vector lines formed on the simulation results that can be seen on the surface current.

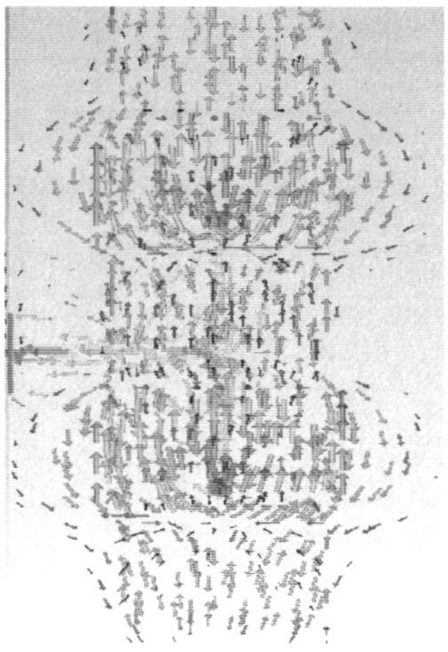

Figure 9. The vector orientation of vertical polarization antenna.

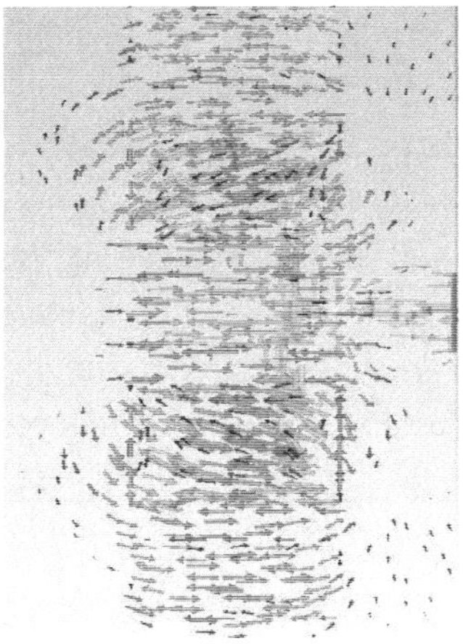

Figure 10. The vector orientation of horizontal polarization antenna.

The radiation pattern of both antennas is directional, but each antenna has different beam width. The beam width of vertical polarization antenna is 45.6 degrees while the horizontal polarization antenna the beam width is 48.2 degrees. The antenna radiation pattern in polar coordinate as shown in Figures 11 and 12.

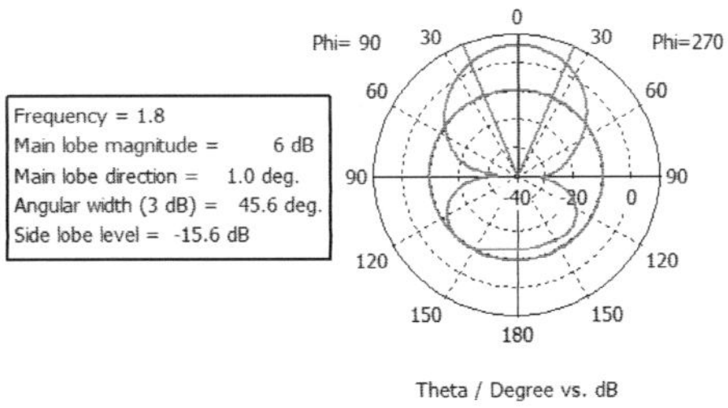

Figure 11. Radiation pattern of vertical polarization antenna.

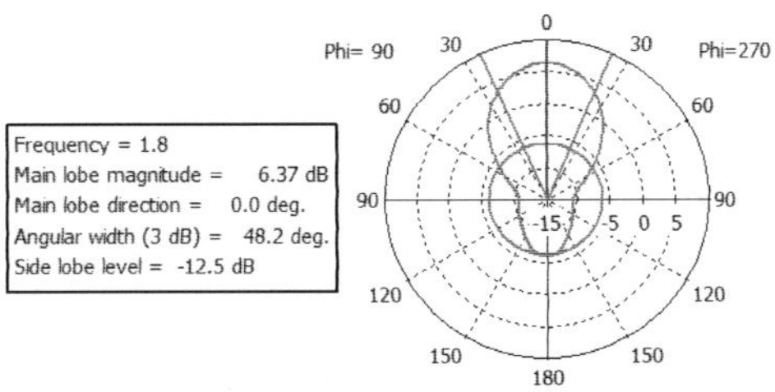

Figure 12. Radiation pattern of horizontal polarization antenna.

4. Conclusions

A 2x2 MIMO antenna with vertical and horizontal polarization has been designed and simulated. This antenna uses FR-4 as substrate with 4.4 permittivity. To enhance the bandwidth, we use coupling feed method that combined with parasitic patch in order to get higher gain. The result from the simulation showed that the antenna operating frequency is 1.73 – 1.90 GHz for the vertical polarization and 1.72 – 1.92GHz for horizontal polarization with gain more than 6 dBi. The simulation results also shown that the mutual coupling of MIMO antenna is very low less than -40 dB. This result can be achieved by using polarization diversity. Regarding this result, this 2x2 MIMO antenna can be implemented for LTE 1.8 GHz application.

IOP Conf. Series: Materials Science and Engineering **384** (2018) 012049 doi:10.1088/1757-899X/384/1/012049

References

[1] Kanwal K, Safdar G A, Ur-Rehman M and Yang X 2017. Energy management in LTE networks *IEEE Access* **5** 4264-4284

[2] Medeiros C R, Lima E B, Costa J R and Fernandes C A 2010 Wideband slot antenna for WLAN access points *IEEE Antennas and wireless propagation letters* **9** 79-82

[3] Soykin O, Kolobov A, Ssorin V, Artemenko A and Maslennikov R 2015 Planar MIMO antenna system with polarization diversity for 2.5–2.7 GHz LTE indoor FemtoCells *In Antennas and Propagation (EuCAP), 2015 9th European Conference* 1-5

[4] ETSI L 2012 Evolved universal terrestrial radio access (e-utra) *User Equipment (UE) radio access capabilities (3GPP TS 36.306 version 10.4. 0 Release 10)*

[5] Vörös Z and Weihs G 2015 Foucault's method for measuring the speed of light with modern apparatus *European Journal of Physics* **36** (3) 035013

[6] Tarokh V, Alamouti S M, Koivunen V and Kim K 2007. MIMO-OFDM and its application *Journal of Communications and Networks* **9** (2) 109-111

[7] Kumar A, Gupta N, Gautam P C 2016. Gain and Bandwidth Enhancement Techniques in Microstrip Patch Antennas-A Review *International Journal of Computer Applications* **148** (7)

[8] Nirmal P, Nandgaonka A B and Nalbalwa S L 2016 A MIMO antenna: Study on reducing mutual coupling and improving isolation. In Recent Trends in Electronics, Information & Communication Technology (RTEICT) *IEEE International Conference* 1736-1740

IOP Conf. Series: Materials Science and Engineering **384** (2018) 012050 doi:10.1088/1757-899X/384/1/012050

Conveyor Prototype of Steam Power Plant

S Sopyantara* and D L Hakim

Department eof Electrical engineering Education, Universitas pendidikan Indonesia, Jl. Dr. Setiabudhi No. 207 Bandung 40154, Indonesia

*sopyantara@student.upi.edu

Abstract. Conveyor is an important tool for moving coal in the steam power plant. But conveyor belts are often broken due to the roller friction is jammed. Based on the problem, this research will make an innovation on conveyor prototype using PLC control system and add a proximity sensor to detect congestion on the conveyor roller. The method used in this research is hardware and software design, tool making, and tool test. The results of this research are expected to be applied to the real conveyor system in the steam power plant so the conveyor operator can monitor the situation through the control room in the steam power plant. It can also notify their family if they are lost.

1. Introduction

The rapid growth of the electricity industry is in tandem with the increasing electricity demand of society, the capacity and large units of coal-fired steam power plants gradually holds the dominant position of power plants [1]. This cannot be separated from the conveyors that play an important role in modern industrial auto off-on, because it can save time effectively in the transport of objects [2], which in this case is coal.

Conveyor drives are driven by a control system. AutoOffc control system for machines in the industry initially only a series of relays. Such a control system becomes less effective because in addition to a large cost to make changes to the system also has a high level of work complexity. The solution to solve the problem is to use control system using PLC (Programmable Logic Controller) [3].

By adopting a less effective and efficient manual control system, this research has a goal to create conveyor prototype based on PLC. This prototype is a conveyor that can carry coal, besides this prototype also equipped with proximity sensor to detect congestion on roller conveyor.

2. Methods

2.1. Material

This prototype is designed using the tool materials as shown in Table 1.

Table 1. Tool Materials.

Tool Name	Specification	Amount
Push Button	2 pin	5
Relay	24 VDC	4
Cable wires	0,75 mm	10 M
Electric motor	12V	3
Power Supply	12V	1
Indicator Lamp	24 VDC	3
Proximity sensor	4 mm	1

2.2. Research methods

There are several stages described in the flowchart in Figure 1.

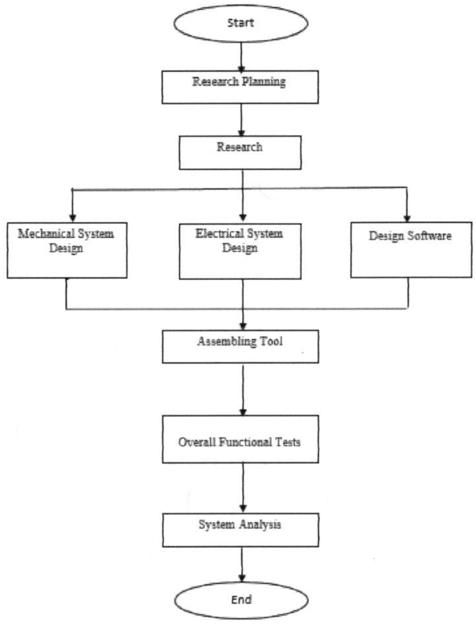

Figure 1. Flowchart Research Methods.

In Figure 1 it can be seen that this research has several stages, this research started with research design phase. At this stage is the stage of designing and looking for what will be the theme and title of the research to be conducted.

After that, the research begins with the design stage of mechanical design starting from the collection of materials to be used and then mechanically assembled according to the initial design. Then, this research continued with the design of electricity and design software that will be used which then assembled with the mechanical part. Further testing is done every function of the tool that has been assembled and analyzed the system used in this tool.

2.2.1. Design. Conveyor prototype based on PLC in design resembles the actual conveyor. There are three conveyors with the height of each of the different conveyors. The first conveyor serves as a start, where the conveyor is the one that distributes coal to the next conveyor. The second conveyor serves as a liaison between the first and third conveyors. This conveyor has two-way rotation, the first is to distribute coal to the next conveyor, while the second round is to distribute the second coal storage area.

IOP Conf. Series: Materials Science and Engineering **384** (2018) 012050 doi:10.1088/1757-899X/384/1/012050

In this conveyor installed sensors to detect congestion on the roller. The third conveyor serves to distribute coal to the first storage area.

2.2.2. Program. The software used for this prototype is Cx-Programmer which serves as a tool for making program design, testing and program improvement related to Omron PLC. Creating a ladder diagram on CX-Programmer is an important part of this research because it involves critical thinking [4], where critical thinking is needed to decide the step by step program that will be applied to this prototype.

2.2.3. Components. For the manufacture of prototype is required tools and materials as follows:

2.2.3.1. PLC (Programmable Logic Controller). PLC are electronic devices that work digitally using "Programmable Memory" for internal instruction storage to implement special functions such as logic, sequencing, time measurement, calculation and arithmetic, to control analog / digital input / output modules, various certain types of machines or processes [5]. The PLC used in the manufacture of this prototype is the Omron PLC with CP1L type as shown in Figure 2. This PLC requires an input voltage of 24 volts DC [6].

Figure 2. PLC (Programmable Logic Controller) Omron CP1L.

2.2.3.2. DC motor. Dc motor is a tool that converts direct electric energy (DC) into mechanic energy in the form of rotation or rotation on the rotor. Motor generated torque can be used to move a tool, its usage is for motor starters, conveyors, elevators, drill machines and others [7]. The DC motor is shown in Figure 3.

Figure 3. DC Motor.

2.2.3.3. Proximity sensor. A sensor is a device that has the function to convert power into another power. Proximity sensor is a sensor that has a working principle based on the object's distance to the sensor that can detect a target with a metal type without any physical contact. Proximity sensor used is inductive proximity which can detect metal, this proximity sensor use power supply equal to 12-24 VDC [8]. One of the proximity sensor is shown in Figure 4.

Figure 4. Proximity Sensor.

IOP Conf. Series: Materials Science and Engineering **384** (2018) 012050 doi:10.1088/1757-899X/384/1/012050

2.2.3.4. Conveyor. In this study, the conveyor is made using iron raw materials and in the form according to the design made. While the material used for conveyor belt is Teflon material. In this study conveyor is used as a tool to transport the sand as a substitute for coal in the process that actually occurs in the industry.

2.2.4. Program Design.

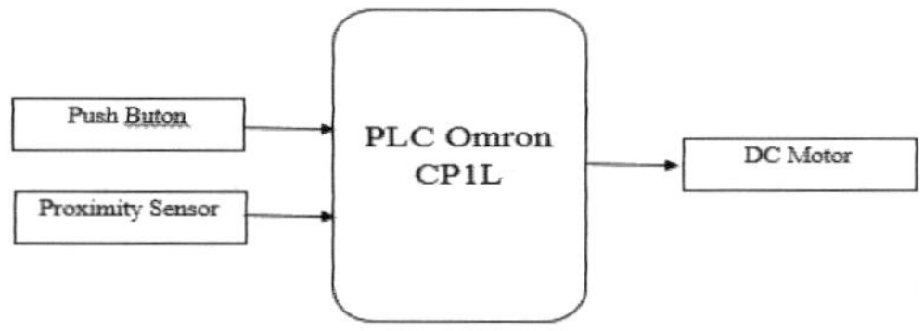

Figure 5. Block Diagram PLC based prototype conveyor in steam power plants.

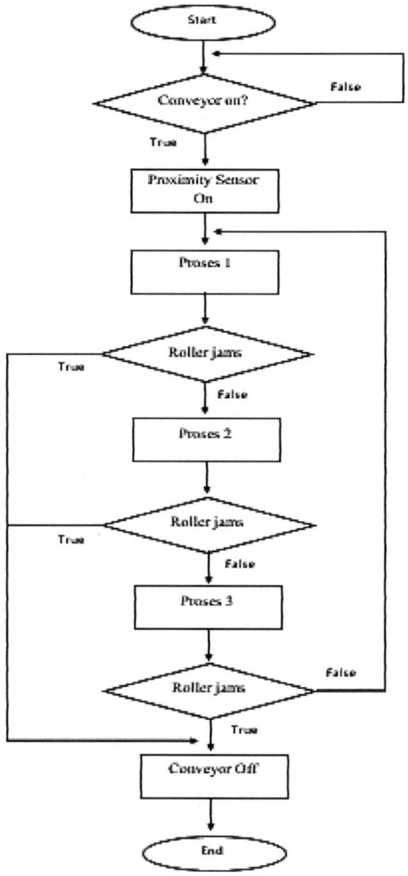

Figure 6. Flowchart of Programming Prototype PLC-based conveyor in steam power plant.

The block diagram and flowchart of programming are shown in Figure 5 and Figure 6, respectively. This conveyor prototype has three coal pipe processing processes. The first is the Direct process where the coal transport flow starts from the first conveyor and ends in the third conveyor to the next to the storage area, in this process the second conveyor revolves in the reverse direction. The second process is called the un-loading process where the coal transport flow begins from the first conveyor and ends

IOP Conf. Series: Materials Science and Engineering **384** (2018) 012050 doi:10.1088/1757-899X/384/1/012050

in the second conveyor with the direction of motor forward rotation which goes to the storage place. Third is the loading process where the conveyor transport flow starts from the second conveyor with the reverse motor rotation direction and ends in the conveyor for the next to the storage area.

At the time of the process of transporting coal using a conveyor, proximity sensor actively detects congestion on the conveyor roller. If there is congestion on the conveyor roller, then the sensor will instruct the PLC to turn off the entire process in order to avoid the breaking of the conveyor belt.

The displacement of each process has a time lag to stop and so is the proximity sensor where the proximity sensor time lag is not working.

3. Results and Discussion

After experiments on PLC-based conveyor prototype in steam power plant, then got some test result as follows:

3.1. Proximity sensor testing

This test is performed on the Proximity sensor to determine the distance the sensor reads to a metal object. Table 2 shows the data taken from the test of proximity sensor.

Table 2. Data from proximity sensor test.

No	testing	distance (mm)		Information
1	Metal testing	**Article I.**	**2 mm**	Detected
2	Metal testing	**Article II.**	**4 mm**	Detected
3	Metal testing	**Article III.**	**6 mm**	Not detected
4	Metal testing	**Article IV.**	**8 mm**	Not detected
5	Metal testing	**Article V.**	**10 mm**	Not detected
6	Metal testing covered paint	Yellow paint		Detected

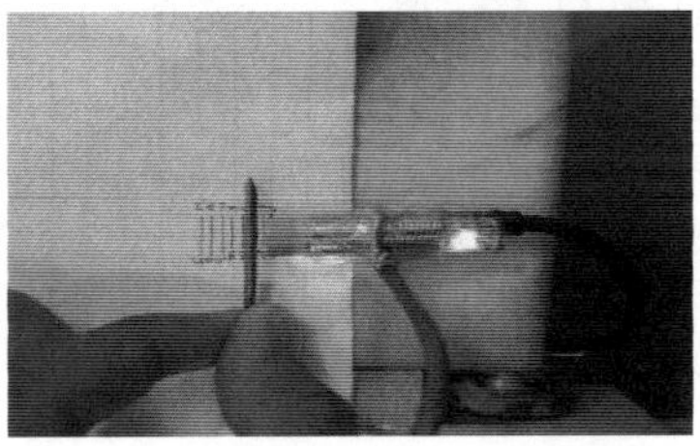

Figure 7. Proximity Sensor Testing.

It can be seen from the Table II that the sensors used in the manufacture of this prototype has a distance of 4 mm readings while the result indicates that the sensor is not reading, it is caused by the sensor readout distance exceeding the distance of 4 mm. As for the experiments conducted on metals coated or covered by paints, there is obtained readings over a distance of 0 to 4 mm.

3.2. Conveyor Testing

Conveyor testing is done by placing the sand above the conveyor, this test is done to know the conveyor can walk well and travel time to deliver the sand to the storage. The data of conveyor test results can be seen in the Table 3.

Table 3. Data from Conveyor Testing.

No	Process	Time
1	Process 1 (Direct)	28 seconds
2	Process 2 (Un-Loading)	22 seconds
3	Process 3 (Loading)	22 seconds

It can be seen from Table III that the conveyor runs according to the PLC program created. For process 1 (Direct) takes 28 seconds to deliver sand to storage. While for process 2 (Un-Loading) takes 22 seconds to deliver the sand to the storage. And for process 3 (Loading) takes the same time with process 2 that is 22 seconds to deliver sand to storage place.

3.3. Congestion testing on the conveyor roller

Roller congestion testing is performed in every process performed by the ongoing conveyor. The data of conveyor congestion test result can be seen in Table 4.

Table 4. Congestion test data on the conveyor roller.

No	Process	Time of congestion	Conveyor
1	Process 1 (Direct)	5 seconds	Off
2	Process 2 (Un-Loading)	5 seconds	Off
3	Process 3 (Loading)	5 seconds	Off

The congestion test is done by holding the 2nd conveyor roller, because the conveyor continues to function. Can be seen in the table above that the results of congestion testing on the roller conveyor on each process run as expected.

4. Conclusions

Based on the test results, it can be drawn some conclusions as follows:

- The proximity sensor used has a detectable distance between 0mm and 4mm.
- From the results of congestion testing on the conveyor roller, it can be concluded that the proximity sensor serves to detect the occurrence of congestion on the roller conveyor.
- From the results of conveyor testing, it can be concluded that the mechanics and system controls are applied to run well and in accordance with the purpose.
- This research has successfully achieved the goal, which is to detect congestion on the conveyor roller and also autoOffc control system using Omron PLC.

This prototype system has the potential to help industry which in this research is steam power plant in detecting congestion on roller conveyor of coal carrier and also autoOffc system which can be applied to actual conveyor system.

References

[1] Jijie Z, Yanbo C and Lihua Z 2011 Power plant coal conveyer coal link PLC control system *Power Electronics Systems and Applications (PESA), 2011 4th International Conference* 1-5
[2] Ning X and Zhao H 2014 A *new testing device for RFID performance factors of Conveyor belt system*
[3] Erickson K T 1996. Programmable logic controllers *IEEE potentials* **15** (1) 14-17
[4] Awalludin A F A 2015 Monitoring and controlling double carriage elevator system using CX-designer and CX-programmer. *Control and System Graduate Research Colloquium (ICSGRC), 2015 IEEE 6th* 151-157
[5] Yulistyawan D 2007 Mesin Vacuum Metalizer Untuk Process Coating *Gematek* **9** 1 99–118
[6] U S B Port, "CP1H / CP1L Wide Lineup of CPU Units with A Wide Variety of Built-in Functions Expand Application Capabilities and Shorten the Design Time Required for the Growing."
[7] Nurhadi M I, Siregar S and Hendrarini N 2015 Desain Mekanik Sistem Pemilah Sampah *eProceedings of Applied Science* **1** (3)
[8] Cylindrical Type DC 3wire 48002

IOP Conf. Series: Materials Science and Engineering **384** (2018) 012051 doi:10.1088/1757-899X/384/1/012051

Performance Simulation of Fiber to the Home (FTTH) Devices Based on Optisystem

W Awalia* and A B Pantjawati

Departement of Electrical Engineering Education, Universitas Pendidikan Indonesia, Jl. Dr. Setiabudhi No. 227 Bandung, Indonesia

*winiawalia@student.upi.edu

Abstract. High speed data rate is a very imperative requirement today and it is increasing according to the customer requirements for various applications such as live broadcasting, video on internet, video conferencing etc. This research is based on how we can design and analysis the optical fiber communication system. The network topology is simulated using Optisystem which is a simulation system used for designing, testing and performance of the optical network by compensate minimum bit error and improve quality factor of the network. This model is designed from the data that has been obtained from PT. Telkom. The value of Bit Error Rate (BER) achieved for the transmission of optical fiber (length=17 Km) for downstream transmission is $2.99513e^{-012}$ and the upstream is 0. So it can be concluded that both values meet the minimum value of BER specified for optical is 10^{-9}. For the optical length (L) = 17 km the maximum Q-factor for downsteam obtained is 6.87953. The Q factor meets the standards because the upstream and downstream values are above 6. The sensitivity of the device is -28 dBm, the calculation using Optisystem of receive shows the number -26.304 dBm so it can be said that this implementation test is feasible.

1. Introduction

After the first low-loss fibers were produced in 1970, fiber optic communications developed very quickly. The fiber optic communication systems has been common, new installations and applications appear continually [1]. With the increasing development of optical fiber communication technology, the needs of higher transmission rates, longer transmission distance, and greater transmission bandwidth are much stronger [2]. PT. Telkom tried to fulfill those needs by applying Fiber Access Network System.

Fiber to the home is one of the fiber optic network configurations that are distributed to the customer's home or to the terminal's location. FTTH supports Triple Play services that provide fixed telephone, data and interactive TV [3]. In FTTH implementation, Provider uses GPON technology [2]

Different from the ITU-T G. 984.1 about physical distance an access network i.e. 20 km. On the standard PT. Telkom, the Access is allowed no more than 17 km. However FTTH on STO Bandung Centrum, just 4 km with coverage of the newly implemented.

In our paper, main focus is to testing performance a GPON network in software "Optisystem" of the optical network by the minimum amount of BER, Q-factor, and power budget to achieve the requirement of the customers of the network [3]. This paper is organised as follows: Section II explain and discuss about the simulation setup of projected network, Section III shows simulation setup results and performance diagram and Section IV shows conclusion.

1.1. Optical Fiber Communication Systems
The application of optical fiber communications is in general possible in any area that requires transfer of information from one place to another. However, fiber-optic communication systems have been developed mostly for telecommunications applications [4]. Arranged as in figure 1, a basic communications system consist of a transmitter, a fiber optic as communications channel or transmission lines and a receiver.

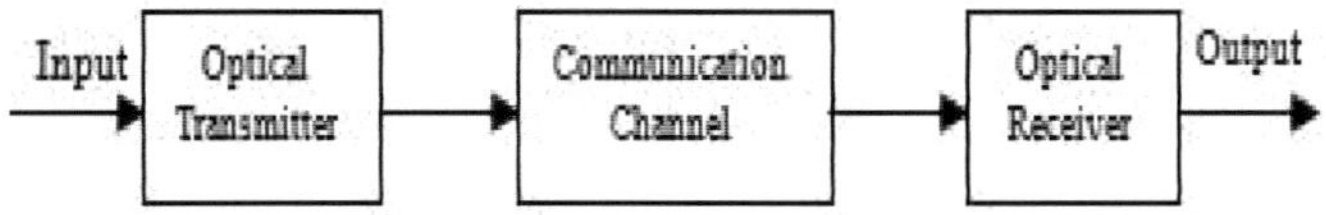

Figure 1. A generic block diagram of an optical communication system [4].

1.2. Fiber to the Home
FTTH is an optical signal transmission format from provider center to user area using optical fiber as channels. The development of this technology is inseparable from the advancement of fiber optic technology that uses copper wires, as well as services known as Triple Play Service [5]. In an OLT it consists of several ODNs that work for transport and distribution of data from OLT to ONU [6]. Another supporting component is Passive Splitter which serves to distribute optical power to all branches. In figure 2 the use of one optical fiber can move from OLT to a customer of about 32 homes.

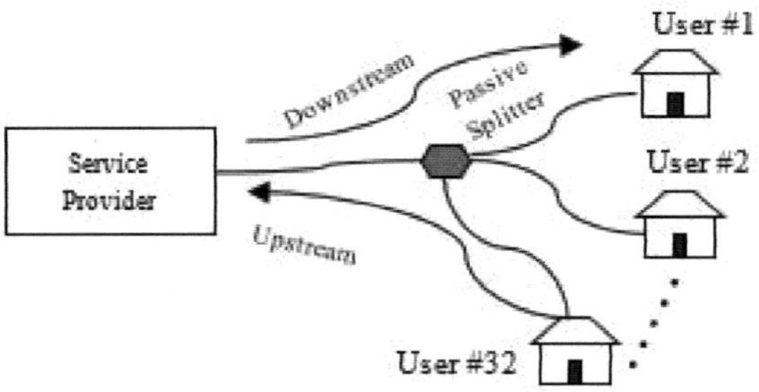

Figure 2. The network connecting the Provider uses a passive splitter [7].

1.3. Gigabit Passive Optical Network
As said above, G-PON systems are characterized, in general, by an optical line termination (OLT) system and an optical network unit (ONU) or optical network termination (ONT) with a passive optical distribution network (ODN) interconnecting them [8]. GPON has a specificity with passive splitter to divide the light signal, so that from one core can be divided into several cores (2, 4, 16, 32, 64) [9]. The principle of Passive Optical Network is a point-to-multipoint system. GPON uses TDMA as a multiple access technique with the nominal bit rate of OLT signals to ONU (downstream) specified in ITU-T G.984.2 Recommendation is 2488.32 Mbit/s. The nominal bit rate of ONU signal to OLT (upstream) is 1244, 16 Mbit/s.

1.4. Optsystem
OptiSystem is a software for simulating fiber optic communications systems and fiber optic sensing systems that are widely used. There are various types of function. Eg: design, test and optimize [1]. Optisystem software is based on modeling a real optical communication system. Optisystem is equipped with Graphical User Interface (GUI) which consists of project layout, component list, component model and graphical display. Library Optisystem consists of active and passive components with wavelength parameters [10].

1.5. Power Link Budget

The power link budget is calculated as a requirement for the link of network, we design the power beyond the threshold of the required power [9]. The power link budget is implemented based on ITU-T G.984 standardization and applied by Industry regulation. The distance is not more than 20 km and the receiving power is not more than -28 dBm. The form of the equation for total damping calculations on the power budget link is:

$$a_{t} = L.a_S + N_c.a_c + N_s.a_s + S \quad (1)$$

$$P_R = P_T - (a_{t} + S) \quad (2)$$

$$M = (P_T - P_{S}) - (a_{t} + S) \quad (3)$$

Where a_{t} the total attenuation is occurs along the fiber-optic cables. Equation 2 [11] to find the incoming power obtained from transmit power or outlet power reduced by total attenuation and safety margin is 6 dB. Equation 3 [11] is the equation to find the power margin values obtained with the emissivity value sensitivity receiver devices then reduced damping and a safety margin. The values of the parameters in the equations 1, 2 and 3 on the basis of the specification of each device's network builders FTTH.

1.6. Bit Error Rate (BER)

Bit error rate occurring in transmitting a digital signal is BER. Sensitivity is the minimum optical power of the received signal at the required BER. The need for different BERs for each application, for example in a communication app requires BER 10^{-10} or better, on some data communications requiring BER equal or better than 10^{-12}. BER for an optical communication system of 10^{-9}. Factors affecting BER include noise, interference, distortion, bit synchronization, attenuation, multipath fading, etc. [3]

2. Simulation Setup

2.1. Network Testing Standards

The analysis performed is testing FTTH network performance using Optisystem. Where the configuration used in the simulator is a real configuration on STO Bandung Centrum. FTTH implementation with a maximum of 13.248 cores is only installed at a distance of 4 km, so the provider needs to perform an analysis of network conditions and field needs. After the simulator generates some parameters, then we calculate the value of Link Power Budget and BER value. In the analysis of power budget link value there are 3 (three) parameters analyzed are total attenuation, receptivity and power margin. The standard used in both simulation and calculation of power budget link is standard from PT. Telkom. Table 1. is a device specification in FTTH implementation at STO Bandung Centrum.

Table 1. Devices Characteristic.

No	Devices	Specification
1.	Fiber Optic cables	Max 0.35 dB/km
2.	Connector loss	0.25 dB (Refer IEC 61300 3-34 Grade B attenuation)
3.	Splice loss	0.1 dB
4.	Splitter 1:4	Max. 7.25 dB
5.	Splitter 1:8	Max. 10.38 dB
6.	Distance	Max. 17 km
7.	Rx Sensitivity	-28 dBm

2.2. Design Simulation

The following table 2. is a List of Materials (LoM) required in the design of FTTH networks divided by segment. Where in the table represent a configuration FTTH especially downstream configuration.

Table 2. List of Material.

Segment	Material Type	Material
Feeder Segment	Temination Material	Optical Distribution Frame
		ODC
	Cable	FO Duct G.652.D
		FO Aerial G.652.D
	Other	Splitter 1:4
		Pathcored
		Connector
Distribusi Segment	Temination Material	ODP
	Cable	FO Duct G.652.D
		FO Aerial G.652.D
	Other	Splitter 1:8
		Connector
Drop Segment	Temination Material	OTP
	Cable	FO Duct G.657.A
		FO Aerial G.657.A
	Other	Connector
		Klam
Indoor Segment	Temination Material	ONT/ONU
	cable	Drop Indoor G.657.A
	other	Connector
		Patchcord
		UTP, PVC, Coaxial cables

The maximum length of feeder cable for RING configuration is 20 km. Max 2 stage splitter with 1 core feeder. Maximum to 32 Home Pass / More in accordance with the link budget obtained. The type of connector used per element is SC-UPC. The type of pole used for aerial systems can use concrete or steel poles along with their respective accessories. In the FTTH configuration, there are several main devices. The first is the Optical Line Termination (OLT) that serves as the end-point of the passive optical network service. OLT provides an interface between Passive Optical Network (PON) systems with service providers (data service, video, or voice / phone). The second device is ODC (Optical Distribution Cabinet). ODC is a passive device that is installed outside STO (outdoor) and can also be indoors or in MDF (indoor). ODC serves as a splitter or distribution point of the feeder cable into a distribution cable.

The third device is the ODP (Optical Distribution Point) which is a device that divides the distribution cable into a drop cable using a 1: 8 splitter. And the fourth device is the ONU (Optical Network Unit) which is the device installed at the customer side.

The distance used in this research reaches the maximum distance that the PT. Telkom Access standard is 17 Km. shown in figure 3. Feeder cable from STO minimum 288 core capacity for both duct and aerial system with cable type G.652.D. While the type is loose tube (max 2x288 core) or ribbon cable (> 288 cores). The configuration used is downstream and upstream with one way simulation. For downstream, passive splitter used is a ratio of 1: 4 on ODC so that from 1 pieces OLT cores can distribute light signals up to 32 ONT / ONU at once. The light signal from 1 ODP is transmitted to 8 ONT by using a passive splitter of 1: 8. The configuration of upstream is shown in Figure 4.

IOP Conf. Series: Materials Science and Engineering **384** (2018) 012051 doi:10.1088/1757-899X/384/1/012051

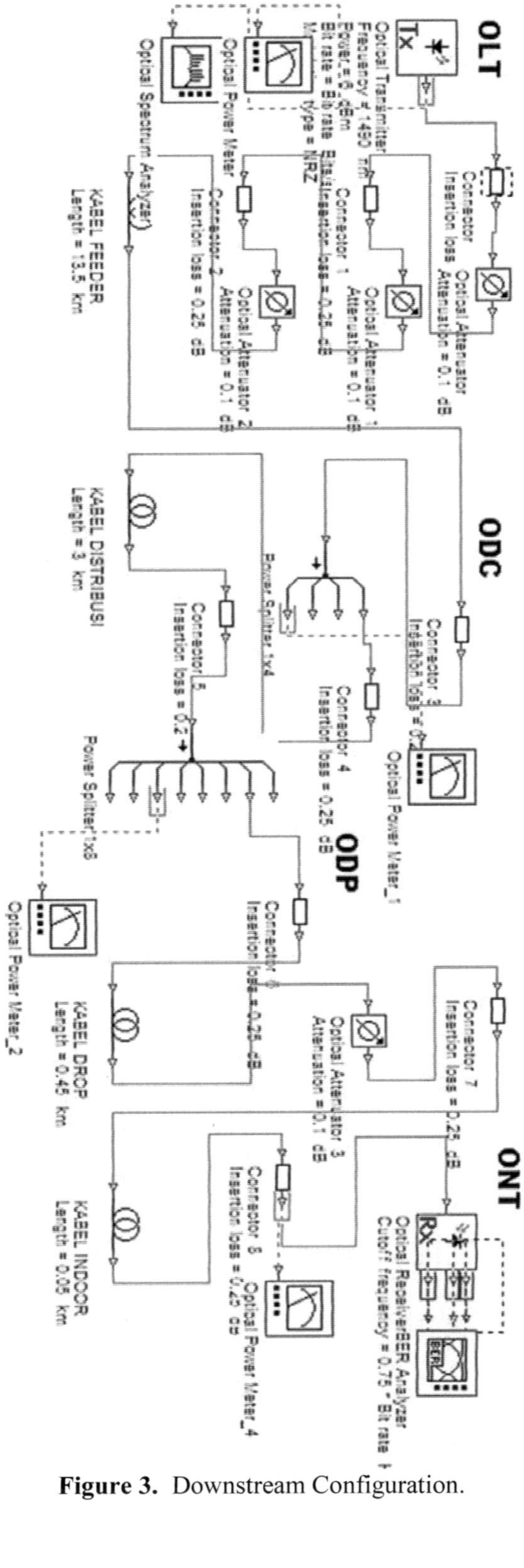

Figure 3. Downstream Configuration.

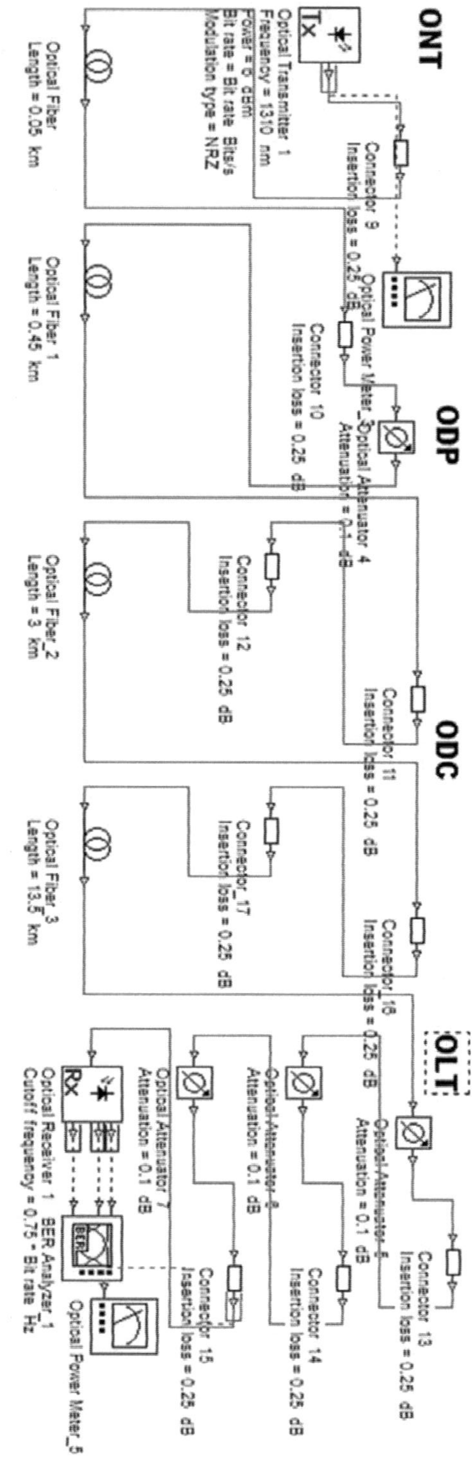

Figure 4. Upstream Configuration.

3. Performance Analysis

3.1. Power Budget Link

The analysis performed is testing FTTH network performance using Optisystem. Where the configuration used in the simulator is a real configuration on STO Bandung Centrum. FTTH implementation with a maximum of 13.248 cores is only installed at a distance of 4 km, so the provider needs to perform an analysis of network conditions and field needs. After the simulator generates some parameters, then we calculate the value of Link Power Budget and BER value. In the analysis of power budget link value there are 3 (three) parameters analyzed are total attenuation, receptivity and power margin. The standard used in both simulation and calculation of power budget link is standard from PT. Telkom. Table 1. is a device specification in FTTH implementation at STO Bandung Centrum.

Link budget calculation uses data on simulation. The data used is the measurement of power by using Optical Power Meter is original to end. Power budget link analysis is used to determine the performance level of new cable network installation before it is operated into the device. This analysis aims to adjust whether the system is running well or not. Tx and Rx power retrieval is done from Optisystem. At power margin the value of Power Margin must be greater than or equal to 0 dB (Power Margin> 0 dB or Power Margin = 0 dB). The receive power value is greater than or equal to the receiver's sensitivity of -28 dBm (PRx> Psensitivity or PRx = Psensitivity). The total attenuation value must be less than or equal to 25 dB. Table 3. is the result of simulation and calculation of power budget link that has been done.

Table 3. Result of Power Budget Link.

Distance (Km)	T_X (dBm)	R_X (dBm)	α_{t_t} (dB)	P_{RX} (dBm)	M
5	3.347	-16.103	19.45	-22.103	5.897
6	3.347	-16.454	19.801	-22.454	5.546
7	3.347	-16.804	20.151	-22.804	5.196
8	3.347	-17.154	20.501	-23.154	4.846
9	3.347	-17.504	20.851	-23.504	4.496
10	3.347	-17.854	21.201	-23.854	4.146
11	3.347	-18.204	21.551	-24.204	3.796
12	3.347	-18.554	21.901	-24.554	3.446
13	3.347	-18.904	22.251	-24.904	3.096
14	3.347	-19.254	22.601	-25.254	2.746
15	3.347	-19.604	22.951	-25.604	2.396
16	3.347	-19.954	23.301	-25.954	2.046
17	3.347	-20.304	23.651	-26.304	1.696

Simulation with range of 5 - 17 km. The distance is sampled because the FTTH network that has been installed on STO Bandung Centrum is 4 km. The total damping generated at a distance of 5 km is 19.45 dB. Damping increases along the configuration when distance is added per 1 km. Until it reaches the maximum FTTH standard of 17 km with a total attenuation of 23,651 dB. In the standard used PT. Telkom, the total damping value should not be more than 25 dB. So with the simulation of FTTH, the network can be said either from the side of total attenuation. Power on the receiver from the simulation results between -22.103 dB to -26.304 dB at the end point of 17 km. The value of the calculated power is smaller, so if the system is used for transmission under normal circumstances. The value is within the allowable value limit on the FTTH network, i.e. Rx sensitivity -28 dBm.

IOP Conf. Series: Materials Science and Engineering **384** (2018) 012051 doi:10.1088/1757-899X/384/1/012051

3.2. Bit Error Rate

Measurement BER on simulator is one method how we can see fiber optic network performance besides with power budget link. In the simulator, the measurement uses BER Analyzer by displaying the eye diagram. Analysis method using eye diagram is a simple method but able to accurately assess from a digital transmission system. The measurement with the eye diagram is made in the time domain and shows the distortion effect of a wave which in its application can be seen by the oscilloscope.

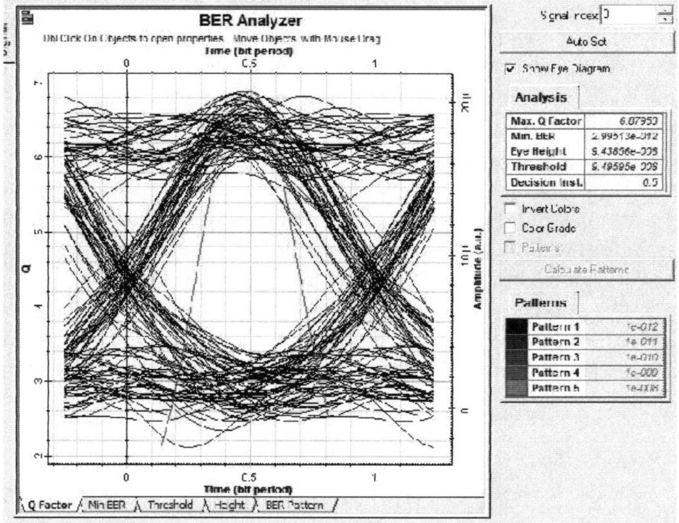

Figure 5. Results BER Analyzer from Downlink configuration.

The eye diagram in figure 5. Shows the result BER Analyzer downlink configuration with a maximum distance of 17 Km. Obtained BER with a value of $2.99513e^{-012}$ and Q-Factor of 6.87953. The result states that within a trillion bits sent, there is 1 wrong bit. The signal waves shown in Figure 6. there is interference but still visible an eye shape indicating the Signal to Noise Ratio (SNR) and the receiver can still read the signal in this configuration. This figure is much lower than the minimum recommended BER for fiber optic communication systems i.e. 10^{-9}. The simulation results prove that with the fiber cable length L = 17 Km (maximum length), the network is still said to be in decent condition.

Good results are also shown in figure 6. The upstream portion of optical fiber communication with the same transmission distance as downstream, but the eye pattern is more clear and open. The resulting BER is zero. This indicates that there is no error in digital transmission.

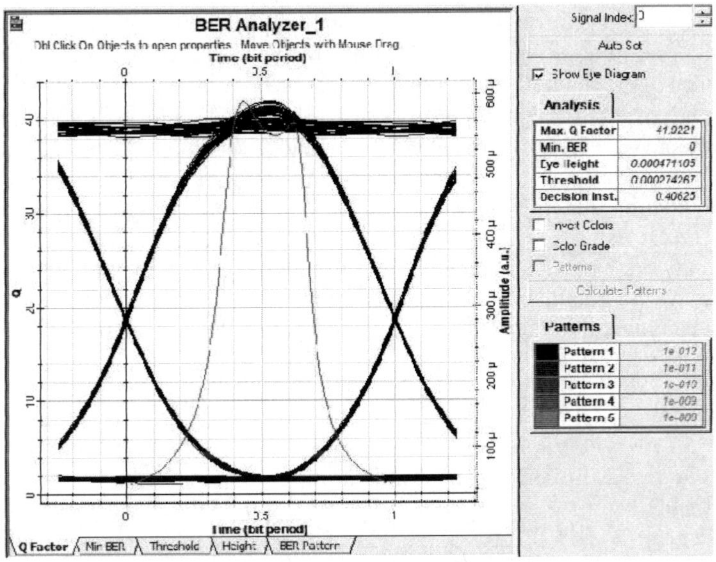

Figure 6. Results BER Analyzer from uplink configuration.

To see the effect of distance on an FTTH network on BER produced, then the simulation by changing the distance with range 5 Km - 16 Km. Seen in Table 4. Comparison of BER to FTTH distance. Changes made is to change the distance of the feeder cable. This is because the maximum distance feeder cable can reach 20 Km.

Table 4. Comparative results.

No	Distance	BER
1.	5 Km	$3.40433e^{-064}$
2.	6 Km	$2.616884e^{-059}$
3.	7 Km	$1.87306e^{-045}$
4.	8 Km	$9.92868e^{-039}$
5.	9 Km	$1.6335e^{-036}$
6.	10 Km	$8.43325e^{-023}$
7.	11 Km	$3.73269e^{-025}$
8.	12 Km	$1.53416e^{-019}$
9.	13 Km	$2.86891e^{-017}$
10.	14 Km	$7.50784e^{-015}$
11.	15 Km	$1.02354e^{-015}$
12.	16 Km	$1.30557e^{-012}$

Changes from BER in configuration can be seen clearly. It states that the length of fiber optic cable used has an influence on performance. The farther the ONU location, the greater the value of BER obtained and the smaller the received power.

4. Conclusions

Based on FTTH device simulation using Optisystem for system performance parameters of BER, Q-factor, and power link budget resulting from OptiSystem simulation. Obtained overall parameters at simulation with length L = 17 km, 1490 nm wavelength for downstream and 1310 nm for upstream. Based on the analysis of FTTH network testing using the PT.Telkom Standard that has been done can be drawn some conclusions as follows:

- Based on the simulation on OptiSystem obtained the value of power budget link on the downstream configuration of 2.99513e^{-012} and for upstream of 0. So it can be concluded that both values can meet the minimum value of BER is 10^{-9}.
- Based on the simulation on OptiSystem got the value of Q-Factor in the downstream configuration of 6.87953 and for upstream of 41.9221. So it can be concluded that both values can meet the minimum value of Q-Factor is 6.
- Based on simulation on OptiSystem we get 23,651 dB total damping. In the standard used PT. Telkom, the total damping value should not be more than 25 dB. So with the simulation of FTTH, the network can be said either from the side of total attenuation. Power on the receiver from the simulation result between -22.103 dBm to -26.304 dBm. So it can be concluded that both values can meet the minimum value of acceptance that has been set by PT.Telkom is -28 dBm.

References

[1] Palais J C 1988 *Fiber optic communications* (Englewood Cliffs: Prentice Hall)
[2] Wang Y and Guan Y 2014 Performance simulations for a high-speed optical transmission system based on OptiSystem *Image and Signal Processing (CISP), 2014 7th International Congress* on 2014 Oct 14 907-911 IEEE
[3] Verma S, Kakati A and Bhulania P 2016 Performance analysis of Q-factor and polarization for GPON network using optisystem *InInformation Technology (InCITe)-The Next Generation IT Summit on the Theme-Internet of Things: Connect your Worlds International Conference* on 2016 Oct 6 138-141 IEEE
[4] Mesh M, Porat Y, Kollmann M and Nachman Y 2002 inventors; Packetlight Networks Ltd, assignee *Fiber optic communication system. United States patent application US* 09/753513 Jul 4
[5] Agrawal G P 2012 Fiber-optic communication systems *John Wiley & Sons* Feb 23
[6] Lokhande M and Singh A., Design and Implementation of FTTH
[7] Razavi B., 2012 Design of integrated circuits for optical communications *John Wiley & Sons* Sep 14
[8] Effenberger F, Cleary D, Haran O, Kramer G, Li R D, Oron M and Pfeiffer T, 2007 An introduction to PON technologies [Topics in Optical Communications] *IEEE Communications Magazine* Mar 45(3):S17-25
[9] Su K, Li J and Fu H 2011 Smart city and the applications *In Electronics, Communications and Control (ICECC), 2011 International Conference on 2011* Sep 9 1028-1031 IEEE
[10] Kramer G, Mukherjee B, Pesavento G 2001 Ethernet PON (ePON): Design and analysis of an optical access network *Photonic Network Communications* Jul 1;3 (3):307-19
[11] Hecht J and Long L 1993 Understanding fiber optics *Prentice Hall* Jan

IOP Conf. Series: Materials Science and Engineering **384** (2018) 012052 doi:10.1088/1757-899X/384/1/012052

Prototype of Water Turbidity Monitoring System

Y Mulyana and D L Hakim*

Departement of Electrical Engineering Education, Universitas Pendidikan Indonesia, Jl. Dr. Setiabudhi No. 207 Bandung 40154, Indonesia

*dadanglh@upi.edu

Abstract. Along with the increasing use of water by the community, the quality of water needs to be considered, especially for human consumption. Water quality is determined based on the value of turbidity of water. This project develops an Arduino Uno-based water turbidity monitoring system, which is aimed to measure water level and water turbidity levels. The turbidity sensor consists of a photodiode and infrared LED. Measurements of water turbidity using principles of light scattering in the water. The turbidity sensor to detect the intensity of light passing through the scattering of particles and the results of the analog signal voltage. The signal entering Arduino Uno of the ultrasonic sensor and the turbidity sensor will be processed. The measurement results is converted to NTU (Nephelometric Turbidity Units) and liter, then are shown on the LCD display 2x16. The turbidity measurement limits on this system are designed between 0 NTU to 40 NTU. This system can potentially help PDAM in monitoring water tanks and it is hoped that water quality monitoring system can be applied online.

1. Introduction

Turbidity is one of the parameters in determining water quality. Turbidity is caused by organic materials and inorganic particles that are suspended in the water [1]. High water turbidity levels can cause health problems if the water is continually consumed. Water turbidity standards are set between 5-25 NTU (Nephelometric Turbidity Unit) [2].

Water-treatment plant station as a provider of clean water in Indonesia needs to pay attention to water quality that will be distributed to society [3]. Monitoring of water turbidity is required in the water treatment process to ensure water quality, so that water is suitable for use in subsequent processes [4].

The research aim to design a prototype system of monitoring the level of water turbidity. Water turbidity is measured based on the principles of light scattering [5]. Water that has a high turbidity levels will be added chemicals to assist in the deposition of particles in water [6].

Turbidity sensors are designed by using components a photodiode as detector and infrared led as light sources. This system is controlled using Arduino Uno microcontroller. The measurement results of the turbidity sensor in the form of ADC value is converted to units of NTU and is shown on the LCD display 2x16.

2. Methods

The prototype design of this turbidity monitoring system begins with the system general design. The system design consists of interconnected devices that is hardware design and software design.

2.1. General System Design
General design of this system include turbidity sensor and ultrasonic sensor as a signal provider, microcontroller Arduino UNO as the signal processing of the measurement and LCD 2x16 for display output.

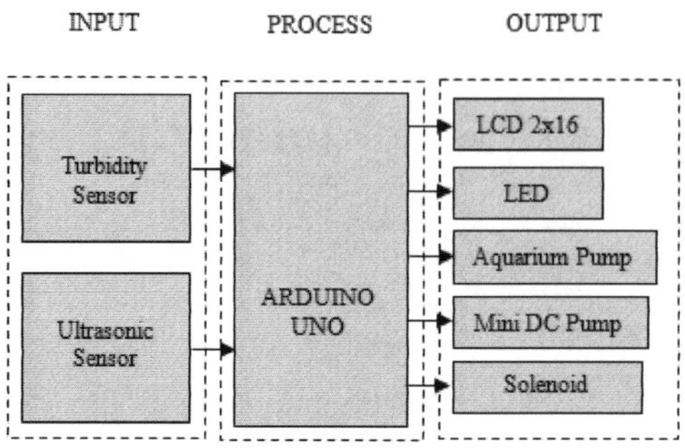

Figure 1. System Block Diagram.

This system is used to monitor water volume and water turbidity in tank. The ultrasonic sensor is used to measure the volume of water and turbidity sensor is used to measure of water turbidity. If the water was turbid be added chemicals such as calcium hypochlorite (Ca $(ClO)_2$) as the lowering of the water turbidity adjusted with the volume of water in tank. This condition is done to know and produce a decent water quality. The measurement results of ultrasonic sensors and the turbidity sensor is processed by a microcontroller Arduino UNO, then shown on the LCD display 2x16.

2.2. Hardware Design
The hardware design consists of pump, ultrasonic sensor, turbidity sensor, mini dc pump, selenoid valve, lcd, and led. Design of this hardware is simplified into the design of ultrasonic sensors, design of turbidity sensor and 2x16 LCD design.

2.2.1. Ultrasonic Sensor Design. SRF05 ultrasonic sensor is a sensor measuring the distance an object has a range that can be detected between 3cm to 300cm. This sensor has 5 pins Vcc pin, Trigger, Echo, Out, and Gnd, and 2 transducers that function as transmitter and receiver [7]. The interface circuit to connect the ultrasonic sensor to the arduino UNO can be seen in Figure 2.

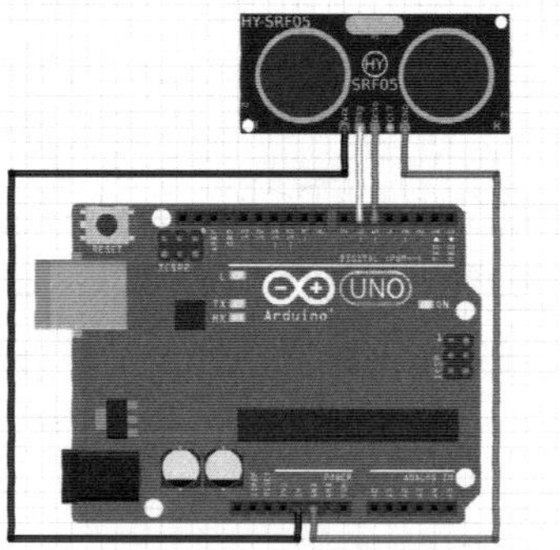

Figure 2. Ultrasonic Sensor Interface Series Srf05 with Arduino Uno.

How ultrasonic sensor work on this tool to measure the volume of water in the tank and a pump switch. Distance measurement is converted to volume, so that the output of the unit water volume (liters).

2.2.2. Design of Turbidity Sensor. Turbidity sensor circuit is made of a photodiode and infrared led. A photodiode is used as detector [8] and infrared LED as a light source. Infrared led has a wavelength of 960 nm [9]. Photodiode and an infrared LED arranged in parallel, so that forming a 90 ° angle position between the two components.

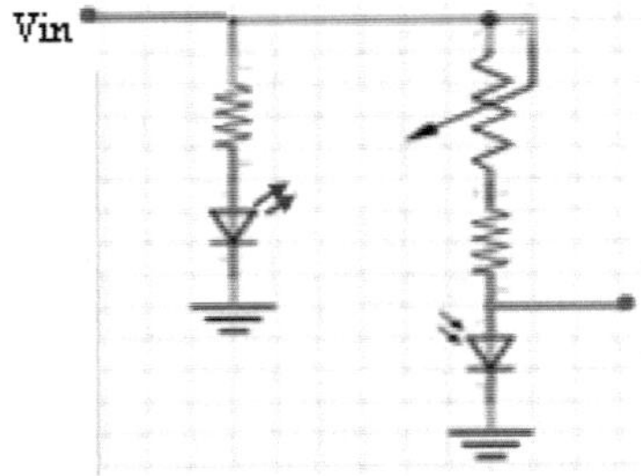

Figure 3. Turbidity Sensor Circuit Scheme.

This sensor utilizes the principle of light scattering by passing water between the detector and the light source. Infrared led will emit light that is partially transmitted and partially dissipated by particles in the water. Photodiode will detect the intensity of light scattered by the particles and converted into a voltage signal [10].

Turbidity sensors are used to measure the turbidity of the water in the tank. The level of turbidity> 25 NTU then mini dc pump will flash to add chemicals and levels of ≤ 25 NTU turbidity then open the solenoid valve to drain the water.

2.2.3. LCD 2x16 Design. LCD 2x16 connection with Arduino UNO using the I2C module to reduce the use of LCD pin on arduino UNO. I2C has a contrast adjust that can be used to adjust the brightness and contrast of characters displayed on the LCD. I2C consists of 2 channels that is SDA (Serial Data) and SCL (Serial Clock) [11]. SDA and SCL pins on the I2C associated with analog pins 4 and 5 on the Arduino UNO.

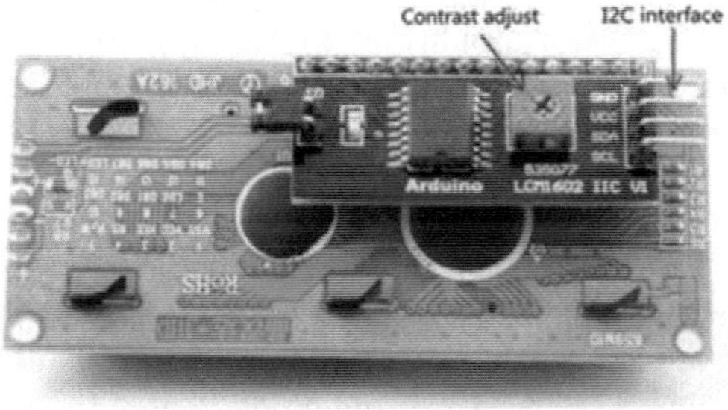

Figure 4. 2x16 LCD Connected With I2C Module.

2.3. Software Design

Design and manufacture of software is used to process the signal changes of the measurement of ultrasonic sensors and the turbidity sensor. Analog signal processing of the sensor is converted into digital data in the form of ADC value. The ADC value is converted to the value of voltage to be processed on the Arduino UNO microcontroller [12]. The final result of the Arduino UNO processing a value of volume and a value of turbidity that shown on the LCD display.

Arduino IDE application is used as a software in the design and manufacture of programs. A software that is used to program the Arduino is application Arduino IDE. Application Arduino IDE is used programming language C ++ [13].

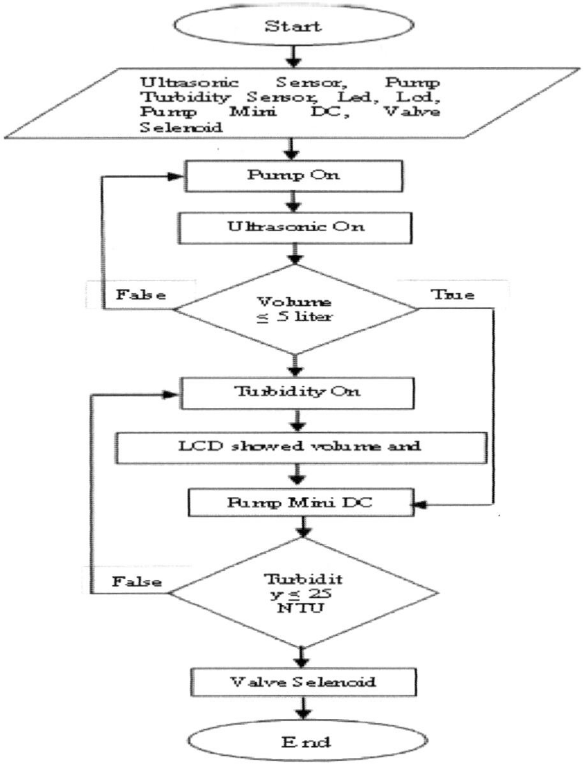

Figure 5. Flowchart Software Design.

3. Results

After designing and manufacturing water turbidity monitoring system prototype is completed, then taken the data with tool testing methods. Tests conducted to determine the ability of tools that have been made to work and function in accordance with the designed. Testing is divided into 2, namely ultrasonic sensor testing and turbidity sensor testing.

3.1. Ultrasonic Sensor Testing

The test is performed on the ultrasonic sensor to determine the height of the sensor against the barrier and to know the response to changes in water volume. The data taken in the form of the height value from the sensor measurement that is converted to water volume unit (liters).

Table 1. Ultrasonic Sensor Test Result Data.

Water Level (cm)	Water Volume (liter)
2,75	1
5,5	2
8,3	3
11	4
13,8	5
16,5	6
19,3	7
22	8
24,8	9
27,5	10

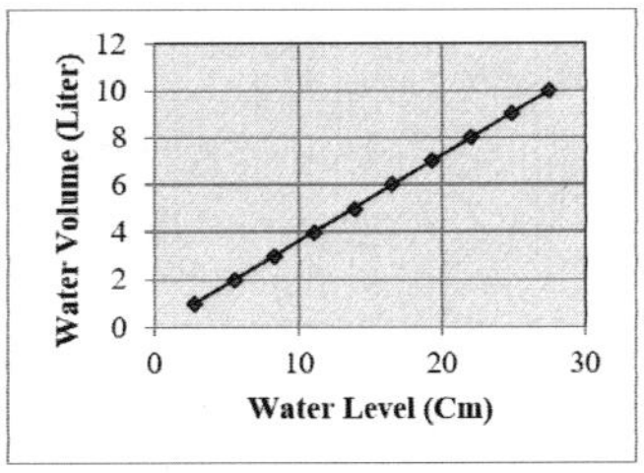

Figure 6. Graphic of Ultrasonic Sensor Characteristics.

3.2. Turbidity Sensor Testing

The test is performed on the ultrasonic sensor to determine the height of the sensor against the barrier and to know the response to changes in water volume. The data taken in the form of the height value from the sensor measurement that is converted to water volume unit (liters).

Table 2. Turbidity Sensor Test Result Data.

Turbidity (NTU)	Voltage (mV)
5	136,4
10	272,8
20	545,6
25	682
30	818,4
40	1091,2

Samples of water used is mineral water, toren, and water mixed with soil. Variations of turbidity concentration in water samples can be a reference to observe the effect of turbidity on the sensor output voltage.

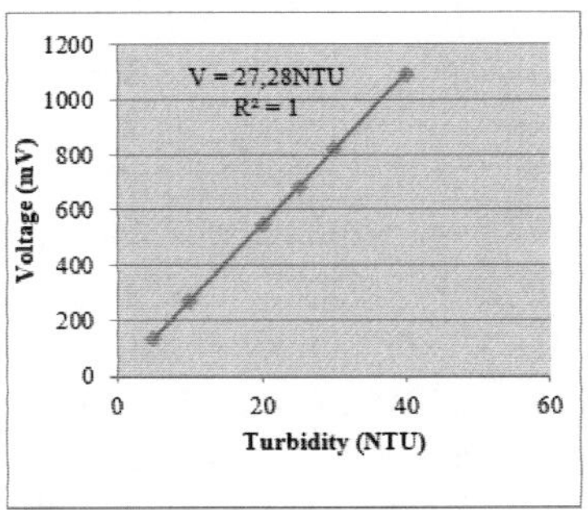

Figure 7. Graphic of Turbidity Sensor Characteristics.

4. Discussion

Figure 6 shows that the volume of water approaching the linear value of the value of water level as measured by the ultrasonic sensor. The higher the water, the greater the volume of water and the less water the lower the volume of water. Ultrasonic sensor measurement results are used as a comparator in adding calcium hypochlorite chemicals (Ca (Cl0) 2) as a decrease in turbidity of water. Ultrasonic sensors are designed to measure the volume of water with a maximum limit 12 liter.

Figure 7 shows that the sensor output voltage is increasing and turbidity increases. In contrast, the sensor output voltage decreases and the turbidity value decreases. From the graphic of the relationship between of turbidity sensor output voltage with the turbidity level obtained the linearity equation is V = 27.28NTU. The value of the output voltage of turbidity sensor represented V. The water turbidity level represented NTU. linear regression coefficient (R2) that is 1. The sensitivity of turbidity sensor as big as 27.28 mV / NTU. The sensor output voltage is converted to water turbidity value of the Arduino UNO using the linearity equation. This turbidity sensor measurement limit is between 0 NTU to 40 NTU. Turbidity sensor has a stable measurement capability without being affected by light around the room.

5. Conclusions

Based on the results of tests that have been done can be concluded ultrasonic sensors used ultrasonic sensors SRF05 which has a range of distances that can be detected between 3cm to 300cm. The chemicals used to decrease water turbidity are calcium hypochlorite (Ca (Cl0) 2). Turbidity sensor consists of a component designed as a photodiode detector and infrared LED as a light source. Infrared led has a wavelength of 960 nm. The turbidity sensor uses light scattering principle. Limit measurement of turbidity sensor between 0 NTU to 40 NTU. The sensitivity of this turbidity sensor is 27.28 mV / NTU. Measurement of turbidity sensor that has been made is not affected by the light conditions that exist around the room. This system can potentially help PDAM in monitoring water tanks and it is hoped that water quality monitoring system can be applied online.

References

[1] Borok A 2014 Turbidity Technical Review: Summary of Sources, Effects, and Issues Related to Revising the Statewide Water Quality Standard for Turbidity *Oregon Department of Environmental Quality*

[2] Edition F 2011 Guidelines for drinking-water quality *WHO chronicle* **38** (4) 104-8

[3] Filgueira A, Seijo M A, Munoz E and Iglesias M M 2007 Project of execution and viability study of photovoltaic facility in a water-treatment plant station *Clean Electrical Power, ICCEP'07* 666-669

[4] Vijayakumar N and Ramya R 2015 The real time monitoring of water quality in IoT environment *Circuit, Power and Computing Technologies (ICCPCT)*

[5] Wiranto G, Hermida I D P and Fatah A 2016 Design and realisation of a turbidimeter using TSL250 photodetector and Arduino microcontroller *Semiconductor Electronics (ICSE), 2016 IEEE International Conference* 324-327

[6] Ibrahim S N, Hakim M L, Asnawi A L and Malik N A 2016 Automated water tank filtration system using LDR sensor *Computer and Communication Engineering (ICCCE), 2016 International Conference* 195-199

[7] Yu K H, Yoon M J and Jeong G Y 2009 3d detection of obstacle distribution and mapping for tactile stimulation *Mechatronics, ICM 2009 IEEE International Conference* 1-6

[8] OptoElectronics O S I *Photodiode characteristics and applications*

[9] Toledano M F J 2012 *Diseño y construcción de una maqueta para el control semafórico con Arduino*

[10] Mylvaganaru S and Jakobsen T 1998 Turbidity sensor for underwater applications *OCEANS'98 Conference Proceedings* **1**158-161 IEEE

[11] Gay W 2017 *I2C LCD Displays In Custom Raspberry Pi Interfaces* (Berkeley: Apress) 35-54

[12] Made I S, Wibawa I K P and Santiyasa I W 2017 Prototype Development of Water Turbidity Measuring Device with ARDUINO UNO and LCD LMB16A Display *Int. J. Sci. Res.* **6** 6 2015–2017

[13] Evans B 2011 *Beginning Arduino Programming* (Berkeley: Apress)

IOP Conf. Series: Materials Science and Engineering **384** (2018) 012053 doi:10.1088/1757-899X/384/1/012053

Design of A 1:12 Power Divider at 5 GHz for Ground Surveillance Radar Application

Y A Rayisiwi* and T Hariyadi

Department of Electrical Engineering Education, Universitas Pendidikan Indonesia, Jl. Dr. Setiabudhi No. 207, Bandung, Indonesia

*yolitarayisiwi2009@student.upi.edu

Abstract. Radio detection and ranging (radar) is a system that can be used to detect and map an object by using electromagnetic waves and the echoes. Radar technology is a system that can help state institutions and apparatus in mastering and securing a country. One of the most important components in a radar is an antenna. The function of antenna is to transmit and receive the signals that will detect the presence of objects around the radar. The antenna commonly used in a radar usually has the high gain, and one of the techniques to enhance the antenna gain is using an antenna array. An antenna array has an important part that is the power divider. In this paper, we will discuss regarding the 1:12 power divider for ground surveillance radar at C-band frequency. We use CST Studio Suite to design and simulate the power divider. In this design and simulation, we use FR-4 with a dielectric constant of 4.5, and the thickness is 1.6 as a substrate. According to the simulation results, it can be seen that this power divider has the bandwidth 200 MHz with the center frequency 5 GHz, VSWR is less than 1.5, and all the output ports are in phase.

1. Introduction

Radio Detection and Ranging (radar) is a useful system for detecting, measuring distances, altitude and mapping of an object. Can be used to detect, measure distances and create maps of objects, such as airplanes, motor vehicles and weather. Basically the radar uses the electromagnetic waves emitted by the transmitter to detect objects from a distance. Radar works by emitting electromagnetic waves in a certain direction, then object reflecting the emitted waves. The reflected signal is then captured by the radar [1]. One of the most important components on the radar is an antenna. The antennas used on the radar usually have high gain, and one of the techniques to increase the antenna gain is using antenna array. How to increase the dimensions of an antenna, without having to increase the size of individual elements, is an element of element assemblies that radiate from electrical and geometric configurations. This new antenna, formed by multielements, is called array [2]. An array of antennas has an important part that is the power divider.

Power divider is a passive microwave component used for power sharing or power merging [3]. The power divider doubles the input signal power into two or more output signals. Three port dividers can one T-junction independently four network ports can be directional coupler and hybrid. Power divider are often used in telecommunications, and radar technology. The Wilkinson method is a commonly used technique in this regard. Figure 1 shows the common Wilkinson power divider [4].

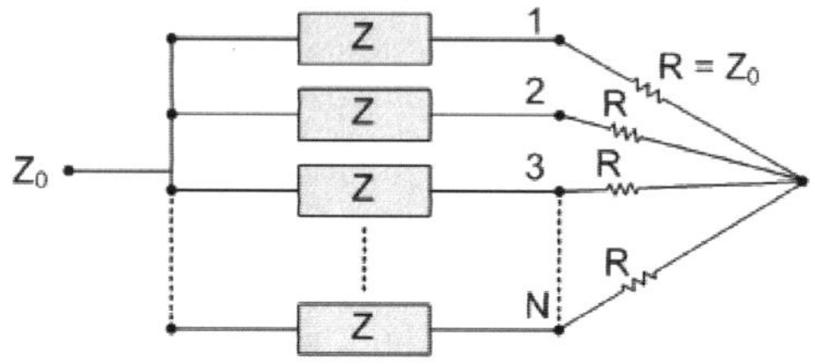

Figure 1. N-Way Wilkinson Divider/Combiner.

In the Wilkinson method, the value for impedance Z is given by the following equation:

$$Z = Z_0\sqrt{N} \tag{1}$$

Where *N* is the number of branching points and Z0 is the initial input impedance. The common power divider technique used in antenna array is T-junction. There are 2 types of T-Junction 50 Ω which can be used as a power divider as shown in Figure 2 [4]:

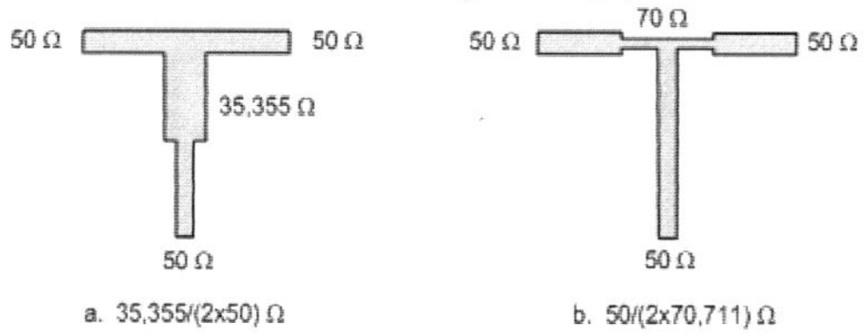

Figure 2. T-Junction 50 Ohm.

In this study, the design a 1:12 power divider using. The power divider is designed to work at C-Band center frequency of 5 GHz. The first stage is to create one element T-junction like in figure (a), then developed up to 12 outputs. This power divider is designed using a FR-4 substrate with dielectric constant of 4.5 and thickness 1.6 mm.

2. Design of power divider

The power divider designed and simulated by using CST Studio Suite [5] with FR-4 material as substrate. The input impedance of our power divider design is 50 Ω. Before designing we do the calculation to get the expected impedance, then the width (W) of the feeder line is done with the following equation (3):

$$\mathrm{B} = \frac{60\pi^2}{\mathrm{Z0}(\varepsilon r)^{1/2}} \tag{2}$$

$$W = \frac{2h}{\pi}B - 1 - \ln(2B-1) + \frac{\varepsilon_r - 1}{2\varepsilon_r}\left[\ln(B-1) + 0.39 - \frac{0.61}{\varepsilon_r}\right] \tag{3}$$

Where B is the magnitude of the impedance on the channel. With the value Zo = 50 Ω; and ε_r = 4.5, then the channel width of 50 Ω is:

$$B = \frac{60\pi^2}{Z0\sqrt{\varepsilon r}} = \frac{60(3.14)^2}{50\sqrt{4.5}} = 5.58\ mm \tag{4}$$

$$W = \frac{2(1.6)}{3.14}\left\{5.58 - 1 - \ln(2x5.58 - 1) + \frac{4.5-1}{2x4.5}[\ln(5.58-1) + 0.39 - \frac{0.61}{4.5}\right\}$$
$$W = 3.01\ mm \tag{5}$$

W is the feeder width with 50 Ω of impedance. After calculating the width of the feeder we must calculate the length of the feeder having a value of 50 Ω and 35.35 Ω. To calculate channel length (L) using the following equation:

$$L = \frac{c}{4fr\sqrt{\varepsilon_r}} \tag{6}$$

$$w_1 = \frac{c}{2fr\sqrt{\frac{\varepsilon r+1}{2}}} = \frac{3x10^8}{2x5x10^9\sqrt{\frac{4.5+1}{2}}} = 19.09\ mm \tag{7}$$

$$\varepsilon_{reff} = \frac{\varepsilon r + 0.3}{2} + \frac{\varepsilon r - 1}{2}\left[\frac{1}{\sqrt{1 + 12\left(\frac{h}{w_1}\right)}}\right]$$

$$\varepsilon_{reff} = \frac{4.5 + 0.3}{2} + \frac{4.5 - 1}{2}\left[\frac{1}{\sqrt{1 + 12\left(\frac{1.6}{19.09}\right)}}\right] \tag{8}$$

$$\varepsilon_{reff} = 3.64\ mm$$

$$L = \frac{3x10^8}{4x5x10^9\sqrt{3.64}} = 7.86\ mm \tag{9}$$

After we obtain the dimension of the T-junction, the next stage is simulating it on CST Studio Suite. To make it easier, we create the first 2 outputs, then 4 outputs, 6 outputs, and so on up to 12 outputs.

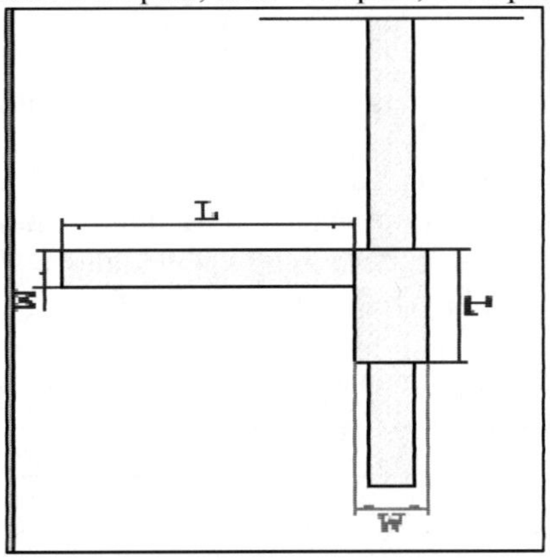

Figure 3. Parameters of 1:2 power divider.

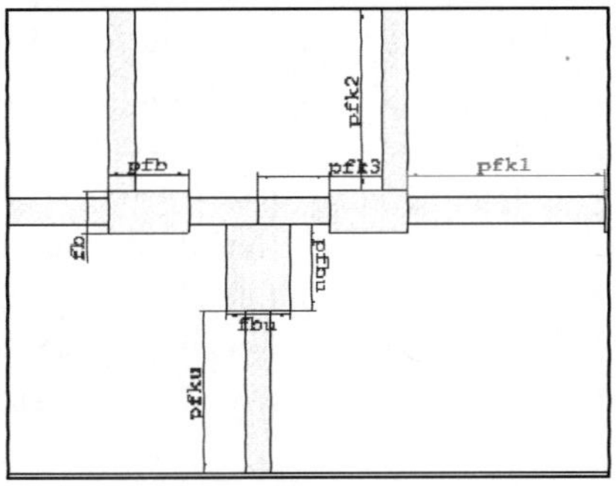

Figure 4. Parameters of 1:4 power divider.

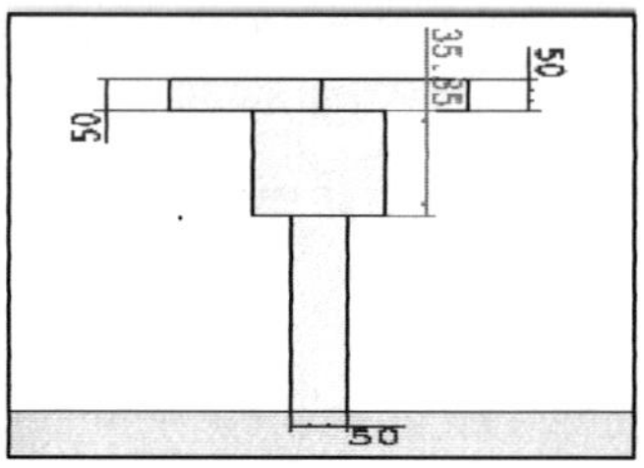

Figure 5. Parameters of T-junction.

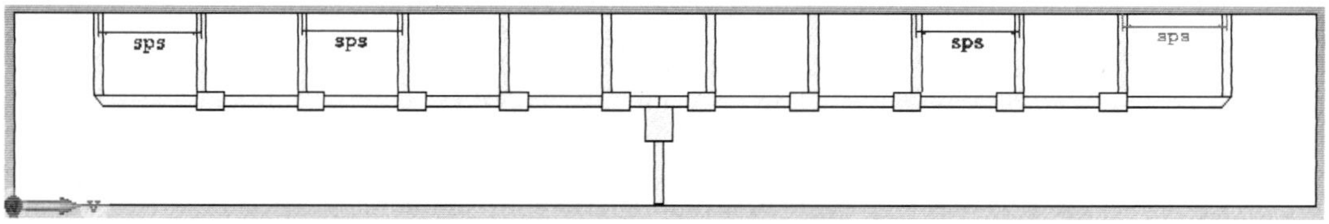

Figure 6. Front view of 1:12 power divider.

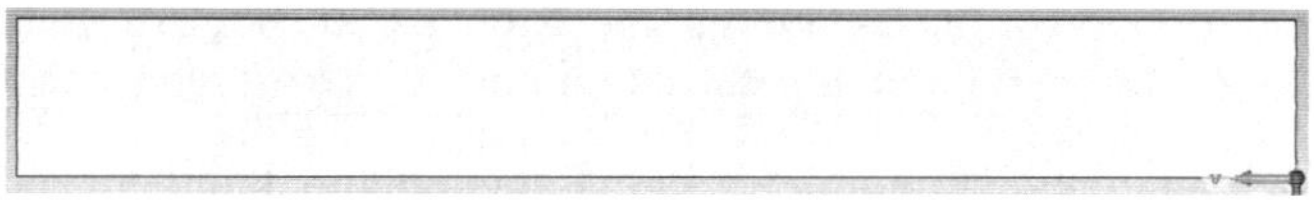

Figure 7. Back view of 1:12 power divider.

Figure 3 shows a 1:2 power divider with a description of W and L calculations. Figure 4 shows a 1:4 power divider with a description that can be seen in table 1. Figure 5 shows the parameters of T-junction power divider with impedance 50 Ω and 35.35 Ω as shown in Figure 2. Figures 6 and 7 show the antenna dimensions of 12 front and back 1:12 power dividers, and the table 1 is the optimization result to get the desired result.

Table 1. Dimension list of antenna design.

Parameter	Value (mm)	Description
ht	0.035	Thick of Ground & Patch
fk1	1.5	Width Small Feeder 1
fk2	1.5	Width Small Feeder 2
fk3	1.5	Width Small Feeder 3
hs	1.6	Thick of Substrat
fb	5	Width Big Feeder
pfb	8.8	Length Big Feeder
pfku	21.5	Length Small Feeder Prime
pfk1	23.95	Length of Small Feeder 1
pfk2	23.5	Length of Small Feeder 2
pfk3	9.1	Length of Small Feeder 3
s	19	Width Substrat
g	19	Width Ground
ps	105	Length of Substrat
pg	105	Length of Ground
pfbu	8.65	Length of Big Feeder Prime
sps	32.7	Space of element to element

After optimizing the dimensions obtained power divider with 12 outputs is 418.32 mm x 53.95 mm x 1.6 mm. The center frequency is 5.04 GHz, with impedance 50 Ω and VSWR <1.5. The optimization is done by changing pfk1, pfk2, and pfku to obtain the similar phase difference at 5 GHz. Changing the length or width can affect the frequency and phase of the power divider itself. The distance between elements is sps, where the distance between elements must be the same around 0.55λ-0.75λ. The power divider geometry is based on the T-junction model, but there are few modifications.

3. Results and Discussion

After optimized, the simulation result are return loss less than -10 dB with 50 Ω of impedance at input port and the phase difference on all output port are similar. The final design of 1:12 power divider can be seen in figure 8.

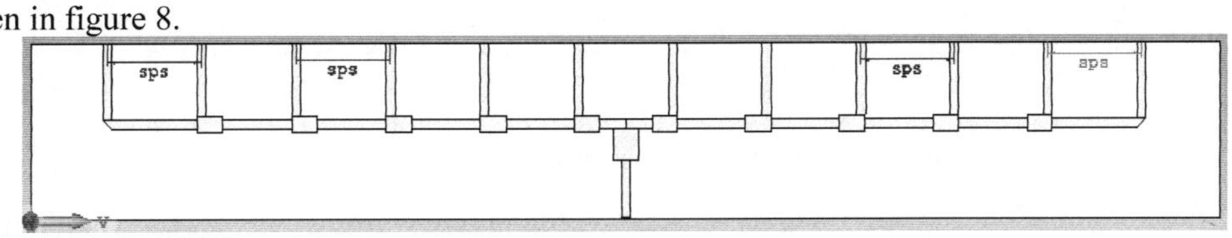

Figure 8. Final design of 1:12 power divider.

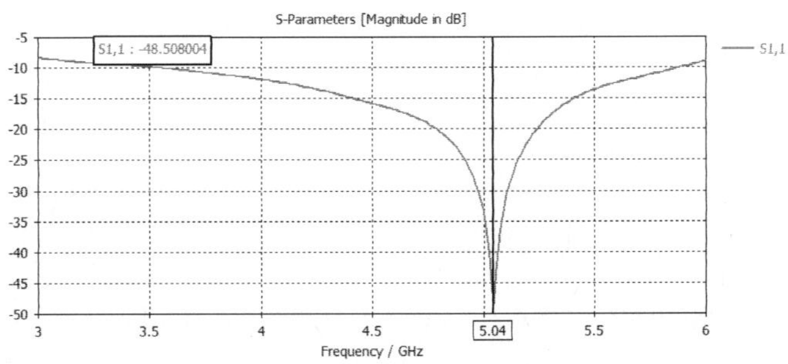

Figure 9. The result of 1:12 power divider shows a frequency of 5.04 GHz.

The power divider that is designed is in accordance with the desired result, where the frequency to be achieved is 5 GHz. At frequency 5 GHZ return loss is -48,51 dB. The result can be seen in figure 9.

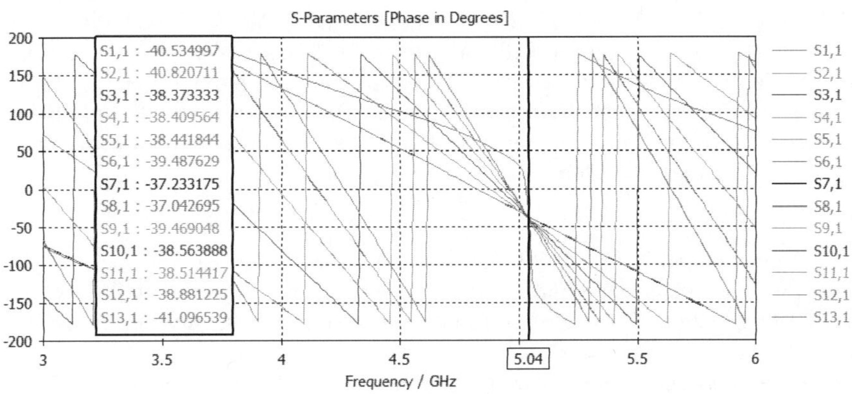

Figure 10. The value of phase of 1:12 power divider.

Figure 10 is a 1 phase divider of 1:12, where at a frequency of 5 GHz there is an intersection point of equal value. This phase indicates that the power distributed to each element is the same.

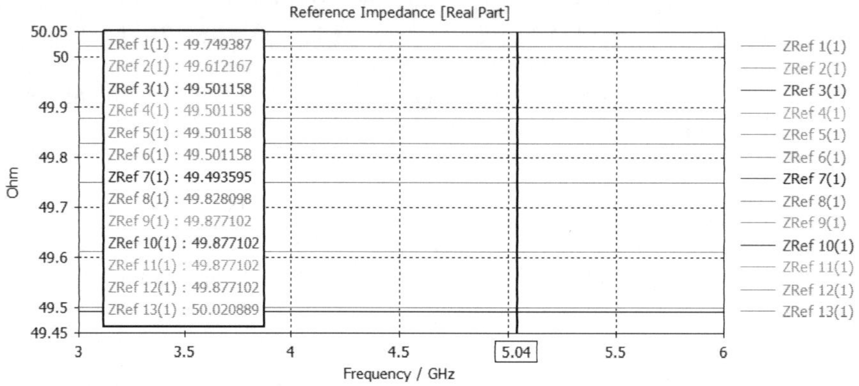

Figure 11. The impedance value of 1:12 power divider.

Figure 11 shows the impedance value for 1:12 power divider that already meets the value of 50.

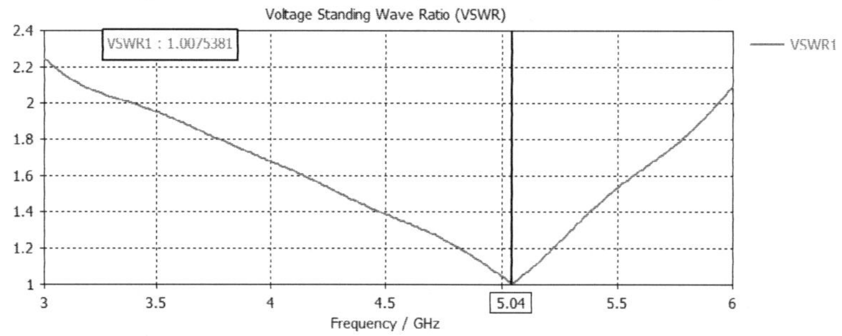

Figure 12. VSWR value of 1:12 power divider.

It can be seen from Figure 10 that at the 5.04 frequency shows the same phase value. Just as in Figures 11 and 12 shows impedances close to 50 Ω and VSWR 1 indicates that the power divider has matching.

4. Conclusions

This power divider design is made with 12 elements at a C-band frequency. The target of 12 elements at a frequency of 5 GHz has been reached with almost the same phase values. The length and width of each element will affect the frequency and phase. Things to consider in the design of power divider with 12 elements is the distance must be the same between elements, because it will greatly affect the results. The materials used in this power divider design are customers using FR4 with 4.5 dielectric constant, with size 418.32 mm x 53.95 mm.

References

[1] Nicolas A, Wijanto H and Wahyu Y 2015 Perancangan Dan Realisasi Antena Mikrostrip S-band Susunan Linier Untuk Radar Kapal *eProceedings of Engineering* **2** 2 2942–2950
[2] Balanis C A 1982 *Antenna Theory: Analysis and Design* (Hoboken, New Jersey: John Wiley & Sons, Inc.)
[3] Pozar D M 2009 *Microwave engineering* (New Jersey: John Wiley & Sons)
[4] Alfadil P A and Rambe A H 2014 Studi Perancangan Saluran Pencatu Untuk Antena Mikrostrip Array Elemen 2x2 Dengan Pencatuan Aperture Coupled *Singuda ENSIKOM* **6**(1) 19-24
[5] Computer Simulation Technology 2017 *CST STUDIO SUITE Student Edition* (online) available at: https://www.cst.com/Academia/Student-Edition

IOP Conf. Series: Materials Science and Engineering **384** (2018) 012054 doi:10.1088/1757-899X/384/1/012054

Smart Cane Using ESPectro with GPS Tracking System

Y Retnowati and A H S Budi*
Department of Electrical Engineering Education, Universitas Pendidikan Indonesia, Jl. Dr. Setiabudhi 207, Bandung 40154, Indonesia.

*agusheri@aol.com

Abstract: A blind person has a physical limitedness on the sense of vision. Therefore to facilitate their mobility, they usually use a cane. However, the cane has a deficiency in terms of function. The deficiencies of the cane are it cannot detect an out of range obstacle and blind's family is unable to track the blind's location. Based on these problems, this research will develop a smart cane for a blind person using ESP8266 as a microcontroller and it has sensors and buzzer to facilitate and provides assistance to its user. To assist in tracking the location, this smart cane utilizes GPS to determine the location and send it via SMS. The method used in this research is the design of hardware and software, tool making, and testing tool. The expected results of this research are to create a smart cane for a blind person which can assist their daily activities. A blind person can avoid a collision because this smart cane will notify the person through the buzzer and it can also notify their family if they are lost.

1. Introduction

Eyes are one of the senses which is very important for humans because it serves to record the circumstances or conditions that are all around, so people can find out things that he saw. Anyone can do a variety of activities more easily if they have normal eyesight. However, not everyone was born with a normal eye, there are also experiencing a vision disorder since birth [1].

A very common eye defect among communities around the world is blindness. According to the World Health Organization (WHO), there are a total of 285 million people in the world are experiencing vision problems i.e. as many as 39 million people are blind and 246 million more who have a low vision [2]. One of the problems often encountered by the blind is the limitation of activity and mobility. Blind people desperately need tools to detect obstruction so they could walk anywhere without worry of going to crash into objects in front of them. Therefore to facilitate their mobility, they usually use a cane [3][4].

The blind cane is one of the important facilities for the blind who used to walk. The sticks used will give the response when an object is exposed to the other end of the stick. However, the stick could not provide information to the user if the object is outside the range of the cane, so users could not prepare themselves when faced with objects that are before them. The other deficiencies of the cane are blind's family is unable to track the blind's location [3][5][6].

Based on these problems, this research will develop a smart cane for the blind use ESPectro which is this board using ESP8266, and then it will be installed the ultrasonic sensor, touch sensor, GPS+GPRS module, and a buzzer to facilitate it's the user. To assist relatives to the blind's location, the smart cane will use the GPS system as the determination of the location and it will send through SMS and can also be seen in real time on the makestro cloud.

2. Methods

2.1 Research Methods

The methods used in this research with the title "Smart Cane Using ESPectro with GPS Tracking System" has several stages described in the flowchart in Figure 1.

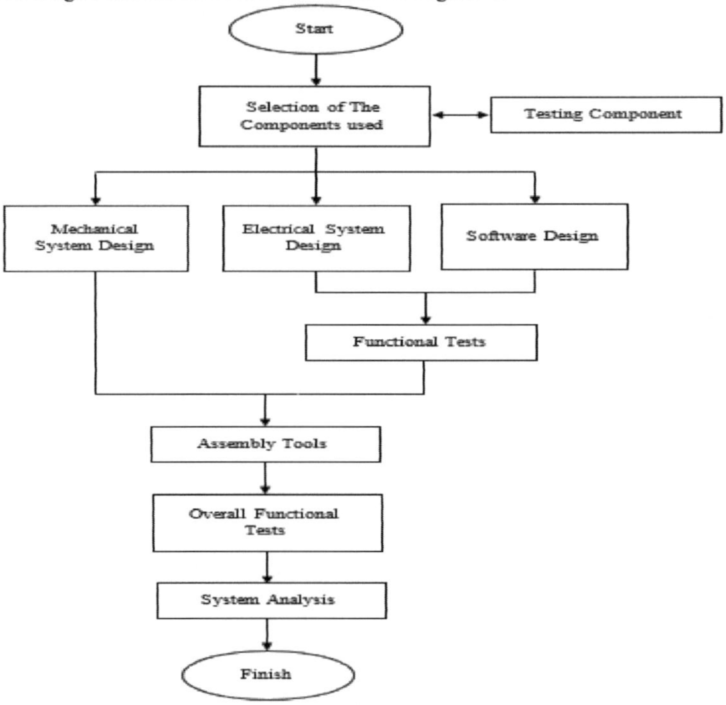

Figure 1. Flowchart of Research Methods.

In this first stage, the selection of components to be used in smart cane is to use modules such as:

2.1.1. ESPectro V3 Board. ESPectro V3 board in Figure 2 is one of the microcontroller board made by PT. DycodeX Nusantara Technology which is this board uses ESP8266 as the core [7]. ESPectro V3 has come the ability to auto-and manual flashing-flashing [8]. ESPectro also has a programmable LED, RGB LED, and button.

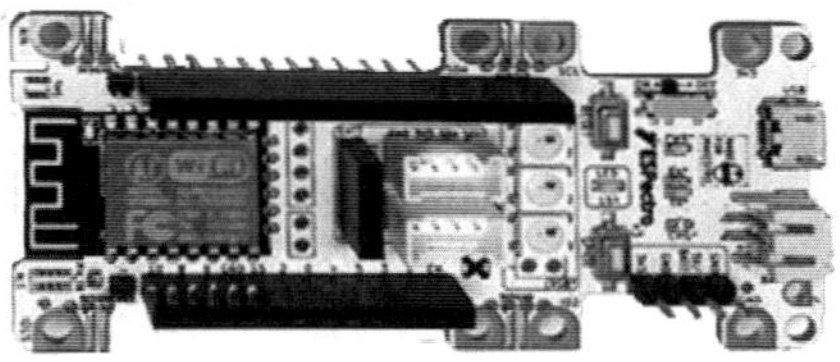

Figure 2. ESPectro V3 board.

2.1.2. Ultrasonic Sensor HC-SR04. The ultrasonic sensor is a sensor that works based on the principle of the reflected sound waves, the sensor produces a sound wave that is then reflected back by the time difference as the basis for measuring distances. Ultrasonic signal emitted by the transmitter. The signal has a frequency above 20 kHz, a frequency usually used to measure the distance of objects is 40 kHz [3]. Figure. 3 is the Ultrasonic sensor HC-SR04 used in this research.

Figure 3. Ultrasonic Sensor HC-SR04 [5].

2.1.3. Touch Sensor TTP223B. This module is a capacitive touch switch module. Under normal circumstances, low-output module; Low power consumption. When a finger touches the corresponding position then the module output becomes high and if not touched for 12 seconds then the module will switch to low power mode. Figure 4. It is the physical appearance of the touch sensor module TTP223B.

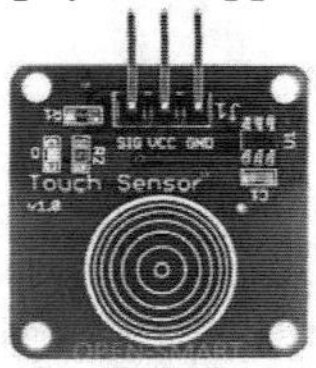

Figure 4. Capacitive Touch Sensor MPR121.

2.2 System Specifications
The design and realization of the system of "Smart Cane Using ESPectro with GPS Tracking System" have the following specifications:

- ESPectro with the ESP8266 chip as the main controller microcontroller.
- Ultrasonic HC-SR04 sensor for detecting obstacles.
- Indicators buzzer sounds voltage 3.3 Volt.
- TTP223B touch sensors are used as sensors when smart cane grip.
- GPS as the user's location signal transmitter.
- A Push button that serves as an emergency button to send SMS when the blind gets lost.
- SMS (Short Message Service) to provide news when the blind get lost in the form of text to the intended number.
- Makestro cloud, the cloud to store locations for visually impaired relatives to see the blind spot in real time.
- LiPo battery 3.7 V as a voltage source.
- Source code with C language for ESPectro board.

2.3 Program Design
When the condition of the smart cane is turned on, the system will immediately detect the distance so that at that time also will respond to value to the distance in accordance with the distance measured by the sensor. If the distance is detected more than 100 cm then the buzzer will not sound. However, if the distance is less than 100 cm is detected then the buzzer will sound. When the smart cane is detached and dropped then the buzzer will sound, it is useful for the visually impaired no trouble to look for smart cane when dropped or detached. For the GPS system used, the GPS module will continuously update the smart cane position and this data will be sent to the cloud so that the relatives can track the location in real time. If the blind person feels lost, the smart cane will automatically send an SMS requesting help by pressing the emergency button provided. This system has a block diagram that refers to Figure. 5 and for simplicity, the system is full of flowcharts that can be seen in Figure. 6.

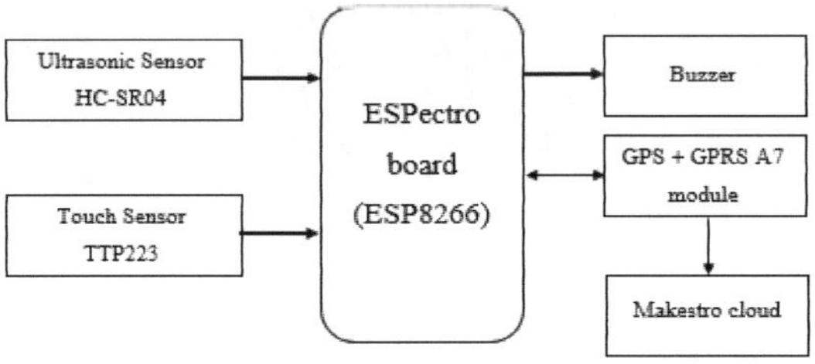

Figure 5. Block Diagram of Smart Cane Using ESPectro with GPS Tracking System.

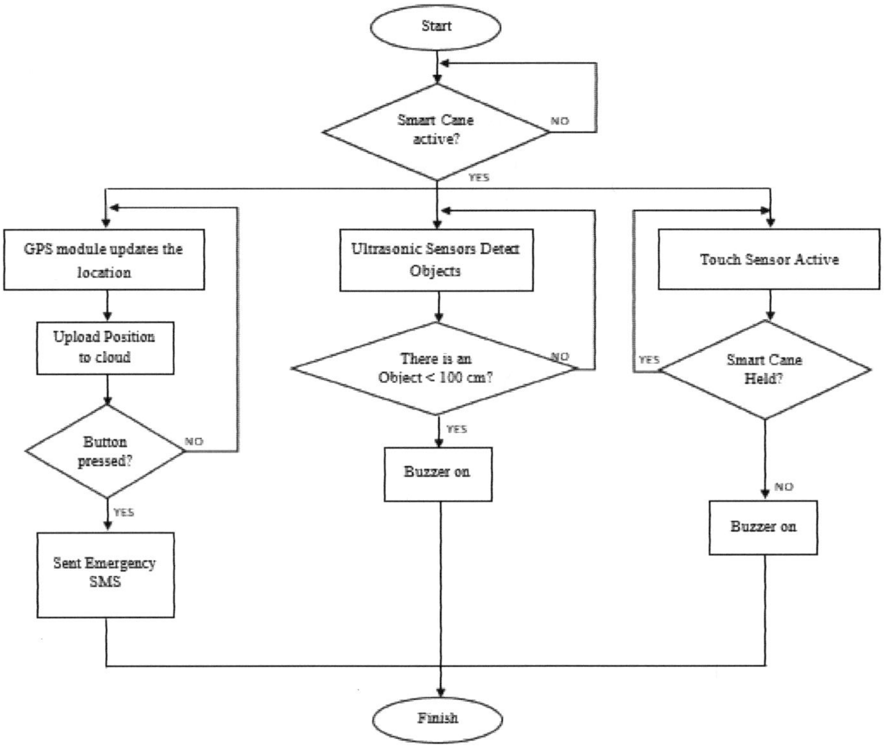

Figure 6. Flowchart of Smart Cane Using ESPectro with GPS Tracking System.

3. Results and Discussion

3.1. Research Results

Smart cane design was created for the visually impaired which is able to detect obstacles as shown in figure 7. The shape of the stick is made as a blind stick in general, with a white stick and red colours.

Figure 7. Physical smart cane.

The design of the casing or storage components and sensors used a black box with a plastic material and in this box had two push button such as one as the power on/off button to turn on the entire system of smart cane and one of the had the function to send emergency SMS if blind people feel lost and need help. Figure 8 was the photographs looked physical storage components and buttons on the smart cane.

Figure 8. The black box to store the components and buttons on the smart cane.

Smart grip cane shows the presence of copper in which copper is associated with the touch sensor. This is useful when the smart cane out of the grasp users automatically buzzer will sound. This meant that blind people have no trouble to find out the smart cane when falling.

3.2. Ultrasonic Sensor Testing

Ultrasonic sensor has been mounted on a black case on a smart cane bottom. These sensors are used to detect objects in front of the smart cane. The buzzer will automatically go off when the ultrasonic sensors to detect objects that are less than 50 cm in front of the smart cane.

Ultrasonic sensor HC-SR04 is used as a sensor that detects objects or obstacles in front of smart cane with a distance of less than 50 cm in front of the sensor then the buzzer will sound automatically. The difference in sound from a buzzer also depends on the distance of the object detected by the sensor. Can be seen in table 1 the results of ultrasonic sensor testing of objects that are in front of it.

Table 1. Ultrasonic Sensor Testing.

No.	Distance (cm)	Buzzer	Note
1.	154	Off	Buzzer off because the distance is more than 50 cm
2.	34	On	Buzzer on because the distance is less than 50 cm
3.	46	On	Buzzer on because the distance is less than 50 cm
4.	75	Off	Buzzer off because the distance is more than 50 cm
5.	23	On	Buzzer on because the distance is less than 50 cm
6.	134	Off	Buzzer off because the distance is more than 50 cm
7.	200	Off	Buzzer off because the distance is more than 50 cm
8.	78	Off	Buzzer off because the distance is more than 50 cm
9.	80	Off	Buzzer off because the distance is more than 50 cm
10.	10	On	Buzzer on because the distance is less than 50 cm
11.	120	Off	Buzzer off because the distance is more than 50 cm
12.	22	On	Buzzer on because the distance is less than 50 cm
13.	35	On	Buzzer on because the distance is less than 50 cm
14.	12	On	Buzzer on because the distance is less than 50 cm
15.	25	On	Buzzer on because the distance is less than 50 cm
16.	150	Off	Buzzer off because the distance is more than 50 cm
17.	340	Off	Buzzer off because the distance is more than 50 cm
18.	42	On	Buzzer on because the distance is less than 50 cm
19.	200	Off	Buzzer off because the distance is more than 50 cm
20.	59	Off	Buzzer off because the distance is more than 50 cm

IOP Conf. Series: Materials Science and Engineering **384** (2018) 012054 doi:10.1088/1757-899X/384/1/012054

From the test results of ultrasonic sensors that do, it can be seen that the ultrasonic sensors are successfully working as expected because when the obstacle is within a distance of less than 50 cm from the smart cane then the buzzer sounds, and if the obstacle is within a distance of more than 50 cm then the buzzer will not sounds.

3.3. GPS Testing

At this stage of validation trials connection range and accuracy of GPS tracking using GPS + GPRS Thinker Ai A7 conducted at 2 different locations with each condition inside and outdoors. Table 2 is the result of data obtained while testing the validation range GPS connection.

Table 2. Testing range of connections GPS.

No	Location	Range	Time Required	Explanation
1.	Jl. Jatiwangi, Margaasih	Affordable	7 minutes 40 seconds	Outdoor
2.	Jl. Jatiwangi, Margaasih	Affordable	10 minutes 18 seconds	Indoor
3.	FPTK UPI	Affordable	13 minutes 23 seconds	Outdoor
4.	FPTK UPI	Affordable	18 minutes 10 seconds	Indoor

Can be seen in Table 2 test range GPS connection, that the time required by the module Ai Thinker GPS + GPRS A7 to obtain the latitude and longitude of data vary depending on location and signal strength are common in the area. To get the accuracy of data in areas with good signal strength, it takes less time compared to areas that have poor signal strength [9]. In addition, the factor of the carrier used in this module also greatly affect the speed of the module to get an accurate location [10].

3.4. Cloud Computing Testing

When testing cloud computing tools, long time required is equal to the current range of testing a GPS connection which can be seen in Table 2. This is because the program has been designed to automatically download data upload latitude and longitude of the position of the tool obtained by the module AI Thinker GPS + GPRS A7 to the makestro smart cane cloud so that location can be viewed in real time. Figure 15 is a view of the makestro cloud while performing GPS tracking.

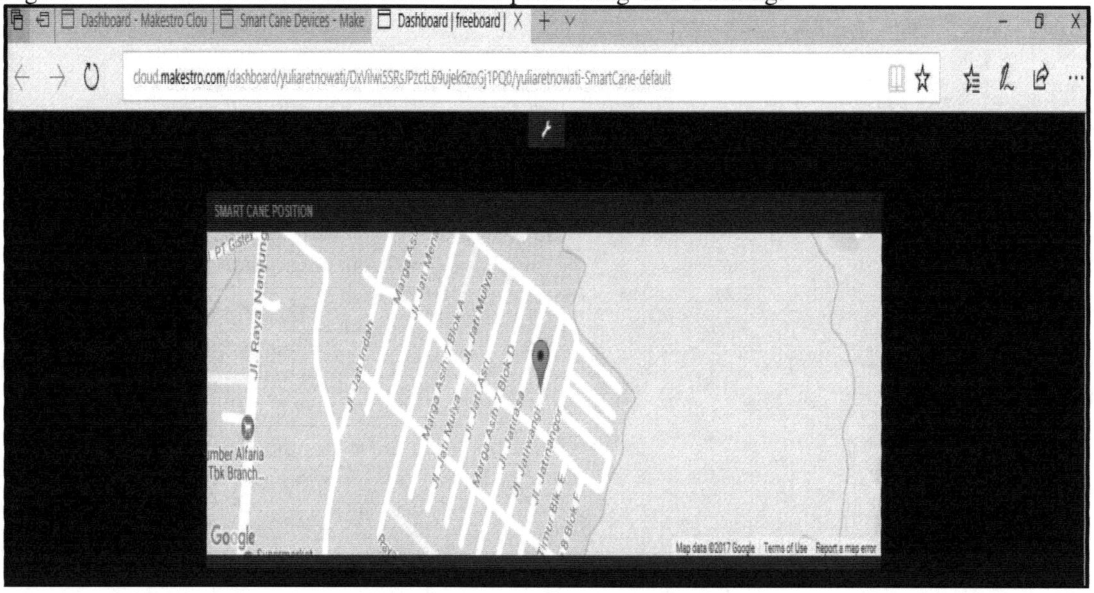

Figure 9. Display the makestro cloud when making GPS tracking.

IOP Conf. Series: Materials Science and Engineering **384** (2018) 012054 doi:10.1088/1757-899X/384/1/012054

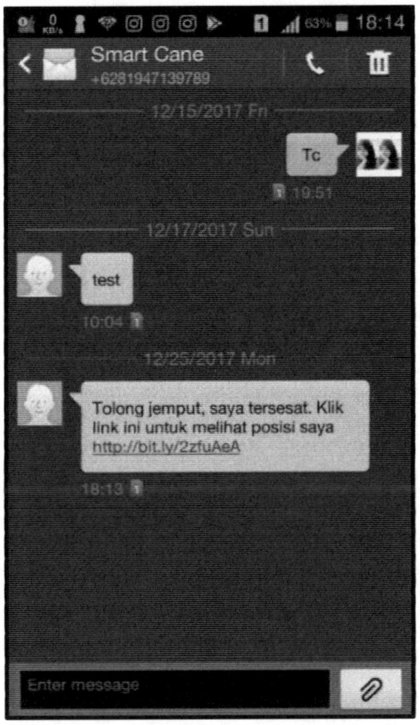

Figure 10. Display SMS requests for assistance (Emergency).

3.5. SMS Testing

Emergency SMS testing is done by pressing the emergency button that is on the smart cane. Then, the module AI Thinker A7 GPS + GPRS will automatically send an SMS request for aid to a prearranged number on the program. As fast or SMS length can be up to the recipient depending on the signal strength in the smart cane and mobile receiver, other influencing factors also strongly influenced by the operator used. Figure 16 shows the contents of the SMS received requests for assistance. The content of the SMS is also can be set as desired.

4. Conclusions

Based on the study "Smart Cane Using ESPectro with GPS Tracking System" has been done, it can be concluded that this study develops a walker for the visually impaired in the form of a stick of intelligent (smart cane) that can detect obstacles in front of him by way of sounding the buzzer as a warning if the obstacle is less than 50 cm. Smart cane can help users when smart cane falls and loss of grip then automatically buzzer will sound to let users know where the smart cane because cane smart paired touch sensors. To help relatives trace the location of the blind in the blind, smart cane is using GPS as determining the location of which can be seen by relatives in real time in the makestro cloud.

Disadvantages owned by cane's smart of them if they are crowded place then the resulting sound less audible buzzer. Then, GPS is used require relatively little long time to get an accurate position of the smart cane because of many factors affecting among other things used in the module carrier and signal strength exists at that location.

Smart cane is still not perfect so that improvements are needed in order for the system have complex functional value. Some suggestions that could be developed include the use of technologies such as the use of cameras to record in front of the object can then inform the user through earphones and their waterproof features so smart cane can still be used even if it is in rainy conditions. GPS module development necessary to make the time to track the position can accurately be shorter than the previous one.

References

[1] Pamungkas T B 2013 *Design of Ultrasonic Detector Sticks And Road Barriers For People With Visual Impairment Based Perforated ATmega16*

[2] Nahar V V, Nikam J L and Deore P K 2016 Smart Blind Walking Stick *International Journal of Modern Trends in Engineering and Research* 2349 645-652

[3] Junfithrana A P and Ruhiyat U S 2015 Design Tool for Road Base with Blind Arduino Uno *J. Engineering Nusaputra* **1** 1 1-5

[4] Suhaeb S 2016 Stick Design Electronics For Blind And Ultrasonic Sensor Based Microcontroller Atmega8535 *J. Sci. pinisi* **2** 2 131-136

[5] Ahmad F and Tjut A Z *Design Tool For The Visually Impaired Ultrasonic Sensor And System Using The Location-Based Mobile* 1-11

[6] Sheth R, Rajandekar S, Laddha S and Chaudhari R 2014 Smart White Cane - An Elegant and Economic Walking Aid *Am. J. Eng. Res.* **3** 10 84-89

[7] Saputra A, Saputra E, Nandar I F and Informatics J T *Sticks Blind Guide Using Ultrasonic Sensor Microcontroller Arduino* 1-11

[8] Subandi 2009 Aid to the Blind Based Mobility Electronics *J. Teknol.* **2** 28 29-39

[9] Farhan A A, Sunarya U and Nur D 2015 Designing and Implementing of A Blind Tool Using Ultrasonic Sensors and Global Positioning System (GPS) **1** 21569-1576

[10] El-aziz M M A and Khalifa W M 2015 Smart Blind Guidance System **11** 1 6-9

IOP Conf. Series: Materials Science and Engineering **384** (2018) 012055 doi:10.1088/1757-899X/384/1/012055

The Development of Walking Pattern and Stabilization of Humanoid Soccer Robot DARDROID (DARWIN-ODROID)

Y F Hestrio and P Prajitno*

University of Indonesia, Depok, Indonesia

*prawito@ui.ac.id

Abstract. One of the main problems on humanoid robot is motion, which is related to the stability of the robot when walking. This problem can be solved by using static walking and dynamic walking method, and the application of zero moment point and inverse kinematics supported by inertial measurement unit (IMU) with also the application of Kalman filter method and Proportional Derivative (PD) controller. The PD controller uses feedback from IMU to improve the servo position obtained from the calculation of zero moment point and inverse kinematics during the robot walking. This research would be done by comparing results from previous research and direct examination of a humanoid robot to get its stability margin and velocity. The result of humanoid robot stability margin is 5.367 mm in X-axis and 10.567 mm in Y-axis, and its minimum velocity 0.0524 m/s and maximum velocity 0.0633 m/s on grass with altitude 3 cm and minimum velocity 0.0933 m/s and maximum velocity 0.1048 m/s on grass with altitude 1.5 cm.

1. Introduction

There are problems and challenges that often arise on humanoid robot which are the design and control of the stability of walking movement of the humanoid robot, to prevent the robot from falling due to the instability, it's required a control instruction from accurate calculations, which it's get from an IMU.

Research on humanoid robot has been done in the University of Indonesia, but the research focus is only on the mechanical design of the bracket so there are some problems arise in the autobalance system due to the imprecise servo position on the bracket resulting obstructed of servo movement[1]. Due to this problem, further research is needed from the previous research to optimize joint trajectory control, balancing control, and motion planning[2].

In order for the robot to move properly, it takes a kind of cooperation between mechanical system, electrical, and artificial intelligence algorithm or program of robot with each system has its own role that supports the overall robot movement system[3][4]. The walking model of the humanoid robot can be divided into two main motion phases namely Single Support Phase (SSP), which only one foot is in contact with the ground to sustain the robot, and Double Support Phase (DSP), which both feet make contact with the ground[5][6][7]. This research will discuss about the zero moment point control using forward and inverse kinematics.

2. Theoretical Review and Research Methodology

2.1. Artificial Intelligence Motion Robot System

In order to design a capable artificial intelligence system in controlling the robot motion well, a method is needed to define the relation between each robot part as the basis of movement. This method is later referred to as Denavit-Hartenberg (DH) Parameters. This method can only be used if the right and left sides of the robot are symmetrical so that the equations in one part of the robot can be used in the symmetrical pair by mirroring the matrix [8].

Table 1. Parameter 6 DOF (Degree Of Freedom) right foot DH parameter [9].

i	α_{i-1} (deg)	a_{i-1} (mm)	d_i	θ_i
1	0	0	0	$\theta_7 + 90°$
2	90°	0	0	$\theta_9 - 90°$
3	-90°	0	0	θ_{11}
4	0	L_{13}	0	θ_{13}
5	0	L_{15}	0	θ_{15}
6	90°	0	0	$\theta_{17} + 90°$

By knowing the sixth parameter of DOF, DH parameter, and angle limitation on the right foot, so to get the DH parameter value and the angle limitation on the left leg only need to do the reflection matrix to the rotation axis [9].

Meanwhile, using the matrix equation on the right foot ${}^{C}_{f}T$, calculating the angle values $\theta_1, \theta_2, \theta_3, \theta_4, \theta_5, \theta_6$ can use partial geometry and partial algebra. Because of ${}^{6}_{f}T$ and ${}^{C}_{0}T$ are constant, so it can be simplified into:

$$ {}^{0}_{6}T = [{}^{C}_{0}T]^{-1}\,{}^{C}_{f}T\,\left[{}^{6}_{f}T\right]^{-1} \tag{1}$$

2.2. Kalman Filter

In order to make the humanoid robot walks well and efficiently, the humanoid robot requires a system that calculates servo movement to keep the robot stable when switching from SSP to DSP and vice versa. One of the systems is Kalman filter method. Kalman filter method is a technique of mathematical calculation (algorithm) which provides an efficient calculation method in the process of estimating the validity of a data by minimizing the mean squared error (MSE)[10].

Kalman Filter is used to know the zero moment point (ZMP) and center of mass (CoM) on the robot when moving dynamically or statically. Kalman filter is used in a process that can be expressed in form of linear state equation as in the following equations.

$$x_{k+1} = A_k x_k + B_k u_k + w_k \tag{2}$$

The above equation can be observed with a measurement model that maps state x to y output as in the following equation.

$$y_k = H_k x_k + v_k \tag{3}$$

Process noise (w) and measurement noise (v) are not bound noise, they are mutually independent noise. The estimated state value $\dot{x_k}$ in the Kalman filter is determined from the posteriori estimate $\dot{x_k}$ and the difference between the actual measurement $\dot{y_k}$ and the estimated measurement $H_k x_k$ as in the following equation.

$$\dot{x} = \dot{x_k} + K_k(y_k - H\dot{x}_k) \tag{4}$$

$$\dot{x} = \dot{x_k} + K_k(H_k x_k + v_k - H\dot{x}_k) \tag{5}$$

The difference in value between y_k actual measurement and estimated measurement is referred to as residual value. If the residual value is zero, then it indicates that the estimation result is the same as the measurement result. The value of K_k is the gain factor on the Kalman filter [10].

2.3. PID Controller

Besides using the Kalman filter to filter the data used, a control system is also required to adjust the servo movement. A PID controller which is a control system that utilizes data feedback to maintain the stability of the system by continuously calculating the value of error as the difference between the desired setpoint value and the measured process variables, is used in the system[11]. PID controller will minimize the error value in any time with adjustment of control variable. The value of PID controller can be calculated using the following equation model:

$$u(t) = P(t) = K_p e(t) + K_i \int_0^t e(\mathrm{T}) d\,\mathrm{T} + K_d \frac{de(t)}{dt} \tag{6}$$

with u is the value of PID controller, Kp is the proportional gain, Ki is the integral gain, Kd is the derivative gain, e is the error of the setpoint minus the process variable, t is time, T is the integration variable from time zero to t. PID controller model is required to keep the robot balance parameter value either static or dynamic [12] [11].

Then, to control the servo on the foot so that the CoM position does not exceed the robot zero moment point, an acceleration equation is used:

$$\frac{d}{dt}\ddot{x} = u_x \tag{7}$$

By using the basic equation of ZMP [13]

$$Tzmp = mg(x - p_x) - m\ddot{x}z_c = 0 \tag{8}$$

So, the equation of an optimal control with zero moment point can be expressed as follows

$$J = \sum_{i=k}^{\infty} Q_e\big(p_d(i) - p(i)\big)^2 + \Delta X^T Q_x \Delta X + R\Delta u(i)^2 \tag{9}$$

Q_e is the multiplier of the ZMP error reference, Q_x is the error multiplier of the system condition and R is the input error multiplier. By following the law of the control system, it is obtained that

$$u(k) - K_s \sum_{i=0}^{k}\big(p(i) - p_d(i)\big) - K_x X(k) - \sum_{i=k}^{N_L} G(i) p_d(k+i) \tag{10}$$

K_s is gain error, K_x contains the gain of system condition and $G(i)$ is a function variable that can show optimal preview gain for N_L preview steps [14].

3. Results and Discussion

3.1. Results of Walking Parameter Testing

This test is done with five parameters i.e. *step forward, foot height, Y offset, Hip pitch offset and time period.* The test is done by comparing the inverse kinematics calculation with the result that occurs when the robot is walking, the given parameters is foot height of 45 mm.

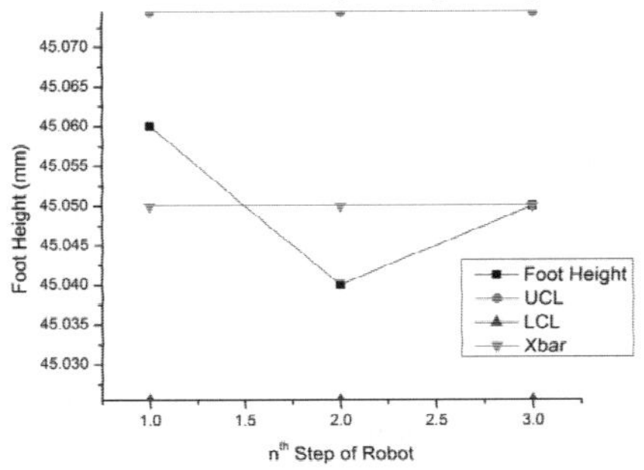

Figure 1. Testing on Height Right Foot Step on Grass 3 cm in This Research.

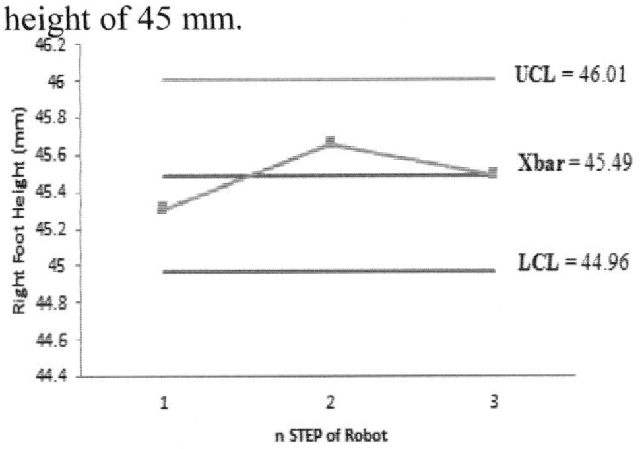

Figure 2. Testing on Height Right Footstep in Previous Research.

The test on the robot height footsteps in Fig. 1 uses a control chart with a range between UCL (Upper Control Limit) and LCL (Lower Control Limit) using 2 times the standard deviation of data. The data taken are the footsteps as much as 6 steps on grass with height 3 cm. In Figure 1, the result can be seen that there is a deviation of 0.00667 mm.

Figure 2 shows that there is a deviation of 0.49 mm. Even figure 1 and figure 2, That value are the average of 3 times the test on the position of the SSP, which the proper height of the foot is 45mm, but this is not so on the test. Therefore, it needs to be analyzed every angle and compare with the calculation of inverse kinematics angle.

3.2. Results of Robot Trajectory Testing

Trajectory is a pattern of a robot trace in the walking process. Trajectory testing is done to be able to know the mathematical calculation of robot velocity when walking.

$$X(t) = (b + f)\left(\frac{t-t1}{t2-t1} - \frac{1}{2\pi}\sin\left(2\pi\frac{t-t1}{t2-t1}\right)\right) - b \tag{11}$$

b is the initial footstep (m) and f is the final footstep (m), t is the time variable, t1 is the start time of the first step walking (s), t2 is the end time of the last step walking (s) and X is trajectory direction of the x axis.

In addition, the required trajectory support is the trajectory direction of y axis to know the pattern of y axis direction movement on the robot by using the following equation:

$$Y(t) = \left(2 - \left(\cos\left(2\pi\frac{t-t1}{t2-t1}\right) + 1\right)\right)\frac{h}{2} \tag{12}$$

h is the height of the robot footstep when walking (m) [14]. From the above equations, the following graphs are obtained.

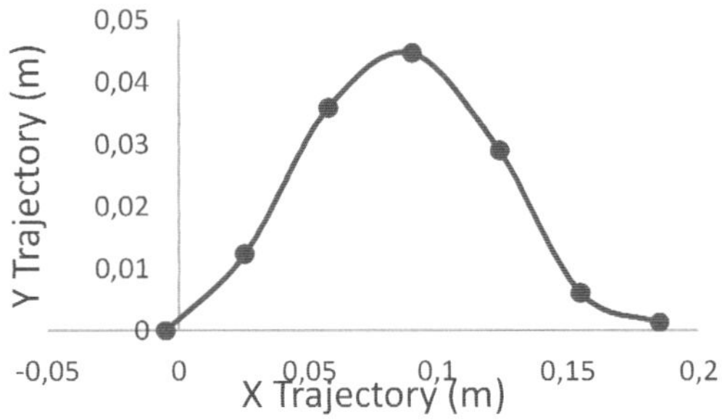

Figure 3. Trajectory of Robot Ankle Steps with 5 mm step size in This Research.

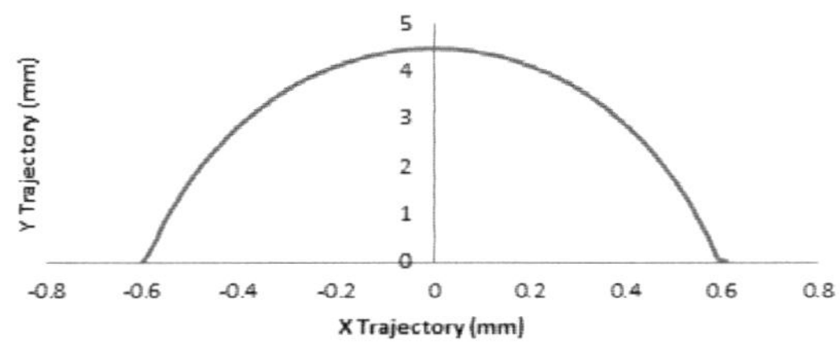

Figure 4. Trajectory of Robot Ankle Steps step size in Previous Research.

From this research, if testing the distance traveled is 3.5 m with a distance of 1 step is 0.03 m, it will require 117 steps. However, if a distance of 1 step is 0.01 m it will require 350 steps. From previous research, if testing the distance traveled is 0.8 m with a distance of 1 step is 0.012m, it will require 67 steps. So, to get the initial velocity of a robot requires a period of one phase walking, and in order to get the required acceleration, it takes the distance of step of the overall walk.

3.3. Results of Robot Balance Testing

In testing the robot stability, the measurement is using a stability margin on the robot. The measurement of stability margin value is done in walking mode. By using Zero Moment Point method, the value of (Xcom, Ycom, Zcom) obtained are (10, 7, 26.5) cm. The value of Xzmp and Yzmp occur during the initial position of the robot, so there is no effect due to acceleration and velocity of work.

However, when at a stage where acceleration in different condition resulting the spread of ZMP points as in the following graph:

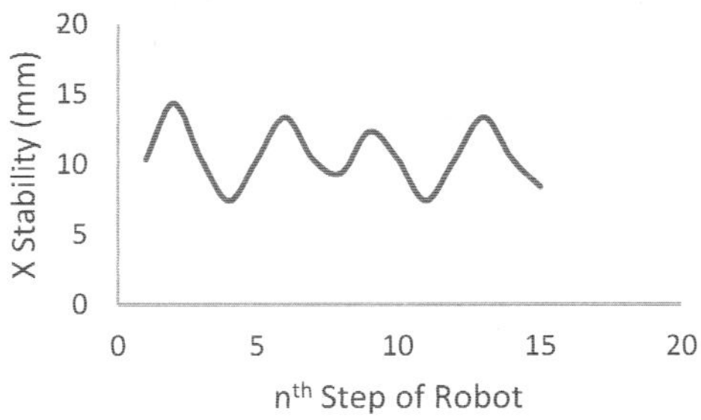

Figure 5. The Robot Stability when walking, on X axis in This Research.

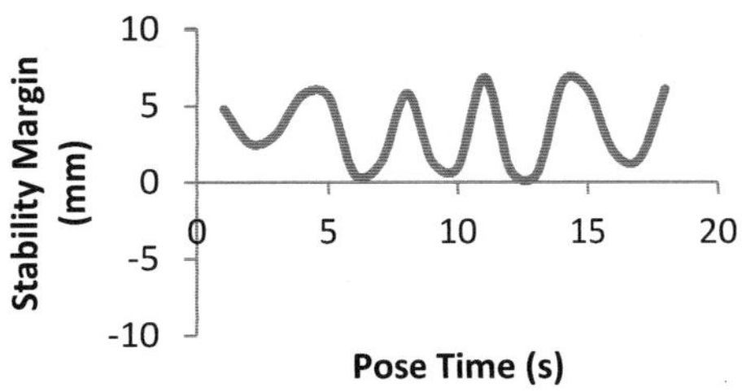

Figure 6. The Robot Stability when walking in Previous Research.

From Figure 5, we can see the zmp points when the robot is moving its feet without stepping forward which means it is in the s*upport polygon*, indicating that the robot is still in a stable state. But from Figure 6, we can see the zmp points. They have been reduction because it has unstable transition Single Support Phase (SSP) into Double Support Phase (DSP) and otherwise.

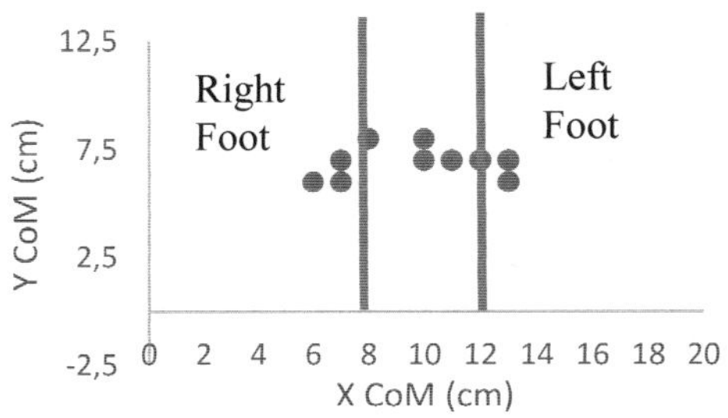

Figure 7. The Distribution Points of Center of Mass at Walking Time in This Research.

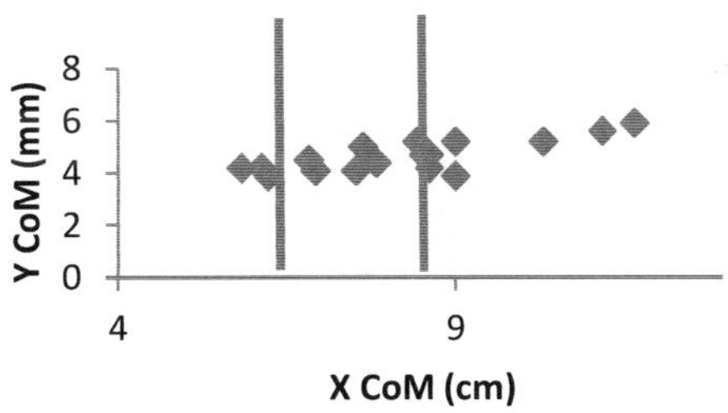

Figure 8. The Distribution Points of Center of Mass at Walking Time in Previous Research.

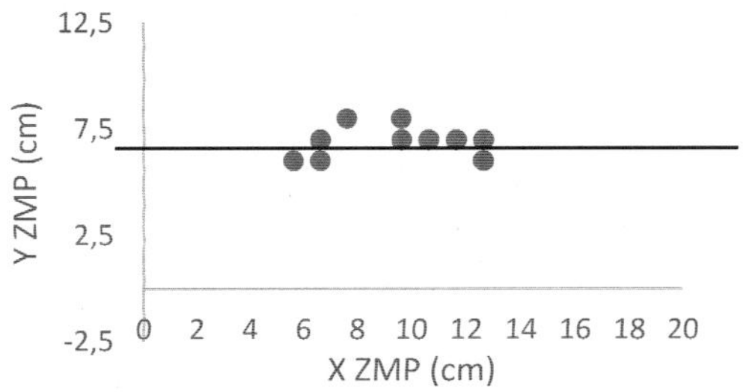

Figure 9. The Distribution of ZMP points at walking time in This Research.

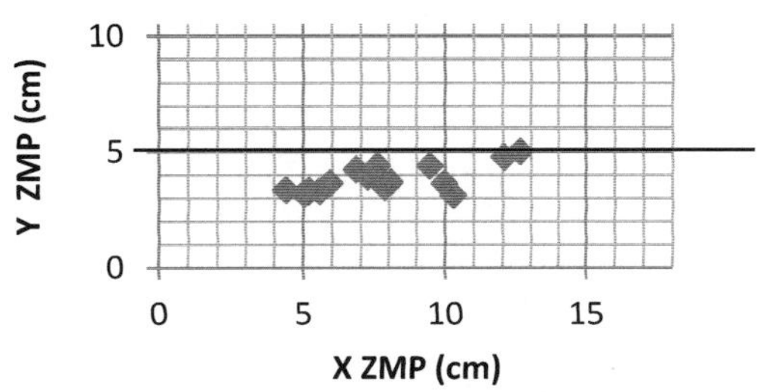

Figure 10. The Distribution of ZMP points at walking time in Previous Research.

Figure 7 and figure 9 shows that the ZMP points are in the front of the robot feet. This is because the position of walking tends to be humpback, by purpose when the robot is in a condition move moment inertia at the top of the robot body is at ZMP point and com robot when it is going nowhere. It can still be categorized as safe because ZMP is still on support of polygon robot. Therefore, it is necessary to view the distribution of ZMP during Single Support Phase (SSP) and Double Support Phase (DSP) separately. However, in figure 8 and 10 shows that the robot have been unstability because the ZMP points have been unstable condition between Single Support Phase (SSP) into Double Support Phase (DSP) and otherwise. Sometimes, the distribution of ZMP during SSP and DSP isn't on support of polygon robot.

3.4. Results of Robot Velocity Testing

The Robot Velocity Testing is done with 2 velocity approaches which are the maximum velocity and the minimum velocity with certain distance, and 2 big steps that are maximal and minimal. For the velocity of the robot can be stated by the following equation:

$$V = 2\,\frac{b}{T} \tag{13}$$

V is the velocity of Robot (m/s), b is the footstep size (m) and T is the period of the robot foot step (s) [4]. From the testing of walk parameter, it is known that the robot walking period is 0.6s, with minimum distance of 0.005m and maximum distance of 0.015m, so it is obtained that the minimum velocity is 0.01667 m/s and the maximum velocity is 0.05m/s.

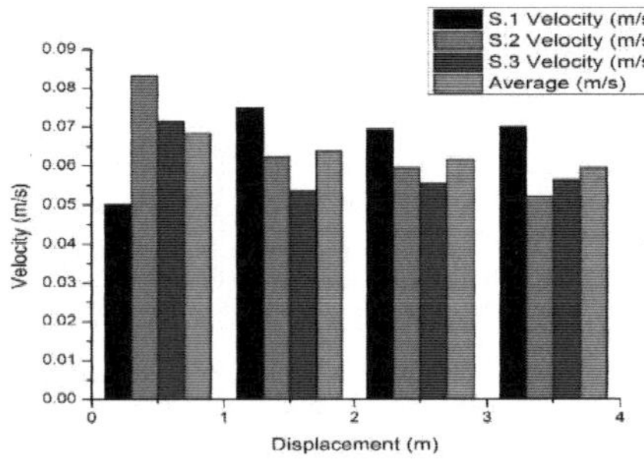

Figure 11. The Graph of Maximum Velocity of the Robot on 3 cm Grass in This Research.

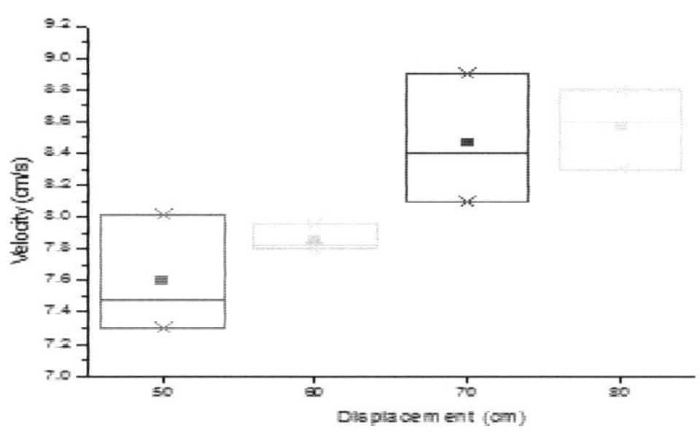

Figure 12. The Graph of Velocity of The Robot in Previous Research.

Figure 11 and 12 above represent the velocity graphs on the distance travelled. If at the previous research maximum velocity calculation of 8.11 cm/s and in this research maximum velocity of 0.05m/s, then in this test obtained the average maximum velocity is 3.809cm/s from previous research, while the average maximum velocity is 0.0633m/s from this research. If it is analyzed further, this can happen due to an increase in velocity, thus we can know the acceleration that occurs with the following equation [14]:

$$a = \frac{V(t)\,T - 2b}{T\,t} \tag{14}$$

From the equation and results of this test, it is obtained the maximum acceleration average is -5.51x10-5m/s^2 from previous research and the maximum acceleration average is 0.000995m/s^2 from this research. This is because at a short distance the robot still has a considerable momentum of velocity due to the start of the walking phase.

4. Conclusions

In this research shown, the average robot stability is 10mm by using auto-balance system from IMU and Kalman filter method and least square data acquisition with maximal error margin of 16%. The average robot stability can be 10mm cause the designed Robot Stability. The designed Robot Stability has average stability margin of 5.367 mm on the X axis and 10,567 mm on the Y axis during walking. For the sensor of robot, the authors use gyro as the feedback servo balance control, with Kp, Ki and Kd Knee values of 0.3, 0.0 and 0.1, Kp, Ki and Kd ankle pitch values of 0.9, 0.0 and 0.1, Kp, Ki and

IOP Conf. Series: Materials Science and Engineering **384** (2018) 012055 doi:10.1088/1757-899X/384/1/012055

Kd hip roll value of 0.5, 0.0 and 0.1 and Kp, Ki and Kd ankle roll value of 1, 0.0 and 0.1. Whereas, in the previous research did not use Gyro as its balance control and robot DARWIN-OP only use Kp on balance. The result from the gyro as sensor stability and designed Robot stability, the authors have got the minimum average of the robot velocity is 0.0524 m/s on 3 cm grass. Moreover, the maximum average of the robot velocity is 0.0633 m/s on 3 cm grass. Meanwhile, in the previous research is 8.11 cm/s. The average minimum acceleration of the robot is -1.206x10-5 m/s^2 on 3 cm grass. Next, the average maximum acceleration of the robot is 0.000995 m/s^2 on 3 cm grass. Meanwhile, the previous research amounted to 0.551 cm/s^2.

References

[1] CUI X and LI X 2015 *Kinematic Analysis and Control for Darwin Humanoid Robot*

[2] Indrajit, W 2013 *Rancang Bangun Sistem Motion Capture pada Biped Robot dengan Pengendalian ZMP* (Depok: Universitas Indonesia) 18-19

[3] Liu J 2016 Bipedal Walking with dynamic balance that Involves three-dimensional Upper Body Motion *Robotics and Autonomous Systems* **77** 39-54

[4] Liu R 2015 *Walking Pattern Generation for Humanoid Robots*

[5] Priyono M G 2016 *Rancang Bangun Stabilitas dan Kecepatan Pergerakan Humanoid Robot dengan Kombinasi Material pada Bracket* (Depok: Universitas Indonesia) 1 75-76

[6] Rahmdianto A, Adiprasetiano H and Priyono M G 2014 Analisis Stable Walking Humanoid Robot Soccer Berbasis Zero Momen Point dengan Metode Praktik *Indonesian Symposium Robot Soccer Competition*

[7] Veselý V R 2011 Robust PID Controller Design *Proceedings of the 18th International Conference on Process Control* 565-570

[8] Saputra A A, Takeda T and Kubota N 2015 Efficiency energy on humanoid robot walking using evolutionary algorithm *IEEE Congress on Evolutionary Computation (CEC)* 573–578

[9] Putra G N 2016 Analisis kecepatan Humanoid Robot dengan metode elemen hingga dan kinematika terbalik *Indonesian Symposium Robot Soccer Competition*

[10] Wang Z 2015 Slow Walking Model for Children With Multiple Disabilities *Mechanical Systems and Signal Processing*

[11] Feng Z Y 2014 Dynamic Modeling and Analyzing of A Walking Robot *The Journal of China Universities Post and Telecomunication*

[12] Saputra A A 2015 Pengembangan Pola Berjalan pada Humanoid Robot Soccer *EEPIS Robotic Research Centre*

[13] Serway J 2010 *Physics for Scientists and Engineers 7th Edition* (Thomson Brooks/Col)

[14] Bouhajar S 2015 Trajectory generation using predicitive PID control for stable walking humanoid robot *Procedia Computer Science* **73** 86-93

IOP Conf. Series: Materials Science and Engineering **384** (2018) 012056 doi:10.1088/1757-899X/384/1/012056

Automation Testing System for Energy Efficiency of Electric Fan Based on SNI IEC 60789:2013

Q Lailiyah*, P Bakti, N Kusnandar, I Kasiyanto, H Tjahjono and I Muchlis

Research Center for Quality System and Testing Technology, Indonesian Institute of Science, Puspiptek Region Building 417, Setu, Tangerang Selatan, Banten, Indonesia

*quds001@lipi.go.id

Abstract. Automation system testing is needed by testing laboratory to reduce testing time. This paper attempt to examine the developed automation testing procedure to the effectivness testing duration. For this purpose, SNI IEC 60789:2013 was applied as testing procedure and the effectivness of testing duration was determined by comparing the duration test of developed automation procedure against manual procedure. The mechanical system component consists of a rail that can drive an anemometer buffer according to the desired distance. The process of moving an anemometer is electronically controlled by relay and timer. An anemometer sensor produces digital data to be read and recorded by the computer. The data of wind speed and electric power from the fan are calculated manually to obtain the percentage value of energy efficiency. This system can significantly reduce test time.

1. Introduction

Electric is the final form of energy that gives effectiveness and ease of use. In Indonesia, electric consumption at the end of 2015 increased by 2.13% from the previous year, and the largest consumption was in the household sector [1]. Many household appliances are switching to using electrical energy sources, one of which is a fan. Fan became one of the household appliances that are popular in the community considering Indonesia is a tropical country [2]. This condition will certainly increase the load of national electricity consumption if there is no effort to suppress the growth. Therefore an increase in fan efficiency energy is a strategic solution without suppressing the growth in demand for such household appliances.

Determination of minimum energy consumption level is very important to be done in an effort to increase energy efficiency of household appliance fan [3]. Currently the ESDM ministry is preparing a draft related to the Energy and Mineral Resource Regulations on Minimum Performance Standards (SKEM) and Inclusion of Energy Saving Label Labels for Fan. To support this, need for readiness in testing system for energy efficiency of electric fan. Therefore, it is necessary to conduct research and development in the field of measuring methods for fan energy consumption. In development of energy testing methods, test procedures must can be repeated and producing accurate data [4]. National standard that regulates the electrical fan performance measurement method that is SNI IEC 60879: 2013. Based on these methods the energy-saving rate is determined by the airflow generated by the fan compared to the electric power consumed [5].

Energy efficiency test method of electric fan requires realtime data monitoring and distance position settings with many shifts in measuring point. This has an impact on the length of test time and the high involvement of operators on each measurement. Both of these problems have the potential to decrease the accuracy of the test results. The solution to this problem is create of a test system that can work automatically. Automation test system can speed up the completion of testing and improve the accuracy and precision of the test results.

2. Test Setup According SNI IEC 60789:2013

Testing method of energy efficiency for electric fan is based on SNI IEC 60879: 2013. The electric fan used in this research is a stand fan. The positioning of the fan is set in the standard. The center of the fan blade is placed 1.5m from the floor and the front of the blade is at least 1.2m from the rear wall, 1.8m from the side wall and 4m from the front wall [5].

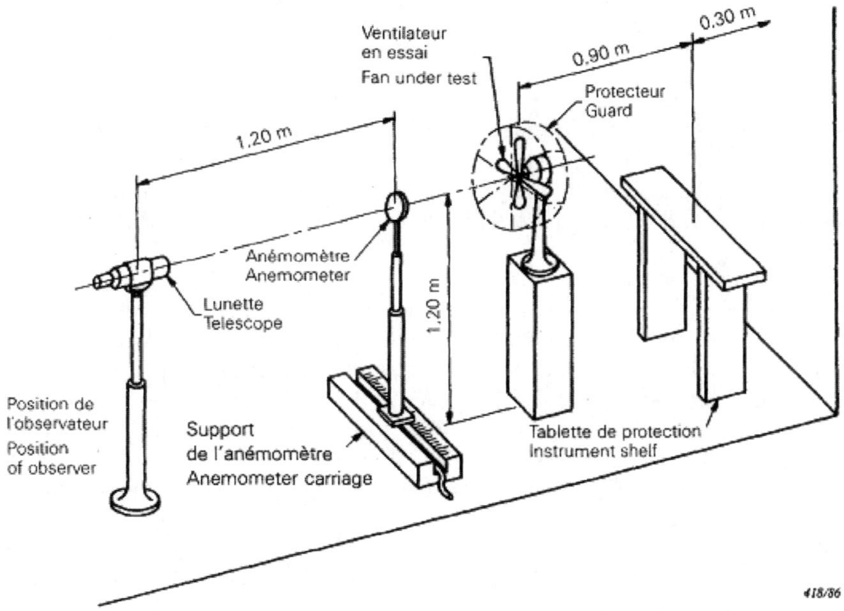

Figure 1. Test setup based on IEC 60879:1986 [6].

Before performing the test, the fan has operated and steady at the test voltage. Measurements are made on the fan operating at full speed. The air velocity reading starts at a point 2 cm from the axis of the fan blade, and should move forward along the horizontal line, with an increase of 4 cm wide and observed for 120 seconds each increase of 4 cm wide. Data retrieval is carried out until air speed falls below 24m / min.

3. Initial System

Several laboratories in Indonesia are receiving equipment assistance from the BRESL-UNDP program for energy efficiency testing of several household products [7]. For testing of efficiency energy for electric fan, there is a mechatronic system for moving an anemometer as shown in the figure 2. The mechanical system consists of anemometer supporting pole, DC rail and motor. There is an electronic control system consisting of a mechanical stopper, a dc voltage source and an anemometer direction button.

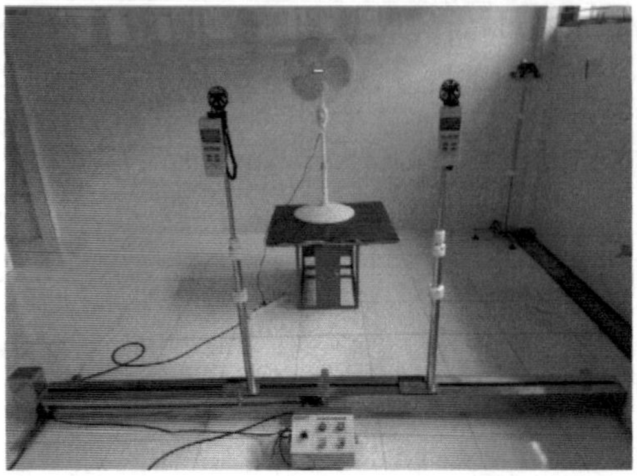

Figure 2. Mechatronic system of anemometer sensor driver.

The operation of the system to get a distance of 4 cm is still through direct observation that is by looking at the ruler that is available on the system. If the movement has reached 4 cm, the system is switched off manually by disconnecting the source voltage contact via switch on / off. To move it back is also done by turning the switch on / off and pressing the green or red direction button. This process is quite time-consuming and effort, especially when the test volume is overwhelming.

4. Automation System Design

Automation system that will be made is a system based on the rules of on - off control. According to Ryniecki, "the on-off control is the most common type of control used in industry because discontinuous actuators, like relays or solenoid valves, are simpler and cheaper than actuators used in the continuous type control systems"[8]. Therefore, in this study the authors want to try to make the design of automation systems using relays and timers that will be incorporated into existing systems.
The timer used is a H3CR-A8 timer that has 8 pins. There are 5 types of operation mode that isA (ON-delay), B (Flicker OFF start), B2 (Flicker ON start), E (Interval) dan J (One-shot). The time limit output type is relay output (DPDT)[9].

The MK2P-I relay has 8 pins with a working voltage of 250 VAC. The maximum current is 10 A at resistive load and 7 A for inductive load with power factor 0.4. Operation time while in AC voltage is 20 ms and release time at 20 ms maximum [10].

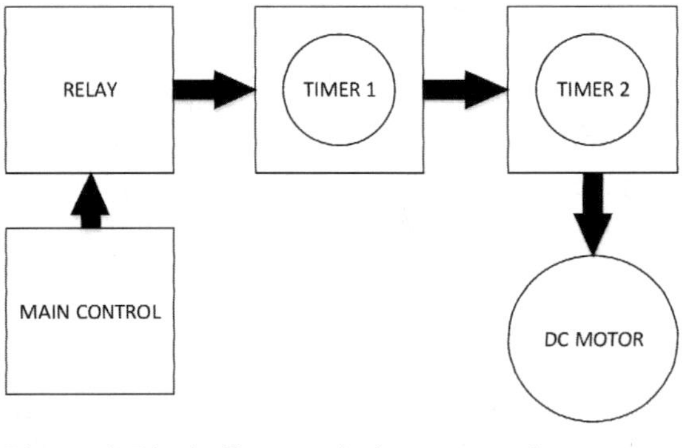

Figure 3.Block diagram design automation system.

Based on the figure. 3, the existing system will be integrated with 2 timers and relays. The main control consists of a switch on / off and directional keys, the red button to move away from the midpoint, and the green button is the opposite. Timer 1 is a timer that will be used to delay during a certain time so that the anemometer can record the airflow from the fan. Timer 2 delays for a certain time, assuming the anemometer support pole can move as far as 4 cm during the timer 2 works.

System workflow based on the Figure. 3 is the main control system provides a signal to the relay where the received signal is a voltage that will determine the direction of motion of the system. The signal from the relay will go to timer 1 and activate the timing, along with it timer 2 will also work and start the timing calculation. When timer 2 does the calculation, the dc motor will move and will stop when timer 2 finishes the calculation. Timer 2 will be active again after timer 1 finishes the calculation.

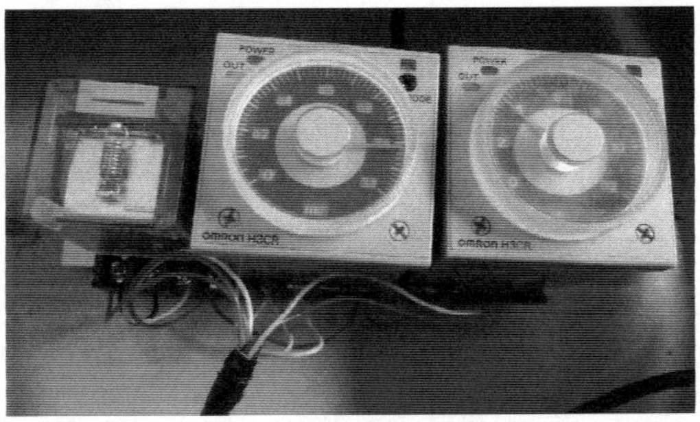

Figure 4. Automation system design with relays and timers.

5. Results and Discussion

Adjustment is performed to obtain an accurate value for both anemometer readings and anemometer pole movement. The result of the adjustment is quite close to the standard requirement which the timer setting for 2 minutes, resulting in a measurement delay of approximately 2 minutes. Setting timer 2 for 4 seconds, resulting in a movement approaching 4 cm. The results more clearly can be seen in table 1.

Table1. Measurement of points shift and times of data collection.

Data Collection	points shift (cm)		difference between points shift (cm)		Time of data collection between points (s)	
	automatic	manual	automatic	manual	automatic	manual
1	4	4,15	-	-	-	-
2	8	8,1	4	3,95	124	151
3	12	12,05	4	3,95	124	155
4	16	16,2	4	4,15	124	150
5	20,2	20	4,2	3,8	124	145
6	24,4	24,3	4,2	4,3	124	144
7	28,7	28	4,3	3,7	124	143

Based on table 1 it is seen that for shifting point of measurement in automated system has value of difference approaching standard requirement while the biggest difference value exists on taking data to 7 with value 4,3 cm, more 0,3cm than specified standard. As for the manual system obtained the same value, there is a difference of 0,3cm from the desired standard on the collection data to 6 and to 7 that is equal to 4,3cm and 3,7cm. Viewed as a whole, the difference between point shifts in automated systems is more precise than the manual system.

The difference in movement occurring in the automation system is due to the different friction between the anemometer with the rails at each point. For manual systems there are also differences between points because the reading of the measuring instrument is not ideal because the position of the ruler is at the bottom of the rail.The data retrieval time for the automatic system is quite precise due to the settings on the 1st and 2nd timers that are consistent. The average data retrieval time for the automatic system is 124 seconds. For manual measurement the average data retrieval time is 148 seconds. There is a 24 second difference where data retrieval is automatically faster.

6. Conclucion

Automated systems that have been designed are close to the standard desired specifications. This system is able to reduce the data retrieval time of 16.22% from manual system. The precision in shifting between points, both automatic and manual systems is still lacking, so the need to develop control methods to improve its precision.

Acknowledgement
Our gratitude goes to UNDP through the BRESL program which has donated equipment. In addition, we also thank Mr. Suyut and Mr. Heri who have helped with this research.

References

[1] Electrical Power Statistics 2015 *Directorate General of Electricity of the Ministry of Energy and Mineral Resources* 29 Year Budget 2016
[2] Citra I A and Hari P 2017 Desain Kipas Angin Neodymium Menggunakan Metode Quality Function Deployment (QFD) *Seminar Nasional IENACO* ISSN 2337-4349
[3] Mahlia T M I, Masjuki H H and Choudhury I A 2002 Theory of energy efficiency standards and labels *Energy Convers. Manag.* **43** 6 743–761
[4] Meier A K and Hill J E 1997 Energy test procedures for appliances *Energy Build* **26** 23–33
[5] SNI IEC 2013 *Standard SNI IEC 60879: Kinerja dan konstruksi kipas angin listrik dan regulator*
[6] CEI IEC 1986 *Performance and constructionof electric circulating fans and regulators*
[7] UNDP Indonesia 2017 *Barriers Removal to the cost-effective development of energy Efficiency Standards and Labelling (BRESL)* (online) available at: http://www.id.undp.org/content/indonesia/en/home/operations/projects/environment_and_energy/barriers-removal-to-the-cost-effective-development-of--energy-ef.html accessed on September 29nd 2017 at 4.16pm
[8] Ryniecki A 2015 Basics of Process Control:the On-Off Control System *Testing Technology Precess Control System*
[9] OMRON 2017 *H3CR-A* (Online) available at: https://www.ia.omron.com/products/family/193/specification.html accessed on 29/9/2017, at 5:31pm
[10] OMRON 2017 *MK2P-I Datasheet (PDF) - Omron Electronics LLC* (Online) available at: http://pdf1.alldatasheet.com/datasheet-pdf/view/146669/OMRON/MK2P-I.html accessed on 29/9/2017, at 5:52pm

Preliminary Comparative Study of Innovation Process Model for Startup in Indonesia

A Asmoro*, L E Nugroho and S Selo

Electrical Engineering and Information Technology Department, Gadjah Mada University, Yogyakarta, Indonesia

*anung.s3te14@mail.ugm.ac.id

Abstract. The startup is the catalyst that transforms ideas into products. Innovation is widely recognized as an essential enabler of a competitive startup, organizational, academic, and enterprise to survive and thrive. However, many of innovation initiatives do not generate satisfactory profit or competitive advantage. The problem does not rest in the invention part or the generation of innovative ideas, but more in the profitable management of the innovation process from an approach to a successful product in the market. This paper puts forward a comparative study of customer-oriented innovation process model Blank, Ries, and Maurya. This paper quarry is the most appropriate model and best fit with startups samples, and looking for rational reasons to do further research to obtain the innovation process model that more appropriate and specific to startups. This research used an explanatory sequential research design in which qualitative and quantitative studies. These data were gathered from 5 startups in Indonesia through FGD, interviews, and observations. Substantial evidence supports that model Maurya is the most appropriate model and best fit with these startups and additional research is needed to formulate a new model that fully supports creative ecosystem industry in Indonesia.

1. Introduction

Innovation is defined as the invention of new products, processes, knowledge or services which provide a degree of novelty either to the developer, the industrial sector, the nation or the world and succeeds in the marketplace. Innovation is widely recognized as an indispensable competitive enabler by organization, academic, and company to survive and grow [1].

Startup's customer development model begins with a simple premise: learn and find out who will be targeted as early customers, and where the segment is them on the market, is a separate and distinct process from product development. The activities were then called as customer development. Most beginners do not have the startup process to find a market, find the first customer, validate their assumptions, and develop its business.

While some of that success is that it managed to do all things. The difference is perhaps that the fortunate people who managed to create a model of customer development [2]. In essence, the startup is the catalyst that transforms ideas into products. When customers interact with the product, they generate feedback and data.

Qualitative feedback (like what they like and do not like) and quantitative (such as how many people use it and find it worthwhile). Lean startup method can be used to build companies with capital-efficient because it allows startups that avoid wasteful of time and money [3].

The classic product-centric approach allocates customer involvement during the requirements-gathering phase but leaves the customer validation until after the software is released. There is a delay when the startup disengages from customers for weeks or months while they build and test their solution. During this time, it's entirely possible for the startup to either create too much or be led astray from making what customers want [4].

However, although most enterprises fully realize the importance of innovation and they continue to spend more and more on innovation, many of these initiatives do not generate satisfactory profit or competitive advantage. The problem does not lie in the invention part or the generation of innovative ideas, but more in the profitable management of the innovation process from an idea to a successful product in the market [5-11].

This paper puts forward a comparative study of three customer-oriented innovation process models, best known today, which is a model of innovation process Blank, Ries, and Maurya. This study aims to answer the following research questions:

- What is an appropriate model and best fit from the three models with startups in Indonesia, and
- Is it necessary to do further research to obtain the innovation process model that more appropriate and specific to startup in Indonesia?.

2. Method and Data

2.1. Method

This research used an explanatory sequential research design in which qualitative and quantitative studies conducted using the model explanations. Interpretation and analysis of data obtained through the union to get models of the innovation process.

This research conducted two different research phases: quantitative followed by qualitative [12] as illustrated in Figure 1.

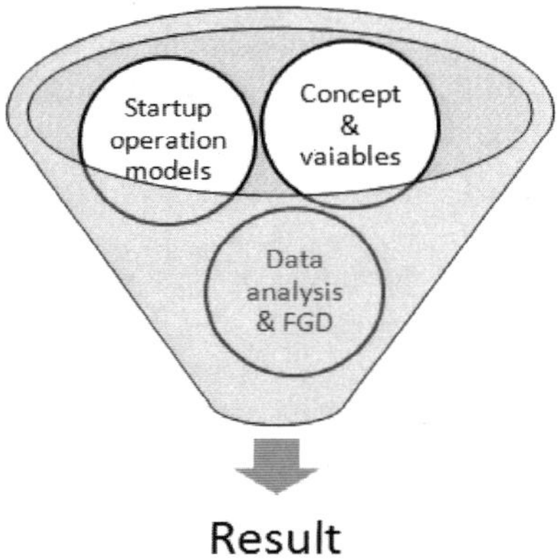

Figure 1. Research Design of Explanatory Sequential Mixed Method

2.2. Data

Sources of data and information that have been used in this study were obtained from two sources, namely the direct result of research in the field as the primary data and data already available as secondary data.

The variables that affect the performance of the creative industries that have been measured to analyze the data models of existing innovations [13] and interpret the startup staging model [2] are illustrated in Figure 2.

IOP Conf. Series: Materials Science and Engineering **384** (2018) 012057 doi:10.1088/1757-899X/384/1/012057

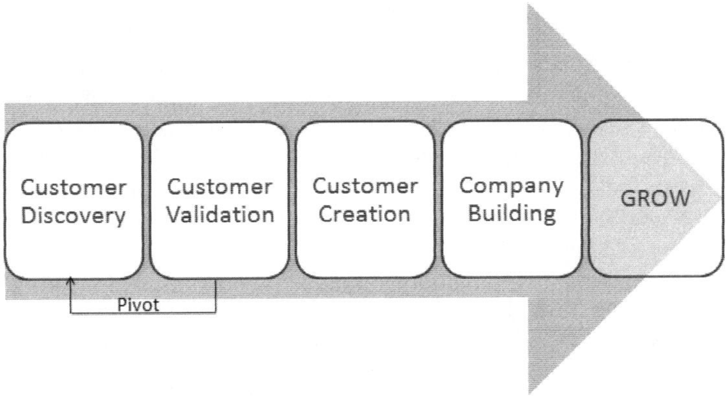

Figure 2. Startup Staging Model.

The data required are the primary and secondary data. These data were gathered through FGD, interviews, and observations. The data sample is taken from the number of startups in several major cities in Indonesia, with a composition as showed in Table 1.

Table 1. Data Model Comparison.

No	Phase	Customer Development	Lean Startup	Running Lean
1	Ideation		-	-
2	Filtering	Customer discovery	-	-
3	Product Development	Customer validation Customer creation Customer development by search execute business models. Product development: continuous iteration and pivots. Use actionable metrics	MVP validation. Product development cysle: build-measure-learn. Innovation accounting: Registration rate, acquisition, activation, retention, revenue, referral.	Customer development: problem/solution fit (improvement by use Lean Canvas Business Model Generation) Product development: Product market fit by validated learning and pivots. Innovation accounting (acquisition, activation, retention, revenue, referral)
4	Company building	Company building	Engine of growth	Engine of growth
5	Grow	Grow	-	-

In interpretation data, the results of quantitative and qualitative research are interpreted and analyzed its impact on business performance of creative industries. This process is necessary to ensure that choose a model that is more appropriate startup innovation to improve the performance of innovative businesses. Also to decide vitally or not, to conduct additional research to obtain the innovation process model that more appropriate and specific to startup.

3. Result

The result obtained from FGD process analysis data from the data samples that success collected shown in table 2.

Table 2. FGD Data Analysis.

No	Phase	Needed by Startup	FGD Results
1	Ideation	Pre-seed capital funding capability sharing academic program community development working space & supporting facilities	80% respondent need pre-seed capital funding. 100% respondent working space and supporting facilities.
2	Filtering	Success Prediction for early stage startup creativity for monetizing measurement	100% respondent need success prediction for early stage startup and monetizing measurement.
3	Product Development	Costumer validation: Lean canvas business model customer deveopment by problem/solution and delivering Product validation: Product-market fit by validated learneing and pivots Business model validation: Innovation accounting using actionable metrics (Acquisition, Activation, Retention, Revenue, Referal)	30% respondent us3 running lean and 20% use lean statup
4	Company building	Market validation: Enine of growth market access to customers	10% use market validation before launch the product
5	Grow	Value validation: Venture capital funding	All respondent not implement yes this method

From Table 1 and Table 2 above, it is seen that all of three models not cover ideation stage, even though 80% respondent need pre-seed capital funding and all respondents need working place and supporting facilities. Only customer development model covers the filtering ideation. Unfortunately, no responsdent use this model.

4. Discussion

In general, the results of research from Table 2 are that these models have a lot of advice about the importance of creative innovation in the IT world. In particular, the average value of Maurya model shows that this model is discussed more fully in business performance and competitive strategy. This model is also equipped with the tools of lean canvas and validation board to prepare a viable plan, customer validation and startup performance measure.

Lean canvas and validation board tools from Maurya can help determine the competitive strategy and measure performance validation process. But both the special tools can be used on stage at the customer validation until validation market. Both tools have not reached the stage of nurturing creativity, creativity selecting, and value validation.

5. Conclusion

Substantial evidence from result and discussion supports that Model Maurya is the most appropriate model and best fit (from the three models) with startups in Indonesia. Lean canvas and validation board tools from Maurya can be used at the customer validation stage until validation market only. Because these three models have not been fully supported creative industries and ecosystem management in the creative digital industries that can build a reliable startup, so further research is needed to formulate a new model that can address this issue. .

References

[1] Drucker P 1985 *Innovation and Entrepreneurship: practise and principles* (London: Heineman)

[2] Blank S and Dorf B 2012 *The Startup Owner's Manual The Step-by-Step Guide for Building a Great Company* K&S Ranch
[3] Duncan S P 2012 The Lean Startup *Software Quality Professional* **14** (4) 48
[4] Maurya A 2012 *Running Lean* (Newton: O'Reilly Media)
[5] Paternoster N, Giardino C, Unterkalmsteiner M, Gorschek T, and Abrahamsson P 2014 Software development in startup companies: A systematic mapping study *Information and Software Technology* **56** (10) 1200-1218
[6] Sarel G, Martie-Louise V and Tim K 2012 The Role of Networks in Small and Medium-Sized Enterprise Innovation and Firm Performance *Journal of Small Business Management* **50** (2) 257–282
[7] Chun-Hsien W, Chih-Cheng L and Pei-Yu C 2014 Exploring High-technology Firms' Open Innovation Strategy and Performance from Knowledge-based View *Proceedings of PICMET '14: Infrastructure and Service Integration*
[8] Da Silva F Q, Santos A L, Soares S, França A C C, Monteiro C V and Maciel F F 2011 Six years of systematic literature reviews in software engineering: An updated tertiary study *Information and Software Technology* **53** (9) 899-913
[9] Kemp R G, Folkeringa M, De Jong J P and Wubben E F 2003 *Innovation and firm performance* (The Netherlands: EIM)
[10] Lööf H and Heshmati A 2006 On the relationship between innovation and performance: A sensitivity analysis *Economics of Innovation and New Technology* **15** (4-5) 317-344
[11] Van der Panne G, van Beers C and Kleinknecht A 2003 Success and failure of innovation: a review of the literature *International Journal of Innovation Management* **7** 3 309-38
[12] Creswell J W and Creswell J D 2017 *Research design: Qualitative, quantitative, and mixed methods approaches* Sage publications
[13] Best R J 2009 *Market-based Management: Strategies For Growing Customer Value and Profitability* (Upper Saddle River, New Jersey: Prentice Hall)

IOP Conf. Series: Materials Science and Engineering **384** (2018) 012058 doi:10.1088/1757-899X/384/1/012058

Robot Operating System (ROS) Compatible Low Cost Rotating Light Detection and Ranging (Lidar) Design

N S Widodo[1], S A Akbar[1] and A Rahman[2]

[1]Program Studi Teknik Elektro, Universitas Ahmad Dahlan, Jl Prof Dr Soepomo SH, Yogyakarta
[2]Program Studi Sistem Informasi, Universitas Ahmad Dahlan, Jl Prof Dr Soepomo SH, Yogyakarta

*nuryono.sw@ee.uad.ac.id

Abstract. Robot Operating system (ROS) is collection of software frameworks for robot software development. Researchers that work with ROS will have an opportunity to use or to be exact to reuse any software framework from another researcher as long as they use a similar robot platform, or similar device that has been include in the ROS. This paper present a design for a low cost rotating Lidar that compatible with ROS. As all we know distance measurement is the basic for building an autonomous capability for robots. Rotating lidar is among one the most popular device used to acquire this kind of information. However most of them are not suitable for low budget project because of their high cost. The design presented here tries to build a low cost rotating lidar based on Lidar Lite device with the compatibility with the ROS. The design incorporate a control board based on a microcontroller.

1. Introduction

LIDAR (Light Detection and Ranging) is an optical remote sensing technology that measure the diffused property of the light to calculate the distance or another quantity from a remote targeted object. In general it can be said that lidar is a method to measure the distance to an object or surface using laser pulse. Just like a radar, that uses a radio wave rather than light, a lidar measure the distance to an object by measuring the time elapsed between light pulse transmission and the reflected pulse detection. LIDAR technologies has been widely applied in the field of geodetic, archeology, geographic, and atmospheric surveys. Another area that benefited by the lidar technology is the area of robotic, with its main function as distance measurement sensor, the lidar has become increasingly popular in robotic, especially in the area of mobile robotic, as presented in [1]. As we all know distance information is the basic information to build autonomous robot behavior, ranging from obstacle avoidance behavior to simultaneous location and mapping, based on distance measurement information. Many researchers use vision sensor to build their robot localization ability, such as [2]-[6]. Others use ranging sensor, such as lidar to build the capability. The most popular lidar used in robotics is the scanning lidar or often mentioned as scanning laser, in which the unit is rotating continuously and performs series of distance measurement simultaneously. However the current scanning laser available on the market are relatively high cost for a small budget robotic. Fortunately, there has been a cheap lidar with the price below $200, but it lack the scanning capability compared to the higher priced lidar. This paper presented a design to build a scanning lidar based on the aforementioned cheap lidar with the compatibility to Robot Operating System (ROS).

Robot operating system or ROS, is actually not an operating system, but it rather serves as a middleware between a computer and robot related hardware or subsystem. ROS enables a robot researcher to use and reuse a package, a collection of program, including a driver for a specific robot hardware, in another robot. It provide a package for many robot platforms, sensors, or actuators commonly used both in industrial robot or research robot. Anyone also possible to build his own robot platform in ROS based on the available sensor, actuator and control algorithm packages in ROS. ROS has been used in many kind of robots, as can be found in [7]-[10]. The design presented on this paper based on a lidar sensor and rotating mechanism controlled by a microcontroller to build a rotating lidar compatible with ROS with a target cost below $400.

2. Research Methodology

The design comprised hardware configuration and software configuration. It is important to note that the design has a targeted budget not more than $400, since it's the lowest scanning lidar available in the market right now.

2.1. Hardware Configuration

The main hardware component is obviously a lidar, a Lidar Lite V3 from Garmin, as shown in figure 1, was used in this design. It has a maximum range of 40 meters, 1 cm resolution, 2.5 cm accuracy for range below 5 m and 10 cm in range more than 5m. The unit has an update rate of 270 Hz in typical, and 650 Hz in fast mode, it means that the unit is able to provide 270 measurement in 1 second in typical an 650 measurements in in 1 second in fast mode. The price tag for the unit is around $150.

Figure 1. Lidar Lite V3.

As for the rotating actuator, a stepper motor is used, stepper motor was chosen because its simplicity to obtain a consistent step of rotation, as long as it's not loaded with heavier load that its torque. The Arduino Nano is used for controlling the lidar operation and stepper rotation. It also works as message generator for the ROS through a serial connection with the computer running ROS. To eliminate a mechanical design complexity, in this case, mechanical design needed to avoid cable tangling especially the cable for the LIDAR due to its scanning/rotating movement, a Bluetooth based connection is used between the Arduino and the Computer. The overall block diagram presented in figure 2.

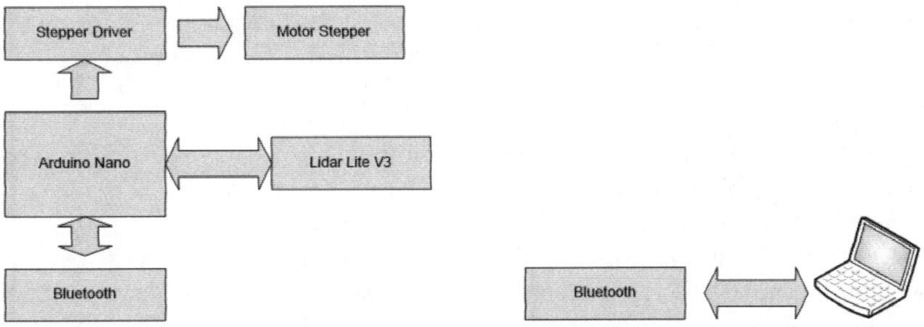

Figure 2. Block Diagram.

2.2. Software Configuration

The software section of the design serves mainly to control the operation of the lidar and the stepper motor, and to form a message containing the information from the lidar that compatible with the ROS laser scan message type. The flowchart of the software section is shown in figure 3.

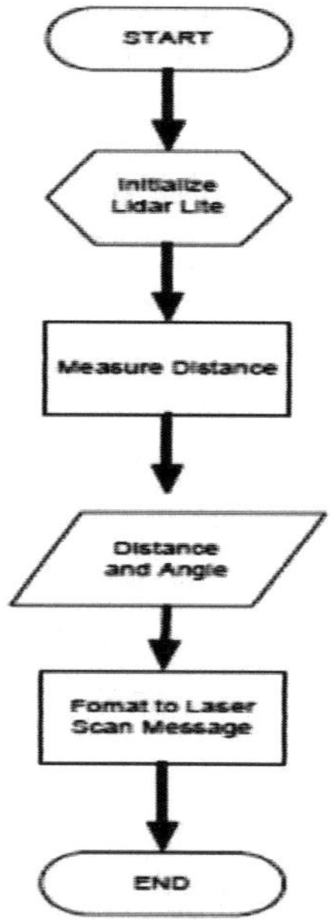

Figure 3. Flowchart.

The Lidar Lite V3 has two interfaces, I2C and PWM. The design presented here uses the I2C interface, and based on the sample code provided in from Garmin as the manufacturer of the unit, https://github.com/garmin/LIDARLite_v3_Arduino_Library [11]. After the measurement the distance information is packed together with the position of the stepper, in the form of angle, that range between 0° to 360°, as a representation of a full rotation.

For laser scanners, ROS provides a special Message type in the sensor_msgs package called LaserScan to hold information about a given scan. LaserScan Messages make it easy for code to work with virtually any laser as long as the data coming back from the scanner can be formatted to fit into the message. Base on this, the next stage conducted in the Arduino Nano is to ensure that the message its generate is has the same format with the LaserScan message in ROS so it can be processed further in ROS.

On the ROS side we need at least two functions, the first is a driver that enable the ROS to recognize and find the designed rotating lidar module and to receive a message from it, and the second is a publisher that enable another part of the ROS listen to the message and process it accordingly.

3. Result and Discussion

The rotating lidar built in this paper has been tested with ROS, especially ROS Indigo, in an Ubuntu 14.4 LTS Laptop. For verification purpose the result of the scan from the rotating lidar built was visualized using rviz. Rviz (ROS visualization) is a 3D visualizer for displaying sensor data and state information from ROS. Using rviz, user can display live representations of sensor values coming over ROS Topics including camera data, infrared distance measurements, sonar data, and more. It even has the ability to display a robot configuration on a virtual model of the robot. The scan result is displayed in rviz as shown in figure 4.

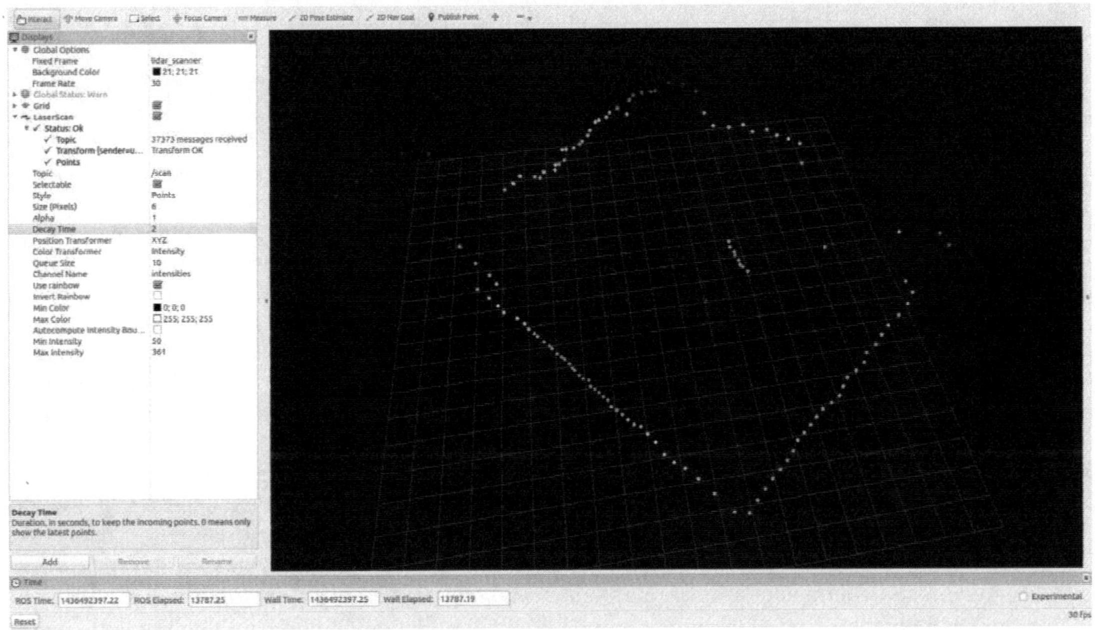

Figure 4. Scan Result in rviz.

The overall cost spent to build the rotating lidar is $195, with the detail presented in table 1.

Table 1. Cost Spent.

Component	Unit price ($)	Quantity needed	Subtotal Cost
Lidar-Lite V3	150	1	150
Stepper Motor	10	1	10
Stepper Driver	3	1	3
Arduino Nano	10	1	10
Bluetooth set	10	1	10
Li ion Battery	2	2	4
3D Printed Housing	8	1	8

4. Conclusions

Presented in this paper a design for a ROS compatible rotating lidar or laser scanner, based on the easy off the shelf part, with the overall cost under $195. As for the improvement, an encoder could be added to ensure that there is no glitch in the rotation that lead to a valid consistent scan result. Another improvement possible is to add an IMU, so with an appropriate software enhancement the unit can be used as a basic sensor to perform simultaneous location and mapping (SLAM).

Acknowledgement

This paper is a result of research conducted using *Penelitian Dosen Pemula* grant, contract number PDP-039/SP3/LPP-UAD/IV/2017.

References

[1] Oßwald S, Görög A, Hornung A and Bennewitz M 2011 Autonomous climbing of spiral staircases with humanoids *Intelligent Robots and Systems (IROS), 2011 IEEE/RSJ International Conference* 4844-4849

[2] Hoermann S and Borges P V K 2014 Vehicle localization and classification using off-board vision and 3-D models *IEEE Transactions on Robotics* **30** (2) 432-447

[3] Minaeian S, Liu J and Son Y J 2016 Vision-based target detection and localization via a team of cooperative UAV and UGVs *IEEE Transactions on systems, man, and cybernetics: systems* **46** (7) 1005-1016

[4] Lategahn H and Stiller C 2014 Vision-only localization *IEEE Transactions on Intelligent Transportation Systems* **15** (3) 1246-1257

[5] Widodo N S and Rahman A 2012 Vision based self localization for humanoid robot soccer *TELKOMNIKA (Telecommunication Computing Electronics and Control)* **10** (4) 637-644

[6] Utomo O K, Santoso D and Nugroho S 2016 Location and Position Determination Algorithm for Humanoid Soccer Robot *TELKOMNIKA (Telecommunication Computing Electronics and Control)* **14** (1) 4-13

[7] Schwarz M, Pastrana J, Allgeuer P, Schreiber M, Schueller S, Missura M and Behnke S 2013 Humanoid teensize open platform nimbro-op *Robot Soccer World Cup* 568-575 Springer, Berlin, Heidelberg

[8] Hellmund A M, Wirges S, Taş Ö Ş, Bandera C and Salscheider N O 2016 Robot operating system: A modular software framework for automated driving *Intelligent Transportation Systems (ITSC), 2016 IEEE 19th International Conference* 1564-1570 IEEE

[9] Wong K W and Kress-Gazit H 2017 From High-Level Task Specification to Robot Operating System (ROS) Implementation *Robotic Computing (IRC), IEEE International Conference* 188-195 IEEE

[10] Allgeuer P, Farazi H, Schreiber M and Behnke S 2015 Child-sized 3D printed igus humanoid open platform *Humanoid Robots (Humanoids), 2015 IEEE-RAS 15th International Conference* 33-40 IEEE

[11] Github 2016 *LIDAR-Lite v3 Arduino Library* (Online) available at: https://github.com/garmin/LIDARLite_v3_Arduino_Library

The Effect of Tutorial Multimedia on the Transformator Learning Outcomes Based on the Students' Visual Ability

S Sriadhi[1*], S Gultom[2], R Restu[3] and J Simarmata[4]

[1]FT Universitas Negeri Medan, Indonesia
[2]FMIPA Universitas Negeri Medan, Indonesia
[3]FIS Universitas Negeri Medan, Indonesia
[4]FE Universitas Negeri Medan, Indonesia

*sriadhi@unimed.ac.id

Abstract. The low learning outcomes have a negative impact on the quality of graduates. This study is aimed at revealing the effect of multimedia exploratory tutorial model on the learning outcomes of Basic Competence (BC) and Advanced Competency (AC) based on the students' visual ability differences in the course of Transformator. The research was conducted in State University of Medan using quasy experiment. Samples of 82 students were divided into three groups: Visual High (V_H), Visual Medium (V_M) and Visual Low (V_L). The comparison of research results was analyzed by Anova. The result shows that at the BC level the students' learning outcomes in the V_M and V_H groups were higher than V_L while the V_M and V_H did not differ significantly. At the AC level the three student groups differ significantly, the V_H group is higher than the V_M and the V_M is higher than the V_L. This study concludes that multimedia exploratory tutorials are effective for V_M and V_H student groups but not for V_L. Therefore exploratory multimedia should be used for V_M and V_T students to obtain optimal BC and only for V_H group to obtain optimal AC. However, this multimedia is not recommended for students with low visual abilities (V_L). It is advisable to prepare a special exploratory media for high visual ability students, while for students with low visual ability need to do further research to find effective learning media used.

1. Introduction

Technology and vocational education in Indonesia have played an important role in the transition of science and technology that enable graduates to prepare themselves before entering the mid-level workplace. Nevertheless, the facts show that vocational technology education graduates have not properly reached the standard competency [1, 2]. Learning outcomes in the Transformator field which is one of the main areas in vocational technology education is considered low so that it affects the low competence of graduates. Several studies revealed that low learning outcomes in electricity field was mainly caused by the lack of learning media [1, 3, 4, 5]. This weakness needs to be addressed so that the decreased quality of education would not trigger the low competence of graduates. The problem now is how to improve students' learning outcomes with the aforesaid problems, lack of learning media and the level of visual ability among different students.

Visual ability is the ability to manipulate an object in a rotational perspective that includes intelligence in rotation operations, manipulation of 2-D and 3-D forms of an object [6]. The instructional medium developed is an exploratory tutorial model [7] with the design as shown in Figure 1.

IOP Conf. Series: Materials Science and Engineering **384** (2018) 012059 doi:10.1088/1757-899X/384/1/012059

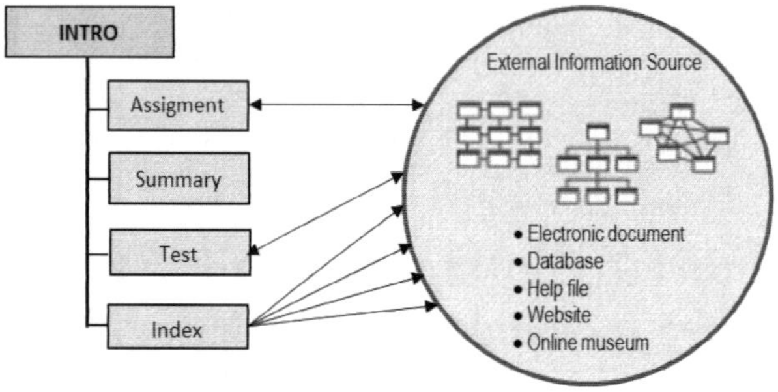

Figure 1. Exploratory tutorial design.

The use of instructional multimedia as an effort to improve learning outcomes is based on problem analysis and research studies that multimedia-based learning is an alternative to improve learning outcomes in electricity [1, 2, 8]. In addition, Kora kakis, et al. [9] found that the use of 3-D multimedia is becoming an enhancement of science learning. Moore [4] also revealed that multimedia learning tutorials become solutions to improve the quality of electromagnetic learning. Similarly, Hsiu [10] who proved the benefits of multimedia-assisted learning, Jenkins and Burtner [5] developed visualization of measurement technology in mechanical engineering using computer-aided learning modules, and Höhne and Henkel [11] who developed multimedia applications in technological education design. The results of these studies proves the contribution of instructional multimedia for learning outcomes improvement in science and engineering, particularly in electricity.

Instructional multimedia has many advantages such as being able to visualize abstract and conceptual events into easily understood concepts [12], and to improve students' learning motivation [1, 13, 14]. Multimedia will be effectively applied if its development follows the rules of multimedia design [15], and is oriented on students' cognitive loads [16]. It is in line with the theory of information processing systems in the human brain [17]. Multimedia-based learning is expected to improve learning outcomes which are distinguished in two levels, namely Basic Competencies (BC) and Advanced Competencies (AC). Basic competencies are competencies used to master theory and usage of Transformator subject, while Advanced Competencies are competencies to planning and problem solving of Transformator subject.

The purpose of this study is analysis the effect of multimedia-based learning on the transformator learning outcomes for students with different visual ability. The data were analyzed by using descriptive statistics and inferential statistics, i.e. one way Anova.

2. Research Methodology

The research was carried out in the form of quasy experiment, by means of multimedia Exploratory Tutorial model in the Transformator subject. The sample of 82 students were divided into three groups of abilities: Visual High (V_H), Visual Medium (V_M) and Visual Low (V_L) determined based on the measurement result using The Purdue Visualization of Rotation Test [6]. Learning outcomes are measured using test instrument in two levels, namely Basic Competencies (BC) and Advanced Competencies (AC). Data analysis for comparative study of learning outcomes was conducted with Anova after the requirements test [18].

3. Results and Discussion

The results showed that most of the respondents had medium-level visual abilities, n=18 (58.54%), the rest were Visual Low, n=48 (21.95%) and Visual high n=16 (19.51%). The ability of students who are generally at the middle level and only a small part of high visual ability becomes an obstacle factor in the learning process of students to understand the subject is abstract and conceptual. This fact also evidenced by the measurement data that show the relevance of the level of visual ability with learning outcomes as shown in Table 1.

Table 1. Learning outcomes of Basic Competencies.

Visual Abilities	n	Learning Outcomes (mean)		
		Basic Competencies	Advanced Competencies	Total
V_H	18	75.89	69.78	72.83
V_M	48	81.25	76.67	78.96
V_L	16	82.13	81.63	81.88
Total	82	80.24	76.12	78.18

Overall learning outcomes of Basic Competencies (BC) is higher than Advanced Competencies (AC). This analysis proves that students' competence is generally only at the level of understanding of teaching subjects while application competence and creativity for further development and problem solving are still low.

Differences in learning outcomes based on group ability were done by comparative analysis with Anova was conducted after passing the requirements test. The data distribution for the Basic Competencies variable is normal, and Levene's Test of Equality of Error Variances produces $F (2.79) = .574$; $p= .565$ so that the variant is homogeneous. Anova for Test of Between-Subject Effects obtained $F (2,79) = 7.83$; $p = .001$ which shows significant differences between sample groups based on visual ability level. Similarly, the Advanced Competencies variable, the price $F = 0.972$; $p = .383$ indicates homogeneous variance, and the Test of Between-Subject Effects yields $F = 19.58$; $p = .00$.

Thus the requirements have been met to perform comparative analysis using Anova.

Table 2. Pairwise Comparison for Basic Competencies.

Visual Abilities (I)	Visual Abilities (J)	Mean Different (I – J)	Std. Error	Significant [b]
V_L	V_M	-5.36*	1.48	.001
	V_H	-6.25*	1.84	.003
V_M	V_L	5.36*	1.48	.001
	V_H	-.88	1.54	1.000
V_H	V_L	6.25*	1.84	.003
	V_M	.88	1.54	1.000

Table 2 indicates that there is a difference in learning outcomes of Basic Competencies between groups of students with low visual abilities (V_L) with students of medium visual skills (V_M) and high visual abilities (V_H). There is no significant difference between V_M and V_H. The result of the data analysis means that the learning result of Basic Competencies of V_L group is lower than V_M and V_H, but is not significantly different between V_M and V_H groups. This implies that visual abilities affect the achievement of learning outcomes in Basic Competencies especially in low visual groups. This finding is in line with previous research results that visual ability determines student learning outcomes [1, 5]. Multimedia that is loaded with visualization of learning content will be easily understood by students with high visual abilities. This is in line with information processing theory, work memory [17], cognitive load theory [16], and multimedia cognitive learning theory [19]. In the dependent variable of Advanced Competencies, there are significant differences between the three sample groups as shown in Table 3.

Table 3. Pairwise Comparison for Advanced Competencies.

Visual Abilities (I)	Visual Abilities (J)	Mean Different (I – J)	Std. Error	Significant [b]
V_L	V_M	-6.89*	1.55	.000
	V_H	-11.85*	1.92	.000
V_M	V_L	6.89*	1.55	.000
	V_H	-4.96*	1.61	.009
V_H	V_L	11.85*	1.92	.000
	V_M	4.96*	1.61	.009

The result of data analysis proves that there are learning outcomes differences in Advanced Competencies between the three group of students which is indicated by significant value $p<0.05$. This result shows that visual abilityis directly proportional to the learning outcomes of Advances Comptencies, while groups mean $V_H > V_M > V_L$.

This findings reinforce that ability to plan and problem solving in transformer subject require visual ability which linearly affects the learning outcomes. Therfore, the result of study is appropriate with the previous studies as well as proving the suitability of processing theory, process of thinking and multimedia learning cognitive [16,17,19]. The graph of learning outcomes is briefly shown in figure 2.

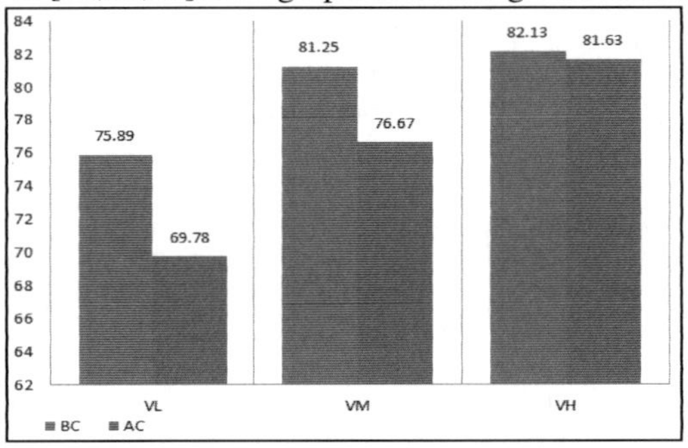

Figure 2. Comparison of Learning outcomes.

In the group of students, the visual abilities of low and medium gaps occur in the achievement of Basic Competencies and Advanced Competencies, but not so in the high visual ability group. This indicates that high visual ability positively influences achievement of Basic Competencies and Advanced Competencies and for students of medium and low visual ability is more difficult to achieve Advanced Competencies than Basic Competencies. This supports the results of previous studies [1, 4].

Multimedia Exploratory Tutorial model developed with the rules of teaching materials structure in parallel gives the opportunity for students to choose the lesson material according to their ability. This parallel pattern will be easily understood by students with high visual abilities but difficult for students with low visual abilities. It supports the theory of principles of multimedia form [15] and the rules of the preparation of multimedia tutorials [7]. At the level of Basic Competencies, students with medium visual abilities are parallel to high-ability visual students, but not so at the Advanced Competencies level, where high-ability visual students are able to achieve higher learning outcomes from the two underlying groups.

The results of this study are in line with previous research that the visual ability has a positive effect on the learning outcomes of engineering [20]. This study also supports research that proves students' learning outcomes of electrical power generation is determined by the level of spatial (visual) abilities [21]. However, the academic service system should also be improved to support learning outcomes. This is in line with Sriadhi's research [22] suggesting improving academic services including the integrated information system to support the teaching-learning process.

4. Conclusions

The development of media in accordance with the rules of design and structure of the content of teaching materials will determine the feasibility of tutorial media. This study proves that multimedia-based learning model Exploratory Tutorial, visual abilities have positive and linear effect on student learning outcomes. At the level of Basic Competencies V_L students achieve learning outcomes lower than the V_M and V_H groups, but the learning outcomes of the V_M group were not different from the V_H group. While at the Advanced Competencies level the visual ability more significantly determines the learning outcomes as evidenced by the learning outcomes of the V_H group of students higher than the V_M group and higher than the V_L group. Therefore exploratory multimedia tutorials are effective for V_M students for the Basic Competencies level and V_H students for the Basic Competencies and Advanced Competencies levels, but are not effective for V_L students for both the Basic Competencies and Advanced Competencies learning achievement levels.

Acknowledments
Acknowledgments are addressed to all those who have assisted this research, such as, heads of study programs, lecturers and students as research samples at State University of Medan.

References

[1] Sriadhi 2016 Pengaruh pembelajaran berbasis multimedia model activity centered tutorial dan exploratory tutorial terhadap hasil belajar bidang elektronika industri *Konaspi* **VIII**
[2] Nisa C and Agung Y A 2014 Pengembangan media pembelajaran berbasis ict menggunakan multisim 10 simulations pada pelajaran teknik elektronika *J.Pendidikan Teknik Elektro* **3** 311-317
[3] Truong M T 2014 Multimedia teaching methods in positive course of engineering *3rd UPI International Conference on Technical and Vocational Education and Training*
[4] Moore J C 2018 Efficacy of Multimedia Learning Modules as Preparation for Lecture-Based Tutorials in Electromagnetism *Education Sciences* **8** (1) 23
[5] Jenkins H and Burtner J 2007 An approach to develop and measure engineering visualization in an introductory mechanics course using computer-aided learning modules *Int. J. of Engineering Education* **23** 1 150-161
[6] Bodner G M and Guay R B 1997 The Purdue Visualization of Rotation Test *The Chemical Educator* **2** 4 1-17
[7] Horton W 2006 *E-Learning by design* (San Fransisco: John Wiley & Sons, Inc.)
[8] Sriadhi 2014 Analisis Karakteristik media pembelakjaran dan motivasi berdasarkan gaya belajar siswa sekolah menengah kejuruan *J. Educandum* **8** 2 37– 47
[9] Korakakis G, Paylatou E A, Palyvos J A and Spyrellis N 2009 3-D visualisation types in multimedia applications for science learning: A case study for 8th grade in Greece. *Computer and Education* **52** 390-401
[10] Hsiu P Y 2012 Effect of student engagement on multimedia-assisted instruction *Knowledge Management & E-Learning: An Int. J.* **4** 3 346 -358
[11] Höhne G and Henkel V 2004 Application of multimedia in engineering design education *European J. of Engineering Education* **29** 1 87–96
[12] Wouters P, Paas F and Merrienboer J J G 2008 How to optimize learning from animated models: A review of guidelines based on cognitive load *Review of Educational Research* **78** 645-675
[13] Malik M and Maharashtra P 2014 Effectiveness of ARCS model of motivational design to overcome non completion rate of students in distance education *Turkish Online Journal of Distance Education-TOJDE* **15** 2 Article 14
[14] Marshall J and Wilson M 2013 Motivating e-learners: Application of the ARCS model to e-learning for San Diego Zoo Global's Animal Care Professionals *The J. of Applied Instructional Design* **3** 2
[15] Mayer R E 2014 *Multimedia Learning* (New York: Cambridge University Press)
[16] Sweller J 2008 *Human Cognitive Architecture* (Sydney: University of New South Wales)
[17] Baddeley A, Eysenck M W and Anderson M C 2009 *Memory* (New York: Psychology Press)
[18] Cohen L M, Lawrence and Morrison K 2005 *Research Methods in Education* (New York: Taylor & Francis e-Library)
[19] Clark R and Mayer R E 2008 *e-Learning and the Science of Instruction* (San Francisco: John Wiley & Sons, Inc.)
[20] Sriadhi 2014. Multimedia Exploratory tutorial learning (ETL) untuk pembelajaran pembangkit listrik *Konferensi Nasional Aptekindo*
[21] Sriadhi 2015 The effect of exploratory multimedia learning towards learning outcomes of electrical power generation based on difference of students' spatial ability *J. Education Science* **1** 1
[22] Sriadhi 2017 Model of the material inventory management using multimedia based information system *IOP Conf. Series: Materials Science and Engineering* **180** 012239

IOP Conf. Series: Materials Science and Engineering **384** (2018) 012060 doi:10.1088/1757-899X/384/1/012060

Simulation of Design and Implementation of Smart Socket Prototype Controlled by Android Application

R A Wahyudi, A Saripudin and A H S Budi

Department of Electrical Engineering Education, Universitas Pendidikan Indonesia, Jl. Dr. Setiabudhi No. 207 Bandung 40154, Indonesia

*rauufashshally@student.upi.edu

Abstract. Internet of Things (IoT) is one of the emerging technologies at the moment. IoT allows use to control and monitor the use of electric power in rental homes via smart phones. One of the IoT support devices for monitoring power consumption is Smart Socket. Because of its usefulness, the author tries to design a prototype Smart Socket to calculate electric power in a rental house. The method used to create a Smart Socket is an experimental method. For communication needs used HTTP protocol and IoT Platform. The results show that smartphones can be used to control the Smart Socket and retrieve information about the total power consumption of each socket. While Smart Socket can do the calculation of total power consumption and send it to IoT Platform. Smart Socket successfully created and works well.

1. Introduction

Internet of Things (IoT) is one of the emerging technologies at the moment. With IoT, one can enable or disable an electronic system remotely [1]. In the IoT system, many variations of data are read, collected, and transmitted in an efficient and secure manner [2]. One of the proper modules to be used as the basis of an IoT device is the ESP-12E due to its economical price, low power consumption capability and integrated with TCP / IP [3]. The device is equipped with WiFi capability with satisfactory function [4]. With the WiFi feature, the ESP8266 module can connect to the internet.

One part of IoT is a smartphone. Smartphones are used to be the control and monitoring center of an electronic device. One of the most widely used operating systems on smartphones is the Android operating system. Android is an operating system for linux-based mobile devices that includes middleware and applications [5].

In general, a rental house has only one kWh-meter, to calculate the power consumption. Meanwhile, the use of electricity from each occupant of the rental house is not the same. However, the fees paid from each rental resident are the same. This causes the injustice and inconvenience of other rental residents. With IoT, rented house owners can monitor the power consumption of each room of the residents so that it can be used as a reference to quote the rental price per month per room. This calculation system will reduce the sense of injustice if payment is generalized.

One of the IoT support devices can be implemented for monitoring power consumption is the Smart Socket. Smart Socket designed to disconnect or connect the flow of electric current. Because of its usefulness, the author tries to design a prototype Smart Socket for the calculation of electrical power in a rental house, which aims to provide a solution to the problem of electricity use inequality between residents of rental homes.

IOP Conf. Series: Materials Science and Engineering **384** (2018) 012060 doi:10.1088/1757-899X/384/1/012060

2. Research Methods

2.1. Circuit and Working

In the schematic design and PCB design described the arrangement of components and pathways used to create a prototype smart socket. The design is done using Eagle software. Schematic design and design of smart socket prototype PCB is shown in Figure 1.

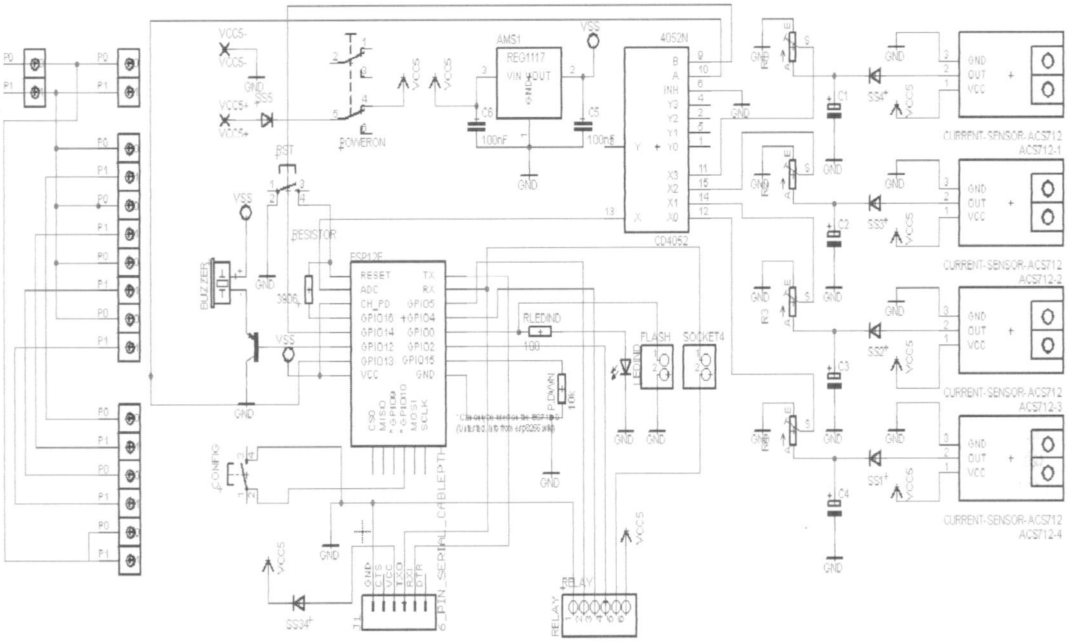

Figure 1. Schematic project.

2.2. Work Principle

The working principle of the smart socket prototype is illustrated by using the block diagram shown in Figure 2. In the block diagram it is explained that the smart socket prototype is divided into 2 modes of work i.e. config mode and running mode. In config mode, local communication is performed between the prototype smart sockets will be access point. In running mode smart socket prototype will retrieve data from Cloud Storage and calculate total power consumption in socket. Android application is used to control relay at smart socket and take the total power consumption information on each socket.

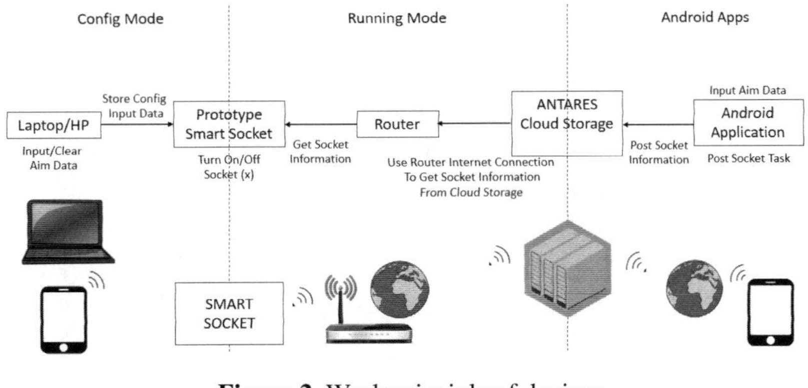

Figure 2. Work principle of devices

2.3. Algorithm

2.3.1. Smart socket. The smart socket prototype algorithm is illustrated using the block diagram shown in Figure 3.

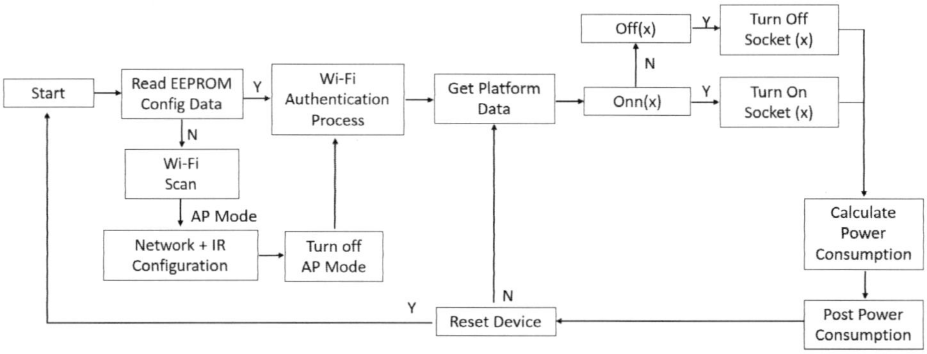

Figure 3. Smart socket prototype algorithm

2.3.2. Android application. The android application algorithm is illustrated using the block diagram shown in Figure 4.

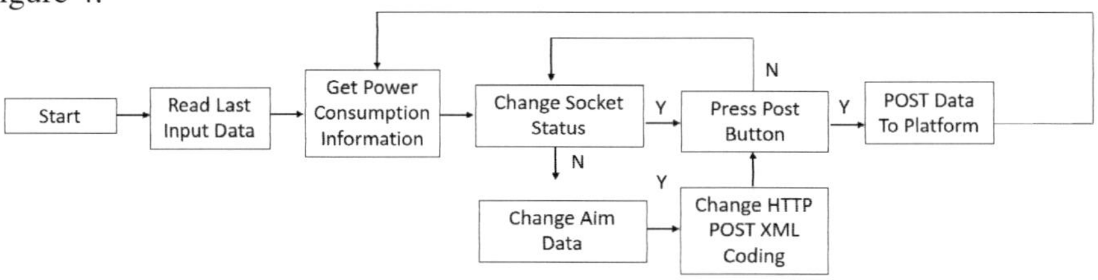

Figure 4. Android apps algorithm.

2.4. Hardware aspect

2.4.1. ESP-12E Wi-Fi Module. The module is best suited for Internet of Things (IoT) at its low cost, low power consumption capability as it requires 3.3V power, built in Wi-Fi module, integrated TCP/IP protocol stack, easy to flash and erase firmware and is usb powered [3]. As the IoT application module it can deploy in home automations, home appliances, industrial wireless network, sensor networks fields [4]. Table 1 shown ESP-12E Wi-Fi module specification.

Table 1. ESP-12E Wi-Fi Module Specification.

Parameters	Specification
Microcontroller	ESP8266
Memory	32 bit
Processor	TenSilica I. 106
Processor Clock	80 ~ 160 MHz
RAM	36 Kb
Storage	4096 byte
Built-in Wi-Fi	2.4GHz supports 802.11 b/g/n
ADC Pin	1 (10bit Resolution)
GPIO pins	10
Operating Voltage	3.0 ~ 3.6 Volt
Operating Current	80nA (Average)
Operating Temperature	-40^0 ~ 125^{0}C

2.4.2. Relay for 220-250 volt, 30A. A relay is a switch that is operated by using electricity. Relay is an electromechanical component consisting of 2 main parts namely electromagnet (coil) and mechanical (switch switch). Relays use the electromagnetic principle to drive the switch contacts, so that a small electric current can conduct higher voltage electricity. In Figure 5 is shown the shape and symbol of the relay.

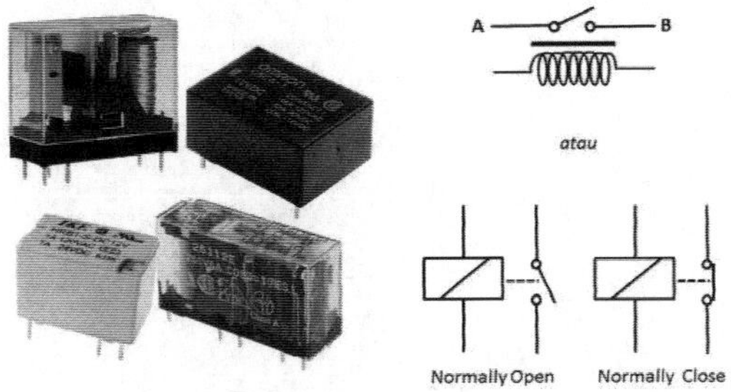

Figure 5. Relay devices and symbol.

2.4.3. Smartphone with OS Android 5.0 (Lollipop). Android is an operating system for linux-based mobile devices that includes operating systems, middleware and applications [5]. Android provides an open platform for developers to create an app. In addition, android is an open source operating system that is freely distributed and used by any vendor. This resulted in Android continues to grow rapidly both in terms of technology and in terms of the number of devices that exist in the world. On Android there is Davilk Virtual Machine (DVM) and Java Virtual Machine (JVM). DVM is a low-cost, specially designed Android-designed low memory machine running on embedded systems. While JVM is a machine that is in a machine that mimics its host machine and is used to run java code written on Android [5]. On android there are four main components on the android of activities, service, broadcast receiver and content provider.

2.5. *Software Aspect*

2.5.1. Arduino IDE software. Arduino IDE software is used to write down the code to run a system. The Arduino IDE software is open-source, it makes it easy to write the code and upload it into the microcontroller board. Coding on Arduino IDE Software written using Javascript language or C / C + language.

2.5.2. Cloud Storage. Cloud storage is an information technology architecture in which storage resources are available as services accessible via the internet. Cloud storage basically uses internet-based services to support business processes. One of cloud storage owned by Indonesia is made by Telkom Indonesia, which is ANTARES Platform. Cloud storage technology provides services to its users in 3 levels: infrastructure as service, Software as service and Platform as service.

2.5.3. Android Studio Software. Android Studio is an IDE for application development on the Android platform. Android Studio is designed in such a way as to make it easier for developers to develop Android apps using Javascript. Android Studio also provides a complete library to access the various features that exist on the smart phone, such as sensors, cameras, GPS and the others. Android and Android Studio Software developed by Google. Android Studio software is a development of the Eclipse IDE Software, and is based on the IDE Java IntelliJ IDEA. Android Studio is planned to replace Eclipse IDE in the future as the official IDE for Android app development.

3. Experimental Result

PCB is made using 2 layers, the top layer and the bottom layer. To connect between top and bottom layers used through holes. After PCB is made, installation of components using soldering equipment is

IOP Conf. Series: Materials Science and Engineering **384** (2018) 012060 doi:10.1088/1757-899X/384/1/012060

done. The components used are components with legs, SMD components and ready-made modules. After all components are installed, the PCB section is connected to a 4-socket terminal using a jumper as shown in Fig 20.

Figure 6. Smart socket prototype hardware

3.1. The Test Results of Android Apps

In android applications that have been designed to be tested. The smart socket android app is designed to monitor total power on the smart socket and disconnect and connect the power supply on each available socket. Figure 7 shows the interface display on main activity and setting activity Android applications that have been created. In the picture is shown some conditions that is when opening the application, at the time of taking the total data power consumption and when turning on the socket.

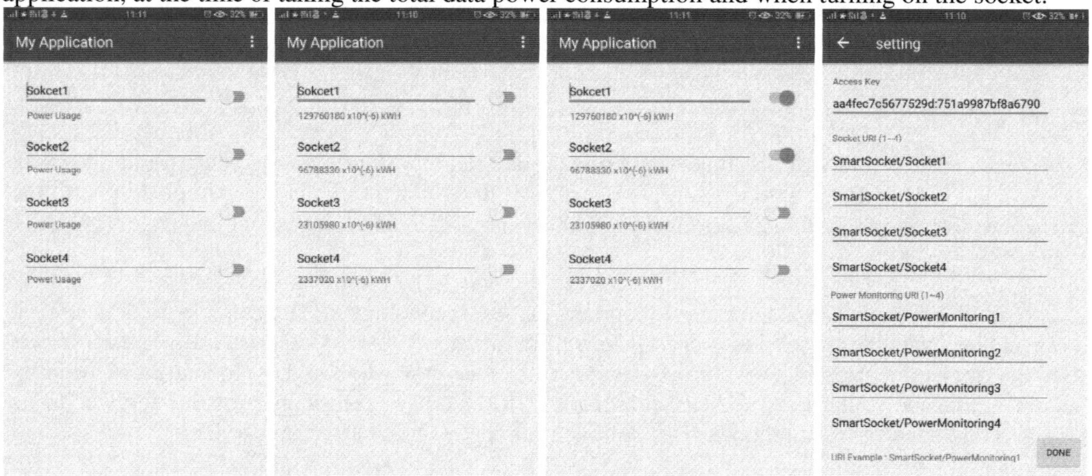

Figure 7. Main activity & setting activity on Android Apps.

To retrieve the total power information and send the socket status information, the destination server address is required. To input the destination server address can be done on the activity settings, by pressing the selection button on the top right of the main activity. Fig 21 shows the interface display activity settings.

3.2. The Test Results of Smart Socket Prototypes

In Running Mode, ESP-12e module will be enabled to Wi-Fi client. Running Mode will run if at the time of Config Mode has been done. At the start of Running Mode, the device will read the data on

IOP Conf. Series: Materials Science and Engineering **384** (2018) 012060 doi:10.1088/1757-899X/384/1/012060

EEPROM memory. The read data is the information ssid and Wi-Fi password, as well as the destination server address. Then the ESP-12E module will authenticate and connect to the specified ssid and password. Figure 8 and 9 shows serial monitor readings in Running Mode.

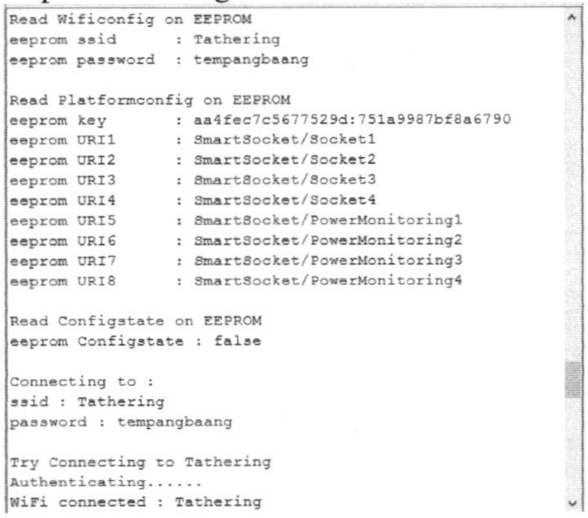

```
Read Wificonfig on EEPROM
eeprom ssid       : Tathering
eeprom password   : tempangbaang

Read Platformconfig on EEPROM
eeprom key        : aa4fec7c5677529d:751a9987bf8a6790
eeprom URI1       : SmartSocket/Socket1
eeprom URI2       : SmartSocket/Socket2
eeprom URI3       : SmartSocket/Socket3
eeprom URI4       : SmartSocket/Socket4
eeprom URI5       : SmartSocket/PowerMonitoring1
eeprom URI6       : SmartSocket/PowerMonitoring2
eeprom URI7       : SmartSocket/PowerMonitoring3
eeprom URI8       : SmartSocket/PowerMonitoring4

Read Configstate on EEPROM
eeprom Configstate : false

Connecting to :
ssid : Tathering
password : tempangbaang

Try Connecting to Tathering
Authenticating......
WiFi connected : Tathering
```

Figure 8. Serial monitor at Running Mode.

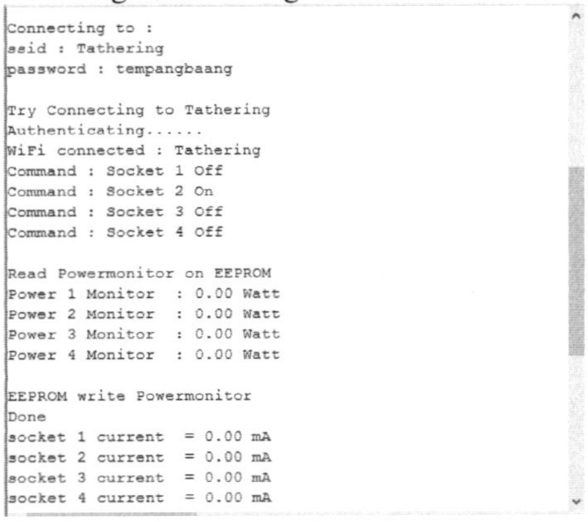

```
Connecting to :
ssid : Tathering
password : tempangbaang

Try Connecting to Tathering
Authenticating......
WiFi connected : Tathering
Command : Socket 1 Off
Command : Socket 2 On
Command : Socket 3 Off
Command : Socket 4 Off

Read Powermonitor on EEPROM
Power 1 Monitor  : 0.00 Watt
Power 2 Monitor  : 0.00 Watt
Power 3 Monitor  : 0.00 Watt
Power 4 Monitor  : 0.00 Watt

EEPROM write Powermonitor
Done
socket 1 current  = 0.00 mA
socket 2 current  = 0.00 mA
socket 3 current  = 0.00 mA
socket 4 current  = 0.00 mA
```

Figure 9. Serial monitor when get server data.

Once connected to an access point, the smart socket prototype has access to connect to the internet. With the internet, the prototype will retrieve the data socket information and send the total power consumption information on each socket. This can be seen in the monitor series shown in Figure 9. Socket status information will be directly applied to each socket. While the power information will be calculated in accordance with the use of the socket. Figure 10 shows one of the conditions in the smart socket prototype after it has been implemented according to the data sent through the server.

Figure 10. Devices when socket 1 & 3 are on and 2 & 4 off.

3.3. Data on ANTARES Cloud Storage Server

The data sent and retrieved on smart android apps and smart socket prototypes are stored on the server. The server used is cloud storage, or server that is used only to store data only. Cloud storage used is made by Telkom Indonesia is ANTARES. The data that is sent and retrieved for the needs of the smart socket prototype is stored in the project. In a project is divided into several devices that function as a division of functions between power monitoring and socket control. Figure 11 shows the socket status data that is sent in the smart phone application to be picked up by the smart socket prototype.

Data

Time	Resource Index (ri)	Value	Unit
2017-12-10 06:12:29	/in-cse/cin-346541903	onn2	socket
2017-12-19 10:19:04	/in-cse/cin-403105487	onn2	socket
2017-12-19 10:22:29	/in-cse/cin-523967401	off2	socket
2017-12-19 10:22:53	/in-cse/cin-90471617	onn2	socket
2017-12-19 10:23:42	/in-cse/cin-879492494	off2	socket
2017-12-19 10:23:43	/in-cse/cin-922206368	off2	socket
2017-12-19 11:04:39	/in-cse/cin-173986241	onn2	socket

Figure 11. ANTARES socket2 device interface

4. Conclusion

Design of smart socket prototype to calculate electric power consumption in rental house has been successfully done. The smart socket prototype works well to monitor power consumption in each rental home room and disconnect or connect the electrical current in each room remotely via a smart phone.

References

[1] Kodali R K and Naikoti A 2016 ECDH based security model for IoT using ESP8266 *Control, Instrumentation, Communication and Computational Technologies (ICCICCT), 2016 International Conference* 629-633 IEEE

[2] Keoh S L, Kumar S S and Tschofenig H 2014 Securing the internet of things: A standardization perspective *IEEE Internet of Things Journal* **1** (3) 265-275

[3] Kodali R K and Mahesh K S 2016 A low cost implementation of MQTT using ESP8266 *Contemporary Computing and Informatics (IC3I), 2016 2nd International Conference* 404-408 IEEE

[4] Manan M 2015 ESP8266: A Breakthrough in wireless sensor networks and internet of things *IJECET* **6** 8

[5] Nazruddin S H 2011 *Android: Pemrograman Aplikasi Mobile Smartphone dan Tablet PC* (Jakarta: Informatika)

The Analysis of Harmonics on LED Lamps

S Adhigunarto*, E Mulyana and W Surya

Department of Electrical Engineering Education, Universitas Pendidikan Indonesia, Jl. Dr. Setiabudhi No. 207 Bandung, Indonesia

*shamadhigunarto@gmail.com

Abstract. The development of the era affects the technological advances that currently appear two types of loads, those are linear and non-linear loads. LED lamp included non-linear loads so causes harmonics. For that reason, the comparison of power consumption between LED lamp load and the combination of TL-D and XL LED lamps to find which one is better to use. Due to the occurrence of harmonics by the LED lamp load, it is necessary to improve the harmonics by designing a Single Tuned Filter passive filter with the help of ETAP Power Station 12.6 applications. This research method with literature study, measurement, load data processing due to harmonics to the comparison of TL LED and TL-D lamps in New FPEB Building, and the design of passive filter design so that can get the desired result. The current and voltage harmonics on the SDP are not compliant with IEEE 519-2014 standard, that is current harmonics on each floor of it ranges above 45.95% to 94.73%, in contrast to voltage harmonics ranging from 2.2% to 2.8% which still within the standards allowed that is under 5%. As a result of harmonics causing total power loss before exposed to harmonics of 7715.11 Watt and total harmonized power of 9639.92 Watt. So the total loss of power due to harmonics of LED lamp loads of 1924.81 Watt. At investment cost, lamp replacement cost based on lifetime, and lower power consumption cost of LED type lamp compared to combination of TL-D and XL LED lamps. In Single Tuned Filter type filter simulation designed to reduce 3rd, 5th, and 7th harmonics can decrease THDi respectively R, S, and T phases of 7.19%, 7.01%, and 10.95%.

1. Introduction

The development of the era has an effect on the technological advancement that nowadays there are various objects and electrical appliances, thus the load on electric power system is classified into two types namely linear load and non-linear load. Linear load is a load having a current waveform equal to its voltage waveform while the non-linear load is a load whose current waveform is not equal to its voltage waveform [1], with both types of load affecting the quality of the Power System (STL). Power System Quality is related to Power Quality. LED lamps can be classified into non-linear load, because in the LED lamp components, there are non-linear components such as diodes, transistors, and others that can affect the quality of Power System harmonics. So the researcher will analyze the harmonics of LED lamps in FPEB new building because in the building all the lighting using LED type of lights and to compare which one is better between LED lights that are small powerless but the price is expensive, or ordinary type of lights with greater power but the price is cheap. In that comparison, the researcher will compare the type of TL LED lamp to TL-D or usual TL.

Non-linear load harmonization especially lighting using LED lamp can affect the power quality at New FPEB Building after harmonic measurement is obtained that the harmonics of voltage on each floor is 2.5% thus the harmonic stress of the building is still in IEEE 519-2014 standard different from current harmonics on the SDP on each floor of the building does not meet the standard that is above 45.95% to 94.73% and on the MDP current harmonics flows over 27.26% to 36.70% where the grounding of the building is 0.1 Ω, hence the improvement of harmonics by using filter is required. Hence, the researcher designs filter. For harmonics reparation, the researcher designs passive filter using a software that is ETAP 12.6.0, the researcher chooses passive filter because it is plentiful on the market and easy to find.

2. Theoretical Basis

2.1. Harmonics

Harmonics of voltage or current waves arising in the alternating current (AC) voltage system are influenced by the non-linear load supplied by the power company. Harmonics is a sinusoidal or current waveform whose frequency is an integer multiple of the frequency at which the supply system is designed to operate (called the fundamental frequency, usually 50 or 60 Hz).

The distorted waveform can be decomposed to a number of fundamental frequencies and harmonic distortion harmonics derived from the non-linear load and load characteristics of the power system [2]. This nonlinearity is that the current is not proportional to the applied voltage. It can be seen in the figure 1.

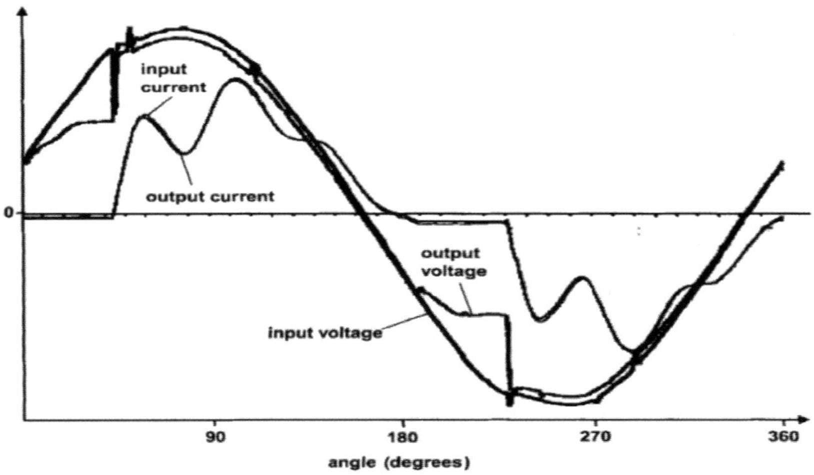

Figure 1. Waves affected by harmonics [3].

2.2. The influence of Hamonics

The effect of harmonics negatively impacts the equipment for instance the emergence of mechanical vibrations in electrical panels, harmonics may cause additional torque on kWh meters of electromechanical type using rotating induction disks, frequency interference on load-bearing telecommunication systems.

The load breaker may work under its rated current or may not work at rated current, reducing the efficiency of the electrical system and its utilities, reducing the isolation life of power supply equipment, increasing resonance of the power system [4]. The impact of voltage and current harmonics on the power system (STL) is increased harmonic level due to resonance parallel relationship, generator power, transmission, and other tools decreased efficiency, errors on the meter of energy measuring plate [5].

2.3. THD (Total Hamonic Distortion)

To measure the effective value of the harmonic components of the defective wave (distorted) the amount of THD (Total Harmonic Distortion) is used is the ratio of the rms value of the harmonic component of a quantity (current or voltage) to the rms value of the current or voltage at its fundamental frequency and is usually calculated in percent that can be seen in equation 1.

$$THD = \frac{\sqrt{\sum_{h>2}^{hmax} {M_h}^2}}{M_1} \quad (1)$$

Where M_h is the rms value of the harmonic component to h of a magnitude of M. The rms value itself can be obtained if it is known that the rms value of the first harmonic component and its THD value can be seen in equation 2:

$$RMS = \sqrt{\sum_{h=1}^{hmax} {M_h}^2} = M_1 \sqrt{1 + THD^2} \quad (2)$$

For the THD value of current in equation 3:

$$THD_I = \frac{\sqrt{{I_2}^2 + {I_3}^2 + {I_4}^2 + \cdots + {I_n}^2}}{I_1} \quad (3)$$

For the THD value of the voltage in equation 4:

$$THD_V = \frac{\sqrt{{V_2}^2 + {V_3}^2 + {V_4}^2 + \cdots + {V_n}^2}}{V_1} \quad (4)$$

THD currents vary from a few percent to over 100%. THD voltages are usually smaller than 5%. Below 5% is generally acceptable but if above 10% then this is unacceptable and will cause problems for sensitive loads and equipment [6].

At the IEEE 519-2014 harmonic standard there are two criteria used to evaluate harmonic distortion ie the limits for current harmonics and constraints for voltage harmonics. For current harmonic standards, is determined by the ratio of Isc / IL. Isc is the maximum short-circuit current available on the PCC (Point of Common Coupling) or at the measuring point, whereas IL is the maximum load current at the fundamental frequency. Meanwhile for the standard harmonic voltage is determined by the voltage of the system used.

2.4. LED Light (Light Emitting Diode)

LED (Light Emitting Diode) or often abbreviated as LED is an electronic component that can emit monochromatic light when is given forward voltage. LED is a family of Diodes made of semiconductor materials. Block diagram of LED electronic ballasts can be seen in the figure 2.

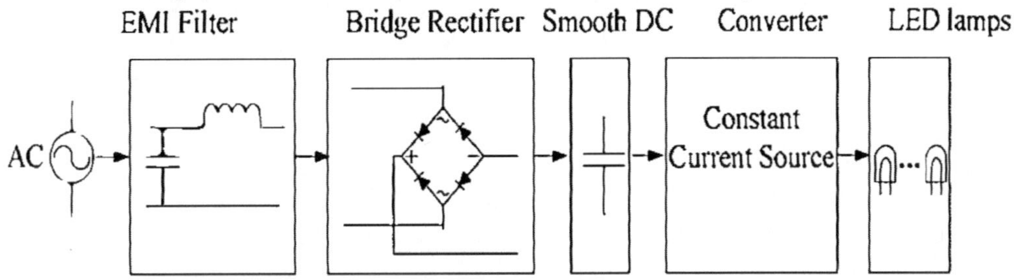

Figure 2. Block diagram of LED electronic ballasts [7].

LEDs require a constant current source from a low DC voltage source, but they should also operate from AC power. Therefore, it is necessary to use the converter to adjust the voltage and control the current applied to the LED [7].

2.5. Harmonic Filters
The main purpose of a harmonic filter is to reduce the amplitude of a particular frequency of a voltage or current. With the addition of a harmonic filter on a power system containing harmonic sources, the spread of harmonic current throughout the network can be minimized. In addition, harmonic filters at fundamental frequencies can compensate for reactive power and are used to improve the system power factor [8].

2.6. Passive Filters
Passive filters are an effective and economical solution for harmonic problems. Passive filters consist of R, L, and C sequences. Passive filters are
;mostly designed to provide a special section for diverting undesired harmonic currents in the power system. Passive filters are widely used to compensate for reactive power losses due to harmonics in the installation system. Passive filters are used as common and easy-to-find applications in the market for power system and designer techniques to reduce voltage distortion and harmonic current through electrical installations [9].

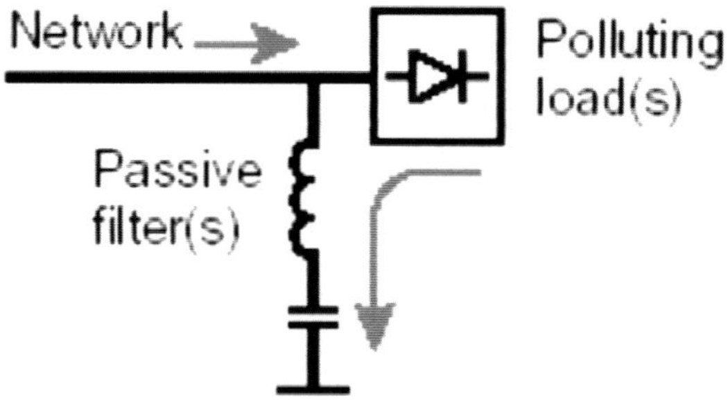

Figure 3. Passive Filters [10].

From the figure 3 the installation of passive filters are located after the source, in the circuit aims to harmonic flow that does not flow to the source but through the filter and grounded, there are some harmonic passive filter topology used, those are single tuned filter and high pass filter. Passive filter configuration can be seen in figure 4.

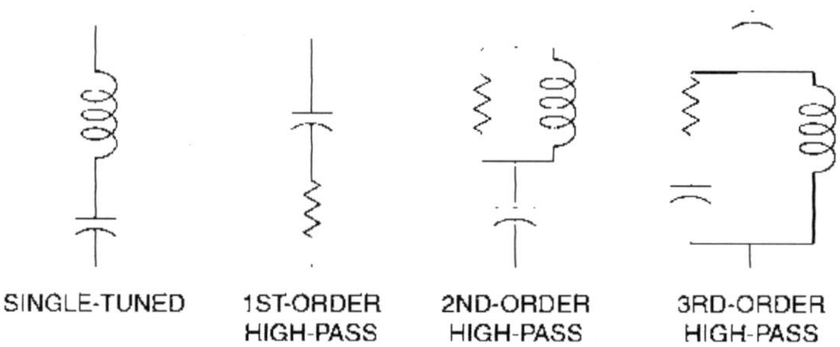

Figure 4. Passive Filter Configuration [2].

3. Methods

3.1. Harmonic Measurement Procedures
A series of harmonic testing on the installation of LED lights on each floor of the New FPEB Building can be seen in figure 5.

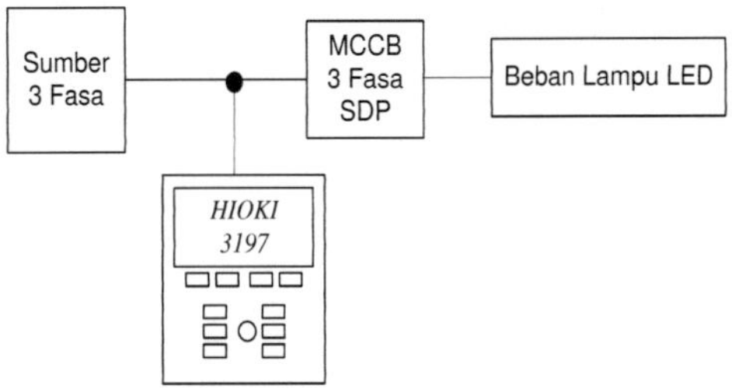

Figure 5. Series of harmonic measurements on LED lamp installations.

At this harmonic measurement is taken when all the LED lights are turned on, the measurement of harmonics on the LED Lights in FPEB Building. The harmonic measurements on the SDP on each floor use HIOKI 3197 POWER QUALITY ANALYSER. The measured data are voltage (V), current (A), active power (Watt), reactive power (VAR), pseudo power (VA), power factor (PF), THDv voltage (%), and THDi current (%). The harmonic measurement circuit on MDP uses Clamp on Power Hitester HIOKI 3286 - 20 can be seen in figure 6.

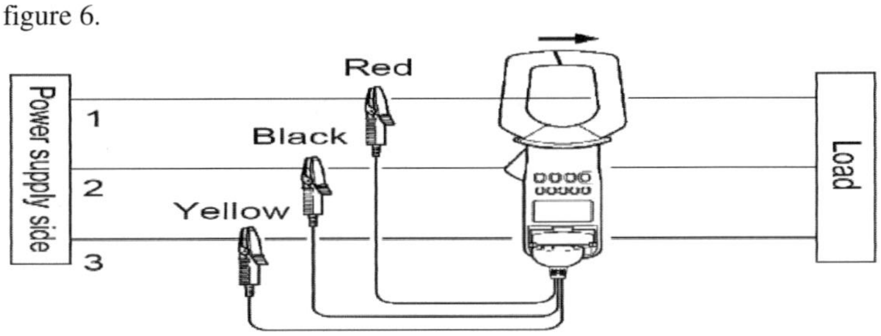

Figure 6. Series of harmonic measurements on MDP.

On MDP uses Clamp on Power HIOKI 3286 - 20 Hitester. The measured data are voltage (V), current (A), active power (Watt), reactive power (VAR), pseudo power (VA), power factor (PF) THDv voltage (%), and THDi current (%). All measured data are recorded.

3.2. Design of Passive Filter Design Single Tuned Filter Type

From the result of measurement in Table 3.6 that the current harmonics value at each of the 3, 5, and 7 phase exceeds 10.7% and below 22%, and the harmonic voltage is still within the allowed standard. So the passive filter will be designed to reduce the current harmonics. The filter designed will improve the power factor to 0.9.

After determining the order of harmonics to be repaired then the next step is to determine the capacitance capacitor, in the calculation for the filter design taken data from phase R, then performed the following calculation:

R Phase Filter

$$\begin{aligned} Q_C &= 16800.\ (\tan(51{,}9)) - (\tan(25{,}84)) \\ &= 16800\ (1{,}275 - 0{,}48) \\ &= 16800\ (0{,}795) \\ &= 13356 \text{ VAR} \end{aligned} \tag{5}$$

The result of the calculation shows that to improve the power factor to 0.9, reactive power of 13356 VAR is needed, for single tuned filter to be installed that is 3rd order harmonic filter.

Furthermore, the capacitor reactance (X_C) can be found.

$$Xc1 = \frac{V^2}{VAR} = \frac{380^2}{13356} = 10{,}81\ \Omega \tag{6}$$

Furthermore, when the capacitor reactance (Xc) results can be found, so is the capacitance of the capacitor (C).

$$C = \frac{1}{2\pi f Xc} = \frac{1}{2(3{,}14)(50)(10{,}81)} = 0{,}00029\ F \tag{7}$$

After obtaining the capacitance of the capacitor, then reactance from the inductor (XL) can be sought. The inductive reactance required at harmonic h is, in this case as follows:

$$\begin{gathered} XLh = Xch \\ XL3 = Xc3 \\ XL1 = \left(\frac{1}{h}\right) XLh = \left(\frac{1}{h}\right) XL3 = \frac{3{,}6}{3} = 1{,}2\ \Omega \end{gathered} \tag{8}$$

$$I1 = \frac{(1{,}05)(VLN)}{(Xc - XL)} = \frac{(1{,}05)(219{,}39)}{(10{,}81 - 1{,}2)} = 23{,}97\ A \tag{9}$$

$$I3 = \frac{1}{h}\frac{KVAload}{\sqrt{3}\,(13{,}8)} = \frac{1}{h}\left(\frac{27{,}2}{\sqrt{3}\,(13{,}8)}\right) = 0{,}379\ A \tag{10}$$

$$Xc3 = \left(\frac{1}{h}\right) Xc1 = \frac{10{,}81}{3} = 3{,}6\ \Omega \tag{11}$$

The peak voltage (Vpeak) and rms voltage (Vrms) on the capacitor,

$$\begin{aligned} Vc\ peak &= \sqrt{2}\,(Vc1 + Vc3) \\ Vc\ peak &= \sqrt{2}\,(Xc1.I1 + Xc3.I3) \\ Vc\ peak &= \sqrt{2}\,\big((10{,}81\ x\ 23{,}97) + (3{,}6\ x\ 0{,}379)\big) \\ Vc\ peak &= \sqrt{2}\,(259{,}16 + 1{,}37) = 368{,}44\ Volt \end{aligned} \tag{12}$$

$$Vc\ rms = \sqrt{Vc1^2 + Vc3^2} = \sqrt{259{,}16^2 + 1{,}37^2} = 259{,}16\ V \tag{13}$$

If 219.39 V (neutral phase voltage to 380 V on the system).

$$\frac{Vc\ rms}{Vc\ peak\ rated} = \frac{259{,}16}{219{,}39} = 1{,}181\ p.u$$

(Limit of IEEE-18 below 1.1 p.u.)

$$\frac{Vc\ peak}{Vc\ peak\ rated} = \frac{368{,}44}{\sqrt{2}\,(219{,}39)} = 1{,}1867\ p.u$$

In the calculation for 3rd order harmonics, is still not in accordance with IEEE-18 standards. After the inductive reactance is known, then the inductance of the inductor (L) can be sought.

$$L = \frac{XL}{2\pi f} = \frac{1{,}2}{2(3{,}14)(50)} = 0{,}0038\ H \tag{14}$$

Obtaining the inductor, and the capacitor, then the resistance (R) can be found.

$$R = \frac{\sqrt{\frac{L}{C}}}{80} = \frac{\sqrt{\frac{0{,}0038}{0{,}00029}}}{80} = 0{,}045\ \Omega \tag{15}$$

Having known the value of each component for 3rd order filter then the filter specification for the order of 5 and 7 can be seen in table 1 is a single tuned filter model specification used.

Table 1. Specification Model Single Tuned Filter Design Results.

Fasa R			
0rde	3	5	7
Qc	13356	13356	13356
Xc1	10,81	10,81	10,81
Xc3	3,60	2,16	1,54
C	0,00029	0,00029	0,00029
XLh=Xch	XL3=Xc3	XL5=Xc5	XL7=Xc7
XL1	1,20	0,43	0,22
I1	23,97	22,19	21,75
I3	0,379	0,228	0,163
Xc3	3,60	2,16	1,54
Vc1	259,16	239,96	235,16
Vc3	1,37	0,49	0,25
Vpeak	368,44	340,05	332,92
Vrms	259,16	239,96	235,16
p.u limit	1,181	1,094	1,072
p.u limit IEEE-18	1,187	1,096	1,073
L	0,0038	0,0014	0,0007
R	0,045	0,027	0,019
Fasa S			
0rde	3	5	7
Qc	12684	12684	12684
Xc1	11,38	11,38	11,38
Xc3	3,79	2,28	1,63
C	0,00028	0,00028	0,00028
XL=Xch	XL3=Xc3	XL5=Xc5	XL7=Xc7
XL1	1,26	0,46	0,23
I1	22,76	21,08	20,66
I3	0,261	0,156	0,112
Xc3	3,79	2,28	1,63
Vc1	259,16	239,96	235,16
Vc3	0,99	0,36	0,18
Vpeak	367,90	339,86	332,83
Vrms	259,16	239,96	235,16
p.u limit	1,181	1,094	1,072
p.u limit IEEE-18	1,095	1,073	0,000
L	0,0040	0,0015	0,0007
R	0,047	0,028	0,020
Fasa T			
0rde	3	5	7
Qc	3605,8	3605,8	3605,8
Xc1	40,047	40,047	40,047
Xc3	13,35	8,01	5,72
C	0,00008	0,00008	0,00008
XL=Xch	XL3=Xc3	XL5=Xc5	XL7=Xc7
XL1	4,45	1,60	0,82
I1	6,47	5,99	5,87
I3	0,215	0,129	0,092
Xc3	13,35	8,01	5,72
Vc1	259,16	239,96	235,16
Vc3	2,87	1,03	0,53
Vpeak	370,56	340,82	333,31
Vrms	259,17	239,96	235,16
p.u limit	1,181	1,094	1,072
p.u limit IEEE-18	1,194	1,098	1,074
L	0,0142	0,0051	0,0026
R	0,167	0,100	0,072

IOP Conf. Series: Materials Science and Engineering **384** (2018) 012061 doi:10.1088/1757-899X/384/1/012061

Table 1 is the result of calculations to find passive filter specifications. The calculations of order 5 and 7 are the same as calculating the filter specifications for the order of 3. The filter specifications for the 5th and 7th order can be found in table 1.

4. Results and Discussion

4.1. THD Value on LED Lights Installation of New FPEB Building

At this harmonic measurement by turning on all LED lights that are in FPEB Building. Measurement is carried out to determine the current harmonics value and the voltage generated by the LED light and the effect on the power consumption due to harmonics produced by the non-linear component on the LED lamp. Here are the measurements of each floor using a HIOKI 3197 Power Quality Analyzer.

Table 2. THDi Measurement on LED Lamp Installation.

Floor	**THDi %**			**THDv %**		
	R	S	T	R	S	T
1	58,00	78,97	87,36	2,50	2,50	2,30
2	93,75	73,70	75,27	2,50	2,60	2,40
3	80,40	75,36	59,07	2,70	2,80	2,30
4	82,67	73,70	90,55	2,30	2,50	2,30
5	84,06	78,69	60,85	2,40	2,50	2,30
6	68,37	94,74	45,95	2,40	2,40	2,20

From Table 2 the measurement data of each floor can be seen harmonic current and voltage, on the harmonic measurement data that is not in accordance with IEEE 519-2014 standard that is harmonic current in each floor harmonic current ranges above 45.95% to 94.73%, unlike the harmonic voltage ranging from 2.2% to 2.8% is still within the allowed standard of under 5%. The influence of the harmonics can have an impact on power losses.

4.2. Effect of Harmonics on Power Loss

From the harmonic measurement results can be seen that the influence of harmonics can cause loss of power. It can be seen from the current measured after exposure to harmonics and before harmonics. The current which is not affected by harmonics can be seen on the harmonic value of 1st order current. In this case we can know the loss of power generated due to harmonics and comparison of power consumption before exposed and after exposed to harmonics with real power calculation (Watt).

Calculation of real power (P) for 1st floor before exposed to harmonics:

$P = (V)\,(I_{orde\ 1})\cos\varphi$

$P = (228{,}8)\,(5{,}1)\,0{,}817$

$P = 953{,}34$ Watt

Table 3. Total Power before Affected by Harmonics.

Power (W)	**Lt 1**	**Lt 2**	**Lt 3**	**Lt 4**	**Lt 5**	**Lt 6**	**Amount of P Fasa**
R	953,34	410,11	380,24	523,06	308,08	234,75	2809,57
S	271,11	239,59	233,25	235,56	508,13	305,33	1792,97
T	430,96	417,09	452,35	423,09	199,30	1189,78	3112,57
Amount of P	1655,4	1066,78	1065,84	1181,71	1015,50	1729,86	7715,11
Amount of P kW	1,66	1,07	1,07	1,18	1,02	1,73	7,72

Calculation of real power (P) for 1st floor before affected by harmonics:

$P = (V)\,(Irms)\cos\varphi$

$P = (228{,}8)\,(5{,}9)\,0{,}817$

$P = 1102{,}88$ Watt

Table 4. Total Power after Affected by Harmonics.

Power (W)	Lt 1	Lt 2	Lt 3	Lt 4	Lt 5	Lt 6	Amount of P Fasa
R	1102,8	560,48	482,61	676,90	406,10	288,92	3517,89
S	348,57	306,14	298,04	300,99	651,04	427,46	2332,25
T	568,87	525,22	572,98	573,22	234,47	1315,02	3789,78
Amount of P	2020,3	1391,8	1353,6	1551,1	1291,6	2031,4	9639,92
Amount of P kW	2,02	1,39	1,35	1,55	1,29	2,03	9,64

From the data table 3 and 4 can be known the difference in power values before and after affected by harmonics. The impact of the harmonics on the loss of power from the actual known value on the calculation of the power of 1st floor phase R before harmonized of 953.34 Watt and the value of power calculation after harmonization of 1102.88 Watt. The difference in power loss at the R phase of 149.54 Watt, so that harmonics can be detrimental in terms of power consumption and costs borne.

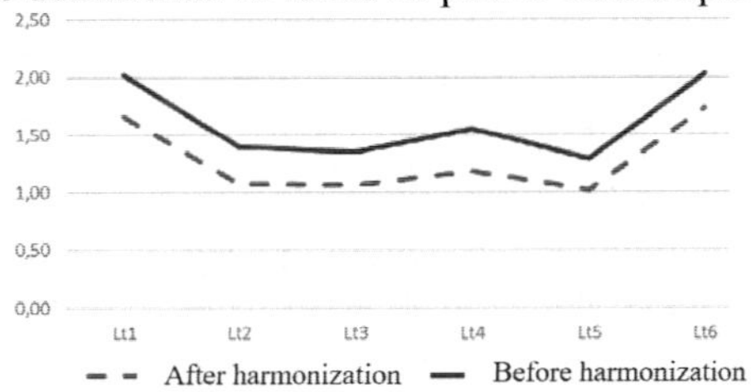

Figure 7. Graph of comparison of power consumption before and after harmonization.

Figure 7 shows the difference in power consumption of LED lamp installations before and after affected by harmonics. So the consequences of harmonics can affect the loss of power of a system.

4.3. Design of Single Tuned Filter Design Using ETAP Power Station 12.6

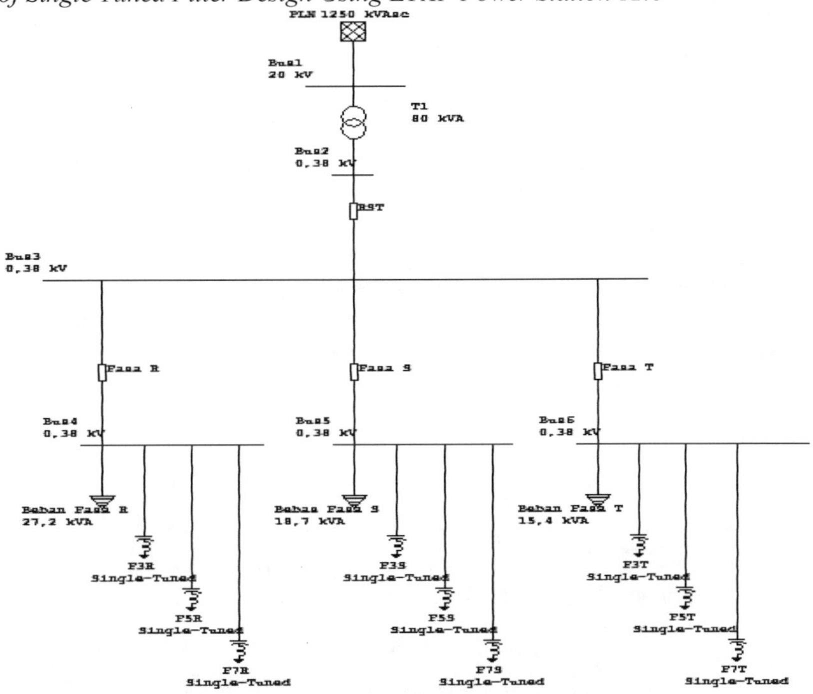

Figure 8. Installing 3, 5, and 7 single tuned filters.

In figure 8 installation of 3rd, 5th, and 7th order filters to reduce current harmonics (THDi) at 3rd, 5th, and 7th order, after the installation of 3rd, 5th, and 7th order filters, the results are as follows.

Table 5. Data of Harmonic Flow of Fitting Results of 3rd and 5th Order Filter.

Orde	Before Filter			Filter Orde 3,5,7		
1	0	0	0	0	0	0
2	2,8546	3,88196	2,40053	4,03874	4,3971	3,07648
3	20,152	18,147	17,3623	0,81638	0,7495	3,02102
4	0,93637	1,1907	1,96136	0,87579	1,3449	2,03651
5	22,312	16,6641	15,461	1,29543	0,89781	4,20688
6	0,88395	4,59412	2,46625	0,49778	2,23365	3,03624
7	18,261	10,7968	11,3327	0,924848	0,519492	2,59537
8	0,5268	2,83449	1,56918	0,314125	0,865434	1,61355
9	2,0558	9,50373	4,90581	0,898116	2,10395	4,43736
10	0,85666	4,26437	2,06841	1,03711	1,55832	1,74055
11	5,4471	5,90634	4,08292	3,98322	2,67024	4,37819
12	0,53288	0,993437	0,858778	0,341195	0,290031	0,82335
13	4,1474	1,86278	2,48619	2,79565	1,40272	2,54694
14	0,71168	1,08882	0,913395	0,632037	0,412226	1,11784
15	1,8958	1,23151	1,9366	1,09724	0,594073	2,07045
16	0,4012	1,30009	0,361039	0,336099	0,558437	0,91033
17	2,1622	1,74735	1,70422	1,41866	1,30041	1,7164
18	0,52975	1,2776	0,576837	0,309872	0,581673	0,81876
19	1,912	1,11662	1,38765	1,21489	1,06422	0,85557
20	0,50612	1,36017	0,954547	0,269866	0,771836	1,07875

In table 5 it can be seen that the current harmonic value at 3rd, 5th, and 7th order has decreased from the previous value. So THDi at each phase decreases THDi value on phase R, S, and T that is 7,20%, 7.01%, and 10,95%. Although the THDi value on each phase is decreasing, the THDi value still has not reached the 5% IEEE 519-2014 standard.

Table 6. Power Comparison Before and After Filter Installation.

Daya	P (kW)	S (kVA)	Q(kVAR)	I (A)
Sebelum di filter	37,3	61,3	47,5	102,9
Setelah di filter	33,79	37,54	16,37	98,8
Selisih	3,51	23,76	31,13	4,10

Table 6 summarizes the power ratio before and after the installation of passive filters in the simulation. The pre-filtered power of the filter is taken from the sum of the values recorded on the gauge, and the after-power filter is generated from the calculation of power with a voltage of 380 V, a power factor of 0.9, and a current of 98.8 A. The voltage, power, and current data Obtained from the simulation. For a clearer comparison of power can be seen in Figure 9.

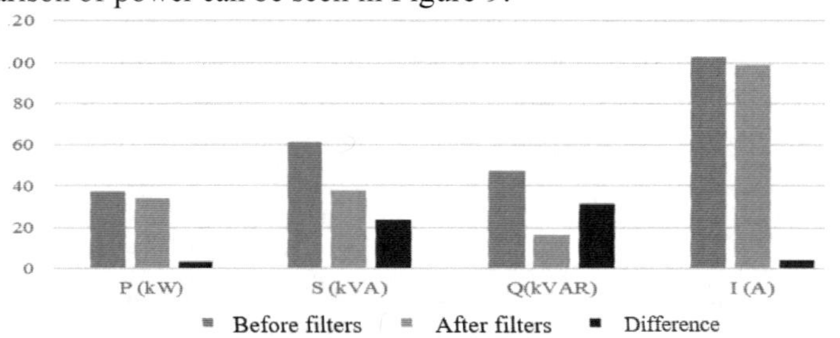

Figure 9. Power Comparison Before and After in Filters.

Figure 9 is a power comparison graph before and after the installation of passive filters. There is a difference between power consumption when harmonics are not filtered and when harmonics are filtered. The difference for real power (P) is 3.51 kW, apparent power (S) 23,76 kW, and reactive power (Q) 31,13 kW. So, for the filters installation can reduce power consumption.

5. Conclusions

From the results of the research that has been conducted can be drawn conclusion as follows:

- The use of LED lamps resulted in harmonics in the New FPEB Building is not meeting the IEEE 519-2014 current harmonic standards in the SDP of each current harmonic floor ranging above 45.95% to 94.73%, and at MDP the current harmonic over 27.26% up to 36.70%, in contrast to the voltage harmonics of the SDP and the MDP is still within the allowable standard of 5%,
- The results of large harmonic measurement data and not in accordance with the IEEE 519-2014 standards will have an effect on the quality of power resulting in loss of power, since the total power before the harmonics hit by 7715.11 Watt and the total harmonized power of 9639, 92 Watts. So the total loss of power due to harmonics of LED lamp loads of 1924.81 Watt.
- Installation of Single Tuned Filter type filter at each load is more effective for reducing current harmonics and filters can work on any perfume load, in Single Tuned Filter type filter simulation designed to reduce 3rd, 5th, and 7th harmonics can decrease THDi respectively - respectively R, S, and T phases are 7.19%, 7.01%, and 10.95%, since passive filters can attract harmonic currents so that harmonic currents do not enter the system resulting in loss of power.

References

[1] De La Rosa F 2006 *Harmonics and power systems* (Boca Raton: CRC press) 1-184

[2] Dugan R C, McGranaghan M F and Beaty H W 1996 *Electrical power systems quality* (New York: McGraw-Hill)

[3] Ewald F F and Mohammad A M 2008 *Power quality in power systems and electrical machines* (London, UK: Elsevier Academic)

[4] Irawan H, Hermawan H and Sukmadi T 2011 *Analisis pengaruh pemasangan filter pasif terhadap penurunan harmonik pada sistem multi mesin 9 bus IEEE* (Semarang: Diponegoro University) 1-11

[5] Elih M, Surya W, Dadang L and Hasbullah 2008 *Analisis Harmonisa Tegangan dan Arus Listrik Di Gedung Direktorat TIK Universitas Pendidikan Indonesia* (Bandung: Universitas Pendidikan Indonesia) 3976

[6] Grady W M and Santoso S 2001 Understanding power system hannonics *IEEE Power Engineering Review* **21** (11) 8-11

[7] Uddin S, Shareef H, Mohamed A and Hannan M A 2012 An analysis of harmonics from dimmable LED lamps *Power Engineering and Optimization Conference (PEDCO) Melaka, Malaysia, 2012 Ieee International* 182-186

[8] Basu S 2006 *Single Phase Active Power Factor Correction Converters-Methods for Optimizing EMI, Performance and Costs* (Göteborg: Chalmers University of Technology)

[9] Cho Y S, Kim B R and Cha H 2010 Transfer function approach to a passive harmonic filter design for industrial process application *Mechatronics and Automation (ICMA), 2010 International Conference* 963-968

[10] Rinas I W 2011 Analisis Perbandingan Penggunaan Filter Pasif fan Filter Aktif Untuk Menanggulangi THD Pada Sistem Kelistrikan Di Ruang Puskom Jurusan Teknik Elektro Fakultas Teknik Universitas Udayana *Jur. Tek. Elektro, Fak. Tek. Univ. Udayana Kampus Bukit Jimbaran, Bali* **10** 1

IOP Conf. Series: Materials Science and Engineering **384** (2018) 012062 doi:10.1088/1757-899X/384/1/012062

The Effect of Temperature to the Formation of Optically Active Point-defect Complex, the Carbon G-centre in Pre-amorphised and Non-amorphised Silicon

D D Berhanuddin[1], M A Lourenço[2], R M Gwilliam[2], and K P Homewood[2]

[1]Institute of Microengineering and Nanoelectronics (IMEN), Universiti Kebangsaan Malaysia, Bangi, 43600 Selangor, Malaysia
[2]Advanced Technology Institute, Faculty of Engineering and Physical Sciences, University of Surrey, Guildford, Surrey GU2 7XH, United Kingdom

*dduryha@ukm.edu.my

Abstract. The effect of the temperature ranging from cryogenics to room temperature were investigated on the formation of the optically-active point defect called the G-centre. The G-centre as an emissive point defect gained a lot of attention recently due to its sharp zero phonon luminescence peak at a wavelength of 1.28 μm (0.97 eV) with the evidence of lasing occurred in the structure. The emission of the G-centre is attributed to the carbon substitutional–carbon interstitial (CsCi) complex which interacts with silicon interstitials during the damage event. This complex is generated by implantation of carbon and followed by proton irradiation. Prior to the carbon implantation, two of the samples were pre-amorphised by germanium. Photoluminescence (PL) measurements were carried out at temperature ranging from 80 K up to room temperature to observe the intensity of the main peaks. The results confirm that the main peaks of point-defect centre in all of the samples including the G-centre suffer from the temperature quenching. However, the peak intensity for some of the wavelength especially the ones with high FWHM, do perform better at high temperature. The temperature quenching phenomena observed in the point-defect technique is the main problem that needs to be addressed and solved before realizing the method in the all silicon photonic system.

1. Introduction

There is a crucial requirement to create a silicon-based optical emitter, which is compatible with standard ULSI technology used in fabricating integrated circuits [1]. Despite silicon dominance in the electronics industry, it is still lacking key properties in optoelectronics. This is because of the indirect band gap which limits the radiative recombination efficiency. However, in the past decade, various approaches have been introduced to try circumventing the band gap limitation, thus transforming silicon as an optically active material. Among the successful and promising methods of emitting light from silicon are point defect centres [2-5], Raman conversion [6-8] strained silicon i.e. dislocation engineering [9,10] and nanocrystals with or without rare earth doping such as erbium(Er) [11-14]. The first silicon laser was reported in 2004 by Jalali et al using the Raman conversion techniques, however by its nature can only ever be optically pumped [6].

The G-centre as an emissive point defect gained a lot of attention recently due to its sharp zero phonon luminescence peak at a wavelength of 1.28 μm (0.97 eV) with the evidence of lasing occurred in the structure [15]. The emission of the G centre is attributed to the carbon substitutional–carbon interstitial (CsCi) complex which interacts with silicon interstitials during the damage event. The G centre emission at 1280 nm is in the spectral range which is vital in long-haul fiber-optic networks and is ideally suited to silicon intra-chip and inter-chip low-power data interconnects. A new method fully compatible with CMOS (Complementary Metal Oxide Semiconductor) technology to efficiently produce the lasing G-centre has been recently reported [16-18]. It combines the implantation of carbon, subsequent heat treatment followed by high-energy proton implantation on bulk silicon substrates to form CsCs complexes and Si interstitials.

The luminescence intensity of the G-centre emission varies with the implant conditions such as the carbon and proton doses. Ion implantation damages the target and displaces many atoms. In CMOS technology, the electrical behaviour after implantation is dominated by deep-level electron and hole traps, which capture the carriers and make the resistivity high. Annealing or rapid thermal process (RTP) is needed after the implantation to electrically activated the implanted atoms while removing the radiation damage. RTP is usually favored among other conventional heat treatment methods because of the shorter processing times and higher electrical activation of the implanted dopants [19]. In the case of pre-amorphized samples, an annealing process is required to allow solid phase epitaxial regrowth (SPER) where the orientation of crystalline region is replicated into the volume that evolves from the amorphous solid with the crystalline region of the wafer serving as the 'seed'.

In this paper, we report the temperature dependence study of both pre-amorphised and non-amorphised silicon that have been implanted with carbon and irradiated with protons to form the optically active centres especially the G-centres. The amorphisation of the silicon lattice was done by bombarding samples with germanium. Photoluminescence measurements have been carried out at different temperatures to investigate the G centre's peak and other related luminescence intensity in bulk silicon samples. The measurements confirm temperature quenching problem suffered by employing the point-defect technique to gain luminescence in silicon.

2. Sample preparation and analysis

The experiment was performed using N-type silicon wafers (100) with resistivity of ~ 10 Ωcm^{-1} and manufacturer-specified residual carbon content of 2.5×10^{16} cm-3. One of the samples were pre-amorphised with 2×10^{15} cm^{-2} of germanium at energy 180 keV. Samples were then implanted with different doses of carbon, first at 30 keV and followed by implantation at 10 keV. The double implantations were done to construct a reasonably flat carbon profile thus ensuring better uniformity of carbon concentration along the depth of samples. After the C implantation, samples were annealed at 1000 °C for 20 seconds in N2 ambient. This is to allow the solid phase epitaxial regrowth (SPER) of the silicon lattice, repair the lattice damage and to activate while introducing carbons into the substitutional sites. Proton irradiation of the annealed samples was subsequently carried out at different doses. Details of the samples are given in Table 1.

Table 1. Sample details. The first carbon implant was at 30 keV and the second implant at 10 keV. All samples were annealed at 1000 °C for 20 s after the C implants. The average carbon volume concentrations were calculated using SUSPRE [20]. Proton irradiation was performed at 2 MeV after the C post-implant anneal. No further sample processing was done after the proton irradiation.

Sample	Ge dose (cm^{-2})	C volume concentration (cm^{-3})	Proton dose (cm^{-2})
1	-	2×10^{19}	5×10^{14}
2	-	2×10^{20}	5×10^{15}
3	2×10^{15}	2×10^{19}	5×10^{13}

3. Results and Discussion

Figure 2 shows the photoluminescence spectra of sample 1, measured at different cryogenic temperature ranging from 80 to 120 K, of bulk silicon implanted to 2×10^{19} cm-3 volume concentration of carbon, followed by proton irradiation at 5×10^{14} cm^{-2} at energy of 2 MeV. Two optically active point-defect centre observed in these samples, which are *CsCs* complex, the G-centre at ~ 1.28 μm and C-centre at ~ 1.57 μm. The latter is another zero phonon line luminescence feature attributed to the carbon-oxygen (CiOi) complex intensity while developing a broad peak in the ~ 1.4 to 1.6 μm range. The broader structured luminescence from 1.3 to 1.4 μm observed in both spectra is a familiar feature of the G-line spectra resulting from local phonon modes [21]. At 80 K, the sharp G-line peak is dominating the spectrum but it is quickly decreasing to quarter of its value when temperature is increase to 90 K. The G-line peak intensity continues to drop at higher temperature denoting the temperature quenching problem suffered by the optically active point-defect. At 120 K, the G-centre peak is no longer observed in the spectrum. As for the C-line, the peak intensity can be seen to drop gradually and finally absence at 120 K. It has been suggested that temperature quenching problem by employing point-defect technique is due to non-radiative process by unspecified energy barrier and deep level trap [22].

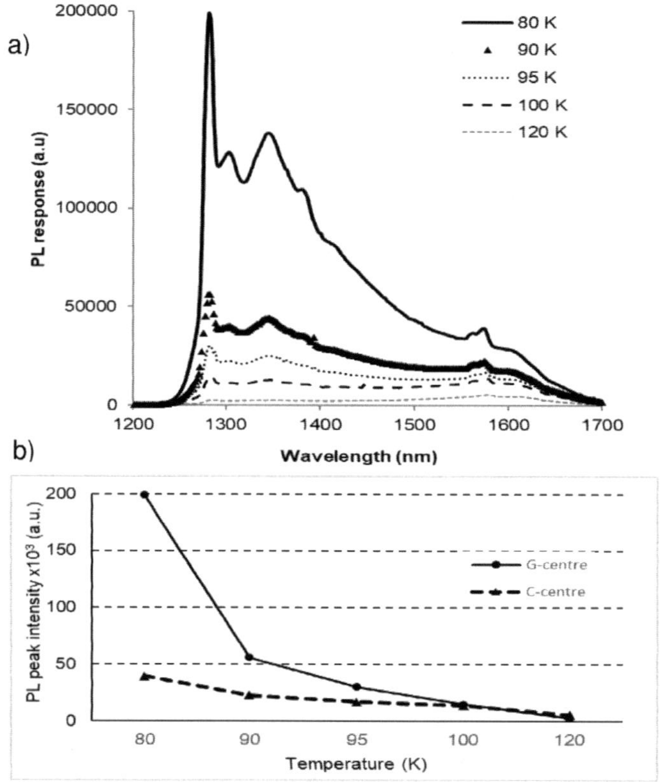

Figure 1. Shows the a) PL spectra of the sample with the different measurement temperatures from 80 to 120 K and b) luminescence intensity of the G and C-centres' peaks for non-amorphised sample with $2\times10^{19}cm^{-3}$ of carbon and 5×10^{14} cm^{-2} protons. The luminescence especially of the G-centre peak quenches rapidly when the temperature increased from 80 to 90 K.

Sample 2 is a bulk silicon wafer that has been implanted to 2×10^{20} cm^{-3} volume concentration of carbon, followed by proton irradiation at 5×10^{15} cm^{-2} at energy of 2 MeV. The photoluminescence spectra of this sample at different temperature is shown in Figure 3. There are three main peaks observed in the sample which is the G-line, W-line attributed to self-interstitials defect clusters formed after ion irradiations and a broad peak at ~ 1.55 μm [23].12 The presence of the W-line indicates the existence of an excess of silicon interstitials created after the proton irradiation, which compete to form various silicon-related radiative and non-radiative centres, including the W- and G-centres. The broad peak at 1.55 μm is clearly dominating the spectrum with a gradual drop of peak intensity from 80 to 120 K. As before, the G-line peak intensity decrease rapidly with the increase in temperature and completely absence at 120 K. Similar to the G-centre, W-line peak has also drop rapidly with temperature and cannot be observed at temperature higher than 95 K. Higher carbon and proton doses in this sample as compared to sample 1 does not give a higher peak intensity of the G-centre. This is because; the excessive amount of the carbon and silicon interstitials in the lattice will tend to agglomerate and forming other radiative and non-radiative centres [17].

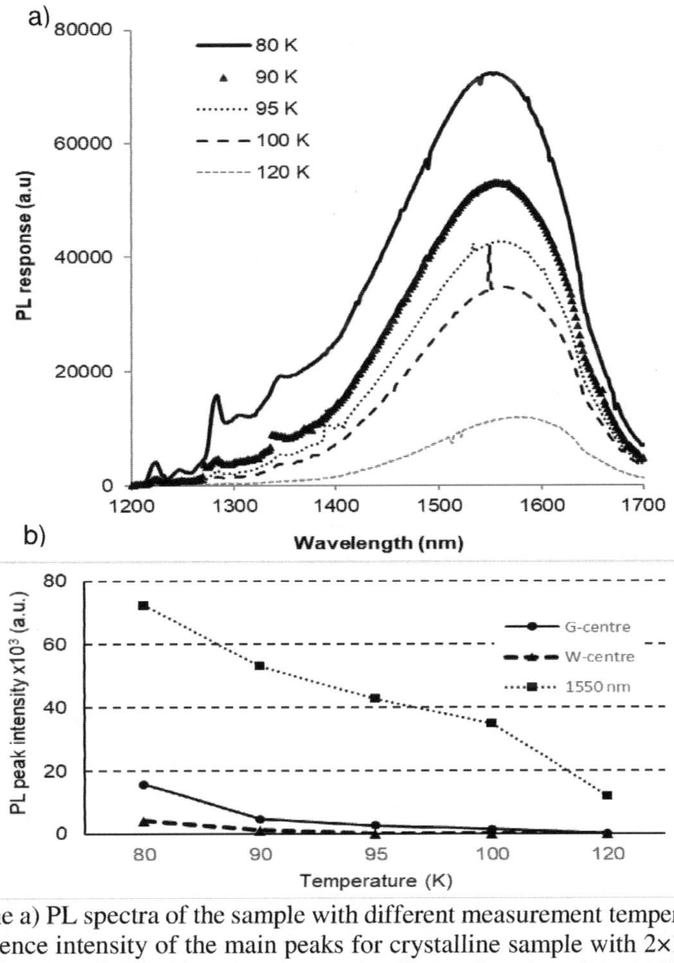

Figure 2. shows the a) PL spectra of the sample with different measurement temperature from 80 to 120 K and b) luminescence intensity of the main peaks for crystalline sample with 2×10^{20} cm^{-3} of carbons and 5×10^{15} cm^{-2} protons. The highest peak of this sample is a broad peak at 1550 nm whose intensity decreases uniformly with temperature.

Figure 3 shows the photoluminescence spectra for sample 3 which has been pre-amorphised with ge prior to carbon implantation with volume concentration of 2x1019 cm-3 and proton irradiation dose of 5x1013 cm-2. Three main peaks the G-line, C-line and broad peak at ~1.6 μm are observed in this sample with the C-line at ~ 1.57 μm dominating the spectrum. End-of-range (EOR) defects are common occurrence in a preamorphised sample and observed at a former amorphous/crystalline (a/c) interface after solid phase epitaxial regrowth (SPER) [17]. EOR defects might trap the majority of silicon interstitials which are vital in the formation of the G-centre thus resulting in the domination of the C-centre luminescence in this sample. We also suggest that the formation of the broad peak at 1600 nm is due to the formation of the interstitials after the SPER. The C-line peak can still be observed at 120 K giving it strong presence and have the potential to emit at room temperature.

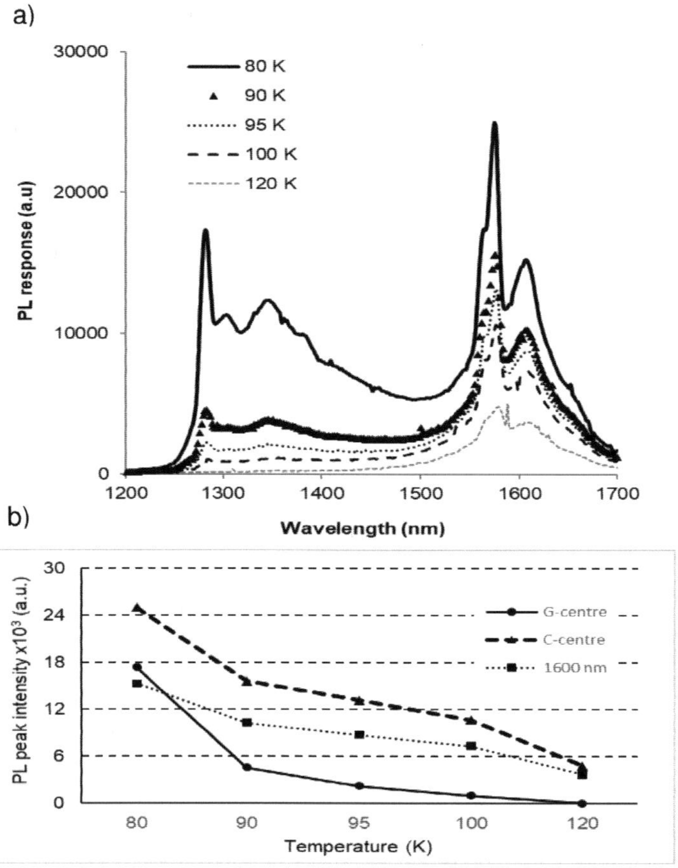

Figure 3. shows a) the PL spectra of the sample with different measurement temperature from 80 to 120 K and b) the luminescence intensity of the main peaks for the pre-amorphised sample with 2×10^{19} cm^{-3} of carbon and 5×10^{13} cm^{-2} protons. The highest peak of this sample is at 1574 nm (C center) whose intensity decreases gradually with temperature but started to decrease rapidly at 100K.

4. Conclusion

We have investigated the temperature effect on the formation of the optically active point defect while focusing at the G-centre. As predicted, most of the luminating centres suffers from the temperature quenching problem and usually disappear at higher temperature. However, there are some potential point-defect centre such as the carbon-oxygen (CiOi) C-centre when the sample was first pre-amorphised with ge.

Acknowledgments

We acknowledge the European Research Council for financial support under the FP7 for the award of the ERC Advanced Investigator Grant SILAMPS 226470. The author would also like to thank Universiti Kebangsaan Malaysia for the grant GGPM-2016-028.

References

[1] Homewood K P and Lourenço M A 2005 Light from Si via dislocation loops *Materials Today* **8**(1) 34-39

[2] Bao J M 2007 Point defect engineered Si sub-bandgap light-emitting diode *Optics Express* **15**(11) 6727-6733

[3] Rotem E, Shainline J M and Xu J M 2007 *Stimulated emission and emission efficiency enhancement in nanopatterned silicon - art. no. 67750I*, in *Active and Passive Optical Components for Communications Vii* (Bellingham: Spie-Int Soc Optical Engineering) I7750-I7750

[4] Rotem E, Shainline J M and Xu J M 2007 Enhanced photoluminescence from nanopatterned carbon-rich silicon grown by solid-phase epitaxy *Applied Physics Letters* **91**(5) 3

[5] Canham L T, Barraclough K G and Robbins D J 1987 1.3-Mu-M Light-Emitting Diode from Silicon Electron-Irradiated at Its Damage Threshold *Applied Physics Letters* **51**(19) 1509-1511

[6] Boyraz O and Jalali B 2004 Demonstration of a silicon Raman laser *Optics Express* **12**(21) 5269-5273

[7] Rong H S 2005 A continuous-wave Raman silicon laser *Nature* **433**(7027) 725-728

[8] Rong H S 2005 An all-silicon Raman laser *Nature* **433**(7023) 292-294

[9] Ng W L 2001 An efficient room-temperature silicon-based light-emitting diode *Nature* **410**(6825) 192-194

[10] Kittler M 2005 Dislocation engineering for a silicon-based light emitter at 1.5 /spl mu. in Electron Devices Meeting *IEDM Technical Digest, IEEE International*

[11] Pavesi L 2000 Optical gain in silicon nanocrystals *Nature* **408**(6811) 440-444

[12] Pacifici D 2003 Erbium-doped Si nanocrystals: optical properties and electroluminescent devices *Physica E: Low-dimensional Systems and Nanostructures* **16**(3-4) 331-340

[13] Lourenco M A, Gwilliam R M and Homewood K P 2007 Extraordinary optical gain from silicon implanted with erbium *Applied Physics Letters* **91**(14) 141122-1

[14] Castagna M E 2003 Si-based materials and devices for light emission in silicon *Physica E-Low-Dimensional Systems & Nanostructures* **16**(3-4) 547-553

[15] Cloutier S G, Kossyrev P A and Xu J 2005 Optical gain and stimulated emission in periodic nanopatterned crystalline silicon *Nature Materials* **4**(12) 887-891

[16] Berhanuddin D D 2012 Co-Implantation of Carbon and Protons: An Integrated Silicon Device Technology Compatible Method to Generate the Lasing G-Center *Advanced Functional Materials*

[17] Berhanuddin D D Structural analysis of silicon co-implanted with carbon and high-energy proton for the formation of the lasing G-centre *AIP* 103110

[18] Berhanuddin D D 2016 Photoluminescence study of the optically active, G-centre on pre-amorphised silicon by utilizing ion implantation technique 2016 *IEEE International Conference on Semiconductor Electronics (ICSE)* 256-259

[19] Hart M J and Evans A G R 1988 Rapid thermal processing in semiconductor technology *Semiconductor Science and Technology* **3**(5) 421

[20] Webb R 2001 *Surrey University Sputter Profile Resolution from Energy deposition,* (SUSPRE: IBC, University of Surrey)

[21] Thonke K 1981 New model of the irradiation-induced 0.97-eV (G) line in silicon: A CS-Si* complex *Physical Review B (Condensed Matter)* **24**(10) 5574-86

[22] Davies G 1989 The Temperature-Dependence of the 969 Mev G-Optical Transition in Silicon *Semiconductor Science and Technology* **4**(3): 200-206

[23] Davies G 1989 The Optical-Properties of Luminescence-Centers in Silicon *Physics Reports-Review Section of Physics Letters* **176**(3-4) 83-188

IOP Conf. Series: Materials Science and Engineering **384** (2018) 012063 doi:10.1088/1757-899X/384/1/012063

Utilization of Arduino Uno on DikaTron TV Trainer Kit

A Syaputra and A B Pantjawati

Department of Electrical Engineering Education, Universitas Pendidikan Indonesia, Jl. Dr. Setiabudhi No. 227 Bandung, Indonesia

*andika17@student.upi.edu

Abstract. This study aims to design Android-based TV trainer with DikaTron app on smartphone is an innovation from previous TV trainer. TV trainer is designed by connecting TV circuit path through the relay to the system that will be controlled, such as a sound circuit, image, color, power supply, tunner, and more. This TV trainer combines Arduino UNO, relay, Bluetooth modules, and TV circuit, with DikaTron application software that installed in smartphone. The Arduino Uno, relay and Bluetooth modules are located in a box separate from the TV set, so it can be developed to control other devices. Through the trial obtained results in accordance with the desired, this means that the function on the television set can work properly and can be controlled from the smartphone via Bluetooth network. From the research conducted can be concluded that Android-based TV trainer with DikaTron app on smartphone can work in accordance with the desired specifications and can be used to support the lab activities.

1. Introduction

The widely used TV Trainer kits are still manually operated through many switches as controllers. To be more easily understood and operated with more efficient use of time, required innovation TV trainer models [1]. This research provides a new innovation in the form of DikaTron application on smartphone combined with Arduino Uno module, relay module, and Bluetooth module are located in a box separate from the TV set, so it can be developed to control other devices [2]. With this innovation, TV trainer can work in accordance with the desired specifications and can be used to support the teaching activities [3].

2. Material and Methods

This research was conducted by experimental method through 4 stages, that is to create DikaTron apps, create Arduino coding, control systems, and test runs. DikaTron application creation using App Inventor by online. Making this application begins by designing the appearance of the application as shown in figure 1.

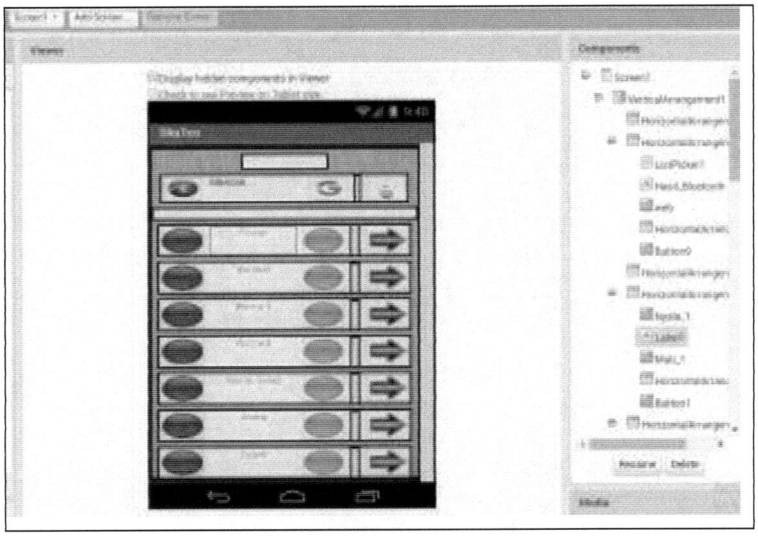

Figure 1. DikaTron application display creation process.

When finished creating the DikaTron app, proceed to create the arduino coding and match the coding with coding on App Inventor because it is interconnected as shown in figure 2 [4].

```
arduino_TA_maybe
char val;
void setup() {
  pinMode(3,OUTPUT);
  pinMode(4,OUTPUT);
  pinMode(5,OUTPUT);
  pinMode(6,OUTPUT);
  pinMode(7,OUTPUT);
  pinMode(8,OUTPUT);
  pinMode(9,OUTPUT);
  Serial.begin(9600);
}

void loop() {
if( Serial.available() >0 ) {
    val = Serial.read();
    Serial.println(val);
}
  if( val == '1' ) {
    digitalWrite(9,HIGH); }
  else if( val == '2' ) {
    digitalWrite(8,HIGH); }
  else if( val == '3' ) {
    digitalWrite(7,HIGH); }
  else if( val == '4' ) {
    digitalWrite(6,HIGH); }
  else if( val == '5' ) {
    digitalWrite(5,HIGH); }
  else if( val == '6' ) {
    digitalWrite(4,HIGH); }
  else if( val == '7' ) {
    digitalWrite(3,HIGH); }

  else if( val == 'A' ) {
    digitalWrite(9,LOW); }
  else if( val == 'B' ) {
    digitalWrite(8,LOW); }
  else if( val == 'C' ) {
    digitalWrite(7,LOW); }
  else if( val == 'D' ) {
    digitalWrite(6,LOW); }
  else if( val == 'E' ) {
    digitalWrite(5,LOW); }
  else if( val == 'F' ) {
    digitalWrite(4,LOW); }
  else if( val == 'G' ) {
    digitalWrite(3,LOW); }
}
```

Figure 2. Coding on Arduino Uno.

In DikaTron application there are 16 buttons that are designed as needed. Before changing lines on a TV machine, first learn the parts on the TV machine, how it works, function, and take measurements [5]. If you already understand it can be continued by changing the path as needed. The disconnected circuit line is connected to the relay. After all connected, it can be tested by connecting the TV machine to the control box via the connector cable [6]. Then create various paths in the Proteus software to make the control box neater and set the control box using acrylic as shown in figure 3 and 4.

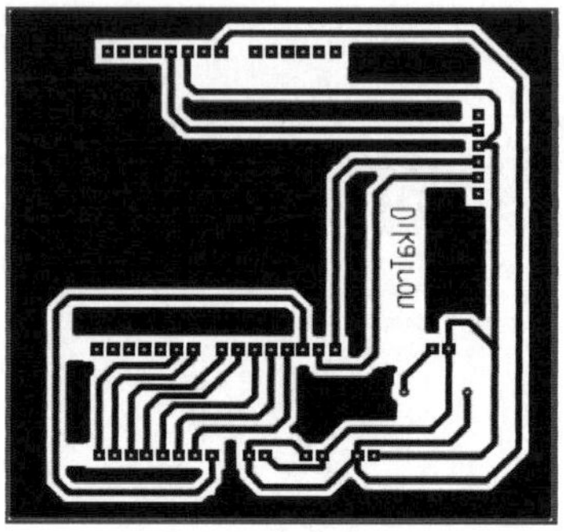

Figure 3. PCB board.

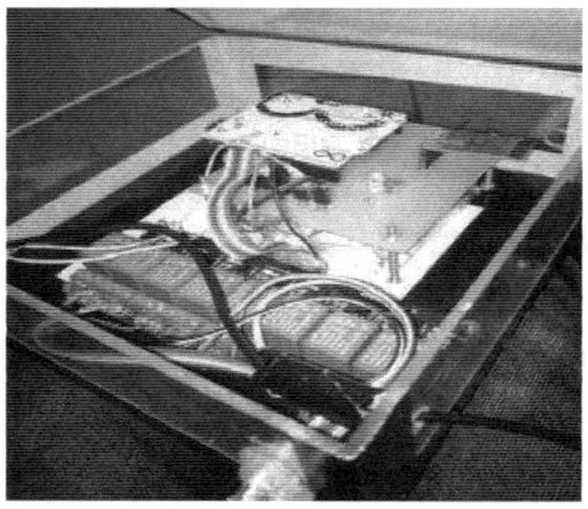

Figure 4. Control box.

The TV trainer design is based on the following specifications:

- Control system via bluetooth network on smartphone;
- Available a content of learning about colour TV and troubleshoot of colour tv
- One trainer can be accessed by more than one user via smartphone;
- Instructors are only required when in complex circumstances;
- The material on color TV troubleshoot is obtained directly from experienced TV technicians.

3. Result and Discussion

Experimental results show that TV trainers can work according to the desired specifications, that is combination of TV machines, control boxes, and connected smartphones. Table 1 is a comparison between conventional TV trainer and DikaTron TV trainer.

Table 1. Comparison.

No	Spesification	Conventional	DikaTron
1	Control system	Using the switch	Using the Smartphone
2	Material content	Not available	Available on smartphone
3	Access trainer	One control many users	One control one users
4	Instructur	Need at all times	No need at all times

In table 1 point 1, the results of the system control section show the performance in accordance with the desired specifications as in table 2.

Table 2. Performance in table 1 point 1.

No	Variable	Function	Does not work
1	Power	✓	-
2	Vertikal	✓	-
3	Colour 1	✓	-
4	Colour 2	✓	-
5	Dark Image	✓	-
6	Sound	✓	-
7	Antenna Tunner	✓	-

In table 1 point 2, the results of the system control section show the performance in accordance with the desired specifications as in table 3.

Table 3. Performance in table 1 point 2.

No	Variable	Function	Does not work
1	Power	✓	-
2	Vertikal	✓	-
3	Colour 1	✓	-
4	Colour 2	✓	-
5	Dark Image	✓	-
6	Sound	✓	-
7	Antenna Tunner	✓	-

In table 1 point 3, the results of the system control section show the performance in accordance with the desired specifications as in table 4.

Table 4. Performance in table 1 point 3.

No	Variable	Function	Does not work
1	Power	✓	-
2	Vertikal	✓	-
3	Colour 1	✓	-
4	Colour 2	✓	-
5	Dark Image	✓	-
6	Sound	✓	-
7	Antenna Tunner	✓	-

In table 1 point 4, the results of the system control section show the performance in accordance with the desired specifications as in Table 5.

Table 5. Performance in table 1 point 4.

No	**Variable**	**Function**	Does not work
1	Power	✓	-
2	Vertikal	✓	-
3	Colour 1	✓	-
4	Colour 2	✓	-
5	Dark Image	✓	-
6	Sound	✓	-
7	Antenna Tunner	✓	-

The displays of TV machine, control box and DikaTron application can be seen in figure 5, 6 and 7 respectively.

Figure 5. TV machine.

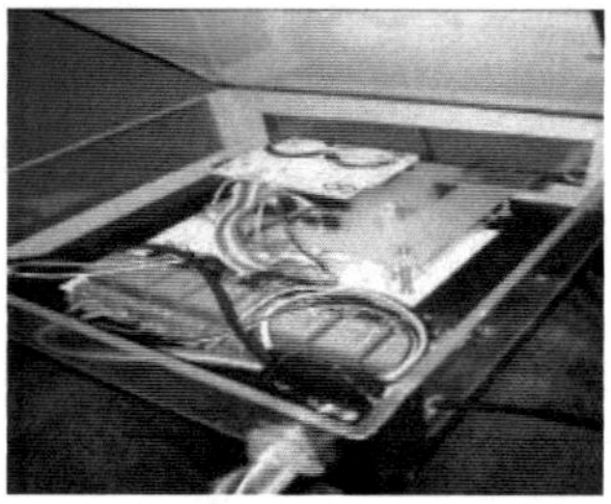

Figure 6. Control box.

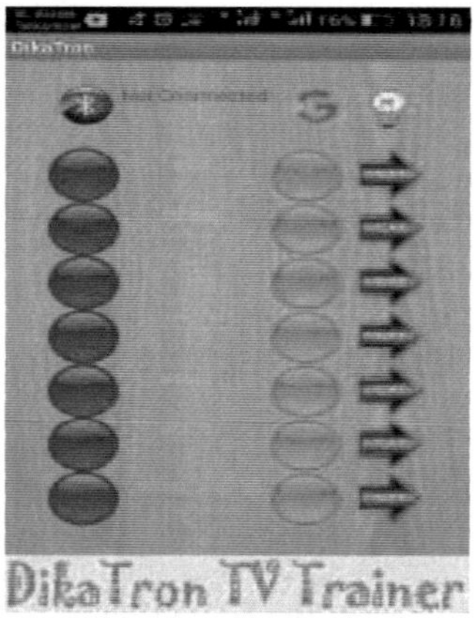

Figure 7. DikaTron App.

DikaTron application has 7 ON / OFF switch, 7 hyperlink buttons to color TV troubleshoot content, 1 piece of bluetooth button, 1 piece google button, and 1 piece of content and matter content. Each has a function in accordance with the needs. Here is the DikaTron application key function shown in figure 7.

- The Bluetooth button functions to connect the smartphone with the control box;
- Google button serves to look for other references;
- The light button contains the material;
- Red and green buttons as control switches;
- The arrow keys contain specific material / explanations according to the key pressed.

Figure 8 shows DikaTron application code and can be downloaded at link: http://ai2.appinventor.mit.edu/b/15qg.

Figure 8. DikaTron App barcode.

At the time of the test, the TV trainer had little disturbance. When the TV is still in normal condition, the TV can work optimally without any noise. After the TV is used as a trainer there is a little noise due to the waste of current flowing on the cable connector to the control box. This can happen because the flowing current is damping along the length of the cable [7].

This research can be developed to be simpler, does not require TV trainer and control box, only need one smartphone only. TV trainer is only shaped audio visual on the smartphone screen.

4. Conclusion

From the research conducted it can be concluded that Android-based TV trainer kit with DikaTron app on smartphone can work in accordance with the desired specifications and can be used to support practical activities. This DikaTron TV trainer kit can be developed again to be more complex, and the use of the control box can be developed its utilization to control other devices.

References

[1] Pi WG, Saher R and Daud S 2017 *Led Tv Trainer Use As Teaching Aids For Dee 3052 Electronic Equipment Repair Courses*
[2] Kaur J and Reddy S R N 2016 Design & Development of My SmartPhone Kit *National Conference on Product Design (NCPD 2016)*
[3] Radio T and Servicing G *National Board For Technical Education*
[4] Onibonoje M O, Umeh U N and Kehinde L O 2015 Development of an Arduino-based trainer for building a wireless sensor network in an undergraduate teaching laboratory *International Journal of Electrical and Electronic Science* **2**(3) 64-73
[5] Fireman J 1977 *TV book: The ultimate television book* (Workman Publishing Company)
[6] Mukai M and Kobayashi R 1988 *TV Trouble-Shooting Manual, Volume 1* (Geneva: ILO Publications, International Labour Office)
[7] Buchsbaum W H 1975 *Color TV servicing* Prentice Hall

IOP Conf. Series: Materials Science and Engineering **384** (2018) 012064 doi:10.1088/1757-899X/384/1/012064

Wireless Sensor Network on LPG Gas Leak Detection and Automatic Gas Regulator System Using Arduino

L Dewi* and Y Somantri

Department of Electrical Engineering Education, Universitas Pendidikan Indonesia, Bandung, Indonesia

*laksmitadewi@student.upi.edu

Abstract. Most of LPG explosions are caused by undetected gas leakage in the pre-detection condition. So that, LPG detection system is needed. The purpose of this system is to detect gas leakage, neutralize it, and prevent the explosion. Gas leakage could happen due to improper regulator installation or the hose is broken. This detection should not work in just one location because gas can leak at the gas regulator and its hose. Therefore, Wireless Sensor Network (WSN) is one of the methods that suitable for detecting gas leakage in the wider area. This method uses two or more gas sensors to detect leakage in two or more locations around the gas tube and its distribution line. WSN system works based on gas sensor MQ-6 and wireless module Bluetooth HC-05. Explosion prevention system works based on alarm/buzzer, exhaust fan, and automatic gas regulator. If the gas leaks, the sensor will send its data wirelessly to Arduino. Then, explosion prevention system will be activated. The system will turn the alarm/buzzer on, automatically releases gas regulator, and neutralizes the air with the exhaust fan. Both systems will be fully controlled by Arduino platform.

1. Introduction

There are still many cases of LPG explosion that happen in the society. Despite, the number of it is not as many as formerly when the first LPG is applied in massive. However, it frequently causes fire forest. Forest fires may cause some relevant losses towards residential structures, infrastructures, industrial facilities, farms, and vegetation, and they are a relevant threat to human life [1].

Almost all of the explosion happen caused by the undetected leakage. In the beginning, LPG has no scent. So it will make LPG hard to detect. Be aware of that, Pertamina added mercaptans gas to make it has a unique and nose piercing smell [2]. Then the LPG will be easier to detect. But sometimes the leakage occurs when there is nobody home. If that happens, gas will spread widely with no warning. This is so risky and the explosion will happen in anytime.

The leakage can occur with some factors. In some cases, it happens because of incorrect regulator installation [3], the tube rubber is worn out, brittle hose, the tube is already obsolete [3], and many others. In addition, BSN present the comprehensive data about under standard gas stove. They are 20 percent of the regulator, 50 percent of the gas stove, 66 percent of valve tube, and 100 percent of hose that out of standard[3]. From all that factors, most of it occurs around the tube mouth and the hose. Because in the tube mouth, gas can slip in and out easily. But, the leakage can also occur around the connection between the hose and the stove. It will happen when the hose is obsolete and/or the connector is gradually loose.

Based on that background, desire to create explosion preventing action appear. This action needs a system that detects early leakage. Weigh the leakage factors can occur about the tube and the hose, then detection will be done in two areas. In addition, the system needs quick action to release the regulator to the tube. So, it will use WSN (Wireless Sensor Network) and automatic regulator.

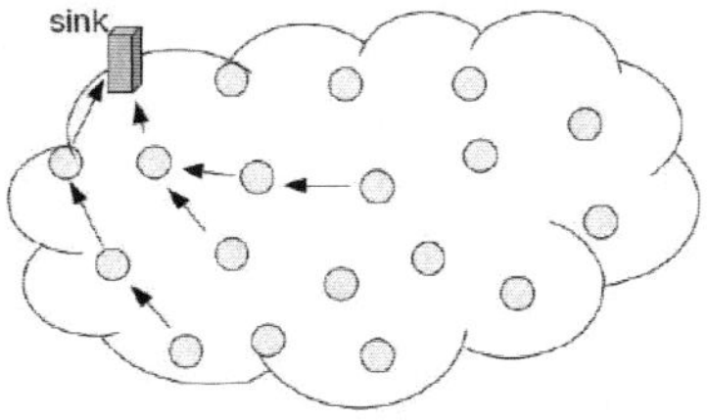

Figure 1. Wireless Sensor Network System [2].

WSN [2] is a wireless ad-hoc multi-hop network which consists wireless sensor node and sink as the main data receiver like figure 1. Environmental information collected by sensor nodes is routed from the ad hoc network to the sink node, where the data can be analyzed and aggregated [4].

From the background explanation above, then we can propose this research goal is to create a tool. A tool for wireless detection and prevention system. The research is still underway when this paper is made. This research hopefully will answer the problems of explosion prevention system concept which still have a gap. Also hopefully this will help leakage detection problems.

2. Methods

2.1. System Design

The system is divided into two. Main system and supporting system. The main system consists of the Arduino Uno [5], gas sensor MQ-6 [6], Bluetooth HC-05 [7], buzzer, LED, LCD, exhaust fan, and synchronous motor [8]. This system is the main system for the research. In addition, the supporting system consists of the Arduino Nano [9] which will operate the MQ-6 and Bluetooth HC-05. The system will be placed separately with the main system and connected wirelessly to it. WSN will connect detection in two areas. The first sensor connects straightaway to the main system and placed in the gas tube. The second sensor connects to the second system and placed near the stove and the hose connection.

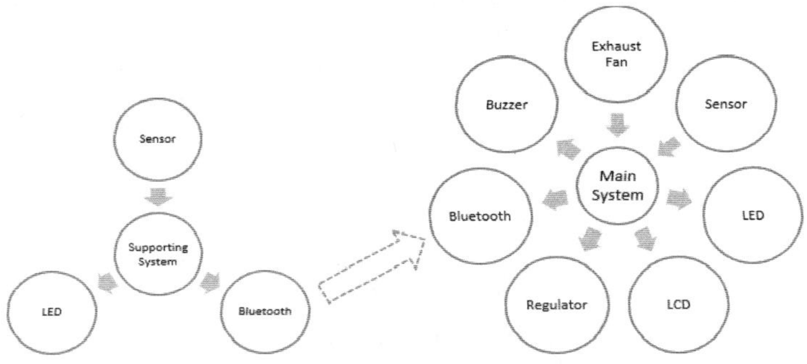

Figure 2. Hardware Design.

2.2. Hardware Design

The research uses some hardware. Arduino Uno and Nano are used as the microcontroller for both systems. The gas sensor is MQ-6. MQ-6 has high sensitivity to natural gas and LPG compositions like Propane and Butane [10]. MQ-6 can detect 300 – 10.000 ppm of gas concentration. Bluetooth HC-05 is

IOP Conf. Series: Materials Science and Engineering **384** (2018) 012064 doi:10.1088/1757-899X/384/1/012064

used as the wireless module that connects both systems. HC-05 has 3 Mbps data rate [7] and its range approximately 10 m. The warning system is provided by the buzzer and the LED. The LCD displays the level of gas leakage. The exhaust fan can neutralize the air when the leakage occurs. Meanwhile, the automated regulator is driven by one phase synchronous motor. The synchronous motor is usually used as rotary on the fan.

Hardware design for the main system is designed with a plastic container. The first container will hang on the gas tube. The supporting system that placed near the hose is also designed with a plastic container. But, this two container have different plastic materials. The second container can be placed anywhere because of its flexibility. It is flexible because it doesn't need wire to be connected to the first container. The second container can be placed 10 m maximum range from the first container.

The main system uses AC as the power supply. Because there are so many components that are used, AC is chosen to replace the battery. However, the supporting system still uses the battery since it just consists of Arduino Nano, HC-05, MQ-6, and LED. The battery even uses to support its flexibility.

2.3. Software Design

Software design is all the work out of programming design. How it makes a system that works based on the required instructions. Because Arduino is used, then the programming design is fully done with Arduino IDE.

```
void setup() {
  digitalWrite(FAN, HIGH);
  digitalWrite(REG, HIGH);

  pinMode(LED_YELLOW, OUTPUT);
  pinMode(LED_GREEN, OUTPUT);
  pinMode(LED_RED, OUTPUT);
  pinMode(sensor, INPUT);
  pinMode(BUZZER, OUTPUT);
  pinMode(FAN, OUTPUT);
  pinMode(REG, OUTPUT);
  pinMode(limswitch, INPUT);

  lcd.setBacklightPin(BACKLIGHT_PIN,POSITIVE);
  lcd.setBacklight(HIGH);
  lcd.begin(16, 2);
  digitalWrite(LED_YELLOW, HIGH);
  tone(BUZZER, 2500);
  delay(100);
  noTone(BUZZER);
  delay(100);
  tone(BUZZER, 2500);
  delay(100);
  noTone(BUZZER);
  delay(100);
```

Figure 3. Arduino Ide Program.

2.4. Simulation

The first thing to do in simulation process is sensor calibrating. Specific tools for calibrating the sensor will cost so much. Also, it will take too much time and exact knowledge will be needed. But, if the calibration is done with the proper test, it will show as precise as it should be. So the easier way to calibrate the sensor is using a simple Arduino program. The calibration program is provided by one of the online electronics design platforms. Calibration should be done in 1000 ppm of LPG. So we could get the value of MQ-6 reading.

```
Calibrating...
LPG = 0ppm
CH4 = 0ppm
Ro = 60.85kohm
MQ = 32.00
```

Figure 4. MQ-6 Calibration.

According to the data that released by NIOSH [3], the threshold for LPG is 2000 ppm (10% LEL). LEL is an abbreviation from Lowest Explosion Limit [11]. There are two explosion mechanisms that need to be considered when evaluating the inflammable gas incident, they are detonation and deflagration. Detonation is shocked reaction where the fire moves faster at supersonic speed (faster than sound). While deflagration is where the fire moves at subsonic speed [3].

After sensor calibration is done, the simulation can be started. In this simulation, the gas match is used for giving a gas leakage simulation around the sensor. The sensor will detect the gas and will send its data to the main system Arduino. Then Arduino will activate all the warning systems, display its data to LCD, turn on the exhaust fan, and release the regulator.

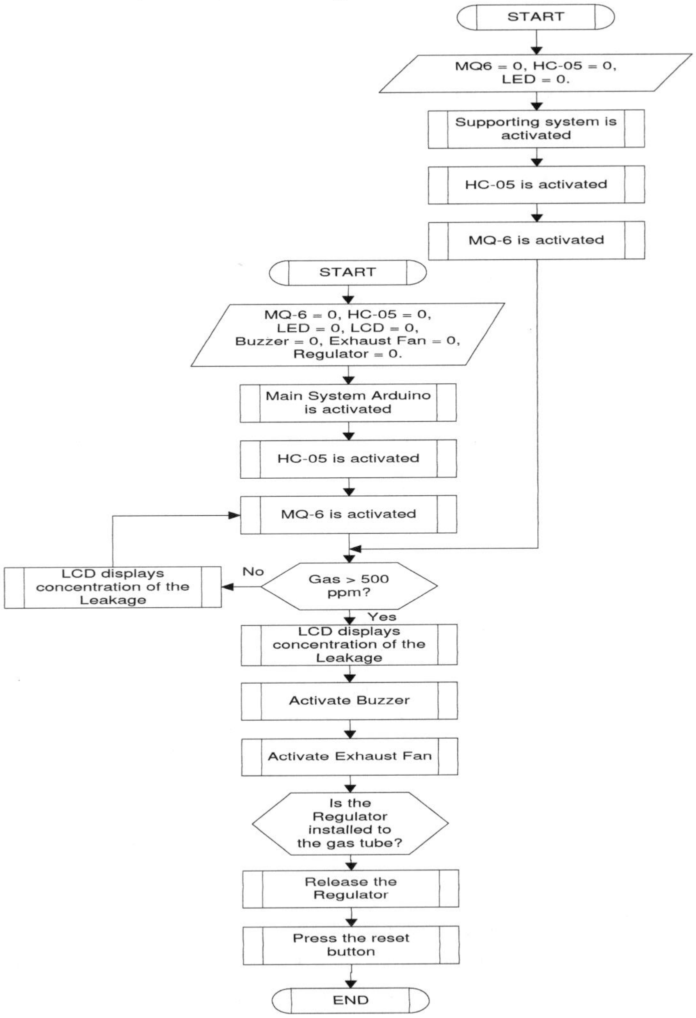

Figure 5. Simulation Flowchart.

3. Results

As the result, this research should detect the leakage around both sensors. Then as the feedback, the system should give the response. The responses are should activate the warning systems, the exhaust fan, and the regulator.

In fact, calibrate the sensor needed a lot of efforts because it was hard to get the value 1000 ppm and the value of Ro precisely. Calibrating the sensor still need a comparison tool. So in this simulation, MQ-6 would be set at the value of 300. The buzzer will sound when the sensor reaches the lowest limit 300 based on the value presented by MQ-6.

Arduino state three conditions. They are Initialize, Stand-By, and On Fire. First, Initialize state is the condition where Arduino recently turns on. It shows the opening message and the opening sound. Second, Stand-By state is the condition where the sensors start sensing. The condition is marked by the blinked green LED. In this condition, the sensors sense over and over again. Third, On Fire state is the condition when the sensor is detecting gas leakage. In this condition, the buzzer is keep ringing, the red LED and the exhaust fan are on, and also, the regulator is released by the system. The table below is showing all the data from this research. The table is showed according to the component's response based on Arduino's condition.

Table 1. Supporting System.

Arduino	MQ-6	Yellow LED	Green LED	Red LED	HC-05
Initialize	-	On	Off	Off	-
Stand By	<300	Off	On	Off	-
On Fire	≥300	Off	Off	On	-

Table 2. Main System.

Arduino	MQ-6	Yellow LED	Green LED	Red LED	Buzzer	LCD	Exhaust Fan	Regulator	HC-05
Initialize	-	On	Off	Off	On	==WELCOME!== SISTEM ACTIVATED	Off	Off	-
Stand By	<300	Off	Blink	Off	Off	Kadar LPG: … Kadar LPG Aman	Off	Off	-
On Fire	≥300	Off	Off	On	On	Kadar LPG: … Bocor Pak Bu!!	On	On	-

4. Discussion

Unfortunately, based on both tables, the wireless system is still not working yet because it should need some troubleshooting. The problem is the Bluetooth module couldn't communicate each other even they are connected. This is still a homework job for this research. One more thing to be considered from this research, the DC supply for the supporting system needs to be more assessed because it could be out of power at any time. In this case, a battery 9V can only power supporting system up during 15 hours approximately.

5. Conclusions

The conclusions that can be drawn from this research are:

- Calibration process still needs comparison process.
- MQ-6 has high sensitivity to the contents of LPG.
- All components of the main system have a successful response except the Bluetooth module.
- The main system and the supporting system are ready to use.
- Need some wires to connect the supporting system to the main system since the Bluetooth module didn't work yet.

The research is still needed evaluation to be a successful research. It has some problems and needs some repair. But, the system is ready to be analyzed and continued by anyone who has interest in solving matters about this LPG explosion.

References

[1] Scarponi G E, Landucci G, Heymes F and Cozzani V 2017 Experimental and numerical study of the behavior of LPG tanks exposed to wildland fires *Process Safety and Environmental Protection*

[2] Sabiq A and Alfarisi T 2017 *Sistem Wireless Sensor Network Berbasis Arduino Uno dan Raspberry Pi untuk Pemantauan Kualitas Udara di Cempaka Putih Timur, Jakarta Pusat* 301–305

[3] Dian S K 2012 *Analisis Konsekuensi Dispersi Gas, Kebakaran, dan Ledakan Akibat Kebocoran Tabung Lpg 12 Kg Di Kelurahan Manggarai Selatan Tahun 2012 Dengan Menggunakan Breeze Incident Analyst Software Selama* (Depok: Universitas Indonesia)
[4] Tian Y, Lv Y and Tong L 2013 Design and application of sink node for wireless sensor network *COMPEL-The international journal for computation and mathematics in electrical and electronic engineering* **32**(2) 531-544
[5] Uno A and Front R 2012 *Full-Text*
[6] Kim 2014 MQ-6 LPG LNG Gas Sensor Module _ Sandbox Electronics *Sandbox Electronics* [Online] Available: http://sandboxelectronics.com/?p=191 [Accessed: 01-Jan-2017]
[7] Studio I 2010 *HC-05-Bluetooth to Serial Port Module* Datasheet
[8] Bedell F and Ryan H J 1895 Action or a single-phase synchronous motor *Journal of the Franklin Institute* **139**(3) 197-214
[9] Arduino LLC 2017 Arduino Nano **9210** 662 1
[10] Data T 2014 *MQ-6 Semiconductor Sensor for LPG* 2–4
[11] Dhawale D S, Salunkhe R R, Patil U M, Gurav K V, More A M and Lokhande C D 2008 Room temperature liquefied petroleum gas (LPG) sensor based on p-polyaniline/n-TiO2 heterojunction *Sensors and Actuators B: Chemical* **134**(2) 988-992

IOP Conf. Series: Materials Science and Engineering **384** (2018) 012065 doi:10.1088/1757-899X/384/1/012065

Analysis of Transformer Conditions using Triangle Duval Method

S Permana*, S Sumarto and W S Saputra

Department of Electrical Engineering Education, Universitas Pendidikan Indonesia, Dr. Setiabudhi Street No. 207, Bandung, Indonesia 40154

*permanasidik1408@gmail.com

Abstract. Electrical or thermal failure that occurred in the transformer oil can produce a dangerous gas called gas fault. By indicating the content of dissolved gases in the transformer oil, can be obtained information about what happened to the oil so it can be estimated indication of the type and cause of the failure in the transformer. The method used to identify and analyze the dissolved gases in oil is called Dissolved Gas Analysis (DGA). In this study conducted by the DGA Duval Triangle method for transformer oil used in geothermal power plant UPJP Kamojang PT. Indonesia Power. The results using Duval Triangle method produces thermal failure and partial discharge.

1. Preliminary

In the electric power transformer system is one of the vital parts, which serves to distribute electrical energy to the low voltage and high voltage to, the delivery takes place within the same frequency. This function is known as the term step-up and step-down,As electrical devices, transformers must never be separated from failures, especially the failure of thermal and electrical failure or what we call electrical breakdown. The failure of the type that usually occur in the transformer, namely: Arcing, Sparking, heat disorders and Partial Discharge [1].

Therefore, the reliability of the transformer also need to be maintained, and the necessary preventive measures to determine the condition of the transformer. One method often used is the method of DGA (Dissolved Gas Analysis). Methane (CH_4), Ethane (C_2H_6), Ethylene (C_2H_4), acetylene (C_2H_2) and Air (H_2O) [2].

2. DGA Testing

2.1 Sampling Process Oil

The sampling process oil in the transformer is divided into stages, due to prepare the equipment for oil sampling and sampling procedures of the oil in the transformer. For the first part is a tool that is used for sampling are: Syringe, oil flushing units, buckets, gloves, wipes and vials.[3] In the next stage of the sampling process transformer oil and gas extraction process to determine the gas content in the transformer.

2.2 Gas Extraction Process

a. Gas Chromatograph is a technique for separating substances - certain of a joint compound, usually substances - are separated by evaporation rates (volatility).[4]

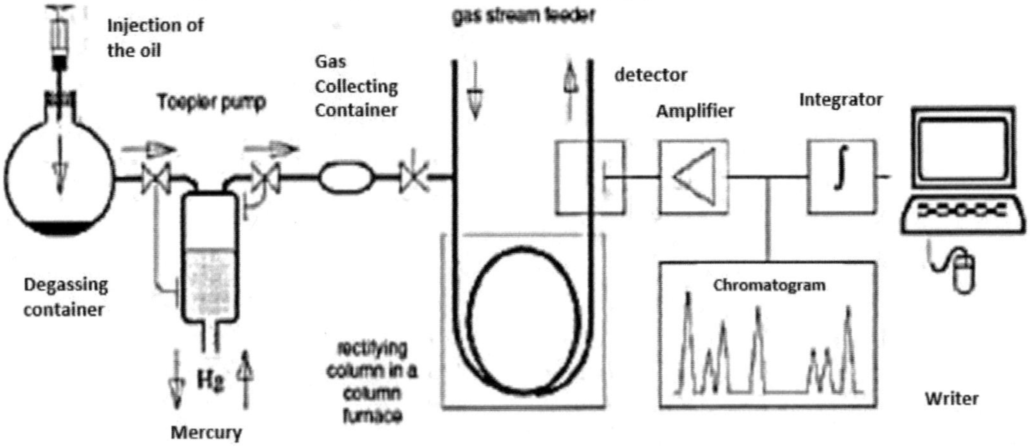

Figure 1. Method Chromatograph

b. Photoacoustic Spectroscopy (PAS), (hydrogen, methane, oxygen, and others) basically have the ability to absorption of electromagnetic radiation that is unique and distinctive. This capability is usually applied to the infra-red spektokopis method to produce an effect of photo acoustic. Absorption of electromagnetic radiation by gases will increase the gas.[5]

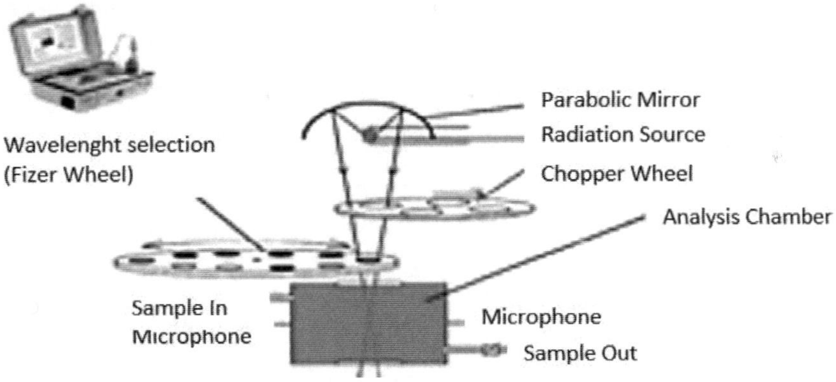

Figure 2. Photoacoustic Spectroscopy Method

Results obtained from the extraction of gas will be used as input to the gas content analysis or DGA. DGA analysis of the calculation results can be obtained an indication of the failure of the transformer is likely to emerge.[6]

2.3 Analysis of Data From Extraction

After the gas is separated from the oil sample through the extraction process, it can be data in the form of gas contained in the sample, then the data is analyzed using Dissolved Gas Analysis. The method used is:

IOP Conf. Series: Materials Science and Engineering **384** (2018) 012065 doi:10.1088/1757-899X/384/1/012065

2.3.1 Limit IEEE / Total Dissolved Combustible Gases. TDCG is one type of Dissolved Gas Analysis method to classify the type of combustible gas, the higher the gas content obtained it must do further tests. [7]

Table 1. Standar TDCG

level	Concentration (Ppm)	Description TDCG
1	<= 720	Indications that the normal transformer operation
2	721 - 1920	Indications high starting a gas composition, there arises the possibility of failure, prevention of the symptoms do not continue.
3	1921 - 4630	Indications decomposition of a high level of insulation. The failure may have occurred. Make prevention of disorder does not continue.
4	> 4630	Indications of deterioration are very high and the decomposition / damage to the insulator is already widespread. Imminent damage to the transformer. Soon

2.3.2 Key Gas. Key gas is defined as gas - gas that is formed in the transformer oil coolers are quantifiable and can determine the type of failure occurs yan, based on the type of gas that is typical or dominant formed at various temperatures. [7]

Table 2. Standar Key Gas

Fault	Key Gas	Criteria	Gas Percent Amount
Arcing	Acetylene (C_2H_2)	Large amount of and (and minor quantities of (, (CO), and may Also exist if cellulose is Involved$(H_2)(C_2H_2)(C_2H_4)(CO_2)$	(H_2) : 60% (C_2H_2): 30%
Corona (Low Energy PD)	Hydrogen (H_2)	Large amount of some, with small quantities of (and (. CO and may be comparable if cellulose is Involved.$(H_2)(CH_4)C_2H_6)C_2H_4)(CO_2)$	(H_2) : 85% (CH_4) : 13%
Overheating of Oil	Ethylene (C_2H_4)	Large amount of (less amount of (, some quantities of and$C_2H_4)C_2H_6)(CH_4)(H_2)$	(C_2H_4): 63% (C_2H_6): 20%
Overheating of Cellulose	Carbon Monoxide (CO)	Large amount of CO and Hydrocarbon gases may exist(CO_2)	CO: 92%

2.3.3 Roger Ratio. Roger Ratio method is one of the DGA methods by comparing the value / amount of gas that is different by dividing the gas with other gases. [8], Comparison of the gas used by themselves have some reference standards starting from IEC Standards, CEGB Standards and ASTM Standards. Type of gas used as a comparison, namely: Hydrogen (H_2), Methane (CH_4), Ethane (C_2H_6), Ethylene $(C_2H_4)$and acetylene (C_2H_2) . [9]

2.3.4 Triangle Duval. This method uses a triangular plot for analyzing interference, Methane (CH_4), Ethylene (C_2H_4) and acetylene (C_2H_2). When using the Triangle Duval necessary to determine the existence of a problem or not, how to determine the existence of a problem or not the gas is contained in the transformer oil is to ensure at least one of hydrocarbons or hydrogen should be at the level of L1. [10]

In addition, Duval triangle method was created to help method - other analytical methods. This method is a closed system (closed system) is different from the method of Key Gas and Gas Ratio which is (open system), so that this method can reduce the percentage of cases outside the analytical criteria.[11]

Table 3. Limit L1

Gas	Limit L1 ppm
H_2	100
CH_4	75
C_2H_2	3
C_2H_4	75
C_2H_6	75
CO	700
CO_2	7000

Now that we know that we obtain the data entered in L1 criteria, then the data previously obtained we calculated by using a calculation formula Duval Triangle, namely:

$$\%CH_4 = \frac{CH_4}{CH_4+C_2H_4+C_2H_2} \text{ x } 100\%$$

$$\%C_2H_2 = \frac{C_2H_2}{CH_4+C_2H_4+C_2H_2} \text{ x } 100\%$$

$$\%C_2H_4 = \frac{C_2H_4}{CH_4+C_2H_4+C_2H_2} \text{ x } 100\%$$

Function formula above is to determine the value of (X, Y, Z) which represent the value of the proportion of CH_4, C_2H_4 and C_2H_2, from 0% to 100%, the first assumption in all three grades, namely: (CH4 + C2H4 + C2H2) = S in ppm, then calculation of the third value in percent of the gas is X = CH_4% CH4 = 100 (A / S), Y = C_2H_4% C2H4 = 100 (B / S), Z = C_2H_2% C2H2 = 100 (C / S). So the value in (CH4 + C2H4 + C2H2) must equal 100%.C_2H_4[12]

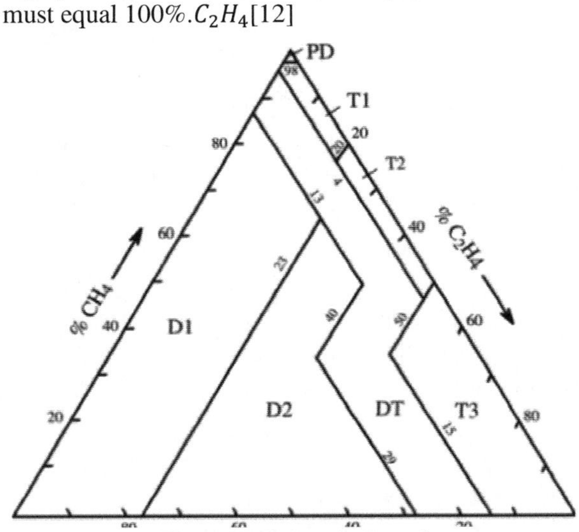

Figure 3. Triangle Duval

Table 4. Types of Disorders of the Duval Triangle

Symbol	Fault	examples
PD	Partial discharge	Discharge of the cold plasma (corona) type in gas bubbles or voids, with the possible formation of X-wax in paper
D1	Discharges of low energy	Partial discharges of the sparking type, inducing pinholes, punctures in carbonized paper. Low energy arcing carbonized inducing perforation or surface tracking of paper, or the formation of carbon particles in oil.
D2	Discharges of high energy	Discharges in paper oil, with the power of follow-through, the resulting in extensive damage to paper or large formation of carbon particles in the oil, metal fusion, tripping of the equipment and gas alarms.
T1	Thermal Fault T <300°C	Evidenced by turning brownish paper (> 200°C) or carbonized (> 300°C).
T2	Thermal Fault 300 <T<700°C	Carbonized of paper,the formation of carbon particles in oil.
T3	Thermal fault T> 700°C	Extensive formation of carbon particles in oil, metal coloration (800°C) or metal fusion (1000°C)

3. Result and Discussion

This study uses data derived from geothermal power plant UPJP Kamojang PT. Indonesia Power. The data in the form of content of the extracted gas in transformer oil, gas content in transformer oil is used as a sample calculation is the gas content in transformer oil T31 during the period 2012 - 2017. The sample is what will be used as the calculation in Duval Triangle methods of analysis.

Prior to the Duval triangle analysis on the data obtained, first do data matching with the limit L1, because not all types of data acquired is at the limit L1. Adjustment function here is the first step in the analysis process, it is undeniable that the limit L1 is one of the conditions required in the analysis of Duval Triangle as the minimum limit indication of failure on the transformer.

Table 5. Data Transformer T31

date`	H_2 ppm	C_2H_4 ppm	CO ppm	CO_2 ppm	C_2H_4 ppm	C_2H_6 ppm	C_2H_2 ppm
11/27/12	32	5	243	2405	1	8	0
11/12/2012	32	1	243	2300	6	10	0
03/07/2013	45	6	511	5944	1	10	0
12/16/2013	34	5	379	6188	7	8	0
03/20/2014	5	2	4	64	1	1	0
07/24/2014	48	6	336	3406	1	6	0
25/09/2014	54	7	462	4460	1	5	0
11/06/2014	90	13	525	5588	1	7	21.6
07/11/2014	33	11	348	4257	1	69	0
11/12/2014	42	7	500	5384	1	8	0
01/22/2015	44	10	551	6744	2	7	0
07/04/2015	49	9	548	7476	5	8	0
16/10/2015	55	14	641	10395	5	6	0
12/04/2015	48	11	633	11025	5	13	0
01/19/2016	52	9	633	11819	5	12	0
03/08/2016	51	10	601	11 865	3	16	0
26/07/2016	39	11	641	13835	13	5	0
19/12/2016	22	1	36	209	3	3	0.5
29/05/2017	38	4	132	419	2	3	0.5

Table 6. Data transformer T31 is at L1 Limit

date`	H_2 ppm	CH_4 ppm	CO ppm	CO_2 ppm	C_2H_4 ppm	C_2H_6 ppm	C_2H_2 ppm
11/06/2014	90	13	525	5588	1	7	21.6
07/04/2015	49	9	548	7476	5	8	0
16/10/2015	55	14	641	10395	5	6	0
12/04/2015	48	11	633	11025	5	13	0
01/19/2016	52	9	633	11819	5	12	0
03/08/2016	51	10	601	11 865	3	16	0
26/07/2016	39	11	641	13835	13	5	0

Table 6 shows the data transformer T31 is at the limit L1, can be seen on the vulnerable 04-07 - 2015 to 26 - 07-2016 gas CO_2 content in transformer oil contains very high, CO_2 exceeding the limit on Limit L1 of 7000 ppm, because it should be noted rise on CO_2 can also affect the temperature rise in cellulose. On April 11 - 06-2014, gas content increased in C_2H_2 the transcend limit L1. The increase in C_2H_2 the type of failure can cause Arcing.

3.1 Triangle Duval

Duval triangle method using three gas in the process are: CH_4, C_2H_2, and C_2H_4. The third gas present value (X, Y and Z) on. X for the value% CH4, C2H4% Y for value, and Z value% C2H2. If it is assumed further that the value - the value of determination serves as a meeting point in determining the type of interference on Duval Triangle.

Data on April 11 - 06-2014 indicates that the value of H_2 = 90 ppm CH_4 = 13 ppm, CO = 525 ppm CO_2 = 5588 ppm C_2H_4 = 1 ppm C_2H_6 = 7 ppm, and C_2H_2 =21.6 ppm , said to be worth using because the Duval triangle method has a C_2H_2 value that exceeds the limit L1, the final result of these data are:

$$\%CH_4 = \frac{13}{13+1+21.6} \text{ x } 100\% = 35\ 519\%$$

$$\%C_2H_2 = \frac{21.6}{13+1+21.6} \text{ x } 100\% = 59.02\%$$

$$\%C_2H_4 = \frac{1}{13+1+21.6} \text{ x } 100\% = 5{,}464\%$$

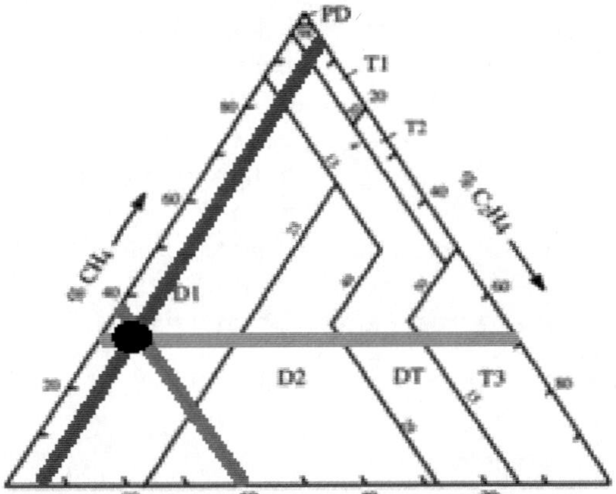

Figure 4. Failure on the type transformer

Table 7. Analysis on Transformer T31using Triangle Duval Method

date`	H_2 ppm	CH_4 ppm	CO ppm	CO_2 ppm	C_2H_4 ppm	C_2H_6 ppm	C_2H_2 ppm	Fault
11/06/2014	90	13	525	5588	1	7	21.6	D1
07/04/2015	49	9	548	7476	5	8	0	T2
16/10/2015	55	14	641	10395	5	6	0	T3
12/04/2015	48	11	633	11025	5	13	0	T2
01/19/2016	52	9	633	11819	5	12	0	T2
03/08/2016	51	10	601	11 865	3	16	0	T2
26/07/2016	39	11	641	13835	13	5	0	T3

The results of the analysis can be seen in Table 7, that the transformer T31 has the possibility of symptoms - symptoms of failure are diverse ranging from Partial Discharge, Thermal Fault 300 $< T <$ 700°C, and Thermal fault $T >$ 700°C. This is because in some cases have increased activity in the C_2H_2 and CO_2, where the gas is a gas trigger of failure - the failure.

References

[1] R. N. Digdayanti, W. Martiningsih, and S. Wardoyo, "Aplikasi Fuzzy Logic Pada Metode Dissolved Gas Analysis Untuk Mengklasifikasikan Tipe Fault Pada Minyak Trafo," vol. 1, no. 1, pp. 1–8, 2012.

[2] A. R. L Demmassabu and F. Patras, S Lisi, "Analisa Kegagalan Transdformator Daya Berdasarkan Hasil Uji DGA Dengan Metode TDCG, Key Gas, Roger's Ratio, Duval's Triangle pada Gardu Induk" 2014.

[3] B.A.Ismanto, T.Haryono, Suharyanto, "Metode DGA (*Dissolved Gas Analysis*) Untuk Perawatan Minyak Transformator Daya Di PT. Pembangkitan Jawa Bali Unit Pembangkitan Paiton. 2010" .

[4] R. Hardityo, "Deteksi dan Analisis Indikasi Kegagalan Transformator dengan Metode Analisis Gas Terlarut," 2008.

[5] D. Bates, "For Training Purpose Only DGA in a Box a Utility's Perspective By DGA in a Box."

[6] G. Yulisusianto, H. Suyono, and R. Nurhasanah, "Diagnosis Kondisi Transformator Berbasis Analisis Gas Terlarut Menggunakan Metode Sistem Pakar Fuzzy," vol. 9, no. 1, pp. 1–6, 2015.

[7] T. Committee of the IEEE Power Engineering Society, *IEEE Std C57.146TM-2005, IEEE Guide for the Interpretation of Gases Generated in Silicone-Immersed Transformers*, vol. 2008, no. February. 2006.

[8] D. Bhalla, R. K. Bansal, and H. O. Gupta, "Transformer incipient fault diagnosis based on DGA using fuzzy logic," *India Int. Conf. Power Electron. IICPE 2010*, 2011.

[9] R. Hooshmand and M. Banejad, "Application of Fuzzy Logic in Fault Diagnosis in Transformers using Dissolved Gas based on Different Standards," *Int. J. Mech. Aerospace, Ind. Mechatron. Manuf. Eng.*, vol. 2, no. 17, pp. 157–161, 2007.

[10] S. S. Desouky, A. E. Kalas, R. A. A. El-Aal, and A. M. M. Hassan, "Modification of Duval triangle for diagnostic transformer fault through a procedure of dissolved gases analysis," *EEEIC 2016 - Int. Conf. Environ. Electr. Eng.*, vol. 1, no. 3, 2016.

[11] S. Singh and M.N. Bandyopadhyay,"Duval Triangle : A Noble Technique for DGA in Power Transformator. 2010" .

[12] M. Duval, "Dissolved Gas Analysis and the Duval Triangle," *TechCon Asia Pacific, Sydney, Aust.*, pp. 1–20, 2006.

International Symposium on Materials and Electrical Engineering (ISMEE) 2017 IOP Publishing
IOP Conf. Series: Materials Science and Engineering **384** (2018) 012066 doi:10.1088/1757-899X/384/1/012066

Control System Flatness Fabric on Double Folding Machine

Z M Zahri* and Y Somantri

Program Studi Pendidikan Teknik Elektro FPTK UPI, Universitas Pendidikan Indonesia, Bandung, Indonesia

*mohammadzahri_@student.upi.edu

Abstract. Increasing the production in the industry can bring the wrong fabric quality on the process of flat rolling cloth. Needs a simulation of prototype that is used as a simulation media that has the same shape and function as the tool or unit. Jointly to know how to work control on flatness of fabric in double folding machine, so as to give description and as media of simulation in industry. Multiply doubled, with this simulation is expected to be a reference and learning in the development of tools that can help the industry to be able to process cloth at the time of production. In application can improve the quality and quantity of production. The research method used in this research. As explained diagram block diagram used is a fabric sensor that is not right, whether the position of the fabric in a tilted / uneven then the electric photo sensor will give info to the arduino to force the servo motor and then the lever to fit the position of the fabric is tilted / uneven to flat back in position. At the conclusion for the future expected to be the tools that will be generated through research will show whether through this simulation can help and develop the process of fabric in the industry.

1. Introduction

Modern era as it is necessary to know the world of industry, especially the textile industry must produce quality and quality of rolls on the fabric. Textile industry in Indonesia has long been developed in line with the development of cloth industry business. This idea can be accommodated by designing a new custom machine or introducing the reconfiguration process [1]. Reconfiguration can be obtained using a robot and it has been used in various ways to fold fabric. In addition, there is a desire associated with the fabric folding machine. When designing a fabric-side control system there is a need to simulate the engine and its interaction with scrolls. The degree of automation in the non-rigid materials industry is still low, as concluded in a survey 96 published research papers on the manipulation of flexible materials, performed by Saadat and Nan. Most of these publications (76%) are related to sheet material and only 58% of them are industry-related. Few of these publications deal with folded fabrics, while the results have not yet been applied to the apparel industry. A study of folding depending on fabric modeling has been guided by Eischen and Kim. The problems that often occur to date make us to continue to think and innovate how to make the results of quality products and quality to give the impression of mutual benefit between the industry and consumers [2]. One of them for the flatness of the fabric is still a lot in the industry that does not care about the quality of fabric rolls result from the final process for double folding machine that serves as folding cloth. The large demand of the market that wants the quality of the neat fabric roll is still a problem today in the industry. Therefore, where the design is done in the form of prototype control system flatness of the fabric on the double folding machine so that the results of the fabric rolls to be neat on the left and right.

2. Methods

As for several stages of the planned project. The system will be shared with two common parts of hardware and software [3]. The software designed at this initial level is only for testing. In the hardware section, consideration of many aspects is very important in terms of size, characteristics, goals, and results. After completing these two important parts, and before integrating the two, some testing should be done on each of the components used. After confirming the perfect circuit, simple program will be tested to microcontroller and all ports are tested and verified.

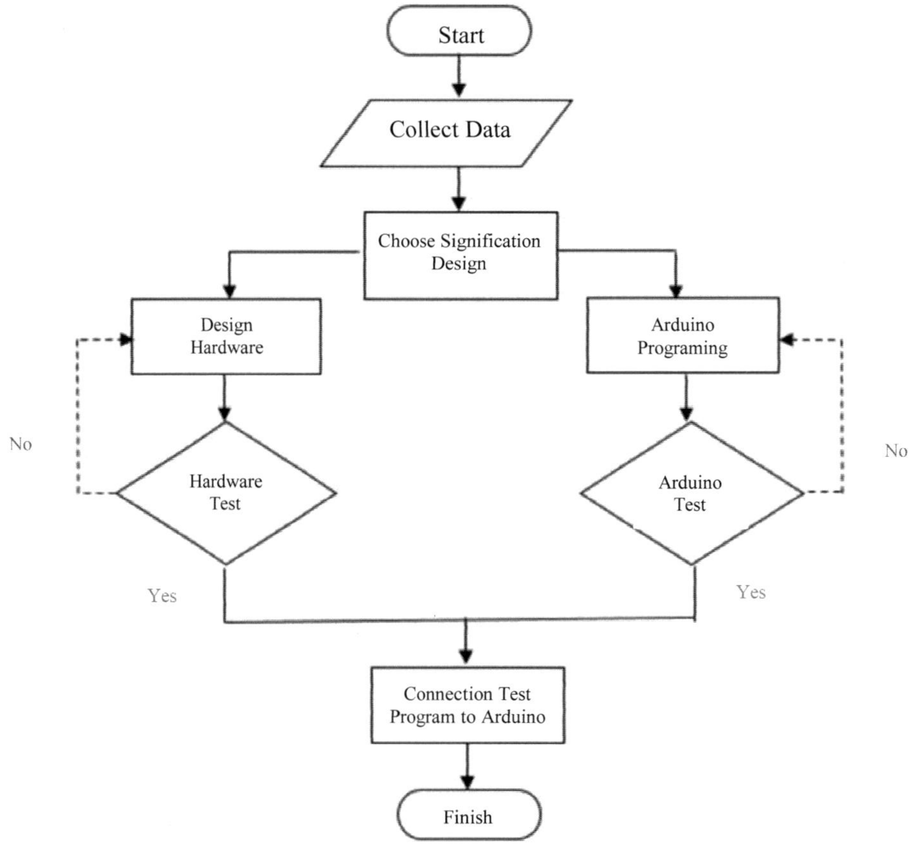

Figure 1. Flowchart method

This prototype process will start as soon as the appliance is turned on the fabric will start rolling up, then when the fabric passes through the electric photo sensor then the sensor will detect whether the fabric is in a tilted state, if the roll position of the fabric is tilted then the photoelectric sensor will give info to arduino MEGA2560 to move the stepper motor, then push the lever / cylinder to adjust the uneven position of the fabric to roll the fabric back flat in position [3].

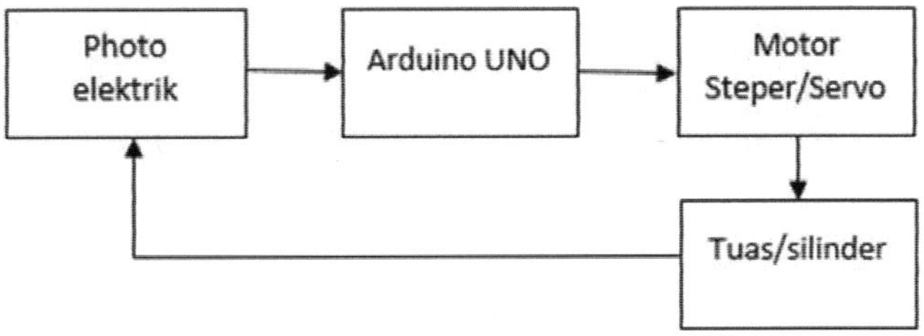

Figure 2. Block Diagram

The first step taken on the design of the prototype is the identification of equipment and materials used. The following table lists the equipment and materials used.

Table 1. Equipment and materials used

No	Name of equipment	Explanation
1.	Arduino UNO	ATMEGA 2560
2.	*Personal Computer*	Toshiba AMD A8
3.	*Photo Eletric*	1 unit
4.	*Tuas/Silinder*	1 unit
5.	*Motor Stapper*	2 unit

The first stage is the design stage schematic circuit using software fritzing. The second stage is the stage of making the program. In the design of software programs used arduino software. The third stage is to perform performance testing on the prototype whether it is in accordance with the program and control system that has been made.

3. Result and Discussion

The design results are obtained from testing every part of the tool that has been made. The test is performed in accordance with the procedure of using the tool. Testing hardware (hardware), done by doing a repeated testing of each component used. After testing has been obtained hail that all components contained in the tool works well. Software testing (software), this is done by testing to test the program on the tool that has been made whether in accordance with the design. From the results of the experiments conducted concluded that the program arduino run according to the initial plan of the program. Testing overall. At this stage of testing, the steps performed are arranging the control system in the arduino program using the jumper and testing whether the program and the control system is running according to the program command or not. After testing the program and the control system runs according to program commands and tools runs well. However, the problems that often occur when the tool is not running that is wrong in assembling a control system that is not in accordance with the output inpu of the program created, or there are cables that do not fit the installation.

However, the problems that often occur when the tool is not running that is wrong in assembling a control system that is not in accordance with the output inpu of the program created, or there are cables that do not fit the installation [3]. Initially there were some problems encountered when first making a prototype model of the fabric flatness control system on an easy folding machine. The most difficult part is choosing the right motor. The motor must be strong enough to lift the folding material. The weight after the cloth is placed on the folding material should also be considered when choosing the motor. Motor also does not have to be high in price as well as easily controlled by microcontroller. Problems also occur when selecting folding materials. The material should not be heavy because the motor will have difficulty in lifting it. The surface of the material must also be slightly coarse so the fabric will not

IOP Conf. Series: Materials Science and Engineering **384** (2018) 012066 doi:10.1088/1757-899X/384/1/012066

slip off of it as soon as its wings are rotated. So once tested with a little material, polystyrene has finally been selected as a suitable material because it is very light and also has a rough surface that can keep the fabric fixed on the roll. Another problem encountered is how to attach the folding material to the motor because the motor needs something that attaches to the shaft like a gear system.

4. Conclusion

A new yet simple project was created. The conventional way home hold chorus, in terms of folding the fabric is now done more neatly and nice views. Because this machine is completely automatic and proven with a complete fabric cycle with the push of a button. This machine can help users reduce their load and is very useful for buyers with many fabrics to fold. The project is also useful for some industries, fabric manufacturers and others who need fabrics to fold in a much more tidy way than expensive and expensive CNC machines.

References

[1] Mullineux, G., & Matthews, J. 2010. Constraint-based simulation of carton folding operations. *CAD Computer Aided Design*, **42**(3), 257–265. https://doi.org/10.1016/j.cad.2009.12.003

[2] Wulfhorst, B., Gries, T., Veit, D., Wulfhorst, B., Gries, T., Veit, D., & Elsasser, N. 2006. *Processes and Machines for Textile Finishing. Textile Technology*, 216–242. https://doi.org/10.3139/9783446433472.009

[3] Gomesh, N., Daut, I., Kumaran, V., Irwanto, M., Irwan, Y. M., & Fitra, M. 2013. Photovoltaic powered T-Shirt Folding Machine. *Energy Procedia,* **36**, 313–322. https://doi.org/10.1016/j.egypro.2013.07.036

IOP Conf. Series: Materials Science and Engineering **384** (2018) 012067 doi:10.1088/1757-899X/384/1/012067

Preliminary Study of Micro-hydro Power Plant (MHPP) in The Rural Area

I W Ratnata*, W S Saputra, M Somantri, E Mulyana and A Ardhika

Faculty of Vocational and Technology Education, Universitas Pendidikan Indonesia, Bandung, Indonesia

*iwayanratnata@upi.edu

Abstract. Pakenjeng village is one of the villages in the subdistrict Pamulihan, Garut. Garut regency has the potential of natural resources is quite large and varied, namely water resources, energy, and mineral resources. Energy resources available in the form of the potential of renewable energy such as geothermal energy (geothermal), micro-hydro, solar, wave, and wind. Judging from potential micro hydro, Pakenjeng village crossed by two rivers with a flow of water large enough, the river Cikondang and Cibatarua. Currently, both the river Cikondang and river Cibatarua been used for hydroelectric power plant, with a capacity> 100 kW (scale mini-hydro), ie power plants Kombongan (river Cikondang) with a capacity of 165 kW and a power plant Panyairan (river Cibatarua) with capacity of 3 x 200 kW. The result of the potential study (pre-feasibility study) showed that Sungai Cibatarua have the quality and quantity of water that is eligible to be a source of hydroelectric energy for water discharge stable and constant reaches 5100 liters / sec. Cibatarua River has a waterfall that has a head that is high enough that Niagara Panyairan with head up to 63 meters. Hydraulic power that can be generated to reach 3248 kW and a power that can be awakened by the turbine reaches 2404 kW. Potential analysis (pre-feasibility study) of the mechanical and electrical indicates that the turbines are suitable for the MHP in the River Cibatarua namely Cross Flow Turbine Type TC 15 BO 650 with power up to 1 MW.

1. Introduction

Garut regency is one of regencies in West Java province which has the potential of natural resources is quite large and varied, namely water resources, energy, and mineral resources. Energy resources available in the form of the potential of renewable energy such as geothermal energy (geothermal), micro-hydro, solar, wave, and wind. From this potential, already done the research, either in the form of an initial survey (survey's review), a preliminary survey and exploration, to investigate the existence of potential energy and mineral resources by various parties, such as the central government, private companies, other institutions, or by Garut regency government itself. Even in some locations already implemented the exploitation and utilization of mineral resources (minerals) and exploitation activities (development) for the resource development of renewable energy, namely geothermal, micro hydro power (GMHP), and solar power plants *(Pembangkit Listrik Tenaga Surya*/ PLTS).

Although there is a power plant that is capable of supplying electricity to Java, Madura and Bali of geothermal electricity, yet very ironic with the state of the region and society Garut regency very shortage enjoy network services of electricity, which until now electrification ratio in Garut regency reached 62.16%, with the number of households that have not had electricity networks 37.84% (data

2013) to overcome the above problems, Garut district government has sought to do development of infrastructure of electrification through rural electricity development program that has pushed the development of public economic activities [1]. Nevertheless, the development of electricity infrastructure is still in need of support from various parties in view of the budget allocation for the program is still very low, so that the achievements of the electricity network services for citizens are still not able to able to increase the electrification ratio [2].

Pakenjeng village fed by rivers Cikandang which in this river there is a waterfall known as Sanghyang Taraje waterfall, located in Kampung Kombongan. The area around the waterfall is 5000 m2 and a height of about 82 m waterfall. In the waterfall area Sanghyang Taraje has been built micro hydro power plant (MHPP) with an initial capacity of 85 kW (2006) [3]. In 2009, the Center for Research and Technological Development of Electricity and Renewable Energy (Pusat Penelitian dan Pengembangan Teknologi Ketenagalistrikan dan Energi Baru Terbaukan (P3TKEBT) has developed interconnection MHPP Kombongan by upgrading the Automatic Voltage Regulator (AVR) and the addition droopkits, as well as the replacement mainboard Electronic Load Controller with digital system for easy operator in the operation of the interconnection. From the research and development of MHP Kombongan interconnection is obtained an increase in electric power of 165 kW from the previous power by 85 kW. But for people Kombongan supplied electrical power of 65 kW for free and the 100 kW done grid so that it becomes non-tax revenue, from the optimization of the grid connection required the addition of ballast load to 120% of the maximum output capacity [4].

In addition Cikandang River, in the village Pakenjeng Cibatarua River also flows in the river where there are waterfalls Panyairan. Based on preliminary information, in this river are the MHPP Panyairan with power capacity is large enough that 3 x 200 kW. This research will be carried out a study on the possibility of making another MHPP Cibatarua River flow with a smaller capacity (<100 kW). Consideration of smaller capacity is that the investment costs required to be smaller, but the electricity produced can fulfill local needs. Will have several locations that allow it to be built PLTMH and further analyzed to compare the potential and opportunities for each of these locations to be recommended to the next step.

2. Literature Review

Micro hydro power plants are generating electrical energy utilizing water energy as an energy source initially, with output power in the range of 5-100 kW[5]. Mechanical energy is the energy produced by the water is used to turn turbines; this rotating turbine connected to a generator so that is produced electrical energy. Based on the operation of power plant, This MHPP within the classification of the hydroelectric plant run off river type. Simply, a run off river hydropower utilizing the flow of river water that is partially passed on a channel, turn a turbine and then discharged back to the main river. The components contained in plants run off river are: dam or weir diversion and intake (diversion weir and intake), like sedimentation (settling basin), the channel carrier (headrace), like tranquilizers (forebay), the pipeline exploded (penstock), turbines and generators, home generator (power house) and sewer (tail race)[6][7].

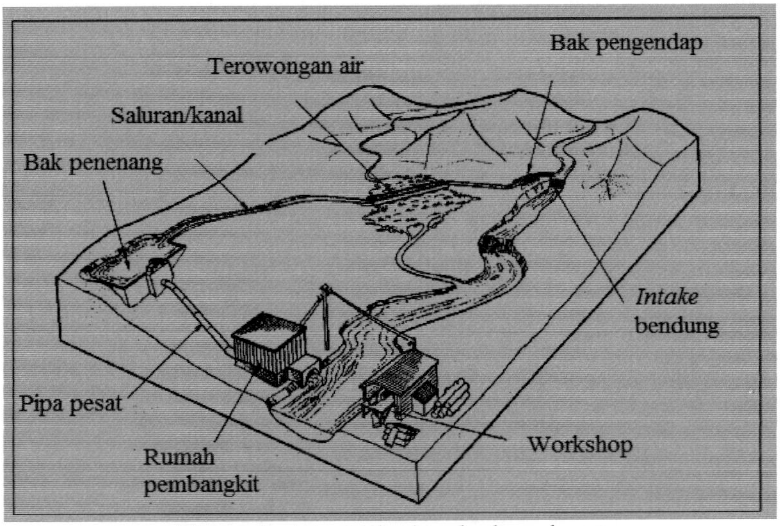

Figure 1. A typical micro hydro scheme

In order the planned MHPP according to the target, giving a clear benefit for the community, and the cost of economic development, it should be done a feasibility study of technical and non-technical. Technical feasibility study is a study to identify the potential popularity barometer (parameter) technical quantitatively to determine whether the potential locations meet the criteria requirements (standards) viable technical aspects. Based on the requirements (standards) are feasible, an MHPP development plan submitted by the related people can be evaluated so that their eligibility can be expressed in the technical aspects[8].

While non-technical feasibility study of the parameters (qualitative) that determines whether the location of the potential to meet the criteria requirements (standards) are non-technical aspects. Based on experience, if these studies neglected the work projects to be financed may result requires funds expensive, less achieve their goals, or less beneficial to the community.

Technical feasibility study includes some essential aspect, namely the feasibility study hydrology, civil, mechanical and electrical. While non-technical feasibility studies include a feasibility study covering aspects of economic / financial, social, cultural, and environmental sustainability.

3. Methods

This research is a descriptive study that aims to describe the data and information and the results of its analysis with respect to electricity is the potential of micro hydro power plant (MHPP) in river flow Cibatarua, Pakenjeng village. The data collection is done by documentary study, observation and interviews. The design and research stage briefly formulated as follows:

1. Assessment of the literature pertaining to the analysis of potential (pre-feasibility study) to study the feasibility of micro hydro power plant (MHPP);
2. The collection of basic reference materials include: location maps, geological maps, availability of construction materials, the data flow;
3. Selection of Potential Location:
   - Selection using topographic maps
   - Selection of the Kemi-light Consideration River / Debit channel and river / channel
4. Data and Information Technical Feasibility location;
5. Study Search Location;
6. Making Lay Out Initial System MHPP;
7. Non-Technical Data Collection: Socio-Economic Profile, Financial Analysis and Inventory Potential Electrical Supplies;

8. Analysis for Feasibility Study Recommendations.

4. Results and Discussion

4.1. Calculation of River Water Flow Cibatarua

Calculation of flow rates Cibatarua River starts with the determination of water cross-sectional area (A) and the water flow rate (v). To determine the cross-sectional area required several variables river width (W) and the average depth of the river (drata average), while the water flow velocity (v) is measured using a float method; water flow rate is obtained by calculating the time data (t) and distance (d) the travel buoy. Once the data is obtained cross-sectional area and the flow rate of the water, then the water flow rate can be calculated.

Table 1. The Calculation of Water Flow Debit

Month	A (m^2)	v (m/s)	Q m^3/s	ltr/s
January	9	0,42	3,78	3780
February	8	0,26	2,08	2080
March	11	0,54	5,10	5100

4.2. Power Potential Analysis

After getting data the river flow and an average elevation (head), then do the calculation of potential hydraulic power and data in the field then the next stage of the analysis of potential hydraulic power of a waterfall or river Panyairan Cibatarua. To get how much power can be generated from the potential of the river. It can also be a criterion in the selection of turbine type. Here is the analysis of the calculation of potential hydraulic power and power awakened.

Table 2. The Calculation of Hydraulic Power Potential

Month	Q (ltr/s)	G (m/s^2)	H_g (m)	P (kW)
January	3.780	9,8	65	2.407
February	2.080	9,8	65	1.324
March	5.100	9,8	65	3.248

Based on data from the hydraulic power potential, can have the type of turbine what is appropriate for use in the realization of the MHP. In the research of Cross Flow turbines have to have an efficiency of up to 74%. With an efficiency of it, then by referring to Table 2, can be calculated awakened power potential, that is equal to 1,781 kW (January), 980 kW (February), and 2,404 kW (March).

4.3. Analysis Pipeline Planning Rapid (Penstock)

Planning rapid pipe or penstock is very dependent on the cost and loss of head on the pipe rapidly. Because the cost of procurement and installation of pipe rapidly relatively expensive, then planning should take into account type of materials to be used, the length, diameter and thickness of the pipe. In planning the design of the pipeline rapidly to MHP in this Cibatarua river, the authors calculate the diameter of the pipe by considering the optimum speed and maximum discharge. From these calculations, pipe specifications rapidly obtained as shown in Table 3.

Table 3. Specifications of Rapid Pipe

Coordinate point upstream	7°23'19.09"LS dan 107°41'17.08"BT
Coordinate point downstream	7°23'19.30"LS dan 107°41'12.95"BT
Length Intake	126 meter
Wide cross-section	77 cm
Depth	3 meter

4.4. Intake Planning Analysis
Intake was planned as far as 126 meters upstream and downstream coordinates on the coordinates. For a wide cross-section of the pipe diameter intake equated to rapidly and to a depth of 2 times the intake taken from the water level in the river.

Table 4. Designing of Intake

Coordinate point upstream	7°23'19.09"LS dan 107°41'17.08"BT
Coordinate point downstream	7°23'19.30"LS dan 107°41'12.95"BT
Length Intake	126 meter
Wide cross-section	77 cm
Depth	3 Meter

4.5. Soothing Swimming Planning Analysis
Swimming sedative function to contain the water flow and channel it into high ideally penstock sedative pool is 2-4 times the maximum height of the water to obtain the height (h) of the pool is 8 meter. Location of planning tranquilizer sedative not far from the mouth of the river that drains adjacent to the river, this is due so that when the volume of water exceeds the capacity of the pool volume sedative wasted water can be poured into the river.

Table 5. Design Swimming Soothing

Long Swimming Soothing	20 meter
Wide	3,6 meter
Maximum depth	8 meter
The volume of water	396 m^3 / 396.000 Liter
Coordinate point	7°23'19.30"LS &107°41'12.95"BT

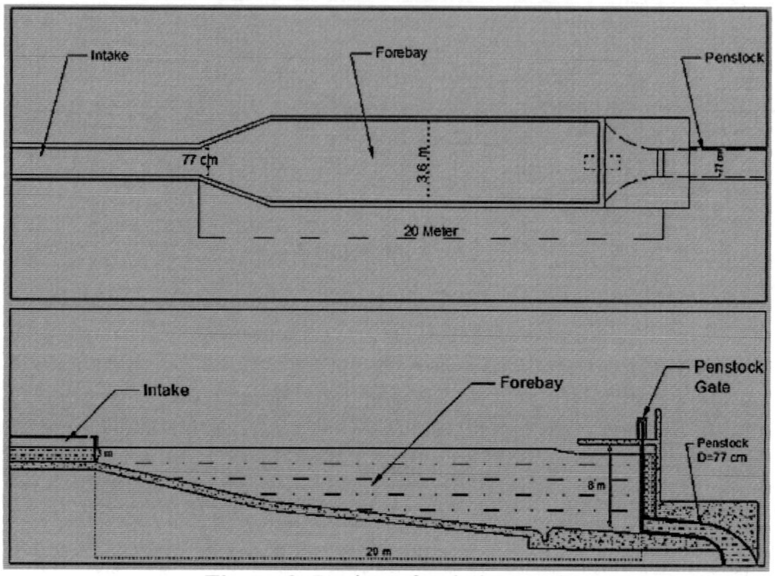

Figure 2. Design of sedative pool

4.6. Design Planning MHPP
After planning penstock, intake and a sedative, then the next step is the design of MHPP using data and variable results prior planning.

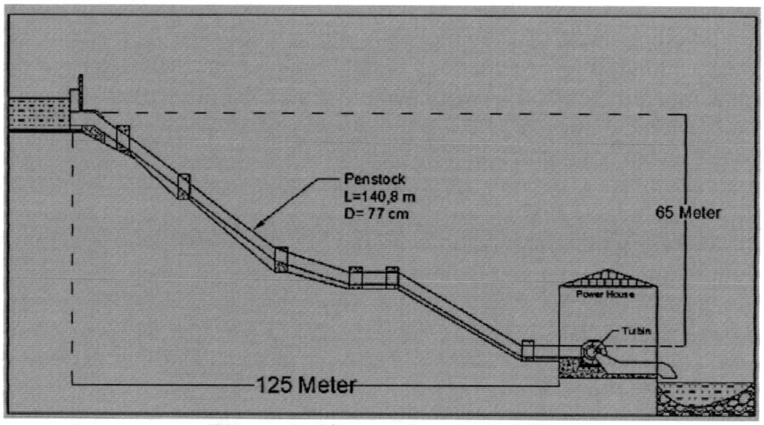

Figure 3. Skema planning MHP

4.7. Mechanical and Electrical Analysis Planning
With reference to prior the data the calculation and analysis, the turbine and generator specifications proposed in this study are as shown in Table 6 and Table 7.

IOP Conf. Series: Materials Science and Engineering **384** (2018) 012067 doi:10.1088/1757-899X/384/1/012067

Table 6. Turbin Specification

Magnitute	*Specipication*
Type of turbine	Cross Flow TC 15 BO 650
Diameter runner	ϕ500
Rotational speed turbine	1100 rpm
The maximum efficiency of the turbine	74%
Manufacture	CV. Cihanjuang Inti Teknik, Bandung

Table 7. Generator Specification

Magnitude	*Spesifikasi*
Type of generator	Generator Synchronous
Drive	Gearbox Speed increaser 600-1500
The number of phasa	3
Power (active power)	1000 Kw
Apparent Power	1200 kVA
Frequensi	50 H_z
Speed of Generator	1500 rpm
Maximum Efisiensi of generator	96%
cosϕ	0,83

4.8. Distribution Network Analysis

The distribution network is required to supply power to consumers. Ideally, the location of the power house to the consumer is not more than 5 km, this is to minimize power losses in the distribution network. Here is the planning system in the village Pakenjeng distribution network.

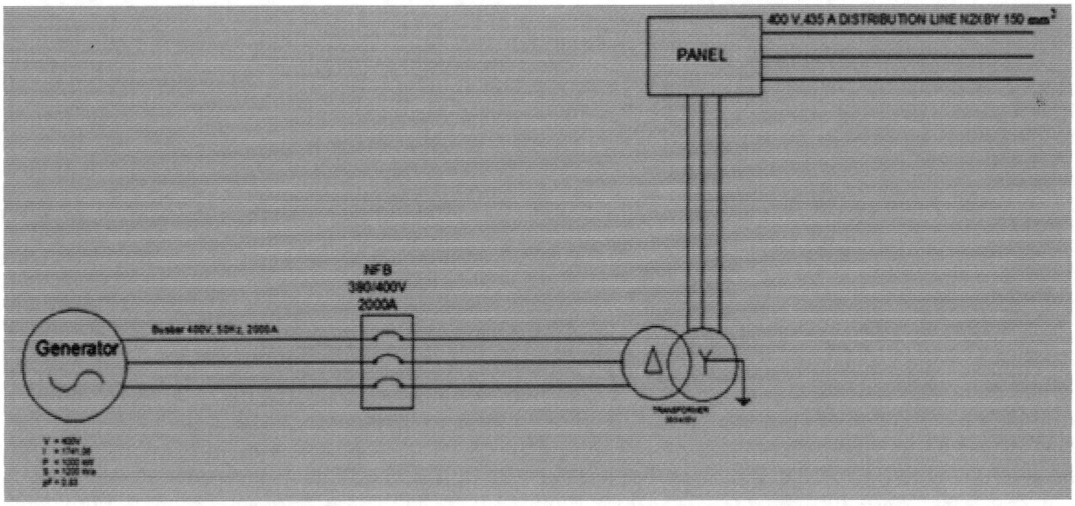

Figure 4. Diagram Distribution of electrical energy

Tabel 8. Electrical Mechanical Components

Component		Remarks
Generator	V	400 V
	I	1741,06 A
	P	1000 kW
	Q	1200 kVA
	pF	0,83
	F	50 Hz
Busbar	Size	80×5 mm^2
	Cable cross-sectional area	400 mm^2
	Number of core wires	2
	Loading Continue	1800 A
No Fuse Breaker		2000 A
Trafo	Tegangan	380/400 V
	Kapasitas	1200 kVA

Referring to SPLN 43-8: 1994, the type of cable used for the distribution network to the consumer is N2XBY 1 x 150 rm 0.6 / 1.2 kV, the code letter stating the type of cable used is a cable XLPE insulated and PVC sheathed, steel tape armored, single-core, twisted copper conducted round cross-sectional area of 240 mm2 with a rated voltage of 0.6 / 1.2 kV. This cable is used for distribution using 4 wire that is 3 phase and one neutral.

Tabel 9. Distribution Network Specifications

Distribution network	
Cable type	N2XBY
cable length	4,1 km
Total cable	4 (R, S ,T, N)
Number of core wires	Tunggal (1 Core)
Cable cross-sectional area	150 mm^2
reactance DC	0,124 Ω/km
AC reactance	0,157 Ω/km
Inductive reactance	0,084 Ω/km
The maximum current strength	479 A
high pole	7 meter

5. Conclusion

The results of the study concluded that the natural conditions around the River Cibatarua feasible for the construction of MHP, because of the location on the slopes of the mountain that has a hard ground contour and contains rocks so that the level of abrasion in this area is relatively small. This area also has a stable precipitation ranging from 1500-3000 mm annually. The feasibility study carried out showed that Cibatarua River has water quality and quantity eligible for discharge of water a stable and constant reaches 5100 liters / sec. Cibatarua River has a waterfall that has a head that is high enough that Niagara Panyairan which has a head of up to 63 meters. This shows that the development potential of the River Cibatarua can be beneficial as the MHP to the surrounding community. Hydraulic power that can be generated to reach 3248 kW and a power that can be awakened by the turbine reaches 2404 kW. The feasibility study mechanical and electrical are done in CV. Cihanjuang showed that the turbines are suitable for the MHP in the River Cibatarua namely Cross Flow Turbine Type TC 15 BO 650 with power up to 1 MW. The distance from the site to the village of Pakenjeng which is about 4.11 km from the location of the point of planning the MHP, the generated power is distributed to the conductor

N2XBY 1 x 150 rm 0.6 / 1.2 kV with a voltage drop of 12.01 V. Thus, location MHP is feasible to be built because it was close to the MHP.

References

[1] Departemen ESDM. *Pedoman dan Pola Tetap Pengembangan Industri Ketenagalistrikan Nasional 2003 -2020.*

[2] Ramdhani, Aji Saka Dwi. 2011. *Studi Perencanaan PLTMH 1 x 12 kW sebagai Desa Mandiri Energi di Desa Karangsewu, Cisewu, Garut, Jawa Barat.* Tugas Akhir. Teknik Elektro Institut Teknologi Sepuluh November, Surabaya.

[3] DJLPE. *Kepmen Rencana Umum Katenagalistrikan Nasional (RUKN) 2005-2025.*

[4] Jumadil Akhir, D. 2013. P*ertumbuhan Listrik Nasional Meningkat 10,1%.* [Online]. Tersedia di: http://economy.okezone.com/read/2013/02/20/19/764478/pertumbuhan-listrik-nasional-meningkat-10-1. Diakses 24 Februari 2014.

[5] Layman's Guidebook. *On How to Develop a small Hydro Site.* European Small Hydropower Association.

[6] K.V. Alexander, E.P. Giddens, A.M. Fuller, Axial-flow turbines for low head microhydro systems, Renewable Energy,Volume 34, Issue 1, 2009,Pages 35-47, ISSN 0960-1481,https://doi.org/10.1016/j.renene.2008.03.017.

[7] Joseph Kenfack, François Pascal Neirac, Thomas Tamo Tatietse, Didier Mayer, Médard Fogue, André Lejeune, Microhydro-PV-hybrid system: Sizing a small hydro-PV-hybrid system for rural electrification in developing countries, Renewable Energy, Volume 34, Issue 10, 2009, Pages 2259-2263, ISSN 0960-1481

[8] A. Fuller, K. Alexander, 6.16 - Simplified Generic Axial-Flow Microhydro Turbines, Editor(s): Ali Sayigh, Comprehensive Renewable Energy, Elsevier, 2012, Current as of 17 July 2015, Pages 435-466, ISBN 9780080878737, https://doi.org/10.1016/B978-0-08-087872-0.00617-X.

IOP Conf. Series: Materials Science and Engineering **384** (2018) 012068 doi:10.1088/1757-899X/384/1/012068

Implementation of Instructional Media Trainer Fiber Optic Sensor System on Sensor and Actuator Learning in SMKN 1 Katapang Bandung

Y M R Faozan*, B Mulyanti, and D Wahyudin

Department of Electrical Engineering Education, Universitas Pendidikan Indonesia, Bandung, West Java, Indonesia

*yuski.maolid@student.upi.edu

Abstract. Instructional media trainer of fiber optic sensor system is media that utilize optical fiber as props in the lessons sensors and actuators. The purpose of this research is the study of students learning outcomes on subjects sensor and actuator by use trainer fiber optic sensor system and student's responds about media has used. The method used in this research is experimental design, with the pre-experimental design. This research was conducted in class XI technique electronics industry (TEI) 1 in SMKN 1 Katapang Bandung with 38 students. The research results show that there has been increasing study results students who demonstrated by the strengthening (the gain) study results students on the cognitive domain an average of 0,64 in the medium category. In the domain affective and in the domain psychomotor the average value on the good predicate. All of students gives a very positive responds about implementation of this media. From the results of this research can be stated that using fiber optic sensor system could improve student learning outcomes and students's response is very positive to the media used. Thus we can conclude that fiber optic sensor system on the subjects of sensor and actuator can improve learning outcomes students and very use full to be used in vocational school.

1. Introduction

One of the productive subject that is needed to be mastered by the students of electrical engineering vocational high schoolers is sensors and actuator subject. The Sensors and Actuator subject cannot be conveyed only in theory without any media to prove the theory. Media usage in teaching could give a realistic experience and giving the foundation for the development to give a better study result using teaching media could boost students' motivation to be more active participating and giving more room to be more creative, innovative, and self-dependent [1].

After doing an observation in industrial electronics major in SMK Negeri 1 Katapang about sensor and actuator subject, researcher conclude that the teaching has not use teaching media or trainer in teaching the students, so the students still find it hard to understand the concept of the subject.

The Sensors and Actuator teaching only explaining the theory to the students without using any media to teach them, making them puzzled with the concept. After doing an interview with one of the teacher of Sensors and Actuator subject, it is revealed that the school doesn't have the teaching media for the subject.

Researcher have found out that without using the teaching media of Sensors and Actuator subject, SMKN 1 Katapang Bandung has not been able to implement the concept of working principles of sensors, which is very important in industrial electronics.

Based on the above finding, the researcher is interested in implementing fibre optic teaching media to help students understanding the material and helping them learning more skills in Sensors and Actuator. With that, the writer proposes a research named "Implementation of Instructional Media Trainer Fibber Optic Sensor System on Sensor and Actuator Learning In SMKN 1 Katapang Bandung"

2. Methods

This research is using experiment method. The experiment method is used to compare the result before and after experimental treatment [2]. The design of the experiment that will be used in this research is pre-experimental design. This design is chosen because the samples that is chosen are not random samples. And then the shape of the pre-experimental design that will be used is one-group pretest-posttest design. One group of pretest-posttest is chosen because this design will be used to compare the situation before and after the experimental treatment, so the result of the treatment could be observed effectively [3]. This research is done in two sessions.

This research is done in SMK Negeri 1 Katapang Bandung, located in Jl. Ceuri Terusan Kopo KM 13,5 Kabupaten Bandung. Sample that is used is 11th graders of industrial electrical engineering of SMKN 1 Katapang Bandung. The instrument used is tested first against the students of class XII TEI 1. The test instrument used in the research must be tested prior to a study [4]. The cognitive instruments performed should be tested first Testing of testing instrument performed to determine the validity, reliability, difficulty level, and the differentiator of the problem [5]. The validity test for the instrument used using Product Moment correlation formula from Preason [6]. In addition to the validity of the instrument also tested its reliability using the formula Kuder-Richardson 21 (K-R.20) [7]. Then the instrument also tested the level of difficulty and test of its power dissipation.

Observation instruments used include affective and psychomotor spheres. In the affective domain, the assessment includes Receiving with discipline, responding with enthusiasm, and Valuing with a sense of responsibility. While the psychomotor domain includes Skills and Problem Solving.

Students response are measured using questionnaires instrument that had been used in the Department of Electrical and Electronics, Neihu Vocational High School, Taipei, Taiwan to test the students' response to the LED photometer application to detect enzyme coupling reactions [8]. Response test conducted using Likert scale with scale 1 to 5.

This research was conducted in three stages, namely the preparation stage, the implementation stage and the stage of processing and data analysis. At the preparatory stage, the researcher conducts literature study, studying the curriculum to know the purpose and basic competence to be achieved in the learning process, determining the research sample, preparing the lesson plan with the learning model of project based learning, preparing the research instrument, observation, conducted by interviewing with subject teachers, tested the test instrument, and analyzed the test results of the test instrument and then determined the feasible questions to be used to obtain the cognitive, affective, and psychomotor student learning outcomes. In the implementation stage, the researcher performs pretest test on the students, giving treatment that is by implementing the learning with the project based learning model on the students, observing the students on the affective and psychomotor aspects, doing the final test (posttest) on the students, and giving a questionnaire in the students to know the response of students to the learning media provided. In the final stages, the researcher analyzes the research data, draws conclusions, and reports.

Data analysis performed on the cognitive domain using the N-gain test. To see an improvement in concept understanding is calculated by the N-Gain <g> formula [9]. The data of the research should also be seen as normality test. Normally distributed data will show that the students' ability to spread evenly [10]. Test the normality of data using chi squared (x2). Hypothesis in this research is as follows:

Ho: Learning media of optical fiber sensor system cannot improve student learning outcomes on Sensor and Actuator subjects in Vocational High school if the average N-gain of cognitive learning achievement has scores less than moderate criteria.

Ha: Learning media of optical fiber sensor system can improve student learning outcomes on Sensor and Actuator subjects in Vocational High school if the average N-gain of cognitive learning achievement has score more than or equal to medium criterion.

The research hypothesis test is based on the data of student learning outcomes, namely the difference between pretest and posttest values. T-test is used for two independent samples with interval data types. To perform T-test conditions the data should be homogeneous and normal.

Student response test with Likert scale is classified into ordinal scale [11]. If the Likert scale is considered an ordinal scale, the average calculation and standard deviation are not correct [12]. Therefore, the Likert scale score calculation in this study did not use the average score or standard deviation. Likert scale data analysis step using median or mode for central tendency (central tendency) and data variation using frequency of score [13]. In this research, the first order to do to find the value of each statement mode. The mode value of each statement is searched for its frequency. The result of the mode and frequency values is shown in percentage [14].

3. Result and Discussion

3.1. *Results of The Test Validity*

The validity is validity of questions. The number of questions tested as many as 55 questions. The validity test was conducted on every questions with 5% of level of significance and degree of freedom (df) is df = n-2= 30-2 = 28, and obtained of t table is 1,701. When t > t table, then the question is valid. The result of the test validity represent that 33 questions is valid and 22 questions is invalid.

3.2. *Result of The Test Reliability*

The test of reliability was conducted to 33 questions valid. The reliability calculation was conducted using the Kuder-Richardson formula (K-R.20) at the 5% significance level and the degrees of freedom (df) = n - 2. The value of reliability is 0,878 and the questions are reliable.

3.3. *Result of The Difficulity Test*

The test of reliability was conducted to 33 questions valid. The result shows at the table 1.

Table 1. Result of The Difficulity Test

Criteria	Number of Questions
Easy	10
Middle	17
Difficult	6
Total	33

3.4. *Result of The Test Discrimnations Power*

The test discrimnations power was conducted to know level of ability students. The test discrimnations power was conducted to 33 questions valid. The result shows at the table 2.

Table 2. Hasil Uji Daya Pembeda

Kriteria	Number of Question
Excellent	15
Good	9
Middle	9
Total	33

3.5. Result of Pretest and Posttest

Pretest and posttest result data gives an overview of students' initial ability before treatment and after treatment. The number of students given treatment as many as 38 people. The result of the mean pretest and posttest values of the students is shown in Figure 1.

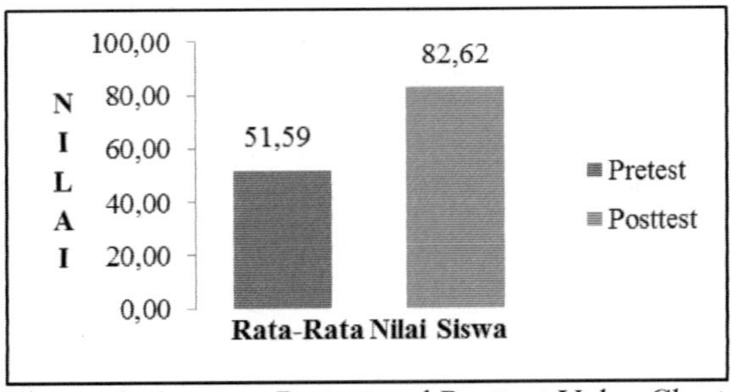

Figure 1. Average Pretest and Posttest Value Charts

From the test results at the beginning of pretest is divided into three groups of students based on ability. The result of the mean score of the student group is shown in Figure 2.

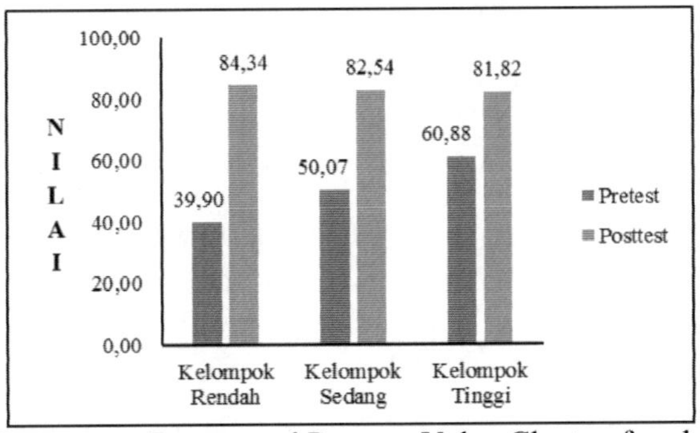

Figure 2. Average Pretest and Posttest Value Charts of each Criteria

3.6. Result of The Test N-Gain

From the result of student's learning value was obtained N-gain test data of each student shown in Figure 3.

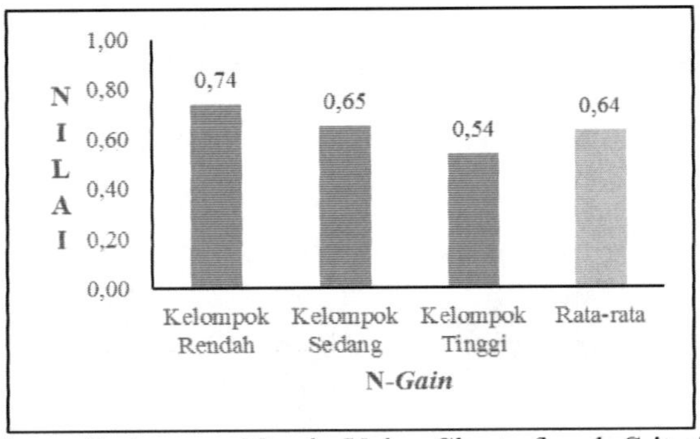

Figure 3. Average N-gain Value Chart of each Criteria

3.7. Result of Affective and Psychomotor

The result of the data test in the affective domain is shown in Figure 4.

IOP Conf. Series: Materials Science and Engineering **384** (2018) 012068 doi:10.1088/1757-899X/384/1/012068

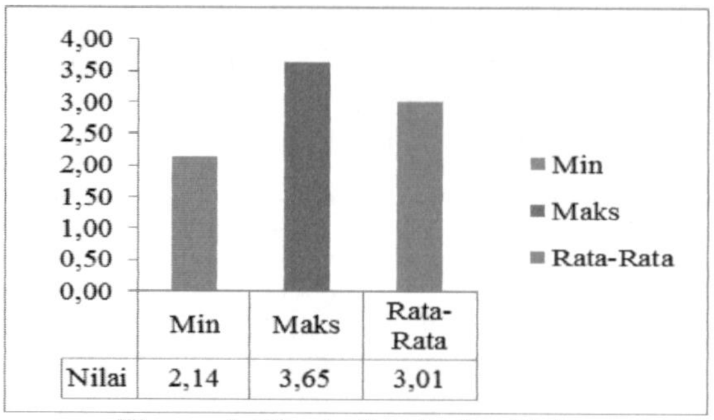

Figure 4. Chart of Affective Value

The affective data test results of each student group shown in Figure 5.

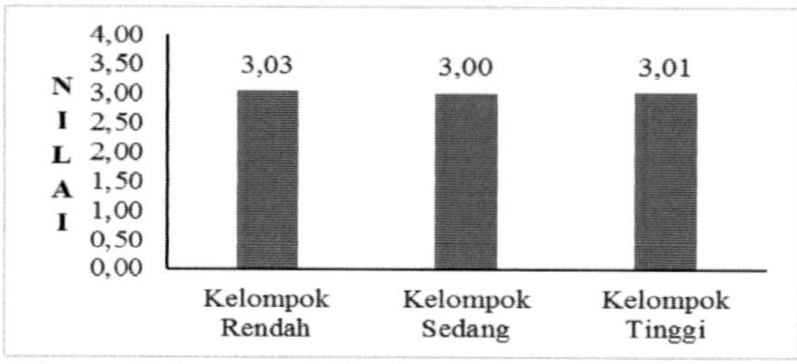

Figure 5. Average Affective Value of Each Criteria

The results of the data test in the psychomotor domain are shown in Figure 6.

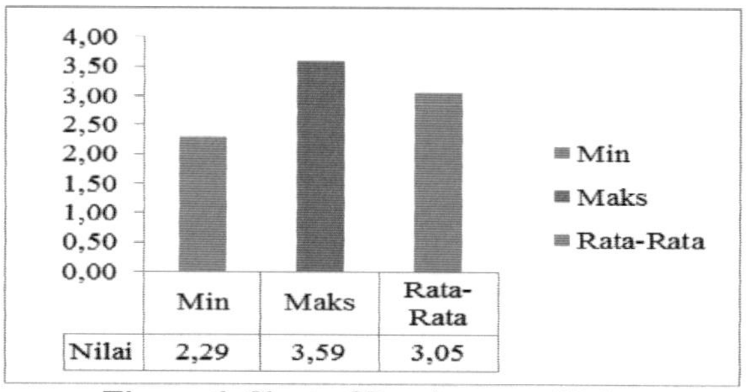

Figure 6. Chart of Psychomotor Value

The psychomotor data test results of each student group shown in Figure 7.

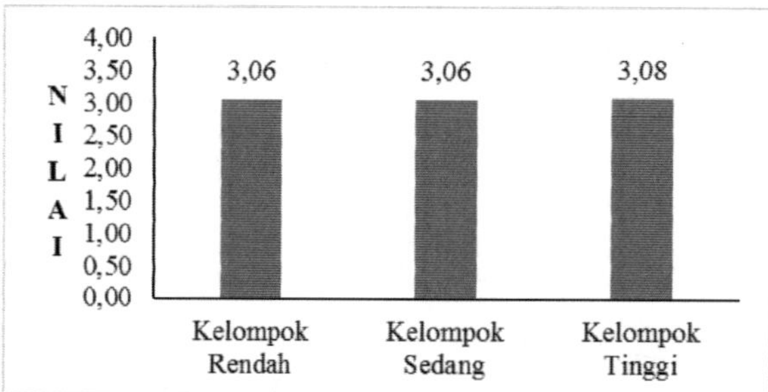

Figure 7. Average Psychomotor Value of Each Criteria

3.8. Result of The Test Data Normality

Data is said to be normal when χ2count <χ2table. The data normality test is performed on degrees of freedom (df) = k - 1 = 6 - 1 = 5 and with a significance level of 5% to obtain χ2table of 11.07. The results of normality data shown in Table 3.

Table 3. Result of the Test Data Normality

Data	χ2 Count	χ2 Table	Criteria
Pretest	7,11	11,07	Normal
Posttest	10,25		Normal

From table 3, can be concluded that:

The result of normality test of pretest shows that the average pretest data is normally distributed because χ2count (7,11) <χ2table (11,07). This means that the average value of pretest is normally distributed.

The result of normality test of posttest shows that the average posttest data is normally distributed because χ2count (10,25) <χ2table (11,07). This means that the average value of posttest is normally distributed.

3.9. Result of The Test Hypothesis

Hypothesis test was conducted to find out whether the hypothesis proposed in this research accepted or rejected. Test the comparative hypothesis (in this case the right-hand test because Ho sounds smaller or equal to (≤) and Ha sounds bigger (>).

Ha accepted and Ho rejected if tcount> ttable. Hypothesis testing was conducted on degrees of freedom (df) = N1 + N2 - 2 = 38 + 38 - 2 = 74, df 74 is between df 60 with significance level 5% obtained ttable = 1.671 and df 120 with significance level 5% obtained ttable = 1.658 and df. Difference df (Δdf) = 60 and ttable difference (Δt) = 0.013. From both of df on ttable, hence obtained ttable df 74 = 1,668. Result of hypothesis test is tcount = 18,995. Because t count (18,995)> ttabel (1,668), Ha is accepted and Ho is rejected. So that instructional media trainer fiber optic sensor system can improve student learning outcomes on sensors and actuators learning in Vocational High School.

3.10. Result of Student Responses

Response data test was conducted to know attitude and student's response to instructional media trainer of fiber optic sensor system. A list of student response statements is shown in Table 4.

Table 4. Questions of Response

Symbol	Questions
Q1.	This implementation (the practice of measuring light sensors with fibre optic systems) enhances my understanding of light sensors in Sensors and Actuators learning
Q2.	I learned something useful about fiber optic sensors
Q3.	Through this implementation, I was build the light sensor circuit and evaluate its performance
Q4.	From my experience with the implemented system, I gained more knowledge than I would have from just theory from textbook or whiteboard teaching.
Q5.	I can practice the use and measurement of fiber optic systems as light sensors
Q6.	This implementation encouraged me to integrate between software and hardware and assemble light sensor circuits
Q7.	This implementation is very interesting
Q8.	I like the concept of fibre optic sensor system
Q9.	This implementation will inspire me to explore fibre optic sensor system further
Q10.	I will recommend this implementation to other students.

The result of the student's response to this instructional media is shown in Figure 8.

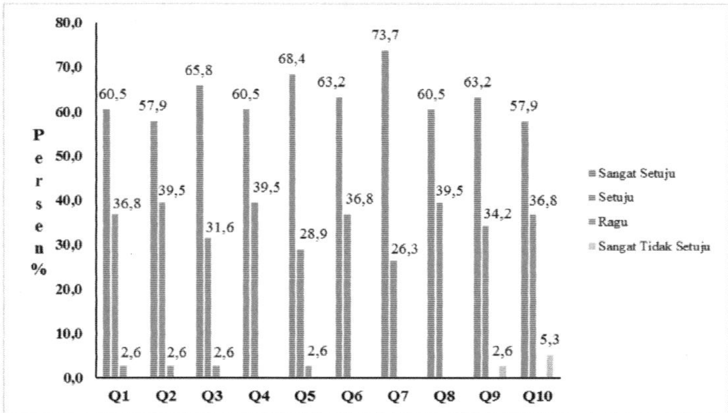

Figure 8. The student's response

Overall the students gave the response shown in Figure 9.

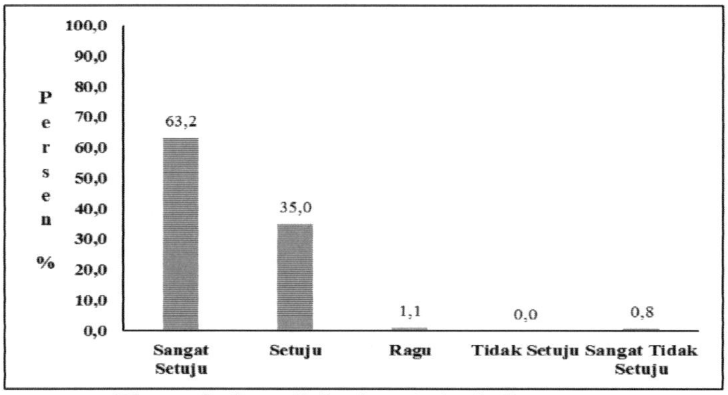

Figure 9. Overall Student Attitude Response

3.11. Discuss

Based on the result of research of Implementation of Instructional Media Trainer Optical Fiber Sensor System on Sensor and Actuator learning at SMK Negeri 1 Katapang Bandung in cognitive domain has increased learning outcomes with average N-gain of 0.64. Affective domain obtained an average value of 3.01 and psychomotor domain obtained an average value of 3.05.

In the group of students with low ability criteria had N-gain 0.74 greater than the high criteria had N-gain of 0.54 and the medium criteria N-gain is 0.65. This can be influenced by the affective value of each group criterion. Affective value of low ability criteria group is greater than 3.03 compared with high criterion group having affective value 3,01 and the medium criteria group affective value is 3,00. This is in accordance with the theory that the higher the use of instructional media and motivation, the higher the learning achievement obtained by the students, but if the use of instructional media and motivation is low then student achievement is low [15]. The motivation is the drive to achieve certain goals. The impulse may take the form of: enthusiasm, hope, and spirit [16].

In the cognitive domain as a whole the students had an improvement that was shown with an average value of N-gain of 0.64. In the affective domain there are 63.16% which has a value above the average class value and 36.84% has a value below the average class value. In the psychomotor domain 44,74% of total all student have value above the average class value and 55,26% student who have psychomotor value below the average class value. This shows that in the psychomotor domain, the number of students who have above average value is less than in the affective domain.

In the affective and psychomotor domain have a good value that the average value of students above 3 or on the predicate B. In the psychomotor domain seen that 55.26% of students have values below the average grade. This shows that skills in practice and problem solving are minus.

There is a stimulation function in the implementation of instructional media trainer fiber optic sensor system. It is marked by 73,7% students give statement that implementation of instructional media trainer of fiber optic sensor system is very interesting on student response instrument with symbol of Q7. The instructional media of optical fiber sensor trainers can be mediated between teacher and student on sensor and actuator learning shown by 60.5% of students giving statements acquire more knowledge than just studying from textbooks and blackboards on statement with Q4 symbols. This instructional media also has an information function proven by a number of 60.5% of students giving a statement that the implementation of this media allows students to gain more knowledge (Q7), 68.4% of students can practice the use of sensors (Q5), and 57 , 9% of students learn something useful (Q2). Overall, 63.2% of students showed a very positive attitude response and 35% of students showed positive attitude response, also 1.1% of doubtful students and 0.8% of students who strongly disagree. Therefore, the implementation of Optical Fiber Sensor System on Sensor and Actuator Learning is beneficial in teaching bioelectronics in vocational high schools.

4. Conclusion

Based on the results of research, the implementation of Optical Fiber Sensor System on Sensor and Actuator Learning at SMKN 1 Katapang Bandung can improve students' cognitive learning outcomes with medium N-gain category. In the affective and psychomotor domain the average value of students has a good value. The attitude of students showed 63.2% of students gave very positive attitude response and 35% gave positive attitude response. So it can be concluded that the implementation of fiber optic sensor system on Sensor and Actuator learning can improve the learning outcomes of students of Industrial Electronics Engineering Department at SMK Negeri 1 Katapang Bandung. Also, the implementation of fiber optic sensor system on Sensor and Actuator learning is very useful in vocational high school.

References

[1] Wisnu, T. N. 2015. Pengembangan Triner Kit Fleksibel Untuk Mata Pelajaran Teknik Mikrokontroler Dan Robotik. *Jurnal Teknik Mekatronika Vol.**5** No.2*, Yogyakarta.

[2] Creswell, J. W. 2012. Educational Research: Planning, Conducting, and Evaluating Quantitative and Qualitative Research. (4rd ed). Bolyston Street, Boston: Pearson Education.

[3] Mertens, D. M. 2010. Research and Evaluation in Education and Psychology Tirhd Edition. California: SAGE Publications.

[4] Copo, A. R. 2015. Students' Initial Knowledge State And Test Design: Towards A Valid And Reliable Test Instrument . *Journal of College Teaching & Learning* .

[5] Osabede, P. U. 2014. Standardization of Test for Assessment and Comparing of Students' Measurement. *International Education Studies*.

[6] Best, J. W. 1981. *Research In Education.* Michigan: Prentice-Hall.

[7] Gravetter, F., & Forzano, L. 2012. *Research Methods for the Behavioral Sciences (4th Edition).* Canada: Cengage Learning.

[8] Chen, H. -Y., Nieh, H. -M., Yang, M. -F., Chou, Y. -K., Chung, J. -H., & Liou, J. -W. 2015. Implementation of a Low-Cost Automated LED Photometer for Enzymatic Reaction Detection to Teach Basic Bioelectronics Technologies in Vocational High Schools. *IEEE Transactions On Education* , **8**

[9] Meltzer, D. 2002. The Relationship between Mathematics Preparation and Conceptual Learning Grains in Physics : A Possible "Hidden Variable" in Diagnostice Pretest Score. *American Journal Physics Vol **70** (12)*, 27.

[10] Ghasemi, A., & Zahediasl, S. 2012. Normality Tests for Statistical Analysis: A Guide for Non-Statisticians. *Int Journal Endocrinol Metab Spring.*

[11] Budiaji, W. 2013, Desember. Skala Pengukuran dan Jumlah Respon Skala Likert. *Jurnal Ilmu Pertanian dan Perikanan, 2.*

[12] Budiaji, W. 2013, Desember. Skala Pengukuran dan Jumlah Respon Skala Likert. *Jurnal Ilmu Pertanian dan Perikanan, 2.*

[13] Boone, D. A., & Boone, Jr, H. N. 2012. Analyzing Likert Data. *Journal Of Extension.*

[14] Warmbrod, J. R. (2014). Reporting and Interpreting Scores Derived from Likert-Type Scale. *Journal of Agricultural Education.*

[15] Slameto. 2003. *Belajar dan Faktor-faktor yang Mempengaruhinya.* Jakarta: Rineka Cipta.

[16] Robbins, S. 2002. *Perilaku Organisasi Jilid I. Tim Indeks (terj).* Bandung: PT. Indeks Kelompok Gramedia.

International Symposium on Materials and Electrical Engineering (ISMEE) 2017 IOP Publishing
IOP Conf. Series: Materials Science and Engineering **384** (2018) 012069 doi:10.1088/1757-899X/384/1/012069

Development of Fiber Optic Chemical Sensor for Monitoring Acid Rain Level

B Mulyanti[1], Y M R Faozan[1], W S Putro[2,*], R E Pawinanto[1], A H S Budi[1], G Sugandi[3] and A B Pantjawati[1]

[1]Department of Electrical Engineering Education, Universitas Pendidikan Indonesia, 40154, Bandung, West Java, Indonesia.
[2]Department of Atmospheric and Planetary Science, Institut Teknologi Sumatera (ITERA), 35365, Jati Agung, Lampung, Indonesia.
[3]Department of Electronics and Telecommunication Research Centre, Lembaga Ilmu Pengetahuan Indonesia (LIPI), 12710, Jakarta, Indonesia.

*wahyu.putro@staff.itera.ac.id

Abstract. Acid rain event is very dangerous for human activity especially in five cities of Indonesia. Here, acid rain has consistently attacked over five cities in Indonesia such as Jakarta, Manado, Pontianak, Bogor, and Surabaya with average pH level 4.22 to 6.34 in July 2010. Thus, in this study aim to develop Fiber Optic Chemical Sensor (FOCS) for monitoring acid rain level over five cities in Indonesia. The development of FOCS is used evanescent wave type and photodiode FDS10X10 to monitor acid rain quality. In this study, the Optical Power Meter (OPM) is used as a validator equipment to capture FOCS capability. The result shows absorption loss was successfully detected using OPM and photodiode FDS10X10 in FOCS evanescent wave type. Here, the maximum value ammonia mass over acid rain level has successfully detected with value range 7.12 dBm to 8.34 dBm and 7.51 dBm to 7.71 dBm using photodiode FDS10X10 and OPM, respectively. The good result over validation process shows FOCS has successful working with strong correlation 0.78 to monitor acid rain level.

1. Introduction

Fiber Optic Chemical Sensor (FOCS) is categorized as a chemical sensor based on fiber optic to monitor acid rain level over aerosol and environmental. In this study, the FOCS evanescent field type was used to assess chemical condition in acid rain level. Here, FOCS evanescent field type measured chemical material involved hydrocarbon oxides, ammonia, and nitrogen (NO_x) over acid rain level [1]. In current study of acid rain measurement level using chemical material sensor, the NO_x gases sensor was used to calculate precipitation index and chemical emission index [2]. However, calculation result doesn't good due to lack of precipitation index during measurement and processing data. To improve precipitation index of calculation acid rain level, Cai *et al* propose to develop a new sensor using stabilized zirconia and MoO_3–In_2O_3 Nano composites [3]. The highly sensitive using stabilized zirconia and MoO_3–In_2O_3 were given improvement over NO_x gases sensor and calculation result. However, the stabilized zirconia and MoO_3–In_2O_3 sensor does not safety due to have chemical reaction over electrochemically material. Thus, Mulyati *et al* has successful to develop Micro-ring resonator to monitor ammonia substance over acid rain event [4, 5]. Here, the Micro-ring resonator is a dielectric ring sensor with smallest refractive index based on optical principal was developed to measure chemical reaction over electrochemically material [6, 7]. In addition, the Micro-ring resonators have a capability to change ammonia substance from acid rain to optical wavelength 5000 nm to 10000 nm (as a chemical transducer based optical sensor) [8, 9]. However, the developing cost of Micro-ring resonator it's very expensive if the farmer and fish breeder used this sensor to measure acid rain level. Thus, in this study we develop FOCS and prototype ammonia detector to assess ammonia substance during acid rain level based on fiber optic. The optical field from FOCS has highly capability to transmit ammonia substance over acid rain level. Finally, we successful to develop FOCS and ammonia detector over acid rain level and calculate the FOCS performance using statistical and correlation analysis.

2. Methodology

2.1. Fiber Optic Chemical Sensor (FOCS)

Today, fiber optic function expanded as a Fiber Optic Chemical Sensor (FOCS) to monitor acid rain level. The two designs of FOCS are suggested to obtain ammonia mass over acid rain level monitoring. Here, the fiber optic was modified into two type such as distal type probe and Evanescent wave type to monitor acid rain level (see Figure 1).

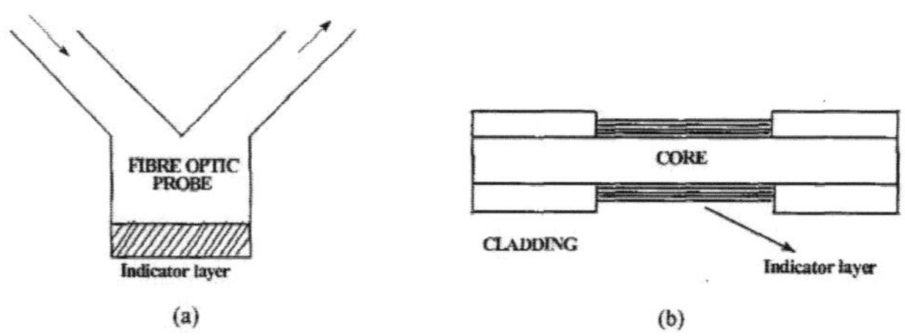

Figure 1. The common FOCS design (a) Distal type probe and (b) Evanescent wave type.

As can be seen in Figure 1, the distal type probe and evanescent wave type had different design to assess acid rain level. Here, the differences design of distal type probe FOCS and evanescent wave type FOCS located on indicator layer (sectional area). In order to monitor acid rain level, the evanescent field type was selected due to have large sectional area compared Distal type probe FOCS. During development process, we assess five sample of ammonia mass to obtain bias index (n) over experimental method using evanescent wave type FOCS. Here, the absorption loss (α) from ammonia bias index obtained using prototype ammonia detector. In addition, we develop this system in two steps design such as Fiber Optic Chemical Sensor (FOCS) and prototype ammonia detector using FDS10X10 (photodiode sensor) to obtain α value. In the first step, we choose 10 cm single mode fiber and etched using hydrogen fluoride (HF) 40% over 30 minutes (etching rate 1.52 μm/minute). During 30 minutes etching process the temperature, pressure, and RF power setting are reached 40°C, 60 mTorr, and 300 Watt, respectively. After that, we obtain the FOCS diameter $\pm$ 79.30 μm in 30 minutes etching process. In second steps, the developing prototype ammonia detector system using FDS10X10 (photodiode sensor) was applied in this study. Here, the photodiode FDS10X10 integrated with FOCS and microcontroller Atmega328 system to obtain α value. In addition, the Optical Light Source (OLS) with 1310 nm is use as a light source over input FOCS while photodiode FDS10X10 was used as a detector output FOCS (see Figure 2).

IOP Conf. Series: Materials Science and Engineering **384** (2018) 012069 doi:10.1088/1757-899X/384/1/012069

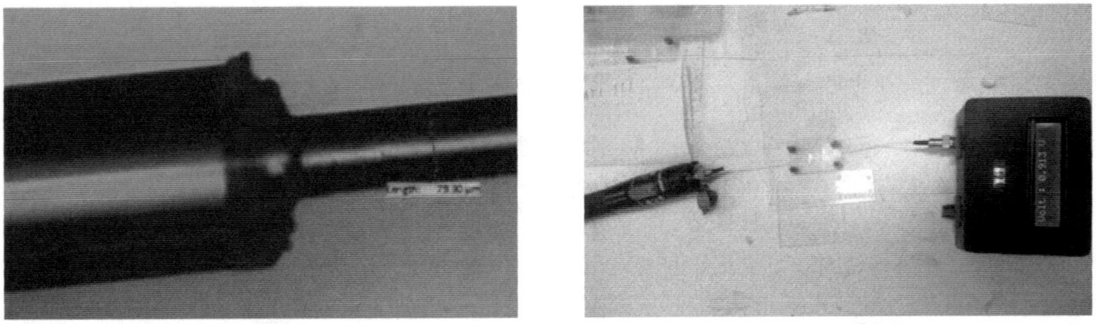

(a) **(b)**

Figure 2. The system development FOCS in (a) first and (b) second steps to monitor acid rain level.

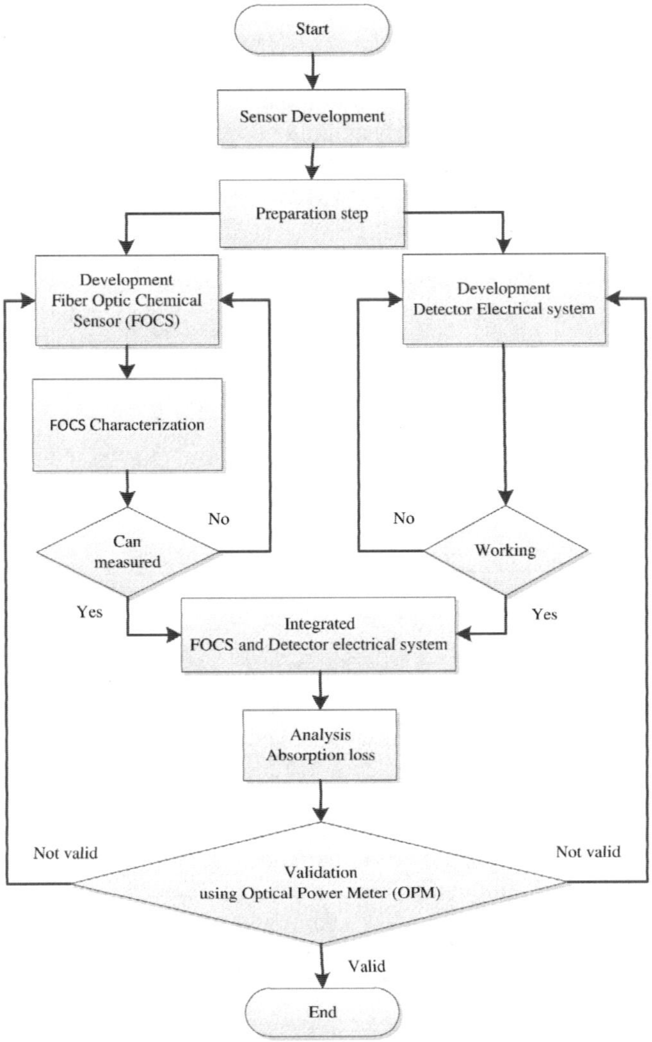

Figure 3. Flowchart testing process of FOCS with prototype ammonia detector and OPM.

As can be seen in Figure 2, the OLS light source displaced into ammonia bias index from acid rain.

Here, FOCS sensitive layer (in indicator layer) captured the differences bias index over acid rain level. The absorption of OLS light source by ammonia bias index detected FDS10X10 (photodiode sensor). In addition, the prototype ammonia detector has equipped one backlight negative LCD 16X2 to monitor α value in real time measurement to assess variation ammonia over acid rain level. After the system has successful developed, the Optical Power Meter (OPM) was used to validate capability FOCS and prototype ammonia detector during testing process. The comparison results between OPM and ammonia detector prototype was analyzed to obtain Root Mean Squared Error (RMSE), Mean Absolute Error (MAE), and Percent True (PT). Here, we have three steps during tested process to obtain α value between OPM and ammonia detector prototype such as developing process, integrated process, and validation (see Figure 3). In order to obtain minimum α value over FOCS, the material tested has placed between substrate and sensitive material. Here, we use variation of ammonia mass (1 ~ 5%) to test capability FOCS capture minimum α value.

2.2. Acid rain level

Acid rains are containing Ammonia (NH_3), Sulphur Dioxide (SO_2) and Nitrogen Oxides (NO_x) in low atmosphere. Here, acid rain will be contaminated rainwater and brings chemical pollutant trapped in the low atmosphere back to earth when the changing chemical composition of the surface soil and water [10]. The contaminated rainwater with acid was going down over lake, river, and water resources especially in five big cities, Indonesia. The pollution index was increased due to car and motor cycle with fossil fuel increasing 9.8% per-years over five big cities in Indonesia. Thus, the pH level of water resources was increased with average range 4.22 to 6.34 over five big cities in Indonesia (Jakarta, Manado, Pontianak, Bogor, and Surabaya). This condition means water resources over five cities in Indonesia are contaminated ammonia substance (acidic) with average range 4.22 to 6.34. Table 1 shows pH quality over July 2010 over Jakarta, Manado, Pontianak, Bogor, and Surabaya during acid rain event.

Table 1. pH quality over July 2010 over five big cities in Indonesia during acid rain event.

No.	pH quantity	Indonesia Province
1.	4.52	Jakarta
2.	4.22	Manado
3.	4.29	Pontianak
4.	4.40	Bogor
5.	6.34	Surabaya

As can be seen in Table 1, the pH quality over acid rain level is very harmful for environmental and human life. Here, the increasing acid rain level over five big cities in Indonesia are indicated environmental status is unhealthy for human life. The unhealthy environmental effect was disturbed physical condition human during life activity. Thus, in this study we use fiber optic technology to assess ammonia substance using FOCS with prototype ammonia detector

3. Results and discussion

In order to achieve the objective, we obtain a minimum Fiber Optic Chemical Sensor (FOCS) α value using substrate and sensitive material. Here, the variation ammonia mass (1 ~ 5%) are added over sensitive material to obtain minimum FOCS α value. During laboratory experiment, we collect seven sample measurements to see FOCS responses. By using variation ammonia mass (1 ~ 5%), we assess FOCS responses using Optical Power Meter (OPM) and prototype ammonia detector system using FDS10X10 (photodiode sensor) with light source 1310 nm (see table 2 and 3).

Table 2. Performance FOCS to capture ammonia mass using OPM.

Ammonia mass (%)	Absorption loss using OSA with light source 1310 nm						
	I (dBm)	II (dBm)	III (dBm)	IV (dBm)	V (dBm)	VI (dBm)	VII (dBm)
1	7.510	7.510	7.510	7.440	7.510	7.710	7.510
2	7.510	7.550	7.990	7.550	7.860	7.550	7.550
3	7.550	7.550	7.760	7.550	7.550	7.830	7.550
4	7.570	7.620	7.570	7.570	7.550	7.570	7.660
5	8.000	7.710	7.710	7.710	7.710	7.710	7.710

Table 3. Performance FOCS to capture prototype ammonia detector system using FDS10X10 (photodiode sensor).

Ammonia mass (%)	Absorption loss using FDS10X10 (photodiode sensor) with light source 1310 nm						
	I (Volt)	II (Volt)	III (Volt)	IV (Volt)	V (Volt)	VI (Volt)	VII (Volt)
1	0.908	0.908	0.908	0.910	0.908	0.908	0.908
2	0.997	0.908	0.908	0.908	0.908	0.952	0.915
3	0.913	0.913	0.913	0.940	0.900	0.932	0.900
4	0.918	0.913	0.990	0.918	0.918	0.918	0.918
5	0.977	0.990	0.977	0.977	0.980	0.977	0.977

As can be seen in Table 2 and 3, the FOCS performance assessment using variation ammonia mass (1 ~ 5%) has successful developed. Here, we obtain minimum and maximum FOCS performance assessment (α values) over seven measurements started between 7.440 until 8.000 dBm using OPM. However, in the second assessment using prototype ammonia detector system we obtain FOCS performance between 0.900 to 0.997 Volts. The differences measurement result between OPM and prototype ammonia detector based on set point unit over both of this device (dBm and Volt). Thus, the statistical method (modus) was performed in this study to see ammonia mass effect over voltage output value between OPM and prototype ammonia detector, respectively (see Table 4).

Table 4. The modus method to see variation ammonia mass and voltage value between OPM and prototype ammonia detector.

Ammonia mass (%)	Modus OPM (dBm)	Modus prototype ammonia detector (Volt)
1.	7.51	0.908
2.	7.55	0.908
3.	7.55	0.913
4.	7.57	0.918
5.	7.71	0.977

After all the data obtained, the regression fitting is applied in this study to shows the relation between voltage (V_{out} from prototype ammonia detector), OPM loss value, and ammonia mass, respectively. Here, the calculation result was performed to see capability prototype ammonia detector and OPM devices to assess ammonia mass during acid rain event (see Figure 4).

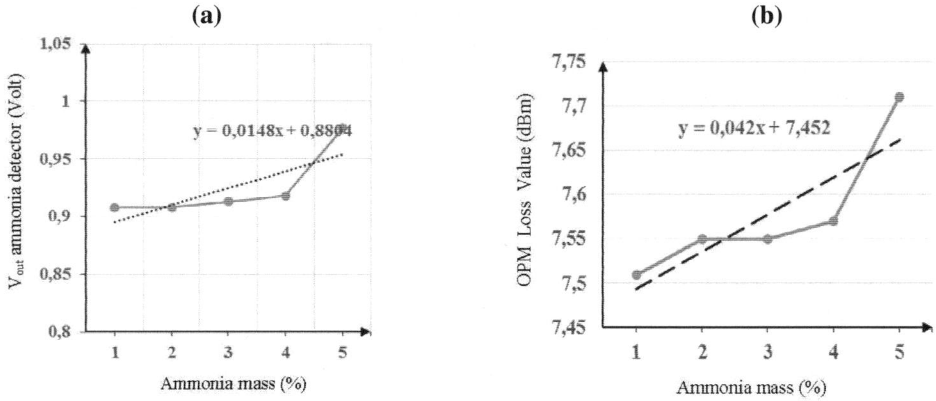

Figure 4. The regression fitting between (a) V_{out} from prototype ammonia detector and (b) OPM loss value over ammonia mass during acid rain event.

As can be seen in Figure 4, the regression fitting gives the model status between Figure (4a) and (4b) to obtain correlation for each parameter in conjunction with ammonia mass. Here, we find two models regression for each parameter in Figure (4a) and (4b). However, we cannot find the correlation between prototype ammonia detector and OPM due to different unit (dBm and Volt). Thus, during correlation analyses, we use all the measurement data from OPM and prototype ammonia detector after ammonia detector unit converted into dBm unit in this study. Here, we obtain R^2 value between OPM and prototype ammonia detector we obtain 0.78 (strong correlation) with value range 7.12 dBm to 8.34 dBm and 7.51 dBm to 7.71 dBm using prototype ammonia detector and OPM, respectively. The advantages of prototype ammonia detector are cheapest compared by Micro-ring resonator. However, the Micro-ring resonator is more sensitive to assess acid rain level [4, 5] compared prototype ammonia detector due to the material of both sensors is different [5]. In addition, the principal of both sensor is use optical concept to assess acid rain level [11]. Finally, the development FOCS to assess acid rain was successful developed using prototype ammonia detector.

4. Conclusions

The development Fiber Optic Chemical Sensor (FOCS) with Evanescent wave type to assess acid rain level was successful in this study. The FOCS system is capable to applied over big five cities in Indonesia such as Jakarta, Manado, Pontianak, Bogor, and Surabaya during acid rain event. The development of FOCS is equipped with prototype ammonia detector to monitor α value. The tested result of FOCS sensor with prototype ammonia detector was successful and validated with Optical Power Meter (OPM). The correlation analysis shows FOCS sensor with prototype ammonia detector and OPM have strong correlation in the 0.78 over 7.12 dBm to 8.34 dBm and 7.51 dBm to 7.71 dBm, respectively. Thus, based on the result we were successful developed FOCS with prototype ammonia detector to assessed acid rain level.

Acknowledgments

We would like to thank the Ministry of Research, Technology and Higher Education of the Republic of Indonesia (Ristekdikti) for research funding. Professor Dr. Sahbuddin Shaari and the other parts were supporting this research activity and all staff from Institute of Microelectronic and Nano Electronic (IMEN), Universiti Kebangsaan Malaysia (UKM).

References

[1] Seinfeld J H and Pandis S N 2016 *John Wiley & Sons* **1** 3.
[2] Seppälä A, Verronen P T, Clilverd M A, Randall C E, Tamminen J, Sofieva V, Backman L, and

Kyrölä E 2007 *Geophysical research letters* **34** L12810.

[3] Cai H, Sun R, Yang X, Liang X, Wang C, Sun P, Liu F, Zhao C, Sun Y, and Lu G 2016 *Ceramics International* **42** 12503.

[4] Mulyanti B, Menon P S, Shaari S, Hariyadi T, Hasanah L, and Haroon H (2014) *Sains Malaysiana* **43** 247

[5] Mulyanti B, Hasanah L, Hariyadi T, Novitasari R, Pantjawati A B, and Yuwono H (2015) *Advanced Materials Research* **1112** 32

[6] Gassensmith J J, Kim J Y, Holcroft J M, Farha O K, Stoddart J F, Hupp J T, and Jeong N C 2014 *Journal of the American Chemical Society* **23** 8277.

[7] Zeng W, Shu L, Li Q, Chen S, Wang F, and Tao X M 2014 *Advanced Materials* **31** 5310.

[8] Zhang M, and Wu G 2017 *Optics Communications* **403** 68.

[9] Simbula A, Rodriguez G A, Menotti M, Galli , Bajoni D, Weiss S, and Liscidini M 2016 *In Transparent Optical Networks (ICTON)* 18th International Conference on (pp. 1-4). IEEE.

[10] Park C C 2013 Acid Rain (Routledge Revivals): Rhetoric and Reality *Routledge* **30** 57.

[11] Mulyanti B, Ramza H, Pawinanto R E, Rahman JA, Ab-Rahman M S, Putro W S, Hasanah L, Pantjawati A B 2017 *IOP Publishing* **852** 43.

Graphical User Interface (GUI) of GPS Data (Dumy Data)

M Kholil[1,2,*] and D Wahyudin[1]

[1]Faculty of Technology and Vocational Education, Universitas Pendidikan Indonesian
[2]Indonesian Institute of Sciences

*kholiltunggal@gmail.com

Abstract. Increasingly growing technology also develops software (software) for computers such as mission planner. This mission planner works to remotely control remote with telemetry support and GPS background to use graphical user interface (GUI) so that it can describe mapping of ship position existence and depth which have been passed by the tool. Meanwhile, for the simulation using autopilot (unmanned) which its manufacture using C + + language. What is done in this survey ship experiment is that it can transmit data to the graphical user interface for processing and mapping made in simulation we use dummy data to measure success in the manufacture. The hypothesis of this research is that graphical user interface can display soil contour graph that can show depth of river and use original map taken directly from Bingo maps online for tracking GPS. But in the simulation experiments in Visual Studio to display graphical user interface not in accordance with the hypothesis that should display 3D is still 2D.

1. Introduction

Visual programming at this point is very important for use in prototypes, interactive design and fast analysis. This visual programming is not designed to replace conventional programming, but to convert a pictorial presentation into a structural list, then we create a graphical user interface (GUI) that serves for communication interface that users interact with the computer graphically (not text), graphical user interface can also be concluded as an application that displays menus, icons, navigation, and other tools that replace command prompt commands or shell commands. This makes it easier for a computer user to operate a device than to remember commands that use text in a particular command box. It is based that humans are easier to remember images than writing. In addition to the computer, the use of graphical user interface is also applied to some specific business types. For example is a bank owned ATM. Although text-based, it is considered semi graphical user interface because users do not have to type in commands to transfer money. You just need to press a button parallel to the existing menu. Even today many bank ATM machines are already equipped with touch screen.

Graphical user interface is created to monitor between users with a microcontroller that is connected to the graphical user interface via serial port, Graphical user interface is made to monitor the user with the microcontroller yanyang we will apply to the submarine cabling problems, which installation is divided into several stages' connected with graphical user interface via serial port, The first is a submarine cabling planning survey, usually done within a period of 6 months to 2 years prior to the installation stage, the survey was conducted with the aim of obtaining basic baseline sea level information, existing cable and pipeline, b carcasses, as well as natural hazards that will be ensured once again [1]. Due to the large amount of data required in designing marine cabling routes, marine cabling is one of the most valuable marine engineering works due to the considerable installation costs, so an optimal marine cabling planning is required. Which often runs a risk of 3 to 5 times the risk of failure [2]. The second stage is an installation that aims to check whether the underwater installations are in accordance with the planned. We can assume that the graphical user interface will be used to monitor the state of submarine structures which require some technical requirements, one of which the user graphical user interface understands the morphology of the seafloor. Therefore we will visualize the data obtained from a depth gauge for the mapping.

2. Methods

The method used in the research "Graphical User Interface (GUI) of GPS Data Dummy" is contained in several stages described in the flowchart in Figure 1.

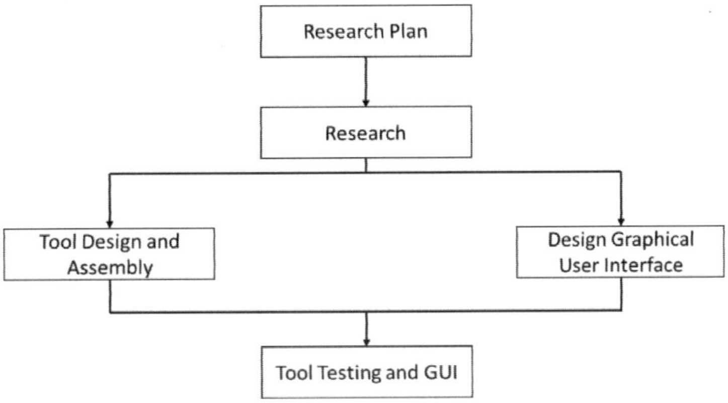

Figure 1. Making GUI.

2.1. Research planning

In this stage there are several things that need to be determined and considered, among others:

- Initial research framework
- Estimated needs
- Possible application of the application to be designed.

2.2. Design and assembling tool

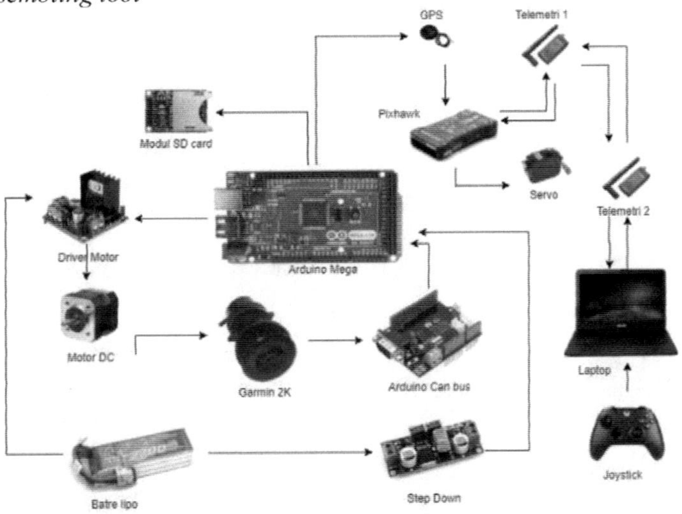

Figure 2. Assembly.

The above block diagram describes the components of the survey vessel designed to measure the depth that is in control with the ATMega microcontroller which has the Atmel AVR (Arduino) processor as shown above, where the arduino here serves to connect all components to execute commands from the microcontroller installed in the RC (Remote Control) small ship. In this ship using Arduino Mega microcontroller which is given a 5 volt supply of 5600 mAh lippo battery which is stabilized by stepdown for the incoming voltage fixed stated 5 volts, This microcontroller is in program to turn the dc motor with the help of motor driver and can be controlled with TTL or microcontroller circuit. DC motors controlled by L293D IC drivers can be connected to ground or to a positive voltage source because in the driver L293D the driver system used is the totem pool. In 1 unit of IC L293D chip consists of 4 pieces of DC motor driver stand alone with the ability to drain current 1 Ampere per driver.

In addition to turning on this microcontroller motor ordered 2000 garmin sensor to measure the depth that requires 12 volt power directly from the battery and transmit data that has been taken and stored into the Sdcard. Pixhawk serves to connect the user with the help of telemetry and GPS that RX and TX is inserted into arduino pin as the receiver of data that sent by GPS and stored in Sdcard other than that also functions for the flight control of the rader (servo) which to turn the ship, where this report deals more with pixhawk communication with arduino.

2.3. Design graphical user interface

Figure 3. Visual studio.

The graphical user interface design was developed using Visual Studio which is used to develop applications which in this software have adopted Object Oriented Programming, which will be used for monitoring. This paper presents a new feedback-based technique for automated testing of graphical users [3].

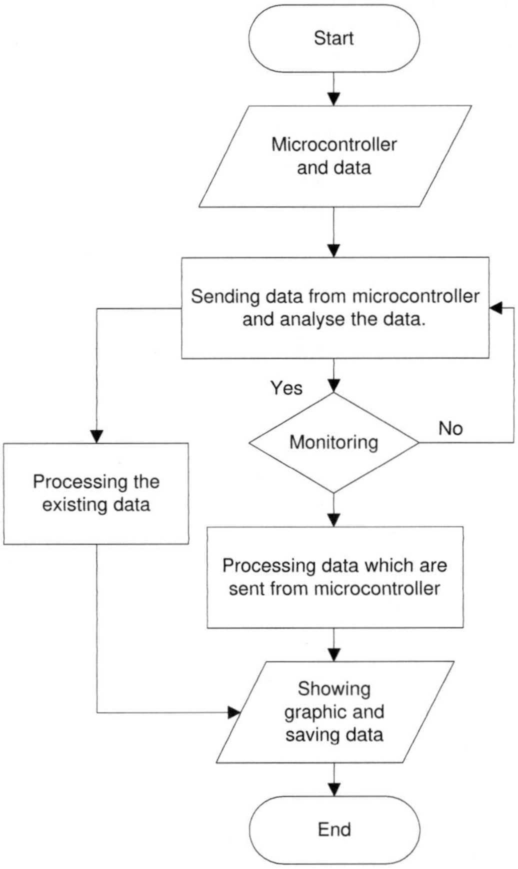

Figure 4. Flowchart GUI.

The use of GUI is like the figure above when we start the graphical user interface it will identify the connected microcontroller, when it is connected then the graphical user interface will monitor the data sent from the microcontroller in the form of text and processed into a graph that is received by the GUI itself or can process data that already exist for the graph process.

3. Results and discussion

3.1. Result tool

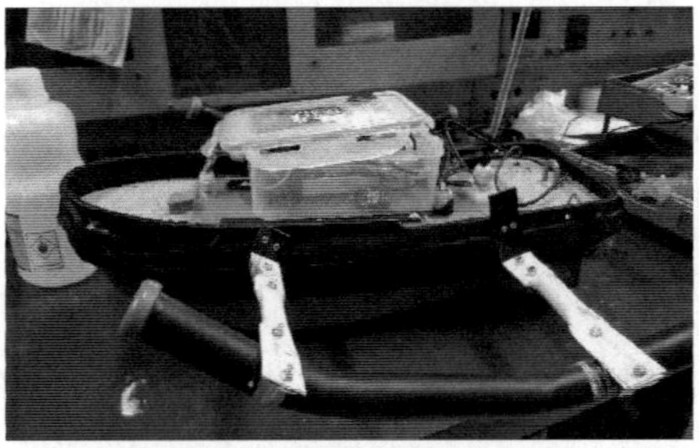

Figure 5. Ship design.

IOP Conf. Series: Materials Science and Engineering **384** (2018) 012070 doi:10.1088/1757-899X/384/1/012070

The figure above is the result of the assembly planning tool, which is used in the simulation at the foot of the river which gets some of the depth data included in this journal taken from around the Saguling reservoir as the table below.

Table 1. Depth data table.

Latitude:	Longitude:	Depth:
-6,83207	107,4686	0,64
-6,83206	107,4686	0,75
-6,83205	107,4686	0,87
-6,83203	107,4686	1,05
-6,83203	107,4686	1,14
-6,83202	107,4686	1,16
-6,83201	107,4686	1,25
-6,832	107,4686	1,28
-6,83199	107,4686	1,34
-6,83198	107,4686	1,37
-6,83197	107,4686	1,37
-6,83196	107,4686	1,51
-6,83196	107,4686	1,57
-6,83195	107,4686	1,54
-6,83194	107,4686	1,57
-6,83193	107,4686	1,63

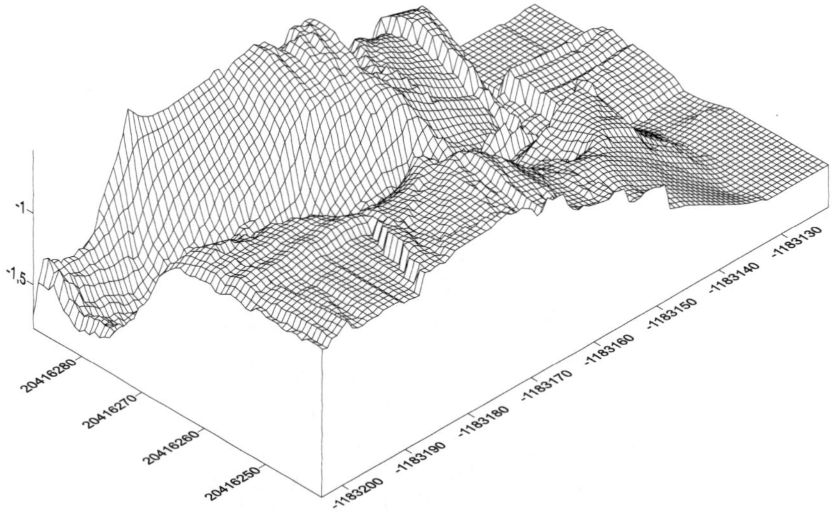

Figure 6. Contour data obtained.

The above figure is processed by using a surfer that shows the depth of the river foot Saguling.

3.2. Result GUI

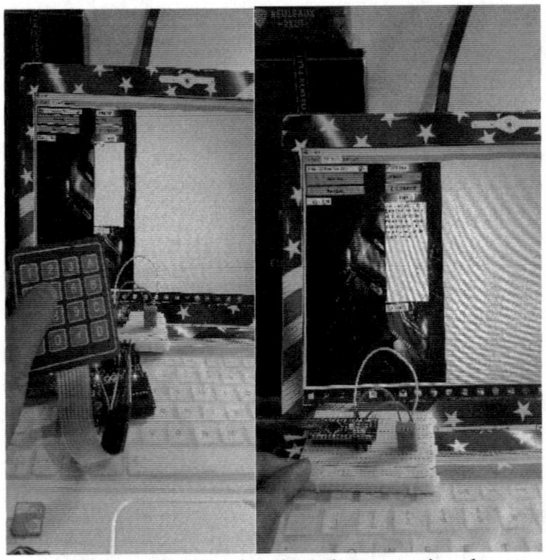

Figure 7. Monitoring data received.

The figure above is an overview of the graphical user interface that serves to send data from microcontroller to gui simulated by looking at temperature monitoring, which graphical user interface will be used for gps dummy data, in addition to monitoring can also process the data received to be made graphics and can display maps to view data latitude and longitude sent by gps but the graph.

4. Conclusions

The creation of this graphical user interface is still in the process of making the mapping (3D graphics) because the lack of clarity of the reference makes it, the better if creating a graphical user interface that smells with algorithms use MATLAB to make it easier to map data in order to be structured.

In the next development I hope this programmed shortage can be improved and hopefully can be better than the previous one.

References

[1] "Geophysical Pre- Engineering Route Survey for Pipeline Instalation.," pp. 5–23.
[2] Mulia S B and Hidayat S 2014 "Analisis kekuatan mekanis dari kabel power bawah laut," J. *ELECTRANS* **13** (2) pp. 181–194.
[3] Yuan X and Memon A M 2010 "Generating event sequence-based test cases using GUI runtime state feedback," *IEEE Trans. Softw. Eng.* **36** (1) pp. 81–95.

International Symposium on Materials and Electrical Engineering (ISMEE) 2017 IOP Publishing
IOP Conf. Series: Materials Science and Engineering **384** (2018) 012071 doi:10.1088/1757-899X/384/1/012071

Smart Building Lighting System

R Firdaus* and E Mulyana

Department of Electrical Engineering Education, Universitas Pendidikan Indonesia, Bandung, Indonesia

*ridwanfirdaus377@student.upi.edu

Abstract. In a building, lighting contributes 20-60% of the total consumption of electrical energy usage. To turn on and off the lamp with the switch manually sometimes people forget to turn it off when leaving the room. This resulted in the inefficiency of energy and cost, where the lamps work not based on need. To overcome this problem designed a lighting control system using internet of things concept so that the lights can be controlled and monitored remotely. This control system is designed using Arduino Mega 2560 as a microcontroller, relays as connectors and electric breakers voltage lamp and motion sensor as a switch that turns off the lights when there is no person in the room and indicator of the activity in the room. In addition, outdoor lights are lit in accordance with a predetermined schedule for more efficient power usage. With lighting control system is expected to reduce consumption of electrical energy usage so more efficient.

1. Introduction

Nowadays, the most popular development of Internet technology is the Internet of Thing. The Internet of things is a global concept that enables information services in an easy way both physical and virtual based on current technological developments [1]. Internet of things can be applied to various things in the environment, among the buildings, the largest contributor to electricity use in many countries is in buildings. More than half of the energy used in buildings lies in electricity consumption and one-third is for lighting [2].

There are still many-storied buildings that have not implemented the lighting control and monitoring system. This resulted in the inefficiency of energy and cost, where the lamps work not based on need. Based on this, the issue of energy saving in lighting facilities is of great importance to research [3]. For example, another paper describes a different intelligent control system research project for the development of intelligent remote-controlled street lighting systems based on enhanced technology [4].

Smart Building is a concept that combines architectural design, interior design, and mechanical-electrical in order to provide the speed of movement / mobility and ease of control from any direction and anytime in the case of automation where all activities that occur in a building or building can occur in the absence of human interference inside it.

In this research is designed a lighting control system on the building using the concept of an internet of things. This control system uses motion sensors as a switch. In addition, this system uses the application and web server so that users can manage and control in real-time from a distance.

2. Methods
The project of this smart parking system is build by some system, starting from hardware architecture, software architecture and system.

2.1. Hardware architecture
Figure 1 shows the design and schematic of the control system circuit. The components used in this circuit are: Arduino Mega 2560, ESP 8266 (wifi module), two PIR sensors, six pushbuttons, 4 - channel relay module, six pushbutton, six 10k ohm resistors, and three pieces 1k ohm resistor.

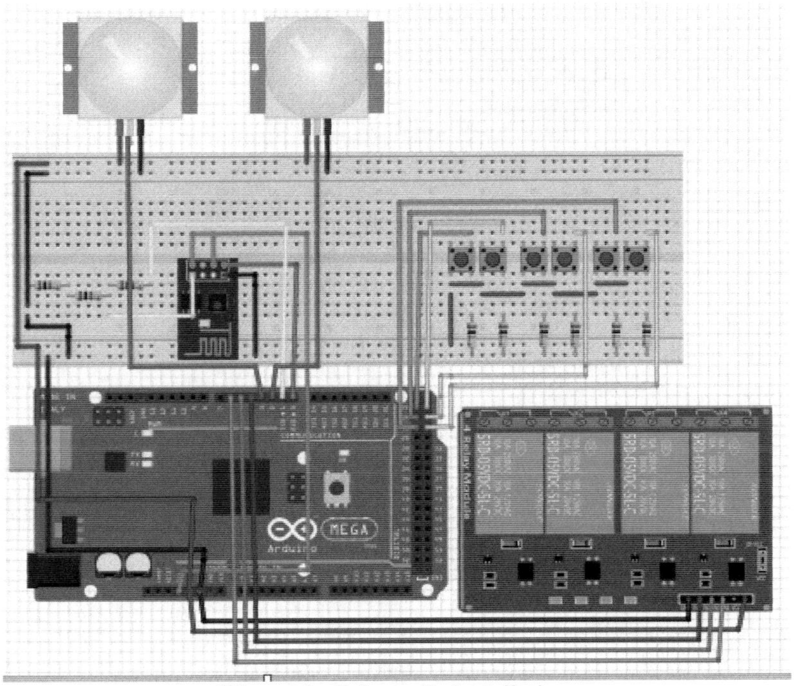

Figure 1. Schematic installation.

2.1.1. Arduino Mega 2560. Arduino Mega 2560 microcontroller is a microcontroller that uses ATmega2560 chip on the board [5]. Arduino Mega 2560 has 54 digital I / O pins (15 pins of which are PWM), 16 pin analog input, 4 pin UART (serial port hardware). The Arduino Mega 2560 is equipped with a 16 MHz oscillator, a USB port, DC power jack, ICSP header, and reset button. This board is very complete, it already has everything needed for a microcontroller.

Figure 2. Arduino Mega 2560.

IOP Conf. Series: Materials Science and Engineering **384** (2018) 012071 doi:10.1088/1757-899X/384/1/012071

2.1.2. ESP8266. ESP8266 is a wifi module that serves as a microcontroller enhancement such as Arduino to connect directly to wifi and create TCP / IP connections. This module requires about 3.3v power with three modes of wifi namely Station, Access Point and Both (tools). This module is also equipped with processor, memory, and GPIO where the number of pins depends on the type of ESP8266 we use. This module can stand alone without using any microcontroller because it already has equipment like the microcontroller [6]. The default firmware used by this device uses AT Command.

Figure 3. ESP8266.

2.1.3. PIR sensor (HC-SR501). Passive infrared sensor (PIR sensor) is an electronic sensor used for motion detection from a distance of about 20 feet, this sensor has an adjustable delay time of about 2 - 4 seconds. This sensor is most often used in motion detectors in a room [7]. In this system control, PIR sensor is used as an indicator of activity in the room and also as a switch that turns off the lights when there is no activity in the room.

Figure 4. PIR sensor.

2.1.4. Relay module. The relay is electrically operated and is an electromechanical component consisting of two main parts namely Electromagnet. Relays use the Electromagnetic Principle to drive the switch contacts so that a small electric current can perform higher electrical voltages. In this control system, relay serves as a connector and breaker voltage of the lamp.

Figure 5. Relay module.

2.1.5. Push button. Push button switch is a simple device that functions to connect or disconnect the flow of electric current with unlock press system. Unlock system here means the switch will work as a connecting device or circuit breaker when the button is pressed, and when the button is not pressed, then the switch will return under normal conditions. In this control system, pushbutton serves as a switch that serves to turn off and turn on the lights.

Figure 6. Push button.

2.2. Software architecture
Software that is used in this research is Arduino IDE, application and web server cayenne.

2.2.1. Arduino IDE. Arduino IDE is software for writing and uploading programs into Arduino and some other microcontrollers. Arduino IDE uses its own programming language that resembles the C language.

2.2.2. Apps and web server cayenne. Cayenne is an application and web server to create project internet of things. Cayenne has features for monitoring and controlling the objects connected to it. Using Cayenne is very easy to use because it is only with visual drag and drop. Use of Cayenne on this project as monitor and control the lights remotely manually and scheduled.

Figure 7. Logo Cayenne.

2.3. System architecture

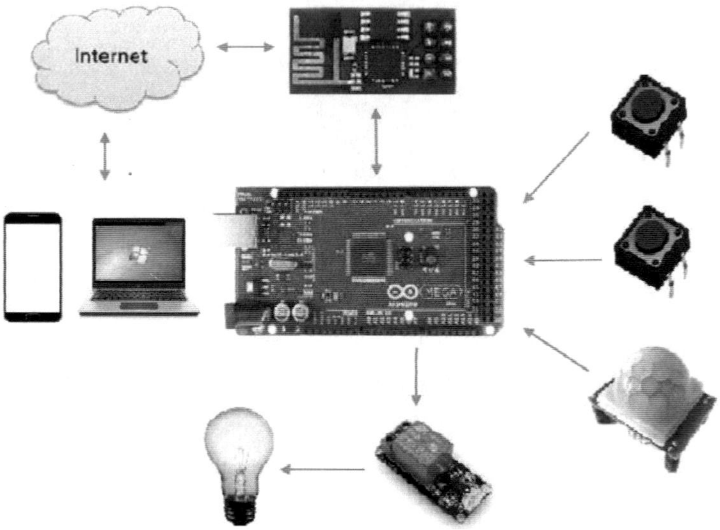

Figure 8. Architecture of smart lighting system.

As shown in Figure 8, the picture explains the architecture of intelligent control building lighting system. This system is divided into two parts hardware and software. In the hardware, the section used a smartphone / personal computer, ESP 8266 wifi module, Arduino Mega 2560, Pushbutton, Motion Sensor (HC-SR501), Relay Module, Lamp, and other supporting components. On software using Arduino IDE, Cayenne Application for smartphone and web server cayenne for a personal computer.

In this control system, pushbutton and sensor function as a giver input digital signal I / O on Arduino. Then Arduino will receive a digital signal so that the relay will set the lights to turn on or off. At the same time, Arduino Mega 2560 will transfer information to app or web server cayenne by using wifi module so it can monitor in real time.

Using the Cayenne app and web server not only as monitoring but also as a remote control so that can turn on and off the lights with automatic or manual. This control system uses a motion sensor that serves as an indicator of activity in the room and as a switch to turns off lights automatically when there is no activity in the room.

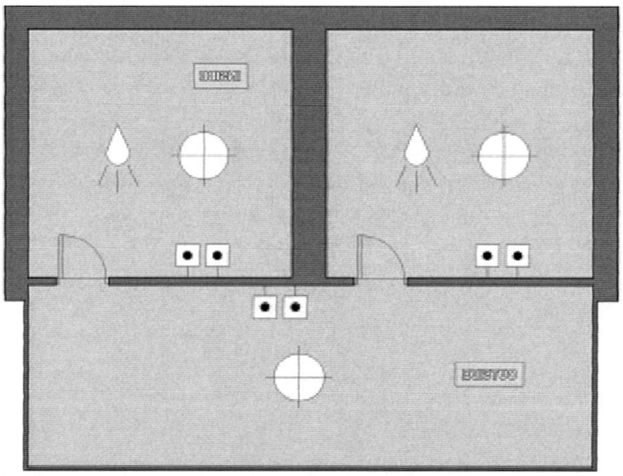

Figure 9. Architecture of smart lighting system.

As shown in Figure 9, the picture describes the design of miniature smart building lighting system. In each room, there is one ON switch and one OFF switch, one lamp and one motion sensor. Outside the room, there is one lamp and two switches to turn off and turn on the light manually.

3. Results and discussion

The result of this project is a building prototype by using a lighting control system connected to the application and web server cayenne so that it can be controlled and viewed remotely in real-time.

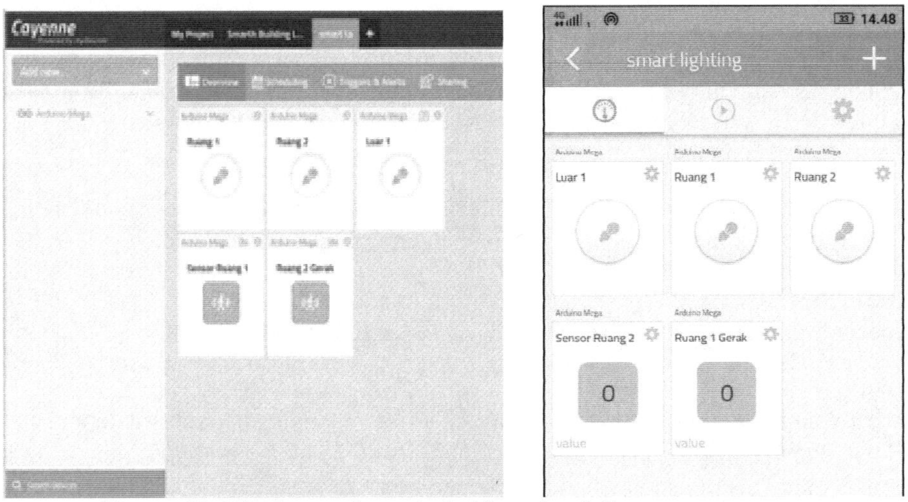

Figure 10. Display on the web dashboard and Cayenne apps.

Based on the prototype can be seen in figure 10, the light will light up when pushbutton ON is pressed so that the relay will be in high position and when pushbutton OFF at press lamp will go off. The lights can also be switched on or off via the internet by pressing the light icon on the web server or mobile application. When there is activity in the room PIR sensor will provide information through ESP8266 which can be seen on the web server and mobile application in the form of motion icon. This sensor also serves as an automatic switch that turns off the lights when there is no activity in the room.

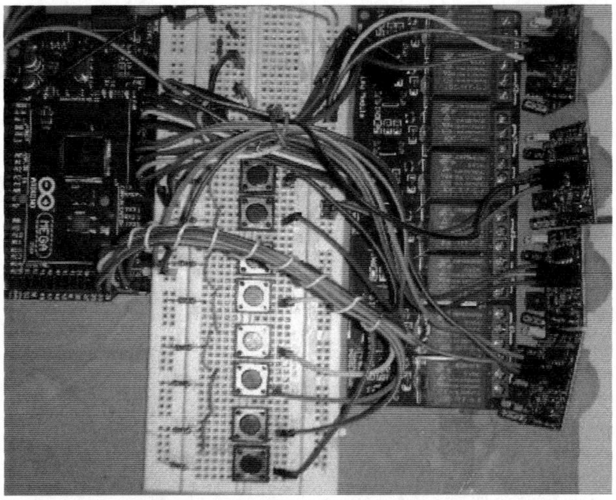

Figure 11. Hardware system.

In figure 11. Showed the hardware for lighting control system in buildings. This prototype has several that is the mismatch between the goals with the results achieved. These constraints include:

- In the planning of this project, the prototype consists of 2 floors each floor has 2 rooms and 1 lamp outside. Due to a hardware error so it is not in accordance with the planning.
- There is still a delay time exchange of information between hardware and web server or application.
- Sometimes hardware is not connected to the Internet network because ESP8266 is not stable
- The sensor used has a problem with sensitivity.

4. Conclusions

The smart lighting system in building in this project isn't too much developing as a whole. Although the project is still many shortcomings, the system has been made in accordance with the purpose of the smart lighting system. The purpose is to monitor and control the lights automatically or manually on the building through the web server and applications using the concept of an internet of things so that it can connect on personal computer or smartphone to efficiently energy and cost.

References

[1] Qi D, Lu G and Dai X 2017 "Design of Urban Greening Intelligent Monitoring System Based on Internet of Things Technology," *2017 9th Int. Conf. Intell. Human-Machine Syst. Cybern.* pp. 317–321.

[2] Roohi M H, Khorsandi A, Setayesh A, Eslamieh A and Saidi H 2015 "Design and implementaion of smart building lighting system," *2015 Int. Congr. Electr. Ind. Autom. ICEIA 2015* pp. 7–10.

[3] Wang Y and Dasgupta P 2015 "Designing an adaptive lighting control system for smart buildings and homes," *ICNSC 2015 - 2015 IEEE 12th Int. Conf. Networking, Sens. Control* pp. 450–455.

[4] Park S, Member S, Seo S, Lee B and Byun J 2014 "An Energy Efficient Smart LED Lighting System for Building Energy Management," *18th IEEE Int. Symp. Consum. Electron. (ISCE 2014)* pp. 1–2.

[5] Kusriyanto M and Putra B D 2017 "Smart Home Using Local Area Network (Lan) Based Arduino Mega 2560," *Proc. - ICWT 2016 2nd Int. Conf. Wirel. Telemat. 2016* pp. 127–131.

[6] Walia N K 2016 "An IOT by Information Retrieval approach : Smart Lights controlled using WiFi," pp. 708–712.

[7] Kitbutrawat N, Yamaguchi H and Higashino T 2017 "Localization of Binary Motion Sensors in House," pp. 1132–1137.

IOP Conf. Series: Materials Science and Engineering **384** (2018) 012072 doi:10.1088/1757-899X/384/1/012072

Reliability Evaluation of Power Distribution System with Reliability Index Assessment (RIA)

T Sucita*, Y Mulyadi and C Timotius

Department of Electrical Engineering Education, Universitas Pendidikan Indonesia

*tasmasucita@upi.edu

Abstract. Distribution network system reliability is a measure of the level of service of the system of electricity supply to consumers. This study aimed to get some reliability index value of the electricity distribution network system at PT PLN (Persero) APJ Bandung, namely the failure rate, the average output time, the average annual unavailability, SAIFI, SAIDI, CAIDI, ENS and then AENS. These results will be used as a reference for determining the level of reliability of the system. Search reliability index value itself is done by way of basic data recap interference on each unit in the network (UPJ) and then calculating the index of reliability by using formulas reliability index. The results of this study, the value of reliability indices SAIFI on the system 2.02 disorders / customer / year with a target of 1.27 disorders / customer / year, while the reliability index SAIDI on the system 0.28 hours / customer / year with a target of 0.21 hours / customer / year, with these results show electric power distribution network system of PT PLN (Persero) APJ Bandung could not reach the target reliability especially SAIFI and SAIDI. The results of this study are also expected to give an idea of the level of system reliability of electric power distribution network PT PLN (Persero) APJ Bandung.

1. Introduction

The amount of electrical energy needs required today shows that electrical energy has become a necessity that is very important to people's lives, need for electricity continues to increase this needs to be accompanied by efforts to improve the quality of service to customers of electricity [1], therefore PT. PLN (Persero) APJ (Area Services Network) Bandung as one part of Distribution West Java and Banten in charge of the management of power distribution in the city of Bandung and its surroundings, also participated in the improvement of service quality, But in improving the quality of service of course there are some problems that must be addressed. The most fundamental problem in the process of distribution of electrical power is on quality, continuity and availability of electric power service to customers. And one of the basic process of overcoming these problems is to determine the reliability (reliability) and availability (availability) system of electric power distribution network.

Reliability in the distribution network system is a measure of the availability or the level of service Provision of electricity from the system to the user within a certain period [2]. To determine the reliability of a service system or an electrical feeder then assigned a reliability index that scale to compare the appearance of a distribution system [1]. The indices of reliability include the failure rate (failures / year), the average time out (hours / failure), the average unavailability of annual (hours / year) there are also indices of reliability that is often used in assessing the reliability a power distribution network system is SAIFI, SAIDI, CAIDI [3, 4].

Therefore, the search value indices of reliability is very important, because the results of the search for these values can be used to determine all the measures to ensure the handling is really a problem, so it can be anticipated disruption of distribution due to the decreased level of reliability beyond the limits of adequate or due to lack of maintenance, which will result in a shortening of the lifespan of the equipment in question and would also affect the quality of electricity service to consumers.

2. Literature review

2.1. Electric power distribution system

The distribution system is part of a power distribution system that is on the side adjacent to the user or the load. The distribution system is useful to deliver electricity from a power source (Bulk Power Source) in the transmission network to the consumer. So the function of power distribution is the distribution or the distribution of electricity to some places (the customer), and a sub power system that directly relate to the customer, because the power supply to the load centers (customers) are served directly through the distribution network.

In the remote power distribution system, always use a voltage as high as possible in order to obtain the value of'losses' lowest of the low, using equipment transformers voltage step-up (step-up). The value of these very high-voltage (HV, UHV, and EHV) raises a number of consequences, among others: dangerous for the environment and the high price of equipment-equipment, in addition to being incompatible with the required voltage value on the load side. Therefore the central regions of the burden of high drain voltage is lowered again by using transformers lowering the voltage (step-down). As a result, when reviewed voltage value, starting from the source point to point load, there are parts of channels that have different voltage values. Besides, the number of connections between the distribution of electricity (transmission) and distribution of electrical power (distribution) and the low voltage network (JTR), resulting in a system often experience considerable interference. With the system configuration broad and complex load on the distribution, then increased the system becomes inefficient, critical interruption or failure affecting services to frequent outage.

Distribution of electricity carried out by PT. PLN is generally carried out in a radial system. But the shape of the distribution network radial has some modified form, among others: (a) Network Radial Type Center expenses, (b) Network Radial With Phase Area, (c) Distribution Network the Web (NET), and (d) Distribution Network spindle.

2.2. The concept of reliability methods reliability index assessment

Reliability (reliability) in general is a possibility the system will be able to function well for a certain period of time. Size reliability can be expressed as how often the system experienced a blackout, how long the outage occurred and how much time is required to recover from the outage occurred. The system has a high reliability will be capable of providing electrical power required at any time, while the system has a low reliability will lead to frequent outage. Research on the reliability of the distribution network system Reliability Index Assessment models are often referred to as RIA, previous researchers conducted through a simple network model is in reconfiguration [5, 6]. Then more specific research a case study by analyzing partially on electric power distribution system to Tuban cement plant to determine the level of reliability of the distribution by the RIA method [5].

The reliability of the distribution system [2] is "A measure of service availability of electricity from the system to the user. Size reliability can be stated how often the system experienced a blackout, how long the outage occurred and how much time is required to recover from a blackout that occurred (restoration) ". Meanwhile to measure the level of reliability of each point of load (load point) can be determined by a reliability index. Reliability index itself is a scale to compare the appearance of the electricity distribution system. According to [7] "To evaluate the reliability of the distribution network analytical techniques using the mathematical formula, the index of basic reliability used the failure rate λ (failures / year), the average time out (outage) r (hours / failure) and average annual unavailability U (hour / year), while the index based systems include the SAIFI and SAIDI ".

Indicators of reliability in the form of basic reliability indexes and several index-based reliability assessment index system reliability include: (a) Failure Rate (λ), (b) Average Time Out Or Repair (r), (c) Average annual unavailability (U), (d) System Average Interruption Frequency Index (SAIFI). This reliability index informs about the average frequency interference experienced by each customer in a given period. Also can be defined as follows: (e) System Average Interruption Duration Index (SAIDI), (f) Customer Average Interruption Duration Index (CAIDI), (g) Energy not Supplied (ENS) and Average Energy not Supplied (AENS).

Whereas in determining the merits of a reliability of the reliability of the electricity distribution system itself other than to note the value of reliability we also need to compare these values with a target of reliability, as from this table standard category reliability index.

Table 1. Table category standard reliability index.

	SAIFI, No. of Interruption/Year			**SAIDI, No. of Interruption/Year**		
	25%	50%	75%	25%	50%	75%
IEEE Std. 1366-2000	0.90	1.10	1.45	0.89	1.50	2.30
EEI (1999) [excludes storms]	0.92	1.32	1.71	1.16	1.74	2.23
EEI (1999) [with storms]	1.11	1.33	2.15	1.36	3.00	4.38
CEA (2001) [with storms]	1.03	1.95	3.16	0.73	2.26	3.28
PA Consulting (2001) [with storms]				1.55	3.05	8.35
IP & L Large City Comparison	0.72	0.95	1.15	1.02	1.02	2.41

(Indianapolis Power & Light, 2000)

As for the various levels of reliability in the service can be divided into 3 (three): (a) The reliability of the system is low (Low Reliability System). (b) Reliability intermediate system (Medium Reliability System), and (c) the reliability of the system is high (High Reliability System).

3. Research methods

In this study the steps that need to be done to get the value of the indices is as follows: (1) Basic data collection period disruption in January 2013 until December 2013 from each unit in the network and data feeders medium voltage feeders, (2) recapitulation and calculation of the basic data for the disorder gets basic reliability indexes on system power distribution network and the feeder, (3) Having obtained the basic reliability indices with recapitulations process data and calculations, then calculating reliability indices SAIFI, SAIDI, CAIDI, ENS and AENS system power distribution, (4) Comparing the results of the calculation by the calculation of the PT PLN (Persero) APJ Bandung that the calculation results as expected, (5) Drawing conclusions about the level of system reliability of electric power distribution network APJ Bandung when viewed from achieving the target reliability index set by PT PLN.

The following flow chart analysis study of system reliability of electric power distribution network:

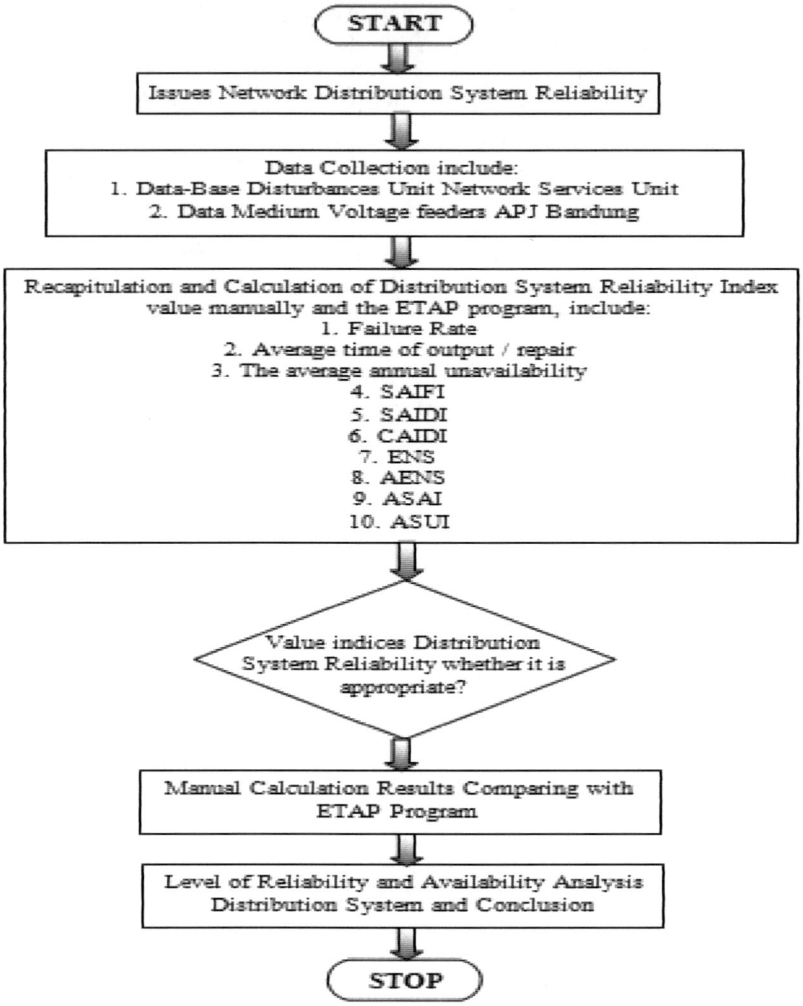

Figure 1. Process flow diagram research.

4. Results and discussion

In the analysis of system reliability of electric power distribution network PT PLN (Persero) APJ Bandung in 2013 will be divided into several basic analysis, reliability analysis on each group of outages due to interference, reliability analysis on each medium voltage feeders, and analysis of the level of the reliability of electric power distribution network system of APJ Bandung in 2013.

System Reliability Analysis of Electric Power Distribution Network APJ Bandung on Each group Outages Due to Interference by 2013.

Analysis of system reliability distribution network in each group extinction due to this interference aimed to compare the performance between parts of the system so that the electrical power distribution network can be seen which part is often impaired, and may also facilitate the work of PT. PLN (Persero) APJ Bandung in making efforts to repair and maintenance to system components that are often impaired.

Table 2. Group outages due to impaired.

Code	Cause Crashes
	Outages Due to Interference
00	Electricity connection group & guides & Measurement Tools
10	Groups of low voltage network
20	Group distribution transformer substation
30	Group of electricity poles TR and TM
40	Medium voltage air channels (SUTM)
50	Medium Voltage cable channels (SKTM)
60	Interruption of transmission and substation
70	Power outages group resources
80	Group of natural disasters

Here is the presentation of data on the calculation of system reliability analysis of electrical power distribution network PT. PLN (Persero) APJ Bandung on each group of outages due to interference.

Table 3. Calculation of reliability index-index based on each group outages due to impaired.

No	Code Disorders	Total Disruption	The average time out (Hours)	The duration Disorders (Hour)
1	A.00	**15994**	0.6649	**10633.8886**
2	A.10	1182	0.5856	692.1936
3	A.20	121	0.7423	89.8238
4	A.30	3	0.6016	1.8049
5	A.40	316	0.4793	151.4611
6	A.50	236	0.1802	42.5191
7	A.60	6	1.1417	6.8499
8	A.70	8	**2.2354**	17.8833
9	A.80	8	0.2313	1.8500
	Total	**17874**	**0.6511**	**11638.2747**

Table 4. Result of reliability index calculation of index-SAIFI, SAIDI, CAIDI, ENS and AENS.

No	Code Disorders	SAIFI	SAIDI	CAIDI	ENS	AENS
1	A.00	0.0244594	0.0163028	0.0080461	**3255839.17**	**4.8668343**
2	A.10	0.0031182	0.0015789	0.0007792	945633.13	1.4135341
3	A.20	0.2579983	0.0481261	0.0237522	71032.60	0.1061797
4	A.30	0.0000045	0.0000027	0.0000013	438.86	0.0006560
5	A.40	**0.8886597**	0.0842517	0.0415816	300147.71	0.4486613
6	A.50	0.8209018	**0.1236873**	**0.0610447**	123839.34	0.1851153
7	A.60	0.0000090	0.0000102	0.0000051	6766.67	0.0101148
8	A.70	0.0000120	0.0000267	0.0000132	1605.00	0.0023992
9	A.80	0.0310127	0.0072301	0.0000132	1.00	0.0000015
	Total	**2.0261755**	**0.2812164**	**0.1352366**	**4705303.48**	**7.0334962**

4.1. Reliability analysis of medium voltage feeder-feeder

Analysis of the reliability of medium voltage feeder-feeder at PT PLN (Persero) APJ Bandung was conducted to determine the extent to which the performance of each feeder. Medium voltage feeders APJ Bandung counted until the end of December 2013 alone amounted to 176 feeders, with a total length *peyulang* SUTM 1,328,481.43 km and a total length of feeder SKTM 330092.71 km. Analysis *keadalan* on this feeder will be very useful in order to keep the electrical energy distribution systems to customers

IOP Conf. Series: Materials Science and Engineering **384** (2018) 012072 doi:10.1088/1757-899X/384/1/012072

in addition to the results of these calculations can also be used to determine feeder-feeder anywhere that requires monitoring, maintenance and even replacement.

Here is the presentation of data on the reliability index calculation results feeder-medium voltage feeders at PT PLN (Persero) APJ Bandung in the Year 2013.

Table 5. Calculation results reliability index values SAIFI.

No.	Feeder	SAIFI
1	KMO	0.1432363954
2	CKK	0.1358834652
3	CKM	0.1165347504
4	UGB	0.1051219385
5	KPO	0.1021801685

Table 6. Results of reliability index value calculation SAIDI and CAIDI.

No.	Feeder	SAIDI	CAIDI
1	CKK	0.0259793319	0.0126796334
2	TBC	0.0179594211	0.0087653862
3	TE	0.0170039936	0.0082990743
4	NAK	0.0155635777	0.0075960560
5	TJSI	0.0135229489	0.0066000941

Table 7. Results of reliability index value calculation ENS and AENS.

No.	Feeder	ENS	AENS
1	NBN	12402.1970	0.0185
2	TJSI	12088.4903	0.0181
3	NPI	10448.8828	0.0156
4	CHM	9225.9170	0.0138
5	CPDU	7522.0977	0.0112

4.2. Achievement of target index system reliability of electric power distribution network

This discussion aims to determine the extent to which performance of the system in servicing and maintaining the continuity of distribution of electrical energy to customers and to determine the level of achievement of the target reliability index during 2013, especially reliability indices SAIFI and SAIDI reliability indices. The results of the reliability analysis can also be used as an evaluation of system performance and as a basis for determining future steps for the system to work the electricity distribution network become better or more reliable.

The following graph of the results of the index calculation reliability and reliability index value growth also attainment targets SAIFI and SAIDI reliability of the electricity distribution APJ Bandung during 2013.

IOP Conf. Series: Materials Science and Engineering **384** (2018) 012072 doi:10.1088/1757-899X/384/1/012072

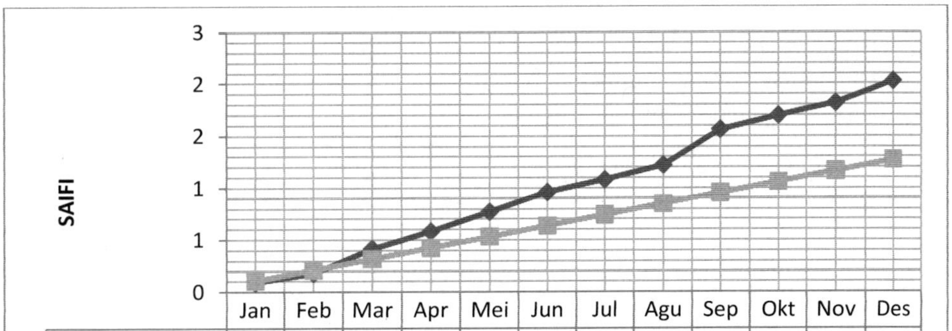

	Jan	Feb	Mar	Apr	Mei	Jun	Jul	Agu	Sep	Okt	Nov	Des
SAIFI	0.090	0.177	0.414	0.580	0.769	0.959	1.080	1.219	1.559	1.695	1.817	2.026
Target SAIFI	0.105	0.211	0.317	0.423	0.529	0.635	0.740	0.846	0.952	1.058	1.164	1.270

Figure 2. Graph of growth and target reliability index values SAIFI system reliability.

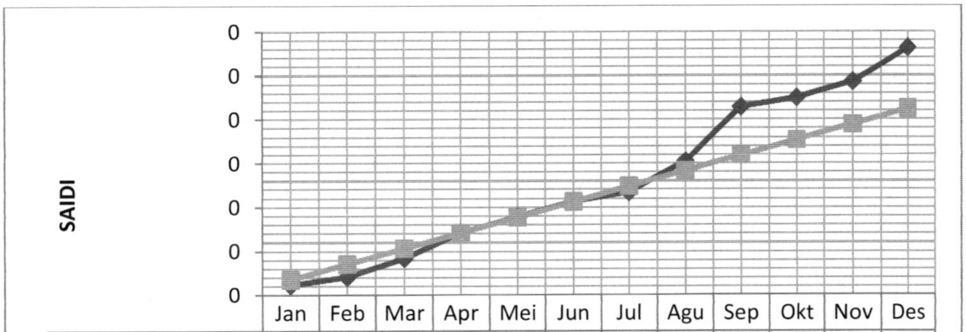

	Jan	Feb	Mar	Apr	Mei	Jun	Jul	Agu	Sep	Okt	Nov	Des
Kumulatif SAIDI	0.011	0.020	0.041	0.071	0.089	0.106	0.117	0.152	0.214	0.224	0.242	0.281
Target SAIDI	0.017	0.035	0.053	0.070	0.088	0.106	0.123	0.141	0.159	0.176	0.194	0.212

Figure 3. Graph of growth values and target reliability index SAIDI system reliability.

From the analysis results obtained electricity distribution network system of PT PLN (Persero) APJ Bandung in 2013 has the reliability indices SAIFI at 2:02 disorders / customer / year while the reliability index SAIDI for 0.28jam / customer / year. Due to the high value of reliability indices SAIFI and SAIDI reliability indices on the system, in the year 2013 PT PLN (Persero) APJ Bandung cannot meet reliability targets that have been set forth. SAIFI reliability index value of 0.75 was higher than the target only 1:27 disorders / customer / year, while the SAIDI reliability indices are higher than the target only 0:07 0:21 hours / customer / year. So it can be said to be based on the achievement of value indices of reliability especially reliability indices SAIFI and SAIDI reliability indices of power system distribution network PT PLN (Persero) APJ Bandung in 2013 is still not reliable.

5. Conclusions

From the analysis of the reliability of the power distribution network system of PT PLN (Persero) Distribution West Java Bandung area network services in 2013 can be summarized as follows:

- The system power distribution network PT PLN (Persero) Distribution West Java area of Bandung has a service network reliability indices as follows: (a) The rate of failure (failure rate) on mencapai17874 system disorders/year; (b) Average time of output or improvements to the system mencapai0.65 hours/ interruption; (c) The average annual unavailability on reach 11,638.27 system; (d) SAIFI reliability indices in the system reaches 2:02 disorders/customer/year; (e) SAIDI reliability indices in the system reaches 0.28jam/customers/year and the system reliability index CAIDI mencapai0.13 hours/ customer outages/year; (f) The amount of energy that is not channeled

(ENS) during 2013 commencing reached 4,714,299.49 kWh / year with an average energy is not channeled to the customer (AENS) commencing reach 7.046943492 kWh / customer / year.

- Due to the high value of reliability indices SAIFI and SAIDI reliability indices on the system, in the year 2013 PT PLN (Persero) Distribution West Java Bandung area network services can not meet the reliability targets that have been set forth. SAIFI reliability index value of 0.75 was higher than the target only 1.27 disorders / customer / year, while the SAIDI reliability indices are higher 0.07 than the target only 0.21 hours / customer / year.

References

[1] PT. PLN (Persero) 1985 *SPLN 59: Keandalan pada Sistem Distribusi 20 kV dan 6 kV* (Jakarta : Departemen Pertambangan dan Energi Perusahaan Umum Listrik Negara).
[2] Rukmi S H dkk. 2007 "Penentuan Angka Keluar Peralatan Untuk Evaluasi Keandalan Sistem Distribusi Tenaga Listrik" *Jurnal Teknologi Elektro* **VI** (2).
[3] IEEE 1998 *IEEE Std 1366-1998: IEEE Trial-Use for Electrics Power Distribution Reliability Indices* (New York : The Institute of Electricals and Electronics Engineers, Inc).
[4] Wiwied P P dkk. 2009 "Evaluasi Keandalan Sistem Tenaga Listrik Pada Jaringan Distribusi Primer Tipe Radial Gardu Induk Blimbing" *Jurnal EECCIS* **III** (1).
[5] Herdianto P dkk. 2013. "Studi Analisis Keandalan Sistem Distribusi Pabrik Semen, Tuban Menggunakan Metoda Reliability Index dan Program Analisis Kelistrikan" *Jurnal ITS* **1** (1).
[6] Li F 2005 "Distributed Processing of Reliability Index Assessment and Reliability–Based Network Reconfiguration in Power Distribution System" *IEEE Transaction on Power Systems* **20** (1) pp. 231.
[7] Wayan S 2007 *Evaluasi Keandalan Jaringan Distribusi Tegangan Menengah 20 KV dan 6 KV* (Jakarta: PT. PLN).

Power Transformer Interruption Analysis Based on Dissolved Gas Analysis (DGA) using Artificial Neural Network

A Muthi, S Sumarto and W S Saputra

Electrical Engineering Education Department, Universitas Pendidikan Indonesia, Jl. Dr. Setiabudhi 229 Bandung, West Java Indonesia

*ahmad.muthi@student.upi.edu

Abstract. The power transformer is an important component in the power system, as it is directly related to the reliability of the electric power system operation. Therefore, the diagnosis of disturbances in power transformers is important for device safety as well as electrical system stability. Dissolved Gas Analysis (DGA) is a disruptive diagnostic technique in power transformers that has been recognized effectively, because it provides knowledge of the state of the transformer based on the dissolved gas content in the transformer oil. The DGA test results can be represented by different methods such as Doernenburg ratio, Rogers ratio, IEC ratio, Duvals triangle, and key gases. The problem presented here is that two methods namely Doernenburg ratio and Rogers ratio, for the same data inputs give two different results from the error diagnosis to know the actual state of the transformer. In this paper, a combination method is proposed to solve the problem of conflict between Doernenburg ratio and Rogers ratio by utilizing multi-layer artificial neural network perceptron to localize and identify the error on the transformer and to select the most appropriate method.

1. Introduction

The power transformer is an important component in the electric power system [1], because it is directly related to the reliability of power system operation [2]. Therefore, the diagnosis of disturbances in power transformers is important for device safety as well as electrical system stability [3].

Dissolved Gas Analysis (DGA) is a disruptive diagnostic technique for power transformers that has been acknowledged effectively, because it provides knowledge of the transformer state based on dissolved gas content in transformer oil [4]. The DGA test results can be represented by different methods such as Doernenburg ratio, Rogers ratio, IEC ratio, Duvals triangle, and key gas [5].

However, the DGA method still has limitations mainly in the analysis and interpresentation of dissolved gases which are not an exact science, but the subject of art for variability [6], and does not involve mathematical formulations and interpresentations based on the method heuristics that vary so reliably [7]. To overcome these weaknesses, various computational models of artificial intelligence have been used such as artificial neural networks, expert systems, and fuzzy logic [8].

In a previous study using the neuro-fuzzy approach, it was concluded that most of the transformer faults could be attributed to the damage to its accessories. These accessories include bushing, load tap chargers, and cooling systems [9].

The problem presented here is that two methods namely the ratio of Doernenburg and Rogers ratios, for the same data inputs give two different results from the error diagnosis to know the actual state from transformers [10]. In this paper, a combination method is proposed to solve the problem of conflict between the ratio of Doernenburg and Rogers ratios by utilizing multi-layer artificial neural network perceptron to localize and identify errors in the transformer and to select the most appropriate method.

2. Dissolved Gas Analysis

Transformer oil analysis is a useful and predictive maintenance to determine the health of the transformer. Along with the quality test of oil samples, the dissolved gas analysis (DGA) of the insulating oil is used to evaluate the health of the transformer. Details of electrical insulation materials and related components inside the transformer produce gas in the transformer. The resulting gas identity can be very useful information in the treatment program as well as the prevention of damage to the transformer [11].

There are several techniques for detecting dissolved gases in transformer oils and DGA has been recognized as the most informative method [4]. This method involves oil sampling and sample testing to measure the concentration of dissolved gases. Associations related to sampling standards, testing, and analysis of the results are ASTM D3613, ASTM D3612, and ANSI / IEEE C57.104.

The two main causes of gas formation in the transformer are electrical disturbance and thermal decomposition. All transformers produce gas to some extent at normal operating temperatures. The insulating oil for the transformer is a mixture of many different hydrocarbon molecules, and the decomposition process for these hydrocarbons in thermal or electrical failure produces complex chemical reactions involving breakdown of carbonhydrogen and carbon-carbon bonds. During this process, active hydrogen atoms and hydrocarbon fragments are formed. These fragments may be combined with each other fragment to form gases such as: hydrogen (H2), methane (CH4), acetylene (C2H2), ethylene (C2H4), and ethane (C2H6). Furthermore, on insulation of cellulose / kraf paper in case of thermal decomposition or failure of electricity produces methane (CH4), hydrogen (H2), carbon monoxide (CO), and carbon dioxide (CO2). The gases listed above are generally referred to as key gases [11].

3. Artificial Neural Network (ANN)

Basically ANN is a machine designed to work in the same way as the human brain in performing certain tasks or functions of interest. Generally implemented by electronic components or simulation software on the computer. ANN generates computing power first through massively distributed parallel structures and the ability to learn and generalize. Generalization basically refers, when a reasonable output is generated for some input by ANN and input not found during the training process (Learning). Both capabilities of processing information for ANN allow to solve large-scale complex problems that are not currently trackable. MLP (Multi-layer perceptron) is used as one of the most widely used ANN structures for problem classification that has back propagation (BP) algorithm [10].

3.1. Multi-layer Perceptron (MLP) neural network

Multi-layer perceptron (MLP) basically consists of a finite number of consecutive layers or an organized network of layers, each of which has a limited number of processor units called neurons. Uniform groups of neurons with no connection to each other make the transformation vector. MLP is composed by an input layer, a number of hidden layers, and an output layer. In Figure 1 the network is fully connected using the sigmoid output function that has been considered because it is known that the number of these levels allows any decision-making area of any kind. In addition, MLP is the most popular method of ANN in pattern recognition [10].

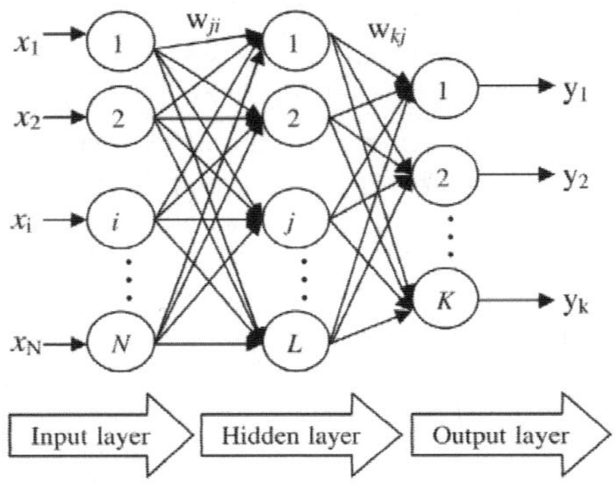

Figure 1. MLP ANN architecture.

MLP training often uses back-propagation algorithms consisting of forward pass and backward pass. Forward pass has a fixed value and equation (1) is a recurring thing used to get the output of all input layers. During the backward pass all values refer to the error correction equation.

$$y_j^{(l)} = \Phi\left(v_j^{(l)}\right) = \Phi\left(\sum_{i=0}^{p} w_{ij}^{(l)} x_{ij}^{(l)}\right) \tag{1}$$

Where:

l = the layer number ($l > 0$),
$y_j^{(l)}$ = the output of the jth neuron lth layer,
$v_j^{(l)}$ = the weighthed sum of neuron's input,
$x_{ij}^{(l)}$ = the ith input of the neuron,
$w_{ij}^{(l)}$ = the contribution weight of the ith input to the neuron,
$\Phi(\ldots)$ = the activation function of the neuron.

The activation function is a fine non-linear function and can have various forms, such as the logistics function in equation (2) and the hyperbolic tangent function in equation (3).

$$\Phi(v) = \frac{1}{1 + \exp(-av)} \tag{2}$$

$$\Phi(v) = a \tan h(bv) \tag{3}$$

During the learning process (training), sample data must be represented to the network at random. Presents all the sample data to the network once called epoch. The number of epochs is usually required to train the network. The training ends when the quadratic error or system error is on average less than the predetermined value as the previous reference [3].

4. Fault Classification of Transformator by MLP ANN

4.1. Source of Training and Testing Date Collection

DGA method is implemented base on Multi-Layer Perceptron Neural Network with database, which consist of 37 samples actual gas taken from transformer Unit-2 in geothermal power plant Kamojang. The diagnostic outputs to the available proposed MLP ANN are: (1) no fault (NP), (2) partial discharge (PD), (3) discharge of low energy discharge (D1), (4) discharge of high energy discharge (D2), (5) thermal fault T<700°C (T1), (6) high temperature thermal fault T>700°C (T2).

4.2. Transformer Fault Diagnosis by Proposed MLP ANN Approach
The MLP ANN algorithm process is divided into two phases:
- Step 1: Training data generation: To simplifying the input variables for the MLP network.
- Step 2: Multi-Layer Perceptron Artificial Neural Network: The artificial neural network basically contains three processes that is training, validation and testing.

4.3. Implementation of Proposed MLP ANN Model
In this study, MLP ANN is used for fault identification of transformers on the MATLAB platform. While constructing MLP ANN, it is important to choose several essential parameters and a suitable transfer function for the correct classification. Selection of parameters and their corresponding values is shown in table 1.

Table 1. Parameters and Values for MLP ANN Model.

Name of parameters	Value of parameters
Neurons of input layer	4, 5, and 6
The amounts of hidden layers	1-2-3
Neurons of hidden layer	10
Neurons of output layer	6
Transfer function	Tangent sigmoid
Epochs	1000

In this process, the Doernenburg ratios, the Rogers ratios, and combined ratios are implemented using their corresponding input ratios. The tangent sigmoid function has been widely used in the past for solving complex problems. Structure of the MLP ANN consist of an input layer of which the number of neuron is equal to ratios used in the network, hidden layer (varied as one, two, and three) and an output layer for each output types.

5. Test Result and Discussions of MLP ANN Models

5.1. Rogers Ratios MLP ANN Model
The rogers ratios MLP ANN training model is developed using three inputs and all essential parameters is selected. Result obtained using this model at different hidden layers is shown in table 2.

Table 2. Percentage classification accuracy of MLP ANN combination ratios.

Number of hidden layer amounts	Output accuracy (%)
Single	
Two	
Three	

5.2. Doernenburg Ratios MLP ANN Model
The four inputs, Doernenburg ratios MLP ANN training model is developed and all essential parameters is selected. Result obtained using this model at different hidden layers is shown in table 3.

Table 3. Percentage classification accuracy of MLP ANN combination ratios in doernenburg ratios MLP ANN model.

Number of hidden layer amounts	Output accuracy (%)
Single	
Two	
Three	

5.3. Proposed Combined Ratios MLP ANN Model

To improve the performance of MLP ANN base DGA method, a proposed combined ratio of two methods that is Rogers and Doernenburg ratio used. Thus, the five ratios proposed combination models and essential parameters is selected. Result obtained using this model at different hidden layers is shown in table 4.

Table 4. Percentage classification accuracy of MLP ANN combination ratios in proposed combined ratios MLP ANN model.

Number of hidden layer amounts	Output accuracy (%)
Single	
Two	
Three	

6. Conclusions

In this paper, for the purpose of incipient faults classification of power transformer, MLP ANN has been adopted. MLP ANN based DGA methods have been developed for the fault diagnosis on MATLAB platform. After comparing the performance of the proposed combined model of Rogers and Doernenvurg, it has been found that proposed combination ratio of MLP ANN model detects faults more accurately.

References

[1] Zakaria F, Johari D and Musirin I 2013 Artificial neural network (ANN) application in dissolved gas analysis (DGA) methods for the detection of incipient faults in oil-filled power transformer *Proc. - 2012 IEEE Int. Conf. Control Syst. Comput. Eng. ICCSCE 2012* 328–332

[2] Song L, Xiu-ying L. and Wen-xu W 2011 Fault Diagnosis of Transformer Based on Probabilistic Neural Network *2011 Fourth Int. Conf. Intell. Comput. Technol. Autom.* 128–131

[3] Salami A and Pahlevani P 2008 Neural network approach for fault diagnosis of transformers *2008 Int. Conf. Cond. Monit. Diagnosis* 1346–1349

[4] Seifeddine S, Khmais B and Abdelkader C 2012 Power transformer fault diagnosis based on dissolved gas analysis by artificial neural network *First Int. Conf. Renew. Energies Veh. Technol.* 230–236

[5] Malabanan D R and Nerves A C 2014 *Power Transformer Condition Assessment using an Immune Neural Network Approach to Dissolved Gas Analysis*

[6] Barbosa F R, Almeida O M, Braga A P S, Amora M A B and Cartaxo S J M 2012 Application of an Artificial Neural Network in the use of physicochemical properties as a low cost proxy of power transformers DGA data *IEEE Trans. Dielectr. Electr. Insul.* **19** 1 239–246

[7] Muhamad N A, Phung B T, and Blackburn T R 2011 Dissolved gas analysis for common transformer faults in soy seed-based oil *IET electric power applications* **5** (1) 133-142

[8] Yang F and Liang Z 2004 Comprehensive method detecting the status of the transformer based on the artificial intelligence *Int. Conf. Power Syst. Technol. (PowerCon 2004)* **2** 1638–1643

[9] Msiza I S, Szewczyk M, Halinka A, Sowa P and Marwala T 2011 *Evaluating the Relevance of the Input Space Parameters* 1–6

[10] Siddique M A A and Mehfuz S 2015 Artificial neural networks based incipient fault diagnosis for power transformers *2015 Annu. IEEE India Conf.* 1–6

[11] Hamrick L 2009 Dissolved gas analysis for transformers *Neta World* 01-04

IOP Conf. Series: Materials Science and Engineering **384** (2018) 012074 doi:10.1088/1757-899X/384/1/012074

Radio Altimeter (Radalt) Simulator on Aircraft using Arduino UNO

A R Ramdhani and A Saripudin

Department of Electrical Engineering Education, Universitas Pendidikan Indonesia, Jl. Dr. Setiabudhi No.229, Bandung, Indonesia

Arif.rahmat.ramdhani@student.upi.edu

Abstract. This research proposes the architecture simulation radio altimeter (radalt) on aircraft using the Arduino UNO. This research is a method of experimentation through the process of designing the software and hardware configuration, components, implementation tools and the evaluation of the test results of simulation of the radio altimeter (radalt) on aircraft using the Arduino UNO. On the design of the simulation tool is the main set of 6 blocks. Arduino Uno as a microcontroller to process input from the sensors and results will be displayed on the display panel that contains the 16x6 LCD, LED, and Buzzer. Meanwhile, dc motor and servo motor used for the mechanism of the movement of the model plane. The results of the research of the LCD will show the distance indicating system goes according to the design.

1. introduction

In the era of globalization such as technology evolves very quickly. So is technology on a means of transport. The current aircraft is one of the means of transport which is quite sophisticated. One of them is shown with components that are best in class. The most important component in the plane is a navigation system that showing the position of the aircraft at the time of broadcast.

The use of inertial navigation systems (INS) in collaboration with the global navigation satellite system (GNSS) is one of the most typical methods that have been used to reach the aircraft navigation solutions because it works complementary to integrate their respective strong points. . Although the data is considered the main initial INS to achieve aircraft navigation solutions, disability INS that matters such as the drifting time is still unresolved. Thus, the use of the INS is assisted by other sources to complement the shortcomings. Because of the GNSS can keep the vehicle positioning errors are limited in extent, GNSS is used as a method of supporting the standard INS [1].

The radar Altimeter is one tool to measure the height of the most widely used by aircraft on landing. This measures the height above the Earth's surface based on difference frequency (beat frequency) between signals that are transmitted and received [2]. This included in Altimeter an instrument landing system. Instrument Landing System provides information guide terminal to the aircraft during the final approach to the airport. Remote control and monitoring of the operations of the station in the system is done with solid state circuits that are connected to a cable that is buried connecting stations in a centralized location field (control tower) [3].

IOP Conf. Series: Materials Science and Engineering **384** (2018) 012074 doi:10.1088/1757-899X/384/1/012074

However valuable an Altimeter is quite expensive, therefore it needs more affordable simulation tools in the issue price but should still be effective in how it works. Simulation tool rated very effectively to address this problem.

It takes a few components to build this simulation tools such as ultrasonic sensor, Arduino UNO, dc motor, 16x2 LCD, led, and buzzer. This tool is used in ultra-sonic sensors to determine the distance between objects with other objects. Sensors ultrasonic (US) are widely used to detect objects in various areas of applications such as parking sensors in the car, help navigation UAV and others [4].

Ultrasonic sensor is a sensor that serves to change physical quantities (sound) into electric quantity and vice versa. The workings of this sensor is based on the principle of the reflection of a sound wave so it can be used to interpret the existence (distance) of an object with a specific frequency. Referred to as Ultrasonic sensors because the sensor using ultrasonic waves (ultrasonic sound).

Ultrasonic waves are sound waves that have a very high frequency of IE 20,000 Hz. ultrasonic Sound cannot be heard by the human ear. Ultrasonic sounds can be heard by dogs, cats, bats, and dolphins. Ultrasonic sounds could propagate through solid, liquid and gas. Ultrasonic sound reflectivity on the surface of solid substances similar to ultrasonic sound reflectivity on the surface of the liquid. However, ultrasonic sound waves will be absorbed by the textiles and foam.

Ultrasonic sensor transmits ultrasonic waves 40 kHz from the ends of the trigger far outside the range of human hearing and received a 40 kHz ultrasonic waves back in the late reverberation. This moves away from the pulse transmitter in the shape of a cone at the speed of sound (340 m/s). The wave is reflected in the case of the existence of objects. Flight time sent and received is the wave of the calculated distance between sensor and object [4].

2. Methods

Aircraft components are certainly not sold freely in the market. The radar altimeter is an instrument in an aircraft that measures distance two-way between radar and the Earth's surface, which depends on the return signal is transmitted from the aircraft to the ground [5].

The design of this tool made for media in schools vocational learning which refers to the technology of aircraft. Of course this tool must be easy to operate and easily understood by its users, so it is in the pupils of the school.

Using components that are easily obtainable and the price is relatively cheap does not mean to precision functions of tools are ignored. It is precisely with a simple component expected students will more readily understand the workings of each component.

This tool will light up when the switch on/off is turned on. All the programming is done by the Arduino. Meanwhile the distance measurements of the miniature aircraft against the object below was done by Ultrasonic sensors. Ultrasonic sensor reading of results will be displayed on the LCD 16x2 in the form of letters and numbers. The distance will be read in units of centimeters. Each zone height will be distinguished by the sign of the sound of the buzzer and the LEDs of different colors. Meanwhile, the movement of the miniature aircraft driven dc motor in gear give miniature can move forward, backward, up, and down. Motor dc here have different sources with Arduino, because dc motors require 12 volts.

2.1. Block Diagram

The block diagram in this system can see in figure 1.

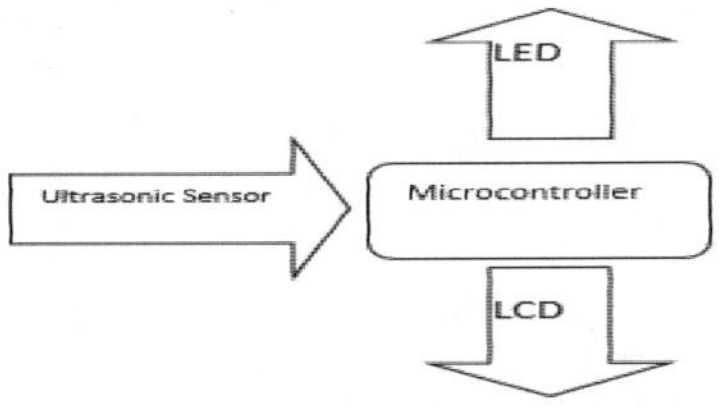

Figure 1. Block Diagram of design tool.

2.2. Selection of Components use, Such as:

2.2.1. Ultrasonic Sensor. An Ultrasonic sensor is a device that can measure the distance to an object by using sound waves [6]. It measures distance by sending out a sound wave at a specific frequency and listening for that sound wave to bounce back. By recording the elapsed time between the sound wave being generated and the sound wave bouncing back, it is possible to calculate the distance between the sonar sensor and the object. Ultrasonic sensor can see in figure 2.

Figure 2. Ultrasonic Sensor.

2.2.2. Microcontroller using Arduino. The microcontroller functions as center coltrolling in the whole system and will be displayed on the LCD. Arduino UNO can see in figure 3.

Figure 3. Arduino UNO.

*2.2.3. LCD 16*2.* LCD function to display data that has been processed, the installation LCD with Arduino UNO can see in figure 4.

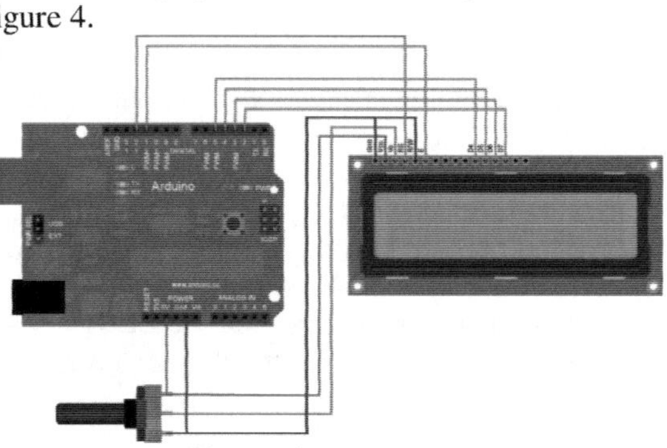

Figure 4. LCD 16*2.

3. Results and Discussion

3.1. System Analysis
The decomposition of a complete information system into its component parts with a view to identifying and evaluating the problems, opportunities, constraints and expected needs so that proposed improvements can be proposed.

3.2. Problem Analysis
The airplane movement mechanism part is still manually driven and not yet digitized. It's just switch which directly through the DC motor.

3.3. Discussion results
Because the tool designed by the author is still in the stage of the design process, then the results of the tool cannot be exposed.

4. Conclusions and Future Work
Like inhr problem we have, the airplane model movement mechanisis still driven manually. It's it would be better if the movement of the model of the aircraft was digitized to make it easier for users to move the model of the airplane in accordance with the wishes.

References
[1] Park J, Kim Y and Bang H 2017 A new measurement model of interferometric radar altimeter for terrain referenced navigation using particle filter *Navigation Conference (ENC), 2017 European* 57-64 IEEE
[2] Thulasi S and Kashwan K R 2016 Implementation of Frequency Modulated Continuous Wave Radar Altimeter *Green Engineering and Technologies (IC-GET), 2016 Online International Conference* 1-4 IEEE
[3] Singh N A and Borschbach M 2017 Effect of external factors on accuracy of distance measurement using ultrasonic sensors *Signals and Systems (ICSigSys), 2017 International Conference* 266-271 IEEE
[4] Huddleston G K and Bush G G 1975 Lightning Protection for Status and Control Lines of the Mark III Instrument Landing System *Electromagnetic Compatibility, 1975 IEEE International Symposium* 1-3 IEEE
[5] Hua Z and Xiulin H 2010 Pulse radar altimeter terrain return model and its simulation *Future Computer and Communication (ICFCC), 2010 2nd International Conference* **3** V3-549 IEEE
[6] Carullo A and Parvis M 2001 An ultrasonic sensor for distance measurement in automotive applications *IEEE Sensors journal* **1**(2) 143-147

IOP Conf. Series: Materials Science and Engineering **384** (2018) 012075 doi:10.1088/1757-899X/384/1/012075

Learning Media Based on Augmented Reality Applied on the Lesson of Electrical Network Protection System

F Aditya* and B Trisno

Universitas Pendidikan Indonesia, Jl. Dr. Setiabudhi No. 229 Bandung, Indonesia

*febrian.aditya.ramadhan@student.upi.edu

Abstract. Technology is created and developed to help people work more efficiently. One of them is Augmented Reality technology. Augmented Reality is a technology that combines two or three dimensional virtual objects into the real three-dimensional environments, then it is projecting the virtual objects in real time, in this case it was possible to help ease the learning process for displaying props and also to increase students attention. In this study, the learning media based on Augmented Reality applied to the subject of Electrical Network Protection System.

1. Intoduction

1.1. Bacground

It is undeniable that today's technological advances are growing very rapidly. This is evidenced by the many innovations that have been made in this world. Technology is created and developed to help humans in doing work to be more efficient [1]. Technology today is still growing rapidly and inspire innovation and new works. One of them is Augmented Reality technology. Augmented Reality is a technology that replaces two-dimensional virtual objects and three-dimensional dimensions, then projecting virtual objects in real time [2]. This technology uses a system integrated into Smartphone or other gadgets. This is intended to help the perception and interaction with the real world.

In Indonesia, especially in the world of self-education, learning with the concept of Augmented Reality is still new, because this technology is only popular with the military world, Augmented Reality Technology in education to be a learning tool for fun learning process [3].

For that Augmented Reality technology in Indonesia to become an up to date teaching method and make this Augmented Reality technology as a fun learning media, it is necessary supporting tools such as Smartphone that can be integrated with this technology.

Smartphone users in Indonesia are increasing every time. Indonesia icated ranked fifth as Smartphone users in the world. About 80% of people in Indonesia have Smartphone [4], with the rapid increase of Smartphone in Indonesia, it can be used to integrate Augmented Reality technology into Smartphone.

1.2. Problem Formulation

Based on the above background, the issues raised in the writing of this Final Project are:

- How to facilitate displaying props on the subjects of Electrical Network Protection?
- How to follow the development of technology in teaching and learning activities?

- What is responsible for this planning?
- How to improve student learning attention?

1.3. Purpose
In accordance with the formulation of the above problems, then to be achieved in writing this Final Project is:
- To facilitate the students know the tool in question in the learning process of Electrical Network Protection
- To make teaching and learning activities up to date with the development of the times
- Use more effective gadgets
- Improving student's learning attention

2. Theoretical Review

2.1. Electrical Network Protection Subject
Electrical Network Protection Subject is one part of the productive learning group at Vocational High School, where in the Curriculum 2013 the subjects of the Basic Network are included in the C3 (Apply) group for class XI. Curriculum 2013 is a new curriculum applied by the government to replace the Education Unit Level Curriculum that has been valid for approximately 6 years. The scope of Basic Network material includes the concept of power system, network protection equipment, etc [5].

2.2. Teaching Module
Module is a way of organizing learning materials that pay attention to the function of education. That is, through a module of learning is expected to bring learners the basic competencies expected to bring learners to the basic competencies expected.

2.3. Augmented Reality
Augmented Reality as the incorporation of real and virtual objects in real environments, runs interactively in real time, and there is integration between the three dimensions, the virtual object integrated into the real world [6]. The incorporation of real and virtual objects is possible with appropriate display technology, interactivity is possible through certain input devices, and good integration requires effective tracking.

2.4. Augmented Reality Book
Augmented Reality Book (AR-Book) or in Indonesian means Augmented Reality based book is a merger between a regular books with AR technology. AR-Book outline has two main components, a book with a Quick Response Code (QRC) type marker on almost every page, and the second is an apparatus for capturing markers and displaying the results [7].

2.5. Unity 3D
Unity 3D is an integrated tool for creating 3-dimensional object shapes in video games or for other interactive contexts such as Architectural Visualization or real-time 3D animation. The environment of Unity 3D development runs on Microsoft Windows and Mac OS X, as well as applications created by Unity 3D can run on Windows, Mac, Xbox 360, PlayStation 3, Wii, iPad, iPhone and do not miss the Android platform.

2.6. Vuforia
Vuforia is a software library for augmented reality, which uses a consistent source of computer vision that focuses on image recognition. Vuforia has many features and capabilities, which can help developers to realize their thinking without any technical limitations. With support for iOS, Android and

Unity3D, the Vuforia platform supports developers to create apps that can be used across virtually every smartphone and tablet [6].

2.7. Marker
The AR application creation process required a marker as a marker to display an object. In making marker required an image file with extension .JPG which will be uploaded to QCAR official site. Markers that have been uploaded will be rated quality by the system, the more rating with an asterisk the better marker quality will be. Markers used should tend to have a contrasting color to get the best rating, bad markers will be difficult to detect device or even not work.

2.8. Android
Android is a Linux-based operating system for mobile phones such as smart phones and tablet computers [8]. Android provides an open platform for developers to create their own applications for use by various propulsion devices. Android is an open source operating system, and Google is releasing its code under the Apache License.

3. Results and Discussion

3.1. Tool Design
Tool design can see in figure 1.

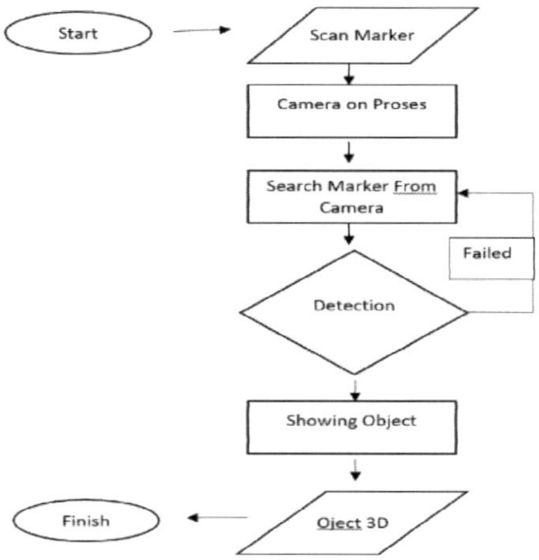

Figure 1. Tool Design.

3.2. Implementation and Unit Testing

3.2.1. Software Requirements Based on the Analysis Performed. The software needed in the development of the Augmented Reality Based Interactive Learning Module for this Basic Network Subject is as follows:
- Unity 3D and Vuforia as library software to develop applications.

*3.2.2. Hardware Requirement.*This application will be developed on laptops with the following specifications:
- 15.6 "HD (1366 x 768) Glossy with integrated camera
- 8GB DDR4 2133 MHz memory
- Intel® Core ™ i7-3520M Processor CPU @ 2.90 GHz
- Android Smartphone with Android 5.1 Operating System Specification, 1.2GHz Eight Core Processor, 3.0 GB RAM

3.3. Integration and System Testing
The testing phase is the stage of testing to determine the application of the teaching module that has been made in accordance with the design and there is still an error or not. The purpose of this stage is to look for as many errors as possible before the unexpected during the manufacture until the application is complete

3.4. Operation and Maintenance
In this research, the maintenance phase is not done by the researcher because the researcher only makes prototype application only and has not set the application environment is operated.

4. Discussion

4.1. Software Implementation
Implementation of an Augmented Reality Interactive Learning Interactive software application for Basic Network Subjects consists of software implementation environments, software implementation constraints, implementation of software architecture, software process implementation, and implementation of software interface screens.

4.2. Application implementation environment
Application implementation environment of Augmented Reality Based Interactive Learning Module for Basic Network Subjects is done on the following software and hardware environment:

4.2.1. Software Specification.
In the software environment, the Augmented Reality-Based Interactive Learning Module application for Network Electrical Subjects runs on the following environments:
- Microsoft Windows 7 Professional Operating System.
- Android Smartphone operating system specification Android 5.1, 1.2GHz Eight Core Processor, 3.0 GB RAM
- Android Tools SDK.
- Vuforia Qualcomm Augmented Reality.
- Unity4.6.1

4.2.2. Hardware Specification
In the hardware environment, applications the Augmented Reality Based Interactive Learning Module for Basic Network Subjects runs on the following environments:
a) Computers
- 15.6 "HD (1366 x 768) Glossy with integrated camera
- 8GB DDR4 2133 MHz memory
- Intel® Core ™ i7-3520M Processor CPU @ 2.90 GHz
- Equipped with input and output tools.

b) Android devices
- Smartphone OPPO F1S Selfie Expert
- Resolution 480 x 800 pixels, 5.0 inches.
- 1.2GHz Eight Core Processor, 3.0 GB RAM.
- RAM 3.0 GB.
- 8 MP primary camera

4.3. Implementation of Markers (Markers)
In accordance with functional model design which has been created, required markers placed in the teaching module to be tracked by a smartphone camera that already contained the Augmented Reality

application. The Augmented Reality Interactive Learning Module software uses a data base device that is ready-to-use bookmark images uploaded to official vuforia sites, then the marker image is processed into a Module AR. Unity package library file for download

4.4. Implementation Screen Software Interface
Implementation of software interface screen is implemented using the features contained in Unity 3D.

5. Conclusions
Interactive Based Trial Module Augmented Reality is an application that contains basic network learning materials in the form of augmented reality that displays 3D objects. Steps for making AR-based lab modules:

- Preparation of the application development environment in the form of hardware and software settings
- Creation of 3D models to be displayed
- Marker creation and configuration d) 3D model settings to be displayed AR in the AR library environment
- Testing

Support hardware and software in making AR-based mobile software needs to be considered carefully because error messages that appear often caused by problems support system.

References
[1] Burdea G C and Coiffet P 1994 *Virtual reality technology* (London: Wiley-Interscience)
[2] Bimber O and Raskar R 2005 *Spatial augmented reality: merging real and virtual worlds* CRC press
[3] Kesim M and Ozarslan Y 2012 Augmented reality in education: current technologies and the potential for education *Procedia-Social and Behavioral Sciences* 47, 297-302.
[4] Putra E Y, Wahyudi A K and Dumingan C 2016 A proposed combination of photogrammetry, Augmented Reality and Virtual Reality Headset for heritage visualisation *Informatics and Computing (ICIC), International Conference* 43-48
[5] Utami I A A, Arthana I K R and Darmawiguna I G M 2015 Pengembangan Modul Ajar Interaktif Berbasis Augmented Reality Untuk Mata Pelajaran Jaringan Dasar Di Smk Negeri 3 Singaraja *KARMAPATI (Kumpulan Artikel Mahasiswa Pendidikan Teknik Informatika)* ISSN: 2252-9063 **4** (5)
[6] Azuma R T 1997 A survey of augmented reality *Presence: Teleoperators & Virtual Environments* **6** (4) 355-385
[7] Hutahaean R, Isnanto R R and Martono K T 2015 Perancangan Aplikasi Multimedia Untuk Pembelajaran Gerbang Logika Menggunakan Augmented Reality *Jurnal Teknologi dan Sistem Komputer* **3**(1) 125-131
[8] Nosrati, M., Karimi, R., & Hasanvand, H. A. (2012). Mobile computing: principles, devices and operating systems. World Applied Programming, 2(7), 399-408.

Load Shedding Analysis Because of Contingency Damage (N-2) at Transmission Lines 150 kV Subsystem Cirata

T Sucita*, Y Mulyadi and W S Saputra

Department of Electrical Engineering Education, Universitas Pendidikan Indonesia, Bandung, Indonesia

*tasmasucita@upi.edu

Abstract. Cirata subsystem is part of the 150 kV interconnection system 150 kV West Java which is supplied by two units of IBT 500/150 kV with a power capacity of each IBT is 500 MVA. It has 15 transmission line connecting the substation 13 (GI), while the generation units that exist in the subsystem Cirata is PLTP Patuha with a power of 55 MW and 180 MW hydropower Jatiluhur. The reliability of the power system must be maintained in order to supply power from the generator to the consumer to keep going well. At peak load conditions and contingencies, an electric power system will undergo changes in value, both current and voltage. In this study, which analyzed disorder is a contingency (N-2), a discharge of the two components of transmission line system and generator. The aim is to evaluate and improve the reliability of electric power systems of 150 kV Subsystem Cirata, analyzing the effects of disturbance contingency (N-2) to changes in current and voltage, identify the bus voltage is outside the operating limits, identifying the transmission line that experienced the imposition of critical / exceeds limit operation, and perform load shedding (load sheeding) to overcome the weaknesses of the power system due to disruption contingency N-2. The study was conducted by collecting data required for power flow simulation using Newton-Raphson method in ETAP 12.6.0. The simulation results show the power flow interruption due to voltage profile under the standard contingency SPLN 150 kV + 5% and -10%, and becomes critical channel loading over 50% of the nominal current conductor. While after the release of the load (load sheeding) voltage profile in Subsystem Cirata back normal or standardized SPLN and *loading the channel becomes lighter.*

1. Introduction

Electric Power System is a set of Electricity Center and Substation (Center of Burden) which each other connected by Transmission network, so it is an interconnection unit [1]. The main role of the electric power system provides and distributes electrical energy reliably and continuously to the load. This means that the reliability of the power system is determined by the ability of the system to supply sufficient electrical energy to satisfied customers satisfactorily in order for the system to survive abrupt disturbances such as short circuit or unanticipated system element loss [2].

One of the things related to the reliability of the system is the release of generating units or transmission lines that need to be taken into account in the security of the system. If one of the generators is off, the current system will experience a lack of power to service the load. If a transmission line is loose, its load will be transferred to another remaining channel, so that the remaining channel will be heavier and may result in overload followed by channel release [3].

Contingency analysis is a study of the release of tissue elements such as transmission lines, transformers and generators, and the result is the post-contingency effect on the power flow and bus voltages in system [1]. Contingency analysis will see limitations such as bus stress, channel loading, and the availability of spinning reserves. The N-2 contingency is the contingency that results from the release of two system elements such as one transmission line and one generator.

Based on the report of PLN West Java, there are 24 interruptions during the range of 2013 - 2015 in the form of contingency (N-1) on the channel 150 kV West Java [4]. In other words, the disruption of one channel in the system has often happened. It indicates that the level of reliability of the system still needs to be improved because of the impact of the electrical system's unreliableness can be widespread and result in the release of the system in stages.

To provide reliable service, the power system must remain intact and able to overcome various kinds of disturbances that may occur. Thus it is very important that the system should be planned and operated in order to be in a contingency state or the release of two elements of the system be it the generator or transmission line will not lose the load.

Load Shedding (Load Shedding) is a form of automatic or manual load release action to secure the operation of generating units from the possibility of black out. Load discharges are automatically performed because the amount of power supply is reduced. Load discharges are automatically performed by detecting the frequency or by observing the condition of an inoperable power plant source [1]. Load discharges can occur due to a decrease in the frequency of imbalance between the generated power generator and the load consumption [5].

The reliability of the power system is determined by adequacy assessment and security assessment. Assessment of adequacy relates to the ability of the system to supply electrical energy to satisfied customers in a satisfactory manner. Security assessments relate to the ability of the electric power system to remain resilient due to sudden disturbances such as short circuit or loss of unanticipated system elements. This includes system responses resulting from the release of generators and transmission lines [2]. One important factor in the operation of electrical power systems is the desire to maintain system security (system security). System security includes activities planned to maintain system operation in case of system component failure For example, a generating unit may have to be off-line due to auxiliary equipment failures. By maintaining a number of appropriate rotary spare plants, the remaining generating units in the system can overcome the power shortage without falling too low or without the need for load shedding. In the generation and delivery of electric power, if a transmission line is damaged by a storm causing the channel to break, the remaining transmission channel will carry a larger load but still within the allowable range [2,6].

2. Research Methods

After the data has been collected, the design and framework of this study are as follows: (a) Model the power system of 150 kV South Bandung subsystem in ETAP; (b) Performing data inputs: generators, loads, buses and channel parameters; (c) Conducting power flow analysis on the Cirata subsystem power system under normal circumstances under ETAP 12.6; (d) Choosing the type of contingency; (e) Conducting a power flow analysis on the Cirata subsystem power system in a contingency state with the ETAP 12.6 program; (f) Identify weak power system (bus and transmission line) elements of the power flow analysis during normal and contingency states (N-2); (g) Performing system improvements, ie load shedding until the return voltage at the limit is permitted; (h) Checking the effectiveness of system reliability improvements as measured by voltage and channel loading. Flow of data analisys research can see in figure 1.

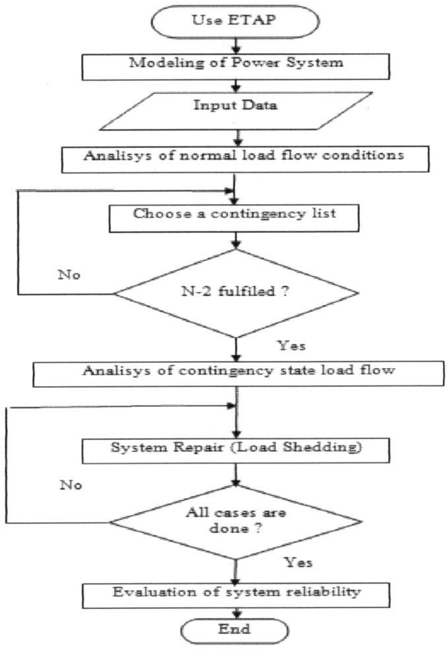

Figure 1. Flow of data analysis research.

3. Results and Discussions

3.1. Peak Load Condition

The magnitude of the system bus voltage at the time of no interference (normal) is at least 127.58 kV at Dawuan load bus and the greatest voltage is 140.81 kV on the bus Cirata. Figure 2 shows that the system voltage at peak load condition is below the standard SPLN seven rails that meet SPLN standards, namely Cirata rail (140.81 kV), Cikumpay (135.65 kV), Pbran (135.60 kV), Kamojang (135.56 kV), and Darajat (135.58 kV). The system bus voltage profile under normal circumstances is as follows:

Table 1. Channel Load Charge Data on Simulation Results.

NO	CHANNEL (From To)	$I_{NOMINAL}$ (A)	LOADING (%)
1	CIRATA - CKPAY	1620	26.95
2	CKPAY - PBRAN	1620	0.98
3	CIRATA - PWKT	1620	35.68
4	CKPAY - PWKT	1620	8.20
5	CIRATA - PDLRNG	1400	90.04
6	PDLRNG - LGDAR	580	12.39
7	PTUHA - LGDAR	1600	4.87
8	BDUTR - PDLRNG	580	46.80
9	CBBRU - PDLRNG	1200	15.80
10	CBBAT - PDLRNG	1200	36.48
11	CBBRU - CBBAT	580	6.23
12	PDLRNG - JTLHR	580	40.77
13	JTLHR - TTJBR	1160	43.52
14	TTJBR - KSBRU	1160	28.62
15	KSBRU - DWUAN	580	31.58

IOP Conf. Series: Materials Science and Engineering **384** (2018) 012076 doi:10.1088/1757-899X/384/1/012076

The minimum transmittal (%) of the transmission line (%) is 0.98% on the largest CKPAY-PBRAN transmission line and the loading (%) is 90.04% on the CIRATA-PDLRNG transmission line segment. From Table 1 it can be seen that under peak load conditions, one channel is overloaded.

3.2. Contingency Conditions

Contingency 1 is the state of the system when there is interference on two transmission lines ie CKPAY-CIRATA & BDUTR1-PDLRNG1 buses which result in the release of two channels.

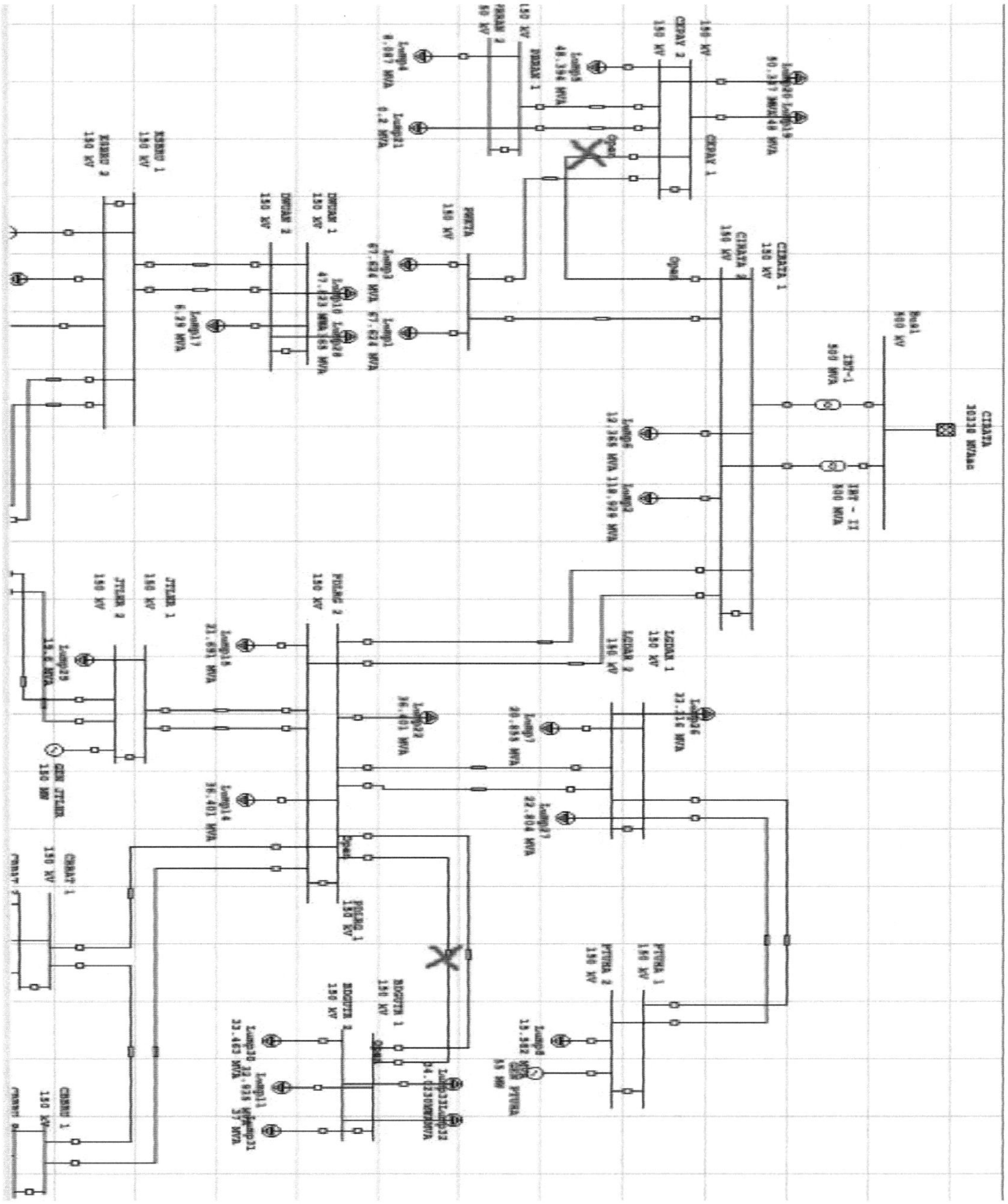

Figure 2. Condition of Contingency system.

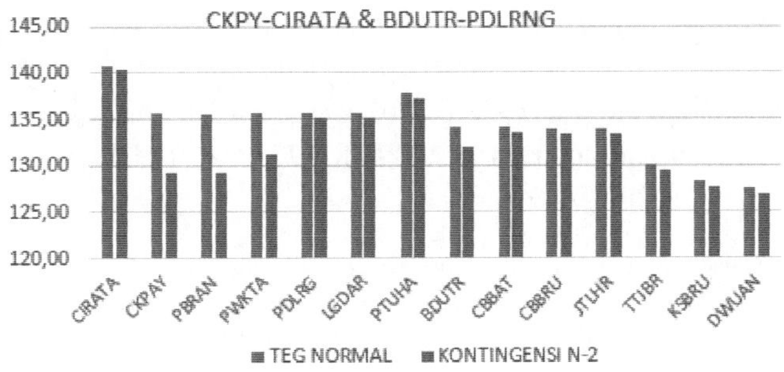

Figure 3. Bus Voltage Profile in Normal and Contingency Conditions.

Figure 3 is bus voltage profile in normal and contingency conditions. In the event of disruption will affect the occurrence of changes in bus voltage, especially buses adjacent to the location of the disturbance. Several buses experienced considerable voltage drop, even some buses decreased to below the defined limit of 135 kV. In the case of contingency, the release of the Cikumpay-Cirata and North Bandung-Padalarang transmission lines made three buses with considerable drops in value, ie Cikumpay bus (129.32 kV), Pbran (129.28 kV), and Purwakarta (131.26 kV). Almost all buses in the Cirata subsystem have decreased to the lower limit of the rated system nominal voltage of 135 kV, except for four buses still within the limits of the SPLN operations ie Cirata buses (140.35 kV), Padalarang (135.16 kV), Lagadar (135.15 kV), and Patuha (137.27).

Table 2. Loading Transmission on Normal Conditions and during Contingency.

Conditions	Channel (From To)	$I_{nominal}$ (A)	$I_{operating}$ (A)
Normal	CKPAY - CIRATA	1620	476.91
	CIRATA – PWKT	1620	689.18
	BDUTR 1 – PDLRNG 1	580	271.46
	BDUTR 2 – PDLRNG 2	580	271.46
Failurer	CKPAY - CIRATA	1620	-
	CIRATA - PWKT	1620	1166.09
	BDUTR 1 – PDLRNG 1	580	-
	BDUTR 2 – PDLRNG 2	580	542.92

In table 2 when there is disturbance on the transmission line CKPAY-CIRATA and BDUTR1-PDLRNG will cause the transmission loading changes especially in the buses are connected and close to the channels that come out of the system. In the CKPAY-CIRATA and BDUTR1-PDLRNG1 channels there are two channels that each channel is loaded at 476.91 A and 271.46 A under normal conditions. Based on the power flow analysis, when there is a disturbance on one channel, the other channel has a load of 1166.09 A and 542.92 A. The carrying current carrying capacity of CIRATA-PWKT is 1620 A while the current carrying current of BDUTR2-PDLRNG2 is 580 A Thus, the CIRATA-PWKT channel has a load of 1166.09 / 1620 x 100% = 71.98%, and BDUTR 2-PDLRNG 2 has 542.92 / 580 x 100% = 93.60%. The price of the CKPAY-CIRATA and BDUTR2-PDLRNG2 channel loading is smaller than 125% in conductor then in this condition the channel does not exit the system, but only the critical loading.

3.3. Load Shedding Solution

Contingency system repair solution (N-2) there are two ways that is by adding the channel to the disruption and do load shedding.

IOP Conf. Series: Materials Science and Engineering **384** (2018) 012076 doi:10.1088/1757-899X/384/1/012076

Table 3. Load Release of Each Case.

No	Name of Solution	Bus load in *-Load Shedding*	Information
1	Load Shedding	CKPAY, CIRATA, PWKTA, BDUTR, PDLRNG, CBBAT, CBBRU,JTLHR, TTJBR, KSBRU, DWUAN	Contingency Condition Solution

Load Shedding Simulation is a solution for contingency conditions [7]. The data from Table 3 can be concluded that by the release of 30% load on each bus whose voltage is less than the SPLN standard of 135 kV when the contingency state 1, causing the CKPAY-CIRATA and BDUTR2-PDLRNG2 ducts to decrease and the tension on the buses whose voltage is less than 135 kV has changed its value to normal, which is at the limit of SPLN.

Table 4. Load Shedding of Simulation Results.

Condition	Bus	Bus Voltage (kV)	Line (Channel)	I_{nom} (A)	$I_{operating}$ (A)
Before Load Shedding	CIRATA	140.35	CKPAY-CIRATA	1620	-
	CKPAY	129.32	CIRATA-PWKT	1620	1166.09
	PWKTA	131.26	-	-	-
	CBBAT	133.64	-	-	-
	CBBRU	133.45	-	-	-
	PDLRNG	135.16	BDUTR1-PDLRNG1	580	-
	BDUTR	132.02	BDUTR2-PDLRNG2	580	542.92
	JTLHR	133.45	-	-	-
	TTJBR	129.55	-	-	-
	KSBRU	127.75	-	-	-
	DWUAN	127	-	-	-
After Load Shedding	CIRATA	144.82	CKPAY-CIRATA	1620	-
	CKPAY	137.84	CIRATA-PWKT	1620	775.86
	PWKTA	139.16	-	-	-
	CBBAT	141.09	-	-	-
	CBBRU	140.97	-	-	-
	PDLRNG	142.08	BDUTR1-PDLRNG1	580	-
	BDUTR	139.72	BDUTR2-PDLRNG2	580	431.26
	JTLHR	143.19	-	-	-
	TTJBR	140.59	-	-	-
	KSBRU	139.39	-	-	-
	DWUAN	138.9	-	-	-

Table 4 is load shedding of simulation results, the impact of doing Load Shedding is a power loss. Power discharged in the event of a contingency disruption. This is done to keep the system safe and secure. Total loss of power at the time of LS load discharge (30%) is 285.51 MW.

The results of system testing performed with the analysis of power flow on the system that has been improved both in normal conditions and contingency conditions, indicating that the system improvements made to fix the weaknesses of existing systems. This is demonstrated by the results of tests performed with the analysis of power flow on the system that has been improved. Both under normal conditions and contingency conditions indicate that system improvements made by the Load Shedding method successfully fix the existing system weaknesses. As shown in Figure 4 below:

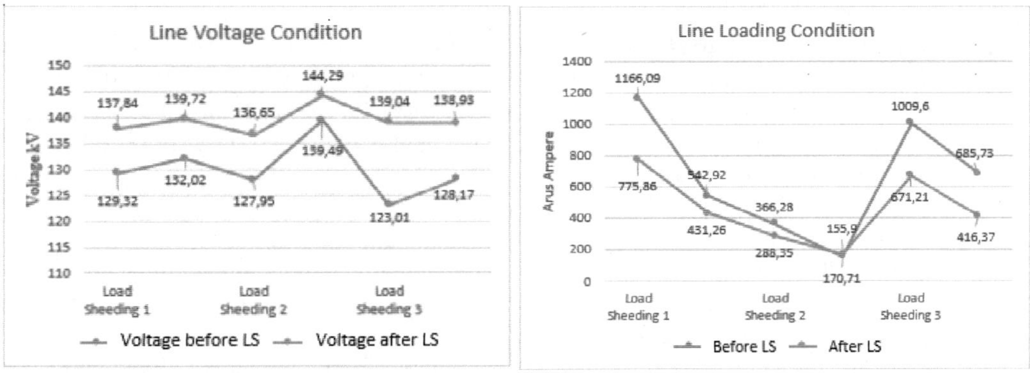

Figure 4. Improved system with Load Shedding method.

- Once the system is repaired there is no bus voltage that violates the limits of operation either in normal circumstances, contingency state.
- System improvements have successfully overcome the loading that exceeds the nominal current of the carrier on the transmission line.

Thus, the N-2 security reliability criteria are met, ie, in the contingency state (N-2), there are no system quantities that violate the specified operating limits. It shows that the system improvement is effective.

4. Conclusions and Recommendation

4.1. Conclusions

Based on data analysis and discussion can be concluded as follows:

- The voltage values on the substations relay in all contingency case studies (N-2) decreased below the lower limit of the SPLN, ie -10%. So at the time of contingency (N-2) the stress on the Cirata 150 kV subsystem is not reliable.
- There are some critical transmission channels that are CIRATA-PWKT transmission line, BDUTR2-PDLRNG2, JTLHR2-TTJBR2, PDLRNG-CBBRU
- System improvements due to contingency disorder can be done by way of load loading (load shedding).
- The discharge load loss due to contingency 1 is 285.51 MW.

4.2. Recommendations

The recommendations of this study are as follows:

- The results of this study should be used as input for PT. PLN to develop the system in order to improve the reliability of Cirata 150 kV subsystem.
- To improve the system during contingency, in addition to using a load-release method, it can also be done by adding channels from either new generator / GI or old generator / GI.
- The need for continuous electrical system operation planning in anticipation of the development of load and the development of transmission network configuration in order to improve the safety and reliability of power system.
- Calculation analysis techniques in this study using ETAP program, for it can also be developed by using other programs as a comparison.

IOP Conf. Series: Materials Science and Engineering **384** (2018) 012076 doi:10.1088/1757-899X/384/1/012076

References

[1] Aulia U and Putranto L M 2014 Analisis Kontingensi Generator Pada Sistem Transmisi 500 kV JAMALI *Jurnal Penelitian Teknik Elektro dan Teknologi Informasi* **1** (03) 100–104

[2] Hartoyo 2006 *Perbaikan Keandalan (N-1) Sistem Tenaga Listrik PLN Jawa Tengah dan DIY*

[3] Rahmawati N 2012 Analisis Kontingensi Sistem Tenaga Listrik dengan Metode Bounding *Jurnal Rekayasa Elektrika* **10** (2) 92-97

[4] PLN 2016 *PT.PLN (Persero) APB Jawa Barat*

[5] Nugraheni A 2011 *Simuasi Pelepasan Beban dengan Menggunakan Rele Frekuensi pada Sistem Tenaga Listrik CNOOC SES LTD*

[6] Dong F, Xu X and Zhang X 2014 Parallel Contingency Analysis Solution based on OpenMP *North American Power Symposium (NAPS)* 1-4 IEEE

[7] Mokhlis H, Karimi M, Member S, Shahriari A, Halim A and Bakar A 2011 A New Under-Frequency Load Shedding Scheme for Islanded Distribution Network *Innovative Smart Grid Technologies (ISGT), 2013 IEEE PES* 1–6

IOP Conf. Series: Materials Science and Engineering **384** (2018) 012077 doi:10.1088/1757-899X/384/1/012077

Integrated Management Information System for Curriculum in University

I R I Astutik[1]* and I Widiaty[2]

[1]Departemen Informatic, Universitas Muhammadiyah Sidoarjo, Jl. Majapahit 666 B, Sidoarjo 61215, Jawa Timur, Indonesia.
[2]Universitas Pendidikan Indonesia, Jl. Dr. Setiabudhi 229 Bandung 40154, Indonesia.

*ikaratna@umsida.ac.id

Abstract. In university, a curriculum is one of the factors that play an important role in preparing graduated student that have High quality and qualified to compete in the era of globalization. The curriculum is also used as a reference in determining the learning achievement and also used as implementation guidelines in the faculty. Learning outcome in the curriculum Described in the courses that are used as material for study or teaching materials for lecturers. This study aims to create a management information system for curriculum that can assist the faculty, Academic Administration Bureau and Institute of Study and Development of education in Managing and evaluation a curriculum, especially when there is a new development in the curriculum. This research uses the Waterfall Software Development Life Cycle (SDLC) model for analyzing the flow of information system such as: (1) Requirement and Analysis, (2) System Design, (3) Implementation, (4) Integration and Testing, (5) Deployment of the system, (6) Maintenance. The system is also supported by twitter bootstrap framework technology that makes the system simple, easier and responsive.

1. Introduction

The higher education curriculum is a program to produce graduates, so the program should ensure that its graduates have qualifications equivalent to the agreed qualifications. [1] Some colleges and universities have developed their curriculum based on educational policy, learning outcome or discussions of faculties. [2] Graduate quality is based on the level of ability expressed in the learning achievement competencies (learning outcomes). From the formulation of learning achievement can be done by using the subject that used as teaching material in learning in college. Good curriculum management can improve the quality of college graduates. Where the quality of graduates or human resources can be seen from the achievement of learning from the curriculum formulated by each faculty at the University.

At the University The current curriculum was evaluated, and sometimes a new concept curriculum was written that make continuous challenge in higher education or in faculties. [3] The curriculum evaluated by the Academic Administration Bureau and expert supervision (Institute of Study and Development of education in Managing and evaluation a curriculum) discusses with the teachers the necessary adjustments needed to make the curriculum relevant to their settings. [4]

The faculty prepares a new curriculum based on applicable government regulations and based on their respective associations of faculties. Which Curriculum changes require strategic and operational management of wide-ranging adjustments within academia. This kind of change might include introducing new teaching and learning, setting out a vision, academics developing new skills and expertise and changing philosophy of faculties. [5]

Here, the purpose of this study was to create a web-based responsive curriculum management information system using PHP programming language, database Postgresql and twitter Bootstrap, where the faculty can activate the course, determine the course prerequisites and do the scheduling of the course without having to submit the file to expert supervision (the Institute of Study and Development of education in Managing and evaluation a curriculum) and also can do monitoring and evaluation if the courses are activated and the schedule in the plotting is not appropriate. The curriculum expert supervision can also check, control and validate the data, especially the teaching load of lecturers.

2. Experimental Method

The development of this application using the Waterfall approach from Software Development Life Cycle Model (SDLC). Which has several stages, such as: Requirement and Analysis, System Design, Implementation, Integration and Testing, Deployment of the system and Maintenance. The first step is Requirement and Analysys information system which is collecting data and information that needed through literature studies, observations and interviews with the head of faculties and expert supervision (head and staff of Development of education in Managing and evaluating a curriculum) and Academic Administration Bureau.

Then analysis of hardware requirements and software requirements required in building information systems. Where this stage determines the minimum hardware specifications needed to run the system as well as to know the software needed in the manufacture of the system. After all the information is obtained, the information system will be designed in accordance with the results of the requirement analysis. The design of the system includes database design, interface, and the flow of the system. The results of the next draft will be implemented through coding or scripting program with the PHP programming language, [6] Twitter Bootsrap and the Postgres database management system. After the system is completed, then the system tested to know if system according what is expected or not.

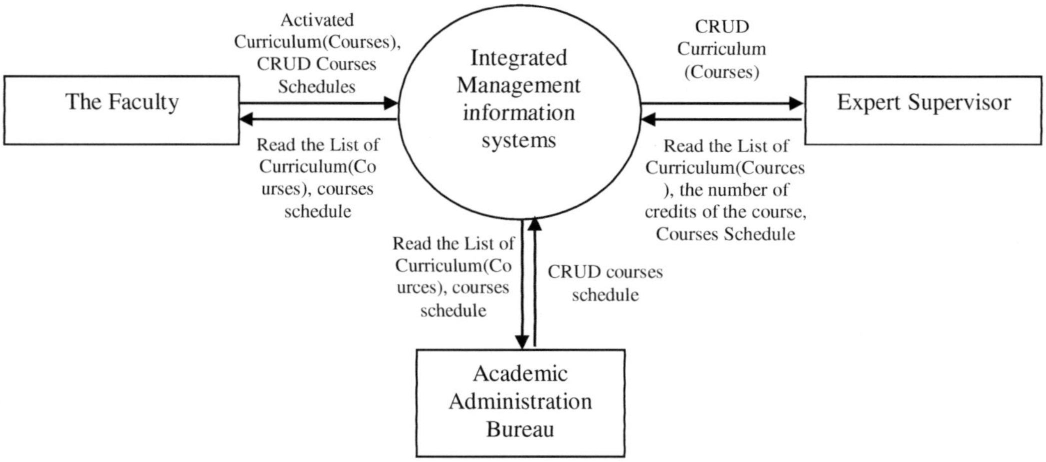

Figure 1. Data Flow Diagram Integrated Management Information System in university

Figure 1 shows that the university has an integrated information system for Program Study, Expert Supervisor and Academic Administration Bureau. Each user has different access and responsibility. The First user is Faculty, not only can activated courses in a curriculum that uses in the academic year in progress. But also have the list of the curriculum either the old or the new. Faculty also can arrange a college schedule for teachers and students. The second user is expert supervisor that can create,read,update, and delete curriculum that is not in accordance with the guidebook prepared by the faculty, also can read the list of all curriculum in university and can read the course schedule that arranged by faculty. The last user is Academin Administration Bueau, this user can read all courses in the curriculum and can create, read, update and delete courses schedule if there is teachers exchanges because overloaded teaching according to the credit load that the university policy. Despite having a different information system, but data storage remains in the same database, because the information system is already integrated.

3. Results and Discussion

Figure 2 shows the first appearance of the curriculum information system. To be able to access the system user information must identify by signing in form sign in system. After signing in successfully, then the user can access menu in the system include: home, table courses list, input data curriculum courses in form, form validation courses that activated by faculty, form input courses schedules, form validation of the teachers learning plan.

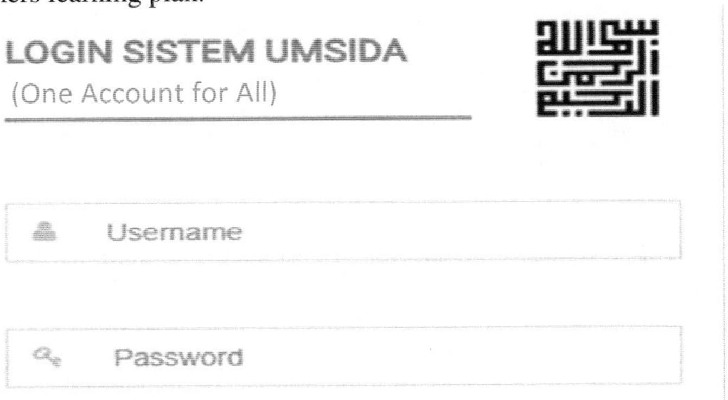

Figure 2. Form Sign in information system.

Figure 3 shows the main page view of the system dedicated to the expert supervisor. This page is useful for displaying data on the number of credits of each course of study active in the current academic year.

Program Studi	Sks	Status
AHWAL AL-SYAKSHIYAH		approved
PENDIDIKAN AGAMA ISLAM		approved
PENDIDIKAN BAHASA ARAB		approved
PENDIDIKAN GURU MI		approved
PERBANKAN SYARIAH		approved
AGROTEKNOLOGI		approved
TEKNOLOGI HASIL PERTANIAN		approved

Figure 3. Form Home information system for curriculum

Figure 4 shows to process the course (a) view and search the course data is done on the page read the course data. (b) As for entering and changing the subject data using the course form. In this form the user input data course code, the name of the course, department, credits of course, semesters of the course will be taught, subject group, the type of course, the status of the course and whether the subject is the concentration course or not.

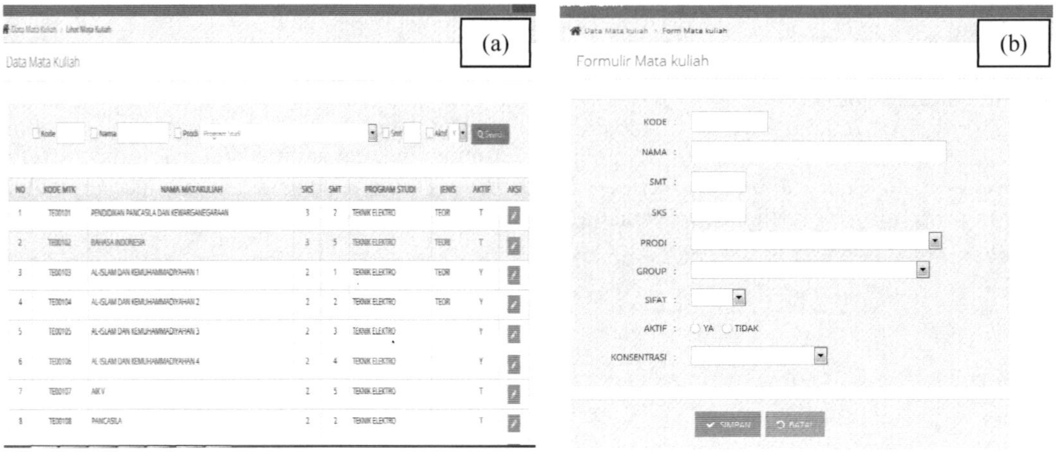

Figure 4. The Form Read the list of courses and form input the course

The curriculum information system for faculty is explained in **Figure 5**. Faculty can view the data of the existing curricula in information systems for the faculty. (a) In addition the system makes it easy to search data curriculum based on code courses, the name of the course and semester. (b) The faculty can also enable the courses that will be used during the current academic year.

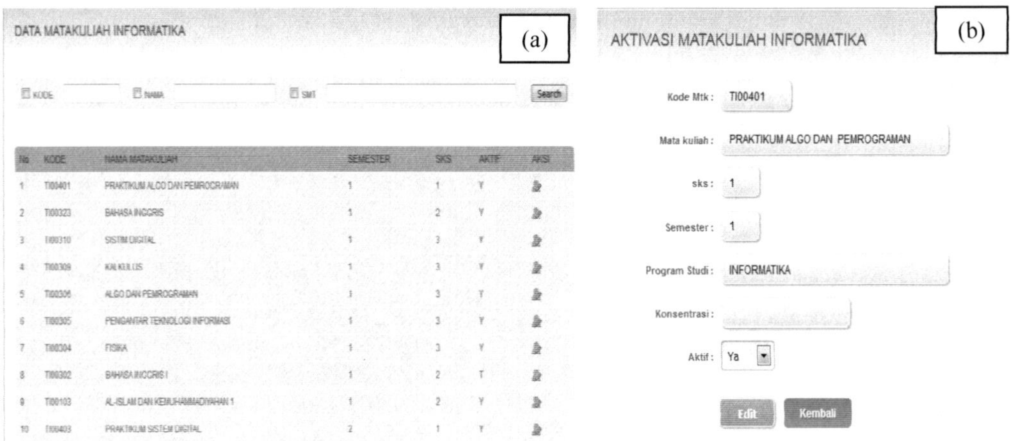

Figure 5. (a) Form List courses and (b) activated courses in the information system Faculty.

After faculty enable courses to use in the current academic year the supervisor will validate the data subjects that later can be used for scheduling courses, this is explained in Figure 6.

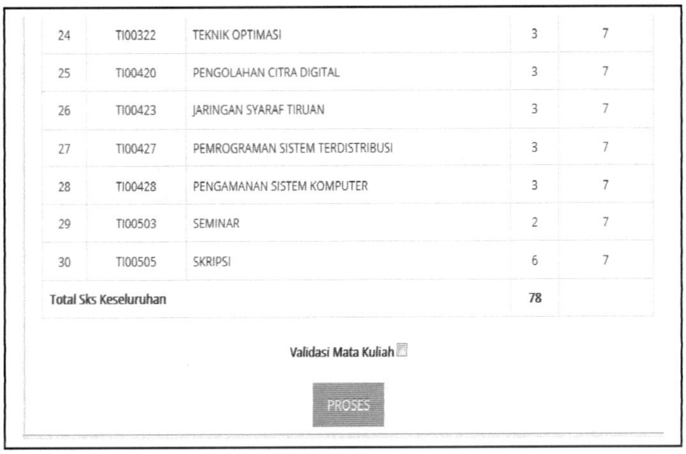

24	TI00322	TEKNIK OPTIMASI	3	7
25	TI00420	PENGOLAHAN CITRA DIGITAL	3	7
26	TI00423	JARINGAN SYARAF TIRUAN	3	7
27	TI00427	PEMROGRAMAN SISTEM TERDISTRIBUSI	3	7
28	TI00428	PENGAMANAN SISTEM KOMPUTER	3	7
29	TI00503	SEMINAR	2	7
30	TI00505	SKRIPSI	6	7
Total Sks Keseluruhan			**78**	

Figure 6. Form validation courses by expert supervisor.

4. Conclusion

The responsive web-based integrated information system proposed in this research can make the curriculum setting process, especially the activation of the course when the student will do the study plan card can be done more efficiently, without having to send data to expert supervisor. From the test results got the conclusion that the proposed system can also be a media search and management curriculum in universities, especially private colleges.

Pre-built systems only provide information for curriculum only institutions. While the proposed integrated system can provide information not only the curriculum management, but also the faculty as the executor and the head of the university can also monitor and evaluate the current curriculum whether in accordance with government regulations and what guidebook does not. To make it easier to know the data of certain courses in show search feature so that the time needed to search certain course becomes shorter and efficient.

Acknowledgements

We acknowledged Muhammadiyah University of Sidoarjo (Grant: Penelitian Internal).

References

[1] RISTEKDIKTI, "Paduan Penyusunan Kurikulum Pendidikan Tinggi," p. 3, 2016.
[2] D Zhao, X Ma, and S Qiao 2017 What aspects should be evaluated when evaluating graduate curriculum: Analysis based on student interview *Stud. Educ. Eval.*
[3] V Gooder and S Cantwell 2017 Student Experiences With a Newly Developed Concept-Based Curriculum *Teach. Learn. Nurs.*
[4] S F Shawer 2017 Teacher-driven curriculum development at the classroom level: Implications for curriculum, pedagogy and teacher training *Teach. Teach. Educ.*, vol. **63**, pp. 296–313
[5] L Newman Lior 2013 *Chapter 2. Design and Development Models and Processes*
[6] T Dan, H Implementasi, P Uang, E Di, M T Informasi, F I Komputer, U Indonesia, J Salemba, and R No, 2017 *Jurnal Sistem Informasi (Journal of Information Systems).* 1 / 13 (2017), 38-48 DOI : http://dx.doi.org/10.21609/jsi.v13i1.465," vol. **13**, pp. 38–48,

IOP Conf. Series: Materials Science and Engineering **384** (2018) 012078 doi:10.1088/1757-899X/384/1/012078

Very Sort Term Load Forecasting Using Interval Type - 2 Fuzzy Inference System (IT- 2 FIS) (Case Study: Java Bali Electrical System)

J Jamaaluddin[1]*, D Hadidjaja[1], I Sulistiyowati[1], E A Suprayitno[1], I Anshory[1], S Syahrorini[1] and A G Abdullah[2]

[1]Program Studi Teknik Elektro, Universitas Muhammadiyah Sidoarjo, Jl. Raya Gelam no 250, Sidoarjo, Jawa Timur, Indonesia.
[2]Sekolah Pascasarjana, Universitas Pendidikan Indonesia, Jl. Dr. Setiabudhi 229 Bandung 40154, Indonesia.

*jamaaluddin@umsida.ac.id

Abstract. One of the important things to do an electric power system operation is load forecasting. Load forecasting consists of short-term forecasting and short-term forecasting. The very short term load forecasting are required for regulating electrical energy generation, maintenance arrangements and regulating the labor involved. This forecasting is done to decide which plant to operate. The capacity of the plant to be operated adjusts to the load plan to be supplied the next day. The very short-term load forecasting is predicting electrical loads with time intervals every 30 minutes for the next day. In this study using Interval Type-2 Fuzzy Inference System (IT-2FIS) because it delivers a high flexibility that can be developed using other methods (hybrid). Laying out the footprint of uncertainty (FOU) membership function of the Interval Type-2 Fuzzy Inference System (IT-2FIS). This method has been applied for short-term load forecasting and will be employed for very short-term forecasting. In very short-term load forecasting IT-2 FIS has Mean Average Percentage Error (MAPE) arround 0,729%.

1. Introduction

Electric power is needed by the world community at this end. In every side of his life always requires electrical energy[1]. Generation, transmission and distribution system in Indonesia also operates this electricity from generated to customer. Handling of this electric power operation, must be professional, meet the quality standards of electrical services and economic factors [2] [3]. On the distribution of electric power that is implementing is PT. PLN through P2B, the agency will generate power generation and load sharing, in which it includes both short loading and very short term loading plans[4]. Very short term load forecasting is required to conduct the generation planning in accordance with the estimated load requirement on the next day. Forecasting a very short term load is to plan the load every 30 minutes from 00.00 until 23.30 the next day. This should be done for economic considerations so that will get the planning of power generation, which should operate and which are not. This problem is related to generating efficiency, maintenance implementation and labor management. In these 2 decades to do load forecasting, have used computing. In this computation has been using Fuzzy Logic for short-term forecasting (Daily)[5] [6]. Both in development using Interval Type 1 - Fuzzy Inference System and Interval Type 2 - Fuzzy Inference System. In forecasting short-term IT-1 FIS and IT-2 FIS loads have resulted in a pretty good MAPE result. Therefore, at this time will try to do very short term forecasting [7].

2. Experimental Method

This research is done several stages of the process, are: 1. Stage before process, 2. Process, 3. After process with more detail explanation as follows:

2.1. Preprocessing

In this process is the stages grouping daily load data every 30 minutes for 24 hours on a predictable day [8], and three hours before the hour predicted at the same time started 3 years ago. In this study, an example is taken on the first Friday of October. This process is required to find Variation Load Difference (VLD_{MAX}). This is done to get the X, Y and Z value [1].

1. For example, to get a very short term forecasting value at 12:00 on the first Friday of October 2015, using data on the first Friday of October 2013, 2014 and 2015.
2. Identify the search for the load on the 4 days before to its at the same hour of the loaded time being analyzed

$$MaxWD_{(i)} = \frac{WD_{(i)d-4} + WD_{(i)d-3} + WD_{(i)d-2} + WD_{(i)d-1}}{4} \quad (1)$$

3. Calculate the difference in load sought (Load Difference) on the clock to be predicted.

$$LD_{MAX}(i) = \frac{MaxSD(i) - MaxWD(i)}{MaxWD(i)} x100 \quad (2)$$

4. Looking for load difference characteristics on typical load hours or TLDMAX (i) by averaging the same LDMAX peak load (i) in previous years.
5. Search for Variation Loads on the clock to be predicted (Variation Load Difference) at each hour

$$VLD_{max}(i) = LD_{max}(i) - TLD_{max}(i) \quad (3)$$

2.2. Processing

At this stage of the process, entering the very short-term forecasting forecast on the first Friday of October into *IT-2FIS* is as follows[1]:

1. The first stage is to build input Membership Function X and Y. The Z is the output membership function for the hour predicted, with the following explanation:

   X: VLDmax (i) Variable Load Difference time that will be predictable

   Y: VLDmax (i) Is a time load at second Friday on October that is adjacent to the same type of time that will be predictable.

   Z: Forecast VID_{max} (on) Variable Load Difference of a time load that would be predicted.

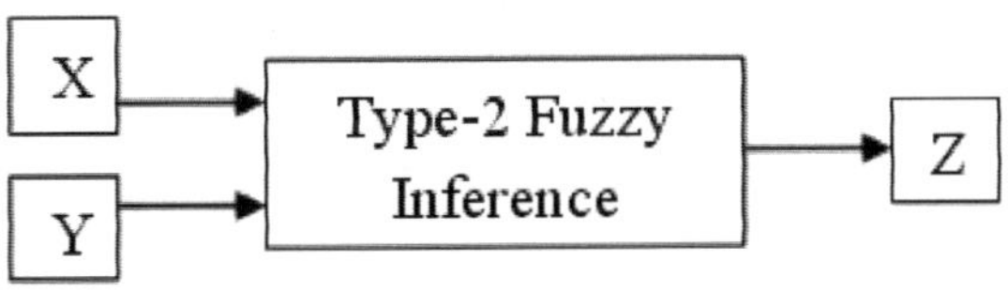

Figure 1. Process of *IT-2FIS*

2. Creating fuzzy rules as follows [6]:

   IF X is Ai AND Y is Bi THEN Z is Ci

IOP Conf. Series: Materials Science and Engineering **384** (2018) 012078 doi:10.1088/1757-899X/384/1/012078

3. Applied on the *(IT-2 FIS)*.
4. Applied the *MIN* function.
5. Applied *MAX* on each fuzzy, implication results.
6. Get the value Forecast VLD_{max}.

2.3. Post Processing

The In the post-processing stage calculated result of Very Short Term Load Forecasting for the time predict as follows[1]:

1. Calculating the time load forecasting :

$$ForecastLD_{MAX}(i)=ForecastVLD_{MAX}(i)-TLD_{MAX} \quad (4)$$

2. Calculating the difference of time load of forecast

$$P'_{MAX}(i) = MaxWD(i) + \frac{(ForecastLD_{MAX} x MaxWD(i))}{100} \quad (5)$$

3. Comparing the value of forecasting results with the actual situation in the already running, namely in 2013, 2014 and 2015. By making the average value Mean Percentage Error (MAPE) on each year forecasting. The formula used is as follows :

$$Error\% = \frac{P_{forecast} - P_{actual}}{P_{actual}} x100$$

$$Error\% = \frac{P'_{MAX}(i) - MaxSD(i)}{MaxSD(i)} x100 \quad (6)$$

2.4. Flowchart of Forecasting by Using IT-2

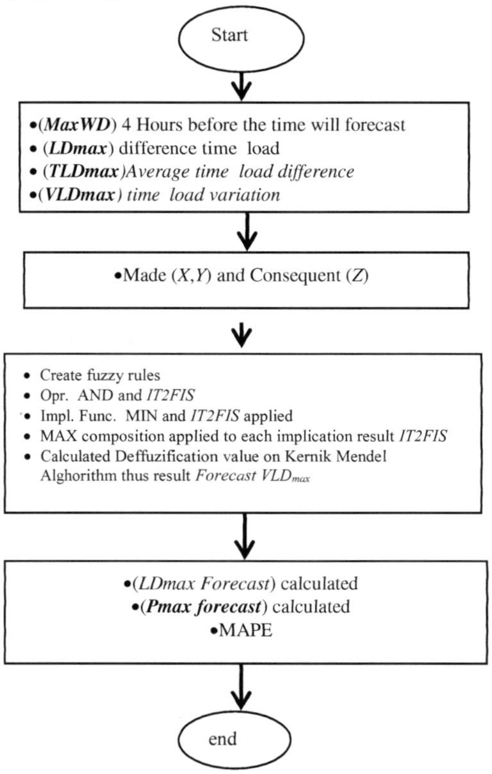

Figure 2. Flowchart of Forecasting by Using IT-2

2.5. Calculation X Value

X value at 2013, obtained from the calculation as below:

Electrical load at 12.00 First Friday in October 2013 $(R_1\text{-}2013)$

$MaxWD(R_1\text{-}2013)_{d-4} = 19.823$ MW

$MaxWD(R_1\text{-}2013)_{d-3} = 20.008$ MW

$MaxWD(R_1\text{-}2013)_{d-2} = 19.763$ MW

$MaxWD(R_1\text{-}2013)_{d-1} = 19.539$ MW

$MaxSD = 19.021$ MW

$$MaxWD(R_1\text{-}2013) = \frac{19.823 + 20.008 + 19.763 + 19.539}{4} = 19.738 \text{ MW}$$

$$LD_{MAX}(R_1\text{-}2013) = \frac{MaxSD(R_1\text{-}2013) - MaxWD(R_1\text{-}2013)}{MaxWD(R1\text{-}2013)} \times 100\%$$

$$Max_{WD}(R1\text{-}2013) = \frac{19.021 - 19.738}{19.738} \times 100\% = \mathbf{-3.854}$$

With the same calculation we can find Max_{WD}(R1-2013) , $LD_{MAX}(R_1\text{-}2014)$, $TLD_{MAX}(R_1\text{-}2014)$, $VLD_{MAX}(R_1\text{-}2014)$, Then the results obtained as the table below :

Table 1. Calculation Forecasting First Friday on October 2013

	day / Hour	d-4	d-3	d-2	d-1	d	WDMAX 2013	LDMAX 2013
1	12.00	19.823	20.008	19.763	19.539	19.021	19.783	-3,854
2	11.00	20.768	20.754	20.892	20.465	20.938	20.720	1,055
3	10.00	20.459	20.566	20.526	20.096	20.551	20.412	0,679
4	09.00	19.812	20.087	20.022	19.350	19.963	19.818	0,736

Table 1, above shows the calculations of WD_{MAX} and LDMAX for the first Friday of 2013, after which the calculations for the first Friday of 2014, 2015 with the results as table 2 below:

Table 2. Calculation Forecasting First Friday on October 2014

	day / Hour	d-4	d-3	d-2	d-1	d	WDMAX 2014	LDMAX 2014	TLDMAX 2014	VLDMAX 2014
1	12.00	23.564	19.369	20.127	21.763	20.536	21.206	-3,160	-3,507	0,347
2	11.00	21.334	20.275	22.423	23.477	21.437	21.877	-2,013	-0,479	-1,534
3	10.00	18.675	19.563	21.897	22.843	21.288	20.745	2,619	1,649	0,970
4	09.00	19.002	19.343	21.453	24.322	20.932	21.030	-0,467	0,134	-0,601

Table 3. Calculation Forecasting First Friday on October 2015

	day / Hour	d-4	d-3	d-2	d-1	d	WDMAX 2015	LDMAX 2015	TLDMAX 2015	VLDMAX 2015
1	12.00	20.566	19.872	20.556	22.329	21.360	20.831	2,538	-1,492	4,030
2	11.00	22.800	19.879	22.121	22.899	22.214	21.925	1,320	0,120	1,199
3	10.00	24.101	20.544	22.017	23.243	22.175	22.476	-1,342	0,652	-1,994
4	09.00	21.637	18.564	19.324	22.887	21.613	20.603	4,904	1,724	3,179

2.6. Calculation of Value of Input Variable Y

To get the value of Y, then selected second Friday in each year starting 2013, 2014 and 2015, in the same way to get the value of X, then get the results as below:

Table 4. Calculation Forecasting second Friday on October 2013

	Day / Hour	d-4	d-3	d-2	d-1	d	WDMAX 2013	LDMAX 2013
1	12.00	19.112	18.973	22.145	19.570	19.348	19.950	-3,018
2	11.00	18.783	18.122	20.312	20.562	19.892	19.445	2,300
3	10.00	20.973	19.887	20.332	22.477	21.700	20.917	3,742
4	09.00	20.431	19.878	21.334	22.769	21.760	21.103	3,113

Table 5. Calculation Forecasting second Friday on October 2014

	Day / Hour	d-4	d-3	d-2	d-1	d	WDMAX 2014	LDMAX 2014	TLDMAX 2014	VLDMAX 2014
1	12.00	20.562	19.879	20.354	21.336	20.239	20.533	-1,431	-2,224	0,793
2	11.00	20.563	20.361	21.782	22.397	21.421	21.276	0,683	1,491	-0,809
3	10.00	20.113	21.785	22.766	23.578	22.488	22.061	1,938	2,840	-0,902
4	09.00	23.444	21.357	22.899	23.765	22.133	22.866	-3,207	-0,047	-3,160

Table 6. Calculation Forecasting second Friday on October 2015

	Day / Hour	d-4	d-3	d-2	d-1	d	WDMAX 2015	LDMAX 2015	TLDMAX 2015	VLDMAX 2015
1	12.00	19.378	19.700	20.765	21.113	20.456	20.239	1,072	-1,125	2,198
2	11.00	19.567	20.878	21.598	22.133	21.271	21.044	1,079	1,354	-0,275
3	10.00	19.789	20.846	22.539	23.782	22.349	21.739	2,806	2,829	-0,023
4	09.00	21.658	22.783	25.786	24.442	23.211	23.667	-1,928	-0,674	-1,254

IOP Conf. Series: Materials Science and Engineering **384** (2018) 012078 doi:10.1088/1757-899X/384/1/012078

2.7 Calculation of Value of Input Variable Z

With the same calculation for the second Friday of October between 2013-2015 in get the value (VLD_{MAX}) which results can be seen as table 7 below:

Table 7. Value of WD_{MAX}, LD_{MAX} and VLD_{MAX} 2013-2015

	pukul	WDMAX	LDMAX	WDMAX	LDMAX	TLDMAX	VLDMAX	WDMAX	LDMAX	TLDMAX	VLDMAX
		2013		2014				2015			
	First Friday on October										
1	12.00	19.783	(3,854)	21.206	(3,160)	(3,507)	0,347	20.831	2,538	(1,492)	4,030
2	11.00	20.720	1,055	21.877	(2,013)	(0,479)	(1,534)	21.925	1,320	0,120	1,199
3	10.00	20.412	0,679	20.745	2,619	1,649	0,970	22.476	(1,342)	0,652	(1,994)
4	09.00	19.818	0,736	21.030	(0,467)	0,134	(0,601)	20.603	4,904	1,724	3,179
	Second Friday on October										
1	12.00	19.950	(3,018)	20.533	(1,431)	(2,224)	0,793	20.239	1,072	(1,125)	2,198
2	11.00	19.445	2,300	21.276	0,683	1,491	(0,809)	21.044	1,079	1,354	(0,275)
3	10.00	20.917	3,742	22.061	1,938	2,840	(0,902)	21.739	2,806	2,829	(0,023)
4	09.00	21.103	3,113	22.866	(3,207)	(0,047)	(3,160)	23.667	(1,928)	(0,674)	(1,254)

3. Results and Discussion

IT-2FIS to forecasting of peak load at the time which forecasting, that the membership function variable input and output of Interval Type-2 Fuzzy Inference System, as follows :

3.1 Membership Function for Input and Output Variable

The set of Interval Type-2 Fuzzy, fuzzy sets similar to the type-1. Interval Type-2 Fuzzy, done twice a fuzzy, membership function type-1. Input variables *(X, Y)* and output variables *(Z)* consists of 11 fuzzy sets are described as follows:

Negative Very Big (NVB) range of values -12 s/d -8
Negative Big (NB) range of values -10 s/d -6
Negative Medium (NM) range of values -8 s/d -4
Negative Small (NS) range of values -6 s/d -2
Negative Very Small (NVS)range of values -4 s/d 0
Zero (ZE) range of values -2 s/d 2
Positive Very Small (PVS) range of values 0 s/d 4
Positive Small (PS) range of values 2 s/d 6
Positive Medium (PM) range of values 4 s/d 8
Positive Big (PB) range of values 6 s/d 10
Positive Very Big (PVB) range of values 8 s/d 12

Translation of antecedent membership functions (*X, Y*) and consequent (*Z*) is used for the manufacture of the *Rules Base Fuzzy Inference System.* Making the basic rules of Fuzzy *(Fuzzy Rule Base)* very short-term load forecasting in 2014-2015 is shown table 8. Through table 12.

Table 8. Input (*X, Y*) and output (*Z*) By VLD_{MAX} in 2014 and 2015

	Hour	*VLDMAX*	*VLDMAX*	Input		output
		2014	*2015*	X	Y	Z
1	12.00	0,347	4,030	0,347	2,198	4,030
2	11.00	-1,534	1,199	-1,534	-0,275	1,199
3	10.00	0,970	-1,994	0,970	-0,023	-1,994
4	09.00	-0,601	3,179	-0,601	-1,254	3,179

Table 9. Process Rules for Input *X* in 2016

			Membership function										
	Hour	X	NVB	NB	NM	NS	NVS	ZE	PVS	PS	PM	PB	PVB	X
1	12.00	0,347						0,21	0,79					PVS
2	11.00	-1,534					0,13	0,87						ZE
3	10.00	0,970						0,32	0,68					PVS
4	09.00	-0,601					0,42	0,58						ZE

Table 10. Process Rules for Input *Y* in 2016

			Membership Function										
	Hour	Y	NVB	NB	NM	NS	NVS	ZE	PVS	PS	PM	PB	PVB	Y
1	12.00	2,198							0,08	0,92				PS
2	11.00	-0,275					0,24	0,76						ZE
3	10.00	-0,023					0,07	0,93						ZE
4	09.00	-1,254					0,09	0,91						ZE

Table 11. Process Rules for Output *Z* in 2016

			Membership Function										
	Hour	Z	NVB	NB	NM	NS	NVS	ZE	PVS	PS	PM	PB	PVB	Z
1	12.00	4,030							0,93	0,07				PVS
2	11.00	1,199						0,83	0,17					ZE
3	10.00	-1,994					0,72	0,28						NVS
4	09.00	3,179							0,85	0,15				PVS

Table 12. Basic Rules table (*fuzzy rules*) for forecasting the year 2016

X/Y	NVB	...	ZE	PVS	PS	...	PB
NVB							
...							
NS							
NVS							
ZE			ZE/PVS				
PVS			NVS		PVS		
PS							
....							
PVB							

If there is a fuzzy rule is the same for input values *X* and, but different *Z* output value, whichever is the greater the value of its output being more removed.

Table 13. Conversion Table Basic Rules Forecasting the Year 2016 for Matlab Software Code

no	Antecendent		Cons	no	Antecendent		Cons
rules	X	Y	Z	rules	X	Y	Z
1	PVS	PS	PVS	1	5	6	6
2	ZE	ZE	ZE	2	5	7	8
3	PVS	ZE	NVS	3	7	6	5
4	ZE	ZE	PVS	4	7	6	8

Membership Function for Input and Output Variable

3.2 Implementation forecasting of Very Short Term Load On Electrical Systems Java Bali using Method Interval Type-2 Fuzzy Inference System (IT2FIS) at 3 Years of data taking into account the Year Actual Data Forecasting

Very Short-term load forecasting using the *Interval Type-2 Fuzzy Inference System (IT2FIS)* executed through m.file program in Matlab using the given function in the Toolbox *IT-2FLT*, to obtain the value of forecasting VLD_{max}. *Value of* VLD_{max} *forecasting* results continued (post processing) using software MS.Excell to get the time load forecasting and forecasting error value. The results of short-term load forecasting error method *IT-2FIS* in 2013 through 2015 can be seen in Table. 14 below.

Table 14. Comparison of Forecasting and Actual load on 2016

	pukul	h-4	h-3	h-2	h-1	h	WDMAX	LDMAX	TLDMAX	VLDMAX	Output Forecast	Forecast LDMAX	Forecast P'MAX	Actual (MW)	Error (%)
							2016								
	First Friday on October 2016														
1	12.00	19.823	20.008	19.763	19.539	19.021	20.831	2,538	(1,492)	4,030	-5,763	-7,254	19.319,61	19.020,72	1,571
2	11.00	20.768	20.754	20.892	20.465	20.938	21.925	1,320	0,120	1,199	-5,003	-4,882	20.854,30	20.938,33	0,401
3	10.00	20.459	20.566	20.526	20.096	20.551	22.476	(1,342)	0,652	(1,994)	-8,668	-8,016	20.674,54	20.550,66	0,603
4	09.00	19.812	20.087	20.022	19.350	19.963	20.603	4,904	1,724	3,179	-3,728	-2,003	20.190,24	19.963,47	1,136
												IT-1FIS	MAPE average		0,928
	First Friday on October 2016														
1	12.00	19.823	20.008	19.763	19.539	19.021	20.831	2,538	-1,492	4,030	-6,763	-8,254	19.111,31	19.020,72	0,476
2	11.00	20.768	20.754	20.892	20.465	20.938	21.925	1,320	0,120	1,199	-5,128	-5,008	20.826,85	20.938,33	0,532
3	10.00	20.459	20.566	20.526	20.096	20.551	22.476	-1,342	0,652	-1,994	-8,068	-7,416	20.809,40	20.550,66	1,259
4	09.00	19.812	20.087	20.022	19.350	19.963	20.603	4,904	1,724	3,179	-4,200	-2,475	20.093,01	19.963,47	0,649
												IT-2FIS	MAPE average		0,729

While for the error value can be seen in the picture. Fig. 3 below :

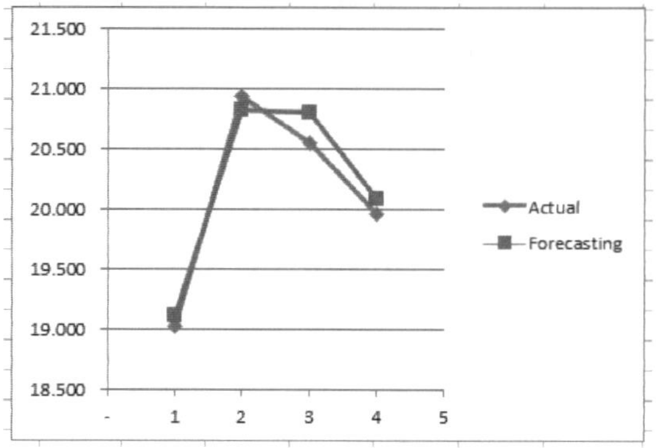

Figure 3. Actual and forecast Load by using *IT-2 FIS*
(First October – 2016 ; 09.00 am-12.00 am)

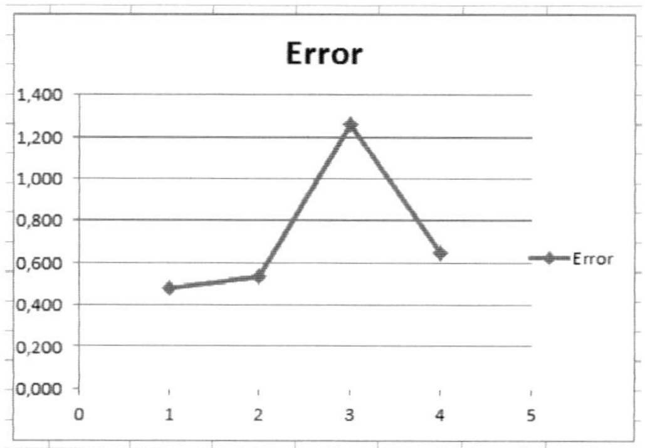

Figure 4. The value of error *IT-2 FIS*

4. Conclusions

After doing this research, then get the results forecasting very short-term daily load every hour by using Interval Type-2 Fuzzy Inference System has better results compared if using Interval type-1 Fuzzy Inference System. With MAPE value if using IT1-FIS equal to 0,928%, whereas if using IT-2 FIS get MAPE 0,729%. It can be concluded that IT-2 FIS can be used to perform very short-term forecasting forecasting of Java Bali system by making in the form of simple and concise software.

Acknowledgements
We acknowledged P3I Universitas Muhammadiyah Sidoarjo.

References

[1] Jamaaluddin, Imam Robandi 2016 Short Term Load Forecasting of Eid Al Fitr Holiday By Using Interval Type – 2 Fuzzy Inference System (Case Study : Electrical System of Java Bali in Indonesia) in *2016 IEEE Region 10, TENSYMP* vol. **0**, no. x, pp. 237–242.
[2] Damayanti R, Abdullah AG, Purnama W, Nandiyanto AB. Electrical Load Profile Analysis Using Clustering Techniques. InIOP Conference Series: Materials Science and Engineering 2017 Mar (Vol. 180, No. 1, p. 012081). IOP Publishing.
[3] N Amral, C S Ozveren, and D King 2007 *Short term load forecasting using Multiple Linear Regression* 2007 42nd Int. Univ. Power Eng. Conf., pp. 1192–1198
[4] P P (PERSERO) P. – B 2015 *Perencanaan, Evaluasi Operasi Sistem Tenaga Listrik Jawa Bali*

2015, 1st ed. Jakarta: PT PLN (PERSERO) P2B – Bidang Perencanaan

[5] M F I Khamis, Z Baharudin, N Hamid, M Abdullah, and F Nordin 2011 *Short term load forecasting for small scale power system using fuzzy logic* 2011 Fourth Int. Conf. Model. Simul. Appl. Optim., 2011

[6] I A Dharma, Robandi 2008 *Aplikasi Metode Fuzzy Inference System (FIS) dalam Perramalan Beban Jangka Pendek Untuk Hari-hari libur (Study Kasus di Pulau Bali).No Title* Procesing 9 thSeminar Intell. Technol. Its Appl., no. 2008, p. 57, 2008.

[7] K Kim, H Youn, S Member, and Y Kang 2000 Short-term load forecasting for special days in anomalous load conditions using neural networks *IEEE Trans. Power Syst.,* vol. **15**, no. 2, pp. 559–565

[8] Mulyadi Y, Farida L, Abdullah AG, Rohmah KA. Anomalous STLF for Indonesia power system using Artificial Neural Network. InScience and Technology (TICST), 2015 International Conference on 2015 Nov 4 (pp. 1-4). IEEE.

IOP Conf. Series: Materials Science and Engineering **384** (2018) 012079 doi:10.1088/1757-899X/384/1/012079

Real Time Facial Recognition Using Principal Component Analysis (PCA) And EmguCV

S Sultoni[1]* and A G Abdullah[2]

[1]Pendidikan Teknologi Informasi, Universitas Muhammadiyah Sidoarjo, Jl. Mojopahit No 666 B, Sidoarjo 61215, Jawa Timur, Indonesia
[2]Departemen Elektro, Universitas Pendidikan Indonesia, Jl. Dr. Setiabudhi No 229, Bandung 40154, Jawa Barat, Indonesia.

*sulton@umsida.ac.id

Abstract. Facial recognition is a challenging research in the field of image processing and computer vision, especially for security systems, weight determiner, and emotional determination based on the face image recognition. Some of the methods that can be used in facial recognition are holistic, feature extraction, hybrids and intelligent systems. This paper used the method of characteristic extraction that used Principal Component Analysis (PCA) which was built using EmguCV application. The purpose of this research is to assess the accuracy of Principal Component Analysis (PCA) method when combined with Emgu CV in face recognition in real time. Based on the results of training and testing, it can be concluded that the PCA method combined with EmguCV has better accuracy, if the data used has the same characteristics, PCA and EmguCV can also be developed to make image processing application especially for security system, because it applies simple statistic method and easy-applied algorithm.

1. Introduction

Facial recognition is a challenging field of research in the field of image analysis and computer vision. Facial recognition is widely used for security systems [1], to determine a person's weight [2] and also to know one's emotions [3]. Face recognition is a biometric system used to identify or verify a person using a digital image [1]. Frequently used facial recognition methods are holistic, feature extraction methods and hybrid methods. Some studies using holistic methods are [4] [5] [6] facial recognition research that uses image data as training data and test data. Feature extraction methods is used in [7] [8] [9] [10] research using only parts of the face that are considered to have the most discriminant features used as training data and test data and which use hybrids are [11] [12] [13] [14] [17] is a study that combines holistic and feature extraction methods. In line with the development of science and technology, facial recognition research is now using intelligent systems, such as those done by [15] [16].

The most widely used facial recognition method is Principal Component Analysis (PCA) which is a feature extrusion technique with the aim to find the eigen value and egien vector aimed to find the feature value of the most discriminant face [1], the result of extraction feature between training data and test data then compared using euclidean distance to know the measurement level of similarity. Research using PCA is a study conducted [15] [16] based on test results using PCA or PCA combined with Genetic Algorithm and Artificial Neural Networks, which showed good accuracy in the number of 92% to 93%. The data used are still images, both test data and training data, making it easier when compared with moving image data, because the still image is not affected by the movement, motion or distance.

This research used the moving image as the tested data, that is face image recognition directly from the research object by using Principal Component Analysis (PCA) and EmguCV method. EmguCV is a multiplatform application that runs on Windows, Linux, Mac OS X, iOS, Android and Windows Phone operating systems as well as open sources, while EmguCV is built for C-based image processing systems that are easy to understand.

2. Experimental Methods

The method applied in this research is displayed in the following chart.

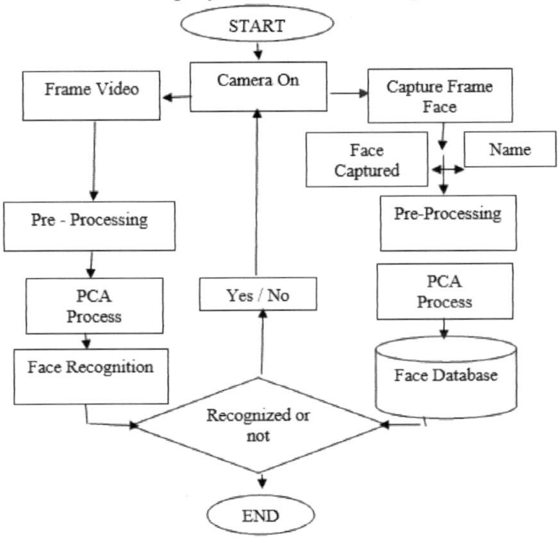

Figure 1. Research Method Diagram.

In this study the data used is the image of faces of students majoring in Pendidikan Teknologi Informasi as much as 10 students who each - each student will be taken 10 facial image from various positions as exercise data.

Then as a test data, will use a webcam that has been connected with the application. From 10 Students each will be matching to 100 frames of webcam. With matching distance divided into three types, namely at a distance of 50 cm, 100 cm and 150 cm. Here is the database of facial image of information technology education students.

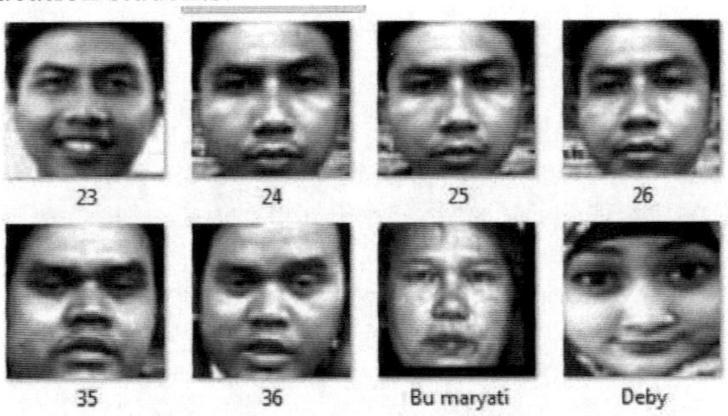

Figure 2. Face Image Database.

3. Results and Discussion

3.1. Image

Image is a representation (image), likeness, or imitation of an object. The image as output of a data recording system may be optical in the form of photographs, analogous to video signals such as images on television monitors, or digital ones that can be directly stored to a storage medium [18].

In general, a rectangular-shaped digital image of dimensions is stated as the x width y height [18]. Digital images have spatial coordinates, with brightness or light intensity (gray scale) having discrete numeric represented in the form of a mathematical function f (x, y) expressing the intensity of light at the point (x, y) itself.

3.2. Principal Component Analysis (PCA)

Principal Component Analysis is a multivariate analysis that transforms interrelated variables into new variables that are not mutually correlated by reducing the number of variables so that they have smaller dimensions but can explain most of the diversity of their original variables [19].

PCA produces large data reduction, so that it is widely used in image processing. The steps in using Principal Component Analysis (PCA) are as follows [19][5]:

3.2.1. Calculating image average

$$\mu = \frac{1}{n}\sum_{k=1}^{n} xk \tag{1}$$

3.2.2. Calculating image matrix covariate

$$C = \sum_{k=1}^{n} (xk - \mu)(xk - \mu)^T \tag{2}$$

3.2.3. Calculating eigen value dan eigen vector of PCA

$$CU_n = \mu_{nU_n} \tag{3}$$

U = Eigen vector
μ = Eigen value

Suppose an M database of N x N face images is transformed into a dimensionless vector (N2) x 1. PCA is used to find M orthonormal vectors μn that best describe the distribution of data. The kth vector, μk is chosen so that:

$$\mu k = \frac{1}{M}\sum_{n=1}^{M} (\mu k^T \emptyset_n)^2 \tag{4}$$

Maximum when

$$\mu l^T \mu k =_{0\ otherwise}^{1, l=k} \tag{5}$$

The vector μk is the eigen vector and λk is the eigen value of the covariance matrix C, the number of images in the database is much smaller than the size of the vector (M <N2), there will be an important M-1eigen vector. So that we can reduce the eigenvectors of size N2 by finding the matrix Eigenvectors with M x M size.

To determine how the eigenface to be taken, the equation is as follows:

$$\frac{\sum_{i=1}^{M} \mu i}{\sum_{i=1}^{M} \mu k} = A \tag{6}$$

A is the value that describes how the database variance we want. Suppose 0.9 (90%), the database variance is the square value of the standard deviation that describes the sum of the mean deviation squared values. The smaller the A value is, the less accurate eigenvalue, but the required number of eigenvectors decreases (M '<M)

3.2.4. Eulcidian distance. To know the level of similarity between trainer data and test data, his research used Euclidian Distance which has the following equation:

$$d \sqrt{(X1 - X2)^2 + (Y1 - Y2)^2} \quad (7)$$

3.3. EmguCV

EmguCV has two layers, as follows.

- The first or the base layer serves to map the functions that exist on the system.
- The second layer contains the classes and is a development of a NET-based program.

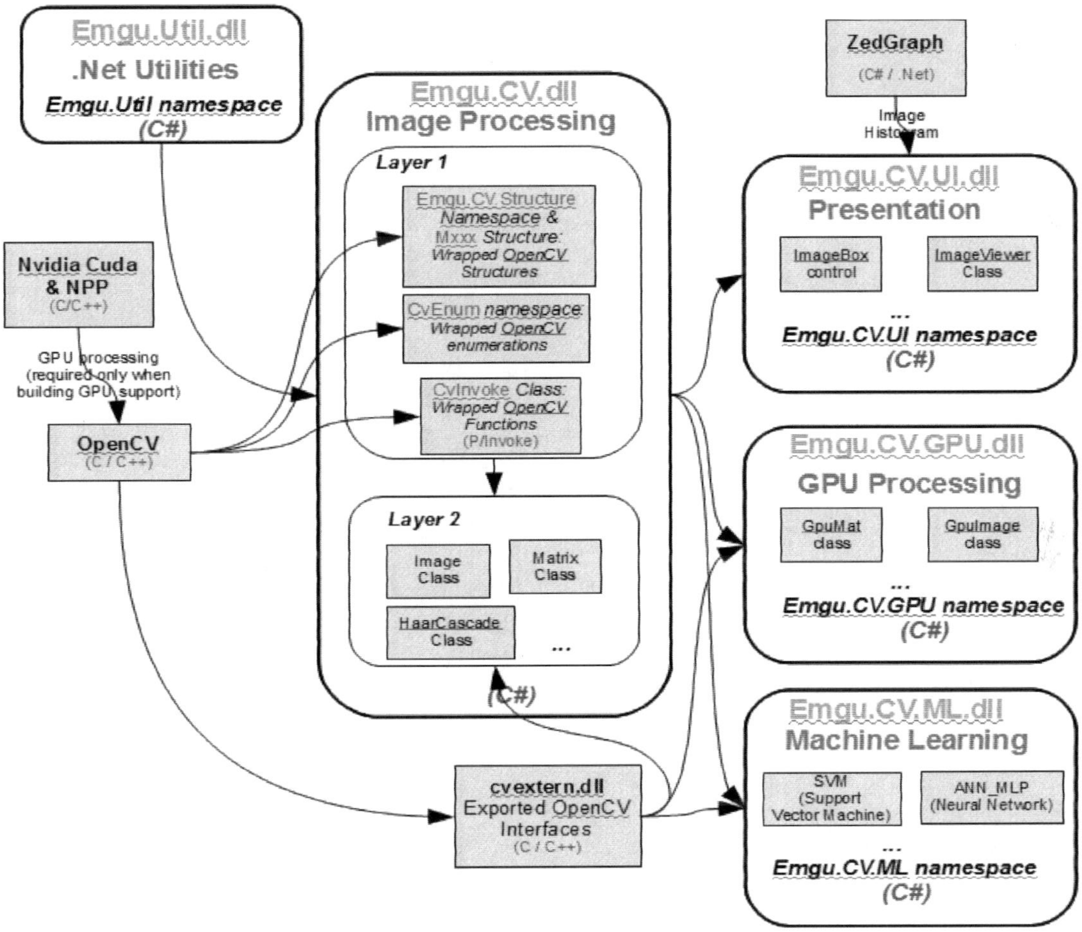

Figure 3. EmguCV Diagram [23].

Based on the results of training and testing related to face recognition using Principal Component Analysis (PCA) and EmguCV, following data are obtained:

3.4. Test Result at 50 cm Distance

In this test, the distance used between the object and the camera is approximately 50 cm. The data obtained are shown in table 1 below.

Table 1. Face Recognition at 50 cm Distance.

NO	Recognition Result		
	Name	*Accuracy*	*Time*
1	Person 1	85	0.96
2	Person 2	88	0.97
3	Person 3	90	0.89
4	Person 4	91	0.99
5	Person 5	89	1.09
6	Person 6	90	1,08
7	Person 7	91	1.08
8	Person 8	88	1.08
9	Person 9	89	1,05
10	Person 10	90	0.90
Average		**89.1**	**1.009**

Based on the data contained in the table above, it can be analyzed that wholistically, the results of the introduction using PCA method with trehsold used between 0.1 to 1, shows recognition accuracy is about 89.1 with the required time is 1.009 seconds.

3.5. Test Result at 100 cm Distance

The next test is facial recognition at a distance of 100 cm. Based on test, the results and training obtained data is shown in table 2 below.

Table 2. The Result of facial recognition on 100 cm distance.

NO	Recognition Result		
	Name	*Accuracy*	*Name*
1	Person 1	87	0.98
2	Person 2	86	1.89
3	Person 3	89	1.89
4	Person 4	88	1.88
5	Person 5	87	0.98
6	Person 6	88	1.009
7	Person 7	86	1.078
8	Person 8	85	1.098
9	Person 9	88	0.987
10	Person 10	87	1.098
Average		**87.1**	**1.289**

Based on the data shown in table 2 above, we can analyze that the average level of accuracy of the facial mask at 100 cm distance is 87.2 with computation time is about 1.289 seconds

3.6. Test Result at 150 cm Distance

The last test is a test at a distance of 150 cm, based on the test results obtained data as shown in table 3 below.

Table 3. Facial recognition result on 150 cm.

NO	Recognition Result		
	Name	*Accuracy*	*Name*
1	Person 1	86	0.788
2	Person 2	84	1.076
3	Person 3	82	1.0876
4	Person 4	85	1.0986
5	Person 5	86	0.987
6	Person 6	87	1.088
7	Person 7	85	0.879
8	Person 8	84	0.879
9	Person 9	86	1.009
10	Person 10	87	1.088
	Average	**85.2**	**0.998**

Based on the data shown in table 3 above, then we can analyze that the average level of recognition accuracy at a distance of 150 cm is 85.2 with a computation time of 0.998 seconds.

Based on some test that has been done at 50 cm, 100 cm and 150. it can be seen that each of them has a different level of accuracy because it is influenced by many factors such as distance and background lighting. Similar things have also been done by [20] [21] where the distance of illumination and background greatly affect the results of the introduction. But overall the accuracy of PCA and EmguCV in face recognition in Real Time is in good category as shown in table 4 below.

Table 4. Acumulation of Face Recognition.

NO	Recognition result		
	Recognition distance	*Result*	*Time*
1	50 cm distance	**89.1**	**1.009**
2	100 cm distance	**87.1**	**1.289**
3	150 cm distance	**85.2**	**0.998**
	Average	**87.13**	**1.987**

Based on the data shown in the above table, we can conclude that the average face recognition in real time using PCA and EmguCV methods has the accuracy of 87.13 with computation time of 1.987. Compared with previous research, this research is still low on accuracy, this is caused by several factors, such as the data used is different and data retrieval techniques are also different. Here is a comparison between the results of current research with previous research.

Table 5. Comparation Result Face Recognition.

NO	Recognition Result	
	Method	*Accuracy*
1	LBP [6]	**94.8**
2	PCA+GA+NN [22]	**81**
3	PCA + EmguCV	**87.13**

Based on the data in the table above, we can see that PCA method has better accuracy when compared with PCA_GA + NN but still lower if with LBP. It is because the difference data are used for LBP method with PCA + EmguCV, where LBP using silent data while PCA + EmguCV uses graded image data. However, when compared to the same PCA + AG + JST using mobile image data, the PCA + EmguCV method has better accuracy

4. Conclusions

Based on the results of the analysis and analysis as described above, it can be concluded that the PCA method combined with EmguCV is very likely to be developed to create security-based system application of image processing, with statistical and algorithm method which is simple but has a good accuracy and computation time. This research can still be combined with other methods to get more optimal results, so that it can be applied in the real world such as for absenteeism and security.

Acknowledgements

The speakers would like to thank all parties involved, especially to the Univeritas Muhammadiyah Sidoarjo who has organized the international seminar and provided financial support for the speakers at the seminar.

References

[1] Divyarajsinh N, Parmar and Brijesh B M 2014 Face Recognition Methods & Applications *IJCTA* **4** (1) 84-86

[2] Nagpal S 2015 Regularized Deep Learning for Face Recognition With Weight Variations *IEEE Special Section On Applying Four Ds Of Machine Learning To Advance Biometrics* **3** 3010 – 3018

[3] Uddin M Z, Hassan M M, Almogren A, Alamri A, Alrubaian M and Fortino G 2017. Facial expression recognition utilizing local direction-based robust features and deep belief network *IEEE Access* **5** 4525-4536

[4] Xu Y, Li Z, Yang J and Zhang D 2017 A survey of dictionary learning algorithms for face recognition *IEEE access* **5** 8502-8514

[5] Sayeed S, Hossen J, Kalaiarasi S M A, Jayakumar V, Yusof I and Samraj A 2017 Real-Time Face Recognition for Attendance Monitoring System *Journal of Theoretical and Applied Information Technology* **95** (1) 24

[6] Vigneau H G 2017 Thermal Face Recognition Under Temporal Variation Conditions *IEEE Special Section On Applying Four Ds Of Machine Learning To Advance Biometrics* **5** 9663 – 9672

[7] Chen Y-C 2015 Dictionary-Based Face and Person Recognition From Unconstrained Video *IEEE Special Section On Applying Four Ds Of Machine Learning To Advance Biometrics* **3** 1783 – 1798

[8] Pan J 2016 Single-Sample Face Recognition Based on LPP Feature Transfer *IEEE Special Section On Applying Four Ds Of Machine Learning To Advance Biometrics* **4** 2873 – 2884

[9] Wan M And Lai Z 2017 Multi-Manifold Locality Graph Embedding Based on the Maximum Margin Criterion (MLGE/MMC) for Face Recognition *IEEE Special Section On Applying Four Ds Of Machine Learning To Advance Biometrics* **5** 9823 – 9830

[10] Zhihua X, Peng J and Shuai Z 2017 Fusion of LBP and HOG Using Multiple Kernel Learning for Infrared Face Recognition *IEEE Computer Society*

[11] Taher K, Faouzi B and Hamid A 2017 Gabor Feature Based Local Generic Representation For Face Recognition with Single Sample per Person *IEEE Computer Society*

[12] Liang J 2017 Bilateral Two-Dimensional Neighborhood Preserving Discriminant Embedding for Face Recognition *IEEE Special Section On Advanced Data Analytics For Large-Scale Complex Data Environments* **5** 17201 – 17212

[13] Hongjun W, Jiani H and Weihong D 2017 Compressing Fisher Vector for Robust Face Recognition *Journal of IEEE access* 1-9

[14] Jianquan G, Haifeng H and Haoxi L 2017 Local Robust Sparse Representation for Face Recognition With Single Sample per Person *IEEE/CAA Journal Of Automatica Sinica* 1-8.

[15] Riggan S B 2015 Coupled Auto-Associative Neural Networks for Heterogeneous Face Recognition *IEEE Special Section On Applying Four Ds Of Machine Learning To Advance Biometrics* **3** 1620 – 1632

[16] Huang P 2016 Fuzzy Linear Regression Discriminant Projection for Face Recognition *IEEE Translations and content mining are permitted for academic research only*

[17] Yimin Z and Zhifei L 2016 Real-time Gender Recognition based on Eigen-features selection from Facial Images *Industrial Electronics Society, IECON 2016 - 42nd Annual Conference of the IEEE* 1025 – 1030

[18] Sutoyo T, Mulyanto E, Suhartono V N O D and Wijanarto 2009 *Teori Pengolahan Citra Digital,* (Yogyakarta: Andi)

[19] Purnomo M H and Muntasa A 2010 *Konsep Pengolahan Citra Digital dan Ekstraksi Fitur* (Yogyakarta: Graha Ilmu)

[20] Sultoni H S, Dachlan M P and Rahmadwati 2016 Pengenalan Wajah Secara Real Time Menggunakan Metode Camshift, Lapalcain Of Gaussian dan Discrete Cosine Two Dimensional (LoGDCT2D) *Jurnal Ilmiah NERO* **2** 3 153-160

[21] Sultoni 2017 Modifikasi Metode Camshift Untuk Pengenalan Citra Wajah Secara Real Time Berdasarkan Warna Kulit Wajah *Seminar Nasional Aplikasi Teknologi di Industri Seniati* A34 1-7

[22] Yimin Z and Zhifei L 2016 Real-time Gender Recognition based on Eigen-features selection from Facial Images *Industrial Electronics Society , IECON 2016 - 42nd Annual Conference of the IEEE* 1025 – 1030

[23] Emgu CV 2017 *Architecture Overview* (Online) available at: http://www.emgu.com/wiki/index.php/Main_Page #Architecture_Overview (on access dated August 19, 2017)

AUTHOR INDEX

AUTHOR INDEX

Institute of Physics
Dirac House, Temple Back
Bristol BS1 6BE UK

ISSN: 1757-8981
ISBN 978-1-5108-6753-6

CURRAN ASSOCIATES INC.
proceedings.com